T0348561

STRUCTURAL GEOLOGY

STRUCTURAL GEOLOGY

The Mechanics of Deforming Metamorphic Rocks

Volume I: Principles

BRUCE HOBBS
Research Fellow, CSIRO
Adjunct Professor, Centre for Exploration Targeting,
The University of Western Australia

ALISON ORD
Winthrop Research Professor, Centre for Exploration Targeting,
The University of Western Australia.

ELSEVIER

AMSTERDAM   BOSTON   HEIDELBERG   LONDON   NEW YORK   OXFORD
PARIS   SAN DIEGO   SAN FRANCISCO   SINGAPORE   SYDNEY   TOKYO

Elsevier
Radarweg 29, PO Box 211, 1000 AE Amsterdam, Netherlands
The Boulevard, Langford Lane, Kidlington, Oxford, OX5 1GB, UK
225 Wyman Street, Waltham, MA 02451, USA

ISBN: 978-0-12-407820-8

Library of Congress Cataloging-in-Publication Data
Hobbs, Bruce E.
    Structural geology : the mechanics of deforming metamorphic rocks / Bruce E. Hobbs, Alison Ord.
        pages cm
    ISBN 978-0-12-407820-8 (hardback)
    1. Metamorphic rocks. 2. Rock deformation. 3. Rock mechanics. I. Ord, Alison, 1955. II. Hobbs, Bruce, 1936. III Title.
    QE475.H63 2015
    552'.4–dc23
                    2014033461

British Library Cataloguing in Publication Data
A catalogue record for this book is available from the British Library

For information on all Elsevier publications visit our
web site at http://store.elsevier.com/

Dedicated to:

Susan Therese Plant (1964–2013)

Evelyn Ethel Ord (1919–2014)

Smile because she lived.

Contents

Preface

This volume discusses the processes that operate during the deformation and metamorphism of rocks in the crust of the earth with the goal of understanding how these processes control or influence the structures that we observe in deformed metamorphic rocks. The deformation and metamorphism of rocks involves structural rearrangements of elements of the rock by processes such as mass diffusion, dislocation slip and climb, disclination and disconnection motion, grain-boundary migration and fracturing at the same time as chemical reactions proceed. In some instances infiltrating fluids introduce or remove chemical components and may influence mechanical properties through changes in volume, fluid pressure or temperature. At the same time, heat is added or removed from the system depending on the surrounding tectonic environment and the processes operating within the deforming rock mass. The intent is to consider these processes within a framework set by both the individual mechanisms of deformation and the fundamental mechanics that describe and relate these mechanisms. Such mechanisms do not operate independently of each other but are strongly coupled so that each process has strong feedback influences on the other leading to structures and mineral assemblages that do not develop in the uncoupled environment. This leads to nonlinear behaviour and so a goal of this book is to treat deforming metamorphic systems as nonlinear dynamic systems. The general system we consider in this book is a deforming, chemically reactive system in which fluid and thermal transports play significant roles. The volume concentrates on the *principles* that govern the development of deformed metamorphic rocks. A second volume deals with *applications* of these principles.

Over the past 30 years there have been dramatic developments in structural geology, metamorphic petrology, physical metallurgy, nonlinear chemical dynamics, continuum mechanics, nonlinear dynamics and thermodynamics that are relevant to processes that operate within the Earth; but these developments have largely evolved independently of each other and many have not been incorporated into mainstream metamorphic geology. This book attempts to integrate aspects of these developments into a common framework that couples the various processes together.

In this volume we first develop the basic groundwork in Mechanics where thermodynamics plays a fundamental role and then move to apply these principles to rocks where deformation, fluid flow, thermal transport, mineral reactions, damage and microstructural evolution contribute to the evolution of the fabric of the rock mass, and produce the structures we observe.

The processes that operate fall into five inter-related categories: (1) mechanical processes, (2) mineral reactions, (3) fluid flow, (4) transport of heat and (5) microstructural adjustments. A goal of this book is to treat all five of these categories under the same umbrella and integrate these processes within a single framework. The basis of this framework lies in thermodynamics. This is not a book about thermodynamics but it uses thermodynamics as an important tool. What do we mean by thermodynamics? By *thermodynamics* we mean: the study of the flow of physical and chemical quantities through a system under the influence of thermodynamic forces. The physical and chemical quantities involve momentum (per unit area), fluids, heat and mass; the corresponding thermodynamic forces (an unfortunate but useful term) are gradients in deformation, the inverse of the temperature, hydraulic and chemical potentials.

This volume is arranged into two parts. Both parts begin with an overview of the section. Part A is a discussion of modern mechanics and includes chapters devoted to the geometry, kinematics, nonlinear dynamics and thermodynamics of deforming systems. We follow with a consideration of common constitutive relations including non-steady state evolutionary processes such as hardening and softening and the development of anisotropy. In particular, in Section A we introduce the basic principles that govern the *nonlinear* behaviour of deforming, chemically reactive systems. This represents a significant departure from current treatments of the mechanics of geological materials.

Part B considers the common processes involved in the development of geological structures including brittle and visco-plastic flow, heat and fluid flow, damage evolution, microstructural rearrangements such as sub-grain formation and rotation and the nonlinear kinetics of mineral reactions. We take the opportunity to introduce the concepts of disclinations and disconnections which, along with dislocations, are now recognised as playing important roles in deformation and grain growth/reduction processes during the deformation of polycrystalline materials. Of greater importance is the concept of coupled grain-boundary migration in which grain- (or subgrain) boundary migration is coupled to a shearing deformation parallel to the moving boundary. This process has a profound influence on the development of metamorphic fabrics and the concept has a great unifying influence on understanding their evolution.

Of particular importance is the notion of critical systems of various kinds where the Helmholtz energy passes from being a convex to a nonconvex function as a function of some critical parameter or forcing. In order for a deforming system to minimise such energy functions and remain compatible with the imposed deformation, the system cannot remain homogeneous and forms structure at finer and finer scales. This behaviour is fundamental to rock deformation and is responsible for most geological fabrics including jointing, localised folding and boudinage and rotation recrystallisation.

Volume II considers the common geological structures such as foliations, lineations, folds and fracture systems and extends the discussion both to the regional scale and to open flow systems such as hydrothermal systems. In particular, the concept of localised structure development is emphasised in the light of progress in nonlinear mechanics over the past 30 years; this enables the fundamental work of Biot to be seen within a generalised framework. We regard Biot as the father of the application of thermodynamics to deforming, chemically reacting systems.

Both volumes assume a basic knowledge of structural geology and metamorphic petrology and so some familiarity with the books quoted below is necessary. A short list of useful additional references for in-depth reading is given at the end of each chapter. Although there is an emphasis on a mathematical treatment, the mathematics is kept to an elementary level and a basic knowledge of differential calculus and tensor algebra suffices to understand the book.

Short appendices on both these subjects are included. The volume is meant to be read by graduate students and research workers in metamorphic geology. The intent also is to supply a vocabulary to enable readers to follow some of the more involved papers on the mechanics of deforming, metamorphic rocks that are appearing with greater frequency.

In an interdisciplinary book such as this, we have found it too demanding to stick to a unique mathematical notation throughout. We have attempted to standardise on commonly used quantities (Appendix A) but for individual examples have mainly used the notation of the original authors of the example considered. Such terms are always defined where they are first used. Equally, for didactic reasons, we have introduced concepts and principles where they are of the greatest use and not where they would be introduced in a standard treatment of the subject. This leads to some repetition and sometimes an ordering of material that is not sequential. Each chapter is largely self-contained. We trust that readers will report errors to one or both of us at bruce.hobbs@csiro.au and alison.ord@uwa.edu.au.

Much of the content is very much in progress and many questions remain open but we hope it will inspire researchers, especially the younger breed, who more commonly already have a grasp of the concepts involved, to question and revisit entrenched wisdom and especially revisit natural examples in the field and at the microscale where most of the fascinating and interesting problems are yet to be solved.

Bruce Hobbs
Alison Ord
Fremantle, Western Australia,
August, 2014.

Further Reading

Fossen, H. (2010). *Structural Geology.* Cambridge University Press.

Hobbs, B.E., Means, W.D., Williams, P.F. (1976). *An Outline of Structural Geology.* John Wiley.

Jaeger, J.C. (1969). *Elasticity, Fracture and Flow.* Methuen.

Passchier, C.W., Trouw, R.A.J. (1996). *Microtectonics.* Springer.

Ramsay, J.G. (1967). *Folding and Fracturing of Rocks.* McGraw-Hill.

Turner, F.J., Weiss, L.E. (1963). *Structural Analysis of Metamorphic Tectonites.* McGraw-Hill.

Twiss, R.J., Moores, E.M. (2006). *Structural Geology.* W. H. Freeman.

Vernon, R.H. (2004). *A Practical Guide to Rock Microstructure.* Cambridge University Press.

Vernon, R.H., Clarke, G. (2008). *Principles of Metamorphic Petrology. Cambridge University Press.*

Yardley, B.W.D. (1974). *An Introduction to Metamorphic Petrology.* Longman.

Acknowledgements

We thank the following for their influence on our evolving thought processes and for their contributions to a nonlinear view of deformation and metamorphism. Some may not recognise that they made such a contribution, but we acknowledge them for sharing with us their approach to mechanics and of deformation during metamorphism. Thanks to: Elias Aifantis, Mike Brown, John Christie, Peter Cundall, David Griggs, Peter Hornby, Giles Hunt, Win Means, Hans Muhlhaus, Gerhard Oertel, Mervyn Paterson, Roger Powell, Klaus Regenauer-Lieb, Ioannis Vardoulakis, Ron Vernon, Vic Wall, Paul Williams, David Yuen and Chongbin Zhao. We would also like to acknowledge the encouragement, patience and help of Elsevier staff John Fedor, Louisa Hutchins and Sharmila Vadivelan.

Introduction

General Statement. The aim of this book is to discuss the processes that operate during the deformation and metamorphism of rocks in the crust and upper lithosphere of the Earth with the goal of understanding how these processes control or influence the structures that we observe in the field. The intent is to consider these processes within a framework set both by the individual mechanisms of deformation and the fundamental mechanics that describe and relate these mechanisms. Such mechanisms do not necessarily operate independently of each other but may be strongly coupled so that each process has strong feedback influences on the others leading to structures and mineral assemblages that do not develop in the uncoupled environment. The general system we consider in this book is a deforming chemically reactive system in which fluid and thermal transport play significant roles.

Over the past 30 years there have been dramatic developments in structural geology, metamorphic petrology, physical metallurgy, nonlinear chemical dynamics, continuum mechanics, nonlinear dynamics and non-equilibrium thermodynamics that are relevant to processes that operate within the Earth but these developments have largely evolved independently of each other and many have not been incorporated into mainstream metamorphic geology. This book attempts to integrate aspects of these developments into a common framework that couples the various processes together.

Although there is a tendency to separate *structural geology* from *metamorphic petrology* in text books, we attempt to bring these two subjects together in this book under the name *metamorphic geology*. We are not too concerned with mineral phase equilibria but treat the

1

processes that give rise to mineral assemblages and fabrics from the viewpoint that such processes commonly contribute to the production of the *structures* and *patterns* we see in deformed metamorphic rocks. We use the term *fabric* to mean *the internal spatial ordering of geometrical, physical and mineralogical elements in deformed metamorphic rocks* (Cf., Turner and Weiss, 1963, p. 19; Vernon, 2004, p. 7). The term *structure* means the assemblage of folds, boudins, foliations, lineations, grain shapes, mineral phases and so on that make up such rocks. At a microscopic scale the *microfabric* consists of the grain shapes, grain arrangements and spatial orientations and patterns of grain distributions. We use the term *crystallographic preferred orientation* (or just *CPO*) to describe the patterns of lattice orientations of grains. An important additional concept is that of *patterns*. We will see that the behaviour of nonlinear systems is characterised by the formation of patterns of various kinds (Cross and Greenside, 2009). These patterns are typified by metamorphic differentiation, the distribution of porphyroblasts and mineral lineations defined by mineral segregations to name a few. Just as important as the patterns are defects in the patterns and we will discuss such features in Chapter 7.

At least while these processes are operating, the system dissipates energy and hence is not at equilibrium. Thus, the overarching concepts that unify these various processes are grounded in the thermodynamics of systems not at equilibrium. The traditional approach in metamorphic geology is strongly influenced by the work of Gibbs where, for the most part, equilibrium is assumed and so approaches based on departures from equilibrium have been largely neglected or even dismissed as irrelevant in the Earth Sciences, the argument being that geological processes are so slow that equilibrium can be assumed. There has been an additional important deterrent to progress in that developments in non-equilibrium thermodynamics have been dispersed across a large number of disciplines and languages, with apparently conflicting propositions put forward and sterile discussions devoted to the relative merits of minimum and maximum entropy production principles and the so-called Curie principle (Truesdell, 1966). The result has been difficulty in bringing forward a unified approach to the subject as far as Earth scientists are concerned.

In this book we first develop the basic groundwork in *mechanics* where thermodynamics plays a fundamental role and then move to apply these principles to rocks where deformation, fluid flow, thermal transport, mineral reactions, damage and microstructural evolution contribute to the evolution of the rock mass and produce the structures we observe. The processes that operate fall into five interrelated categories: (i) mechanical processes such as diffusion, dislocation motion, twinning, grain-boundary sliding and fracturing that contribute directly to the deformation of the material; (ii) chemical reactions that produce changes in mineralogy, volume changes and grain-rearrangements and consume or produce heat; these processes influence the mechanical properties both by the generation of heat and of new weak or strong mineral phases and the ways in which these phases are spatially related to each other; moreover chemical reactions can act as a deformation mechanism in their own right; (iii) fluid flow which not only introduces or removes chemical components such as $(OH)^-$, H^+ and CO_2 but can also be an agent for dissolution and deposition − processes that also contribute to the deformation; (iv) transport of heat by conduction or by advection in a moving fluid that leads to temperature changes that in turn can influence the rates of mineral reactions and of deformation; (v) structural adjustments such as subgrain formation, recrystallisation, preferred orientation development (in the form of both crystallographic axes and grain-shape), metamorphic differentiation, the development of foliations and lineations, folding, boudinage and localised shear zones.

A goal of this book is to treat all five of these categories under the same umbrella and integrate these processes within a single framework. The basis of this framework lies in thermodynamics. This is not a book about thermodynamics but it uses thermodynamics as an important tool. What do we mean by thermodynamics? By *thermodynamics* we mean *the study of the flow of physical and chemical quantities through a system under the influence of thermodynamic forces*. The dominant physical and chemical quantities of interest to metamorphic geologists are momentum (per unit area), heat, fluids and mass; the corresponding thermodynamic forces are gradients in the deformation, the inverse of the temperature, hydraulic potential and chemical potential. If there are no flows the system is at equilibrium.

All of the processes mentioned above produce or consume entropy while they operate. So thermodynamics can also be defined as *the study of the production of entropy*. We use the concept of entropy in a very simple manner. By *entropy* we mean *the amount of heat produced at an instant at a point in the body divided by the current absolute temperature at that point*. Most deformation processes such as plastic flow or fracturing produce entropy, whereas chemical reactions may produce or consume entropy. Even the flow of fluids through fractures or porous media produces entropy. The second law of thermodynamics tells us that for a given system the sum of all the individual entropy production rates is greater than zero (or equal to zero at equilibrium). This means that the processes that operate to constitute a deforming chemically reacting system do not occur independently of each other but are coupled through the second law of thermodynamics to ensure that the total entropy production is greater than zero. Such a simple overarching law forms the basis for the unifying framework we seek for deformed metamorphic rocks.

In this book we introduce the basic principles that govern the *nonlinear* behaviour of deforming chemically reactive systems. This represents a significant departure from current treatments of the mechanics of geological materials. Of particular importance here is the notion of *critical systems* where the Helmholtz energy passes from being a convex to a non-convex function as a function of some critical parameter. This behaviour is responsible for a large number of geological structures including localised folding and boudinage and rotation recrystallisation. In particular the concept of localised structure development is emphasised in the light of progress in nonlinear mechanics over the past 30 years; this represents an extension of the fundamental work of Biot who we regard as the father of the application of thermodynamics to deforming chemically reactive systems through publication of two seminal papers: Biot (1955, 1984).

This introductory chapter discusses a number of concepts that may be unfamiliar to some metamorphic geologists but which are fundamental tenets of the approach adopted in this book:

- The multiscale nature of metamorphic geology.
- The coupled nature of most processes that operate during deformation and metamorphism.
- The need to describe and analyse the processes involved and their interactions in terms of non-equilibrium concepts.

- The distinction between linear and nonlinear systems and the implications of these two different modes of behaviour for the development of the structures and metamorphic fabrics observed in deformed metamorphic rocks.
- The use of *wavelets* as an overarching technique that enables structures and patterns in deformed rocks to be quantified across a range of length scales and characterised in terms of fractal geometrical concepts.

1.1 THE MULTISCALE NATURE OF DEFORMATION AND METAMORPHISM

   This book is concerned with the processes that operate during the deformation and metamorphism of rocks in the crust and upper lithosphere of the Earth. The deformation and metamorphism of rocks involves structural rearrangements of elements of the rock at the microscale by processes such as mass diffusion, dislocation slip and climb, grain-boundary migration and fracturing at the same time as chemical reactions proceed. In some instances infiltrating fluids introduce or remove chemical components and may influence mechanical properties through changes in mineralogical composition, fluid pressure or temperature. At the same time, heat is added to or removed from the system depending on the surrounding tectonic environment. Such mechanisms do not operate independently of each other but are coupled so that each process has feedback influences on the others leading to structures and mineral assemblages that do not develop in the uncoupled environment. Some of the inter-relations between these processes are illustrated in Figure 1.1.

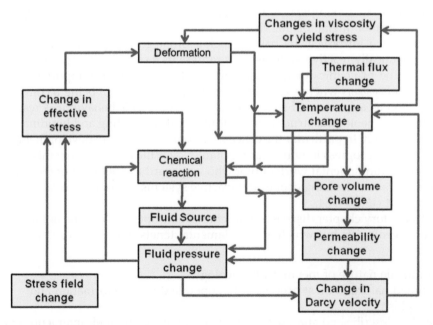

FIGURE 1.1   Coupling and feedback relations for some of the processes that operate during the development of deformed metamorphic rocks.

For instance deformation can induce a pore volume change which in turn produces a change in fluid pressure which alters the effective stress that then influences the deformation. As another example, a chemical reaction can produce fluid and/or a change in temperature both of which induce changes in pore fluid pressure with a consequent change in the deformation. Again, changes in temperature arising from deformation and/or chemical reactions can influence physical properties such as the viscosity or coefficient of friction which in turn influences the deformation. Many feedback relations exist in deforming chemically reacting rock masses and our aim in this book is to discuss the mechanics of these feedback relations and integrate them within a unified framework that describes the formation of the common structures and mineral fabrics we see in deformed metamorphic rocks.

One of the outstanding characteristics of metamorphic processes is the vast range of spatial and temporal scales involved (Figure 1.2). Thus at the atomic scale the forces between atoms need to be modelled with length scales of 10^{-9} m involving timescales for atomic vibrations of 10^{-15} s. Such modelling is becoming common for understanding dislocation dynamics at geological strain rates (Cordier et al., 2012) and some progress has been made in understanding metamorphic mineral reactions (Lasaga and Gibbs, 1990; Gale et al., 2010). Towards the other end of the spectrum is the timescale for regional metamorphism and melting which can be ≈ 100 my (Brown, 2010; Smithies et al., 2011) and involve length scales of 1000 km. The length scales involved in metamorphic geology span ≈ 15 orders of magnitude and the timescales span ≈ 32 orders of magnitude.

Such huge spans in length and timescales have advantages and disadvantages. The advantages lie in the opportunity to uncouple processes operating at different scales thus simplifying the mathematical treatment of the processes involved. For instance we will show in Volume II that for a shear zone 1 km thick deforming at a slow shear strain rate of 10^{-13} s^{-1} the heat dissipated by deformation does not have time to be conducted out of the system and the shear zone heats up thus influencing the deformation and the rates of mineral reactions within the shear zone. For these conditions the thermal, chemical and deformation processes cannot be considered to be independent. On the other hand if the shear zone is 1 m thick then the heat can diffuse out of the shear zone and the deformation can be considered isothermal. Here the thermal and deformation processes can be uncoupled and considered independently of each other. On the other hand if the deformation is fast, say 10^{-1} s^{-1} corresponding to a slow aseismic event, then the thin shear zone also heats up and thermal, chemical and deformation processes need to be coupled.

The disadvantages arising from the huge spans in spatial and timescales involved lie in the present limitations of modern computers and the ways we use them. If we want to model, using finite elements, the deformation of the lithosphere say 150 km thick and 1000 km wide in two dimensions with a resolution of 10 m we need $\approx 1.5 \times 10^9$ finite elements. This is very close to what is possible with a modern computer. If we needed to do the computation in three dimensions the problem is not possible today. However, 10 m is still large compared to the processes such as diffusion and dislocation motion that take place at the atomic to micron scales. The solution adopted is called a *multiscale or homogenisation approach* (Tadmor and Miller, 2011) wherein the results of modelling particular processes at one scale are used to develop a mathematical description of that process which effectively homogenises the details of the processes operating at that scale. This description is then used to model detailed behaviour at the next scale up and so on. Thus dislocation dynamics

(a)

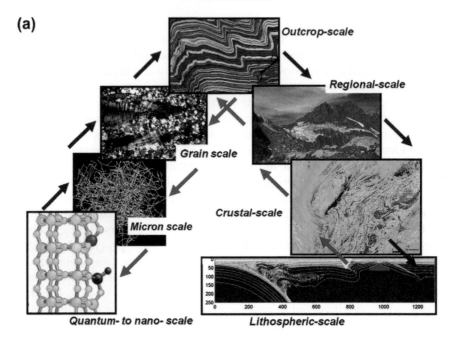

(b)

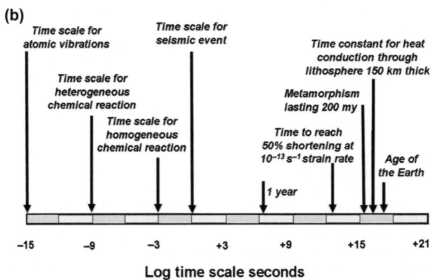

FIGURE 1.2   Spatial and temporal scales associated with deformation and metamorphism in the upper litho-sphere of the Earth. (a) Spatial scale variation leading to a multiscale approach to modelling such systems. *Images in order clockwise from lower left are from Zhu et al., (2005), Patrick Cordier, Ron Vernon, Bruce Hobbs, Jean-Pierre Burg, Catherine Spaggiari and Weronika Gorczyk.* (b) Timescales associated with some metamorphic processes.

modelled at the micron scale can be used to develop a mathematical constitutive relation that links the strain rate to the applied stresses. This relation can then be used to model details of processes at the metre scale and above (Cordier et al., 2012). This *upscaling* approach is shown in Figure 1.2(a) by the blue arrows. On the other hand conditions established by geological observations at a large length scale can be used as boundary conditions for processes operating at smaller length scales. This *downscaling* procedure is shown by the red arrows in Figure 1.2(a).

While these various processes are operating, the system dissipates energy (that is, the system produces or consumes heat) and hence is not at equilibrium. This statement is true no matter what the length and timescales are. As long as dissipative chemical and physical processes are operating the system, by definition, is not at equilibrium. We will show in Chapter 5 and Volume II that for some important metamorphic systems equilibrium is never attained no matter how slow the processes are. We will see that this is particularly true if the system is open so that fluids are entering or leaving the system; examples of such situations are H_2O or H^+ entering a retrograding shear zone, SiO_2 bearing fluids leaving a schist undergoing metamorphic differentiation or melt leaving an anatexite. Thus, the overarching concepts that unify these various processes are grounded in the thermodynamics of systems not at equilibrium.

In this book we first develop the basic groundwork in mechanics where thermodynamics plays a fundamental role and then move to apply these principles to rocks where deformation, fluid flow, thermal transport, mineral reactions, damage and microstructural evolution contribute to the evolution of the rock mass and produce the structures we observe. The overarching principle that unites all these process is the *second law of thermodynamics* which, in a form where the mathematical detail of the processes involved is spelt out, is commonly called the *Clausius−Duhem relation* (Truesdell, 1966).

1.2 MECHANICS, PROCESSES AND MECHANISMS

Mechanics is the study of the ways in which physical and chemical systems respond to imposed forces or displacements at the boundaries of the system; it forms the content of section A: The Mechanics of Deforming Solids of this book. The response is strongly influenced by the nature of the material, the chemical reactions taking place and the physical environment defined by the temperature, deformation rate and whether the deforming system is open or closed with respect to transport of mass including fluids. Deformed metamorphic rocks display a vast array of structures such as dislocations, subgrains, recrystallised microstructures, folds, foliations, mineral lineations, boudins, metamorphic layering, porphyroblasts, leucosomes, localised shear zones and fractures. The characteristic of most of these structures is that they formed during the deformation of a chemically reacting *polycrystalline solid*. In fact the fundamental common characteristic of all of these structures is that the material comprises grains with well-defined crystal structures for most if not all of its evolution. Thus the material resembles a deforming metal rather than a viscous syrup that possesses no structure or a viscous polymer that may develop some of these structures but with no inherent granular structure where the grains have well-defined crystal structures. Thus the response of deforming metamorphic rocks is more akin to that of a solid rather than that of a fluid. From the point of view of mechanics, what do we mean by a solid rather than a fluid?

The common response to this question (Rice, 1993) is that *a solid can support a shear stress over finite periods of time, whereas a fluid cannot.* A fluid will flow under the influence of shear stresses but a solid may remain undeformed (except for elastic deformations) for periods of time that are long compared to the time involved in the observations or experiments being conducted. We discuss other important differences in Chapter 6. This difference has lead to the development of two different streams of mechanics known as *solid* and *fluid mechanics.* These two streams utilise different mathematical tool boxes and have different conceptual approaches. It has been, and still is, common for deforming rocks to be treated as fluids because often the concepts involved and the mathematical treatments are simpler than dealing with solids. Thus the dominant part of the theory of folding in the geological litera-ture is based on fluid mechanics (Johnson and Fletcher, 1994; Ramberg, 1963; Smith, 1977) as are also almost all models of convection within the mantle of the Earth (Davies, 2011). How-ever, the fact that the concepts and mathematics of fluids (especially that of ideal gases or linear fluids) are commonly simpler than dealing with solids is no excuse for proposing that metamorphic rocks actually are fluids or ideal gases. In this book we treat metamorphic rocks as solids and develop the theory of metamorphism within the context of solid mechanics.

The fundamental differences between a fluid and a solid are illustrated in Figure 1.3. Figure 1.3(a) shows a fluid with both viscosity and elasticity subjected to forces on its bound-ary. No matter what these external forces are the fluid, given a relatively short time, experi-ences only the forces normal to the boundary and these forces generate a hydrostatic pressure, P, within the fluid; P depends on the density of the fluid so that if the material is

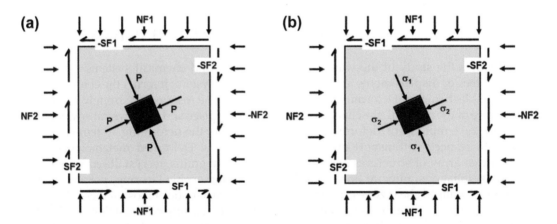

FIGURE 1.3   The difference between a fluid (a) and a solid (b). In both (a) and (b) the external force field is identical with normal forces NF1 and NF2 and shear forces SF1 and SF2. For any orientation of a fluid element in (a) the element feels only the hydrostatic pressure P. For a particular orientation of the solid element in (b) the element feels unequal normal stresses σ_1 and σ_2 and the pressure within the element is defined as the mean of these normal stresses.

compressible, P will be different for different values of the elastic bulk modulus. No matter what the orientation of a volume element within the fluid it feels only this hydrostatic pressure on its faces and the magnitude of P is fixed by the magnitudes of the normal forces on the boundaries of the fluid and, once equilibrium has been achieved, is independent of the fluid viscosity. We will see how to calculate P in Chapter 6. Figure 1.3(b) shows an identical situation for a solid. The solid feels both shear forces and normal forces and the stress state within the body reflects this situation. For a particular orientation of an element within the solid the hydrostatic pressure is now replaced by unequal normal stresses (forces per unit area) labelled σ_1 and σ_2. The orientations and magnitudes of σ_1 and σ_2 are defined by the mechanical properties of the solid and are different for different solids. The pressure within the element is the mean of the stresses, $-\frac{1}{2}(\sigma_1 + \sigma_2)$, and so also depends on the mechanical properties of the solid, being different for different solids. For the solid in Figure 1.3(b) deformation is driven locally by the stress difference, $(\sigma_1 - \sigma_2)$. Elastic solids and fluids always deform elastically even for small boundary forces but elastic–plastic solids only deform inelastically (that is, permanently) once a critical stress (the yield stress) is exceeded. Thus for a fluid the pressure within the fluid is imposed by the *external* normal forces; for a solid the pressure is defined *internally* by the mean stress.

Processes of deformation: As we have seen the *processes* that operate during deformation and metamorphism fall into five interrelated categories: (i) mechanical processes; (ii) chemical reactions; (iii) fluid flow; (iv) transport of heat by conduction or by advection; and (v) structural adjustments at the subgrain or grain scale.

Mechanisms of deformation: The term *mechanism* is used here as a subset of the term *process* and the distinction between the terms is to some extent scale dependent. As an example, a deformation process may be *solid state diffusion* but the mechanisms involved may be vacancy motion, interstitial motion or some more complicated set of cooperative atomic motions. Similarly the deformation process might be *dislocation creep* but the mechanisms involved may be dislocation glide, athermal kink drift, climb, interaction of individual dislocations with the forest dislocations or more complicated dislocation, dislocation array, impurity or vacancy interactions. Or the deformation process may be *crack propagation* where the deformation mechanism is diffusion of impurity atoms to the crack tip and the accumulation of dislocations. The term *deformation process* is a more general term than *deformation mechanism* and a number of deformation mechanisms may act together to produce a given deformation process. The deformation process generally operates at larger length scales than the contributing deformation mechanisms. Similarly a chemical reaction may be written as a process, $A + B \rightarrow C$, but may consist of a number of coupled mechanisms such as $A \rightarrow X$, $B \rightarrow 2Y$, $X + 2Y \rightarrow C$. Again, at a length scale large with respect to the pore scale, fluid flow in a porous medium may be described by a process governed by Darcy's law but at the pore scale the mechanism is described by Stokes' law (Coussy, 2010; Phillips, 1990).

For the most part little direct knowledge is available on the mechanisms that operate during the deformation and metamorphism of rocks. Analogues are continuously drawn with what is known in metals and physical chemistry but the necessity for experiments at elevated temperatures and pressures makes experimental observations on mechanisms within mineral systems very difficult. If one adds the additional complexity of making observations at geological strain rates then the problem becomes intractable. We can of course make experimental observations on the processes involved and try to extrapolate this information to geological conditions but in general the direct establishment of mechanisms is very difficult. A way forward that is growing in momentum is the computer simulation of mechanisms and the establishment of the energy necessary for that mechanism to operate under geological conditions. Examples include pioneer work by Lasaga (Xiao and Lasaga, 1996) on the dissolution of quartz, models of the molecular structure of water at quartz–quartz grain boundaries (Adeagbo et al., 2008), models of the dislocation structure of olivine (Castelnau et al., 2010), models of the dissolution of carbonates (Gale et al., 2010), models of the influence of water on the stress solubility of quartz (Zhu et al., 2005) and models of 'power-law creep' of MgO at geological strain rates and mantle pressure–temperature (P–T) conditions (Amodeo et al., 2011; Cordier et al., 2012). These studies are shedding new light on the mechanisms of deformation and metamorphism. We present two examples here, namely, that of the stress solubility of quartz in the presence of H_2O (Zhu et al., 2005; Figure 1.4) and the development of dislocation structures associated with crack development in a face-centred cubic (FCC) crystal (Abraham, 2003; Figure 1.5)

1.3 LINEAR AND NONLINEAR PROCESSES

A stroll through any deformed metamorphic terrain will soon convince an observer that there is a bewildering array of structures developed in such rocks ranging from folds with an endless variety of shapes and sizes to layering and lineations developed solely by metamorphic processes to fractures and veins that pre- and postdate the metamorphism (Figure 1.6).

In addition a large array of different metamorphic mineral fabrics exists including simple foliations, crenulated foliations, porphyroblasts and foliations/lineations defined by melting processes. Relevant questions are: *Are there guiding principles behind all this complexity or does each structure form independently of each other? If there are guiding principles what are they and what causes the complexity?* The aim of this book is to explore these questions and to propose a systematic basis for the complexity. Part of the answer to these questions lies in the nonlinear response of deforming chemically reacting solids to the deformation conditions and it is important from the outset to understand the implications of such nonlinear behaviour.

A linear system is one that behaves in such a way that the response is proportional to the input. This means that small changes to the input result in small changes to the response. Examples are linear (Hookean) springs that extend in proportion to the load on the spring, linear (Newtonian) fluids that flow at a rate that is proportional to the instantaneous fluid pressure gradient and simple uncoupled isothermal chemical reactions such as $A \rightarrow B$ where the reaction rate is proportional to the *affinity* or driving force for the reaction. One of the characteristics of such linear systems is *stability* which means that if we perturb the system by a small amount from its current position it will return to its initial configuration, although it may undergo some oscillatory motion as it returns. Thus if we pull on the weight at the end of a linear spring and then release the weight the system will oscillate but return to its initial position. This initial position is an equilibrium state where no motion occurs. Such systems possess only one equilibrium state and that state is stable.

For linear systems the mathematical *principle of superposition* holds (Boyce and DiPrima, 2005). The system involved may be a deforming mechanical system or a chemical system

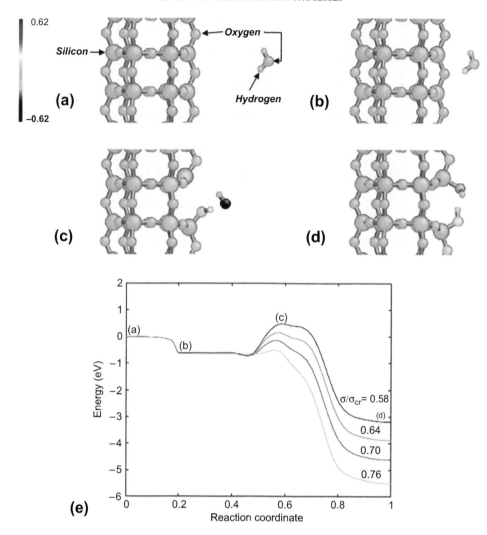

FIGURE 1.4 Molecular simulation of the mechanisms involved in the stress solubility of quartz corresponding to a stress $\sigma/\sigma_{cr} = 0.58$ where σ_{cr} is the maximum stress at fracture for a pure quartz rod (*After Zhu et al. (2005)*). (a) A rod of α-quartz in the absence of H_2O. Medium-sized spheres are oxygen and large spheres are silicon. An H_2O molecule is nearby with hydrogen atoms denoted by the small spheres. Corresponding energy is shown in (e). (b) The H_2O molecule approaches the SiO_2 rod. Corresponding energy is shown in (e). (c) Saddle point configuration where the energy is a maximum. Corresponding energy is shown in (e). (d) Final chemisorbed state. Corresponding energy is shown in (e). (e) Energy versus the reaction coordinate at high and low stresses measured by the ratio σ/σ_{cr}. Atoms are colour coded by charge variation relative to the initial configuration (a).

undergoing a simple uncoupled reaction. This principle says that if functions f_1 and f_2 are two possible solutions describing the behaviour of such a system then $\alpha f_1 + \beta f_2$ is also a solution where α and β are arbitrary constants. Thus if $f_1 = \sin(x)$ and $f_2 = \sin(2x)$ then the principle of superposition says that another solution describing the behaviour is $f_3 = \sin(x) + 2\sin(2x)$ as

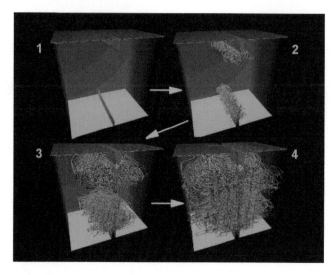

FIGURE 1.5   Computer simulated time sequence of the propagation of dislocations from opposing crack tips. *From Abraham (2003).*

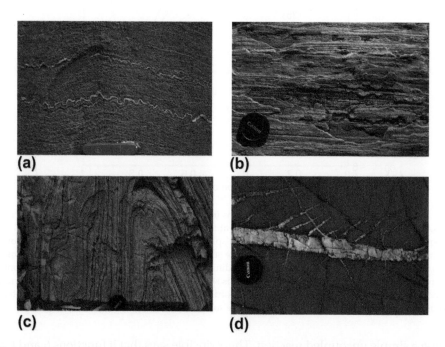

FIGURE 1.6   Typical structures in deformed rocks. (a) Single layer localised folding in schists from Cap de Creus, Spain. (b) Mineral lineation approximately parallel to bedding/foliation intersection in banded iron formations, Tropicana, Western Australia. (c) Folding in Chewings Range quartzites, Alice Springs, Australia. (d) Quartz veins in deformed turbidites, Bermagui, Australia.

shown in Figure 1.7(c); this behaviour is periodic. If the ratio β/α is irrational then the behaviour is *quasi-periodic* as shown in Figure 1.7(e) where $\frac{\beta}{\alpha} = \sqrt{2}$. This means that in a linear system solutions describing the behaviour of the system evolve independently of each other with no mechanical or chemical interference between each other and the final solution can be expressed as a Fourier series of the various independent solutions. One might expect that this could lead to quite complicated geometrical or chemical patterns but for many mechanical and chemical systems of interest it turns out that one solution grows much faster than all others so that a dominant solution appears finally as the only (sinusoidal) solution. This is the process that operates in the classical Biot theory of folding (Biot, 1965) which is rigorously true in two dimensions for a layer composed of any material embedded in a linear matrix. An infinite number of wavelengths start to grow but one grows fastest to become the dominant sinusoidal wavelength.

The behaviour of nonlinear systems, however, is quite complicated and at present there is no comprehensive overarching theory that enables general principles to be stated. We will explore many aspects of nonlinear behaviour in the remainder of this book. However, as an example the buckling behaviour of a single layer embedded in nonlinear material is summarised in Figure 1.8. Here we see that the response may be homogeneous, with no buckling behaviour, or a range of buckling behaviours depending sensitively on initial conditions, on

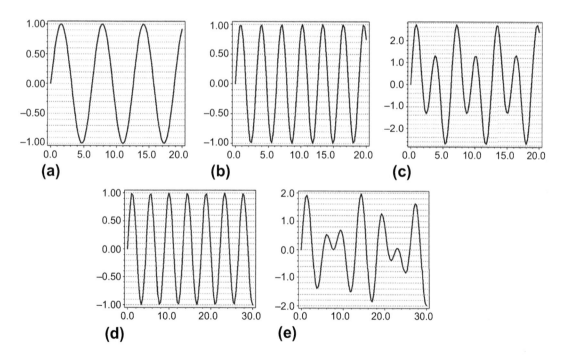

FIGURE 1.7   The principle of superposition. If $f_1 = \sin(x)$ shown in (a) is a function that describes the evolution of some linear system and so also is $f_2 = \sin(2x)$ shown in (b), then another solution is a linear combination of these two solutions: $f_3 = \sin(x) + 2\sin(2x)$ shown in (c); this solution is periodic but not sinusoidal. A quasi-periodic solution is shown in (e) which is given by $f_5 = \sin(x) + \sin(\sqrt{2}x)$. $f_4 = \sin(\sqrt{2}x)$ is shown in (d). A quasi-periodic function is close to periodic but never repeats itself.

FIGURE 1.8 The various
responses to the deformation of
a layer embedded in a
nonlinear material. The homogeneous and sinusoidal responses are characteristic of
linear embedding materials.
The periodic unlocalised
response is characteristic of
linear embedding materials for
very small deformations; such
structures ultimately evolve to
a sinusoidal response. See Volume II for detailed discussions.

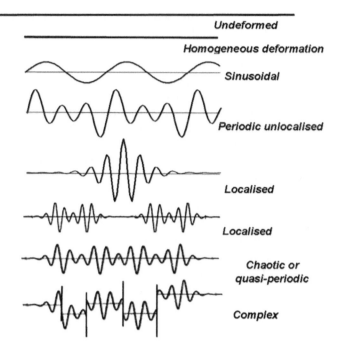

the boundary conditions and on the evolution of geometry and mechanical properties of both the layer and the embedding material during buckling. The inhomogeneous responses may be sinusoidal, periodic but nonlocalised, localised and periodic, localised and non-periodic and quasi-periodic or chaotic. The localised packages are also referred to as solitary waves. These various forms of buckling behaviour are discussed in Volume II.

The responses illustrated here are characteristic of many nonlinear systems including nonlinear chemical systems, shear zone localisation and the development of more permeable pathways in a reacting porous medium with fluid flow. The development of homogeneous, sinusoidal and nonlocalised periodic behaviour is characteristic of many linear systems. We discuss the details of these behaviours in later sections of this book. The essential differences in behaviour between linear and nonlinear systems are summarised in Table 1.1. It is impor-tant to note that the responses for both linear and nonlinear systems are characteristic of sit-uations where there is an initial random distribution of small-amplitude geometrical, physical or chemical heterogeneities. Larger heterogeneities influence the final result for nonlinear systems but are not necessary to produce localised or aperiodic behaviour. The in-fluence of initial imperfections is discussed in Volume II. The localised and chaotic response of nonlinear systems derives solely from nonlinear geometrical, physical and/or chemical interactions.

The nonlinear behaviour may be material or geometrical in nature and in this introductory discussion we concentrate on geometrical nonlinearity and give two examples to emphasise that even the simplest of mechanical behaviour coupled with simple geometrical nonline-arity can lead to localised and nonintuitive behaviour. A more extensive discussion is

TABLE 1.1 The Different Behaviours of Linear and Nonlinear Systems

Linear Systems	Nonlinear Systems
Linear systems obey the superposition principle.	Nonlinear systems do not obey the superposition principle.
Response is homogeneous or sinusoidal or a linear combination of sinusoidal responses.	Response can be homogeneous, sinusoidal, periodic, localised, quasi-periodic, aperiodic or chaotic.
Individual wavelengths grow independently of each other with no physical or chemical interference.	Individual wavelengths interfere with each other physically and/or chemically.
Spatial patterns grow simultaneously throughout the system.	Spatial patterns can be localised both in space and time and develop sequentially; some system responses may be chaotic.
There is a wavelength selection process so that a limited number of wavelengths (commonly just one) eventually grow to dominate the spatial pattern. In many systems there is a single dominant wavelength characterising the final pattern.	No wavelength selection process operates and localised responses grow and/or decay at various places during the evolution of the system. In some systems the response is a sequential development of elements of the pattern. The concept of a dominant wavelength does not exist.
The dominant ultimate spatial pattern is sinusoidal or, for some systems, periodic with two wavelengths.	The system may evolve through a series of different spatial responses. In some systems the initial response is sinusoidal only to be replaced with ever more localised or chaotic patterns; in other systems the initial response may be localised only to grow into a sinusoidal pattern.
In general there is only one solution that describes the evolution of the system. This can commonly be obtained by a linear perturbation analysis. An analytical solution commonly exists.	In general there are many, if not an infinite, number of solutions that describe the evolution of the system. Analytical solutions commonly do not exist. A sinusoidal solution might be obtained by linearization procedures but it is not unique and commonly is not stable. Chaotic and fractal geometries are commonplace.

presented in Chapter 7. As a first example, slight complications in the geometrical arrangements of systems of linear springs can soon lead to complexity. Consider the system shown in Figure 1.9 where two linear springs are arranged initially as shown in Figure 1.9(a). A displacement under the influence of a vertical force leads to the configuration shown in Figure 1.9(b). In doing so the force increases as shown in Figure 1.9(d) from O through D to A. However, this configuration is unstable; a further small displacement results in a jump to the configuration shown in Figure 1.9(c). In doing so the displacement jumps from A to B in Figure 1.9(d). This behaviour is referred to as *snap-through instability*. Continued downward displacement continues on the path indicated by the black arrow from B. In reverse, indicated by the blue arrows in Figure 1.9(d) the system unloads to C and then jumps to D to continue unloading back to the initial configuration.

In the force–displacement diagram in Figure 1.9(d) the branches OA and BC represent stable configurations such that a small increase or decrease in force results in a small change in

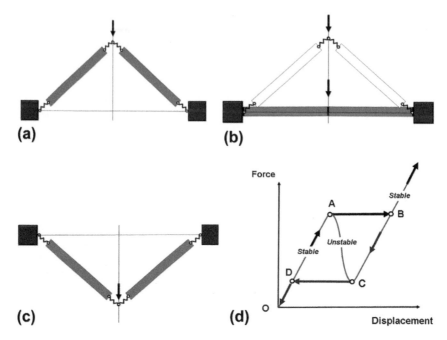

FIGURE 1.9   A simple system of linear springs that displays nonlinear behaviour arising from changes in geometry under force-controlled loading. (a) Initial unloaded condition. This corresponds to the point O in (d). (b) An
unstable configuration corresponding to point A in (d). (c) A new stable configuration that the system jumps to after
(b). This corresponds to point B in (d). (d) Force–displacement diagram showing a stable loading path from O to A,
an unstable path which is never accessed from A to C and a stable path corresponding to a loading path past B and
an unloading path BC. On loading the system follows the path OAB and beyond represented by the black arrows
and on unloading follows the path BCDO represented by the blue arrows. The fact that the loading and unloading
paths are different is referred to as hysteresis. If the same system is subjected to constant displacement rate loading
conditions the system follows the complete force–displacement curve. That is, it follows the unstable branch as well
as the stable branches.

displacement. The branch AC is unstable in that a small change in force can result in a large
displacement. The system behaves differently under loading and unloading conditions
resulting in what is called *hysteresis*. This behaviour results in an otherwise linear system
becoming unstable when a critical geometrical configuration is reached. Hysteresis is typical
of many systems, particularly mechanical systems where some mechanical parameter softens
due to thermal or chemical feedback, of many chemical reactions such as a simple exothermic
reaction, A → B, and of many chemically reacting systems with open fluid flow. One should
appreciate that the above scenario is true for force boundary conditions. If the system is
loaded at constant velocity the complete force–displacement curve is followed including
the unstable branch with no snap-through instability.

As a second example consider a number of balls arranged in layers and connected by
linear springs as shown in Figure 1.10. The packing of the balls is initially close. The nonlinearity in this system is geometrical and results solely from the fact that, during deformation,
balls need to ride over each other, producing a dilation, especially as the rotation of an

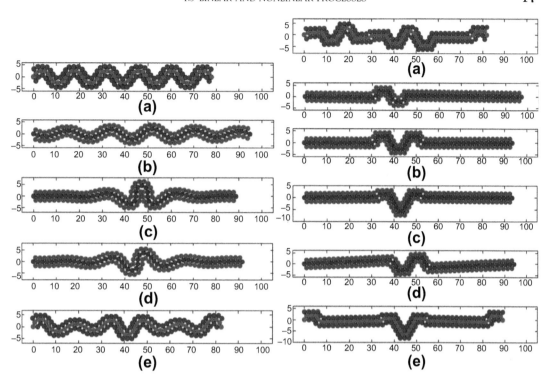

FIGURE 1.10 A three-layer system consisting of initially close packed balls connected by linear springs. The nonlinearity in the system arises from large rotations of groups of balls. In the left-hand panel no work is done against dilation as the balls slide over each other. In the right-hand panel work is done against dilation and hence dilation is minimised and the initial close packing tends to be preserved. Dilation sites are represented by small white circles. (a) One of the two initial solutions which evolves to become (e). (b) The second kind of initial solution which evolves to become (c) and (d). *From Hunt and Hammond (2012).*

individual chain of balls increases. In the two panels shown in Figure 1.10 no energy is associated with the dilation in the left-hand panel, whereas an energy penalty is associated with dilation in the right-hand panel. By *penalty* we mean that some work needs to be done for the material to dilate. The initial length of the system before deformation in both cases is 100 units.

For the left-hand panel of Figure 1.10 even the initial response is complex in that at least two initial solutions are possible. The first shown in (a) is sinusoidal but progresses to strong aperiodic localisation shown in (e). A second solution is shown in (b) which occurs at a slightly lower initial stress than (a), and consists of weak localisation which intensifies by (c) and progressively becomes aperiodic, (d), as deformation proceeds. The lack of any penalty for the formation of voids means that local departures from the initial close packing are common. This enables a rounded waveform to develop so that the fold style could be labelled *concentric*. Differences in the pattern of dilation are responsible for the two different initial responses shown in (a) and (b) in the right and left panels.

In the right-hand panel of Figure 1.10, the response is localised from the start of deformation and the fold style is *kink* or *chevron* like. Again there are two initial solutions (a) and (b).

Solution (a) evolves to become (e), whereas (b) evolves through the stages (c) and (d). The penalty imposed by introducing the work due to dilation forces the packing to remain close packed even for large rotations and any necessary dilation is localised. The result is kink or chevron styles of folds. There is an added element of complexity here in that the width of kink bands can evolve from even numbers of balls to odd numbers of balls or vice versa in quite a complex manner (Hunt and Hammond, 2012). Such behaviour is a common feature of nonlinear systems and will be discussed in greater detail in Chapter 7.

1.4 WAVELET ANALYSIS

Since many deformed rocks have structures that are developed over a range of length scales and (we will show) fractal geometries are ubiquitous, it is natural to use a technology that takes such characteristics into account when attempting to quantify the fabrics of deformed rocks and the associated patterns (and defects in the patterns) that are characteristic of these fabrics. In this book and Volume II we use *wavelet transforms* as a convenient tool to achieve quantification of all deformation/metamorphic fabrics and patterns including the distribution of mineral phases, the CPOs developed during deformation, the shapes of folds at all scales and the geometry of fracture and vein networks. A wavelet transform consists of two parts. The so-called *mother wavelet* is a mathematical function, usually of a wave shape, that is commonly localised in space and designed in order to emphasise some aspect of the pattern being quantified. The second part of the transform consists of scanning the pattern with this function as it is stretched or compressed to become *daughter wavelets*. The amount of stretch or compression is the scaling factor. At each point in the fabric the degree of match between some measure of the fabric with the stretched or compressed wavelet is recorded in the form of a series of coefficients.

The wavelet transform can be viewed as a microscope that enables one to probe a pattern at various scales and hence has a close relationship with fractal analysis methods such as box counting (Feder, 1988) since one can think of the wavelet acting as a 'generalised box' (Arneodo et al., 1995). The advantages of wavelet analysis over other forms of analysis are that details of the pattern distribution are available at every point in the data set and one can interrogate the wavelet transform to see if the pattern is fractal or multifractal (Ott, 1993) and characterise the fractal dimension or the multifractal spectrum of the fabric. The construction of the mother wavelet can be put on a thermodynamic basis (Arneodo et al., 1995) and so designed with the physical and/or chemical processes responsible for the pattern in mind. It is also possible to use the wavelet analysis to solve the inverse fractal problem that gives insight into any multiplicative process that was responsible for the pattern formation.

We leave detailed discussions of the use of wavelets to Chapters 7 and 13 and especially Volume II where the technique is used to characterise a number of fabrics in deformed metamorphic rocks. For the moment we present an example of the application of a wavelet transform to a spatial CPO pattern in a deformed quartzite given in Figure 1.11. A map of the spatial distribution of c-axes in quartz is shown in Figure 1.11(a); also shown are traverse lines to be analysed across the map parallel and normal to the main foliation. Figure 1.11(b and c, upper) show the data extracted from these traverses which consist of the colour at each pixel expressed as a spectrum ranging from 1 to 255. These data sets now contain all of the

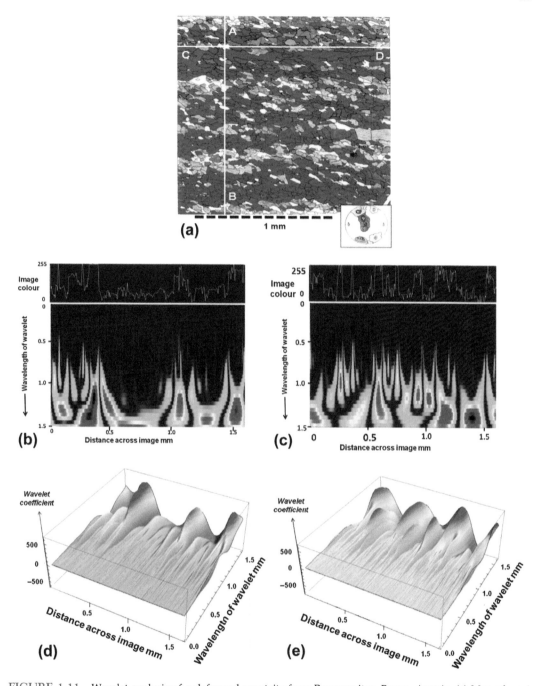

FIGURE 1.11   Wavelet analysis of a deformed quartzite from Rensenspitze, Bozen, Austria. (a) Map of c-axis orientations in quartz. From Sander (1950). Wavelet analyses are presented along the traverses A−B (normal to the lineation and to the foliation) and C−D (parallel to the foliation and lineation). (b) Wavelet scalogram (below) with raw data above. Traverse line is C−D. (c) Wavelet scalogram (below) with raw data above. Traverse line is A−B. (d) Three-dimensional representation of the scalogram in (b). (e) Three-dimensional representation of the scalogram in (c).

information on grain boundary positions and c-axis distributions within the selected traverses. Figure 1.11(b and c, lower) show what are called *scalograms* for these traverses and express the degree of match between the wavelet at various scales. Figure 1.11(d and e) are three-dimensional renditions of the scalograms.

One can see that patterns of correlation exist at scales far larger than the grain size ($\approx 10^{-1}$ mm). For the traverse normal to the foliation this can be related to the foliation but for the traverse parallel to the foliation correlations at wavelengths of 1–1.5 mm are unexpected. We revisit this example and others in Chapter 13 and Volume II but for now we can see that wavelet analysis is a convenient (and fast) way of quantifying fabric geometry.

1.5 NOTATION, CONVENTIONS AND UNITS

We define symbols that are commonly used in this book in Table A1 in Appendix A and when first introduced in the text. Appendix B contains a brief introduction to vectors and tensors. Cartesian coordinates are used throughout and are represented by italic symbols; both the bold tensor notation x is used as well as the component notation x_i. We use the convention that scalar quantities are represented in non-bold font: a, T, Ψ; vectors, tensors and matrices are represented by bold font: $\mathbf{J}$, $\boldsymbol{\sigma}$, $\mathbf{A}$ (the direct notation) or where convenient by component (or indicial) notation, J_i, σ_{ij}, A_{ij}. Vector, tensor and matrix Cartesian components are represented using indices ranging from 1 to 3: J_i for vectors and σ_{ij} for second-order tensors. The Einstein summation convention (Nye, 1957, p. 7) is used so that repeated indices mean summation on that index unless otherwise stated. The various tensor operations are as follows: (i) $\mathbf{a} \cdot \mathbf{b}$ is the scalar product of two vectors $\mathbf{a}$ and $\mathbf{b}$; (ii) $\mathbf{a} \times \mathbf{b}$ is the vector product of two vectors $\mathbf{a}$ and $\mathbf{b}$; (iii) $\mathbf{a}{:}\mathbf{b}$ is the scalar product of two second-order tensors a_{ij} and b_{ij} and is equivalent to $a_{ij}b_{ij}$; (iv) $\mathbf{a} \otimes \mathbf{b}$ is the tensor product of two vectors $\mathbf{a}$ and $\mathbf{b}$ so that

$$\mathbf{a} \otimes \mathbf{b} = \begin{bmatrix} a_1b_1 & a_1b_2 & a_1b_3 \\ a_2b_1 & a_2b_2 & a_2b_3 \\ a_3b_1 & a_3b_2 & a_3b_3 \end{bmatrix} \tag{1.1}$$

The tensor product of two vectors is also known as the dyadic product; the result of the $\otimes$ operation involving two vectors is a rank-1 matrix that transforms as a second-order tensor. This product is important in discussions of crystal plasticity and particularly in establishing compatibility of deformations between two adjacent domains deformed in different manners. Chemical components are distinguished by Arial font: SiO$_2$, A, B, etc. We refer to figures by the total reference 'Figure 1.6', whereas equations are referred to simply by the relevant number in parentheses '(1.6)'; renumbering of both figures and equations begins at the start of each chapter. Sections in the text are referred to as 'Section 1.6'. SI units are used throughout.

Recommended Additional Reading

Cross, M., Greenside, H. (2009). *Pattern Formation and Dynamics in Nonequilibrium Systems.* Cambridge University Press, 535 pp.
   This book covers the basic principles of nonlinear dynamics with an emphasis on spatial pattern formation. The link to systems not at equilibrium is clearly made.

Nye, J.F. (1957). *Physical Properties of Crystals.* Oxford Press, 322 pp.

This book is an excellent introduction to tensors and the Einstein summation convention. Chapters on strain and stress are included. Chapter 10 treats the thermodynamics of equilibrium properties of crystals with coupling between elasticity and thermal effects in particular. The book contains numerous examples of coupling between tensors of different ranks. These are examples of experimentally well-established effects said to be invalid by proponents of the so-called Curie 'principle'.

Rice, J. (1993). *Mechanics of Solids.* In: *Encyclopaedia Britannica,* fifteenth ed., vol. 23, 734–747 and 773.

This article explains the origins and basic principles of solid mechanics and has an excellent summary of the history of the subject with applications.

Tadmor, E.B., Miller, R.E. (2011). *Modelling Materials: Continuum, Atomistic and Multiscale Techniques.* Cambridge University Press.

This book treats the latest developments in the modelling of solids at the atomic scale. A clear link is made to the resulting mechanical properties and behaviour. The techniques involved in multiscale approaches are considered in depth.

Truesdell, C. (1966). *Six Lectures on Modern Natural Philosophy.* Springer-Verlag.

This is an entertaining discourse on mechanics and the way in which thermodynamics is integrated with mechanics. The text contains some interesting discussions on the approaches to thermodynamics that assume linear systems.

The Mechanics of Deforming Solids: Overview of Section A

The subject called *mechanics* is the study of the response of materials to applied forces or applied displacements on the boundaries of the material. Traditionally mechanics has been divided into the *mechanics of fluids* and the *mechanics of solids* with somewhat different mathematical treatments and concepts involved in both. We delay the discussion of the features that distinguish a solid from a fluid to Chapter 6 but our emphasis in this book is on *solid mechanics*. The response referred to is commonly called *deformation* and depends on many environmental factors such as temperature, the rate at which the boundaries of the solid are loaded or displaced, the chemical composition and structure of the solid and its fluid content, and the nature of any chemical reactions that are in progress during the deformation. This book attempts to integrate all of these factors but in Section A, we discuss mostly the physical, as opposed to the chemical, factors involved in the deformation of metamorphic rocks.

Section A of this book is intended to provide the concepts and language of mechanics used to describe the processes involved in deformation and the consequent development of metamorphic fabrics. The language is one of mathematics and our intent is not to emphasise mathematical rigour but to concentrate on giving the various concepts physical meaning of use to metamorphic geologists. Reference is made throughout to more detailed mathematical treatments of the material if the reader intends to delve deeper into the subject.

The fundamental assumptions made in the mechanics of solids and important definitions arising from these assumptions are sixfold:

1. A body of rock, B, exists that is the subject of interest. It may be a micron or many kilometres in size and consists of material particles whose Cartesian coordinates in some reference state are X_i and in a deformed state are x_i where i takes on the values 1, 2 and 3. A *deformation* is described by the *geometrical transformation* that links X_i and x_i.
2. The body B has a mass m and is subjected to external forces (tractions), T, and/or *displacements*, $\mathbf{u}$, on its boundary, S. These boundary conditions induce a system of forces within the body that cannot be established without extra information that is discussed below. The system of forces at each point in the body is measured by a quantity called the stress, σ, which has the units force per unit area. Also an amount of *heat*, q, can be generated within the body by deformation, chemical reactions or other processes such as microstructural rearrangements. In addition an amount of heat q_e can be added or removed from the body by external processes. These processes that produce heat are called *dissipative processes*, and the process of producing heat is called *dissipation*. By definition, if the deformation involves dissipation, the system is not at equilibrium.

3. The laws of thermodynamics are assumed to apply to dissipative processes in keeping with day-to-day experience. These laws are as follows:

4. The zeroth law of thermodynamics is expressed as: *the concept of an absolute temperature scale exists where the absolute temperature is* T *and* T *is always greater than zero.* Once the temperature is defined another useful quantity, the *entropy*, s, is defined at each point within B. The entropy production is defined as the ratio of the rate of heat production at that point divided by the local temperature at that point. The entropy production at a point times the temperature at that point is the *dissipation* at that point. The units of entropy production are Joules per Kelvin per second.

5. The first law of thermodynamics is expressed as: *the body* B *possesses an internal energy, e, which is comprised of the sum of all the energies (dissipative and non-dissipative) derived from the work done by processes operating in* B.

6. The second law of thermodynamics is expressed as: *The total dissipation in the body comprises the sum of the dissipations arising from the individual processes operating within and upon the body. This total dissipation is greater or equal to zero.* If the total dissipation is equal to zero the body is at *equilibrium.*

These six assumptions and the associated quantities that arise from these assumptions are sufficient to describe the evolution of geological bodies as they are deformed (by all mechanisms including plastic flow and fracturing), subjected to fluid infiltration and undergo chemical reactions. The assumptions amount to proposing that the entities mass, force, heat, temperature, entropy and internal energy exist for a given body and need no further definition. In addition the second law of thermodynamics proposes that the dissipation is always greater than zero for a body not at equilibrium.

However one other piece of information is required which is not an assumption and is based on the observation that bodies comprised of different materials, when subjected to identical boundary conditions, behave in different manners. The extra information consists of a description of the relation between the history of deformation and the stress developed within the body and depends on the material and the structure of the material making up B. This description is called a *constitutive relation*. It is also observed that the same materials when subjected to different boundary conditions behave in different manners even though the final deformed shape of the body may be identical. Thus the stresses that develop in a particular material depend not only on the constitutive relation for that material but also on the history of the motions that it has experienced.

In order to flesh out these concepts in greater depth we first in Chapter 2 discuss the changes in the geometry of a body as it is subjected to external influences. These changes are called *deformations*. The concept of a deformation does not consider the history of the particle movements or how the body evolves from one shape to another during the deformation history. For many situations of interest, it is convenient to select some reference state and refer the deformed geometry to that reference state. A deformation involves only the geometrical relation or transformation between the reference state and the deformed state. In Chapter 3 the motions involved during the deformation history are considered. This constitutes the subject called *kinematics*. In Chapter 4 the system of forces developed within the deforming body are considered. This constitutes the subject known as *dynamics*. In Chapter 5 the energy and energy transfers involved in a deformation and associated processes are considered. This constitutes the subject called *thermodynamics*. An important aspect here is the notion that many structures we observe in deformed rocks originate to minimise the deformation induced energy of the system. In Chapter 6 some common constitutive relations for deforming solids are considered. This involves considerations of solids whose mechanical properties are strongly influenced by the rate of deformation as well as those insensitive to the rate of deformation. Chapter 7 considers the different behaviours that result from *nonlinear* as opposed to *linear systems*. This constitutes an introduction to the *nonlinear dynamics* of metamorphic rocks and considers the principles behind the development of the complex structural geometries and fabrics that characterise metamorphic rocks and that make such rocks such fascinating objects to study.

Geometry: The Concept of Deformation

Structural Geology
http://dx.doi.org/10.1016/B978-0-12-407820-8.00002-3

2.1 DEFORMATIONS

The *deformation* of rocks involves the relative *motion* of elements of the rock so that new arrangements of these elements develop. The motions comprise *rigid rotations, rigid translations* and *distortions* (changes in shape) of the elements. The motions are a function of time but it is convenient as a first step to neglect time and consider only the deformed configuration relative to what we imagine to be the undeformed or some reference configuration. The relation between these two configurations is called *deformation*. If the deformation is the same everywhere (that is, *homogeneous*) then it does not really matter if one considers the rotations and translations since they make little difference to what we observe *within* the rock in the deformed state. However, if the deformation varies from place to place (that is, the deformation is *inhomogeneous*) as in folded rocks or rocks with localised shear zones then the local rotations and translations need to be considered since these local adjustments enable variously distorted regions to fit together with no gaps. In much of the geological literature to date the discussion has concerned homogeneous deformations and so the emphasis has been on just the distortional part of the deformation. This is commonly called *strain*.

Thus the study of strain in deformed rocks has occupied much of the literature on geological deformations over the past 45 years or so since the publication of Ramsay's classical book in 1967 (Ramsay, 1967). In emphasising the mechanics of metamorphic rocks it will become apparent that the total deformation picture as described by local rotations, displacements and strains becomes important. For instance, in describing the motions that are developed in rocks by imposed forces the strain plays little role. To paraphrase Truesdell and Toupin (1960, p 233) regarding the behaviours involved in some familiar constitutive relations:

- If the distances between particles do not change this is a *rigid body*.
- If the stress is hydrostatic this is a *perfect fluid*.
- If the stress may be obtained from the rate of stretch alone this is a *viscous fluid* or a *perfectly plastic body*.
- If the stress may be obtained from the strain alone this is a *perfectly elastic body*.

Thus, as far as the mechanics of deformation is concerned, it is only in elastic theory that the strain becomes important. For the constitutive relations that describe the behaviour of deforming rocks (combinations of elasticity, plasticity and viscosity) some measure of the rate of deformation (not the rate of strain) becomes important and much of the development in that area is based on what is defined below as the *deformation gradient*. The strain is a quantity that accumulates during the deformation history; it is not a quantity that controls the behaviour of materials (other than perfectly elastic materials). Hence in this chapter we assume that much of the development in the analysis of strain in deformed rocks is already covered by books such as Ramsay (1967), Means (1976) and Ramsay and Huber (1983) and we concentrate on the subject of *deformation*.

It is also commonplace in the geological literature to begin the study of deformation with *infinitesimal strain theory* and work progressively towards *finite strain theory* (Hobbs et al., 1976; Jaeger, 1969; Means, 1976; Ramsay, 1967). There are two aspects to this approach: (1) The interest in infinitesimal strain theory is based on classical engineering approaches where small deformations of elastic materials were once the common-place application. (2) The approach attempts to progress from the particular to the general situation. First, the interesting deformations in geology are finite in scale and the relations that describe how a material behaves under the influence of imposed forces or displacements involve quantities related to the deformation rather than to the strain so we begin our approach with this in mind. Second, it sometimes is difficult to generalise from the particular to the general; this we will see is particularly relevant in thermodynamics (Chapter 5) where equilibrium thermodynamics appears as a special case of generalised thermodynamics. Interesting deformations in geology are commonly finite and it turns out that a logarithmic measure of deformation is more useful and rigorous than classical infinitesimal measures of strain, although more difficult to calculate. In this chapter infinitesimal strain theory appears as a final approximation only sometimes relevant to the behaviour of geological materials and in some cases (as in the theory of buckling) does not incorporate the geometry of the deforming process.

We consider a body of rock $\mathcal{B}$, with a bounding surface $\mathcal{S}$, and seek to describe the deformation of this body as $\mathcal{S}$ is loaded by forces or is displaced in some manner (Figure 2.1). The subject of *deformation* is commonly dealt with under the heading of *kinematics* in the mechanics literature even though the word *kinematics* carries with it the connotation of *movement* or *motion*. To a structural geologist the distinction between geometry and movement is paramount since the geometry of the deformed state is commonly the only feature that can be directly observed. Hence we prefer to divide the subject into two distinct areas of study, namely, *deformation*, the subject of this chapter and concerned with the *geometrical* features that can be observed in deformed rocks, and *kinematics* (the subject of Chapter 3), to do with the *movements* that a rock mass has experienced. As so defined we emphasise that the subject of deformation is concerned only with the geometrical relations between a deformed

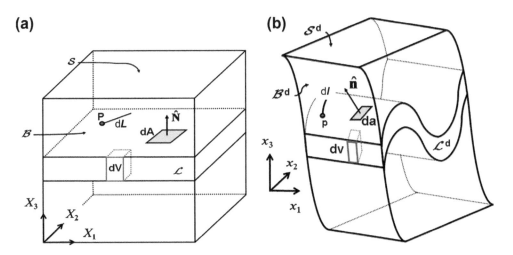

FIGURE 2.1 Reference and deformed states. (a) Reference state for the body $\mathcal{B}$ with bounding surface $\mathcal{S}$. The coordinate axes are X_1, X_2, X_3 and we consider a layer $\mathcal{L}$. (b) The deformed state with coordinate axes, x_1, x_2, x_3. The body has been inhomogeneously deformed to become $\mathcal{B}^d$ with bounding surface $\mathcal{S}^d$. P is deformed to become p, dL is deformed to become dl, the surface area dA with normal $\hat{\mathbf{N}}$ is deformed to become da with normal $\hat{\mathbf{n}}$, the volume element, dV is deformed to dv and the layer to $\mathcal{L}^d$.

body and some reference state. It has nothing necessarily to do with the movements and stresses that the body has experienced and the histories of these quantities. In general much more than just the geometry is required in order to establish these relationships as discussed in Chapters 3 and 4. It is important to understand that simply because one can pre-scribe a particular deformation does not mean that this deformation can actually be achieved for a particular material and still satisfy the boundary conditions. In other words, the defor-mation may be *geometrically admissible* but not necessarily *dynamically admissible* (Tadmor et al., 2012, pp. 247−248).

Figure 2.1 shows a deformation where an initially planar layer, $\mathcal{L}$, is distorted to form a fold system defined by $\mathcal{L}^d$. In the reference state we identify an initial material point, P, an initial line element, dL, an initial surface element, dA, with unit normal, $\hat{N}$, and an initial vol-ume element, dV. In the deformed state these elements become p, dl, da, $\hat{n}$, and dv, respec-tively. We define *surface vectors* in the reference and deformed states as d**A** and d**a** with magnitudes, respectively, of dA and da but oriented in the directions of their unit normals, $\hat{N}$ and $\hat{n}$. Clearly material lines initially coincident with $\hat{N}$ do not remain coincident with $\hat{n}$.

A relevant geometrical question is: How do we calculate the deformed equivalents of these features, p, dl, da, $\hat{n}$ and dv for a given deformation? The answer to this question is the topic of this chapter. The chapter provides a toolbox whereby (1) the *strain* can be calculated from prescribed deformations or from three independent measurements in two dimensions or six in three dimensions; (2) the deformation due to a specified deformation mechanism can be calculated; and (3) the conditions for compatibility of *deformation* across a boundary can be calculated. Such considerations of compatibility are different from the discussion presented by Ramsay (1967) and Jaeger (1969) for compatibility of *strain* (see Section 2.10). We adopt the convention that upper case symbols such as X, dV and $\hat{N}$ refer to the reference configuration, whereas lower case symbols such as x, dv and $\hat{n}$ refer to the deformed configuration.

A *deformation* is a *geometrical transformation* that links the coordinates of material points in a reference state to those of these same material points in a deformed state; commonly the reference state is the undeformed state. Such transformations may be such as to keep all initially straight lines straight, in which case the deformation is an *affine transformation*. Other-wise it is a *non-affine transformation*. From now on we use the terms *deformation* and *transfor-mation* interchangeably. Clearly many deformations in nature are non-affine and some examples are shown in Figure 2.2.

A general deformation is defined by the set of equations:

$$x = f(X) \tag{2.1}$$

where x and X are the coordinates of a material point in the deformed and undeformed (or reference) states (Figure 2.1). If f is a linear function of X the deformation is an *affine transfor-mation* and is said to be *homogeneous*; such a deformation is shown in Figure 2.3(a, c and e). In Figure 2.3(b, d and f) we show representations of the displacements, $\mathbf{u} = x - X$, that are responsible for the deformations in Figure 2.3(a, c and d). These figures and subsequent similar figures are prepared in *Mathematica* (2011) using the *line integral convolution plot* (Cabral and Leedom, 1993) which is a graphical representation of the dispersion of a virtual tracer in the vector field of interest. Such figures show the details (especially when the dis-placements are small) of the map of the displacement field with arrows, which are tangents to the lines, showing the magnitude and direction of the displacements. The map

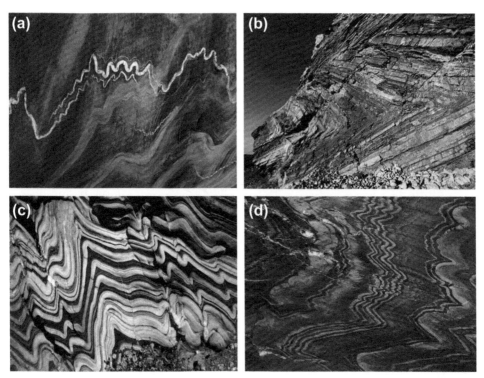

FIGURE 2.2   Some examples of non-affine deformations in metamorphic rocks. (a) Folded quartz veins crossing folded bedding from Cornwall, UK. Image approximately 0.2 m across. (b) Angular folds from Cornwall, UK. (c) Folded multi-layers from Harveys Retreat, Kangaroo Island, Australia. Scale: Outcrop is about 1 m across. (d) Folded bedding in slates from Cornwall, UK. Scale: approximately 1 m across. *(Photo: Tim Dodwell.)*

corresponds to the *material reference frame* with coordinates X. These plots resemble *phase portraits* that are widely used in dynamical systems to describe the trajectories that the system may pass through as the system evolves (Wiggins, 2003). Phase portraits are useful in describing the evolution of many nonlinear dynamical systems including evolving deforming systems (Chapter 3), many nonlinear pattern forming systems (Chapter 7), nonlinear chemical systems (Chapter 14), buckling systems involving nonlinear behaviour (Volume II) and open flow systems (Volume II). Strictly, a phase portrait can only be defined for the evolution of a system with time and figures such as Figure 2.3(b) in this chapter refer to the displacement field at an instant of time. Nevertheless the resemblance to true phase portraits is strong and the principles that govern their geometry are very similar to true dynamical systems. Hence it is instructive to introduce them here and follow with true phase portraits in later chapters.

Figure 2.3(a) shows a general *isochoric* (that is, constant volume), affine transformation in which a rigid body rotation and translation are involved as well as shortening and shearing deformations. If no rigid body rotation is involved the deformation is commonly called a *pure shear* (although *pure stretch* would be a better term) and an example is shown in Figure 2.3(c and d). If the deformation involves only shearing with no deformation normal

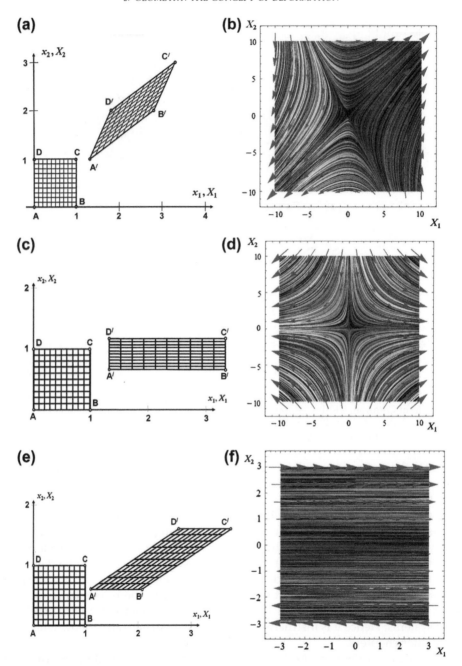

FIGURE 2.3   Undeformed and deformed states for some affine transformations. The way in which the co-ordinates of a material point $[x_1, x_2]^T$ in the deformed state are related to the coordinates of this same point in the undeformed state, $[X_1, X_2]^T$, defines the deformation gradient in (2.3). The deformation, namely, the complete configuration of the deformed state, including the strains and rotations at each point, is defined by the deformation

to the shear planes then the deformation is a *simple shear* and is shown in Figure 2.3(e and f). We will see in Sections 2.3 and 2.5 that this deformation involves a rigid body rotation.

In general f is a nonlinear function of X and the deformation is a *non-affine transformation*; such a deformation is shown in Figure 2.4(a and b) and is said to be *inhomogeneous*. Clearly most natural deformations are inhomogeneous.

From (2.1), using the chain rule of differentiation (see Appendix C), we can derive expressions for the line element, dx, in terms of the undeformed line element, dX;

$$dx_1 = \frac{\partial x_1}{\partial X_1}dX_1 + \frac{\partial x_1}{\partial X_2}dX_2 + \frac{\partial x_1}{\partial X_3}dX_3$$

$$dx_2 = \frac{\partial x_2}{\partial X_1}dX_1 + \frac{\partial x_2}{\partial X_2}dX_2 + \frac{\partial x_2}{\partial X_3}dX_3$$

$$dx_3 = \frac{\partial x_3}{\partial X_1}dX_1 + \frac{\partial x_3}{\partial X_2}dX_2 + \frac{\partial x_3}{\partial X_3}dX_3$$

which can be written as

$$dx = (\nabla_X x) \cdot dX \tag{2.2}$$

This says that the vector dx in the deformed state is produced by the gradient $(\nabla_X x)$ operating on the vector dX in the reference or undeformed state. Thus it is convenient to define the *deformation gradient* as

$$\mathbf{F} = \begin{bmatrix} \dfrac{\partial x_1}{\partial X_1} & \dfrac{\partial x_1}{\partial X_2} & \dfrac{\partial x_1}{\partial X_3} \\[2mm] \dfrac{\partial x_2}{\partial X_1} & \dfrac{\partial x_2}{\partial X_2} & \dfrac{\partial x_2}{\partial X_3} \\[2mm] \dfrac{\partial x_3}{\partial X_1} & \dfrac{\partial x_3}{\partial X_2} & \dfrac{\partial x_3}{\partial X_3} \end{bmatrix} \equiv \nabla_X x \tag{2.3}$$

In (2.1) and (2.3), x_1, x_2, x_3 are the Cartesian coordinates of a material point in the deformed state and X_1, X_2, X_3 are the coordinates of this same material point in the undeformed state as shown in Figures 2.1 and 2.3. Notice that for a general deformation the deformation gradient is not symmetric, so that $\frac{\partial x_i}{\partial X_j} \neq \frac{\partial x_j}{\partial X_i}$ and $\mathbf{F}$ has nine independent components instead of the six independent components of the strain tensor (Means, 1976). This has important implications

◄────────────

gradient, $\mathbf{F}$, as in (2.3). (a) The deformation is $x_1 = 1.5X_1 + 0.5X_2 + 1.25$, $x_2 = X_1 + X_2 + 1$; this deformation is isochoric and consists of a rigid translation through a vector $[1.25, 1]^T$ and an anticlockwise rotation through 11.3° which can be calculated using (2.15). The principal stretches are 2.17 and 0.46. (b) A representation (a pseudo phase portrait) of the displacement field for (a). (c) A pure shear (or better, pure stretch) deformation given by $x_1 = 2X_1 + 1.33$, $x_2 = 0.5X_2 + 0.67$; this deformation is isochoric and consists of an elongation parallel to x_1 of 200% and a shortening parallel to x_2 of 50%. There is no rigid body rotation but a rigid body translation through a vector $[1.33, 0.67]^T$ is involved. (d) A representation of the displacement field for (c). (e) A simple shear deformation given by $x_1 = X_1 + \sqrt{3}X_2 + 1.1$, $x_2 = X_2 + 0.6$. This represents a shear through 60° and a rigid body rotation through 40.9° as is discussed in Section 2.5 or can be calculated from (2.15); also involved is a rigid translation through the vector $[1.1, 0.6]^T$. (f) A representation of the displacement field for (e).

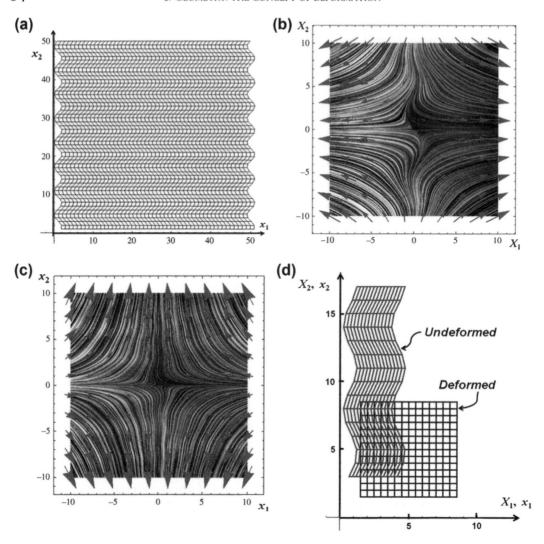

FIGURE 2.4   A non-affine isochoric transformation producing an inhomogeneous deformation. The deformation is expressed by $x_1 = 2X_1 + \sin(X_2)$, $x_2 = 0.5X_2$. (b) A representation of the displacement field in (a) where $-10 \leq X_1 \leq 10$ and $-10 \leq X_2 \leq 10$. This represents a Lagrangian or material view of the deformation. An Eulerian or spatial view of the phase portrait is given in (c) for $-10 \leq x_1 \leq 10$ and $-10 \leq x_2 \leq 10$ and in (d) which shows the grid in the reference or undeformed state that becomes a square in the deformed state and represents the inverse transformation $X_1 = 0.5x_1 - 0.5 \sin(2x_2)$, $X_2 = 2x_2$.

for microfabric development in deforming polycrystals where the deformation field is commonly inhomogeneous. Thus if the *strain* in an aggregate is homogeneous then (Chapter 9) six independent deformation mechanisms must operate in each grain in order to match the six independent components of strain; this number reduces to five if an extra constraint such as isochoric deformation is introduced. However, if the *deformation* is inhomogeneous from one grain to another or within a single grain, then the number of independent deformation mechanisms must increase to nine for a general deformation so that an individual grain, or parts of a grain, can match the imposed deformation with no gaps or overlaps with adjacent, differently deformed parts of the aggregate. We look at this issue again in Section 2.10 and in Chapter 13.

The volume change associated with the deformation is measured by J, the value of the determinant of $\mathbf{F}$ (see Appendix B and Table 2.1); the deformation defined by (2.1) is isochoric for $J = det\mathbf{F} = 1$. The condition $J > 0$ ensures that material does not interpenetrate at the local scale. The condition that an inverse to $\mathbf{F}$, $\mathbf{F}^{-1}$, exists is the condition that the material does not interpenetrate at a global scale. The significance of the deformation gradient is that it completely defines the deformation, including the stretch and rotation, of all line, surface and volume elements at each point within the body that it applies to (Bhattacharya, 2003; Tadmor et al., 2012), whereas various measures of the strain give a subset of this information. If one knows the deformation gradient, $\mathbf{F}$, then the strain may be calculated. However, the converse is not true. For affine deformations it is a matter of convenience from a purely geometrical point of view whether one describes the deformation in terms of the deformation

TABLE 2.1   Expressions for Matrix Operations for Rank Two Matrices

Quantity	Expression
Matrix $\mathbf{A}$	$\mathbf{A} = \begin{bmatrix} A_{11} & A_{12} \\ A_{21} & A_{22} \end{bmatrix}$
Determinant of $\mathbf{A}$	$det\mathbf{A} = \begin{vmatrix} A_{11} & A_{12} \\ A_{21} & A_{22} \end{vmatrix} = (A_{11}A_{22} - A_{12}A_{21})$
Trace of $\mathbf{A}$	$Tr\mathbf{A} = A_{11} + A_{22}$
Cofactor of $\mathbf{A}$	$cof\mathbf{A} = \begin{bmatrix} A_{22} & -A_{12} \\ -A_{21} & A_{11} \end{bmatrix}$
Expression for eigenvalues, λ, of $\mathbf{A}$	$\begin{vmatrix} A_{11} - \lambda & A_{12} \\ A_{21} & A_{22} - \lambda \end{vmatrix} = 0$ or $\lambda^2 - \lambda \cdot Tr\mathbf{A} + det\mathbf{A} = 0$
Eigenvalues of $\mathbf{A}$	$\lambda_1 = \dfrac{Tr\mathbf{A} + \sqrt{(Tr\mathbf{A})^2 - 4det\mathbf{A}}}{2} = \dfrac{Tr\mathbf{A} + \sqrt{\Delta}}{2},$ $\lambda_2 = \dfrac{Tr\mathbf{A} - \sqrt{(Tr\mathbf{A})^2 - 4det\mathbf{A}}}{2} = \dfrac{Tr\mathbf{A} - \sqrt{\Delta}}{2},$ $\lambda_1 + \lambda_2 = Tr\mathbf{A}, \quad \lambda_1\lambda_2 = det\mathbf{A}, \quad \Delta = (Tr\mathbf{A})^2 - 4det\mathbf{A}$
Eigenvectors of $\mathbf{A}$	$v_1 \propto [A_{12}, \lambda_1 - A_{11}], \quad v_2 \propto [A_{12}, \lambda_2 - A_{11}], \quad \|v\| = 1$
$\mathbf{A}$ in terms of the eigenvalues, λ_i, and eigenvectors, Λ_i, of $\mathbf{A}$	$\mathbf{A} = \lambda_1 \mathbf{\Lambda}_1 \otimes \mathbf{\Lambda}_1 + \lambda_2 \mathbf{\Lambda}_2 \otimes \mathbf{\Lambda}_2$
Angle between v_1 and x_1	$\tan^{-1}\dfrac{\lambda_1 - A_{11}}{A_{12}}$
Inverse of $\mathbf{A}$	$\mathbf{A}^{-1} = \dfrac{1}{det\mathbf{A}} cof\mathbf{A} = \dfrac{1}{det\mathbf{A}} \begin{bmatrix} A_{22} & -A_{12} \\ -A_{21} & A_{11} \end{bmatrix}$

In Table 2.1 $\|v\|$ is the *Euclidean norm* of v: $\|v\| = \sqrt{|v_1|^2 + |v_2|^2}$. This is used by some authors to normalise the magnitudes of the eigenvectors. Such a procedure is not universal and others propose that one component of the eigenvector is of unit length. This is the convention adopted in *Mathematica*.

A. THE MECHANICS OF DEFORMED ROCKS

gradient or by some measure of the strain since the rigid body rotation is commonly not of concern. However, for non-affine deformations the deformation gradient is fundamental since it carries all the information on local rotations and rigid body displacements as well as the strains. Specifically the deformation at a point x in the deformed body in terms of the deformation gradient is given by

$$x = \mathbf{F}X + \mathbf{d} \tag{2.4}$$

which says that each vector X in the undeformed state is distorted and rotated by the deformation gradient $\mathbf{F}$ and translated via a rigid motion by the vector $\mathbf{d}$ to become the vector x. Both (2.1) and (2.4) are referred to as a *deformation*. The quantity $\mathbf{F}$ has also been called the *position gradient tensor* (Means, 1982). In such usage the term *position* refers to the spatial coordinates of a material point relative to the coordinates of this same material point in some reference configuration and hence is synonymous with what we have called *deformation*. We continue with the term *deformation gradient* to preserve consistency with the mechanics literature. It is also clear from (2.2) that the deformation gradient defines the deformation of an infinitesimal line length at X.

The deformation expressed by (2.1) and (2.4) constitutes what is known as a *Lagrangian* or *material* description of deformation where coordinates of material particles in the deformed state are related to the coordinates of these same particles in the reference state. This is the natural way of approaching the subject from the viewpoint of a structural geologist who only has the deformed state to study but has in mind what the undeformed state may have been. However, if instead of $x = f(X); x = \mathbf{F}(X)$, we were to write the following:

$$X = \breve{f}(x); \quad X = F^{-1}x$$

where

$$\mathbf{F}^{-1} = \frac{1}{det\mathbf{F}}(cof\mathbf{F})^{\mathrm{T}} \tag{2.5}$$

then the coordinates of material points in the deformed state are used to derive the positions of these same points in the reference state. In (2.5), the superscripts $^{\mathrm{T},-1}$ stand for the transpose and the inverse of the relevant matrix and f is the spatial description of the deformation; $cof\mathbf{F}$ means the cofactor of the matrix $\mathbf{F}$ — see Appendix B and Table 2.1. This constitutes what is called an *Eulerian* or *spatial* description of the deformation; such a description is commonly used in fluid dynamics since for a fluid the concept of a reference or undeformed state has little significance. When we come to discuss the thermodynamics of deforming, chemically reacting systems with fluid flow (Chapters 5, 10, 11 and 14) it is convenient to use a spatial description so that quantities such as chemical concentration are expressed in terms of the deformed volume and heat or fluid flow is measured in terms of the deformed area or pore volume. An example of the differences between the Lagrangian and Eulerian descriptions of a deformation is given in Figure 2.4.

2.1.1 Deformation of a Surface Element

Consider two material vectors in the reference state, $\mathbf{V}_1$ and $\mathbf{V}_2$. Then the area of the parallelogram bounded by these vectors is $\mathbf{A} = \mathbf{V}_1 \times \mathbf{V}_2$. Here $\times$ is the cross or vector product of two vectors — see Appendix B. For a deformation with deformation gradient, $\mathbf{F}$, the two vectors become $\mathbf{v}_1 = \mathbf{FV}_1$ and $\mathbf{v}_2 = \mathbf{FV}_2$ so that the surface vector $\mathbf{A}$ becomes $\mathbf{a}$ in the deformed state and is given by $\mathbf{a} = (\mathbf{FV}_1) \times (\mathbf{FV}_2)$. This can be written as (see Appendix C)

$$\mathbf{a} = cof\mathbf{F}(\mathbf{V}_1 \times \mathbf{V}_2)$$

or

$$\mathbf{a} = cof\mathbf{F}(\mathbf{A}) = (det\mathbf{F})\mathbf{F}^{-T}(\mathbf{A})$$

Thus $\mathbf{a} = cof\mathbf{F}(\mathbf{A}) = J\mathbf{F}^{-T}(\mathbf{A})$. This latter expression is called *Nanson's relation*.

As a summary we see that for the deformation of line, surface and volume elements we have

$$\begin{aligned} d\boldsymbol{x} &= \mathbf{F}d\boldsymbol{X} \\ \mathbf{a} &= cof\mathbf{F}(\mathbf{A}) \\ v &= det\mathbf{F}(V) = JV \end{aligned} \qquad (2.6)$$

Thus, line elements are deformed by the deformation gradient, $\mathbf{F}$, surface elements are deformed by the cofactor of $\mathbf{F}$, $cof\mathbf{F}$, and volume elements are deformed by the determinant of $\mathbf{F}$, $det\mathbf{F}$. These are convenient ways of thinking about the physical meanings of the *gradient* of the deformation and the *cofactor* and *determinant* of the deformation gradient matrix. The distinction between d$\mathbf{A}$ and d$\mathbf{a}$ becomes particularly important when we come to discus the concept of *stress* (Chapter 4) or *force per unit* area since we can adopt a material or spatial view for the area across which the force acts. If the area in the undeformed or reference state is selected then the stress is known as the *Piola-Kirchhoff stress*, whereas if the area in the deformed state is selected then the stress is known as the *Cauchy stress*.

2.2 DISTORTION AND ROTATION

Two fundamental quantities that can be derived from the deformation gradient are what are called the *right* and *left Cauchy–Green tensors*, $\mathbf{C}$ and $\mathbf{B}$, given by

$$\mathbf{C} = \mathbf{F}^T\mathbf{F} \quad \text{and} \quad \mathbf{B} = \mathbf{FF}^T \qquad (2.7)$$

The importance of $\mathbf{B}$ is that at each point in the deformed state it can be expressed as an ellipsoid that has principal axes parallel to those of another ellipsoid that was a unit sphere at the equivalent material point in the reference state; the lengths of the principal axes of $\mathbf{B}$ are the squares of those of the deformed unit sphere. Both $\mathbf{B}$ and $\mathbf{C}$ are readily calculated and hence are convenient representations of the distortion associated with the deformation. The angle between the principal axes of $\mathbf{B}$ and $\mathbf{C}$ is given by the rigid body rotation, $\mathbf{R}$, associated with the deformation so that $\mathbf{B} = \mathbf{RCR}^T$. Both $\mathbf{B}$ and $\mathbf{C}$ are symmetrical tensors meaning that $B_{ij} = B_{ji}$ and $C_{ij} = C_{ji}$.

Important properties of both $\mathbf{B}$ and $\mathbf{C}$ are their respective *eigenvalues* and *eigenvectors*. The term *eigen* is German for *self*. An eigenvector of a (square) matrix is such that when that matrix operates upon the eigenvector, the orientation of that vector is preserved, although its magnitude and direction-sense may change. In three dimensions a matrix may have one, two or three eigenvectors. If one begins with a unit vector parallel to an eigenvector then the magnitude of that vector becomes the eigenvalue of the matrix. The eigenvalues may be real or imaginary. As an example, it follows that the eigenvectors of $\mathbf{C}$, $\mathbf{\Lambda}^{\mathbf{C}}$, have the characteristic that

$$\mathbf{C}\mathbf{\Lambda}^{\mathbf{C}} = \lambda\mathbf{\Lambda}^{\mathbf{C}} \Leftrightarrow C_{ij}\Lambda_j^C = \lambda\Lambda_i^C \tag{2.8}$$

which means that when the tensor $\mathbf{C}$ operates on the vectors $\mathbf{\Lambda}^{\mathbf{C}}$ it only changes their magnitudes (by λ) and not their directions. (2.8) can be rewritten

$$(\mathbf{C} - \lambda\mathbf{I})\,\mathbf{\Lambda} = \mathbf{0} \quad \Leftrightarrow \quad \left(C_{ij} - \lambda\delta_{ij}\right)\Lambda_j = 0 \tag{2.9}$$

where $\mathbf{I}$ is the identity matrix (see Appendix B). (2.9) in turn can be rewritten:

$$\begin{vmatrix} C_{11} - \lambda & C_{12} & C_{13} \\ C_{21} & C_{22} - \lambda & C_{23} \\ C_{31} & C_{32} & C_{33} - \lambda \end{vmatrix} = 0 \tag{2.10}$$

Or, since $C_{ij} = C_{ji}$,

$$-\lambda^3 + I_1(\mathbf{C})\lambda^2 - I_2(\mathbf{C})\lambda + I_3(\mathbf{C}) = 0 \tag{2.11}$$

where I_1, I_2 and I_3 are the *principal invariants* of $\mathbf{C}$:

$$I_1(\mathbf{C}) = C_{ii} = Tr\mathbf{C} = C_{11} + C_{22} + C_{33}$$

$$I_2(\mathbf{C}) = \frac{1}{2}\left(C_{ii}C_{jj} - C_{ij}C_{ji}\right) = \frac{1}{2}\left((tr\mathbf{C})^2 - tr\mathbf{C}^2\right) = tr\mathbf{C}^{-1}det\mathbf{C} \tag{2.12}$$

$$I_3(\mathbf{C}) = \varepsilon_{ijk}C_{1i}C_{2j}C_{3k} = det\mathbf{C}$$

Equations (2.10) and (2.11) are known as the *characteristic equations* for $\mathbf{C}$. $Tr\mathbf{C}$ means the trace of the matrix $\mathbf{C}$ and ε_{ijk} is the permutation symbol – see Appendix B.

In general the deformation (2.1) consists, at each point, of a distortion, a rigid body rotation and a rigid body translation. The *right finite stretch tensor*, $\mathbf{U}$, and the *finite rotation matrix*, $\mathbf{R}$, are given (Bhattacharya, 2003; Tadmor et al., 2012) by

$$\mathbf{U} = \sqrt{\mathbf{F}^{\mathrm{T}}\mathbf{F}} = \sqrt{\mathbf{C}} \quad \text{and} \quad \mathbf{R} = \mathbf{F}\mathbf{U}^{-1} \tag{2.13}$$

$\mathbf{U}$ has the same eigenvectors as $\mathbf{F}^{\mathrm{T}}\mathbf{F}$ but the eigenvalues of $\mathbf{U}$ are the square roots of those of $\mathbf{F}^{\mathrm{T}}\mathbf{F}$. If the x_3 axis is the axis of rotation and φ is the angle of rotation then the matrix $\mathbf{R}$ is given by (Nye, 1957, pp. 9 and 43)

$$\mathbf{R} = \begin{bmatrix} \cos\varphi & -\sin\varphi & 0 \\ \sin\varphi & \cos\varphi & 0 \\ 0 & 0 & 1 \end{bmatrix} \tag{2.14}$$

The displacement field for $\mathbf{R}$ is shown in Figure 2.5(a).

In Appendix B we show that for a two-dimensional deformation, $\mathbf{R}$ can be calculated from

$$R_{ij} = \frac{1}{\sqrt{(F_{11} + F_{22})^2 + (F_{12} - F_{21})^2}} \begin{bmatrix} F_{11} + F_{22} & F_{12} - F_{21} \\ F_{21} - F_{12} & F_{11} + F_{22} \end{bmatrix} \tag{2.15_1}$$

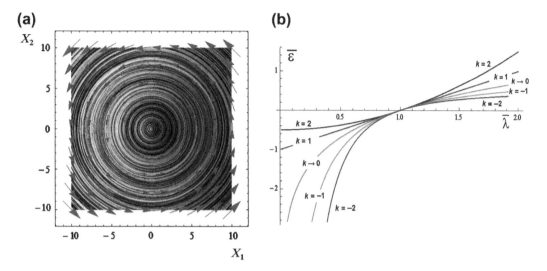

(a)

(b)

FIGURE 2.5 Representations of deformation. (a) Displacement field for the rigid body rotation, **R**. (b) The Seth–Hill family of strain measures.

Thus, from (2.14),

$$\cos \varphi = \frac{F_{11} + F_{22}}{\sqrt{(F_{11} + F_{22})^2 + (F_{12} - F_{21})^2}} \quad \text{or,} \quad \tan \varphi = \frac{F_{12} - F_{21}}{F_{11} + F_{22}} \tag{2.15_2}$$

2.3 DEFORMATION AND STRAIN TENSORS AND MEASURES

The deformation at a point can be conveniently represented by a set of measures that defines the changes in the lengths of lines, the angles between lines and the deformed shapes of initial spheres. If a line of initial length L_0 is deformed to become length L then the *elongation* or *axial strain* is defined as

$$\varepsilon = \frac{L - L_0}{L_0}$$

The *stretch* is $\bar{\lambda} = \frac{L}{L_0}$ while the *quadratic elongation* is $\bar{\lambda}^2 = \left(\frac{L}{L_0}\right)^2$. The *Hencky* or *natural* or *logarithmic strain* is $\bar{\lambda}^H = ln\bar{\lambda} = ln\left(\frac{L}{L_0}\right)$. All three of these measures of distortion, ε, $\bar{\lambda}^H$ and $\bar{\lambda}$ are members of a spectrum of strain measures, $\bar{\varepsilon}$, known as the *Seth–Hill family* of measures given in terms of a parameter, k, by

$$\bar{\varepsilon} = \frac{1}{k}\left(\bar{\lambda}^k - 1\right)$$

For $k=1$, $\bar{\varepsilon} = \varepsilon = \bar{\lambda} - 1$; for $k=-1$, $\bar{\varepsilon} = 1 - \frac{1}{\bar{\lambda}}$; for $k=2$, $\bar{\varepsilon}^L = \frac{1}{2}(\bar{\lambda}^2 - 1)$ where $\bar{\varepsilon}^L$ is known as the *Lagrangian strain*. As $k \to 0$, $\bar{\varepsilon} \to \bar{\lambda}^H$. The Seth–Hill family is shown in

Figure 2.5(b) where one can see that for small stretches $(\bar{\lambda} \approx 1)$ it makes little difference which measure of strain is used. However, we will see (especially in Chapter 3) that for large strains the Hencky strain has special advantages.

Changes in orientations of lines are defined by *shear strains*. If two lines initially at right angles are deformed so that the angle between them becomes ψ then the *shear strain*, γ, is defined as

$$\gamma = \tan \psi \tag{2.16}$$

If we begin with a unit sphere at a point in the reference state this becomes an ellipsoid (known as the *strain ellipsoid*) in the deformed state. The principal axes of this ellipsoid are the principal stretches and we refer to the magnitudes of these axes (in keeping with geological usage) as X, Y and Z with $X \geq Y \geq Z$. An example is given in Figure 2.6 for a two-dimensional deformation involving a shear through $60°$. This deformation is given by

$$x_1 = X_1 + \tan 60° X_2$$

$$x_2 = X_2$$

and we calculate the various measures of strain for this deformation in a worked example in Section 2.5. In Figure 2.6(a) the undeformed circle $P_1P_2'P_1'P_2$ is deformed to become the strain ellipse $p_1p_2'p_1'p_2$. The principal axes of strain in the deformed state, p_1p_1' and p_2p_2', are deformed versions of the principal axes of strain in the undeformed state, P_1P_1' and P_2P_2'. The two sets of axes are related by a rigid body rotation, φ, defined by $\mathbf{R}$ in (2.14). The axes P_1P_1' and P_2P_2' are parallel to the principal axes of $\mathbf{C}$, $\mathbf{U}$, $\mathbf{C}^{-1}$ and $\mathbf{U}^{-1}$. The principal axes of strain, p_1p_1' and p_2p_2', are parallel to the principal axes of $\mathbf{B}$, $\mathbf{V}$, $\mathbf{B}^{-1}$ and $\mathbf{V}^{-1}$ as can be seen by comparing Figure 2.6(a,b and c).

We have seen that a deformation can be represented by a square matrix, $\mathbf{F}$, that in general, is asymmetrical. The *polar decomposition theorem* (see Appendix B) states that any square matrix $\mathbf{F}$ with $det\mathbf{F} > 0$ can be expressed as a rotation, $\mathbf{R}$, and a symmetric matrix $\mathbf{U}$ or $\mathbf{V}$ such that

$$\mathbf{F} = \mathbf{RU} = \mathbf{VR} \tag{2.17}$$

Both $\mathbf{U}$ and $\mathbf{V}$ are representations of the *distortion* or *strain* resulting from the deformation and are called the *right* and *left stretch tensors*, respectively. Since both $\mathbf{U}$ and $\mathbf{V}$ are symmetrical second order tensors they can be expressed as representation ellipsoids. $\mathbf{V}$ is known as the *strain ellipsoid* and is the shape that a unit sphere in the reference state adopts in the deformed state. This means that the semiaxes of $\mathbf{V}$ are the principal stretches, X, Y and Z. $\mathbf{U}^{-1}$ is the *reciprocal strain ellipsoid*; this is the ellipsoid in the reference state that becomes a unit sphere in the deformed state.

The other convenient tensors, the right and left Cauchy-Green tensors $\mathbf{C}$ and $\mathbf{B}$, are given, respectively, by

$$\mathbf{C} = \mathbf{U}^2 \quad \text{and} \quad \mathbf{B} = \mathbf{V}^2$$

The lengths of the principal axes of $\mathbf{C}$ and $\mathbf{B}$ are the principal quadratic elongations X^2, Y^2 and Z^2. The principal axes of $\mathbf{B}$ are parallel to those of $\mathbf{V}$. Summaries of these various tensors are given in Table 2.2 and in Figure 2.6 for a simple shear through $60°$.

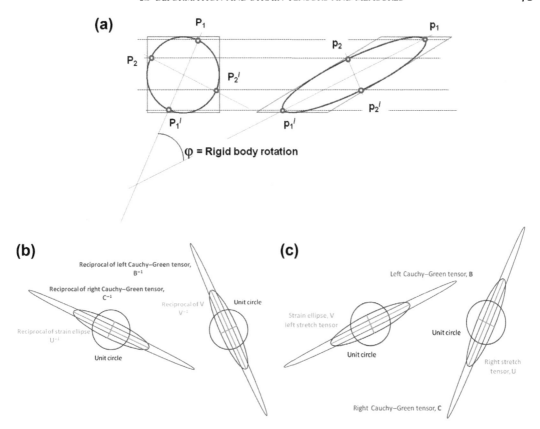

FIGURE 2.6   The various deformation and strain tensors associated with a deformation gradient **F**. (a) A simple shear deformation of a square with an embedded circle. The shear angle in this case is 60°. The square is deformed to become a parallelogram and the circle becomes an ellipse, the *strain ellipse*. The points on the long axis of the ellipse, p_1 and p_1', originated at P_1 and P_1' which define a principal axis of strain in the undeformed state. Similarly p_2 and p_2' originated at P_2 and P_2' which define another principal axis in the undeformed state. The principal axes of strain in the deformed and undeformed states are related by a rigid body rotation, φ, defined by **R** in (2.14) and (2.15). (b) The ellipses which represent the tensors $\mathbf{C}^{-1}$, $\mathbf{B}^{-1}$, $\mathbf{U}^{-1}$ and $\mathbf{V}^{-1}$. (c) The ellipses which represent the tensors **C**, **B**, **U** and **V**. In both (b) and (c) the ellipses are referred to the same coordinate axes as in (a).

For the non-affine deformation shown in Figure 2.4 given by

$$x_1 = 2X_1 + \sin(X_2)$$

$$x_2 = 0.5X_2$$

$\mathbf{F} = \begin{bmatrix} 2 & \cos(X_2) \\ 0 & 0.5 \end{bmatrix}$ and $J = det\mathbf{F} = 1$ so that the deformation is everywhere isochoric. In Section 2.7.1 we calculate the various strain tensors for this deformation and present in Figure 2.7(b) the strain ellipse at a number of places in the deformation. Figure 2.7(a) shows this same deformation but with 33.3% shortening. Calculations of the strain distribution in non-affine

TABLE 2.2   Strain Tensors and Their Significance

Tensor	Expression	Significance
Right Cauchy-Green, **C**	$\mathbf{C} = \mathbf{F}^T\mathbf{F}$	Same eigenvectors as **U** and $\mathbf{U}^{-1}$ but eigenvalues are principal quadratic elongations
Left Cauchy-Green, **B**	$\mathbf{B} = \mathbf{F}\mathbf{F}^T$ $\mathbf{B} = \mathbf{R}\mathbf{C}\mathbf{R}^T$	Same eigenvectors as **V** and $\mathbf{V}^{-1}$ but eigenvalues are principal quadratic elongations
Right stretch, **U**	$\mathbf{U} = \sqrt{\mathbf{F}^T\mathbf{F}}$	Same eigenvectors as **C** and $\mathbf{U}^{-1}$. Eigenvalues are the principal stretches X, Y and Z
Left stretch, **V**	$\mathbf{V} = \sqrt{\mathbf{F}\mathbf{F}^T}$	Strain ellipsoid with eigenvalues equal to the principal stretches, X, Y and Z; eigenvectors same as **B** and $\mathbf{V}^{-1}$
Rotation, **R**	$\mathbf{R} = \mathbf{F}\mathbf{U}^{-1}$ $\mathbf{R} = \mathbf{V}^{-1}\mathbf{F}$	Rotation matrix
Reciprocal of **U**, $\mathbf{U}^{-1}$	$\mathbf{U}^{-1} = \dfrac{1}{det\mathbf{U}} cof\mathbf{U}$	Reciprocal strain ellipsoid; same eigenvectors as **C** and **U**
Reciprocal of **V**, $\mathbf{V}^{-1}$	$\mathbf{V}^{-1} = \dfrac{1}{det\mathbf{V}} cof\mathbf{V}$	Same eigenvectors as **B** and **V**
Green's finite strain, **E**	$\mathbf{E} = \dfrac{1}{2}(\mathbf{C} - \mathbf{I})$	
Almansi's finite strain, $\overline{\mathbf{A}}$	$\overline{\mathbf{A}} = \dfrac{1}{2}(\mathbf{I} - \mathbf{B}^{-1})$	

deformations are given by Jaeger (1969, pp. 245–251), Ramsay and Graham (1970), Hobbs (1971) and Hirsinger and Hobbs (1983).

2.3.1 Physical Significance of C

Consider the deformation of a line element d**X** in the reference state to become d**x** in the deformed state as shown in Figure 2.8(a).

The squares of the lengths of these two lines are

$$dS^2 = dX_I dX_I$$

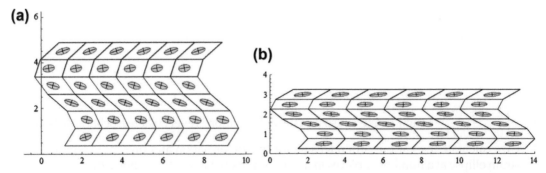

FIGURE 2.7   Strain ellipses for the deformation $x_1 = aX_1 + \sin(X_2)$, $x_2 = X_2/a$. (a) Deformation plus strain ellipses for $a = 1.33$ (33.3% shortening). (b) Deformation plus strain ellipses for $a = 2.0$ (50% shortening).

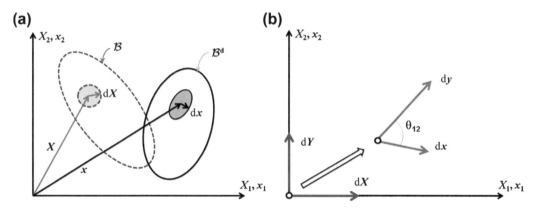

FIGURE 2.8 Deformation of line elements. (a) $d\mathbf{X}$ is deformed to become $d\mathbf{x}$. (b) $d\mathbf{X}$ and $d\mathbf{Y}$ in the reference state are perpendicular to each other and are deformed to become $d\mathbf{x}$ and $d\mathbf{y}$ with an angle θ_{12} between them in the deformed state.

$$ds^2 = dx_i dx_i = (F_{iI}dX_I)(F_{iJ}dX_J) = (F_{iI}F_{iJ})dX_I dX_J = C_{IJ}dX_I dX_J$$

It is convenient to define the *Lagrangian strain tensor* **E** (Table 2.2) as $E_{IJ} = \frac{1}{2}(C_{IJ} - \delta_{IJ})$ where δ_{IJ} is the Kronecker delta (see Appendix B). Then,

$$ds^2 - dS^2 = 2E_{IJ}dX_I dX_J$$

Consider now that $d\mathbf{X}$ is parallel to X_1 so that it is given by $[d\mathbf{X}] = [dX_1, 0, 0]^T$, then the lengths of this line in the reference and deformed states are

$$dS^2 = dX_I dX_I = (dX_I)^2$$

$$ds^2 = C_{IJ}dX_I dX_J = C_{11}(dX_I)^2$$

The stretch along the X_1 direction is therefore

$$\bar{\lambda}^{(1)} = \sqrt{C_{11}}$$

For the X_2 and X_3 directions,

$$\bar{\lambda}^{(2)} = \sqrt{C_{22}} \quad \text{and} \quad \bar{\lambda}^{(3)} = \sqrt{C_{33}}$$

Thus *the diagonal components of* **C** *are the squares of the stretches of material lines initially parallel to the undeformed or reference coordinate axes.*

Now consider two material lines initially parallel to the X_1 and X_2 directions so that these two initially orthogonal vectors are defined by

$$[d\mathbf{X}] = [dX_1, 0, 0]^T \quad \text{and} \quad [d\mathbf{Y}] = [0, dX_2, 0]^T$$

These two vectors become $d\mathbf{x}$ and $d\mathbf{y}$ in the deformed state and the angle between them becomes θ_{12} as shown in Figure 2.8(b) and is given by

$$\cos \theta_{12} = \frac{dx \cdot dy}{|dx||dy|} = \frac{C_{IJ}dX_IdX_J}{[C_{KL}dX_KdX_L]^{\frac{1}{2}}[C_{MN}dX_MdX_N]^{\frac{1}{2}}}$$

$$= \frac{C_{12}dX_1dX_2}{\sqrt{C_{11}}dX_1\sqrt{C_{22}}dX_2} = \frac{C_{12}}{\sqrt{C_{11}}\sqrt{C_{22}}}$$

Thus *the off-diagonal components of C are related to the change in angles between vectors initially parallel to the reference axes in the reference or undeformed state.*

If we return to the Seth–Hill family of strain measures, we can define the spatial and material tensor members, respectively, as

$$\bar{\varepsilon}^{spatial} = \frac{1}{k}\left(\mathbf{U}^k - 1\right) \quad \text{and} \quad \bar{\varepsilon}^{material} = \frac{1}{k}\left(\mathbf{V}^k - 1\right)$$

If the deformation is a pure stretch deformation then there is no difference between these two measures. However, if a material rotation is involved then

$$\bar{\varepsilon}^{spatial} = \mathbf{R}\bar{\varepsilon}^{material}\mathbf{R}^T$$

As a summary, the operations required to derive the various tensors discussed above for the two-dimensional situation are given in Table 2.1. The same operations in three dimensions are more extended, cumbersome and error-prone but mathematical symbolic languages such as *Mathematica* and *MATLAB* have routines that enable such operations to be carried out with relative ease. An extensive tabulation of useful expressions for a large number of matrix operations and relationships is given in Petersen and Pedersen (2008).

2.4 DISTORTION AND VOLUME CHANGE

Many deformed rocks consist of localised regions of volume loss and/or gain. Examples include 'spots' (porphyroblasts), linear regions (lineations) and layers (foliations) where locally the metamorphic mineral reactions have been dominantly volume increasing or decreasing, zones where material has been removed due to 'pressure solution', metamorphic differentiation during the formation of crenulation cleavage (Figure 2.9(a)) and leucosomes where melt has left the system (Figure 2.9(b)). We will discuss the processes involved in these volume changes in Chapter 14 and Volume II but for now we present an example of deformations that might reproduce some of these situations. These consist of inhomogeneous shears with periodic distributions of volume changes (Figure 2.9(c)):

$$x_1 = X_1 + \sin X_2$$

$$x_2 = 0.05 \sin \left(X_2^2\right) + 0.9X_2$$

Various representations of the displacement field for this deformation are given in Figure 2.9(d, e and f) which are to be compared and contrasted with Figure 2.4. The

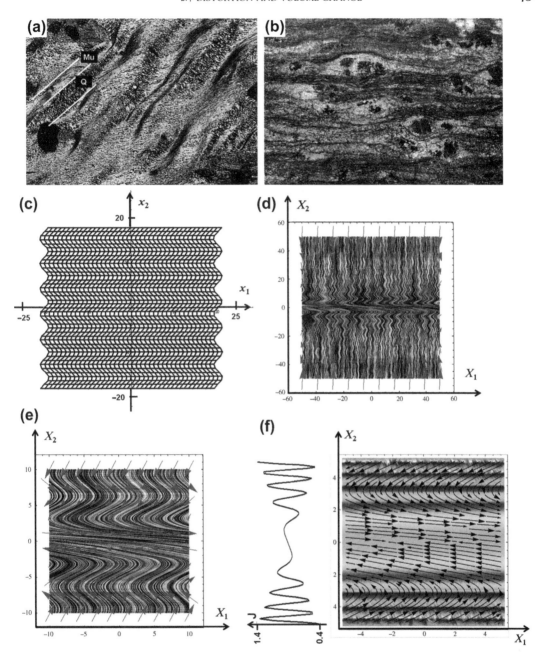

FIGURE 2.9　Non-affine deformations associated with volume changes. (a) Differentiated crenulation cleavage from the Picuris Range, New Mexico, USA. *(Photo: Ron Vernon.)* View about 1.8 mm across. In some models for the formation of this structure, it is proposed that quartz is removed from the muscovite (Mu) rich layers and either removed from the system or added to the quartz (Q) rich layers (Vernon, 2004, pp. 393−389). (b) Differentiated foliation in granulite facies rocks from Round Hill, Broken Hill, Australia. Scale: view about 0.5 m across. Models for the development of this foliation involve removal of melt from the lighter coloured garnet bearing areas (Powell & Downes, 1990). (c) A non-affine deformation with periodic distribution of volume change. (d) The displacement field associated with the deformation in (c). (e) Zoom into part of (d). (f) Divergence of the displacement vector field superimposed on the vector field. On the left hand side the Jacobian of the deformation gradient is plotted as an indication of the volume change.

deformation in Figure 2.4 is a modification of an isochoric pure shear, whereas that in Figure 2.9 is a modification of a pure shear with local volume changes. Both produce deformations that are superficially similar in appearance yet the displacement fields are quite different. For the deformation in Figure 2.9 the volume change is given by the Jacobian:

$$J = \begin{vmatrix} 1 & \cos X_2 \\ 0 & 0.1X_2 \cos \left(X_2^2\right) + 0.9 \end{vmatrix} = 0.1X_2 \cos \left(X_2^2\right) + 0.9$$

The value of the Jacobian is plotted on the right side of Figure 2.9(f) which represents the *divergence* of the displacement field; the volume change ranges from 0.4 to 1.4 times the initial volume. We discuss the concept of the divergence of a vector field in Chapter 3 but for the moment it is sufficient to grasp that zero divergence means that there are no sources or sinks for the vector field, whereas positive or negative values for the divergence mean that there are sources or sinks, respectively. Other examples of non-affine deformations involving volume changes are given in Sections 2.7.2 and 2.7.3.

We note that for finite deformations the Hencky strain tensor is the only strain measure whose trace gives the volumetric strain:

$$Tr\bar{\varepsilon}^H = ln\bar{\lambda}_1 + ln\bar{\lambda}_2 + ln\bar{\lambda}_3 = ln\left(\bar{\lambda}_1\bar{\lambda}_2\bar{\lambda}_3\right) = lnJ$$

and so the Hencky tensor is particularly useful for deformations involving a volume change.

2.5 EXAMPLE 1: THE GEOMETRY OF A SIMPLE SHEAR DEFORMATION

As a worked example we derive the various strain measures and tensors and rotation for a simple shear deformation. Consider the deformation shown in Figure 2.10(a).

The deformation may be written as

$$x_1 = X_1 + \gamma X_2$$

$$x_2 = X_2$$

The deformation gradient and its transpose are

$$\mathbf{F} = \begin{bmatrix} 1 & \gamma \\ 0 & 1 \end{bmatrix} \quad \text{and} \quad \mathbf{F}^T = \begin{bmatrix} 1 & 0 \\ \gamma & 1 \end{bmatrix}$$

One can confirm that the deformation is isochoric since $J = det\mathbf{F} = \begin{vmatrix} 1 & \gamma \\ 0 & 1 \end{vmatrix} = 1$. Thus the right Cauchy–Green tensor is $\mathbf{C} = \mathbf{F}^T\mathbf{F} = \begin{bmatrix} 1 & 0 \\ \gamma & 1 \end{bmatrix}\begin{bmatrix} 1 & \gamma \\ 0 & 1 \end{bmatrix} = \begin{bmatrix} 1 & \gamma \\ \gamma & 1+\gamma^2 \end{bmatrix}$.

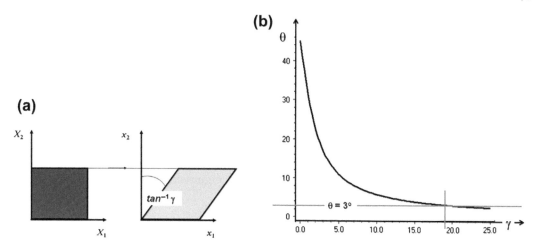

FIGURE 2.10   A simple shear deformation. (a) The angle of shear is $\tan^{-1}\gamma$. (b) Plot of the angle θ, between the maximum principal axis of strain in the deformed state and the shear plane for various values of γ.

The two eigenvalues, λ_1^C and λ_2^C, of $\mathbf{C}$ are given by the solution of

$$\begin{vmatrix} 1-\lambda & \gamma \\ \gamma & 1+\gamma^2-\lambda \end{vmatrix} = 0$$

Or, $\lambda_1^C = 1 - \gamma\beta^-$ and $\lambda_2^C = 1 + \gamma\beta^+$ where $\beta^\pm = \frac{1}{2}(\sqrt{4+\gamma^2}\pm\gamma) \geq 1$. Notice that $\beta^- + \gamma = \beta^+$. One of the eigenvectors, $\mathbf{\Lambda}_1^C$, may be found by solving either of the two equations

$$C_{ij}\Lambda_j^{(1)} = \lambda_1^C \Lambda_i^{(1)}$$

with a similar set of equations for $\mathbf{\Lambda}_2^C$. Here $\Lambda_1^{(1)}$ and $\Lambda_2^{(1)}$ are the components of $\mathbf{\Lambda}_1^C$. There are always two equivalent answers for each eigenvector one corresponding to $i=1$ and another corresponding to $i=2$. We choose $i=2$ and obtain the equation

$$C_{21}\Lambda_1^{(1)} + C_{22}\Lambda_2^{(1)} = \lambda^{(1)}\Lambda_2^{(1)}$$

Or, $\gamma\Lambda_1^{(1)} + (1+\gamma^2)\Lambda_2^{(1)} = (1-\gamma\beta^-)\Lambda_2^{(1)}$
Thus, since $\beta^- + \gamma = \beta^+$ we obtain

$$\Lambda_1^{(1)} = -(\beta^- + \gamma)\Lambda_2^{(1)} = -\beta^+\Lambda_2^{(1)}$$

Now since $(\Lambda_1^{(1)})^2 + (\Lambda_2^{(1)})^2 = 1$ we finally obtain

$$\Lambda_1^{(1)} = -\frac{\beta^+}{\sqrt{1+(\beta^+)^2}} \quad \text{and} \quad \Lambda_2^{(1)} = \frac{1}{\sqrt{1+(\beta^+)^2}}$$

A similar calculation gives Λ_2^C so that the two eigenvectors can be written as

$$\Lambda_1^C = \frac{\left[-\beta^+, 1\right]^T}{\sqrt{1 + (\beta^+)^2}} \quad \text{and} \quad \Lambda_2^C = \frac{\left[\beta^-, 1\right]^T}{\sqrt{1 + (\beta^-)^2}}$$

The orientations of the eigenvectors are

$$\tan \theta^{(1)} = \frac{1}{\beta^-} \quad \text{and} \quad \tan \theta^{(2)} = -\frac{1}{\beta^+}$$

where θ is the angle between the x_1-axis and the eigenvector of $\mathbf{C}$. The right stretch tensor is

$$\mathbf{U} = \sqrt{\lambda_1}\,\Lambda_1 \otimes \Lambda_1 + \sqrt{\lambda_2}\,\Lambda_2 \otimes \Lambda_2$$

and hence can be written as

$$\mathbf{U} = \frac{\sqrt{1 - \gamma\beta^-}}{1 + (\beta^+)^2}\begin{bmatrix}(\beta^+)^2 & -\beta^+ \\ -\beta^+ & 1\end{bmatrix} + \frac{\sqrt{1 + \gamma\beta^+}}{1 + (\beta^-)^2}\begin{bmatrix}(\beta^-)^2 & \beta^- \\ \beta^- & 1\end{bmatrix}$$

Notice that $\mathbf{U}$ for simple shear is singular since $det\mathbf{U} = 0$ and so it is not straightforward to calculate the rotation matrix $\mathbf{R}$ from $\mathbf{R} = \mathbf{F}\mathbf{U}^{-1}$. A way forward is to assume that $\mathbf{R}$ is given by a standard matrix (Nye, 1957) in terms of the rigid body rotation angle, φ:

$$\mathbf{R} = \begin{bmatrix}\cos\varphi & -\sin\varphi \\ \sin\varphi & \cos\varphi\end{bmatrix}$$

Then since $\mathbf{U} = \mathbf{R}^T\mathbf{F}$, $\quad \mathbf{U} = \begin{bmatrix}\cos\varphi & \sin\varphi \\ -\sin\varphi & \cos\varphi\end{bmatrix}\begin{bmatrix}1 & \gamma \\ 0 & 1\end{bmatrix} = \begin{bmatrix}\cos\varphi & \gamma\cos\varphi + \sin\varphi \\ -\sin\varphi & \cos\varphi - \gamma\sin\varphi\end{bmatrix}$

But since $\mathbf{U}$ is symmetrical, $-\sin\varphi = \gamma\cos\varphi + \sin\varphi$ or $\tan\varphi = -\frac{\gamma}{2}$. Thus the rigid body rotation associated with a simple shear is that angle whose tangent is minus one half the shear strain. This can also be calculated using (2.15$_2$).

As a direct application, consider a simple shear through $60°$ as shown in Figure 2.6. We have $\beta^+ = 2.1889$, $\beta^- = 0.4569$ so that $\lambda_1^C = 1 - \gamma\beta^- = 0.2087$ and $\lambda_2^C = 1 + \gamma\beta^- = 4.7913$. Note that $\lambda_1^C = \frac{1}{\lambda_2^C}$ since the deformation is isochoric. We also have $\theta^{(1)} = 65.4°$ and $\theta^{(2)} = -24.6°$ as plotted for the undeformed state in Figure 2.6. The rigid body rotation is $\tan^{-1}(\sqrt{3}/2) = 40.9°$; this can also be calculated using (2.15).

By a similar argument we obtain

$$\mathbf{B} = \mathbf{F}\mathbf{F}^T = \begin{bmatrix}1 + \gamma^2 & \gamma \\ \gamma & 1\end{bmatrix}$$

with eigenvectors $\mathbf{\Lambda}_1^B = \frac{[\beta^+, 1]^T}{\sqrt{1+(\beta^+)^2}}$, $\mathbf{\Lambda}_2^B = \frac{[-\beta^-, 1]^T}{\sqrt{1+(\beta^-)^2}}$ and $\tan\theta^{(1)} = \frac{1}{\beta^+}$, $\tan\theta^{(2)} = -\frac{1}{\beta^-}$ where the θ's now are the angles between the principal axes of strain and the x_1 axis in the deformed state.

We can now plot various items of interest such as the angle between the major principal axis of strain in the deformed state and the shear plane. This is given by

$$\theta = \tan^{-1}\frac{1}{\left[\frac{1}{2}\left(\sqrt{4+\gamma^2}+\gamma\right)\right]}$$

and is plotted in Figure 2.10(b). It can be seen that shear strains of 19 (shear angles of 87°) or so are necessary before a principal axis of strain becomes within 3° of the shear plane.

2.5.1 Deformation of an Arbitrary Line Element in Simple Shear

Consider an arbitrary line element dL in the undeformed state that is deformed to become $\mathbf{F}dL$ so that $dl = \mathbf{F}dL$. For a simple shear the line $[A, B]^T$ becomes $\begin{bmatrix} 1 & \gamma \\ 0 & 1 \end{bmatrix}[A, B]^T$ or $[A+\gamma B, B]^T$ in the deformed state. This allows us to calculate the rotations of all initial lines during the deformation. The angle of rotation, $\overline{\varphi}$, of an arbitrary initial line is given by

$$\cos\overline{\varphi} = \frac{dL \cdot dl}{|dL||dl|} = \frac{A(A+\gamma B) + B^2}{\sqrt{(A^2+B^2)}\sqrt{(A+\gamma B)^2 + B^2}}$$

This means that $\tan\overline{\varphi} = \frac{\gamma B^2}{A(A+\gamma B)+B^2}$

We ask: *what lines in the undeformed state are rotated through the rigid body rotation angle?* This is true for $\tan\overline{\varphi} = \tan\varphi$ which means that

$$\frac{\gamma B^2}{A(A+\gamma B)+B^2} = \frac{\gamma}{2}$$

or,

$$A^2 + \gamma AB - B^2 = 0$$

This means that $A = \frac{B}{2}\left(\sqrt{\gamma^2+4}\pm\gamma\right) = B\beta^\pm$ or that $\frac{B}{A} = \frac{1}{\beta^\pm}$

Hence the only two lines in the undeformed state that suffer the rigid body rotation in a simple shear are the principal axes of strain in the undeformed state; all other lines suffer smaller or larger rotations.

2.5.2 Deformation of an Arbitrary Volume Element

An arbitrary volume element dV at the point P in the undeformed state is deformed according to $det\mathbf{F}$ to become dv in the deformed state so that $dv = (det\mathbf{F}(P))dV$. For a simple shear $det\mathbf{F} = 1$ so all volumes are preserved.

2.6 PSEUDO PHASE PORTRAITS FOR AFFINE DEFORMATIONS

In this chapter we have presented quite a diversity of material representations of displacement fields for two-dimensional deformations. As we indicated these resemble phase portraits to be discussed in the next chapter and should be called *pseudo phase portraits* since they represent static displacement fields. However, as with dynamical systems, the systematics behind this diversity is best analysed in terms of the eigenvectors of the deformation gradient matrix. As a reminder, the eigenvalues, λ_1 and λ_2, of the matrix $\mathbf{F}$ given by

$$\mathbf{F} = \begin{bmatrix} F_{11} & F_{12} \\ F_{21} & F_{22} \end{bmatrix}$$

are obtained as the roots of the quadratic equation $(F_{11} - \lambda)(F_{22} - \lambda) - F_{12}F_{21} = 0$. These roots are $\lambda_1 = \frac{Tr\mathbf{F} + \sqrt{(Tr\mathbf{F})^2 - 4det\mathbf{F}}}{2}$ and $\lambda_2 = \frac{Tr\mathbf{F} - \sqrt{(Tr\mathbf{F})^2 - 4det\mathbf{F}}}{2}$ or $\lambda_1 = \frac{Tr\mathbf{F} + \sqrt{\Delta}}{2}$ and $\lambda_1 = \frac{Tr\mathbf{F} - \sqrt{\Delta}}{2}$ where $\Delta = (Tr\mathbf{F})^2 - 4det\mathbf{F}$ is called the *discriminant* of the quadratic equation. If $\Delta > 0$ both roots are real, if $\Delta = 0$ there is one real root and if $\Delta < 0$ both roots are imaginary. In two dimensions an example of a deformation with two real eigenvectors is a pure shear; the eigenvectors are parallel to the principal axes of strain. An example of a deformation with $\Delta = 0$ and hence one real eigenvector is a simple shear; the eigenvector is parallel to the shear direction. An example of a deformation with imaginary eigenvectors is a pure rotation. The condition $\Delta = 0$ can be written as $det\mathbf{F} = \frac{1}{4}(Tr\mathbf{F})^2$. This is a parabola and is plotted in Figure 2.11(a).

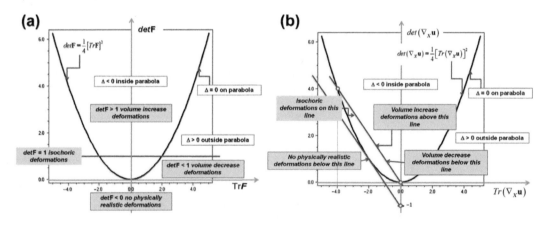

FIGURE 2.11 'Phase portraits' for affine displacement fields. (a) The space $Tr\mathbf{F} - det\mathbf{F}$. The parabola $det\mathbf{F} = \frac{1}{4}(Tr\mathbf{F})^2$ divides the space into deformations with imaginary eigenvectors inside the parabola and real eigenvectors on and outside the parabola. Isochoric deformations are for $det\mathbf{F} = 1$; volume decrease deformations lie below this line and volume increase deformations above. (b) The space $Tr(\nabla_{\boldsymbol{\chi}}\mathbf{u}) - det(\nabla_{\boldsymbol{\chi}}\mathbf{u})$. The parabola $det(\nabla_{\boldsymbol{\chi}}\mathbf{u}) = \frac{1}{4}(Tr(\nabla_{\boldsymbol{\chi}}\mathbf{u}))^2$ divides the space into deformations with imaginary eigenvectors inside the parabola and real eigenvectors on and outside the parabola. Isochoric deformations are for $det(\nabla_{\boldsymbol{\chi}}\mathbf{u}) = -Tr(\nabla_{\boldsymbol{\chi}}\mathbf{u})$ with volume increase deformations above that line and volume decrease deformations below. (c) Plots of pseudo phase portraits for various isochoric and volume increase deformations in $Tr\mathbf{F} - det\mathbf{F}$ space. (d) Plots of phase portraits for deformations involving a 50% decrease in volume. Each insert in (c) and (d) is a plot of displacement vectors in a material representation, (X_1, X_2). Note that the range of volume changes in (c) is unrealistic for geological deformations but is presented to illustrate the trends in pattern development.

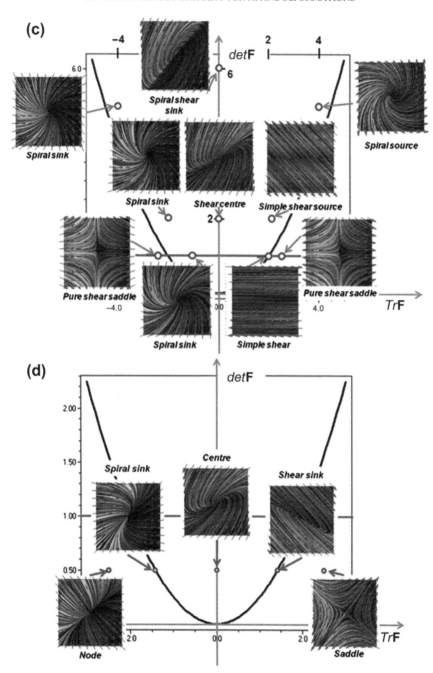

FIGURE 2.11 (*continued*).

A. THE MECHANICS OF DEFORMED ROCKS

The diagram can be divided into a number of regions: $\Delta < 0$ inside the parabola, $\Delta = 0$ on the parabola and $\Delta > 0$ outside the parabola. In addition the deformations involve a volume increase for $det\mathbf{F} > 1$, are isochoric for $det\mathbf{F} = 1$ and involve a volume decrease for $det\mathbf{F} < 1$. No physically realistic deformations exist for $det\mathbf{F} \leq 0$.

Instead of a plot of $det\mathbf{F}$ against $Tr\mathbf{F}$ as in Figure 2.11(a) an argument similar to the above indicates that we could use the displacement gradient matrix derived from $\mathbf{u} = x - X$ and represent these same relations on a plot of $det(\nabla_X \mathbf{u})$ against $Tr(\nabla_X \mathbf{u})$ as shown in Figure 2.11(b). Since $\nabla_X \mathbf{u} = \begin{bmatrix} F_{11} - 1 & F_{12} \\ F_{21} & F_{22} - 1 \end{bmatrix}$, Figure 2.11(b) follows from the relations

$$det(\nabla_X \mathbf{u}) = det\mathbf{F} - Tr\mathbf{F} + 1 \quad \text{and} \quad Tr(\nabla_X \mathbf{u}) = Tr\mathbf{F} - 2$$

The condition for an isochoric deformation is $det(\nabla_X \mathbf{u}) = -Tr(\nabla_X \mathbf{u})$. Physically realistic deformations do not exist below the line $det(\nabla_X \mathbf{u}) = -Tr(\nabla_X \mathbf{u}) - 1$. Clearly the representation of these various fields using the displacement gradient (Figure 2.11(b)) is more complicated than using the deformation gradient (Figure 2.11(a)) so we opt to use the latter below.

Various kinds of patterns are shown in Figure 2.11(c and d) depending on the value of Δ and of $det\mathbf{F}$ and in the study of nonlinear dynamics these are given various labels. For isochoric deformations ($det\mathbf{F} = 1$) the behaviour is either a saddle or a shear. For both $det\mathbf{F} > 1$ and $det\mathbf{F} < 1$ spiral sinks, sources or centres develop. For $\Delta > 0$ other patterns called nodes can form. The outstanding issue that remains is: *What is the range of types of patterns that typify natural deformations, especially those that involve volume changes?* So far this question is open. Perhaps some constraints can be placed on the answer to this question by examining non-affine deformations where we will see that more than one kind of pattern can develop within a single phase portrait.

2.7 EXAMPLE 2: NON-AFFINE DEFORMATIONS

In this section we give some examples of non-affine deformations. First an isochoric deformation is considered, then a non-isochoric deformation which resembles crenulation cleavage with a volume decrease in the strongly sheared zones. This is followed by examples of Type 1C folds of Ramsay (1967), with and without volume changes. Finally we present an example of an isochoric localised shear zone.

2.7.1 Isochoric Non-affine Deformations

Let us first consider the deformation shown in Figure 2.4 and given by

$$x_1 = 2X_1 + \sin(X_2)$$

$$x_2 = 0.5 \sin(X_2)$$

We have $\mathbf{F} = \begin{bmatrix} 2 & \cos X_2 \\ 0 & 0.5 \end{bmatrix}$ and $\mathbf{F}^T = \begin{bmatrix} 2 & 0 \\ \cos X_2 & 0.5 \end{bmatrix}$.

Therefore $\quad \mathbf{B} = \mathbf{F}\mathbf{F}^\mathrm{T} = \begin{bmatrix} 4 + \cos^2 X_2 & 0.5 \cos X_2 \\ 0.5 \cos X_2 & 0.25 \end{bmatrix}$ $\quad$ and $\quad Tr\mathbf{B} = -4.75 - 0.5 \cos(2X_2),$

$det\mathbf{B} = 1.$

Hence, $\quad \lambda_1 = \frac{1}{2}(4.75 + 0.5 \cos(2X_2) - \beta)$ $\quad$ and $\quad \lambda_2 = \frac{1}{2}(4.75 + 0.5 \cos(2X_2) + \beta)$ $\quad$ where

$\beta = \sqrt{-4 + (0.5 \cos(2X_2) + 4.75)^2}.$

We see that $J = 1$ and so the deformation is isochoric. The eigenvectors of $\mathbf{B}$ are

$$v_1 = \begin{bmatrix} 2\{2.125 + 0.25 \cos(2X_2) - 0.5\beta\} \sec(X_2) \\ 1 \end{bmatrix}$$

and

$$v_2 = \begin{bmatrix} 2\{2.125 + 0.25 \cos(2X_2) + 0.5\beta\} \sec(X_2) \\ 1 \end{bmatrix}$$

The eigenvectors of the strain ellipse, $\mathbf{U}$, are those of $\mathbf{B}$ and the eigenvalues of $\mathbf{U}$ are $\sqrt{\lambda_1}$ and $\sqrt{\lambda_2}$. The distribution of strain ellipses is shown in Figure 2.7(a and b).

2.7.2 Non-affine Deformations with Volume Change

Next we consider a non-isochoric, non-affine transformation where the volume change is periodically distributed. The transformation is

$$x_1 = X_1 + \sin(X_2)$$

$$x_2 = X_1 + 1.25X_2 + 0.05 \sin(X_2^2)$$

and is shown in Figure 2.12. The deformation gradient and its transpose are

$$\mathbf{F} = \begin{bmatrix} 1 & \cos(X_2) \\ 1 & 0.1X_2 \cos(X_2^2) + 1.25 \end{bmatrix}; \quad \mathbf{F}^\mathrm{T} = \begin{bmatrix} 1 & 1 \\ \cos(X_2) & 0.1X_2 \cos(X_2^2) + 1.25 \end{bmatrix}$$

and $J = def\mathbf{F} = 0.1X_2 \cos(X_2^2) + 1.25 - \cos(X_2)$ so the deformation is not isochoric; the volume change is plotted to the left of Figure 2.12(d) which is a plot of the divergence of the displacement field. The displacement plots in Figure 2.12(c and d) comprise saddles alternating with spiral sinks embedded in a shear and emphasise that the deformation comprises essentially regions of pure shear with relatively little strain alternating with regions of shear plus rotation, both embedded in a dominant shear deformation with volume change.

If we explore the range of deformations characterised by

$$\begin{aligned} x_1 &= X_1 + \sin(X_2) \\ x_2 &= X_1 + \alpha X_2 + 0.05 \sin(X_2^2) \end{aligned} \tag{2.18}$$

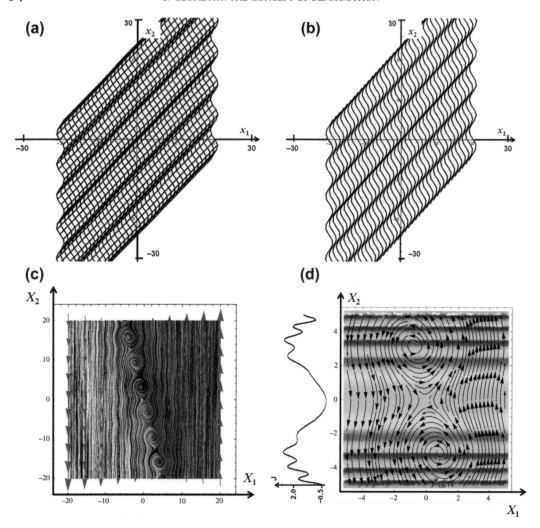

FIGURE 2.12 A nonlinear transformation given by $x_1 = X_1 + \sin(X_2)$ and $x_2 = X_1 + 1.25X_2 + 0.05\sin(X_2^2)$. (a) Deformed grid. (b) Plot of initial lines defined by $X_1 =$ constant to emphasise the resemblance to crenulation cleavage. (c) Displacement field for $-20 \leq X_1 \leq 20$ and $-20 \leq X_2 \leq 20$. (d) Divergence of the displacement field with displacement field superimposed. The dark areas represent volume decreases in the material representation, $-5 \leq X_1 \leq 5$ and $-5 \leq X_2 \leq 5$. On the left-hand side the Jacobian of the deformation gradient is plotted as an indication of the volume change. High volume changes correspond to the boundaries between gold and grey colours for the divergence, whereas low volume changes correspond to the boundaries between grey and light blue for the divergence.

where α varies in the range $1 \leq \alpha \leq 1.5$ the 'phase portraits' vary as shown in Figure 2.13. This variation reflects the increasing role of α in controlling the shear parallel to X_2 together with the increasing volume changes which are given by

$$J = \alpha + 0.1X_2 \cos\left(X_2^2\right) - \cos\left(X_2\right)$$

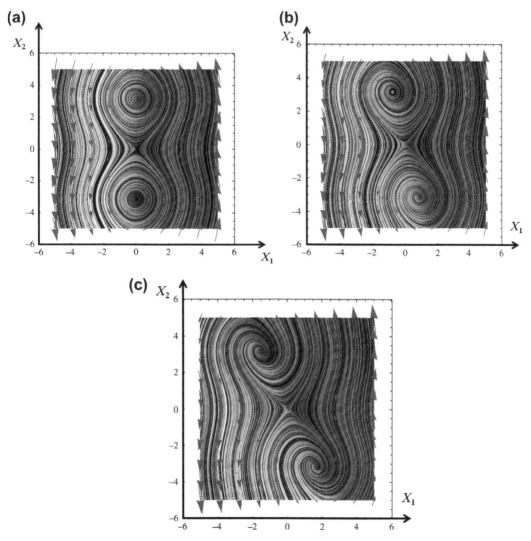

FIGURE 2.13 Pseudo phase portraits for the deformation (2.18) for various values of α. (a) $\alpha = 1$ (b) $\alpha = 1.25$; this is a zoom into Figure 2.12(c) (c) $\alpha = 1.5$.

and hence influenced by α. Figure 2.13(a) represents a deformation comprising a pure shear (a saddle) alternating with regions of almost pure rotation embedded in a simple shear. As α increases from 1 to 1.5 there is a transition pictured in Figure 2.13 from a periodic distribution of close to pure shear saddles alternating with regions of almost pure rotation embedded in a shear deformation (Figure 2.13(a)) to saddles alternating with spiral sources embedded in a shear deformation (Figure 2.13(c)).

We leave as an exercise for the reader the task of calculating the volume changes and plotting the deformed grids and the strain ellipses for parts of these and subsequent

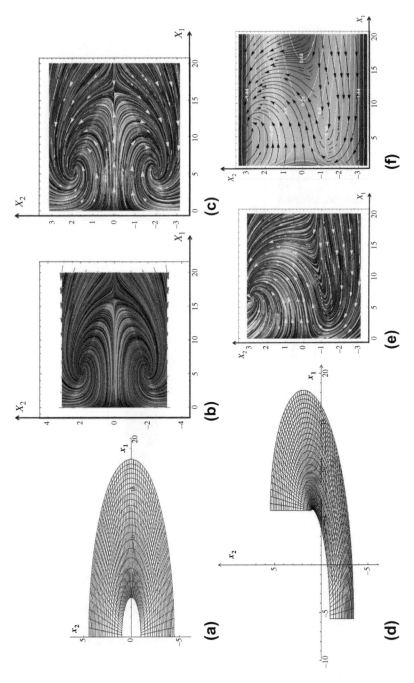

FIGURE 2.14   Deformations that produce Type 1C folds of Ramsay (1969). (a) Isochoric deformation with 50% shortening parallel to x_2. (b) and (c) Phase portraits for the deformation in (a). (d) Type 1C fold with shearing parallel to x_1 and volume changes that depend on both X_1 and X_2. (e) Phase portrait for the deformation in (d). (f) Plot of divergence of the displacement field superimposed on the phase portrait.

deformations. It is to be noted that some of these deformations are restricted in the range of X_2 that ensures $J > 0$.

2.7.3 Type IC Folds

Next we examine the following non-affine transformation that produces the Type IC folds of Ramsay (1967):

$$x_1 = a \cos (CX_2)\sqrt{2AX_1 + B}; \quad x_2 = b \sin (CX_2)\sqrt{2AX_1 + B}$$

This is plotted for $a = 2$, $b = 0.5$, $C = 0.5$, $A = 2$ and $B = 0$ in Figure 2.14 (a). The deformation is isochoric for $a = 1/b$.

The deformed spatial representation is shown in Figure 2.14(a) where the shapes of the deformed initially square grid elements indicate that the principal axes of strain form a convergent fan as proposed by Ramsay (1967). The displacement field is shown in Figure 2.14(b and c).

In Figure 2.14(d, e and f) another deformation is shown for

$$x_1 = 4 \cos (0.5X_2)\sqrt{X_1} + 0.8X_2; \quad x_2 = 0.05X_1 + \sin (0.5X_2)\sqrt{X_1}$$

The deformation consists of a shear of Figure 2.14(a) and volume loss which depends on both X_1 and X_2 as shown by the divergence field in Figure 2.14(f). The divergence field is quite different from those shown in Figures 2.9(f) and 2.12(d). Notice that in Figure 2.14(a) lines that were initially straight and parallel to X_2 remain straight in the deformed state, whereas in Figure 2.14(d) they become curved in parts of the fold. Distinctions of this type become important when we come to consider folding mechanisms in Volume II.

2.7.4 Localised Shearing

In many deformed rocks the deformation is localised into narrow zones commonly referred to as *shear zones* where the deformation is intense and not necessarily homogeneous. The surrounding rocks in many instances are undeformed. An expression that describes such a deformation is

$$x_1 = X_1 + Re\left[2\frac{\exp[i(2X_1 - 1)]}{\cosh[2(X_1 + 3)]}\right], \quad x_2 = X_2, \quad i \equiv \sqrt{-1}$$

where Re stands for the real part of the expression in square brackets and i is defined as the square root of (-1). This expression is a simplified version of a more general expression for solitons given by Bronshtein et al. (2007, p. 549). The deformation is shown in Figure 2.15(a) and the phase portrait in Figure 2.15(b) where the localised nature of the deformation is emphasised.

In this section we have presented a number of examples of phase portraits associated with non-affine deformations. The important point regarding these phase portraits as opposed to phase portraits for affine deformations (Section 2.6) is that two or more attractors such as sinks or saddles can exist for non-affine portraits, whereas only one is ever developed for an affine deformation. This coexistence of two or more attractors in phase portraits from

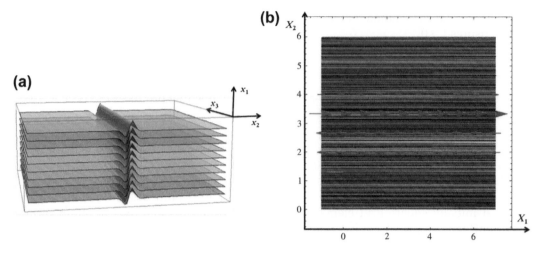

FIGURE 2.15   Localised shearing. (a) Plot of localised, non-affine shearing packet. (b) Phase portrait for (a).

nonlinear dynamical systems is common and we will examine many throughout this book. Figures 2.13 and 2.14 also illustrate some of the ways in which phase portraits can vary as the deformation changes with different combinations of saddles and sinks arranged in various patterns all embedded in a shear.

2.8 THE DEFORMATION ARISING FROM SLIP ON A SINGLE PLANE

As a simple example of a deformation gradient that arises from plastic slip within a single crystal we consider slip in a direction defined by a unit vector, $\mathbf{s}$, on a single plane whose normal is $\mathbf{m}$. We take the situation shown in Figure 2.16(a) where $\mathbf{s}$ is the vector $[1\ 0]^T$ and $\mathbf{m}$ is the vector $[0\ 1]^T$. Then the displacement gradient is $\gamma \mathbf{s} \otimes \mathbf{m} = \begin{bmatrix} 0 & \gamma \\ 0 & 0 \end{bmatrix}$ and $\mathbf{F} = \mathbf{I} + \gamma \mathbf{s} \otimes \mathbf{m} = \begin{bmatrix} 1 & \gamma \\ 0 & 1 \end{bmatrix}$. Here $\otimes$ represents the tensor product (see Appendix B). The deformation is given by $x_1 = X_1 + \gamma X_2$ and $x_2 = X_2$ and so is a simple shear with a shear of γ. Thus $\mathbf{F}^T\mathbf{F} = \begin{bmatrix} 1 & \gamma \\ \gamma & 1 + \gamma^2 \end{bmatrix}$ and the deformation is isochoric since the determinant of $\mathbf{F}$ equals 1. The eigenvalues of $\mathbf{F}^T\mathbf{F}$ are $\frac{1}{2}\{(2 + \gamma^2) \pm \gamma\sqrt{4 + \gamma^2}\}$ and the rotation is anticlockwise through an angle $\tan^{-1}\left(\frac{\gamma}{2}\right)$. Thus if $\gamma = 1$ the principal stretches are 1.618 and 0.618 and the rotation is 26.57°. The total deformation, $\mathbf{F}$, is shown in Figure 2.15(b) and comprises a plastic deformation, $\mathbf{U}^P$, with principal stretches given above and an anticlockwise lattice rotation, $\mathbf{R}^L$, through 26.57° such that $\mathbf{F} = \mathbf{R}^L\mathbf{U}^P$.

In addition a general deformation may also involve an overall rigid rotation $\mathbf{R}$ so that the complete general deformation gradient that arises from single slip is $\mathbf{F} = \mathbf{R}(\mathbf{I} + \gamma \mathbf{s} \otimes \mathbf{m})$.

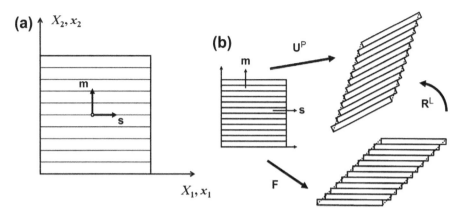

FIGURE 2.16　The deformation of a crystal by slip on a plane whose normal is **m** with a direction of slip, **s**. (a) Initial state. (b) Deformed state. The deformation defined by $\mathbf{F} = (\mathbf{I} + \gamma \mathbf{s} \otimes \mathbf{m})$ can be divided into a plastic strain, $\mathbf{U}^{P}$, and a lattice rotation $\mathbf{R}^{L}$.

This discussion can be extended to include multiple slip within a crystal (Asaro and Rice, 1977; Asaro, 1979; Gurtin et al., 2010, pp. 583–661), climb of dislocations (Groves and Kelly, 1969) and diffusion effects (Wheeler, 2010; Gurtin et al., 2010, pp. 361–414). For instance, if there are α slip systems in a crystal with slip systems defined by $\mathbf{s}^{(\alpha)}$ and $\mathbf{m}^{(\alpha)}$ then the deformation gradient is

$$\mathbf{F} = \sum_{\alpha} \mathbf{R}^{(\alpha)} (\mathbf{I} + \gamma^{(\alpha)} \mathbf{s}^{(\alpha)} \otimes \mathbf{m}^{(\alpha)})$$

If the deformation is by climb of dislocations then the deformation gradient is

$$\mathbf{F} = \sum_{\alpha} (\mathbf{I} + \gamma^{(\alpha)} \mathbf{s}^{(\alpha)} \otimes \mathbf{s}^{(\alpha)})$$

where $\gamma^{(\alpha)}$ is now interpreted as the dilation introduced by the downward climb of an extra half plane associated with the dislocation. If grain boundary sliding occurs the displacement vector in (2.4) needs to be included in $\mathbf{F}$. We explore these areas in Chapter 13 and Volume II.

2.9　INCREMENTAL STRAIN MEASURES

The discussion thus far in this chapter has concerned measures of finite deformations and strains and finds its prime use in describing and exploring large deformations. The main interest in mechanics lies in solving problems where the deformation of the body results from tractions or displacements applied to the boundaries of the deforming body. In some rare instances analytical solutions to these problems are available but for the most part the behaviour of the system has to be explored using some form of computational tool such as the finite element method (Tadmor et al., 2012). In such cases it is necessary to deform the body of interest in suitably small increments and hence it becomes important to understand the incremental forms of the tensors and matrices we have discussed above.

The standard way of proceeding is to deform the body by a small amount and then expand the resulting tensor of interest using a Taylor expansion (see Appendix C); high-order terms that are small in the context of the problem being solved are neglected. Thus if the coordinates of a point, p, are x in a deformed state and p undergoes a further small displacement, u, to become p$'$ (Figure 2.17) then a tensor of interest, $\varphi(x)$ becomes $\varphi(x + u)$. If G is a function of $\varphi(x)$ then G as a function of $\varphi(x + u)$ can be written in a Taylor expansion:

$$G[\varphi + u] = G[\varphi] + \nabla_\varphi G(u) \cdot u + \cdots$$

where $\nabla_\varphi G(u)$ is the gradient of $G(\varphi)$ with respect to $\varphi(x)$.

We give expressions for the incremental forms of some of the quantities discussed above in terms of the incremental displacement, u, in Table 2.3. Note the use of both upper and lower case indices. Derivations of these expressions may be found in Tadmor et al. (2012, p. 92). In this table $\boldsymbol{\varepsilon}$ is the *small strain tensor*. If the undeformed reference configuration ($\mathcal{B}^d$ in this case) is considered to be close to coincident with the deformed configuration (dotted red in Figure 2.17), then $F_{iJ} = \delta_{iJ}$ and we obtain

$$\varepsilon_{ij} = \frac{1}{2}\left(\frac{\partial u_i}{\partial x_j} + \frac{\partial u_j}{\partial x_i}\right) \quad \Leftrightarrow \quad \boldsymbol{\varepsilon} = \frac{1}{2}\left[\nabla u + (\nabla u)^T\right]$$

which is the classical expression for the small strain tensor.

An important issue arises in the expression of the small strain version of the strain tensor for special deformations. In particular we will see in Volume II where we consider the general theory of buckling, that the complete geometry of the problem, namely, the incorporation of curvature, even though the displacements are small, is not addressed unless one considers

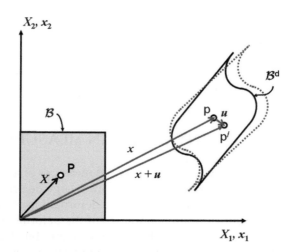

FIGURE 2.17   The initial square, $\mathcal{B}$ undergoes a non-affine transformation to become $\mathcal{B}^d$ so that the point P with coordinates X in the undeformed state becomes the point p in the deformed state with coordinates x. The body $\mathcal{B}^d$ is then deformed incrementally to become the dotted red body and p is displaced incrementally through the vector u to p$'$ with coordinates $(x + u)$.

TABLE 2.3   Incremental Forms of Deformation and Strain Measures

Measure	Incremental Form
Deformation gradient, **F**	$\dfrac{\partial u_i}{\partial X_J} \equiv F_{iJ}$
Right Cauchy-Green tensor, **C**	$\dfrac{\partial u_i}{\partial X_I} F_{iJ} + \dfrac{\partial u_i}{\partial X_J} F_{iI}$
Left Cauchy-Green tensor, **B**	$\dfrac{\partial u_i}{\partial X_I} F_{jI} + \dfrac{\partial u_j}{\partial X_I} F_{iI}$
Lagrangian strain tensor, **E**	$\frac{1}{2}[\mathbf{F}^T\nabla_0 \boldsymbol{u} + (\mathbf{F}^T\nabla_0 \boldsymbol{u})^T]$
Jacobian, **J**	$Tr\boldsymbol{\varepsilon}$ (commonly called the *dilation*)

In Table 2.3 the symbol ∇_0 emphasises that the gradient is calculated in the incremental deformation represented by the right-hand side of Figure 2.17.

second-order terms in $\frac{\partial u_i}{\partial x_j}$. The Lagrangian-Green strain tensor is used for small amplitude buckling and has the following components (Fung, 1965, p. 465):

$$E_{11} = \frac{\partial u_1}{\partial x_1} + \frac{1}{2}\left[\left(\frac{\partial u_1}{\partial x_1}\right)^2 + \left(\frac{\partial u_2}{\partial x_1}\right)^2 + \left(\frac{\partial u_3}{\partial x_1}\right)^2\right]$$

$$E_{22} = \frac{\partial u_2}{\partial x_2} + \frac{1}{2}\left[\left(\frac{\partial u_1}{\partial x_2}\right)^2 + \left(\frac{\partial u_2}{\partial x_2}\right)^2 + \left(\frac{\partial u_3}{\partial x_2}\right)^2\right]$$

$$E_{12} = \frac{1}{2}\left[\frac{\partial u_1}{\partial x_2} + \frac{\partial u_2}{\partial x_1} + \left(\frac{\partial u_1}{\partial x_1}\frac{\partial u_1}{\partial x_2}\right) + \left(\frac{\partial u_2}{\partial x_1}\frac{\partial u_2}{\partial x_2}\right) + \left(\frac{\partial u_3}{\partial x_1}\frac{\partial u_3}{\partial x_2}\right)\right]$$

These components differ from the classical components of the small strain tensor by the addition of second-order terms in $\frac{\partial u_i}{\partial x_j}$.

2.10 COMPATIBILITY OF DEFORMATIONS

Deformation in rock masses is commonly heterogeneous in that the deformation in one part of the rock mass is quite different from other parts. Yet the parts fit together after deformation with no gaps between them. One therefore needs to understand the conditions that enable such compatibility of deformations to be achieved. The issue is not to be confused with discussions of strain compatibility (see Jaeger, 1969; Ramsay, 1967, 1980). Such discussions concern relationships between the infinitesimal strain components in a deformation arising from the fact that six components of strain are derived from just three components of displacement and there are conditions on the strains that define continuity and smoothness of the displacement field in order for this to be possible. Means (1983) addressed the issue of compatibility of deformations across a boundary using off-axis Mohr circle constructions; such a procedure is relevant to two dimensions but there seems to be no equivalent construction available for three dimensions.

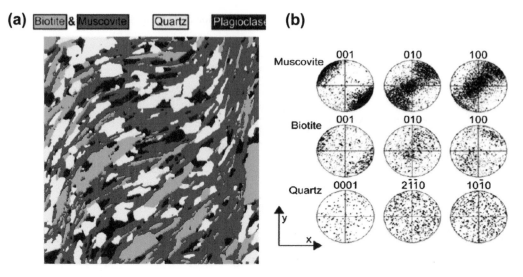

FIGURE 2.18   Grain shape foliation from crenulation cleavage zone. (a) Foliation defined by grain boundaries between biotite, muscovite, quartz and plagioclase. (b) Crystallographic preferred orientations for muscovite, biotite and quartz. *From Naus-Thijssen et al. (2011).*

The problem becomes relevant to polycrystalline aggregates where the deformation in each grain is different but the issue is pertinent at all scales up to the lithospheric scale. Thus in the deformed polycrystalline aggregate shown in Figure 2.18 comprised of biotite, muscovite, quartz and plagioclase, each grain has different strengths and deformation mechanisms and the overall deformation is inhomogeneous yet the grains fit together with no open spaces between grains (at least at the scale of microscopic observation). The issue here is: *What are the mathematical conditions that allow the aggregate to maintain compatibility with no gaps or overlaps between grains?* Below we concentrate on the compatibility of deformation across boundaries between adjacent rock masses or mineral grains. In Volume II a model is presented whereby grain shape foliations such as pictured in Figure 2.18(a) develop by a combination of crystal slip processes, disconnection motion and minimisation of the energy of the surfaces generated between grains. The competition between these processes is responsible for the *grain size and shape* but the crystal slip process leads to a *crystallographic preferred orientation (CPO)* such as depicted in Figure 2.18(b) along with a deformation prescribed for each grain. The shape orientation arises from a combination of the necessity for deformation compatibility across each grain boundary and the fact that CPO's have developed in some or all minerals.

There are two aspects to deformation compatibility in an inhomogeneous deformation. One is local compatibility between deformations in adjacent domains and the other is global compatibility of the array of deformed domains with the imposed deformation. In both instances we require the rock structure to develop with no gaps or overlaps. This requirement also eliminates any local stress concentrations and long-range stresses. The general situation is shown in Figure 2.19 where two initial regions, $\mathcal{B}^{(+)}$ and $\mathcal{B}^{(-)}$, in contact across a surface

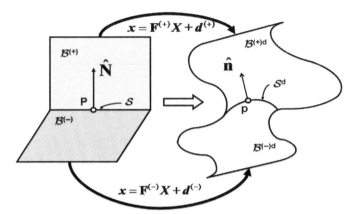

FIGURE 2.19   Two different reference bodies $\mathcal{B}^{(+)}$ and $\mathcal{B}^{(-)}$ separated by a surface $\mathcal{S}$ with normal at P of $\hat{\mathbf{N}}$ suffer two different deformations defined by $\mathbf{F}^{(+)}$ and $\mathbf{F}^{(-)}$ to become $\mathcal{B}^{(+)\mathrm{d}}$ and $\mathcal{B}^{(-)\mathrm{d}}$ separated by a surface $\mathcal{S}^{\mathrm{d}}$ with normal at p of $\hat{\mathbf{n}}$. $\mathbf{d}^{(+)}$ and $\mathbf{d}^{(-)}$ are the displacement vectors relevant to either side of the boundary. The deformations across $\mathcal{S}^{\mathrm{d}}$ are continuous but the deformation gradients are not. The question we ask is: What are the conditions that the deformations should be continuous across $\mathcal{S}^{\mathrm{d}}$?

$\mathcal{S}$ are both inhomogeneously deformed but remain intact across a deformed surface $\mathcal{S}^{\mathrm{d}}$. This means the deformations across $\mathcal{S}^{\mathrm{d}}$ given by

$$x = \mathbf{F}^{(\pm)}X + d^{(\pm)}$$

are continuous but the deformation gradients, $\mathbf{F}^{+}$ and $\mathbf{F}^{-}$, are discontinuous.

We consider two domains labelled $(+)$ and $(-)$ in Figure 2.20(a) separated by a surface, $\mathcal{S}$. x_2 is parallel to $\hat{\mathbf{N}}$ and x_1 is tangent to $\mathcal{S}$ at P. Deformations defined by $x = \mathbf{F}^{+}X + \mathbf{d}^{+}$ and $x = \mathbf{F}^{-}X + \mathbf{d}^{-}$ occur in the two domains at P where $\mathbf{d}^{+}$ and $\mathbf{d}^{-}$ are constant displacement vectors. We are interested in the conditions under which the deformation is continuous across $\mathcal{S}$ at P but the deformation gradients are discontinuous.

For compatibility between two domains characterised by two deformation gradients, $\mathbf{F}^{+}$ and $\mathbf{F}^{-}$, the tangent to the boundary surface separating the two domains before deformation must be equally distorted and rotated by both $\mathbf{F}^{+}$ and $\mathbf{F}^{-}$ during the deformation. This is a way of saying that the boundary surface must be an invariant surface in both deformations. Hence, on the boundary

$$\mathbf{F}^{+}X + \mathbf{d}^{+} = \mathbf{F}^{-}X + \mathbf{d}^{-}$$

or,

$$(\mathbf{F}^{+} - \mathbf{F}^{-})X = \mathbf{d}^{+} - \mathbf{d}^{-}$$

For X to be continuous across the boundary $\mathbf{d}^{+} = \mathbf{d}^{-}$ and hence if l is a line in the boundary

$$\mathbf{F}^{+}l = \mathbf{F}^{-}l$$

or

$$(\mathbf{F}^{+} - \mathbf{F}^{-})l = 0$$

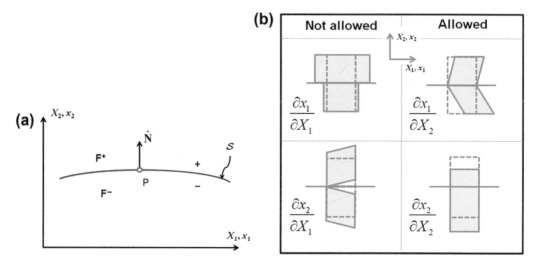

FIGURE 2.20 Compatibility of two deformations across an arbitrary boundary. (a) A curved boundary, S, between two deformations F^+ and F^-. We seek the conditions that the deformation across S with normal $\hat{N}$ at P should be continuous. (b) Two domains outlined by dotted lines in the undeformed state either side of a surface represented by the horizontal full line. These domains are deformed to become the grey boxes. In order for the deformations to be compatible the boundaries of the grey boxes must match across the boundary. $\frac{\partial x_1}{\partial X_1}$ and $\frac{\partial x_2}{\partial X_1}$ must match either side of the boundary while $\frac{\partial x_1}{\partial X_2}$ and $\frac{\partial x_2}{\partial X_2}$ are unconstrained.

which says that F^+ deforms l in the same way that F^- does and hence S is an invariant surface in the bulk deformation.

If X is continuous across the boundary then the following relations hold (see Figure 1.20(b)):

$$\left(\frac{\partial x_1}{\partial X_1}\right)^+ = \left(\frac{\partial x_1}{\partial X_1}\right)^- \quad \text{and} \quad \left(\frac{\partial x_2}{\partial X_1}\right)^+ = \left(\frac{\partial x_2}{\partial X_1}\right)^-$$

There are no restrictions on $\left(\frac{\partial x_1}{\partial X_2}\right)^{\pm}$ or $\left(\frac{\partial x_2}{\partial X_2}\right)^{\pm}$. Hence $(F^+ - F^-)$ must be of the form

$$\left(F^+ - F^-\right) = \begin{bmatrix} 0 & \alpha \\ 0 & \beta \end{bmatrix}$$

where $\alpha = \left(\frac{\partial x_1}{\partial X_2}\right)^+ - \left(\frac{\partial x_1}{\partial X_2}\right)^-$ and $\beta = \left(\frac{\partial x_2}{\partial X_2}\right)^+ - \left(\frac{\partial x_2}{\partial X_2}\right)^-$.

In order to proceed we make use of the identity (Appendix B)

$$(a \otimes \hat{N})l = a(l \cdot \hat{N})$$

where a is a vector yet to be determined. Now l is normal to $\hat{N}$ and so $(l \cdot \hat{N}) = 0$. Hence compatibility is ensured if we can find a vector a such that

$$\left(F^+ - F^-\right) = a \otimes \hat{N}$$

This follows since

$$\mathbf{F}^+ l - \mathbf{F}^- l - \left(a \otimes \widehat{\mathbf{N}}\right) l = a\left(l \cdot \widehat{\mathbf{N}}\right) = 0$$

and hence

$$\mathbf{F}^+ l = \mathbf{F}^- l$$

which means that both $\mathbf{F}^+$ and $\mathbf{F}^-$ deform an arbitrary vector l in the boundary equally. Hence the compatibility condition is

$$\mathbf{F}^+ - \mathbf{F}^- = a \otimes \widehat{\mathbf{N}} \tag{2.17}$$

and a is the vector $[\alpha, \beta]^T$.

These same arguments apply in three dimensions. This relation is known as the *Hadamard jump condition* (Vardoulakis and Sulem, 1995, pp. 39–41); it represents a discontinuity or jump in the deformation gradient but ensures continuity of the deformation across the boundary.

As an example consider the following three-dimensional situation. A grain is divided into two domains A^+ and A^- by a boundary S which is parallel to the $x_1 x_3$ plane and with a normal $\widehat{\mathbf{N}}$ parallel to x_2. These could be two grains undergoing different deformations in an aggregate. The deformation gradients in A^+ and A^- are given by $\mathbf{F}^+ = \begin{bmatrix} 1 & \gamma & 0 \\ 0 & 1 & 0 \\ 0 & 0 & 1 \end{bmatrix}$

and $\mathbf{F}^- = \begin{bmatrix} 1 & -\gamma & 0 \\ 0 & 1 & 0 \\ 0 & 0 & 1 \end{bmatrix}$ so that $\mathbf{F}^+ - \mathbf{F}^- = \begin{bmatrix} 0 & 2\gamma & 0 \\ 0 & 0 & 0 \\ 0 & 0 & 0 \end{bmatrix} = 2\gamma \begin{bmatrix} 1 \\ 0 \\ 0 \end{bmatrix} \otimes \begin{bmatrix} 0 \\ 1 \\ 0 \end{bmatrix} = 2\gamma a \otimes \widehat{\mathbf{N}}$

where it emerges that a is a vector parallel to the x_1 coordinate axis and the region undergoes a simple shear through the combined deformations in A^+ and A^-.

This completes our tour through the geometry of deformation. Next we turn our attention to the *movements* that produce such deformations.

Recommended Additional Reading

Bhattacharya, K. (2003). *Microstructure of Martensite: Why it Forms and How it Gives Rise to the Shape-Memory Effect.* Oxford University Press.
  Although this book is concerned with martensitic transformations, it is a good basic introduction to deformation and contains many useful discussions on the deformation of crystals including subjects such as deformation compatibility, twinning and deformation-induced phase transformations.

Bronshtein, I.N., Semendyayev, K.A., Musiol, G., Muehlig, H. (2007). *Handbook of Mathematics*, fifth ed. Springer-Verlag.
  This is a comprehensive summary of mathematical concepts and expressions that is a useful reference resource.

Fung, Y.C. (1965). *Foundations of Solid Mechanics.* Prentice-Hall.
  A thorough treatment of the principles of solid mechanics.

Gurtin, M.E., Fried, E., Anand, L. (2010). *The Mechanics and Thermodynamics of Continua*. Cambridge University Press.
An up-to-date discussion of mechanics with an emphasis on thermodynamics and the link to the development of constitutive equations. The text includes a good introduction to vector and tensor analysis applied to deformation theory.

Jaeger, J.C. (1969). *Elasticity, Fracture and Flow*. Methuen.
An invaluable, classical treatment of mechanics.

Means, W.D. (1976). *Stress and Strain*. Springer-Verlag.
A discussion of stress and strain concepts with an emphasis on the use of Mohr diagrams.

Petersen, K.B., Pedersen, M.S. (November 14, 2008). *The Matrix Cookbook*. Version. http://matrixcookbook.com.
A comprehensive compilation of matrix manipulations and formulae.

Ramsay, J.G. (1967). *Folding and Fracturing of Rocks*. McGraw-Hill Book Company.
The classical treatment of structural geology with an emphasis on strain and of methods for estimating strain in deformed rocks.

Tadmor, E.B., Miller, R.E., Elliot, R.S. (2012). *Continuum Mechanics and Thermodynamics*. Cambridge University Press.
A modern treatment of mechanics with an excellent introduction to matrices and tensors.

Kinematics — Deformation Histories

3.1 THE MOVEMENT PICTURE AND THE MATERIAL DERIVATIVE

When a body of rock is subjected to forces or displacements on its boundary the particles that comprise the body undergo motions that ultimately result in deformation. It is the history of these motions, especially relative motions between elements of the rock, that results in the deformation fabrics that we observe at all scales in deformed metamorphic rocks and in the patterns of accumulated strain, rotation and displacement that are developed. The map of these motions was called *das Teilbewegung* or *the movement picture* by Sander (1911) who took his inspiration from the extensive treatment of mechanics by Thomson and Tait (1872) and the classic paper by Becker (1893). Thus for nearly 120 years structural geologists have recognised that the *motions* or *flow* involved in deformation are important for understanding the development of rock structures. Moreover, we will see in Chapter 6 that the behaviours of most materials of interest when subjected to boundary forces or displacements are related to these internal motions and not to the deformation or strain. As Truesdell (1966, p. 4) observes: "*The stress in a body is determined by the history of the motion of the body*". Thus the study of deformation history is of fundamental importance. In keeping with geological usage we call this area of study that involves *internal motions* of a deforming body *kinematics*. The subject of kinematics is a *geometrical* description of the flows involved in deformation. Even though one can prescribe a velocity field, there is no guarantee that a particular material can follow these motions and that these motions will be compatible with some set of imposed boundary conditions. This chapter is concerned with some prescribed motions involved in the deformation of rocks and the fundamental features of the flow fields that control the development of deformation fabrics. Of particular importance are the *eigenvectors* of the flow field. The characteristic of a flow eigenvector is that its *orientation* is not changed by continuation of the flow, although the eigenvector may be stretched or may change sense. Thus once material lines or planes become close to or parallel to an eigenvector, that orientation tends to be preserved or reinforced so long as the flow does not change. Hence eigenvectors of the flow play a fundamental role in fabric development.

The *deformation history* is given in terms of the undeformed or reference coordinates, X, as:

$$x = f(X, t) \tag{3.1}$$

where f can be a linear or nonlinear function and t is the time; this is a *material* or *Lagrangian* description of the motion in that it maps the time history of material particles initially at X to the spatial positions, x, of these same material particles at time t. If convenient, one can write the deformation history in terms of the deformed coordinates:

$$X = \breve{f}(x, t) \tag{3.2}$$

where $\breve{f}(x, t)$ is the inverse of $f(X,t)$. (3.2) is a *spatial* or *Eulerian* description of the deformation history. For $\breve{f}(x, t)$ to exist we require, for all values of t, that the Jacobian of the deformation gradient, $\mathbf{F}$, is greater than zero:

$$J = det\mathbf{F} = \left| \frac{\partial x_i}{\partial X_j} \right| > 0$$

In Figure 3.1 we show some examples of features that are commonly used to infer something about the movement pictures for deformed rocks.

When one observes a deformed rock, one is looking at a *spatial* or *Eulerian* representation of material points that defines a deformation. However, a structural geologist has in mind what these rocks looked like in some reference or undeformed state and this is a *material* or *Lagrangian* way of viewing the deformation. Put in another way, a Lagrangian view concentrates on individual material particles that have been deformed relative to some reference state; this is the way structural geologists view the world. An Eulerian view concentrates on the particular volume of space that contains the deformed material. This is the way metamorphic petrologists view the world; concentrations are discussed in terms of the deformed volume at x and metamorphic reactions are considered to have occurred at x with no dependence on the material configuration at X. The various matrices and tensors discussed in Chapter 2 and others that describe the kinematics can be classified as mixed, spatial or

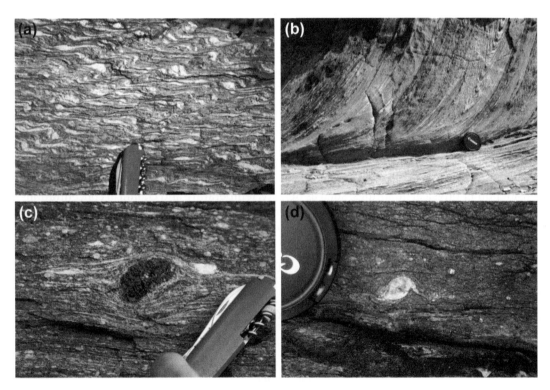

FIGURE 3.1 Some structures in deformed rocks used to say something about the kinematics or movement picture. (a) Deformed granite gneiss from Spencers Gorge in the Chewings Range, Central Australia. The deformed K-feldspars and the asymmetric relation between a horizontal foliation and cross-cutting foliation indicate dextral shearing. (b) Early isoclinal folds deformed by a later zone of intense dextral shearing; Cap de Creus, Spain. (c) Garnet aggregate deformed by dextral shearing; Irindina Gneiss, Lizzy Creek, Central Australia. (d) Porphyroclast of K-feldspar deformed by dextral shearing; Bruna Gneiss, beneath the Mt Ruby Massif, Central Australia.

TABLE 3.1   A Summary of the Operations of Tensors and Matrices Used in Describing Deformation and Flow

Tensor	Name	Type	Operation
$\mathbf{F}, \mathbf{F}^{-T}$	Deformation gradient, inverse transpose of $\mathbf{F}$	Mixed	Map material to spatial vectors
$\mathbf{F}^{-1}, \mathbf{F}^{T}$	Inverse and transpose of $\mathbf{F}$	Mixed	Map spatial to material vectors
$\mathbf{R}$	Rotation matrix	Mixed	Maps material to spatial vectors
$\mathbf{B}$	Left Cauchy–Green tensor	Spatial tensor field	Maps spatial to spatial vectors
$\mathbf{V}$	Left stretch tensor; strain ellipsoid	Spatial tensor field	Maps spatial to spatial vectors
$\mathbf{C}$	Right Cauchy–Green tensor	Material tensor field	Maps material to material vectors
$\mathbf{U}$	Right stretch tensor; reciprocal strain ellipsoid	Material tensor field	Maps material to material vectors
$\boldsymbol{\ell}$	Spatial velocity gradient tensor	Spatial tensor field	Maps spatial to spatial vectors
$\mathbf{d}$	Spatial stretching tensor	Spatial tensor field	Maps spatial to spatial vectors
$\mathbf{w}$	Spatial spin tensor	Spatial tensor field	Maps spatial to spatial vectors

material fields as indicated in Table 3.1. The deformation gradient, $\mathbf{F}$, for instance is a mixed tensor field that maps material vectors to spatial vectors.

However, the Lagrangian viewpoint carries with it some complications when one needs to consider other physical quantities, such as density, temperature and chemical concentrations, that may be identified with a material particle in the undeformed state and that move (or advect) with the particle during deformation. For example, if the spatial description of the temperature is

$$T = T(\boldsymbol{x}, t)$$

the material description is

$$T = T(\boldsymbol{X}(\boldsymbol{x}, t), t)$$

The *material derivative* of T is defined by

$$\frac{DT}{Dt} \equiv \left.\frac{\partial T}{\partial t}\right|_{X}$$

and represents the change in temperature with time while following the particle initially at $\boldsymbol{X}$. The *spatial derivative* of T is

$$\frac{\partial T}{\partial t} = \left.\frac{\partial T}{\partial t}\right|_{x}$$

and represents the change in the temperature at the spatial position x as the material particles move through x. Hence by the chain rule of differentiation the relationship between the two can be found as:

$$\frac{DT}{Dt} = \frac{\partial T}{\partial t} + \frac{\partial T}{\partial x}\frac{dx}{dt}$$

$$= \frac{\partial T}{\partial t} + (\mathbf{v} \cdot \nabla_x T) \tag{3.3}$$

where $\mathbf{v}$ is the velocity of the material point at x. The $(\mathbf{v} \cdot \nabla_x T)$ term in (3.3) arises because the material particles are moving through the spatial coordinates. This term is called the *convective* rate of change of T, whereas the $\frac{\partial T}{\partial t}$ term is called the *local* rate of change of T. $\frac{DT}{Dt}$, the *material derivative* of T, is also known as the *substantial*, *advective* or *convective* derivative of T. The material derivative may be viewed as the time rate of change of a quantity associated with a material particle that would be observed by someone travelling with the material particle.

3.2 VELOCITY AND VELOCITY (OR FLOW) FIELDS

We define the *displacement*, $\mathbf{u}$, of a material particle as

$$\mathbf{u} = x - X$$

Then, since X is independent of time $\mathbf{v} = \frac{dx}{dt} = \frac{d(\mathbf{u}+X)}{dt} = \frac{d\mathbf{u}}{dt}$.

Most of the velocity fields considered in the geological literature produce affine deformations. For such deformations it is convenient to distinguish between those histories where the principal axes of strain remain parallel from one increment to the next and those where the principal axes are not parallel from one increment to the next. These are known, respectively, as *coaxial* and *non-coaxial histories*. Ramberg (1975) considered the following three classes of two-dimensional affine deformation histories (see also Means et al., 1980):

3.2.1 Type 1 Deformation Histories: Non-pulsating

The deformation history is given by:

$$x_1 = [A\beta_{11} + \exp(-\alpha t)]X_1 + A\beta_{12}X_2$$
$$x_2 = -A\beta_{21}X_1 + [\exp(\alpha t) - A\beta_{11}]X_2 \tag{3.4}$$

where $A = \exp(\alpha t) - \exp(-\alpha t)$ and for isochoric deformation histories $\beta_{12}\beta_{21} = \beta_{11}(\beta_{11} - 1)$. The velocity field in spatial coordinates is

$$v_1 = \alpha[2\beta_{11} - 1]x_1 + 2\alpha\beta_{12}x_2$$
$$v_2 = -2\alpha\beta_{21}x_1 + \alpha[1 - 2\beta_{11}]x_2 \tag{3.5}$$

Such deformation histories may be coaxial or non-coaxial and are characterised by situations where lines that were extended in the past always continue to be extended in future increments of deformation but there may also exist lines that were shortened in the past that will be extended in future increments.

3.2.2 Type 2 Deformation Histories: Pulsating

The deformation history is given by:

$$x_1 = \left[\cos\left(\beta t\right) + \frac{a_{11}}{\beta}\sin\left(\beta t\right)\right]X_1 + \frac{a_{12}}{\beta}\sin\left(\beta t\right)X_2$$

$$x_2 = \frac{a_{21}}{\beta}\sin\left(\beta t\right)X_1 + \left[\cos\left(\beta t\right) + \frac{a_{22}}{\beta}\sin\left(\beta t\right)\right]X_2$$

(3.6)

where for isochoric deformation histories $a_{11} = -a_{22}$ and $a_{11}a_{22} = a_{12}a_{21} - \beta^2$. The velocity field in spatial coordinates is

$$v_1 = a_{11}x_1 + a_{12}x_2$$

$$v_2 = a_{21}x_1 + a_{22}x_2$$

(3.7)

These deformation histories are always non-coaxial and are characterised by situations where lines that were shortened in the past are extended in future increments of deformation. These deformations may correspond to those associated with rotating porphyroclasts. Ramberg (1975) and McKenzie (1979) have proposed such histories for packages of rock involved in mantle convection.

3.2.3 Type 3 Deformation Histories: Simple Shearing

The deformation history corresponds to simple shearing and is given by:

$$x_1 = [1 + \beta_{11}t]X_1 + \beta_{12}tX_2$$

$$x_2 = \beta_{21}tX_1 + [1 + \beta_{22}t]X_2$$

(3.8)

where for isochoric deformation histories $\beta_{11} = -\beta_{22}$ and $\beta_{12}\beta_{21} = -\beta_{11}^2$. The velocity field in spatial coordinates is

$$v_1 = \beta_{11}x_1 + \beta_{12}x_2$$

$$v_2 = \beta_{21}x_1 - \beta_{11}x_2$$

(3.9)

Type 3 deformation histories are always non-coaxial.

An important feature of all these spatial velocity fields is that they are independent of time; such vector fields are said to be *autonomous* (Wiggins, 2003). The velocity fields can be plotted in spatial coordinates, **x**, and the patterns illustrate the way in which the velocity field is distributed in space. One can consider the velocity vectors or the curves that are tangent to the velocity vector at each point in the flow. These curves are known as *streamlines* and the maps of the streamlines and of the velocity vectors are called *phase portraits*.

The most general affine, two dimensional, autonomous, isochoric deformation history is given by

$$v_1 = \beta x_2, \quad v_2 = K\beta x_1 \quad \text{for } -1 < K < 1 \tag{3.10}$$

For $K < 0$ the streamlines are ellipses given by $x_2^2 - Kx_1^2 = constant$ (Figure 3.2(a)); these correspond to pulsating histories (Type 2 above). For $K > 0$ the streamlines are hyperbolas with an extension axis inclined at $\theta = \tan^{-1}(1/K)^{1/2}$ to the x_2 axis (Figure 3.2(b)); these correspond to Type 1 histories. For $K = 0$ the streamlines are straight and correspond to simple shearing (Type 3 histories; Figure 3.2(c)).

As well as the velocity field we can define a curve in spatial coordinates that is the trajectory followed by a material particle with initial coordinates, **X**. This is known as the *path line*. Streamlines and path lines are coincident only for autonomous flows. Otherwise they are generally different. Now, in contrast to similar plots in Chapter 2, diagrams such as Figure 3.2 represent the true trajectories of particles in time so long as the velocity field is autonomous. A range of streamline plots for autonomous flows is shown in Figure 3.3 and further discussion is presented by Passchier (1997). Classification of this range of patterns can be made on the basis of the nature of the eigenvectors of the spatial velocity gradient tensor. We define the *spatial velocity gradient tensor* as:

$$\ell = \begin{bmatrix} \dfrac{\partial v_1}{\partial x_1} & \dfrac{\partial v_1}{\partial x_2} & \dfrac{\partial v_1}{\partial x_3} \\[2ex] \dfrac{\partial v_2}{\partial x_1} & \dfrac{\partial v_2}{\partial x_2} & \dfrac{\partial v_2}{\partial x_3} \\[2ex] \dfrac{\partial v_3}{\partial x_1} & \dfrac{\partial v_3}{\partial x_2} & \dfrac{\partial v_3}{\partial x_3} \end{bmatrix} \equiv (\nabla_x \mathbf{v}) \tag{3.11}$$

Following the arguments presented in Chapter 2, the eigenvalues, $\lambda^{\pm}$, for a two dimensional velocity field are given by the solutions to the characteristic equation:

$$\lambda^2 - (Tr\ell)\lambda + (det\ell) = 0$$

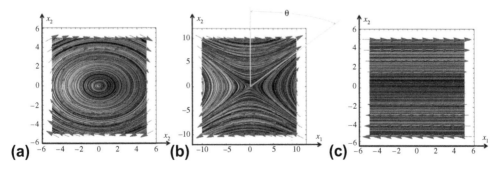

FIGURE 3.2 Phase portraits for the three types of two dimensional, affine, isochoric, autonomous flows. The lines correspond to streamlines which have the velocity vectors (in red) at each point as tangents. (a) $K < 0$, corresponding to Type 2 deformation histories. (b) $K > 0$, corresponding to Type 1 deformation histories. (c) $K = 0$, corresponding to Type 3 deformation histories. K is defined in (3.10).

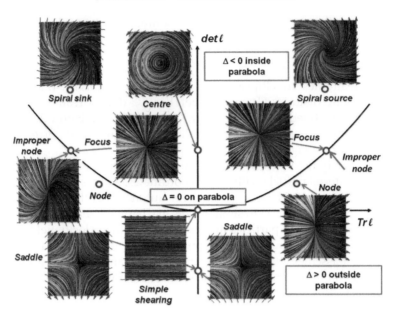

FIGURE 3.3   Classification of two-dimensional, autonomous flow fields. The parabola $det\boldsymbol{\ell} = \frac{1}{4}(Tr\boldsymbol{\ell})^2$ defines fields for which $\Delta = 0$ on the parabola, $\Delta < 0$ inside the parabola and $\Delta > 0$ outside the parabola. These fields define the type of flow pattern.

and the discriminant for this quadratic equation is

$$\Delta = (Tr\boldsymbol{\ell})^2 - 4(det\boldsymbol{\ell})$$

with eigenvalues $\lambda^\pm = \frac{1}{2}\left[Tr\boldsymbol{\ell} \pm \Delta^{\frac{1}{2}}\right]$

Thus $\Delta > 0$ corresponds to real eigenvalues, $\Delta = 0$ corresponds to one real eigenvalue and $\Delta < 0$ corresponds to imaginary eigenvalues. The possibilities are as follows (Ottino, 1989) and can be represented on a diagram (Figure 3.3) similar to Figure 2.11. The parabola $\Delta = 0$ defines four fields of behaviour:

1. All real eigenvalues of different signs.

$\boldsymbol{\ell} = \begin{bmatrix} \lambda & 0 \\ 0 & \mu \end{bmatrix}$ with $\lambda < 0 < \mu$ or with $\lambda > 0 > \mu$. These correspond to *saddles*.

2. All eigenvalues have negative real parts.

$\boldsymbol{\ell} = \begin{bmatrix} \lambda & 0 \\ 0 & \lambda \end{bmatrix}$ with $\lambda < 0$; this corresponds to a *focus*.

$\boldsymbol{\ell} = \begin{bmatrix} \lambda & 0 \\ 0 & \mu \end{bmatrix}$ with $\lambda < \mu < 0$; this corresponds to a *stable node*.

$$\ell = \begin{bmatrix} \lambda & 0 \\ 1 & \lambda \end{bmatrix}$$ with $\lambda < 0$; this corresponds to an *improper node*.

$$\ell = \begin{bmatrix} a & -b \\ b & a \end{bmatrix}$$ with $a < 0$; this corresponds to a *spiral sink*.

3. All eigenvalues have positive real parts.
Same as 2 but with all arrows in Figure 3.3 reversed.

$$\ell = \begin{bmatrix} \lambda & 0 \\ 0 & \lambda \end{bmatrix}$$ with $\lambda > 0$; this corresponds to a *focus*.

$$\ell = \begin{bmatrix} \lambda & 0 \\ 0 & \mu \end{bmatrix}$$ with $\lambda > \mu > 0$; this corresponds to an *unstable node*.

$$\ell = \begin{bmatrix} \lambda & 0 \\ 1 & \lambda \end{bmatrix}$$ with $\lambda > 0$; this corresponds to an *improper node*.

$$\ell = \begin{bmatrix} a & -b \\ b & a \end{bmatrix}$$ with $a > 0$; this corresponds to a *spiral source*.

4. All eigenvalues are pure imaginary.

$$\ell = \begin{bmatrix} 0 & -b \\ b & 0 \end{bmatrix} \text{ and } \ell = \begin{bmatrix} 0 & b \\ -b & 0 \end{bmatrix}$$ correspond to a *centre*.

$$\ell = \begin{bmatrix} 0 & -b \\ 0 & 0 \end{bmatrix} \text{ and } \ell = \begin{bmatrix} 0 & 0 \\ -b & 0 \end{bmatrix}$$ correspond to a *simple shearing*.

This classification of flow patterns has been considered by a number of authors including de Paor (1983), Passchier (1988, 1997), Weijermars (1991, 1993) and Weijermars and Poliakov (1993).

In three dimensions the situation is somewhat more complicated (Chella and Ottino, 1985; Iacopini et al., 2010; Olbricht et al., 1982) and we discuss this situation in Section 3.3.3 below.

3.3 OTHER MEASURES OF THE KINEMATICS

Other than the velocity gradient ℓ it is convenient to define a number of quantities that are measures of the deformation rates, the rates of reorientation of lines and of the relative magnitudes of these quantities.

3.3.1 The Stretching and Spin Tensors

The *velocity gradient tensors,* $\mathcal{L}$ and ℓ, are defined as the material and spatial gradients of the instantaneous velocity field:

$$\mathcal{L} = \begin{bmatrix} \dfrac{\partial v_1}{\partial X_1} & \dfrac{\partial v_2}{\partial X_1} & \dfrac{\partial v_3}{\partial X_1} \\[2mm] \dfrac{\partial v_1}{\partial X_2} & \dfrac{\partial v_2}{\partial X_2} & \dfrac{\partial v_3}{\partial X_2} \\[2mm] \dfrac{\partial v_1}{\partial X_3} & \dfrac{\partial v_2}{\partial X_3} & \dfrac{\partial v_3}{\partial X_3} \end{bmatrix} \qquad \ell = \begin{bmatrix} \dfrac{\partial v_1}{\partial x_1} & \dfrac{\partial v_2}{\partial x_1} & \dfrac{\partial v_3}{\partial x_1} \\[2mm] \dfrac{\partial v_1}{\partial x_2} & \dfrac{\partial v_2}{\partial x_2} & \dfrac{\partial v_3}{\partial x_2} \\[2mm] \dfrac{\partial v_1}{\partial x_3} & \dfrac{\partial v_2}{\partial x_3} & \dfrac{\partial v_3}{\partial x_3} \end{bmatrix}$$

As with the deformation gradient tensor, $\mathcal{L}$ and ℓ are asymmetrical but can be split into symmetrical and skew symmetrical parts:

$$\mathbf{D} = \frac{1}{2}(\mathbf{v}\nabla_X + \nabla_X \mathbf{v}) \quad \text{and} \quad \mathbf{W} = \frac{1}{2}(\mathbf{v}\nabla_X - \nabla_X \mathbf{v})$$

$$\mathbf{d} = \frac{1}{2}(\mathbf{v}\nabla_x + \nabla_x \mathbf{v}) \quad \text{and} \quad \mathbf{w} = \frac{1}{2}(\mathbf{v}\nabla_x - \nabla_x \mathbf{v}) \tag{3.12}$$

so that $\mathcal{L} = \mathbf{D} + \mathbf{W}$ and $\ell = \mathbf{d} + \mathbf{w}$. $\mathbf{D}$ and $\mathbf{d}$ are called the material and spatial *rate of deformation tensors* or more commonly the *stretching tensors* and $\mathbf{W}$ and $\mathbf{w}$ are the material and spatial *spin tensors*. Note that $\mathbf{D}$ and $\mathbf{d}$ *are not strain-rate tensors*. In indicial notation these tensors are:

$$D_{ij} = D_{ji} = \frac{1}{2}\left(\frac{\partial v_i}{\partial X_j} + \frac{\partial v_j}{\partial X_i}\right) \qquad W_{ij} = -W_{ji} = \frac{1}{2}\left(\frac{\partial v_i}{\partial X_j} - \frac{\partial v_j}{\partial X_i}\right)$$

$$d_{ij} = d_{ji} = \frac{1}{2}\left(\frac{\partial v_i}{\partial x_j} + \frac{\partial v_j}{\partial x_i}\right) \qquad w_{ij} = -w_{ji} = \frac{1}{2}\left(\frac{\partial v_i}{\partial x_j} - \frac{\partial v_j}{\partial x_i}\right) \tag{3.13}$$

Both $\mathbf{D}$ and $\mathbf{d}$ are second order symmetrical tensors and so can be represented by ellipsoids with principal axes equal to the principal *logarithmic rates of stretch* as indicated in (3.24). $\mathbf{w}$ represents the instantaneous rotation experienced by the principal axes of $\mathbf{d}$. Two other tensors that measure strain-rates are sometimes useful. These are the material derivatives of the Lagrangian and Euler-Almansi strain tensors given (Tadmor et al., 2012, pp. 98, 99), respectively, by:

$$\dot{\mathbf{E}} = \mathbf{F}^T \mathbf{d}\, \mathbf{F} \quad \text{and} \quad \dot{\mathbf{e}} = \frac{1}{2}\left[\ell^T \mathbf{B}^{-1} + \mathbf{B}^{-1}\ell\right] \tag{3.14}$$

Note that $\mathbf{d}$ in general is not a true rate for a general deformation history but it is for a deformation history that involves no rotation of the principal stretching directions. The stretching tensor and the Lagrangian strain-rate tensor are approximately equal for infinitesimal deformations. The Eulerian strain-rate tensor is approximately equal to the stretching tensor only for infinitesimal deformations with small spins (Eringen, 1962, p. 80). The principal axes of the Lagrangian strain-rate tensor are parallel to the principal axes of the stretching tensor (Eringen, 1962, p. 79). We use the convention that the word termination *-ing* denotes a kinematic quantity. Thus *stretch* and *simple shear* are distortions and are *geometric* terms independent of the deformation history; *stretching* and *simple shearing* are *kinematic* terms and imply an evolution of the flow with time.

It is important to relate **d** and **w** to the rate of change of the stretch tensor **U** and of **R**, the rigid body rotation associated with an instantaneous deformation gradient, **F**. Gurtin et al. (2010, p. 90) show that

$$\mathbf{d} = \mathbf{R}\left[sym(\dot{\mathbf{U}}\mathbf{U}^{-1})\right]\mathbf{R}^{T} \quad \text{and} \quad \mathbf{w} = \dot{\mathbf{R}}\mathbf{R}^{T} + \mathbf{R}\left[skw(\dot{\mathbf{U}}\mathbf{U}^{-1})\right]\mathbf{R}^{T} \tag{3.15}$$

where *sym* and *skw* refer to the symmetrical and skew symmetrical parts of the relevant matrix. This means, from $(3.15)_2$, that the spin is the sum of two terms. The first is the rotational spin $\mathbf{w}_{rot}$ arising from the rotation **R**, and the second is a stretch spin, $\mathbf{w}_{str}$ arising from the stretch **U**:

$$\mathbf{w} = \mathbf{w}_{rot} + \mathbf{w}_{str} \tag{3.16}$$

$$\mathbf{w}_{rot} = \dot{\mathbf{R}}\mathbf{R}^{T} \quad \text{and} \quad \mathbf{w}_{str} = \mathbf{R}\left[skw(\dot{\mathbf{U}}\mathbf{U}^{-1})\right]\mathbf{R}^{T}$$

This expresses the statements commonly made in the structural geology literature (Iacopini et al., 2007; Jiang, 2010; Means et al., 1980) that the spin can be expressed as an 'external' and 'internal' rotation. (3.15) and (3.16) are true for all deformation histories, whereas many of the arguments in the structural geology literature are for specific deformation histories such as the combined simple shearing and rotation discussed by Means et al. (1980). The common interpretation of (3.16) is that $\mathbf{w}_{rot}$ represents a rotation of the fabric with respect to external coordinates, and hence has no influence on fabric evolution, whereas $\mathbf{w}_{str}$ represents rotation of material lines relative to the stretching axes and hence has a fundamental influence on fabric development (Means et al., 1980). It is important to note that such statements are true for affine deformation histories but do not hold for non-affine deformation histories where the local rotations represented by $\mathbf{w}_{rot}$ are an integral part of the deformation at each point in the material and hence also contribute to fabric development (Chapter 11). It is the array of local deformations represented by $\mathbf{w}_{rot}(x)$ in a non-affine deformation history that contributes to the deformed material remaining intact with no gaps throughout the deformation history.

3.3.2 Spin and Vorticity

The *vorticity*, **ω**, associated with a velocity field, **v**, is defined as the vector:

$$\boldsymbol{\omega} = curl\mathbf{v} \equiv \nabla \times \mathbf{v} \tag{3.17}$$

It can be shown (Truesdell, 1954, pp. 3 and 58) that

$$\nabla\mathbf{v} = \mathbf{d} - \mathbf{I} \times \boldsymbol{\omega}/2$$

and

$$\nabla \cdot \boldsymbol{\omega} = \nabla \cdot \nabla \times \mathbf{v} = 0$$

Just as the streamlines are lines tangent to the velocity vectors in a flow, the *vortex lines* are tangent to the local vorticity vectors in the flow.

Another measure widely used to classify flow fields is the *kinematic vorticity number*, W_K, given (Truesdell, 1954; Means et al., 1980) by

$$W_K = \frac{|\omega|}{\left[2\left(d_1^2 + d_2^2 + d_3^2\right)\right]^{1/2}} \tag{3.18}$$

where d_1, d_2 and d_3 are the principal stretchings of $\mathbf{d}$ and $|\omega|$ is the magnitude of the vorticity. An important concept that arises here is that of *objectivity*. A quantity is said to be *objective* if its description is independent of the coordinate frame of reference used (Gurtin et al., 2010, Chapter 20). If we want to compare different flows we want the measuring system to be independent of the reference frame. One can show that the stretching tensor, $\mathbf{d}$, is objective but the vorticity, $\boldsymbol{\omega}$, is not except for special flows (Astarita, 1979; Gurtin et al., 2010; Truesdell, 1977). Thus W_K is not an objective measure of a flow and differs from one flow to another depending not only on the nature of the flow but also on the reference frame. A simple example illustrating the non-objectivity of W_K is that of a simple shearing referred to two different sets of coordinate axes. If the coordinate axes x_1 and x_2 are parallel and normal, respectively, to the shearing plane a possible velocity field is $v_1 = x_2$ and $v_2 = 0$; the curl of this velocity field is 1 and $W_K = 1$. However, if we refer this same velocity field to new axes inclined at φ to the field, the vorticity is $\cos \varphi$ and W_K varies from 1 to 0 as φ varies from 0 to $\pi/2$.

Astarita (1979) defined another measure of a flow that is objective and a version of this is also defined by Jiang (2010). Various other forms of a quantity resembling the kinematic vorticity number have been defined by various authors (Jiang, 2010; Passchier, 1988, 1990, 1991, 1997; Robin and Cruden, 1994). For example, Passchier (1988) defines a quantity W_n called the *sectional kinematic vorticity number* so called because it is relevant to two dimensional flows and depends on the principal instantaneous stretchings in a section which is the plane of flow (defined as that containing the eigenvectors corresponding to d_2 and d_3) so that

$$W_n = \frac{|\omega|}{d_2 - d_3} \tag{3.19}$$

Also defined are the *sectional kinematic dilatancy number*, A_n; a number T_n that measures the magnitude of the out-of-plane stretching to the in-plane stretchings; and V_n, a number that is a measure of relative volume change:

$$A_n = \frac{d_2 + d_3}{d_2 - d_3}, \quad T_n = \frac{d_1}{d_2 - d_3} \quad \text{and} \quad V_n = T_n + A_n$$

This means that $W_n = W_K\sqrt{2T_n^2 + A_n^2 + 1}$.

These measures of the flow are not objective since they either use the vorticity or do not employ invariants of $\mathbf{d}$. In describing or classifying different flows it is fundamental that objective measures are used. One needs to be very careful in using published analyses of flows that use W_K or sectional kinematic numbers. In order to use these kinds of measures to describe or compare different flows three criteria must be met: (1) the measure used must be objective or the coordinate axes must be selected to take the non-objective nature of the velocity field into account in any analysis of the flow; (2) the flows must be autonomous; and (3) the flows must be spatially homogeneous. These criteria are not met for many published studies.

3.3.3 Classification of Three Dimensional Flows

The situation for three dimensional (3D) flows is considered by Iacopini et al. (2007, 2010) who point out that if the discriminant of the characteristic equation for a three dimensional flow is positive then three real eigenvalues exist for the flow. However, if this discriminant is negative there is only one real eigenvalue and the other two are complex-conjugate. Their analysis indicates that although some flows may start as a pulsating flow, they will ultimately evolve into a non-pulsating flow dominated by the real eigenvalue at higher strains as shown in Figure 3.4. Such flow evolution has important ramifications for fabric development and we return to this subject in Section 3.6.2.

Different approaches to the classification of three dimensional flow fields are presented by Tanner and Huilgol (1975), Tanner (1976), Astarita (1979), Olbricht et al. (1982) and Chella and Ottino (1985); these classifications are motivated by attempts to define so called *strong flows* that can be responsible for some kinds of microfabric development. The following discussion is based largely on Olbricht et al. (1982), although their discussion is more general as we will see in Section 3.6.2.

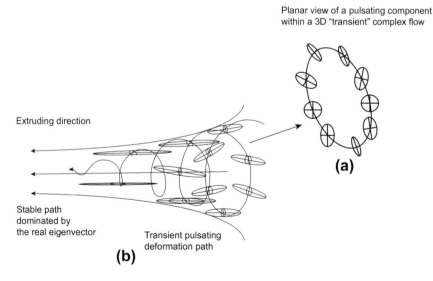

FIGURE 3.4 Illustration of a possible evolution for a three dimensional flow beginning with a transient pulsating history of deformation as shown in (a) and progressing to a stable stretching deformation aligned parallel to a single real eigenvector at later times in the flow (b). *From Iacopini et al. (2010).*

A. THE MECHANICS OF DEFORMED ROCKS

If we orient the spatial coordinate axes parallel to the principal axes of $\mathbf{d}$ we can write for the most general isochoric, affine three dimensional flow:

$$\mathbf{d} = \begin{bmatrix} a & 0 & 0 \\ 0 & b & 0 \\ 0 & 0 & -(a+b) \end{bmatrix} \quad \text{and} \quad \mathbf{w} = \begin{bmatrix} 0 & g & -h \\ -g & 0 & j \\ h & -j & 0 \end{bmatrix}$$

where a, b, g, h, and j are constants. The magnitude of the vorticity vector is $|\omega| = \sqrt{g^2 + h^2 + j^2}$.

The velocity gradient is normalised by dividing by its magnitude and is given by Olbricht et al. (1982) as:

$$\tilde{\ell} = \frac{1}{(2M)^{\frac{1}{2}}} \begin{bmatrix} a & g & -h \\ -g & b & j \\ h & -j & -(a+b) \end{bmatrix} \quad \text{where } M = \sqrt{\ell:\ell} = a^2 + b^2 + ab + \omega^2.$$

The eigenvalues for the flow are given by the roots of the characteristic equation:

$$\lambda^3 - \left(\frac{1}{2} tr\tilde{\ell}^2\right)\lambda - det\tilde{\ell} = 0, \tag{3.20}$$

namely,

$$\lambda_1 = s_1 + s_2$$

$$\lambda_2 = -\frac{1}{2}(s_1 + s_2) + \frac{\sqrt{3}}{2}i(s_1 - s_2)$$

$$\lambda_3 = -\frac{1}{2}(s_1 + s_2) - \frac{\sqrt{3}}{2}i(s_1 - s_2)$$

where $s_1 \equiv \left[\frac{det\tilde{\ell}}{2} + \sqrt{\Delta}\right]^{\frac{1}{3}}$, $s_2 \equiv \left[\frac{det\tilde{\ell}}{2} - \sqrt{\Delta}\right]^{\frac{1}{3}}$ and $i \equiv \sqrt{-1}$. Thus one eigenvector is always real and the other two are complex-conjugate as discussed by Iacopini et al. (2010).

The invariants of $\tilde{\ell}$ are:

$$I_{\tilde{\ell}} = 0$$

$$II_{\tilde{\ell}} = Tr\tilde{\ell}^2 = \frac{1}{M}(a^2 + b^2 + ab - \omega^2)$$

$$III_{\tilde{\ell}} = det\tilde{\ell} = -\frac{1}{(2M)^{\frac{3}{2}}}(ab[a+b] + [g^2(a+b)] - j^2 a + h^2 b)$$

This means that $-1 \le Tr\tilde{\ell}^2 \le 1$ with the minimum of $Tr\tilde{\ell}^2$ corresponding to purely rotational flow ($a=b=0$) and the maximum corresponding to pure straining with $\omega^2 = 0$. Olbricht et al. (1982) show that for each value of $Tr\tilde{\ell}^2$, the maximum and minimum values of $det\tilde{\ell}$ are given by

$$det\tilde{\ell}\Big|_{max/min} = \pm\frac{sgn(b)2(1+3K)}{[6(1+K)]^{\frac{3}{2}}} \quad \text{with } K = \frac{1 - Tr\tilde{\ell}^2}{1 + Tr\tilde{\ell}^2} \tag{3.21}$$

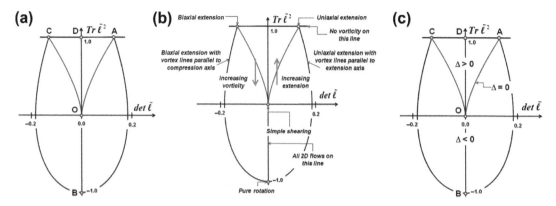

FIGURE 3.5   Classification of three dimensional, affine, isochoric flows. The region ABCD defines the *admissible domain* that contains all possible three dimensional affine isochoric flows. (a) The admissible region ABCD. The red curve COA represents the condition $\Delta = 0$. (b) Characteristics of various parts of the admissible region. (c) The three fundamental parts of the admissible region characterised by $\Delta > 0$, $\Delta = 0$ and $\Delta < 0$. $\tilde{\ell}$ in these diagrams is the normalised velocity gradient tensor.

Thus, any three dimensional, affine flow field can be represented by a point within the region ABCD in Figure 3.5 where the blue line, ABC, is given by (3.21), The black line, CDA, represents the maximum value of the trace of $\tilde{\ell}^2$, $Tr\tilde{\ell}^2 = 1$, and the red line, COA, is the condition that the discriminant of the characteristic equation for the flow field is zero:

$$\Delta = \left[\left(det\tilde{\ell} \right)^2 - \left(Tr\tilde{\ell}^2 \right)^3 / 54 \right] = 0$$

The region ABCD maps out what is called the *admissible domain* that contains all possible three dimensional isochoric affine flows. The various flow fields represented in Figure 3.5 are:

1. All two dimensional flows are represented by the line BOD ($det\tilde{\ell} = 0$) with simple shearing at $Tr\tilde{\ell}^2 = 0$, pure rotation at $Tr\tilde{\ell}^2 = -1$ and plane extension at $Tr\tilde{\ell}^2 = 1$.
2. The point A corresponds to uniaxial 3D extension and C corresponds to biaxial 3D extension (pure flattening).
3. The line $Tr\tilde{\ell}^2 = 1$ represents all flows with no vorticity.
4. If the stretching, **d**, is defined, the value of $det\tilde{\ell}$ defines the orientation of the vorticity vector (or vortex lines) relative to the principal axis of stretching. This angle varies across the diagrams being parallel to the principal axis of stretching on the right hand side of the admissible domain, to 45° to the principal axis of stretching at $det\tilde{\ell} = 0$, to being normal to the principal axis of extension on the left hand margin of the admissible domain. The positions defined by ABC correspond to various forms of monoclinic flow (Iacopini et al., 2007; Jiang, 2010; Jiang and Williams, 1998). For $0 < det\tilde{\ell} < det\tilde{\ell}_{max}$ and $det\tilde{\ell}_{min} < det\tilde{\ell} < 0$ within the admissible domain the flows are triclinic.
5. Within the region OCDA, $\Delta > 0$ and there are three real roots to the characteristic equation (3.20). Outside of this region $\Delta > 0$ and there is one real root and there are two complex conjugate roots.

We return to this diagram in Section 3.6.2 where we consider some aspects of the development of microfabric driven by the flow field and the concept of *weak* and *strong* flows.

3.4 RATE OF CHANGE OF DEFORMATION MEASURES

The following section discusses the time rate of change of various measures of deformation.

3.4.1 Rate of Change of Deformation Gradient

We write (2.2) again:

$$d x = (\nabla_X x) \cdot dX$$

and so

$$\overline{dx} = \overline{FdX} = \dot{F}dX$$

where the overbar means that the material derivative is taken of everything under the bar. Now,

$$\dot{F}_{iJ} = \frac{\partial v_i}{\partial x_j} \frac{\partial x_j}{\partial X_J} = l_{ij}F_{jJ}$$

Thus

$$\dot{F} = \ell F \tag{3.22}$$

From which one may obtain

$$\overline{dx_i} = l_{ij}F_{jJ}dX_J = l_{ij}dx_j \tag{3.23}$$

3.4.2 Rate of Change of Stretch

The stretch, $\bar{\lambda}$, of an initial line element, dS, is $\bar{\lambda} = \frac{ds}{dS}$ where $ds = \sqrt{dx_i dx_i}$ and $dS = \sqrt{dX_I dX_I}$. Using (3.23) the material time derivative of ds is

$$\overline{ds} = \overline{\sqrt{dx_i dx_i}} = \frac{d_{ij}dx_i dx_i}{ds}$$

If $\hat{m}$ is a unit vector along dx then $dx_i = \hat{m}_i ds$ and

$$\frac{1}{ds}\overline{ds} = d_{ij}\hat{m}_i \hat{m}_j$$

Now $\bar{\lambda} = \frac{ds}{dS}$ and $\dot{\bar{\lambda}} = (\overline{ds})/dS$ and so

$$\overline{\ln\bar{\lambda}} = d_{ij}d\hat{m}_i d\hat{m}_j \tag{3.24}$$

This is a relation between the rate of change of the Hencky strain and the stretching tensor and can be used to show that the principal axes of $\mathbf{d}$ are the principal logarithmic rates of stretching (Tadmor et al., 2012, pp. 97, 98).

3.4.3 Rate of Change of Volume

In appendix C we show that

$$\frac{\partial(det\mathbf{A})}{\partial\mathbf{A}} = \mathbf{A}^{-T}det\mathbf{A}$$

Hence

$$\dot{J} = \overline{det\mathbf{F}} = \frac{\partial(det\mathbf{F})}{\partial\mathbf{F}} : \dot{\mathbf{F}} = J\mathbf{F}^{-T} : \dot{\mathbf{F}} \tag{3.25}$$

Using the identity $\mathbf{A}:(\mathbf{BC}) = (\mathbf{B}^T\mathbf{A}):\mathbf{C}$ we obtain

$$\dot{J} = J\mathbf{I} : \boldsymbol{\ell}$$

Now $\mathbf{I} : \boldsymbol{\ell} = \mathbf{I} : \nabla\mathbf{v} = div\mathbf{v}$ and $\mathbf{I} : \boldsymbol{\ell} = Tr\boldsymbol{\ell} = Tr\mathbf{d}$ since $tr\mathbf{w} = 0$.
Hence the useful result:

$$\dot{J} = Jdiv\mathbf{v} = JTr\boldsymbol{\ell} = JTr\mathbf{d} \tag{3.26}$$

If the motion is to be isochoric then $\dot{J} = 0$ which means that $div\mathbf{v} = 0$. For incompressible flows a convenient relation from (3.25) for isochoric flow is

$$\mathbf{F}^{-T} : \dot{\mathbf{F}} = 0 \tag{3.27}$$

3.4.4 Rate of Change of Area

We have from Nanson's formula in Section 2.1 that

$$\overline{d\mathbf{a}} = \left[\dot{J}\mathbf{F}^{-T} + J\overline{\mathbf{F}^{-T}}\right]d\mathbf{A}$$

Since from (3.26) we have $\dot{J} = Jdiv\mathbf{v}$ and from (3.22), $\overline{F_{Ij}^{-1}} = -F_{Ii}^{-1}l_{ij}$ we obtain

$$\overline{d\mathbf{a}} = \left[(div\mathbf{v})\mathbf{I} - \boldsymbol{\ell}^T\right]d\mathbf{a} \tag{3.28}$$

3.5 AN EXAMPLE OF A NON-AFFINE FLOW

Most examples of flows in the literature are affine. As an example of a non-affine flow and the use of the kinematic vorticity number for non-autonomous flows, consider the velocity field:

$$v_1 = \frac{1}{t}[x_1 - \sin(tx_2)]$$
$$v_2 = -\frac{x_2}{t} \tag{3.29}$$

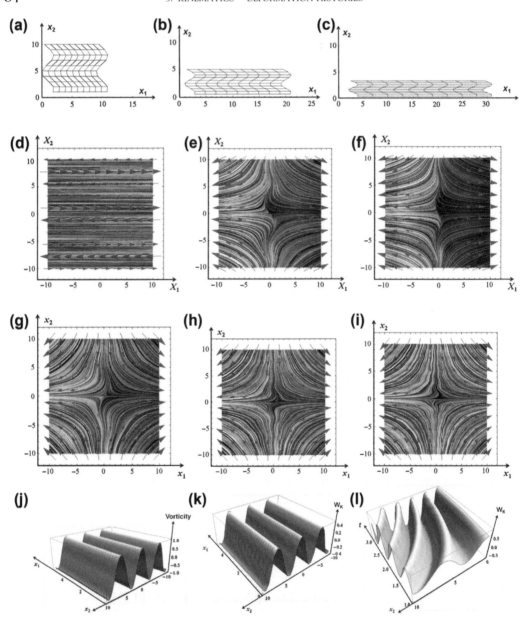

FIGURE 3.6 Deformation, velocity and vorticity fields for the non-affine deformation (3.29). (a), (b) and (c) Deformed grids for the deformation history given in (3.29) for times $t = 1, 2, 3$. (d), (e) and (f) Material representations of the displacement fields for times $t = 1, 2, 3$. (g), (h) and (i) Spatial representations of the velocity fields for times $t = 1, 2, 3$. (j) Distribution of vorticity in spatial coordinates, $t = 1$. (k) The kinematic vorticity number in spatial coordinates, $t = 1$. (l) Time and space variation of kinematic vorticity number. In the plots for W_K the value of the vorticity is used and not its magnitude as in (3.18).

This produces the isochoric deformations shown in Figure 3.6 (a, b, c) for times $t = 1, 2$ and 3 and given by:

$$x_1 = tX_1 + \sin(X_2) \quad \text{and} \quad x_2 = \frac{X_2}{t}$$

The deformations at three times $t = 1, 2, 3$ are shown in Figure 3.6(a,b,c) and consist of non-affine deformations with 0, 50 and 66.7% shortening normal to the axial planes, respectively. The material description of the displacement is shown in Figure 3.6(d,e,f) and emphasises that the deformation in (a) is an inhomogeneous simple shear, whereas subsequent frames (b) and (c) approach pure shear. The spatial description of the velocity field is shown in Figures 3.6(g,h,i). The spatial distribution of vorticity is shown for $t = 1$ in Figure 3.6(j), whilst (k) shows the spatial distribution of W_K also for $t = 1$. The coordinate axes used are the spatial coordinates, x. The basic flow is essentially an inhomogeneous pure shearing as indicated by the mean value of W_K which is zero; departures arising from the non-affine parts of the deformation give maximum and minimum values of $W_K = \pm\frac{1}{\sqrt{6}} = \pm 0.406$. Notice that the *value* of the vorticity has been used here rather than the *magnitude* as is commonly done for affine deformations. The spatial and temporal variation of W_K is shown in Figure 3.6(l) illustrating that at any specified spatial position in a non-affine deformation, W_K varies with time. It hence is meaningless to define a single value for W_K at a particular spatial position.

3.6 KINEMATIC INDICATORS AND FLOW FIELDS

3.6.1 The Development of Microfabric due to Flow

A large number of publications describe structures that indicate a sense of shear movement in deformed rocks. These include Sander (1930, 1950), Simpson and Schmid (1983), Lister and Snoke (1984), Passchier and Simpson (1986), Hammer and Passchier (1991), Olson and Pollard (1991), Passchier and Trouw (1996), Goscombe and Trouw (1999), Goscombe et al. (2004), Cosgrove (2007), and Wagner et al. (2010). Some examples of such indicators are given in Figure 3.1. These kinematic indicators rely for their interpretation on the *symmetry principle of Sander (1911)* according to which *the symmetry of the fabric reflects the symmetry of the movements that produced the fabric.* Many of these indicators show that there is a shear strain parallel to foliation planes (ten Grotenhuis et al., 2003) and to 'stretching' lineations. In fact in many deformed rocks the asymmetry revealed by these kinematic indicators is only exhibited in sections parallel to a 'stretching' lineation or normal to an inferred vorticity vector (Passchier and Simpson, 1986).

Over the past 100 years or so there have been three approaches by structural geologists to linking features of the flow field to the structures observed in deformed rocks. Sander (1911) proposed that the *symmetry of fabric* was all important and that this symmetry reflects that of the *movement picture* or *kinematics* (see Paterson and Weiss, 1961; Turner and Weiss, 1963). Sander proposed that for monoclinic deformation fabrics one could define three mutually perpendicular axes that were reflections of the kinematics: the *a-axis* parallels the direction of *'tectonic transport'*, the *b-axis* is the single symmetry axis of the fabric and is commonly an *axis of rotation* and the *c-axis* is normal to the *ab-plane*. This approach led to considerable controversy (for reasons we will discuss in Volume II) and ultimately fell into disrepute.

The next approach proposed that all structures are related to patterns of strain and so the measurement or estimation of strain in deformed rocks is paramount (Ramsay, 1967; Ramsay and Huber, 1983). This area of research was emphasised in the period 1967–1987 and is commonly rooted in assumptions or propositions that many foliations and lineations form, respectively, parallel to principal planes and axes of strain. Because the existence of the required number of independently distorted objects is rarely available in order to make unique estimates of the magnitudes and orientations of the principal axes of strain without making additional assumptions, most of this work provides some kind of (commonly unspecified) bound on what the strain actually is. Although this approach provides important information on such bounds and what the states of strain are, there remain important issues such as (1) many foliations, although proposed as principal planes of strain, show widespread evidence of shear displacements and shear strains parallel to the foliation so that in their present state they cannot be principal planes of strain since by definition a principal plane of strain is a plane of zero shear strain. There are widespread justifications of such observations in order to protect the initial premise. (2) Many 'stretching lineations' postulated to be parallel to a principal axis of strain show shear parallel to the lineation in the form of rotated porphyroclasts, 'mica fish' or small-scale asymmetric folds (ten Grotenhuis et al., 2003; Passchier and Simpson, 1986; Passchier and Trouw, 1996). Again, by definition, in their present state such lineations cannot represent principal axes of strain.

The third approach is basically a reversion to that of Sander but dressed up in a different terminology. The emphasis is on description and quantification of the kinematics (or movement picture of Sander). One branch of this approach involves describing the nature of the flow field responsible for the deformation by developing criteria for estimation of the *kinematic vorticity number*, W_K, of Truesdell (1952, 1953) or some set of related numbers (Jiang, 2010; Passchier, 1997). W_K was introduced into structural geology by Means et al. (1980) and is given by (3.18); it is a non-objective measure of the relative intensity of vorticity and stretching. Since Means et al. (1980) there have been a number of developments including Passchier and Urai (1988), Passchier (1991, 1994, 1997), Fossen and Tikoff (1993), Robin and Cruden (1994), Jiang and Williams (1998) and Jiang (2010).

From the point of view of this kinematic approach, a direct question that one could ask in trying to understand the nature of the flow responsible for geological deformations is: *What are the eigenvectors for the flow field?* One asks this question initially because the nature of the eigenvectors defines the various classes of flow fields that exist; these distinctions are shown in Figures 3.3 and 3.5 for two- and three-dimensional homogeneous flows. However, a second reason for asking such a question is based on the proposal of Passchier (1997) (see also Iacopini et al., 2007) that the nature of the flow field acts as an *attractor* for passive fabric elements. This means that many fabric elements are advected in the flow to align in some manner with the eigenvectors of the flow. This is particularly relevant to simple shearing deformation histories where there is but one eigenvector (parallel to the shearing direction) and the proposal is that fabric elements tend to align parallel to this direction. In our view the attractor concept represents a fundamental advance in understanding deformation fabrics and we summarise the essential features of the attractor concept below.

1. One way of characterising flow fields is to look at the pattern of streamlines as in Figures 3.2 and 3.3. The parameters that enable a classification of these patterns are the eigenvectors of the spatial gradient, ℓ, of the velocity field as we have seen in Figure 3.3.

2. The classical approach in continuum mechanics is to study the nature of the stretching tensor, $\mathbf{d}$, and the vorticity tensor, $\mathbf{w}$, which are derived as the symmetrical and skew symmetrical parts of the velocity gradient, ℓ. The eigenvectors of $\mathbf{d}$ define the principal stretchings at each point and are directly related to the stresses through the relevant constitutive model for the material (Chapters 4 and 6). Thus the tensor $\mathbf{d}$ plays centre stage in much of continuum mechanics. The same kinds of problems arise in relating $\mathbf{d}$ to deformation fabrics as have been discussed for the strain tensor: for instance, 'stretching lineations' are commonly proposed to form parallel to a principal axis of strain or of straining yet kinematic indicators indicate shear parallel to the lineation.

3. The major advance (Passchier, 1997) has been to propose that *the fabric elements are controlled in their orientations by the eigenvectors of the flow*. The eigenvectors of the stretching tensor, $\mathbf{d}$, in general are not parallel to those of the flow gradient, ℓ. Thus the lineations can reorient to become approximately parallel to an eigenvector of the flow but because they are oblique to the principal axes of $\mathbf{d}$ they still experience shearing parallel to the lineation. This is particularly evident in simple shearing flows where the single eigenvector of the flow is always at $45°$ to the principal axes of stretching.

4. As we have indicated in Chapter 2 the strain is a geometrical feature of deformation that accumulates as the deformation history proceeds. Except for purely elastic materials it plays no single role in controlling the stress or the deformation history. Yet stretches parallel to the principal axes of strain progressively accumulate and at least one of these axes rotates towards an eigenvector of the flow. It is wrong to propose that the flow eigenvector is an attractor for material lines parallel to the principal planes and axes of strain since the asymptotic approach of a principal axis of strain towards a flow eigenvector is purely a geometrical necessity. In general, there is a different line of material particles defining a principal axis of strain at each moment of a (rotational) deformation history and so the asymptotic alignment of a principal axis of strain with a flow eigenvector does not necessarily correspond to the advection of a single line of material particles.

5. In hindsight the *a-axis* of Sander is to be identified with an eigenvector of a monoclinic flow field and the *b-axis* with the vorticity vector, $\boldsymbol{\omega}$, defined as *curl*$\mathbf{v}$.

We illustrate the importance of these points in an example below. Figure 3.7 shows maps of the two eigenvectors of $\mathbf{d}$ for the flow (3.29) at times $t = 1, 2, 3$. Passchier (1997) introduces the concept of *flow eigenvector grids* that can be imagined to be embedded in the deforming rock mass at any particular instant. The diagrams in Figure 3.8(a,b) represent these grids for the flow (3.29) at time $t = 1, 3$. The flow eigenvector pairs are not orthogonal. Nevertheless, in this example, one eigenvector remains constant in orientation throughout the deformation history, as emphasised in Figure 3.8(c), and this is the one parallel to the axial planes of the folds. Thus throughout the deformation history, the principal axes of strain rotate as shown by the deformed grids in Figure 3.6(a,b,c) but one flow eigenvector remains constant in orientation. Throughout the deformation history the principal axes of the stretching tensor are orthogonal but in general are oblique to the eigenvectors of the flow as indicated by comparing

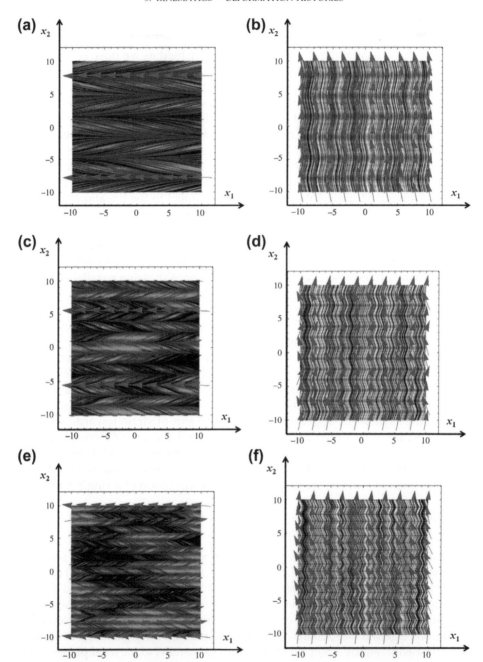

FIGURE 3.7   Examples of stretching eigenvector grids for the non-affine flow (3.29). The lines and arrows represent the orientations of the principal axes of **d** at each point in the spatial representation. $t = 1$ for (a), (b); $t = 2$ for (c), (d); and $t = 3$ for (e), (f). Frames (a), (c), (e) represent the maximum (extensional) eigenvectors for **d** and frames (b), (d), (f) represent the minimum (shortening) eigenvectors for **d**.

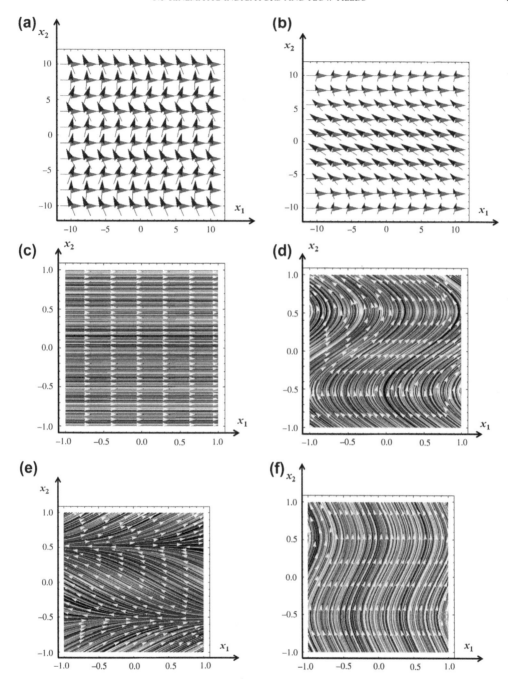

FIGURE 3.8 Comparison between flow eigenvector grids (Passchier, 1997) and a stretching eigenvector grid for the flow (3.29). (a), (b) Plots of the maximum (extensional) flow eigenvector (red) and minimum (shortening) flow eigenvector (blue) for $t = 1$ and 3, respectively. (c), (d) Maps of the flow eigenvectors for $t = 3$. (e), (f), Maps of the eigenvectors for $\mathbf{d}$ at $t = 3$. (e) Map of extensional eigenvector for $\mathbf{d}$ at $t = 3$. (f) Map of the shortening eigenvector for $\mathbf{d}$ at $t = 3$. Notice that (c) and (d) are zooms into the area covered by (b) above and figures (e), (f) are zooms into Figure 3.7(e,f).

A. THE MECHANICS OF DEFORMED ROCKS

Figure 3.8(c,d) with Figure 3.8(e,f). Thus the axial planes of the folds are always aligned parallel to one eigenvector of the flow and although one principal axis of strain approaches parallelism with this eigenvector at high strains, the axial plane always has shearing deformations parallel to it except at positions where the local straining is irrotational.

3.6.2 Microfabric Development and the Concepts of Weak and Strong Flows

The concept of a fabric attractor is made more precise if one is able to propose a model for the rate at which the length scale and/or intensity of a fabric accumulates with different flow fields. Flow fields of sufficient strength (as measured by the form and magnitude of the velocity gradient tensor) that result in significant growth of the spatial scale and/or intensity of the fabric on some specified timescale are said to be *strong*, whereas flow fields that lead to small growth rates are said to be *weak*. The concept has been explored by Tanner and Huilgol (1975), Tanner (1976), Astarita (1979), Olbricht et al. (1982) and Chella and Ottino (1985). Models for the development of microfabric in this context were proposed by Hinch and Leal (1975, 1976) and Leal (2007). The incorporation of such models alters the details of the diagrams in Figure 3.5 depending on the geometrical nature of the microstructure and the constitutive framework (Olbricht et al., 1982) but does not alter the general principles.

It is important to define the growth rate for a microfabric in terms of its tensorial nature. Thus the alignment of a string of non-deforming feldspar porphyroclasts in a flowing matrix may be defined in terms of a vector field, whereas the distortion of initially spherical pebbles to become ellipsoids requires a second-order tensor (Hinch and Leal, 1975, 1976; Leal, 2007). Again if one represents damage as a fabric tensor (Voyiadjis and Kattan, 2006) and links the evolution of that tensor to the velocity gradient or stretching tensor then the growth of the damage microstructure can be described in terms of evolution of the strength of the flow. The distinction is important because those microfabrics that require a vector field to describe their growth can be discussed solely in terms of Figure 3.5, whereas those that depend on a higher order tensor need more specific discussion.

Olbricht et al. (1982) point out that the characteristic equation (3.20) enables contours of the maximum eigenvalue, λ^+, to be plotted within the admissible domain in Figure 3.9(a) as straight lines. These lines have the equations:

$$Tr\tilde{\ell}^2 = \frac{2\left[det\tilde{\ell} - \left(\lambda^+\right)^3\right]}{\lambda^+} \quad \text{within ACOB in Figure 3.5}$$

and
$$Tr\tilde{\ell}^2 = \frac{det\tilde{\ell} + 8\left(\lambda^+\right)^3}{\lambda^+} \quad \text{within COB in Figure 3.5} \tag{3.30}$$

where, as in Section 3.3.3, $\tilde{\ell}$ is the normalised velocity gradient tensor.

A flow is said to be *strong* when the maximum eigenvalue, λ^+, exceeds some critical value so that for $\lambda^+ \geq \lambda^+_{critical}$ the growth of the microfabric is significant on some timescale. The growth is strong and steady (monotonic) to the right of the line $\lambda^+ = \lambda^+_{critical}$ and weak but oscillating to the left (Figure 3.9(b)). For a given value of $Tr\tilde{\ell}^2$ the strongest flows occur on the right boundary of the admissible domain where the vorticity vector is aligned parallel to the principal axis of stretching (extension).

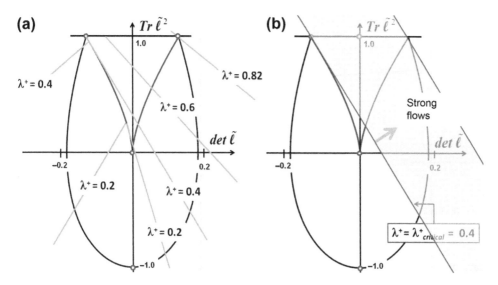

FIGURE 3.9   Flow eigenvalue contours (in grey) and the concept of strong flows. (a) Eigenvalue contours given by [3.30]$_1$ and [3.30]$_2$. (b) The region of strong flows for a critical eigenvalue taken to be 0.4. Strong flows occur to the right of this contour within the admissible domain and weak flows to the left. In these diagrams $\tilde{\ell}$ is the normalised velocity gradient tensor.

If the fabric evolution requires a tensor of higher order than a vector to describe its evolution then the possibility exists that components of the flow field need to be coupled in order to generate a strong flow. For instance, for an evolution equation that involves a second order tensor, **A**, and with the additional constraint that the fabric evolution is isochoric, the situation can become much simpler than for an evolution described by a vector. For instance, in irrotational flows, all elongational flows are of equal strength for such fabric evolution (Olbricht et al., 1982). This arises because the rate of extension in any direction is coupled to the rate of shortening at right angles through $Tr\mathbf{A} = 0$.

Thus in the eigenvector grids shown in Figures 3.7 and 3.8 one expects that microfabric will grow once one of the flow eigenvalues exceeds some critical value that results in a strong flow which drives the increasing length scale or intensity of the fabric. The precise way in which such a fabric evolves depends on the nature of the fabric evolutionary law and the way in which the eigenvectors evolve in orientation and magnitude with time. Such evolutionary laws have not so far been developed for metamorphic fabrics.

3.7 SOME IMPORTANT RELATIONS AND THEOREMS

3.7.1 The Divergence (or Gauss') Theorem

There are many situations in deforming metamorphic rocks where we need to discuss the integral of some vector field over the domain of the deformed body. Examples are establishing the total flux of fluid or heat through a deforming rock mass. If chemical

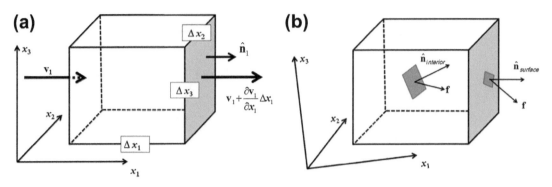

FIGURE 3.10   The concept of divergence and Gauss' theorem. (a) The inflow of a vector, **v**, becomes the outflow $v_1 + \frac{\partial v_1}{\partial x_1}\Delta x_1$. (b) Gauss' theorem.

reactions are proceeding within the rock mass during deformation the individual reaction sites may act as sources or sinks for fluid and heat depending on whether the mineral reactions are exothermic or endothermic and whether the reactions release or consume the fluid phases. Thus it is useful to have a relationship between the total flow into and out of the domain along with details of the sources and sinks within the domain. The divergence theorem supplies this information in that it relates the divergence of the relevant vector field within the body, $\mathcal{B}^d$, at time, t, to the vector field on the boundaries, $\mathcal{S}^d$, of the body.

3.7.1.1 The Divergence of a Vector Field

Consider Figure 3.10(a) where a volume element, dV, at x within the body with total volume, V, has a flux, **v**, through the body. The discussion is general so that **v** may represent a flux of material (fluid, chemical components) or of heat. The flux $\mathbf{v}(x, t)$ is prescribed throughout the body. Let v_1 be the magnitude of the component of the flux across one surface dS of the volume with unit normal $\hat{\mathbf{n}}_1$ parallel to x_1. If the elemental volume has a length parallel to x_1 of Δx_1 then the outward flux in the x_1 direction is $[v_1(x_1, x_2, x_3) + v_1(x_1 + \Delta x_1, x_2, x_3)]\Delta x_2 \Delta x_3$ where Δx_2 and Δx_3 are the edges of the box in the x_2 and x_3 directions.

Similar arguments apply to the x_2 and x_3 directions with associated unit vectors $\hat{\mathbf{n}}_2$ and $\hat{\mathbf{n}}_3$ parallel to x_2 and x_3. Thus the net flux of **v** across the volume is

$$\frac{1}{\Delta V}\int_S \mathbf{v} \cdot \mathbf{n} dA = \frac{[v_1(x_1 + \Delta x_1, x_2, x_3) - v_1(x_1, x_2, x_3)]}{\Delta x_1} + \frac{[v_2(x_1, x_2 + \Delta x_2, x_3) - v_2(x_1, x_2, x_3)]}{\Delta x_2}$$

$$+ \frac{[v_3(x_1, x_2, x_3 + \Delta x_3) - v_3(x_1, x_2, x_3)]}{\Delta x_3}$$

In the limit as $\Delta x_i \rightarrow 0$ the right-hand terms become partial derivatives so that

$$\lim_{\Delta x_i \rightarrow 0} \frac{1}{V} \int_S \mathbf{v} \cdot \mathbf{n} dA = \frac{\partial v_1}{\partial x_1} + \frac{\partial v_2}{\partial x_2} + \frac{\partial v_3}{\partial x_3} \equiv div\mathbf{v} \qquad (3.31)$$

Now consider Figure 3.10(b) where a smoothly varying vector field $\mathbf{f}(x)$ within a closed body $\mathcal{B}^d$ has a bounding surface $\mathcal{S}^d$ with an outward pointing unit normal, $\hat{\mathbf{n}}(x)$, at an element of surface area, dA. If dV is an element of volume within $\mathcal{B}^d$ the divergence theorem states that

$$\int_S \mathbf{f} \cdot \hat{\mathbf{n}} dA = \int_B div\mathbf{f} dV \qquad (3.32)$$

The surface integral is a measure of the flux of $\mathbf{f}$ out of the surface, whereas the volume integral is a measure of the sources and sinks for $\mathbf{f}$ within the body. The divergence theorem says that if we can observe the sum of the components of vector flux normal to the boundary everywhere then we know the divergence of $\mathbf{f}$ within the body. If, for instance, $\mathbf{f}$ corresponds to the displacement or fluid velocity fields then measurement of the surface integral, $\int_S \mathbf{f} \cdot \hat{\mathbf{n}} dA$, constrains the volume change or the sources and sinks for fluid within the body. If $\mathbf{f}$ is the flux of a chemical component, then measurement of $\int_S \mathbf{f} \cdot \hat{\mathbf{n}} dA$ constrains the sources or sinks of that component arising from chemical reactions within the body. An important point to note is that given the integral $\int_S \mathbf{f} \cdot \hat{\mathbf{n}} dA$ we can say something about the divergence of $\mathbf{f}$ within the body but not about the distribution of $\mathbf{f}$ within the body. Thus in Figure 3.11 if the red arrows indicate fluid flux and we measure the total flux on the outside surface of the body

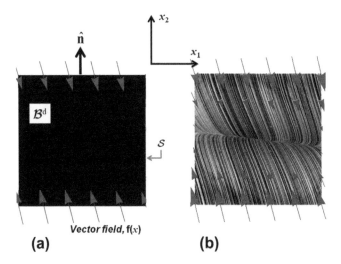

(a)   **(b)**

FIGURE 3.11 An example of Gauss' theorem. (a) shows a 'black box' where measurement of the normal flux across the boundary, $\mathcal{S}$, says that more fluid enters the box than leaves. Gauss' theorem simply says there must be a sink or system of sinks within the body, $\mathcal{B}^d$. It says nothing about what the distribution of sinks is as is revealed in (b). This structure may, for instance, be a layer of sillimanite converting to muscovite and acting as a sink for H_2O.

normal to the surface as in Figure 3.11(a) we see that more material enters the body than leaves. Gauss' theorem simply states that there must be sinks of fluid within the body but says nothing about the distribution of these sinks (Figure 3.11(b)) that may, for instance, represent a layer of sillimanite being converted to muscovite or may be more complicated consisting of many layers.

The divergence theorem applies to tensors of any rank. Thus for a second-order tensor, $\boldsymbol{\sigma}$, we have

$$\int_S \boldsymbol{\sigma}\hat{\mathbf{n}}\,dA = \int_B div\boldsymbol{\sigma}\,dV \tag{3.33}$$

For a scalar, ρ, we have

$$\int_S \rho\hat{\mathbf{n}}\,dA = \int_B grad\rho\,dV \tag{3.34}$$

3.7.2 Reynold's Transport Relation

The *Reynold's transport relation* is a fundamental relationship in continuum mechanics that enables the rate of change of some integral quantity in the reference state to be specified as the body is deformed by the flow, $\mathbf{v}$, and as sources or sinks of that quantity evolve within the body. An example is the rate of change of density within a deforming chemically reacting body where chemical reactions are locally changing the density within the body as it flows and dissolved material is being removed. Our question would be: *How does the time rate of change of mass evolve as the chemical reactions proceed?*

Consider a scalar field whose material and spatial descriptions are $\breve{\varphi} = \breve{\varphi}(\mathbf{X}, t)$ and $\varphi = \varphi(\mathbf{x}, t)$. The element of volume in the undeformed state is dV and in the deformed state dv. Then the time rate of change of φ integrated over the deformed body is $\overline{\int_{B^d} \varphi dv}$.

We want to express this integral in terms of the rate of change of φ and of any contributions to this rate of change arising from sources and sinks within the body. Now since $dv = J\,dV$ and B is constant in time,

$$\overline{\int_{B^d} \varphi dv} = \int_B \overline{\breve{\varphi}J}\,dV$$

$$= \int_B \left(\dot{\breve{\varphi}}J + \breve{\varphi}\dot{J}\right)dV$$

$$= \int_B \left[\dot{\breve{\varphi}} + \breve{\varphi}\left(div\,\breve{\mathbf{v}}\right)\right]J\,dV$$

or

$$\overline{\int_{\mathcal{B}^d} \varphi \, \mathrm{d}v} = \int_{\mathcal{B}^d} [\dot{\varphi} + \varphi(div\mathbf{v})]\mathrm{d}v \tag{3.35}$$

This is known as *Reynold's transport relation*. An alternative and more intuitive form (Tadmor et al., 2012, p. 101) is:

$$\frac{D\varphi}{Dt} = \int_{\mathcal{B}^d} \frac{\partial \varphi}{\partial t}\mathrm{d}v + \int_{S} \varphi \mathbf{v} \cdot \mathbf{n}\mathrm{d}a \tag{3.36}$$

which says that the rate of change of φ is equal to the production of φ inside the deformed body, $\mathcal{B}^d$, plus the net transport of φ across the boundary S.

Recommended Additional Reading

Gurtin, M.E., Fried, E., Anand, L. (2010). *The Mechanics and Thermodynamics of Continua*. Cambridge University Press.
   An up to date discussion of mechanics with an advanced treatment of kinematics.

Ottino, J.M. (1989). *The Kinematics of Mixing, Stretching, Chaos, and Transport*. Cambridge University Press.
   Although this book emphasises chaotic mixing it has some excellent introductory discussions of the kinematics of flow.

Passchier, C.W., Trouw, R.A.J. (1996). *Microtectonics*. Springer-Verlag, New York.
   The definitive treatment of kinematic indicators at the micro-scale. 2nd Edition 2005.

Tadmor, E.B., Miller, R.E., Elliot, R.S. (2012). *Continuum Mechanics and Thermodynamics*. Cambridge University Press.
   A modern treatment of mechanics with an excellent discussion of kinematics.

The Balance Laws:
Forces Involved in Deformation

4.1 GENERAL STATEMENT

In the previous two chapters we have discussed the rules that govern the geometrical transformations and the movements (flows) that are involved in the deformation of rocks. These are both geometrical concepts; one involves the geometrical transformations that constitute deformation and the other the geometry of the flows that result in deformation. These concepts do not require that the causes of the deformation and motion are considered. We now consider the ways in which bodies of rock react to forces or displacements that are applied to the surfaces of the bodies by adjacent parts of the Earth. In order to understand the various responses that may arise due to surface tractions or displacements we need to formulate a set of laws that govern the balance of various quantities of interest. For a metamorphic geologist, there are just four quantities of interest, namely, mass, linear and angular momentum, and energy and hence we need to deal with four laws:

1. *The conservation of mass.*
2. *The balance of linear momentum.*
3. *The balance of angular momentum.*
4. *The balance of energy;* this is commonly known as *the first law of thermodynamics.*

This chapter is concerned with the first three balance laws on the list. Chapter 5 considers the balance of energy.

4.2 CONSERVATION OF MASS

Let us consider a body of deforming rock where the density, ρ, varies both with the spatial position within the body and with time. This might correspond to a deforming/reacting rock where a metamorphic reaction produces denser or less dense phases or where local temperature fluctuations arising from the heat of reaction change the density. Thus we write for the material derivative of the density:

$$\frac{D\rho}{Dt} = \frac{\partial \rho}{\partial t} + \frac{\partial \rho}{\partial x}\frac{dx}{dt} \tag{4.1}$$

and the mass, m, of the body with volume, V, is

$$m = \int_V \rho(x, t)dV \tag{4.2}$$

For mass to be conserved, the Reynold's transport relation gives

$$\overline{\int_V \rho(x, t)dV} = \int_V [\dot{\rho} + \rho\, div v]dv = 0 \tag{4.3}$$

Since (4.3) is to hold for any volume, V, the quantity within the square brackets must be zero so that

$$\dot{\rho} + \rho\, div \mathbf{v} = 0 \tag{4.4}$$

An alternative is

$$\frac{\partial \rho}{\partial t} + div(\rho \mathbf{v}) = 0 \tag{4.5}$$

Both (4.4) and (4.5) are known as the spatial form of the *continuity equation.*
The material form of the continuity equation is

$$J\rho = \rho_0 \tag{4.6}$$

where J is the Jacobian of the deformation and ρ_0, ρ are the densities in the reference and deformed states, respectively (Tadmor et al., 2012).

4.3 BALANCE OF LINEAR MOMENTUM

In this section we are concerned with the causes of deformation in the sense that we want to understand what happens on the boundaries of a body that results in flow being generated within the body. There are two end member ways of loading the surface of a body. One is to apply a system of forces and the other is to displace the surface in some manner. These two loading methods are illustrated in Figure 4.1. In Figure 4.1(a) a dead weight of 100 kg is placed upon the body and this induces a system of forces within the body that are essentially reaction forces in response to the accelerations generated by application of the dead weight. The deformation of the body is a direct response to these forces. If the material is ideally linearly elastic the *displacements* are linearly related to the forces through the elastic constants. If the material is ideally linearly viscous then the *displacement rates* are linearly related to the forces through the viscosity.

In Figure 4.1(b) the body is deformed by the application of a *constant velocity* at the boundary. By Newton's second law of motion there is no acceleration at the boundary of the body and hence no applied force. However, a system of forces is generated in the body arising from the deformation. If the material is ideally linearly elastic the forces generated in the material are linearly related to the current displacement through the elastic constants. If the material is ideally linearly viscous the forces in the material are related to the velocity by the viscosity.

If conducted in a laboratory the first of these experiments is called a creep experiment, whereas the second corresponds to a displacement controlled experiment in an infinitely stiff loading frame. The two loading regimes produce dual responses within the body. In the dead load case the applied force controls the displacements and/or the displacement rates. In the constant velocity case the displacement or the displacement rates control the forces within the material. It is important to distinguish between the two end member loading regimes

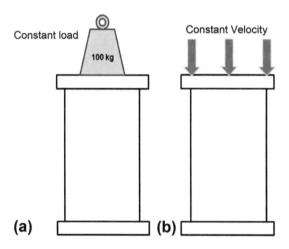

FIGURE 4.1   The two end member ways of loading a body. (a) Dead weight loading. This corresponds to the application of a constant force. (b) Constant velocity loading. This corresponds to loading by an infinitely stiff loading frame.

because entirely different deformation responses arise in the two cases, for instance with respect to buckling of layers (Volume II) and fracture pattern formation (Chapter 8). These two loading regimes are end members of a spectrum of possibilities including mixed force/displacement combinations as well as constant stress and constant strain rate combinations. It is an interesting exercise to speculate what the loading conditions are within the lithosphere of the Earth. Mechanisms of generating a constant force on the boundaries of a body of rock seem more difficult to contemplate except for the dead weight exerted by rocks overlying a system of interest. This contrasts with velocity boundary conditions generated by Plate motions.

However, the result of all loading regimes is the development of a system of forces, *T*, within the deforming material and it is convenient to introduce the concept of *stress at a point*. *Stress* is *the force per unit area* at a point in the material. Immediately, given the discussions in Chapters 2 and 3 the question arises: *Which area do you mean; the spatial or the material representation?* The answer is that both representations are important for different problems. If one uses the spatial (or current, or deformed) area then the stress is called the *Cauchy stress*, **σ**. If the material (or reference or undeformed) area is used the stress is called the *first Piola-Kirchhoff stress*, **P**. There is a *second Piola-Kirchhoff stress* which is formed by transforming *T* using the same affine transformation that links the undeformed (or reference) coordinates to the spatial (or deformed) coordinates and dividing by the material representation of the area (Fung, 1965, Section 16.2). These concepts are illustrated in Figures 4.2 and 4.3. The first and second Piola-Kirchhoff stresses are also known as Lagrangian and Kirchhoff stresses, respectively. It makes sense to refer to the Cauchy stress as an Eulerian stress and this occurs sometimes. The second Piola-Kirchhoff stress is used in discussions of the general three dimensional buckling of a layer in Volume II; unless otherwise specified we use the Cauchy stress.

The dimensions of *stress* are $[ML^{-1}T^{-2}]$ which is the same as the dimensions of *momentum flux per unit area*. Thus one can consider the application of stress to a plane within a body as the result of a momentum flux across that plane. This is the reason that the study of stress is related to the balance of momentum. The *diffusion of momentum* (per unit area) is commonly

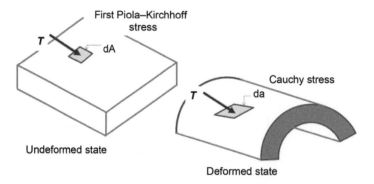

FIGURE 4.2   The first Piola-Kirchhoff and Cauchy stresses. The force, *T*, defined in the deformed state, acts across the area da in the deformed state to define the Cauchy stress. This identical force acts across the equivalent area dA in the undeformed state to define the first Piola-Kirchhoff stress.

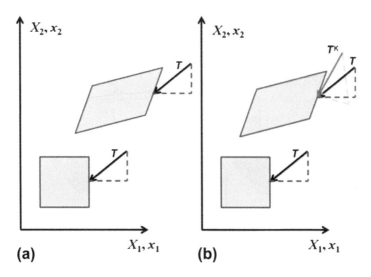

FIGURE 4.3 The first and second Piola-Kirchhoff stresses. (a) The first Kirchhoff stress is defined by the force T, which is a spatial concept defined in the deformed state, acting across the equivalent area in the undeformed state. (b) The force T is transformed by the same affine transformation that links the material coordinates to the spatial coordinates to generate the Kirchhoff force T^K. This force acts across the deformed area to define the second Piola-Kirchhoff stress.

referred to as *stress diffusion*. For metamorphic systems the concept becomes important at large length scales (say of the order of 10's of kilometres) and we consider this in Volume II.

It is also convenient to define the *normal stress* and the *shear stress* across a surface (Figure 4.4). The force T acting on a surface is resolved normal and parallel to the surface of area Δa to produce the normal force T_n and the shear force T_s. The quantities $\sigma_n = T_n/\Delta a$ and $\sigma_s = T_s/\Delta a$ are called the *normal stress* and the *shear stress*.

These concepts can be extended to a general case of a cube at a point P in the body subjected to a set of forces across its faces (Figure 4.5) so that on each face there are two shear

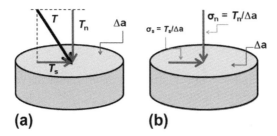

FIGURE 4.4 The concepts of normal and shear stresses. (a) A force applied to the surface (with area Δa) of a body is resolved into two components, one normal to the surface called the normal force and the other parallel to the surface called the shear force. (b) The normal force and the shear force are both divided by the area of the surface to form the normal stress, σ_n, and the shear stress, σ_s. If the area Δa is the material (or reference) representation of the area then the stresses are known as the *Piola stresses*. If the area is the spatial (or current or deformed) representation of the area the stresses are known as the *Cauchy stresses*.

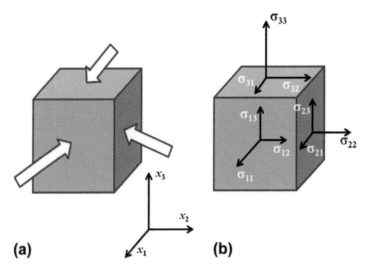

FIGURE 4.5   The concept of stress at a point. (a) A small box with centroid at P (not shown) is subjected to forces on its boundaries. (Forces on the hidden boundaries are not shown). These forces can be resolved into shear forces parallel to the bounding surfaces and normal forces normal to the bounding surfaces. (b) The normal and shear forces divided by the areas of the bounding surfaces give rise to six components of shear stress and three components of normal stress. (Forces on the hidden bounding surfaces are not shown). Each component of stress is labelled in the following manner: $\sigma_{[direction\ of\ surface\ normal,\ direction\ of\ action\ of\ the\ stress]}$. The stresses are shown as positive stresses. They would be negative in the opposite directions. When the box shrinks to a point we have the components of stress at P.

stresses and one normal stress. We do not include a detailed mathematical analysis of the concepts presented above. One can access such discussions in many text books including Tadmor et al. (2012, pp. 113–118) and Gurtin et al. (2010, Chapter 19). The outcome is that both the Cauchy and the two Piola-Kirchhoff stresses are second order tensors. The Cauchy stress is symmetrical if body torques (as would arise from a magnetic material in a magnetic field) do not exist and can be written as

$$\boldsymbol{\sigma} = \sigma_{ij} = \begin{bmatrix} \sigma_{11} & \sigma_{12} & \sigma_{13} \\ \sigma_{21} & \sigma_{22} & \sigma_{23} \\ \sigma_{31} & \sigma_{32} & \sigma_{33} \end{bmatrix} \tag{4.7}$$

with $\sigma_{ij} = \sigma_{ji}$. The first Piola-Kirchhoff stress is asymmetrical while the second Piola-Kirchhoff stress is symmetrical (Tadmor et al., 2012, pp. 123–126).

Since the Cauchy stress, $\boldsymbol{\sigma}$, is symmetrical it can be expressed as a representation ellipsoid (Nye, 1957) — *the stress ellipsoid* — with principal axes (eigenvectors) given by the roots (eigenvalues) of

$$\begin{vmatrix} \sigma_{11} - \lambda & \sigma_{12} & \sigma_{13} \\ \sigma_{21} & \sigma_{22} - \lambda & \sigma_{23} \\ \sigma_{31} & \sigma_{32} & \sigma_{33} - \lambda \end{vmatrix} = 0$$

or,

$$-\lambda^3 + I^\sigma \lambda^2 - II^\sigma \lambda + III^\sigma = 0$$

where I^σ, II^σ, III^σ are the first three *stress invariants* of σ given by

$$I^\sigma = \sigma_{11} + \sigma_{22} + \sigma_{33} = \sigma_1 + \sigma_2 + \sigma_3$$

$$II^\sigma = \begin{vmatrix} \sigma_{22} & \sigma_{23} \\ \sigma_{32} & \sigma_{33} \end{vmatrix} + \begin{vmatrix} \sigma_{11} & \sigma_{13} \\ \sigma_{31} & \sigma_{33} \end{vmatrix} + \begin{vmatrix} \sigma_{11} & \sigma_{12} \\ \sigma_{21} & \sigma_{22} \end{vmatrix} = \sigma_1\sigma_2 + \sigma_2\sigma_3 + \sigma_3\sigma_1$$

(4.8)

$$III^\sigma = \begin{vmatrix} \sigma_{11} & \sigma_{12} & \sigma_{13} \\ \sigma_{21} & \sigma_{22} & \sigma_{23} \\ \sigma_{31} & \sigma_{32} & \sigma_{33} \end{vmatrix} = \sigma_1\sigma_2\sigma_3$$

and σ_1, σ_2, and σ_3 are the principal stresses or eigenvalues of $\boldsymbol{\sigma}$.

If the stress is referred to the coordinate axes parallel to the principal axes of stress then (4.7) becomes

$$\boldsymbol{\sigma} = \begin{bmatrix} \sigma_1 & 0 & 0 \\ 0 & \sigma_2 & 0 \\ 0 & 0 & \sigma_3 \end{bmatrix}$$

4.3.1 Shear and Normal Stresses on an Arbitrary Plane

Some problems arise where we need to calculate the normal and/or shear stresses across an arbitrary plane and in an arbitrary direction in that plane, given a stress referred to arbitrary axes. This problem arises, for instance, in calculating the shear stresses across planes of slip in a crystal during crystallographic preferred orientation (CPO) development (Chapter 13). A way of addressing this problem is to express the components of $\boldsymbol{\sigma}$ in terms of another coordinate frame that has two axes parallel to the plane of interest and one axis parallel to the direction of interest.

In Figure 4.6 we show two sets of Cartesian coordinate axes that are related to each other by the direction cosines shown in Table 4.1. The rotation matrix describing the coordinate transformation is $\mathbf{R}$ so that $\bar{x} = \mathbf{R}x$ and

$$\mathbf{R} = \begin{bmatrix} a_{11} & a_{12} & a_{13} \\ a_{21} & a_{22} & a_{23} \\ a_{31} & a_{32} & a_{33} \end{bmatrix}$$

(4.9)

A vector $\mathbf{v}$ referred to the coordinates x becomes the vector $\bar{\mathbf{v}} = \mathbf{R}\mathbf{v}$ when referred to the axes $\bar{x}$. This transformation is used widely in problems related to crystal reorientation during CPO development (Chapter 13).

A stress tensor $\boldsymbol{\sigma}$ referred to the coordinates x becomes the tensor $\bar{\boldsymbol{\sigma}}$ when referred to the axes $\bar{x}$ and is given by

$$\bar{\boldsymbol{\sigma}} = \mathbf{R}\boldsymbol{\sigma}\mathbf{R}^\mathsf{T}$$

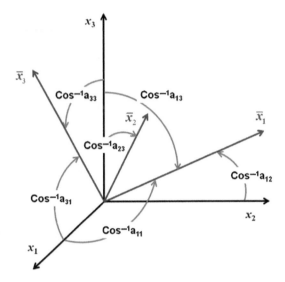

FIGURE 4.6   Two sets of coordinate axes, $\boldsymbol{x}$ (black) and $\bar{\boldsymbol{x}}$ (red) related by a rotation. The rotation is defined by the direction cosines a_{ij} in Table 4.1 and the matrix (4.9).

TABLE 4.1   The Matrix of Direction Cosines Relating a New Set of Coordinates, $\bar{x}_j$, to Another Set, x_i

	x_1	x_2	x_3
$\bar{x}_1$	a_{11}	a_{12}	a_{13}
$\bar{x}_2$	a_{21}	a_{22}	a_{23}
$\bar{x}_3$	a_{31}	a_{32}	a_{33}

The a_{ij} form the components of $\mathbf{R}$ in (4.9).

We illustrate the problem in two dimensions. Consider an arbitrary stress $\boldsymbol{\sigma}$ referred to a set of axes x_1 and x_2 and suppose we want to calculate the stresses $\bar{\boldsymbol{\sigma}}$ relative to a set of axes, $\bar{x}_i$, inclined at φ to x_1 (Figure 4.7).

The rotation is described by a rotation matrix:

$$\mathbf{R} = \begin{bmatrix} \cos\varphi & \sin\varphi \\ -\sin\varphi & \cos\varphi \end{bmatrix} \tag{4.10}$$

The new tensor $\bar{\boldsymbol{\sigma}}$ is obtained as

$$\bar{\boldsymbol{\sigma}} = \mathbf{R}\boldsymbol{\sigma}\mathbf{R}^T = \begin{bmatrix} \cos\varphi & \sin\varphi \\ -\sin\varphi & \cos\varphi \end{bmatrix} \begin{bmatrix} \sigma_{11} & \sigma_{12} \\ \sigma_{12} & \sigma_{22} \end{bmatrix} \begin{bmatrix} \cos\varphi & -\sin\varphi \\ \sin\varphi & \cos\varphi \end{bmatrix} \tag{4.11}$$

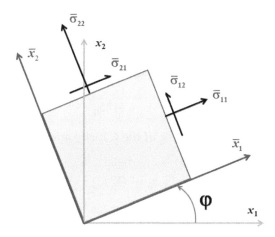

FIGURE 4.7 Transformation of stress in two dimensions. The coordinate axes $\bar{x}$ are related to x by a rotation through φ. The stress $\bar{\sigma}$ referred to the coordinate axes $\bar{x}$ is related to the stress referred to the coordinate axes x by the transformation (4.11) where the rotation matrix is given by (4.10).

This gives

$$\bar{\sigma}_{xx} = \sigma_{xx} \cos^2 \varphi + \sigma_{yy} \sin^2 \varphi + 2\sigma_{xy} \sin \varphi \cos \varphi$$

$$\bar{\sigma}_{yy} = \sigma_{xx} \sin^2 \varphi + \sigma_{yy} \cos^2 \varphi - 2\sigma_{xy} \sin \varphi \cos \varphi$$

$$\bar{\sigma}_{xy} = \left(- \sigma_{xx} + \sigma_{yy} \right) \sin \varphi \cos \varphi + \sigma_{xy} \left(\cos^2 \varphi - \sin^2 \varphi \right)$$

If we use the identities $\sin^2 \varphi = \frac{1}{2}(1 - \cos 2\varphi)$ and $\cos^2 \varphi = \frac{1}{2}(1 + \cos 2\varphi)$ we obtain

$$\bar{\sigma}_{xx} = \frac{\sigma_{xx} + \sigma_{yy}}{2} + \frac{\sigma_{xx} - \sigma_{yy}}{2} \cos 2\varphi + \sigma_{xy} \sin 2\varphi$$

$$\bar{\sigma}_{yy} = \frac{\sigma_{xx} + \sigma_{yy}}{2} - \frac{\sigma_{xx} - \sigma_{yy}}{2} \cos 2\varphi - \sigma_{xy} \sin 2\varphi \tag{4.12}$$

$$\bar{\sigma}_{xy} = -\frac{\sigma_{xx} - \sigma_{yy}}{2} \sin 2\varphi + \sigma_{xy} \cos 2\varphi$$

These expressions are for stress fields referred to general coordinates. The equations reduce to the familiar ones that appear in texts such as Jaeger (1969) and Means (1976) when σ is referred to the principal axes of strain so that σ_{xx} and σ_{yy} are principal stresses and $\sigma_{xy} = 0$; they are the basis of the Mohr circle construction for stress.

One can also note that

$$\tan 2\varphi = \frac{2\sigma_{xy}}{\sigma_{xx} - \sigma_{yy}} \quad \text{when } \bar{\sigma}_{xy} = 0 \tag{4.13}$$

This value of φ gives the orientations of the principal axes of σ.

4.3.2 Stress Deviator and Invariants, the Octahedral Stress

Any tensor can be divided into a *deviatoric* and a *spherical* part (Gurtin et al., 2010, Section 2.7). By definition the deviatoric part of a tensor $\mathbf{A}$ is traceless which means that $Tr\mathbf{A} = 0$ for the deviator. The deviator is constructed by the rule

$$A'_{ij} = A_{ij} - \frac{1}{3}I^A\delta_{ij} \tag{4.14}$$

The spherical part is $\frac{1}{3}I^A\delta_{ij}$. The deviator of the Cauchy stress, the *deviatoric stress*, is

$$\sigma' = \begin{bmatrix} \sigma_{11} - (I^\sigma/3) & \sigma_{12} & \sigma_{13} \\ \sigma_{21} & \sigma_{22} - (I^\sigma/3) & \sigma_{23} \\ \sigma_{31} & \sigma_{32} & \sigma_{33} - (I^\sigma/3) \end{bmatrix}$$

or,

$$\sigma'_{ij} = \sigma_{ij} - \frac{1}{3}I^\sigma\delta_{ij} \tag{4.15}$$

The spherical part of σ is $\frac{1}{3}(\sigma_1 + \sigma_2 + \sigma_3)$, which is the *mean stress*. It is common to identify this quantity with the *hydrostatic pressure*.

The deviatoric stress contains only information on shear stresses and is important because there are many materials where the deformation process depends on the deviatoric stress and is relatively insensitive to the spherical part of the stress (Chapter 6).

4.3.2.1 Invariants of the Deviatoric Stress

The invariants of the deviatoric stress are conventionally labelled J_1, J_2 and J_3; J_2 features prominently in plasticity theory. These invariants are given by

$$J_1 = 0$$

$$J_2 = \frac{1}{3}I^{\sigma 2} - II^\sigma = \frac{1}{2}\sigma'_{ij}\sigma'_{ij} \tag{4.16}$$

$$J_3 = III^\sigma - \frac{1}{3}I^\sigma II^\sigma + \frac{2}{27}I^{\sigma 3} = \frac{1}{3}\sigma'_{ij}\sigma'_{jk}\sigma'_{ki}$$

4.3.2.2 The Octahedral Stress

The octahedral stress, τ_{oct} is an important concept in plasticity theory and is represented using Figure 4.8 which shows, in σ_1–σ_2–σ_3 space, planes equally inclined to the principal stress axes. The direction cosines of these eight planes are $\pm\frac{1}{\sqrt{3}}$. They are known as the *octahedral planes*. The resultant shear stresses on these planes are the *octahedral shear stresses* and are given by

$$\tau_{oct} = \frac{1}{9}\left[(\sigma_1 - \sigma_2)^2 + (\sigma_2 - \sigma_3)^2 + (\sigma_3 - \sigma_1)^2\right] \tag{4.17}$$

The octahedral shear stress can be expressed in terms of the stress invariants as

$$\tau_{oct} = \frac{1}{9}\left[2I^{\sigma 2} - 6II^\sigma\right] \tag{4.18}$$

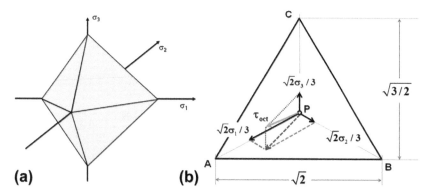

FIGURE 4.8 The octahedral stress. (a) The octahedral planes in principal stress space. (b) Construction on an octahedral plane to find the magnitude and orientation of the octahedral shear stress given the values of the principal stresses. *After Nadai (1950), Figure 10–18.*

Also,

$$J_2 = \frac{3}{2}\tau_{oct}^2 \qquad (4.19)$$

This last relation, introduced by Nadai (1950) plays an important role in plasticity theory (Chapter 6). A useful construction to find τ_{oct} on an octahedral plane (Nadai, 1950, Figure 10–18) is shown in Figure 4.8(b).

4.3.3 Equilibrium Conditions

Many problems require knowledge of the conditions for equilibrium of a deforming body under the influence of a general stress field that is varying with position within the body. By equilibrium here we mean mechanical equilibrium where the body is stationary or moving with constant velocity. In Figure 4.9 we show such an infinitesimal cube with faces parallel to the coordinate axes. All the components of stress that exist in the x_1 direction are shown including that arising from the x_1 component of a body force B per unit volume. The other components comprise, for instance, the stress σ_{11} and its variation through the body, $\sigma_{11} + \frac{\partial \sigma_{11}}{\partial x_1}dx_1$. Similar expressions can be written for the variation of σ_{21} and σ_{31} through the body. In addition there is the component of the body force $B_1 dx_1 dx_2 dx_3$.

For mechanical equilibrium in the x_1 direction all the *forces* in the x_1 direction must balance:

$$\left[\sigma_{11} + \frac{\partial \sigma_{11}}{\partial x_1}dx_1\right]dx_2 dx_3 - \sigma_{11}dx_2 dx_3 +$$

$$\left[\sigma_{21} + \frac{\partial \sigma_{21}}{\partial x_2}dx_2\right]dx_3 dx_1 - \sigma_{21}dx_3 dx_1 +$$

$$\left[\sigma_{31} + \frac{\partial \sigma_{31}}{\partial x_3}dx_3\right]dx_1 dx_2 - \sigma_{31}dx_1 dx_2 + B_1 dx_1 dx_2 dx_3 = 0$$

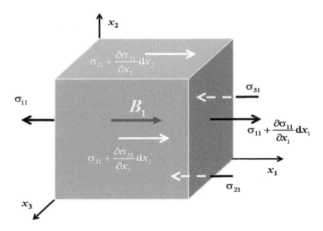

FIGURE 4.9   Components of stress in the x_1 direction for an infinitesimal cube. These components are σ_{11}, σ_{21} and σ_{31} together with the variations of these stresses through the cube. Also involved is the x_1 component of the body force per unit volume, B_1.

If we divide by $dx_1 \, dx_2 \, dx_3$ we obtain

$$\frac{\partial \sigma_{11}}{\partial x_1} + \frac{\partial \sigma_{21}}{\partial x_1} + \frac{\partial \sigma_{31}}{\partial x_1} + B_1 = 0$$

Identical arguments hold for the x_2 and x_3 directions so that the complete set of equations for mechanical equilibrium is

$$\frac{\partial \sigma_{11}}{\partial x_1} + \frac{\partial \sigma_{21}}{\partial x_1} + \frac{\partial \sigma_{31}}{\partial x_1} + B_1 = 0$$

$$\frac{\partial \sigma_{12}}{\partial x_2} + \frac{\partial \sigma_{22}}{\partial x_2} + \frac{\partial \sigma_{32}}{\partial x_2} + B_2 = 0 \qquad (4.20)$$

$$\frac{\partial \sigma_{13}}{\partial x_3} + \frac{\partial \sigma_{23}}{\partial x_3} + \frac{\partial \sigma_{33}}{\partial x_3} + B_3 = 0$$

This set of equations can be obtained more elegantly using the Gauss divergence theorem and we refer the reader to Fung (1965, Section 5.5), Tadmor et al. (2012, pp. 119–120) and Gurtin et al. (2010, Section 19.6). Similar derivations arrive also at the Eulerian form of the *equations of motion* or the *balance of linear momentum*:

$$div\boldsymbol{\sigma} + \rho\mathbf{B} = \rho\mathbf{a} \qquad (4.21)$$

which reduces to (4.20) at mechanical equilibrium when $\frac{Dv_i}{Dt} = 0$. (4.21) is also known as *Cauchy's first law* and is important because it says that given the forces on a body, one can only determine the divergence of the stress not the stress itself (Truesdell, 1966). In order to determine the stress one needs extra information concerning the material that makes up the body in the form of constitutive laws for the material.

One can perform the same exercise as above with respect to the moments of the shear stresses about the centroid of the cube. The result is that

$$\sigma_{ij} = \sigma_{ji} \tag{4.22}$$

which states that the Cauchy stress tensor is symmetrical. This result is known as the *balance of angular momentum* and derives from *Cauchy's second law* (Truesdell, 1966).

Recommended Additional Reading

Fung, Y.C. (1965). *Foundations of Solid Mechanics*. Prentice Hall, New Jersey.
  A readable and extensive treatment of solid mechanics. The treatment of stress is particularly good.

Gurtin, M.E., Fried, E., Anand, L. (2010). *The Mechanics and Thermodynamics of Continua*. Cambridge University Press.
  This text includes a good but advanced discussion of balance laws and stress.

Jaeger, J.C. (1969). *Elasticity, Fracture and Flow*. Methuen.
  An invaluable, classical treatment of mechanics.

Means, W.D. (1976). *Stress and Strain*. Springer-Verlag.
  A discussion of stress concepts with an emphasis on the use of Mohr diagrams.

Nye, J.F. (1957). *Physical Properties of Crystals*. Oxford Press, 322 pp.
  This book is an excellent introduction to stress and coordinate transformations.

Tadmor, E.B., Miller, R.E., Elliot, R.S. (2012). *Continuum Mechanics and Thermodynamics*. Cambridge University Press.
  A modern treatment of mechanics with an excellent treatment of balance laws and stress.

Truesdell, C.A. (1966). *Six Lectures on Modern Natural Philosophy*. Springer-Verlag, Berlin.
  An entertaining and informative view of mechanics, thermodynamics and the role of balance laws.

Recommended Additional Reading

Energy Flow — Thermodynamics

5.1 WHAT IS THERMODYNAMICS?

Thermodynamics is a contentious subject because it means different things to different people. Some quotes http://en.wikiquote.org/wiki/Thermodynamics from prominent scientists and mathematicians make the point:

> Einstein (1903): 'A law is more impressive the greater the simplicity of its premises, the more different are the kinds of things it relates, and the more extended its range of applicability...It is the only physical theory of universal content, which I am convinced, that within the framework of applicability of its basic concepts will never be overthrown.'

Eddington (1915): 'The law that entropy always increases holds, I think, the supreme position among the laws of Nature. If someone points out to you that your pet theory of the universe is in disagreement with Maxwell's equations — then so much the worse for Maxwell's equations. If it is found to be contradicted by observation — well, these experimentalists do bungle things sometimes. But if your theory is found to be against the second law of thermodynamics I can give you no hope; there is nothing for it but to collapse in deepest humiliation.'

These opinions contrast with Truesdell (1969): '...*thermodynamics never grew up'*.

'Thermodynamics...began out of steam tables, venous bleeding and speculations about the universe and has always had a hard time striking a mean between these extremes. While its claims are often grandiose, its applications are often trivial'.

And with respect to some poor soul trying to understand the subject (that many will identify with): *he is told that dS is a differential but not of what variables S is a function; that δQ is a small quantity but not normally a differential; he is expected to believe that not only can one differential be bigger than another but even that a differential can be bigger than something that is not a differential. He is loaded with an arsenal of words like piston, boiler, condenser, heat bath, reservoir, ideal engine, perfect gas, quasi-static, cyclic, nearly in equilibrium, isolated, universe...The mathematical structure...is slight...The examples or exercises consist in no more than calculating partial derivatives or integrals of given functions or their inverses and plugging numbers into the results.*

With such a diverse range of views available we proceed with some trepidation and follow a path that originated with writers such as Bridgman (1943, 1950) and Biot (1955) and was further developed by Truesdell and Toupin (1960), Coleman and Noll (1963), Ziegler (1963), Truesdell (1966, 1969), Kestin (1968), Kestin and Rice (1970), and Rice (1970, 1971, 1975). The issues involved in developing a framework for deforming chemically reacting solids have slowly evolved mainly through the work of the above group of authors.

With this background literature in mind we define *thermodynamics* as *the study of the flow of physical and chemical quantities through or within* a *system driven by thermodynamic forces*. The primary quantities of interest to us are *momentum* (per unit area), *heat, fluid* and *chemical components*. The thermodynamic forces (or *affinities*) that drive these flows are *gradients* in the *deformation*, in the *inverse of the temperature*, in *hydraulic potential* and in *chemical potentials*. Systems are driven away from equilibrium by thermodynamic forces and tend to be driven towards equilibrium by thermodynamic flows (which dissipate energy).

At times the forcing can balance the flow so that one or more *non-equilibrium stationary states* develop. These states may be stable or unstable and can continue to exist or evolve as long as the thermodynamic forcing exists. If there is no thermodynamic forcing the system is at *equilibrium* which is also a stationary state for the system.

Three issues arise in attempting to develop a thermodynamic theory for crystalline solids (such as metals, ceramics and rocks) applicable to non-equilibrium systems. The first issue is to define variables for non-equilibrium systems such as temperature and energy that specify the state of the system and are clearly defined for systems at equilibrium.

The second issue is that the distribution of deformation, temperature, strain rate and chemical reaction rates is commonly non-uniform in a system not at equilibrium so that any overarching theory must take such non-uniformity into account. In many fluids and certainly within the framework of classical chemical equilibrium thermodynamics the temperature, pressure and chemical potential fields are uniform so that such difficulties do not arise.

The third issue concerns the non-uniqueness of the stress that develops in crystalline solids during deformation for given values of temperature and strain. Thus for given conditions of strain and temperature, quite different stresses might develop in a given solid material depending on how fast the deformation is imposed and on the types of microstructural processes that evolve during the deformation. This arises because solids are characterised by internal microstructural adjustments that are history and rate dependent. This is not the case for ideally elastic solids or for ideally inviscid or viscous fluids. It is true for some complicated fluids such as polymers that possess internal structure and might show yield phenomena (Barnes, 1999). Thus a useful theory of non-equilibrium behaviour must address these issues involving the definition and non-uniformity of variables and the rate sensitivity of the behaviour of crystalline solids arising from internal microstructural processes.

The first issue, namely, the definition of variables such as temperature and internal energy, is handled in two ways. One way (due to Truesdell, 1966) is to declare that such variables are *primitive* quantities that exist for any system and need no justification in the same way that concepts such as space, mass, force and time exist for any system with no need for justification. The second way (see Kestin, 1990, 1992 for discussions) is to appeal to the *principle of local action* which postulates that, in a system not at equilibrium, *'all relations between thermodynamic properties which are valid for uniform systems continue to be valid locally at a point and instantaneously, even if the processes occur at finite rates'* (Kestin, 1968). This principle is also commonly (and erroneously, Kestin, 1992) referred to as the *principle of local equilibrium* but because of the use of this concept in a completely different context in metamorphic petrology we prefer *local state* rather than *local equilibrium*.

The second issue, namely, the non-uniformity of the system, is addressed by casting the subject in a continuum framework. The word *continuum* implies that quantities such as temperature, pressure and so on are defined at each point in the system and are allowed to vary from point to point. The term *continuum* does not imply that discontinuities such as fractures and grain boundaries do not exist within the system. We adopt a *spatial representation* of the system of interest so that all quantities are referred to the current or deformed state. Quantities such as energy, temperature, pressure, strain, strain rate and chemical potential are defined at a point and are functions of both the spatial coordinates, x, and time, t. This contrasts with treatments of the thermodynamics of systems at equilibrium where such quantities are taken to be *homogeneous* throughout the system so that no gradients are present, and there is no dependence on spatial coordinates or time. The approach adopted here is consistent with experience where, for instance, one may experience different temperatures in one room and the temperature may change at each point in the room with time.

The third issue, namely, the non-uniqueness of system response, is addressed by allowing a much wider array of variables to define the state of a system than is used in systems at equilibrium. Thus the array of *equilibrium state variables* such as temperature, pressure and chemical potential is expanded to an array of *generalised state variables*. These include *kinematic state variables* such as the plastic strain and *internal variables* that may include, for instance, variables that describe the state of the microstructure such as dislocation or microcrack densities. Such generalised state variables must relax to equilibrium values as the system approaches equilibrium.

The thermodynamics literature that pervades the geosciences is almost exclusively devoted to *equilibrium chemical thermodynamics* developed initially by Gibbs (1906); such an approach is widely used by metamorphic petrologists to calculate mineral phase stability fields (Connolly, 1990, 2005; Powell et al., 1998). The approach as it is widely developed (Kern and Weisbrod, 1967; Powell, 1978; Yardley, 1989; Vernon and Clarke, 2008) is based on the following principles: (1) The system of interest is closed and at equilibrium. This means, as we will see below, that all processes have been driven to completion and hence there is no current dissipation of energy in the system. (2) Following directly from this assertion, the system must be free of all driving forces such as gradients in temperature, pressure or chemical potentials. This means that the system is homogeneous with respect to such variables. (3) The system can be treated as a fluid at rest which means that the only stress in the material is a hydrostatic pressure that can be calculated from the boundary conditions (which essentially means the force due to gravity arising from the overlying rock mass).

From a process point of view there is little interest in examining systems where all processes have been driven to completion. We are interested in the processes that operate during the deformation and metamorphism of the lithosphere and the inter-relationships between them. Hence the emphasis must be on systems not at equilibrium. This does not detract from or threaten the vast amount of effort that has been devoted to the development of chemical equilibrium thermodynamics applied to mineral systems. That body of work forms the fundamental foundation for developments in the thermodynamics of deforming mineral systems far from equilibrium.

5.2 METAMORPHIC SYSTEMS

A *metamorphic system* is that part of the lithosphere whose behaviour we happen to be interested in. The rest of the Earth provides an environment for the system and may exert forces or displacements on the system as well as add or subtract mass and entropy. The system may be weakly coupled with the environment in which case an approximation is an *isolated system* or the system may have strong interactions with the environment in which case it is called an *open system*.

The concept of a metamorphic system resembles that of a *control volume* in the fluid dynamics and engineering literature (Law, 2006, pp. 157–163). A control volume is a convenient spatial (or Eulerian) region of interest, open or closed, in which the state of the control volume is specified by two sets of quantities: (1) *variables* that describe the *state* of the control volume and may include the temperature, the pressure, the fluid pressure, a set of chemical potentials, the volume, the rates of chemical reactions, the stress and the strain; (2) *constraints* that prescribe the mass flux and heat flux through the control volume together with the forces and/or displacements (and their rates) imposed on the boundaries of the control volume. To a large extent a *metamorphic system is defined by the constraints imposed by the external environment*. These constraints control the entropy production in the control volume and whether the system is able to evolve towards equilibrium.

We give two examples of metamorphic systems to illustrate the controls on system evolution. The first is at the lithospheric scale (Figure 5.1(a)) and comprises a system between the Moho and the upper crust outlined by the region ABCD in Figure 5.1(a). The environment consists of the upper mantle, the upper crust and the mid crust that surrounds ABCD. Both

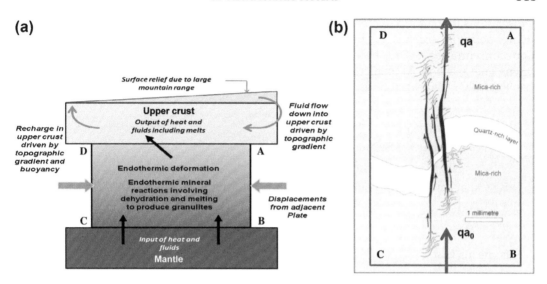

FIGURE 5.1 Examples of metamorphic systems. (a) Lithospheric scale system. The region ABCD is a spatial control volume that remains fixed relative to spatial coordinates. Heat and fluids flow through the boundaries of this control volume so that while the metamorphic processes are operating the system is open. Buoyancy drives fluids (aqueous fluids and melts) upwards. A horizontal component of fluid flow in the mid crust (both for aqueous fluids and melts) is introduced by a topographic induced hydraulic potential gradient (see Chapter 12). Topographically induced gradients in hydraulic potential also allow meteoric fluids to be driven into the upper crust given a favourable permeability structure (see Chapter 12). (b) Thin section scale system. A control volume ABCD has a volumetric influx, q, at the base of the system bearing a SiO_2 concentration, a_0. The outflow is also q bearing a SiO_2 concentration a. Chemical reactions take place to dissolve SiO_2 in the fluid while the control volume is deforming. The rate of dissolution of SiO_2 is controlled by the deformation. *After Worley et al. (1997).*

heat and fluids (perhaps CO_2, H_2O and mafic melts) are added from the upper mantle and heat, together with fluids (perhaps H_2O and granitic melts), are removed from the top of the system and the surrounding mid crust. The presence of a large mountain range introduces horizontal gradients in hydraulic potential (see Chapter 12) that are capable of driving meteoric fluids into the lower crust given a favourable permeability structure, such as might be supplied by steeply dipping shear zones, and also introduces a horizontal component of fluid flow both for aqueous fluids and melts in the mid to lower crust. If we are to examine this system from a thermodynamic point of view the control volume is ABCD and we are interested in the variables that define the non-equilibrium thermodynamic state inside the control volume together with the constraints imposed on the control volume both by fluxes in heat and fluid and by imposed forces or displacement rates on the boundaries of the control volume. These constraints govern whether the system reaches one or more non-equilibrium stationary states, the time scale at which it approaches an equilibrium stationary state and the evolution of entropy production in the control volume.

The second example is at the microscale (Figure 5.1(b)) and is motivated by the model for crenulation cleavage development presented by Worley et al. (1997). We define a control volume ABCD as shown in Figure 5.1(b) which is adiabatic; displacements are applied to the boundaries of the control volume by the surrounding material resulting in localised deformations in the form of crenulation cleavages. A volumetric flow, q (in cubic metres per square metre per second), is applied across the boundary BC and leaves through DA. The input concentration of SiO_2 in the fluid is a_0 and the output concentration is a. Chemical reactions take place in the control volume resulting in the dissolution of quartz. The rate of change of SiO_2 concentration is given by

$$\begin{bmatrix} Rate\ of\ change\ of \\ concentration\ of\ SiO_2 \end{bmatrix} = \begin{bmatrix} Influx\ of\ SiO_2 \\ minus\ outflux\ of\ SiO_2 \\ = q(a_0 - a) \end{bmatrix} + \begin{bmatrix} Rate\ of\ dissolution \\ of\ SiO_2 \end{bmatrix}$$

This system can evolve to one or more non-equilibrium stationary states for

$$\begin{bmatrix} Rate\ of\ change\ of \\ concentration\ of\ SiO_2 \end{bmatrix} = 0$$

This means that

$$\begin{bmatrix} Influx\ of\ SiO_2 \\ minus\ outflux\ of\ SiO_2 \\ = q(a_0 - a) \end{bmatrix} = - \begin{bmatrix} Rate\ of\ dissolution \\ of\ SiO_2 \end{bmatrix}$$

We will see that the behaviour of this non-equilibrium system depends on the form of the rate equation for the dissolution of quartz, which in this case depends on the deformation. If this equation is linear there is just one non-equilibrium stationary state. If the equation is non-linear, multiple non-equilibrium stationary states can exist some of which are stable while others are unstable.

5.3 THERMODYNAMIC SYSTEMS

Four types of thermodynamic systems are distinguished (Niven, 2009; Niven and Andresen, 2010).

(a) *Isolated systems.* The first is an *isolated* system where extensive variables such as the internal energy density, E, the volume, V, and the number of moles of k chemical components, n_k, are kept constant. The system is isolated from other systems and from its environment by an impermeable wall (Figure 5.2(a)). This is also called a *closed system*. The only evolutionary path such systems can adopt is to evolve to equilibrium and the equilibrium state is described by minimising an energy function such as the Helmholtz energy or by maximising an entropy function. If the system is perturbed from equilibrium or begins far from equilibrium the path to equilibrium may be tracked using geometrical methods developed by Gibbs (1906) and elaborated upon by numerous authors including Weinhold (1975a, 1975b, 1975c, 1975d, 1976); for a review see Niven and Andresen (2010). The path from one state to the equilibrium state need not necessarily be steady. We will see in Chapter 14 that in an isolated system even a simple exothermic reaction of the type A → B can produce oscillations in the concentration of A and in the temperature (Gray and Scott, 1994) as the system proceeds to equilibrium. Nevertheless, an isolated system must ultimately proceed smoothly to equilibrium.

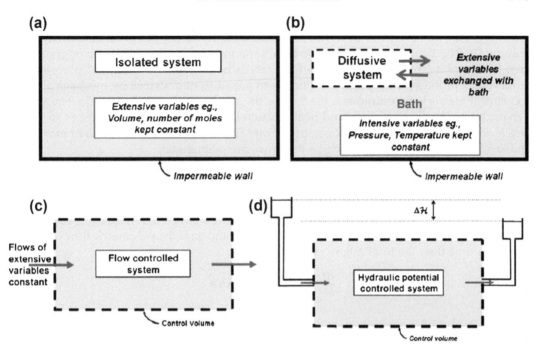

FIGURE 5.2 The four types of thermodynamic systems. (a) Isolated system. Extensive variables are kept constant. The system is isolated from any other system. (b) Diffusive system. The system can communicate with the surrounding bath where intensive variables are kept constant. The bath is isolated from any other system. (c) Flow-controlled system where flows of extensive variables are kept constant within a region known as the control volume. (d) Hydraulic potential, $\mathcal{H}$, control system where the flows of extensive variables are controlled by a gradient in hydraulic potential such as the large mountain range sketched in Figure 5.1(a). Both (c) and (d) can communicate with other systems but the evolution of (c) and (d) is controlled only by flows into and out of the control volume and processes that operate within the control volume. *From Ord et al. (2012) motivated by Niven (2009).*

(b) *Diffusive systems.* The second type of system, a *diffusive* system (Figure 5.2(b)) is also a closed system. The system is embedded in another medium which acts as a *bath* and which itself is isolated from all other systems and its environment by an impermeable wall. Intensive variables such as T and P are held constant in the bath and the diffusive system can interact with the bath by interchange of mass and heat with the bath. Diffusive systems must always evolve to equilibrium and follow the same rules for doing so as do isolated systems. Clearly most considerations of metamorphic mineral reactions assume the system is isolated or diffusive, although the distinction is not commonly well defined or emphasized.

A completely different class of behaviour is that of an *open flow* system (Figure 5.2(c and d)) where the system comprises an open configuration that can communicate with the surrounding environment with constraints on mass and heat flow.

(c) *Flow controlled systems.* One of these systems, the third type of system, is an open *flow controlled* system (Figure 5.2(c)) and is the type of relevance to devolatilising and melting systems and to hydrothermal mineralising systems where the fluid input rate is constrained by the rate of production of the fluid (Phillips, 1991). This is also the case in Figure 5.1 where both metamorphic systems are portrayed as open flow controlled systems. This we propose also is the case for orogenic gold deposits and iron oxide copper-gold (IOCG) deposits (Volume II). The boundaries of the control volume may be fixed spatially or migrate with time. It is important to note that if the flow rate is held constant, the initial porosity and permeability of the system, in general, needs to adjust to new values by mechanical and/or chemical means to accommodate the flow as the system evolves (Chapter 13). The system is characterised by flows of mass and heat through the system and the simplest of such systems is where these flows remain constant in time. These systems evolve to one or more non-equilibrium stationary states so long as the flows are maintained.

The system can be held far from equilibrium for the duration of the flows. Consider the simple open flow system shown in Figure 5.3 where the chemical reaction $A \overset{k}{\to} B$ is progressing inside the control volume. We adopt the convention that chemical components are written as Arial upper case: A, B, while their concentrations are written as Arial lower case: a, b. If qa_0 is the volumetric flow of A into the control volume and qa is the volumetric flow out of the control volume then the total rate of change of A is

$$\frac{\partial a}{\partial t} = q(a_0 - a) - kVa \tag{5.1}$$
$$= qa_0\xi - ka_0V(1 - \xi)$$

where a_0 is the initial input concentration of A, k is the rate constant for the reaction producing B, V is the volume of the system and $\xi = \frac{a_0 - a}{a_0}$ is the *extent* of the reaction that produces B; $0 \leq \xi \leq 1$. The system is at a stationary state when $\frac{\partial a}{\partial t} = 0$, that is, when $\xi = \frac{kV}{q_{ss}+kV}$ (Figure 5.3(b)) where q_{ss} is the imposed volumetric flow rate. One can show that this particular non-equilibrium stationary state is stable against small perturbations in q and a_0 so that the system remains at this non-equilibrium state indefinitely as long as A is added to the system. This simple argument is relevant to the situation portrayed in Figure 5.1(b). The reaction in that example, however, is more complicated since it is exothermic and is a heterogeneous reaction

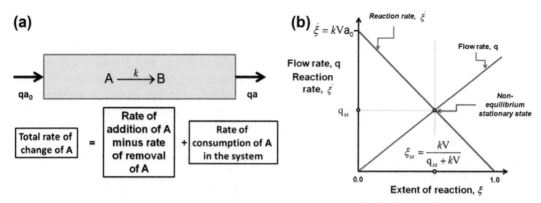

FIGURE 5.3   An open flow controlled system in which the chemical reaction $A \overset{k}{\to} B$ is progressing. (a) The input flow of A is qa_0 where q is a volumetric flow rate and the output flow of A is qa. (b) A flow diagram (Gray and Scott, 1994) showing the non-equilibrium stationary state (for the system shown in Figure 5.3(a)) defined by a balance between the net flow rate and the reaction rate. V is the volume of the system.

taking place at the surfaces of quartz grains: $SiO_2^{solid} \rightarrow SiO_2^{aqueous}$. More complicated versions of the argument presented above exist when other processes operate such as thermal feedback from the heat supplied by the reaction. We examine these cases in greater detail in Chapter 14.

Expressions such as (5.1) are used in Chapter 14 and particularly in Volume II to discuss the behaviour of various types of chemical reactions in flow controlled systems and have been used extensively by Gray and Scott (1994) to produce *flow diagrams* (see Figure 5.3(b)) which are convenient graphic representations of the stationary states of chemical reactions in flow controlled systems.

(d) *Hydraulic potential controlled systems.* The fourth type of system, a *hydraulic potential controlled flow system* (Figure 5.2(d)) is similar to a flow controlled system except that the volumetric flow rate is imposed by a gradient in the hydraulic potential such as the large mountain range sketched in Figure 5.1(a). These systems evolve in a similar manner to flow controlled systems except that if the hydraulic head is kept constant, the fluid velocity within the control volume changes as the permeability within the system changes due to chemical precipitation and/or dissolution (Merino and Canals, 2011). The thermodynamics of such permeability evolution is considered by Coussy (1995) and by Merino and Canals (2011). In order to achieve a stationary state the rates of chemical precipitation and/or dissolution must evolve in order to satisfy $\frac{\partial(\varphi c_i)}{\partial t} = 0$. These kinds of systems are typical of Mississippi Valley-Type (MVT) and Irish lead/zinc deposits as well as uranium unconformity deposits (Anderson and Garven, 1987; Appold and Garven, 2000; Appold et al., 2007; Garven, 1985; Garven and Freeze, 1984a,b; Murphy et al., 2008; Raffensperger and Garven, 1995a,b). Such systems are also characteristic of many upper crustal shear zones where retrograde metamorphic reactions take place involving the influx of meteoric H_2O (Cartwright et al., 1997).

In the mechanics literature, flow controlled systems with constant input flow correspond to what are called *Neumann boundary conditions*, whereas hydraulic potential controlled flow systems correspond to *Dirichlet boundary conditions*. It is of course possible to have *mixed (Robin) boundary conditions*. These latter conditions are relevant to situations such as those shown in Figure 5.1(a) where the flux from the mantle constitutes a flux controlled boundary condition, whereas the topographic gradient constitutes a pressure gradient boundary condition.

5.4 FOUR DIFFERENT STRANDS OF THE DEVELOPMENT FOR THERMODYNAMICS

Over the past 100 years or so, four major lines of thought regarding systems not at equilibrium have developed. These are all compatible (except for some important details) and are converging on a common treatment of nonlinear systems not at equilibrium. These lines of development are summarised below:

1. *Linear thermodynamics.* Onsager (1931a,b), along with Machlup (Onsager and Machlup (1953)), developed a thermodynamically linear theory of non-equilibrium systems with the intent that such systems are 'close' to equilibrium. The treatment was elaborated upon by Prigogine (1955) and De Groot and Mazur (1984) to encompass mainly chemical systems. By *thermodynamically linear* it is meant that the *thermodynamic forces* such as chemical affinity, and gradients in deformation and chemical potential must be linear functions of the *thermodynamic fluxes* such as chemical reaction rate, momentum flux and diffusive flux. The

TABLE 5.1   Thermodynamic Fluxes and Forces for Some Common Processes.

Process	Thermodynamic Flux	Thermodynamic Force
Deformation of Newtonian viscous material	Momentum flux per unit area	Gradient in deformation
Deformation of non-Newtonian viscous material	Momentum flux per unit area	Gradient in deformation
Thermal conduction	Thermal flux, $J^{thermal}$ $J^{thermal} = -k^{thermal}\nabla T$	∇T^{-1}
Chemical diffusion	Mass flux, J^{mass} $J^{mass} = -D\nabla\mu$	$\nabla\frac{\mu}{T}$
Fluid flow	Darcy velocity, u $u = -K^{fluid}\nabla P^{fluid}$	∇P^{fluid}
Chemical reaction	Reaction rate, $\dot{\xi}$ $\dot{\xi} = [r^+ - r^-]$	Chemical affinity $kT \ln\left[\frac{r^+}{r^-}\right]$

$k^{thermal}$ is a thermal conductivity, D is a diffusion coefficient, μ is a chemical potential, K^{fluid} is a permeability and r^+, r^- are the forward and reverse rate constants for a chemical reaction.

Onsager approach is widely used but Ross and co-workers (Hunt et al., 1987, 1988; Ross, 2008; Ross et al., 1988; Ross and Villaverde, 2010; Ross and Vlad, 2005; Villaverde et al., 2011) have pointed out that for *any* isolated or diffusive chemical system (even the simplest and no matter how close to equilibrium) the chemical affinity is *never* a linear function of the chemical reaction rate (see Table 5.1) so that Onsager principles are never applicable to chemical reactions no matter how close the system is to equilibrium (see Chapter 14).

2. *Generalised thermodynamics*. The application of thermodynamics to deforming reacting systems was developed from a non-Onsager (that is, nonlinear) point of view by Truesdell (1966) based on work by Coleman and Noll (1963). This involved the definition of state functions which in turn are functionals of *internal* variables that define the non-equilibrium state of the system. Truesdell referred to these developments as *rational thermodynamics*. Somewhat similar strands of this development were initiated by Bridgman (1943, 1950) and Biot (1955, 1984) who used the concept of the Helmholtz energy for a stressed solid, a concept that had been introduced much earlier by Gibbs (1906), together with the concept of a *dissipation function*. That strand was further developed by Ziegler (1963), Kestin (1968), Kestin and Rice (1970), Rice (1971), Coussy (1995, 2004, 2010), Collins and Houlsby (1997), Collins and Hilder (2002), and Houlsby and Puzrin (2006a). This approach is now commonly known as *generalised thermodynamics*.

3. *Nonlinear chemical thermodynamics*. A completely different line of activity developed independently of the above approaches in nonlinear chemistry (Epstein and Pojman, 1998) and chemical engineering (Gray and Scott, 1994; Liljenroth, 1918; Van Heerden, 1953) with parallel activity in biology (Strogatz, 2001). The concepts involve isolated,

diffusive and flow controlled systems and are based on determining the stationary states for chemical reactions that are coupled. The term *coupled* means that the concentration of a particular component is produced or consumed in more than one reaction (as proposed by Carmichael (1969) for metamorphic reactions) and/or feedback mechanisms exist for the rate of production of that component due to thermal effects (that is, heat production or consumption during the chemical reaction), flow (in the form of diffusion and fluid advection) or deformation. Both coupling and feedback commonly lead to multiple stationary states and the procedure then is to establish which states are stable or unstable. A state is said to be stable if a small perturbation from that state results in the system returning to that state. Unstable states are those where the perturbation continues to grow so that the system moves to another state. The stationary states are further analysed using methods well established in nonlinear dynamics (Epstein and Pojman, 1998; Gray and Scott, 1994; Guckenheimer and Holmes, 1986; Wiggins, 2003) to describe the details of the path or paths the system will follow once perturbed (by a small or large amount). A diverse range of behaviours is possible including oscillatory behaviour in both space or time, travelling compositional waves and chaotic behaviour both in space and time (Epstein and Pojman, 1998; Gray and Scott, 1994; Murray, 1989; Ortoleva et al., 1987a,b; Prigogine, 1955; Ross, 2008; Scott, 1994; Turing, 1952). These processes are fundamental for understanding the development of metamorphic differentiation and the evolution of mineralising systems both from the point of view of spatial pattern formation (compositional zoning, pH and redox fronts) and of temporal oscillatory behaviour (oscillations in redox and pH in time at a given place in mineralising systems). These issues are considered in detail in Chapters 7 and 14 and in Volume II.

4. *Entropy production-based thermodynamics.* The fourth approach derives largely from the work of Jaynes (1963, 2003) and, although its basis is in the statistical physics of systems at equilibrium, it has evolved into general non-equilibrium thermodynamic arguments (Niven, 2009, 2010b; Niven and Andresen, 2010). The approach is characterised by the search for extrema in functions (or functionals) that represent the entropy or the entropy production or that represent some measure of the energy or of the energy dissipation. The correspondence in principle with the search for extrema in the entropy or the Gibbs energy in classical chemical equilibrium thermodynamics (Callen, 1960; Gibbs, 1906) is clear. In equilibrium thermodynamics the use of extremum principles is to predict the *equilibrium state* under constraints imposed by intensive or extensive variables. In non-equilibrium thermodynamics the use of extremum principles is to predict the *stationary states* of a flow system under constraints imposed by thermodynamic fluxes or thermodynamics forces. For details see Niven (2009, 2010a), Niven and Andresen (2010). The great potential of this approach is that it will reveal some general rules for how nonlinear systems evolve with time without investigating all the details of the individual processes involved in the evolution. For metamorphic and mineralising systems this is particularly relevant. The approach at present for mineralising systems is to add more and more detail to models of mineralising systems in an attempt to reproduce and couple all of the processes in Figures 1.1 and 5.4. The result is that one is forced to make simplifying assumptions such as local equilibrium or a lack of coupling between chemical reactions or of mechanical–thermal–chemical–fluid processes in order to

FIGURE 5.4 Some of the feedback processes operating in a
deforming chemically reacting solid
with fluid flow. *From Hobbs, Ord,
and Regenauer-Lieb (2011).*

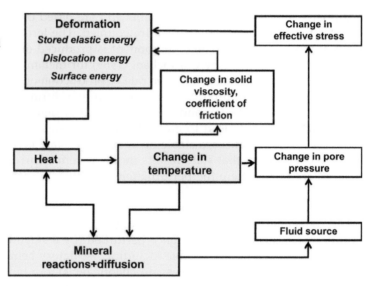

examine models that are tractable on large, modern computers. The situation has its
parallel in modern general circulation models of the climate of the Earth where more and
more detailed mechanisms with associated feedback effects are incorporated into the
models. The spectacular successes of extrema methods in this case are the results of
Paltridge and others (Niven, 2009; Ozawa and Ohmura, 1997; Ozawa et al., 2003;
Paltridge, 1975, 1978, 1981, 2001; Paltridge et al, 2007; Shimokawa and Ozawa, 2001,
2002) who reproduce details of the Earth's climate including detailed distributions of
temperature and cloud cover with a very simple model and the assumption that the
entropy production for the system is a maximum. A similar goal for describing the
evolution of metamorphic and mineralising systems with the bare minimum of detail
and some extremum principle should be the goal of future work. We elaborate upon
these concepts in Section 5.11 and Volume II.

5.5 SUMMARY OF THE NON-EQUILIBRIUM FRAMEWORK

This chapter considers the subject that has come to be known as *Generalised Thermodynamics* (Houlsby and Puzrin, 2006a). This area of study provides a convenient framework
for formulating relationships between coupled processes in a manner that is not *ad hoc*, in
that it ensures compatibility with the laws of thermodynamics, and is applicable to systems
at equilibrium as well as far from equilibrium. Moreover, the framework is a means of integrating and synthesising apparently diverse subjects and approaches. We also consider some
aspects of entropy production. This chapter is more an introduction to the subject than a review of the associated literature and it is intended to provide the basis for future chapters and
as a stepping stone for those who want to delve deeper.

Figure 5.4 shows some of the coupled processes that operate in a deforming rock mass undergoing metamorphism. A thermodynamic approach emphasises that these processes do not operate independently of each other but are strongly coupled through the second law of thermodynamics. Each process dissipates or absorbs energy, that is, it produces or consumes entropy, and so the coupling is dictated by the ways in which the entropy production is partitioned between the processes. By *entropy production* we mean the heat produced at each point in the system per unit time divided by the current temperature at that point (Truesdell, 1966). If $\dot{s}$ is the specific entropy production at a particular point then the specific total dissipation function is defined at that point as $\Phi = T\dot{s}$ where T is the absolute temperature; the *dissipation function*, Φ, is a scalar and has the units joules per kilogram per second; the overdot represents differentiation with respect to time. The total dissipation rate at each point, Φ, is the sum of the individual dissipation rates arising from each dissipative process operating at that point. The dissipation rates that concern us are those arising from plastic deformation, $\Phi^{plastic}$, mass transfer (by diffusion or by advection in a moving fluid), $\Phi^{mass\ transfer}$, chemical reactions, $\Phi^{chemical}$, and thermal transport, $\Phi^{thermal\ transport}$ and so

$$\Phi = T\dot{s} = \Phi^{plastic} + \Phi^{mass\ transfer} + \Phi^{chemical} + \Phi^{thermal\ transport} \geq 0 \qquad (5.2)$$

where the inequality is a direct expression of the second law of thermodynamics for a system not at equilibrium; the equality holds for equilibrium. (5.2) is the fundamental equation that enables various processes to be coupled in a thermodynamically admissible manner (such that the laws of thermodynamics are obeyed); when the individual dissipation functions are expressed in an explicit manner (5.2) is often called the *Clausius—Duhem inequality* (Truesdell and Toupin, 1960). If necessary, additional dissipation functions could be added to (5.2) to represent other dissipative processes such as fracturing, grain size reduction, crystallographic preferred orientation development or microstructural evolution. (5.2) is true independently of the material properties of the material; it is true for homogeneous and inhomogeneous materials and for isotropic and anisotropic materials. Complicated interactions can be incorporated into (5.2) by introducing relations (determined by experiments) between various dissipation functions. Thus chemical softening arising from the formation of weak mineral phases during a metamorphic reaction can be incorporated by writing evolutionary relations between $\Phi^{plastic}$ and $\Phi^{chemical}$.

It turns out that the manner in which the partitioning expressed by (5.2) is achieved across the various processes is dependent on the length and timescales involved; this not only introduces some simplifying aspects but is also the basis of a general principle of scale invariance that emerges from such work; *different processes dominate in producing entropy at different scales but similar structures develop at each scale.* By similar here we mean that structures of identical geometrical appearance are developed at a range of length scales.

Scale issues arise because of the following relations. Many processes that operate during the deformation of rocks can be expressed as diffusion equations. Thus diffusion of momentum, heat, of chemical components and of fluid pressure is governed by equations of the form $\frac{\partial c}{\partial t} = \kappa^{process}\frac{\partial^2 c}{\partial x^2}$ involving a diffusivity for the process, $\kappa^{process}$. c represents the degree of deformation, temperature, chemical potential or fluid pressure. If this process is coupled with deformation then the length scale, $l^{process}$, over which feedback is important is given by the standard diffusion relationship (Carslaw and Jaeger, 1959), $l^{process} = \sqrt{\kappa^{process}\tau}$, where τ is a timescale associated with the deformation. We take $\tau = (\dot{\varepsilon})^{-1}$ where $\dot{\varepsilon}$ is the strain rate; then $l^{process} = \sqrt{\kappa^{process}/\dot{\varepsilon}}$. Values of $l^{process}$ are given in Table 5.2 which shows that the length scale likely to characterise a particular coupled process varies from kilometres to microns depending on the process and the rate of deformation. For thermal–mechanical coupling at tectonic strain rates (say $10^{-12}\,\mathrm{s}^{-1}$) the length scale is the kilometre scale (Hobbs et al., 2008; Regenauer-Lieb and Yuen, 2004). Other scales have been considered in Regenauer-Lieb et al. (2009) and Hobbs et al. (2011) where it is shown that the same fold mechanism (namely, viscosity strain rate softening) operates at outcrop and thin-section scales as does at regional scales, although different physical and chemical processes are involved at the different scales.

As an example, consider a shear zone of thickness h. Then the heat dissipated by mechanical processes at a strain rate $\dot{\varepsilon}$ will be conducted out of the shear zone on a characteristic time scale $\tau = h^2/\kappa^{thermal}$ where $\kappa^{thermal}$ has a value for most rocks of $10^{-6}\,\mathrm{m^2 s^{-1}}$. For a shear zone 1 km thick $\tau = 10^{12}\,\mathrm{s}$, whereas for a shear zone 1 m thick $\tau = 10^6\,\mathrm{s}$ (note that 1 year $\cong 3.1536 \times 10^7$ s). If the shear zones deform at a shear strain rate of $10^{-13}\,\mathrm{s}^{-1}$ then

TABLE 5.2   Length Scales Associated with Various Processes When Coupled to Deformation

Process	Diffusivity, $\mathrm{m^2\,s^{-1}}$	Strain Rate, $\mathrm{s^{-1}}$	Length Scale for Process, m	References
Heat conduction; slow deformations (tectonic deformations)	10^{-6}	10^{-12}	1000	Hobbs et al. (2008)
Heat conduction; fast deformations (slow to fast seismic)	10^{-6}	10^{-2} to 10^2	10^{-2} to 10^{-4}	Veveakis et al. (2010)
Chemical diffusion; slow deformations (tectonic deformations)	Say 10^{-10} to 10^{-16}	10^{-12}	10 to 10^{-2}	Regenauer-Lieb et al. (2009)
Chemical diffusion; fast deformations (slow to fast seismic)	Say 10^{-10} to 10^{-16}	10^{-2} to 10^2	10^{-4} to 10^{-7}	Veveakis et al. (2010)
Fluid diffusion under a pressure gradient	Depends on permeability	10^{-12} to 10^{-2}	Any value from 10^3 to 10^{-4} depending on permeability	Phillips (1991), p. 81.
Chemical reactions	No diffusivity. Coupling depends on chemical dissipation		All scales from 10^3 to 10^{-4}	Hobbs et al. (2009), Regenauer-Lieb et al. (2009), Veveakis et al. (2010)
Diffusion of stress	10^{-4} to 10^{-7}	All strain rates	All length scales	Patton and Watkinson (2010)

the time taken to reach 20% shear strain is 2×10^{12} s. Thus the thick shear zone has the potential to heat up, whereas the thin shear zone remains isothermal. Note that if other endothermic processes such as endothermic mineral reactions occur within the thick shear zone then any heat generated by deformation may be used to enhance reaction rates and the thick zone may remain close to isothermal. If on the other hand the strain rate is 10^{-2} s^{-1} (representing a slow seismic event) then both the thick and thin shear zones heat up. Further discussion on this matter is presented by Burg and Gerya (2005) who point out that for a shear strain rate of 10^{-12} s^{-1} and a shear stress of 10 MPa the resulting dissipation compares in value with radiogenic dissipation in crustal rocks. Another example involving rock pulverisation in fault zones is discussed by Ben-Zion and Sammis (2013). We consider these aspects in Volume II.

The overall outcome is that different processes have strong feedback influences on others at different length scales leading to the scale invariance of structures and to the mineral assemblages that we observe in the crust of the Earth. The general system we consider in this chapter is a deforming, chemically reactive system in which microstructural/mineralogical evolution and mass and thermal transport play significant roles. Some examples of the feedback relations that can exist in such systems are illustrated in Figures 1.1 and 5.4.

While the various processes mentioned above are operating, the system dissipates energy and hence is not at equilibrium. Thus, the overarching concepts that unify these various processes are grounded in the thermodynamics of systems not at equilibrium. The traditional approach within geology as far as thermodynamics is concerned is strongly influenced by the work of Gibbs (1906) where, for the most part, equilibrium is assumed. The outcome is that non-equilibrium approaches have been largely neglected or even dismissed as irrelevant in the Earth Sciences, the argument being twofold: (1) geological processes are so slow that equilibrium can be assumed and (2) within a small enough region we can assume that equilibrium is attained (Korzhinskii, 1959; Thompson, 1959). Although both these points may be excellent approximations in some cases, the important point is that all of the processes mentioned above dissipate energy while they operate no matter how slow the process, or how small the system, and the resulting dissipation must be expressed in some manner.

The subject that is now called *Generalised Thermodynamics* (Houlsby and Puzrin, 2006a) is concerned with the ways in which dissipated energy (entropy production) is partitioned across the various processes operating in a deforming chemically reacting system. However, the system does not need to be far from equilibrium to be treated by generalised thermodynamics. In an adiabatic, homogeneously deforming hyperelastic solid (see Section 6.5) the deformation is reversible (McLellan, 1980); the entropy production for a closed cycle is zero (the processes are isentropic) and the system is an equilibrium system. This is still part of generalised thermodynamics. In this particular case the stresses are non-hydrostatic but the processes are reversible; *non-hydrostatic thermodynamics is not synonymous with non-equilibrium thermodynamics*. Discussions of non-hydrostatic thermodynamics in reversible systems are given by Nye (1957), Paterson (1973) and McLellan (1980). A deterrent to progress in applying non-equilibrium thermodynamics to deforming rocks is that developments in the subject have been dispersed across a large number of disciplines and languages, with conflicting or paradoxical propositions put forward, so that it has been difficult to produce a unified approach to the subject as far as geoscientists are concerned. An important example is the proposition that the entropy production (or the dissipation rate) is a minimum in non-equilibrium systems (Biot, 1955, 1958; Kondepudi & Prigogine,

1998; Onsager, 1931a, 1931b; Prigogine, 1955). In an apparent paradox, others, in particular (Ziegler, 1963, 1983a), have proposed exactly the opposite, that the entropy production is a maximum. We explore these and related concepts in Section 5.11. The outcome is that many of these apparently conflicting views turn out to be an expression of Ziegler's principle (Table 5.3); whether the entropy production is a maximum or a minimum depends on the constraints imposed on the system by its environment (Niven, 2010a). However, neither proposition need be relevant (Ross, 2008; Ross et al., 2012); for instance, in closed chemical systems the only extremum in entropy production is zero which means such systems always evolve to an equilibrium state (Chapter 14 and Ross, 2008).

TABLE 5.3   A Summary of Various Extremum Principles Proposed in Non-equilibrium Thermodynamics

Extremum Principle	Statement of Principle	Status
Ziegler's principle of maximum entropy production rate (Ziegler, 1963)	If the stress is prescribed the actual strain rate maximises the entropy production rate	True for many deforming systems; see Ziegler (1963, 1983b), Rajagopal and Srinivasa (2004).
Onsager's reciprocity relations and principle of least dissipation of energy (Onsager, 1931a,b)	Once the thermodynamic forces are fixed the actual thermodynamic fluxes maximise the dissipation. This is a consequence of the classical Onsager reciprocity relations. See (Ziegler, 1963)	Same as Ziegler's principle. Restricted to thermodynamically linear systems
Biot's principle of minimum entropy production rate (Biot, 1955)	If the stress and the rate of stress are prescribed the actual strain rate minimises the dissipation function	Same as Ziegler's principle but was obtained from Onsager's principle
Prigogine's principle of minimum entropy production rate. (Prigogine, 1955)	If there are n thermodynamic forces and j of these are prescribed then the dissipation function is minimised by equating the k thermodynamic fluxes to zero for $k = j+1, ..., n$	Conceptually of completely different form to Ziegler's principle but follows from Ziegler's principle; relies on the imposition of constraints; see Ziegler (1963) and Rajagopal and Srinivasa (2004). For chemical and thermal conduction processes the minimum entropy production principle does not exist (Ross, 2008)
Taylor's minimum internal work principle (Taylor, 1938)	For a prescribed deformation increment imposed on a polycrystal and a given set of critical resolved shear stresses the shears on the operating slip systems minimise the work	Same as Ziegler's principle
Bishop–Hill maximum external work principle (Bishop and Hill, 1951a,b)	For a prescribed deformation increment imposed on a polycrystal, the stress state that develops is such that the operating slip systems maximise the work. This is the linear programming dual of the Taylor principle	Same as Ziegler's principle

In this book we first develop a groundwork in *generalised thermodynamics* and then move to apply these principles to rocks where deformation, thermal and fluid transport, chemical reactions, damage and microstructural evolution contribute to the evolution of the structure of rock masses. It becomes clear that the principle of minimising the Helmholtz energy is an important principle at all scales. This principle is useful for both equilibrium and non-equilibrium systems and governs the formation of subdomains within the system characterised by differing chemical compositions and/or fabric. Such sub-domains play the roles of energy minimisers and can take on fractal geometries in order to achieve overall compatibility with an imposed deformation.

5.5.1 A Note on the 'Curie Principle'

The dissipation function, Φ, in (5.2) is a scalar quantity given by $\Phi = T\dot{s}$ but comprises the sum of other quantities that involve tensor quantities of higher order. Thus $\Phi^{plastic} = \frac{1}{\rho}\sigma_{ij}\dot{\varepsilon}_{ij}^{plastic}$, where ρ is the density, and so involves the second-order tensors, the Cauchy stress, σ_{ij}, and the plastic strain rate, $\dot{\varepsilon}_{ij}^{plastic}$. Misconceptions concerning the veracity of coupling various processes through the use of (5.2) arise from the work of De Groot (1952) and Prigogine (1955) who proposed a 'principle' which they labelled 'the Curie principle or theorem' that says '*Quantities of different tensorial character cannot interact with each other.*' Truesdell (1966, 1969) points out that there is nothing in the original works of Curie (1894, 1908), that resembles such a statement. Since a general tensor has no symmetry ascribed to it (other than the trivial property of symmetry or lack thereof about the leading diagonal) the De Groot 'Curie principle or theorem' has nothing to do with symmetry. The so-called 'Curie principle' is nothing more nor less than a statement of the most elementary rule of (Cartesian) tensor algebra which says that *tensors of the same order may be added to produce a tensor of the same order* (Eringen, 1962, p. 435). Otherwise an expression which adds Cartesian tensors of different orders makes no sense mathematically, physically or chemically. In (5.2) all of the terms such as $\Phi^{plastic} = \frac{1}{\rho}\sigma_{ij}\dot{\varepsilon}_{ij}^{plastic}$ are scalars with units J kg^{-1} s^{-1} as is the total dissipation. Unfortunately, the De Groot statement is commonly taken to mean *processes involving quantities of different tensorial order cannot be coupled*. This is clearly wrong. Nye (1957, Chapter 10) gives many well documented examples of the equilibrium properties of crystals where properties of different tensorial order are coupled or interact. An example is the elastic strain, ε_{ij}, produced in a piezoelectric crystal such as α-quartz, under the influence of a stress, σ_{ij}, and electric field, E_i, and also subjected to a temperature change, ΔT:

$$\varepsilon_{ij} = S_{ijkl}\sigma_{kl} + d_{kij}E_k + \alpha_{ij}\Delta T$$

This expression represents a coupling between a scalar, ΔT, a vector, E_i, three second order tensors, ε_{ij}, σ_{ij} and the thermal expansion tensor, α_{ij}, a third order tensor, the piezoelectric moduli tensor, d_{ijk}, and the fourth order elastic compliance tensor, S_{ijkl}. Such a relationship is well established both theoretically and experimentally (Nye, 1957).

Part of the misconception here seems to arise from the fact that many authors did not understand that quantities such as $\sigma_{ij}\dot{\varepsilon}_{ij}^{plastic}$ are scalars. After all, the quantity $\boldsymbol{\sigma} : \dot{\boldsymbol{\varepsilon}}^{plastic} \equiv \sigma_{ij}\dot{\varepsilon}_{ij}^{plastic}$ is, by definition, the *scalar* product of two second order tensors.

All of the couplings described by (5.2) are scalars. The De Groot 'Curie theorem' has been called the *'non-existent theorem in algebra'* by Truesdell (1966) since it is merely an expression of an elementary rule in tensor algebra and it certainly has no association with the Curie Principle as enunciated by Curie (1894, 1908) and discussed by Paterson and Weiss (1961).

5.6 STATE VARIABLES AND INTERNAL VARIABLES

A *state variable* is some kind of measure of a system that describes its state. An *equilibrium state variable* is an independent, observable (and hence experimentally controllable) measure of a system at thermodynamic equilibrium. Typical measures are the temperature, pressure and number of moles of a particular chemical component. For systems not at equilibrium independent, observable measures that are relevant to a description of the behaviour of the system are *kinematic state variables*. Typical measures are the stress, the strain and the volume. There are other measures relevant to the behaviour of solids not at equilibrium; these are frequently difficult to monitor during their evolution but have a direct influence on system behaviour. These are *internal variables*; typical examples are the plastic strain, the density of microfractures, the extent of a chemical reaction, the degree and nature of crystallographic preferred orientation and the degree of alignment of grain boundaries. Some examples of state variables of various kinds of use in understanding the evolution of metamorphic rocks are given in Table 5.4 together with what are known as

TABLE 5.4   State or Internal Variables and Conjugate Variables Used in Continuum Thermodynamics Relevant to Metamorphic Systems

State, Kinematic or Internal Variable	Conjugate Variable	Descriptions
$\varepsilon_{ij}^{elastic}$	σ_{ij}	Elastic strain; Stress
$\varepsilon_{ij}^{plastic}$	χ_{ij}	Plastic strain; Generalised stress (Houlsby and Puzrin, 2006a)
T	s	Absolute temperature; Specific entropy
ξ	A	Extent of chemical reaction; Affinity of chemical reaction (Kondepudi and Prigogine, 1998)
δ_{ij}	Y_{ij}	Tensor measure of damage; Generalised damage stress (Lyakhovsky et al., 1997, Karrech et al., 2011a)
m^K	μ^K	Concentration of Kth chemical component; Chemical potential of Kth chemical component (Coussy, 1995, 2004)
d	χ_{ij}^d	Grain size; Generalised stress associated with grain size evolution
ζ	p^{fluid}	Variation of fluid content; Pore pressure of fluid (Coussy, 1995; Detournay and Cheng, 1993)
β_{ij}^{CPO}	B_{ij}	Tensor measure of crystallographic preferred orientation; Generalised stress driving crystallographic preferred orientation development (Faria, 2006a,b, Faria et al., 2006)

TABLE 5.5   Intensive and Extensive Variables

Intensive Variables	Extensive Variables
Specific internal energy, e	Internal energy density, E
	Mass, m
Chemical potential, μ	Number of moles, n
Pressure, P	Volume, V
Temperature, T	Entropy, s
Elastic part of the second Piola−Kirchhoff stress	(Lagrangian) volume strain

conjugate variables. These are variables that when multiplied by their equivalent state variable in the adjacent column in Table 5.4 give the work done by that process. For this reason they are often called *work conjugate.* Some pairs of variables such as stress and strain rate are said to be *power conjugate* because when multiplied they give the power produced during deformation.

It is convenient in what follows to retain the concepts of extensive and intensive variables inherited from equilibrium thermodynamics. We use ϕ_k as the set of *additive* quantities such as volume and entropy density, known as *extensive variables.* The set of quantities, λ_k, are *non-additive* quantities such as pressure, temperature and specific entropy and are known as *intensive variables.* Some of these quantities are given in Table 5.5.

5.7 THE LAWS OF THERMODYNAMICS

We begin, in the spirit of Truesdell (1969), by asserting that quantities such as the internal energy density (the internal energy per unit volume), E, the absolute temperature, T, the entropy density, S, the heat flux vector, $\mathbf{q}$, with magnitude, q, the entropy supply, j (units: joules per unit volume per kelvin per second) and the entropy flux, $\mathbf{J}$ (units: joules per unit volume per kelvin per second), are *primitive quantities* (Truesdell, 1966) and need no definition in the same way that time, mass and force are adopted as primitive quantities in mechanics. Further we have the following definition and relations:

$$T \geq 0 \text{ by definition} \tag{5.3}$$

$$J = \frac{q}{T} \tag{5.4}$$

$$j = \frac{q}{T} \tag{5.5}$$

The laws of thermodynamics are assumed to hold for systems not at equilibrium just as they do for systems at equilibrium. The exception is the so called *Zeroth law* which is relevant only to systems at thermal equilibrium. This states that *two bodies in thermal equilibrium with a third body are in equilibrium with each other.*

The *first law of thermodynamics* is an expression of *energy balance* and states that there exists a quantity called the *internal energy density*, E, such that the rate of change of internal energy density in the body is equal to the sum of all the sources of power, namely, heat flux per unit volume into the system, $\dot{Q}$, and mechanical power input per unit volume, $\dot{W}$:

$$\dot{E} = \dot{Q} + \dot{W} \tag{5.6}$$

If we write the mechanical power input as $\dot{W} = \sigma_{ij}\dot{\varepsilon}_{ij}$ and $\dot{Q} = -\frac{\partial q_i}{\partial x_i}$ then the first law is written

$$\rho\dot{e} = \sigma_{ij}\dot{\varepsilon}_{ij} - \frac{\partial q_i}{\partial x_i} \tag{5.7}$$

where e is the specific internal energy and ρ is the density in the deformed state.

Just as the first law proposes that a quantity called internal energy exists, *the second law of thermodynamics* says there is another quantity, the *entropy density*, S, such that the total rate of change of entropy density of a system is greater than or equal to the sum of all the rates of heat input from various sources divided by the current temperature of those sources and is greater than or equal to zero.

We consider a spatial region, $\mathcal{B}$, that convects with the deforming body, $\mathcal{B}^d$, and that has a bounding surface, $\mathcal{S}$, with outward pointing normal, $\mathbf{n}$, at each point on $\mathcal{S}$ (Figure 5.5). Then we define a *net internal entropy*, $\mathbb{S}(\mathcal{B})$ and an *entropy flow*, $\mathbb{J}(\mathcal{B})$, the rate at which entropy flows into $\mathcal{B}$. Let $\mathbb{H}(\mathcal{B})$ be the *net entropy production* in $\mathcal{B}$. Then the *second law of thermodynamics* says that

$$\mathbb{H} \geq 0 \tag{5.8}$$

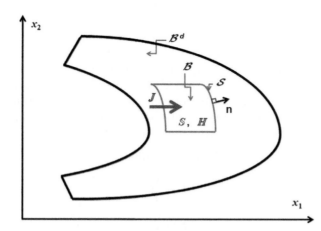

FIGURE 5.5   A spatial region, $\mathcal{B}$, that convects with a deforming body, $\mathcal{B}^d$, is bounded by a surface, $\mathcal{S}$, and has a unit normal, $\mathbf{n}$, at each point on $\mathcal{S}$. $\mathbb{S}$ and $\mathbb{H}$ are the net entropy and net entropy production within $\mathcal{B}$. The entropy flow into $\mathcal{B}$ from the rest of the deforming body is $\mathbb{J}$.

It follows that

$$H = \dot{S} - J \geq 0 \tag{5.9}$$

Or, rearranging (5.9), we have

$$\dot{S} = H + J \geq 0 \tag{5.10}$$

which says that the rate at which the net entropy of $\mathcal{B}$ changes is equal to the rate at which net entropy is produced in $\mathcal{B}$ plus the rate at which entropy flows into $\mathcal{B}$.

We assume that there exists a scalar, s, the *specific entropy* such that

$$\mathbb{S} = \int_B \rho s dv \tag{5.11}$$

We also assume that the entropy flow is comprised of an entropy flux, J, and an entropy supply, j, such that

$$J = -\int_S J \cdot \mathbf{n} da + \int_B j dv \tag{5.12}$$

Using (5.9) and (5.11) in (5.12) we obtain

$$\overline{\int_B \rho s dv} \geq -\int_S J \cdot \mathbf{n} da + \int_B j dv \tag{5.13}$$

Now using (5.4) and (5.5) we arrive at

$$\overline{\int_B \rho s dv} \geq -\int_S \frac{\mathbf{q}}{T} \cdot \mathbf{n} da + \int_B \frac{q}{T} dv \tag{5.14}$$

which is one form of the *Clausius–Duhem inequality*.

From (5.4), (5.5), (5.9), (5.11) and (5.12) we obtain

$$H = \int_B \left(\rho \dot{s} + \mathrm{div}\left(\frac{\mathbf{q}}{T}\right) - \frac{q}{T} \right) dv \tag{5.15}$$

and if we define $\mathbb{T}$ as the net entropy production density at each point in $\mathcal{B}$:

$$H = \int_B \mathbb{T} dv \geq 0 \tag{5.16}$$

Then

$$\mathbb{T} = \rho \dot{s} + \mathrm{div}\left(\frac{\mathbf{q}}{T}\right) - \frac{q}{T} \geq 0 \tag{5.17}$$

(5.17) is another, and commonly used, form of the Clausius–Duhem inequality.

5.7.1 Dissipation

The conversion of mechanical work to heat is called *dissipation* and it is convenient to define a quantity, Φ, called the *dissipation function* such that

$$\Phi = T\dot{s} \tag{5.18}$$

This means that the second law can be written

$$T\dot{s} = \sum_{i=1}^{n} \Phi_i \geq 0 \tag{5.19}$$

where the Φ_i are the dissipations arising from the n different processes that are operating. This is also an expression of the *Clausius–Duhem inequality*. (5.19) is the same as (5.2). Both expressions are simply expressions of the second law of thermodynamics.

5.7.2 Equilibrium or Not?

Most treatments of metasomatism and mineralising processes adopt the concept introduced by Korzhinskii (1959) of *mosaic equilibrium* which has come to be known in the geological literature as *local equilibrium* (Thompson, 1959, 1970). The proposal is that even though the system is not at equilibrium one can identify a large-enough region for the system that can be considered to be at equilibrium. The evolution of the system can then be built up by considering the interaction between these regions. The notion of equilibrium is pervasive in metamorphic and metasomatic petrology, to the extent that the concept is commonly treated as self evident for most geological systems. By definition, however, systems are not at equilibrium during their evolution and there are many situations, which we will discuss, where even the smallest region, in which concepts such as temperature and internal energy can be defined by some kind of averaging of molecular motions, is not at equilibrium. A self evident form of this statement is that mineral reactions cannot be at equilibrium while the mineral reactions are in progress and minerals are growing. There must be chemical affinities driving the reactions and hence by definition the system is not at equilibrium; at equilibrium these affinities drop to zero. The issue is: *In Nature how far from equilibrium are most mineral systems?* This question is not meant as a threat to the vast amount of work that has been devoted to defining mineral systems at equilibrium. If we knew how to handle systems not at equilibrium it would give us an extra toolbox that enables us to investigate the processes that operate during deformation and metamorphism. We return to this question in Chapter 14.

5.8 THE POTENTIALS: HELMHOLTZ AND GIBBS ENERGY FOR DEFORMING SOLIDS

In applying the concepts of equilibrium chemical thermodynamics for small strains to deforming chemically reacting solids, one replaces the mechanical role of the pressure, P, by the Cauchy stress tensor, σ_{ij}, and the mechanical role arising from small changes in the specific volume, $\widehat{V}$, by the small strain tensor, ε_{ij} (Houlsby and Puzrin, 2006a).

The restriction to small strains is not an undue limitation since one is commonly concerned with developing incremental approaches that can be implemented in finite element or finite difference codes.

Taking tensile stresses as positive, the pressure, P, and specific volume, $\widehat{V}$, are now given by

$$P = -\frac{1}{3}\sigma_{kk} \quad \text{and} \quad \widehat{V} = \widehat{V}_0(1 + \varepsilon_{kk}) = \frac{1}{\rho} \tag{5.20}$$

where $\widehat{V}_0$ is the initial specific volume and ρ is the current density. The initial density, ρ_0, is $\frac{1}{V_0}$. In a conceptual mapping of the quantities used in equilibrium chemical thermodynamics, P is replaced by $(-\sigma_{kk}/3)$ and V by $\widehat{V}_0\left(\frac{1}{3}\delta_{ij} + \varepsilon_{ij}\right)$ where δ_{ij} is the Kronecker delta. Thus the term, (PV), is replaced by

$$-\sigma_{ij}\widehat{V}_0\left(\frac{1}{3}\delta_{ij} + \varepsilon_{ij}\right)$$

$$= -\widehat{V}_0\left(\frac{1}{3}\sigma_{kk} + \sigma_{ij}\varepsilon_{ij}\right).$$

$$= P\widehat{V}_0 - \widehat{V}_0\sigma_{ij}\varepsilon_{ij}$$

The expressions for the specific internal energy, e, the specific Helmholtz energy, Ψ, the specific enthalpy, H, and the specific Gibbs energy, G, are then given as in Table 5.6. Also given in Table 5.6 are expressions for the conjugate variables $\{\sigma_{ij}, \varepsilon_{ij}\}$ and $\{s, T\}$ in terms of e, Ψ, H and G.

Gibbs (1906) discussed the solubility of a chemical component, A, in a fluid in contact with a stressed solid comprised of A. In discussing this problem the conclusion of Kamb (1961) is *'it is not possible to usefully associate a Gibbs free energy with a non-hydrostatically stressed solid.'* We discuss this particular problem in Chapter 14 but note that Kamb did not propose that the Gibbs energy could not be defined for a stressed solid. He simply indicated that the concept of a Gibbs energy is not useful for the problem he considered. Similar sentiments are presented by (Paterson, 1973) who discusses the issue in some detail.

TABLE 5.6   Thermodynamic Quantities Used in the Small Strain Formulation of Generalised Thermodynamics

	Specific Internal Energy	Specific Helmholtz Energy	Specific Enthalpy	Specific Gibbs Energy
Potential	$e = e(\varepsilon_{ij}, s)$	$\Psi = \Psi(\varepsilon_{ij}, T)$ $\Psi = e - sT$	$H = H(\sigma_{ij}, s)$ $H = e - \frac{1}{\rho_0}\sigma_{ij}\varepsilon_{ij}$	$G = G(\sigma_{ij}, T)$ $G = H - sT$ $G = \Psi - \frac{1}{\rho_0}\sigma_{ij}\varepsilon_{ij}$
	$\sigma_{ij} = \rho_0\frac{\partial e}{\partial \varepsilon_{ij}}$ $T = \frac{\partial e}{\partial s}$	$\sigma_{ij} = \rho_0\frac{\partial \Psi}{\partial \varepsilon_{ij}}$ $s = -\frac{\partial \Psi}{\partial T}$	$\varepsilon_{ij} = -\rho_0\frac{\partial H}{\partial \sigma_{ij}}$ $T = \frac{\partial H}{\partial s}$	$\varepsilon_{ij} = -\rho_0\frac{\partial G}{\partial \sigma_{ij}}$ $s = -\frac{\partial G}{\partial T}$

After Houlsby and Puzrin (2006).

It is important to realise that the Gibbs energy for a stressed solid has been defined and used by a number of authors over the past 100 years or so. Gibbs (1906, pages 86, 87 and 195–214) himself defines a quantity which is the *enthalpy per unit volume for a stressed solid*. This is equivalent to what we would now call the Gibbs energy per unit volume. Similar quantities have been defined for stressed solids by Truesdell and Toupin (1960, p. 627: the *free enthalpy*), Truesdell (1969, p. 33: the *free enthalpy*), Kestin (1968, p. 180 the *Gibbs function* or *free enthalpy*), Rice (1971, p. 437: the *complementary energy* or the *Legendre transform of the Helmholtz energy*), Rice (1975, p. 32: the *dual potential* to the Helmholtz energy), Shimizu (1997), Ulm and Coussy (2003; the *complementary energy* in many parts of the book), Houlsby and Puzrin (2006a, p. 47 and in many parts of the book), and Plohr (2011) who considers the Gibbs energy for two adjacent elastoplastic solids together with issues involved in defining the Gibbs energy for large deformations. Plohr also considers the differences that arise in defining the Gibbs energy for an elastoplastic solid as opposed to an inviscid fluid. A generalised treatment of this issue is given by Frolov and Mishin (2012).

5.8.1 The Legendre Transform

Many physical and chemical quantities are considered in the deformation of chemically reacting systems and it is useful both conceptually and pragmatically to understand the ways in which some of these quantities are related. Particularly important in this regard is the *Legendre transform* which is useful in expressing the relations between thermodynamic potentials (this chapter), between the yield surface and the dissipation surface in plasticity (Chapter 6) and between the singularity spectrum and a measure of the scaling exponent for the generalised fractal dimension (Arneodo et al., 1995) in the wavelet transform modulus maxima (WTMM) method in wavelet analysis (Chapter 7).

The Legendre transform of a function $y = f(x)$ at the point (x, y) is illustrated in Figure 5.6 and consists of the intercept of the tangent to $y = f(x)$ at (x, y) on the y-axis. For a function $y = f(x)$ the Legendre transform of $f(x)$ at x is given by

FIGURE 5.6   The Legendre transform.

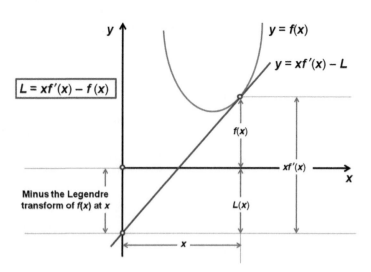

$$y = f(x) - x\frac{\partial f(x)}{\partial x} \tag{5.21}$$

As an example we take $f(x) \equiv \Psi(\varepsilon_{ij}, T)$, then $f'(x, T) \equiv \frac{\partial \Psi}{\partial \varepsilon_{ij}} = \frac{1}{\rho}\sigma_{ij}$. Thus the Legendre transform of Ψ is

$$y = \Psi - \frac{1}{\rho}\sigma_{ij}\varepsilon_{ij} \equiv G$$

Hence the Legendre transform of the Helmholtz energy is the Gibbs energy.

5.9 THE DISSIPATION FOR A MATERIAL WITH INTERNAL VARIABLES

The fundamental difference between the thermodynamics of relatively simple fluids such as those described by Newtonian viscosity or power-law viscosity (Chapter 6) and solids is that the state of stress in fluids is described by variables such as the velocity gradient, the temperature and the pressure, whereas in solids the stress state is also a function of the history (see comment by Kestin (1968), p. 200) of deformation as reflected in microstructural evolution such as the degree and character of crystallographic preferred orientation, the alignment of grain boundaries and the degree and geometry of microfracturing. As pointed out by Kestin (1968, p. 200), the use of the term *history* does not imply that the material has a *memory* of its past; the important point is that the stress is dependent on the *evolution* of microstructural adjustments. Thus extra variables are required to define the state of the body. These additional variables are called *internal variables*. Complicated fluids, such as polymers, which possess internal microstructure also require internal variables to define the state during deformation.

We introduce a *tensor internal variable* as α_{ij}. Then the internal energy is

$$e = e(\varepsilon_{ij}, \alpha_{ij}, s) \tag{5.22}$$

so that

$$\dot{e} = \frac{\partial e}{\partial \varepsilon_{ij}}\dot{\varepsilon}_{ij} + \frac{\partial e}{\partial \alpha_{ij}}\dot{\alpha}_{ij} + \frac{\partial e}{\partial s}\dot{s} \tag{5.23}$$

We write the second law of thermodynamics as

$$\dot{s} + \frac{1}{\rho}\frac{\partial}{\partial x_i}\left(\frac{q_i}{T}\right) = \frac{\Phi}{T} \geq 0 \tag{5.24}$$

where Φ is the total dissipation. The first law of thermodynamics is written as

$$\rho\dot{e} = \sigma_{ij}\dot{\varepsilon}_{ij} - \frac{\partial q_i}{\partial x_i} \tag{5.25}$$

so that combination with the second law, (5.24) gives

$$\rho\dot{e} = \sigma_{ij}\dot{\varepsilon}_{ij} + \rho T\dot{s} - \frac{q_i}{T}\frac{\partial T}{\partial x_i} - \Phi \tag{5.26}$$

Using (5.26) with (5.23) gives

$$\overbrace{\left(\frac{1}{\rho}\sigma_{ij}-\frac{\partial e}{\partial \varepsilon_{ij}}\right)\dot{\varepsilon}_{ij}+\left(T-\frac{\partial e}{\partial s}\right)\dot{s}-\frac{\partial e}{\partial \alpha_{ij}}\dot{\alpha}_{ij}}^{\text{Mechanical dissipation}}-\overbrace{\Phi}^{\text{Total dissipation}}-\overbrace{\frac{1}{\rho}\frac{q_i}{T}\frac{\partial T}{\partial x_i}}^{\text{Thermal dissipation}}=0 \qquad (5.27)$$

As indicated, (5.27) may be divided into terms that represent mechanical dissipation and thermal dissipation. Although the second law requires that the total dissipation be non-negative the assumption is made that the *processes* involved in thermal dissipation are independent of the *processes* involved in mechanical dissipation so that we can write

$$\left(\frac{1}{\rho}\sigma_{ij}-\frac{\partial e}{\partial \varepsilon_{ij}}\right)\dot{\varepsilon}_{ij}+\left(T-\frac{\partial e}{\partial s}\right)\dot{s}-\frac{\partial e}{\partial \alpha_{ij}}\dot{\alpha}_{ij}-\Phi^{mechanical}=0 \qquad (5.28)$$

where $\Phi^{mechanical}=total\ dissipation-thermal\ dissipation\geq 0$. We see from Table 5.6 that $\sigma_{ij}=\rho\frac{\partial e}{\partial \varepsilon_{ij}}$ and $T=\frac{\partial e}{\partial s}$ so that

$$-\frac{\partial e}{\partial \alpha_{ij}}\dot{\alpha}_{ij}=\Phi^{mechanical}\geq 0 \qquad (5.29)$$

A common situation arises where α_{ij} is the plastic strain in which case, as expected, (5.29) states that the mechanical dissipation arises solely from the plastic and not the elastic strain rate. In other applications there may be more than one internal variable in addition to the plastic strain (arising from dislocation motion) such as the strain arising from microfracturing in which case the mechanical dissipation consists of the sum of the dissipations arising from the various processes.

5.10 THE THERMO-MECHANICAL HEAT BALANCE EQUATION

It is important in many applications to define the temperature evolution of a system arising from all the dissipative processes that are operating. The equation that describes the rate of change of temperature with time is called *the thermo-mechanical heat balance equation*. As a summary of many points that have been raised in this chapter it is instructive, in deriving this equation, to begin with the first law of thermodynamics, (5.25), written as $\rho\dot{e}=\sigma_{ij}\dot{\varepsilon}_{ij}-\frac{\partial q_i}{\partial x_i}$ (where $\dot{\varepsilon}_{ij}$ is the total strain rate; internal heat sources have been neglected) and the specific Helmholtz energy (Table 5.6): $\Psi=e-Ts$. Then

$$\dot{\Psi}=\dot{e}-T\dot{s}-\dot{T}s \qquad (5.30)$$

Substitution of the first law into (5.30) gives

$$\rho\left(\dot{\Psi}+T\dot{s}+\dot{T}s\right)-\sigma_{ij}\dot{\varepsilon}_{ij}+\frac{\partial q_i}{\partial x_i}=0 \qquad (5.31)$$

For materials where contributions to the entropy production arise from processes involved in the dissipative strain rate, $\dot{\varepsilon}_{ij}^{dissipative}$ (such as dislocation motion and diffusion), and also from k other processes (such as fluid flow) characterised by the array of scalar, vector and tensor quantities α_k, the Helmholtz energy can also be written

$$\Psi = \Psi\left(\varepsilon_{ij}^{elastic}, \alpha_k, T\right) \tag{5.32}$$

Hence, by the chain rule of differentiation,

$$\dot{\Psi} = \frac{\partial\Psi}{\partial\varepsilon_{ij}^{elastic}}\dot{\varepsilon}_{ij}^{elastic} + \sum_k\left(\frac{\partial\Psi}{\partial\alpha_k}\dot{\alpha}_k\right) + \frac{\partial\Psi}{\partial T}\dot{T} \tag{5.33}$$

Or, since (Table 5.6)

$$s = -\frac{\partial\Psi}{\partial T} \tag{5.34}$$

$$\dot{\Psi} = \frac{1}{\rho}\sigma_{ij}\dot{\varepsilon}_{ij}^{elastic} + \frac{1}{\rho}\sum_k\left(\frac{\partial\Psi}{\partial\alpha_k}\dot{\alpha}_k\right) - s\dot{T} \tag{5.35}$$

Assuming that the behaviour of the material is such that (see Chapter 6)

$$\dot{\varepsilon}_{ij} = \dot{\varepsilon}_{ij}^{elastic} + \dot{\varepsilon}_{ij}^{dissipative} \tag{5.36}$$

substitution of (5.35) and (5.36) into (5.31) gives

$$\rho T\dot{s} = \sigma_{ij}\dot{\varepsilon}_{ij}^{dissipative} - \sum_k\left(\frac{\partial\Psi}{\partial\alpha_k}\dot{\alpha}_k\right) - \frac{\partial q_i}{\partial x_i} \tag{5.37}$$

Now since

$$q_i = -k_{ij}\frac{\partial T}{\partial x_j} \tag{5.38}$$

(5.37) can be rewritten as

$$\rho T\dot{s} = \sigma_{ij}\dot{\varepsilon}_{ij}^{dissipative} - \sum_k\left(\frac{\partial\Psi}{\partial\alpha_k}\dot{\alpha}_k\right) + k_{ij}\nabla^2 T \tag{5.39}$$

From (5.34) we have

$$\dot{s} = -\frac{\partial^2\Psi}{\partial T^2}\dot{T} - \sum_k\left(\frac{\partial^2\Psi}{\partial T\partial\alpha_k}\dot{\alpha}_k\right) - \frac{\partial^2\Psi}{\partial T\partial\varepsilon_{ij}^{elastic}}\dot{\varepsilon}_{ij}^{elastic} \tag{5.40}$$

We write

$$c_p = -T\frac{\partial^2\Psi}{\partial T^2}; \frac{\partial^2\Psi}{\partial T\partial\alpha_k} = \frac{\partial}{\partial T}\frac{\partial\Psi}{\partial\alpha_k} = \frac{\partial\chi_k}{\partial T}; \quad\text{and}\quad \beta_{ij} = -\frac{\partial^2\Psi}{\partial T\partial\varepsilon_{ij}^{elastic}} = \frac{\partial s}{\partial\varepsilon_{ij}^{elastic}} = -\frac{\partial\sigma_{ij}}{\partial T} \tag{5.41}$$

where c_p is the specific heat at constant pressure, χ_k is an array of generalised stresses conjugate to the α_k and β_{ij} is a material coefficient that describes how the entropy varies with the elastic strain; it is a measure of the thermoelastic effect (Nye, 1957, pp. 173–176). Therefore,

$$\rho T \dot{s} = -\rho c_p \dot{T} - T \sum_k \left(\frac{\partial \chi_k}{\partial T}\right) \dot{\alpha}_k + T \beta_{ij} \dot{\varepsilon}_{ij}^{elastic} \tag{5.42}$$

and hence

$$\rho c_p \dot{T} = \widehat{\chi} \sigma_{ij} \dot{\varepsilon}_{ij}^{dissipative} + \sum_k \left(\frac{\partial \chi_k}{\partial T} T - \chi_k\right) \dot{\alpha}_k - T \beta_{ij} \dot{\varepsilon}_{ij}^{elastic} + k_{ij} \nabla^2 T \tag{5.43}$$

where $\widehat{\chi}$ is the Taylor–Quinney coefficient (Stainier and Ortiz, 2010; Taylor and Quinney, 1934) that is the proportion of heat that is stored as energy (in the form of dislocations or other defects) in the deformed material; $\widehat{\chi}$ commonly is of the order of 0.9. (5.43) is *the thermomechanical heat balance equation*. This equation, in many different forms, is used to calculate the temperature rise due to various dissipative processes that occur in a deforming chemically reacting body.

Equation (5.43) says that the change in temperature arises from the various dissipative processes that operate in the body to produce the permanent deformation (the first term on the right-hand side), the dissipation arising from other internal processes such as chemical reactions and fluid flow (the second term on the right-hand side), modified by a thermoelastic effect (the third term on the right-hand side), and thermal dissipation (the fourth term on the right-hand side). In its most general usage the equation is highly nonlinear since both the density and the specific heat are temperature dependent. Various approximations are commonly made such as temperature-independent density and specific heat, $\widehat{\chi} = 1$ and neglect of the thermo-elastic effects.

In practice, it is sometimes possible to simplify (5.43) by considering the time and length scales associated with various processes that operate simultaneously. Thus if the strain rate is slow (say $10^{-12} \, \text{s}^{-1}$) and the system is 1 km^3 in size then Table 5.2 indicates that the system may be considered isothermal as far as mechanical dissipation is concerned so that $\dot{T} = 0$ and $\nabla T = 0$ in (5.43).

5.11 ENTROPY PRODUCTION AND SYSTEM CONSTRAINTS

The extremum principle universally used in equilibrium chemical thermodynamics is that the Gibbs energy is a minimum at equilibrium. This principle has proved very powerful and has allowed exceptional progress in the construction of mineral phase equilibrium diagrams for quite complicated metamorphic reactions and bulk chemical compositions (Powell, 1978; Powell et al., 1998). However, for systems not at equilibrium some other principle, or set of principles, would be useful in describing the evolution of the system as it proceeds towards equilibrium or is maintained far from equilibrium by fluxes of energy and/or mass. The extremum principles that have evolved over the past 30 years all involve a statement concerning the entropy production, the most general of which is the Clausius–Duhem relation.

A well-known example in structural geology is represented by the Taylor (1938) and Bishop—Hill (Bishop and Hill, 1951a,b) extremum principles used in calculating the evolution of crystallographic preferred orientations (Lister et al., 1978). The Taylor principle says: *In a deforming polycrystal, for a given imposed strain-increment, the energy dissipation by the shear strains on the slip systems (with specified critical resolved shear stresses) needed to accomplish that strain-increment is a minimum.* The Bishop—Hill principle is the linear programming dual of the Taylor principle and says that: *In a deforming polycrystal, for a given imposed strain-increment, the stress state that actually exists within the deforming crystal is that which maximises the energy dissipation.* We will see that the Taylor—Bishop—Hill principles are equivalent to the (Ziegler, 1963, 1983a) principle of maximum entropy production. The Bishop—Hill maximum work principle chooses that stress (represented by a vector in stress space) in the deforming crystal which is as close as possible parallel to the imposed strain rate vector subject to the constraint that the imposed strain increment must be achieved only by slip systems that define the yield surface.

For many systems the commonly quoted extremum principle is that the entropy production is minimised (Biot, 1958; Prigogine, 1955). Examples of such systems that are widely quoted are steady heat conduction in a material with constant thermal conductivity (Kondepudi and Prigogine, 1998) and simple, uncoupled chemical reactions (Kondepudi and Prigogine, 1998). However, Ross and Vlad (2005 and references therein) and Ross (2008, his Chapter 12) show that, for these two classical systems, the only extremum in the entropy production corresponds to $\dot{s} = 0$. It is only for systems where the relationship between thermodynamic fluxes and forces is linear that a maximum exists for the entropy production and this corresponds to a stationary state that is not an equilibrium state. If the relationship between thermodynamic fluxes and forces is nonlinear then only one extremum in the entropy production exists and that corresponds to an equilibrium state. The linear situation is that commonly quoted and discussed by Prigogine (1955), Kondepudi and Prigogine (1998) and a host of others. However, reference to Table 5.1 shows that at least for thermal conduction and chemical reactions the thermodynamic fluxes are not linear functions of the thermodynamic forces. For mass diffusive processes the thermodynamic flux is a linear function of the thermodynamic force only for isothermal situations. For further discussion and clarification of what is meant by 'close' and 'far' from equilibrium reference should be made to the interchange between Hunt, Hunt, Ross, Vlad and Kondepudi (Hunt et al., 1987, 1988; Kondepudi, 1988; Ross, 2008; Ross and Vlad, 2005). We return to this subject in Chapter 14 and Volume II.

For some systems (and this involves many mechanical systems; Houlsby and Puzrin, 2006a,b; Rajagopal and Srinivasa, 2004) the entropy production is maximised. Ziegler (1963 and references therein) seems to have been the first to propose this principle, although we will see that the principle of maximum work proposed by Bishop and Hill (1951a,b) is identical to that of Ziegler; Ziegler (1963) gives a good summary of other related principles. The Ziegler principle is a powerful tool for defining the evolution of simple plastic deformations (Houlsby and Puzrin, 2006a,b). In order to understand Ziegler's arguments it is instructive to examine two ways in which the relationship between stress and dissipative strain rate can be graphically represented. The first way, the yield surface, is well known and widely used. This surface was discussed in relation to localisation in rate-independent materials by Hobbs et al.

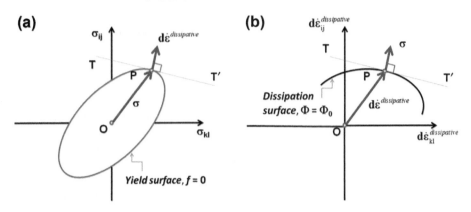

FIGURE 5.7   Yield and dissipation surfaces. One is the Legendre transform of the other. For the yield surface, (a), the incremental strain rate vector is normal to the yield surface where the imposed stress vector touches the yield surface. For the dissipation surface, (b), the stress vector is normal to the dissipation surface where the incremental strain rate vector touches the dissipation surface.

(1990) and is shown in Figure 5.7(a). Detailed discussions of the yield surface are given in Chapter 6. The yield surface is drawn in stress space and stress states within the surface correspond to elastic deformations while stress states on the yield surface correspond to plastic deformations. If the yield surface corresponds to a function $f(\sigma_{ij})$ then the condition for plastic yield is $f = 0$ while $f < 0$ corresponds to elastic deformations. Stress states corresponding to $f > 0$ are physically unrealistic. If no volume change accompanies plastic deformation then the increment of plastic strain rate, $\dot{\varepsilon}_{ij}^{dissipative}$, associated with the stress, σ_{ij}, is normal to the yield surface at the end point of the stress vector (Figure 5.7(a)) and is given by $\dot{\varepsilon}_{ij} = \lambda \frac{\partial f}{\partial \sigma_{ij}}$ where λ is a constant known as a *plastic multiplier*. The dissipation is the scalar product of the stress and the incremental strain rate: $T\dot{s} = \Phi = \frac{1}{\rho}\sigma_{ij}\dot{\varepsilon}_{ij}^{dissipative}$. One can see from Figure 5.7(a) that Φ is always positive in accordance with the second law of thermodynamics. This arises from the assumption that the yield surface is convex.

One can also plot another surface, the *dissipation surface* in strain rate space (Figure 5.7(b)) for a particular value of the dissipation function, $\Phi = \Phi_0$. Ziegler (1963, 1983b) showed that for uncoupled deformations the stress corresponding to a particular strain rate increment is orthogonal to the dissipation surface at the end point of the strain rate vector (Figure 5.7(b)) and is given by $\sigma_{ij} = \nu \frac{\partial \Phi}{\partial \dot{\varepsilon}_{ij}^{dissipative}}$ where ν is a constant. Thus the yield surface and the dissipation surface are complementary to each other (Collins and Houlsby, 1997; Houlsby and Puzrin, 2006a) and in fact, just as the Gibbs energy can be obtained from the Helmholtz energy through a Legendre transformation (Callen, 1960; Houlsby and Puzrin, 2006a; Powell et al., 2005) and *vice versa*, the yield surface can be obtained from the dissipation surface through a Legendre transformation (Collins and Houlsby, 1997; Houlsby and Puzrin, 2006a) and *vice versa*.

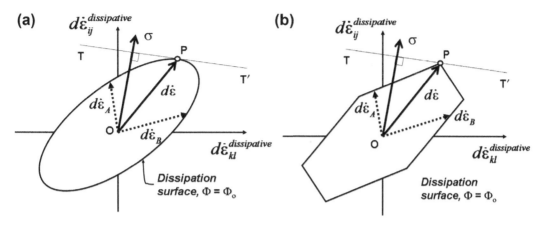

FIGURE 5.8 Illustration of Ziegler's principle of maximum entropy production rate. The dissipation surface is drawn in incremental strain rate space. (a) The stress associated with the imposed strain rate is normal to the dissipation surface at P. Of all other possible strain rate vectors, for example $d\dot\varepsilon_A$ and $d\dot\varepsilon_B$, the projection of the strain rate vector on to the stress vector is smaller than the projection of the imposed strain rate vector. This means that the dissipation arising from the stress and the imposed strain rate is a maximum. The distance of the tangent plane TPT' from the origin, O, is the work done by the stress during the imposed strain rate increment. (b) The equivalent of (a) drawn for multiple slip single crystal rate insensitive plasticity. The argument for (a) follows but there is ambiguity with respect to the orientation of the tangent at the corner of the yield surface. This ambiguity is removed with the introduction of rate sensitivity (Chapter 6). *From Hobbs et al. (2011).*

The importance of the *Ziegler orthogonality relation* is better seen by redrawing Figure 5.7(b) as in Figure 5.8(a). In this figure the strain rate vector touches the dissipation surface, $\Phi = \Phi_0$ at P where the tangent TPT' is drawn. The Zeigler orthogonality relation says that the stress vector associated with the strain rate is normal to this tangent. The dissipation (and hence the entropy production at a given temperature) is the scalar product of the strain rate with the stress (or the projection of the strain rate vector on to the stress vector). One can see from Figure 5.8(a) that, because of the convexity of the dissipation surface, this scalar product is a maximum for the stress and incremental strain rate that obey the orthogonality relation. For all other strain rate vectors the scalar product with the stress is smaller. This is a graphical demonstration of the Ziegler principle of maximum entropy production which says that *of all possible stress states, the stress that is actually associated with a particular strain rate produces a maximum in the entropy production.* Although this argument depends on the convexity of the dissipation surface, Houlsby and Puzrin (2000) point out that the Ziegler principle holds also for weakly non-convex yield surfaces and for non-associated flow (where the incremental strain rate vector is not normal to the yield surface).

The Bishop—Hill theory of crystal plasticity is illustrated in Figure 5.8(b) where a section through the dissipation surface for a crystal undergoing rate-insensitive deformation on multiple slip systems is shown. The same arguments that apply to Figure 5.8(a) hold except now, for rate-insensitive plasticity, there is an ambiguity in the orientation of the tangent to the dissipation surface at a corner. Nevertheless the overall argument concerning entropy production rate holds as in Figure 5.8(a). The ambiguity is removed by the introduction of rate sensitivity into the constitutive equations for slip (Chapter 6, Rice (1970); see also

Lin and Ito (1966) who also discuss the rounding of corners on a yield surface during deformation).

The differences between the Ziegler maximum entropy production principle and the Prigogine minimum entropy production principle are illustrated in Figure 5.9. For a process where the thermodynamic affinity, A, is a linear function of the thermodynamic flux, J, the entropy production, $\dot{s}$, is a quadratic function of the thermodynamic fluxes. A simple example is Newtonian viscosity where the stress is given in terms of the viscosity, η, as $\sigma = \eta\dot{\varepsilon}$. The entropy production is $\rho\dot{s} = \sigma\dot{\varepsilon} = \eta\dot{\varepsilon}^2$. Thus a two-dimensional plot of entropy production versus the fluxes is a paraboloid as shown in Figure 5.9(a) This paraboloid passes through

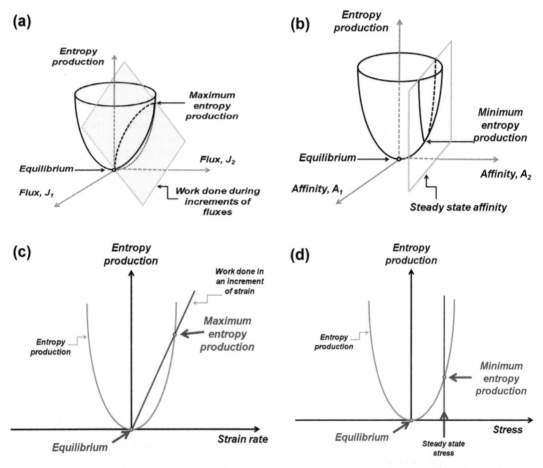

FIGURE 5.9   Effect of various constraints and system behaviour on entropy production. (a) A parabolic dissipation surface where the entropy production is constrained by a planar surface corresponding to the power. This is a linear system where the intersection of the two surfaces corresponds to a maximum in entropy production. This is the Ziegler model. (b) The same system as (a) but constrained by the imposition of steady state conditions. The entropy production is a minimum. This is the Prigogine model. (c) and (d) Sections through (a) and (b) for a deforming Newtonian viscous material. *Motivated by Rajagopal and Srinivasa (2004).*

the origin at $\dot{s} = 0$ which corresponds to equilibrium. The power for a fixed A is given by $A \cdot J$ which is represented by a plane that passes through the origin in Figure 5.9(a) This plane represents the constraint on entropy production; it intersects the paraboloid in an ellipse and the highest point on the ellipse represents maximum entropy production for that particular constraint. This illustrates the Ziegler principle. If on the other hand both A and J are fixed so that the system is at steady state the situation is represented by Figure 5.9(b). Now the constraint is such that the entropy production is a minimum. This is the Prigogine argument. Sections through these two different constraint models for a Newtonian viscous material are given in Figure 5.9(c,d).

5.12 CONVEX AND NON-CONVEX POTENTIALS: MINIMISATION OF ENERGY

In chemical systems it is common for the Gibbs energy to be expressed as a convex function with a single minimum at equilibrium (Kondepudi and Prigogine, 1998; Powell, 1978). At a phase change the function may become non-convex with two minima corresponding to the Gibbs energies of the two phases at equilibrium (Kondepudi and Prigogine, 1998; Powell, 1978). The same situation arises in mechanical (and other) systems where more than one equilibrium or non-equilibrium state exists (Chapter 7; Ortiz and Repetto, 1999; Thompson, 1982).

Deforming nonlinear elastic systems represent perhaps the simplest manifestation of systems with multiple stationary states and we include a brief description to highlight that even in such systems the evolution of the system can be complex, even chaotic. In linear elastic systems undergoing deformation one equilibrium stationary state exists and that is the one that maximises the stored elastic energy (McLellan, 1980, pp. 313–314). This is the only stationary state that can exist for an adiabatically deforming linear elastic system. However, for nonlinear elastic systems where the nonlinearity is introduced via a softening of the elastic modulus arising from some form of elastic damage or from geometrical softening arising from large rotations, multiple equilibrium stationary states commonly exist (Budd et al., 2001; Ericksen, 1998; Hunt, 2006; Hunt et al., 1997a,b; Hunt et al., 2000a; Hunt and Wadee, 1991) and correspond to non-convex forms of the Helmholtz energy function. By non-convex here we mean that a plot of the Helmholtz energy against some measure of the deformation or deformation gradient has a number of 'bumps' or discontinuities in it (see Chapter 7).

The extremum principle involves the development of minima in the stored elastic energy and corresponds to points where the stored elastic energy matches the energy required to drive some form of deformation in the system. Such a principle corresponds to the Maxwell construction common in chemical systems involving two or more phases (Kondepudi and Prigogine, 1998; Powell, 1978) and also represents a lower bound for the force required for the initiation of deformation. Such behaviour is common in buckling systems. However, the first bifurcation in the system occurs after the peak stress is attained (Tvergaard and Needleman, 1980 for plastic buckling; Hunt and Wadee, 1991 for elastic buckling). Sequential bifurcations occur as the system jumps from one equilibrium stationary state to another, although complexity can arise due to mode locking (Everall and Hunt, 1999; Hunt and Everall, 1999). All such behaviour in an adiabatically deformed elastic system represents a progression through a series of equilibrium states and a full description of the system evolution relies on nonlinear bifurcation theory (Guckenheimer and Holmes, 1986; Wiggins, 2003). In addition, softening behaviour can lead to localisation of fold packets in layered materials and a progression to chaos with fractal geometries (Hunt and Wadee, 1991). Multiple stationary states are also well known in the study of microstructural development in martensitic transformations (Ericksen, 1998). Stationary states arising from thermal–mechanical feedback in mantle deformations are discussed by Yuen and Schubert (1979). In nonlinear chemical systems multiple stationary states are common (Epstein and Pojman, 1998; Ross, 2008) and are considered in detail in Chapter 14. The fundamental principle involved in understanding the evolution of these kinds of mechanical and chemical systems is that the Helmholtz energy is no longer convex and the system evolves to minimise the Helmholtz energy by forming a series of finer structures or chemical domains that lead to compatibility with the imposed deformation (Ball, 1977; Ball and James, 1987; Ortiz and Repetto, 1999). We explore such behaviour in detail in Section 7.7 and in Chapter 8. The transition from convex to non-convex Helmholtz energy functions is the hallmark of critical phenomena (Ben-Zion, 2008; Sornette, 2000) so that such behaviour seems to be a fundamental and universal aspect of deforming- reacting systems.

Recommended Additional Reading

Callen, H. B. (1960). *Thermodynamics: An Introduction to the Physical Theories of Equilibrium Thermostatics and Irreversible Thermodynamics*. New York, London: John Wiley and Sons, 376 pp.
   The classical text on thermodynamics with excursions into non-equilibrium thermodynamics.

Coussy, O. (1995). *Mechanics of Porous Continua*. Chichester, UK: Wiley.
   The definitive text on the thermodynamics of fluid flow through deforming porous media with sections on chemical phase changes, dissolution and precipitation and chemical reactions.

Coussy, O. (2010). *Mechanics and Physics of Porous Solids*. Chichester, UK: Wiley.
   An excellent text on thermodynamics applied to flow in porous media.

Gurtin, M. E., Fried, E., & Anand, L. (2010). *The Mechanics and Thermodynamics of Continua*. Cambridge University Press.
   An advanced text with in-depth treatments of thermodynamics applied to deforming materials both solids and fluids.

Houlsby, G. T., & Puzrin, A. M. (2006). *Principles of Hyperplasticity*. London: Springer-Verlag.
   An important text on the thermodynamics of hyperplastic materials with an emphasis on pressure-dependent constitutive behaviour.

Kern, R., & Weisbrod, A. (1967). *Thermodynamics for Geologists*. Cooper and Co: Freeman.
   An excellent text on equilibrium thermodynamics with geological applications.

Lavenda, B. H. (1978). *Thermodynamics of Irreversible Processes*. MacMillan Press.
   An important text devoted to non-equilibrium thermodynamics with a historical perspective.

Nye, J. F. (1957). *Physical Properties of Crystals*. Oxford Press.
   This book is the definitive treatment of the equilibrium properties of crystals with a chapter devoted to the thermodynamics of coupled phenomena.

Powell, R. (1978). *Equilibrium Thermodynamics in Petrology: An Introduction*. Harper and Row, London.
   A clearly written text that acts as an introduction to equilibrium thermodynamics applied to petrological examples.

Rice, J. R. (1975). Continuum mechanics and thermodynamics of plasticity in relation to microscale deformation mechanisms. In Ali S. Argon (Ed.), *Constitutive equations in plasticity* (pp. 23–79). Cambridge, Mass: MIT Press. *A seminal paper that introduces the concept of the stress being defined in terms of a potential and treats the incorporation of internal variables to describe microstructural mechanisms, including diffusion, phase changes and fracturing, during deformation. The treatment involves large deformations.*

Tadmor, E. B., Miller, R. E., & Elliot, R. S. (2012). *Continuum Mechanics and Thermodynamics*. Cambridge University Press. *A treatment of the thermodynamics of deforming materials that is fairly easy to read.*

Truesdell, C. A. (1966). *Six Lectures on Modern Natural Philosophy*. Berlin: Springer-Verlag. *A critical overview of mechanics with a discussion of the role of thermodynamics in mechanics.*

Truesdell, C. A. (1969). *Rational Thermodynamics* (second ed.). New York: Springer-Verlag. *An important work that is devoted to the application of thermodynamics to the prescription of constitutive equations.*

6

Constitutive Relations

Structural Geology
http://dx.doi.org/10.1016/B978-0-12-407820-8.00006-0

6.1 WHAT IS A CONSTITUTIVE RELATION?

When a solid is loaded mechanically and thermally it behaves in different ways depending on environmental factors such as the temperature, the confining pressure, the deformation rate and, if fluids are present, the fluid pressure. Also solids comprising different materials behave in different ways even if the environmental factors are identical. An example is shown in Figure 6.1 where two different materials, a laminated elastic-viscous material in Figure 6.1(c and d) and a laminated elastic-plastic material in Figure 6.1(e and f) are deformed in sinistral simple shearing, both to the same strain, as shown in Figure 6.1(b). The two materials behave in completely different manners. The elastic-viscous material forms gentle folds with axial planes at a *high angle* to the shearing direction, whereas the elastic-plastic material forms localised folds with somewhat sharper hinges and with axial planes *parallel* to the shearing direction. Moreover, the fine scale structure is different for the two materials (Figure 6.1(d and f)). This makes it clear that the material behaviour is paramount in controlling the types of structures that form and also gives some confidence that we may be able to test particular constitutive relations by informed inspection of natural examples.

A *constitutive relation* is a mathematical expression, relevant to a particular material, that relates the stress, the deformation gradient, the stretching tensor and the heat flow during deformation. Such a relation is commonly only useful over a restricted range of environmental factors. These relations are usually greatly idealised and describe materials such as those with ideal elasticity, viscosity and plasticity. Nevertheless, considerable detail can be built into these relations, in the form of internal variables, in order to depict the influence of material heterogeneities, anisotropy, geometrical imperfections, fluid flow, chemical reactions and so on. A constitutive relation takes the same place in mechanics as does an *equation of state* in chemical equilibrium thermodynamics. In this chapter we explore only thermal-mechanical constitutive relations and extend the arguments to materials with greater variations in behaviour in later chapters. The outcome of developing these relations is that we must be able to explain what we see in the field or in the microscope based on what these relations predict. The field and the microscope are our ultimate laboratory in which these constitutive relations can be tested.

There are three aspects to the development of a constitutive relation and these must be mutually compatible and should be carried out hand in hand. First, laboratory experiments give direct information on the form of the constitutive relation for a given material and on what environmental factors are important and the range of variation in these factors where the relation is relevant. For instance, is the relation sensitive to variations in confining pressure, temperature or deformation rate? If so are the relations linear or nonlinear? Such experiments give direct data on constitutive parameters such as activation energy, activation volume, frictional and dilatancy measures and details of any rate dependence. There is now a considerable literature devoted to this approach and reviews and compilations of data can be found in many publications including Ranalli (1995), Evans and Kohlstedt (1995), papers in Karato and Wenk (2002), Paterson and Wong (2005), Karato (2008) and Kohlstedt and Mackwell (2009).

Second, the proposed relations that are derived from experiments must be compatible with the laws of thermodynamics and obey certain simple rules that we will discuss in detail

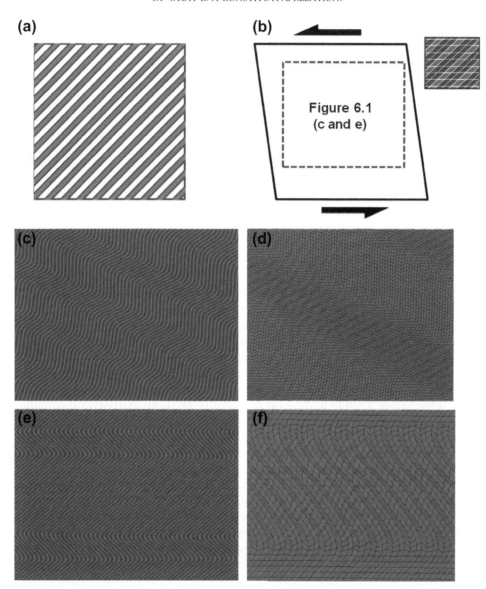

FIGURE 6.1 Control of constitutive behaviour on the development of structures. A laminated material shown in (a) undergoes sinistral simple shearing, (b), through 8.5°. A zoom into the undeformed computational grid (white lines) is shown in the inset. Figures (c) and (e) are zooms into the area outlined in (b). (c) and (d) Elastic-viscous material with identical elastic properties throughout but viscosity varies by a factor of 100 between adjacent layers (red and cream). (d) is a zoom into (c) and shows details of the deformed computational grid. (e) and (f) Elastic-plastic material with identical elastic properties throughout but yield stress varies by a factor of 10 between adjacent layers (red and cream). (f) is a zoom into (e) showing details of the deformed grid.

in Section 6.4. The outcome of the experimental development of a constitutive relation may be a simple mathematical relation that fits all the experimental data but which does not obey the laws of thermodynamics or that changes its form when the material reference coordinate frame is changed. Some rules that ensure experimentally derived constitutive relations are thermodynamically admissible are given in Section 6.4.8 and result in the *Coleman-Noll procedure*. Surprisingly even the simplest of constitutive relations, that for ideal (Cauchy) elasticity, does not obey the laws of thermodynamics for some situations (see Section 6.5) and therefore cannot be a general relation. A classic example of an experimentally produced constitutive relation that is thermodynamically inadmissible is the original formulation of the Cam Clay model for fluid bearing clays (Collins and Kelly, 2002). A third example is the commonly used form of the Mohr-Coulomb relation. We look at this relation in Section 6.6.4 and show that for certain combinations of constitutive parameters no plastic work is done or even negative plastic work is done during deformation thus violating the first law of thermodynamics.

The rules that constitutive relations must obey are discussed in Section 6.4; an important one is that the relation must be independent of the material coordinate frame that is used to express the relation. Simple outcomes of this are that the relation must involve objective measures of the deformation, deformation rate and stress rate. Thus variables such as the stretching tensor, d, must be used rather than the commonly used strain rate. These considerations become particularly important as research workers increasingly have access to computer codes that are capable of producing results at large deformations that may involve large rotations. Unless the constitutive relations that are used in these codes, and the codes themselves, are expressed in ways to obey the rules laid out in Section 6.4 the results of computer simulations at large deformations must be suspect (see Karrech et al., 2011c).

Third, it is becoming feasible to develop constitutive relations starting at the quantum or molecular level and upscaling such relations to coarser scales (Amodeo et al., 2011; Tadmor and Miller, 2011). This approach is now commonplace in biology and physical metallurgy and new developments for mantle mineralogies and conditions are given by Cordier et al. (2012). Again, constitutive relations developed by molecular or dislocation dynamics modelling must be thermodynamically admissible and obey the rules for the development of constitutive relations. Fundamentally they must also predict what we see in experiments under laboratory conditions. This approach is particularly important in the geosciences where geologically realistic deformation rates are not accessible in the laboratory. We address this particular issue in Section 6.7.3 and in Chapters 9 and 13 with some thought provoking repercussions.

Finally, in the geosciences we face an important dilemma, as do biologists, in formulating constitutive relations for open flow systems and for systems that involve chemical reactions. Most thermodynamically admissible constitutive laws for solids are based on the concept of a material reference state from which deformations are measured. The procedure is to map material points in the reference state to new spatial coordinates in the deformed state so that the approach is Lagrangian. Conceptually this is straight forward in a closed thermal-mechanical system with no internal physical or chemical rearrangements during the deformation. The situation becomes blurred in open systems with fluid flow-through and the possible addition or removal of material particles during deformation. In such systems and in chemically reacting deforming systems the definition of a *material particle* needs careful attention when the nature (or even the existence) of this material particle may change during deformation.

The solutions to these problems may lie in the development of an Eulerian approach to constitutive relation development. Attempts in this direction are considered by Rajagopal and Srinivasa (2007). The development of constitutive relations for such systems is an important area of future research.

6.2 WHY ARE CONSTITUTIVE RELATIONS IMPORTANT?

This chapter completes the toolbox that we need to explore the ways in which various bodies deform when subjected to boundary forces and/or displacements. In Chapter 2 we looked at various forms of *deformation* followed by Chapter 3 with a description of the *motions* or *flows* that result in deformation. Chapter 4 provided us with the driving force for flow, namely, the *internal stress field* that is generated in a body by external loadings or displacements. However, we noted in Chapter 4 that Cauchy's first law (4.21) says that all we can discover about the stress field given a set of external loadings is the divergence of the stress; we need more information about the material itself to go the last step and define the stress field. That is, *bodies made of different materials with identical loads or displacements on their boundaries will have different internal stress fields*. All of the principles developed in Chapters 2 through 5 are applicable to any material. We need extra information and that is supplied in the form of *constitutive relations* (or response functions) which are equations that link the response or flow of a body comprising *a specific material* to mechanical and thermal loads placed on the body. We do, however, have two constraints on what the stress field may be: first, Cauchy's second law says that in the absence of internal torques, $\boldsymbol{\sigma} = \boldsymbol{\sigma}^{\mathrm{T}}$, that is, the stress tensor is symmetrical. Also Chapter 5 supplies us with the Clausius-Duhem equation (a specific form of the second law of thermodynamics) that places constraints on any material behaviour that results in the internal stress, namely, the behaviour must obey the laws of thermodynamics.

The need for constitutive relations is made clear by considering the information we have available to us from previous chapters and the way in which this information is put together to solve a problem in mechanics. In this chapter we consider the simplest of systems, namely, thermo-mechanical systems with no internal fluid flow or chemical reactions. Such features are left to Section B. As such, the problems we are interested in mainly comprise an initial undeformed body of a given material in which the undeformed or reference coordinates, X, of each material point are known along with the initial temperature, T, at each point. The body is loaded or displaced at its boundary where the temperature is known. The question is: What are the coordinates, x, and the temperature of each material point after time, t? We assume that the rate, r, of any internal heat production, arising say from radiogenic sources, is known along with the body force, B, per unit volume. In Table 6.1 we list the various equations available to us from previous chapters together with the number of unknown variables that appear in these equations for problems where the initial and/or boundary conditions are prescribed. As we have seen these may be in the form of prescribed forces or displacement rates or combinations of these. There are eight equations (all balance laws) that govern the behaviour of a thermo-mechanical system, namely, one for the conservation of mass, (4.4 or 4.5), three equations for the balance of linear momentum, (4.21), three for the balance of angular momentum, (4.22), and one, (5.7), for the conservation of energy (which is

TABLE 6.1   Governing Equations for Boundary Value Problems; Knowns and Unknowns

Name of Governing Equation	Equation	Number of Independent Equations	Unknown Dependent Variables	Number of Unknown Dependent Variables
Conservation of mass	$\dot{\rho} + \rho(div\mathbf{v}) = 0$ (4.4)	1	Scalar density, ρ	1
Balance of linear momentum	$div\boldsymbol{\sigma} + \rho\mathbf{B} = \rho\mathbf{a}$ (4.21)	3	Components of stress tensor, σ_{ij}	9
Balance of angular momentum	$\boldsymbol{\sigma} = \boldsymbol{\sigma}^T$ (4.22)	3	0	0
Conservation of energy (first law)	$\boldsymbol{\sigma}:\mathbf{d} + \rho r - div\mathbf{q} = \rho\dot{e}$ (5.7)	1	Heat flux vector, $\mathbf{q}$; Specific internal energy, e	3 1
GEOMETRIC AND KINEMATIC RELATIONS				
Deformation gradient	$\mathbf{F} = \nabla_X x$	0	Components of position vector, x_i	3
Spatial velocity	$\mathbf{v} = \dot{x}$	0		
Spatial acceleration	$\mathbf{a} = \dot{v}$	0		
Stretching tensor	$\mathbf{d} = \frac{1}{2}(\nabla\dot{x} + [\nabla\dot{x}]^T)$ (3.12)$_3$	0	Components of stretching tensor, d_{ij}	0
OTHER THERMODYNAMIC RELATIONS				
Clausius–Duhem inequality (second law)	$\dot{s} \geq \frac{r}{T} - \frac{1}{\rho}div\frac{q}{T} \geq 0$ (5.17)	0	Temperature, T; Specific entropy, s	1 1
Total		8		19
Known independent variables	Material coordinates, X			0
	Time, t			0
	Internal heat production rate, r			0
	Body force, B			0

an expression of the first law of thermodynamics). However, there are 19 independent unknowns, namely, the density (1), the spatial position (3), the stress (9), the internal energy (1), the temperature (1), the heat flux (3) and the specific entropy (1). All other quantities are related through equations in Table 6.1. The time is prescribed. The extra 11 equations required to solve a problem for a given material are supplied by the mechanical and thermal *constitutive equations* for that material and by the second law of thermodynamics expressed as the Clausius-Duhem equation, (5.17). These constitutive relations are for

- Internal energy, e (1 equation)
- Stress, $\boldsymbol{\sigma}$ (6 equations)
- Temperature, T (1 equation)
- Heat flow, $\mathbf{q}$ (3 equations)

In addition, some quantities are readily derived from the spatial coordinates, x, as discussed in Chapters 2 and 3. Thus the deformation gradient, $\mathbf{F}$ (and hence the strain), the velocity, $\mathbf{v}$, the acceleration, $\mathbf{a}$, and the stretching tensor, $\mathbf{d}$, are given by the geometric and kinematic relations:

$$\mathbf{F} = \nabla_X x, \mathbf{v} = \dot{x}, \mathbf{a} = \dot{\mathbf{v}} \quad \text{and} \quad \mathbf{d} = \frac{1}{2}\left(\nabla \dot{x} + [\nabla \dot{x}]^{\mathrm{T}}\right)$$

There is a tradition in much of the literature, particularly the engineering geomechanics literature, to invent constitutive relations based on empirical curve fitting to experimental data. However, constitutive relations must conform to certain rules not the least of which is that they should be compatible with the laws of thermodynamics as expressed, for instance, by the Clausius-Duhem relation; Chapter 5 gives us the tools to proceed in that direction.

6.3 THE FOUR BASIC FORMS OF CONSTITUTIVE BEHAVIOUR

There are four fundamental forms of constitutive behaviour for thermo-mechanical deformation. One relates the heat flow in a material to the thermal properties of the material and the other three relate the mechanical response of a given material to the stress imposed on that material. We discuss heat flow in Chapter 11, although some constraints on this constitutive relation are discussed in Section 6.4.8 (6.13). This chapter is concerned essentially with the three forms of mechanical response, namely:

1. Materials where the stress is solely a function of the deformation gradient and the temperature and has no dependence on deformation rate; these materials do not dissipate energy (however, see Figure 6.4(b)). These are *elastic* materials, although the only class of the family of elastic materials that is always thermodynamically admissible is the *hyperelastic* or *Green elastic* material.
2. Materials where the stress is insensitive to the deformation rate and that dissipate energy. These are *plastic* materials. We are interested in a *hyperplastic* formulation for such materials, that is, where the stress is defined in terms of a potential, namely, the Helmholtz energy. Plastic materials may or may not show a strong dependence of the deviatoric stress on the confining pressure.
3. Materials where the stress is sensitive to the deformation rate and dissipate energy. These are *viscous* materials. Viscous materials commonly show only a weak dependence of stress on the confining pressure and a strong dependence on temperature.

Combinations of these three types of behaviour are the common situation.

6.4 THE RULES FOR CONSTRUCTING CONSTITUTIVE RELATIONS

For the materials we will consider in this book there are seven rules that ensure any proposed constitutive relation is robust. Details of these rules may be found in Truesdell (1966, 1969), Gurtin et al. (2010) and Tadmor et al. (2012).

6.4.1 Determinism

This principle says that: *The stress in a body is determined by the history of the motion of the body* (Truesdell, 1966). For many materials the stress in the system is defined by constitutive equations that depend on the initial configuration of the material, that is, on X and for some materials, on the history of states since deformation began. This means that heterogeneous systems can be fully described by an array of constitutive equations that may be different for every point, X. In addition, a constitutive equation at each material point may depend on time so that the material has a memory of past configurations or stress states. The obvious application of this part of the principle is the dependence of stress on some reference state that the material 'remembers'. As indicated below we do not consider materials where the current stress state depends on the history of stress states.

6.4.2 Local Action

This principle states that the behaviour of a material depends solely on events that occur spatially close to X and that events far from X have no influence on the stress. This principle is followed in this book but relatively recently many constitutive relations that involve gradients in X have been developed. These are part of a wider group of *nonlocal constitutive relations* and are known as *gradient constitutive relations* (Aifantis, 1984, 1987, 1999). The advantage of nonlocal relations is that they automatically involve a length scale such as the grain size or a layer thickness, whereas classical constitutive relations that strictly follow the principle of local action involve no intrinsic length scale. The lack of an internal length scale in some of the constitutive relations we discuss in this book has its disadvantage in that when one models deformation using finite element or finite difference computer codes, the only length scale introduced is that of the computational grid. Thus the results of such modelling (such as the width and spacing of shear zones) are commonly sensitive to the grid size. This effect is known as *mesh sensitivity.*

6.4.3 Thermodynamic Admissibility

This means that any proposed constitutive relation must obey the laws of thermodynamics or to be a little more precise, the Clausius-Duhem relation, (5.17). This principle was extended by Coleman and Noll (1963) and is now expressed as the *Coleman-Noll procedure* which we consider in some detail in Section 6.4.8.

6.4.4 Material Frame Indifference or Objectivity

This principle simply says that a constitutive relation must be independent of the material coordinate frame from which it is viewed. This can be expressed as: *Any two observers of the motion of a body find the same stress* (Truesdell, 1966). This in turn means that the relation must be expressed in terms of objective quantities, that is, quantities that remain of the same form with respect to any change in a material coordinate frame. In Table 6.2 we list many of the quantities that are used to describe deforming systems

TABLE 6.2   Objective and Non Objective Quantities

Quantity	Objective or Not?
Scalar quantities such as mass density, temperature and entropy density	Invariant
Displacement, **u**	Objective
Finite strain tensors, **B**, **V**, **E**	Objective
Finite strain tensors, **C**, **U**, **E**	Invariant
Deformation gradient, **F**	Non objective
Velocity, **v**	Non objective
Acceleration, **a**	Non objective
Velocity gradient tensor, ℓ	Non objective
Spatial stretching tensor, **d**	Objective
Cauchy stress, σ	Objective
First Piola–Kirchhoff stress, **P**	Objective
Second Piola–Kirchhoff stress, **S**	Invariant
Vorticity, ω	Non objective
Kinematic vorticity number, W_K	Non objective
Sectional vorticity number, W_n	Non objective
Vorticity numbers of Astarita (1979) and Jiang (2010)	Objective

and classify them as to whether they are objective or not. Some of these quantities are identical for all material coordinate frames and are called *invariant* (Gurtin et al., 2010, pages 150 and 275).

Although one can be rigorous in the use of objective measures in expressing constitutive relations, one still needs extreme care in the use of such relations in studying deformations that involve high shear strains where large rotations are important in the deformation. Many objective measures of stress and of deformation result in severe oscillations of stress at shear strains larger than ≈ 1 unless one uses logarithmic measures of strain (the Hencky strain tensor) and special measures of the stress (Karrech et al., 2011c).

6.4.5 Memory

In this book we consider only materials that remember a reference state. We do not include materials that are influenced by the history of past stress states. We do, however, consider materials where the stress depends on the *evolution* of internal variables. Such behaviour is characteristic of solids. Notice, however, that such dependence does not constitute a memory of past events (see Kestin, 1969, p. 200).

6.4.6 Restrictions on the Internal Energy

Only materials where the internal energy depends on the entropy, the deformation gradient and admissible internal variables are considered. This restriction is sufficient to allow inclusion of a large array of material behaviours of interest in metamorphic geology. This means we can write

$$e = e(s, \mathbf{F}) \tag{6.1}$$

If we can write (6.1) for a particular material then that material is known as a *simple material*. If we include the internal variables, α_i, then we can write $e = e(s, \mathbf{F}, \alpha_i)$.

Notice that a further step needs to be considered in arguments that follow from the Coleman-Noll procedure in order for the constitutive relations such as (6.1) to be material frame indifferent; that is, the form of the dependence on non objective quantities such as $\mathbf{F}$ needs to be spelt out. This means that (6.1) might be written in greater detail as $e = e(s, \mathbf{F}^T\mathbf{F}) = e(s, \mathbf{C})$.

6.4.7 Material Symmetry

Any constitutive relation must take into account the symmetry of the material involved.

6.4.8 Compatibility with Thermodynamics

6.4.8.1 The Coleman-Noll procedure

In order to ensure that a given constitutive law is compatible with the laws of thermodynamics a procedure has been developed based on Coleman and Noll (1963). This procedure consists essentially of ensuring that the constitutive law can be derived from the Clausius-Duhem relation, (5.17). Below we follow Tadmor et al. (2012). For an open system the Clausius-Duhem relation can be written:

$$\dot{s}^{internal} = \dot{s} - \dot{s}^{external} \geq 0 \tag{6.2}$$

where $\dot{s}$ is the total specific entropy production, $\dot{s}^{internal}$ and $\dot{s}^{external}$ are the specific internal and external entropy production. We have

$$\dot{s}^{external} = \frac{r}{T} - \frac{1}{\rho} div \frac{\mathbf{q}}{T} \tag{6.3}$$

and hence,

$$\dot{s}^{internal} = \dot{s} - \frac{r}{T} + \frac{1}{\rho} div \frac{\mathbf{q}}{T}$$

$$= \dot{s} - \frac{r}{T} + \frac{(div\mathbf{q})T - \mathbf{q} \cdot \nabla T}{\rho T^2}$$

$$= \dot{s} - \frac{1}{\rho T}[\rho r - div\mathbf{q}] - \frac{1}{\rho T^2}\mathbf{q} \cdot \nabla T \geq 0 \tag{6.4}$$

Rearranging and using the first law of thermodynamics, (5.7), we obtain,

$$\rho T \dot{s}^{internal} = \rho T \dot{s} - \rho \dot{e} + \boldsymbol{\sigma} : \mathbf{d} - \frac{1}{T} \mathbf{q} \cdot \nabla T \geq 0 \tag{6.5}$$

If we take the material time derivative of the internal energy given in (6.1) we have

$$\dot{e} = \frac{\partial e}{\partial s} \dot{s} + \frac{\partial e}{\partial \mathbf{F}} \dot{\mathbf{F}} \tag{6.6}$$

Substitution of (6.6) into (6.5) gives

$$\rho \left[T - \frac{\partial e}{\partial s} \right] \dot{s} + \left[\boldsymbol{\sigma} : \mathbf{d} - \rho \frac{\partial e}{\partial \mathbf{F}} : \dot{\mathbf{F}} \right] - \frac{1}{T} \mathbf{q} \cdot \nabla T \geq 0 \tag{6.7}$$

We also have $\boldsymbol{\sigma} : \mathbf{d} = \boldsymbol{\sigma} : \boldsymbol{\ell}$, since $\boldsymbol{\sigma}$ is symmetric, and $\dot{\mathbf{F}} = \boldsymbol{\ell}\mathbf{F}$(from (3.22)), so that $\boldsymbol{\ell} = \dot{\mathbf{F}}\mathbf{F}^{-1}$. Hence

$$\rho \left[T - \frac{\partial e}{\partial s} \right] \dot{s} + \left[\boldsymbol{\sigma}\mathbf{F}^{-T} - \rho \frac{\partial e}{\partial \mathbf{F}} \right] : \dot{\mathbf{F}} - \frac{1}{T} \mathbf{q} \cdot \nabla T \geq 0 \tag{6.8}$$

The proposal of Coleman and Noll (1963) is that every process that is compatible with the laws of thermodynamics must satisfy (6.8). Such processes are said to be *thermodynamically admissible* and the application of (6.8) is called the *Coleman-Noll procedure*. As examples we apply such a procedure below to find the general forms of constitutive relations for the temperature, heat flux and stress for certain steady conditions.

6.4.8.2 Constitutive Relation for the Temperature

Consider a steady deformation, so that $\dot{\mathbf{F}} = 0$, and where the temperature is uniform so that $\nabla T = 0$. Then (6.8) becomes

$$\rho \left[T - \frac{\partial e}{\partial s} \right] \dot{s} \geq 0 \tag{6.9}$$

Since $\dot{s}$ takes on arbitrary values by changing r and since $\dot{s}$ can be positive or negative, (6.9) can only be satisfied for every process if the constitutive law for temperature takes the form

$$T = T(s, \mathbf{F}) = \frac{\partial e}{\partial s} \tag{6.10}$$

6.4.8.3 Constitutive Relation for the Heat Flux

We substitute (6.10) into (6.8) then (6.8) becomes

$$\left[\boldsymbol{\sigma}\mathbf{F}^{-T} - \rho \frac{\partial e}{\partial \mathbf{F}} \right] : \dot{\mathbf{F}} - \frac{1}{T} \mathbf{q} \cdot \nabla T \geq 0 \tag{6.11}$$

Again if we consider a steady deformation so that $\dot{\mathbf{F}} = 0$ we obtain:

$$-\frac{1}{T} \mathbf{q} \cdot \nabla T \geq 0 \tag{6.12}$$

Our experience says that $\mathbf{q}$ changes sign when ∇T changes sign and so $\mathbf{q}$ must depend on ∇T. Thus the constitutive law for heat flux must be of the form

$$\mathbf{q} = \mathbf{q}(s, \mathbf{F}, \nabla T) \tag{6.13}$$

6.4.8.4 Constitutive Relation for the Stress

Consider the case where the temperature is uniform so that $\nabla T = 0$. Then (6.8) becomes

$$\left[\boldsymbol{\sigma} \mathbf{F}^{-T} - \rho \frac{\partial e}{\partial \mathbf{F}} \right] : \dot{\mathbf{F}} \geq 0 \tag{6.14}$$

This relation can only be satisfied for all $\dot{\mathbf{F}}$ if

$$\left[\boldsymbol{\sigma} \mathbf{F}^{-T} - \rho \frac{\partial e}{\partial \mathbf{F}} \right] = 0 \tag{6.15}$$

This conclusion introduces an issue with respect to (6.11) since if (6.15) is true then the inequality in (6.11) arises solely from the heat flux term and not from dissipation arising from deformation. This is contrary to experience. To proceed we partition $\boldsymbol{\sigma}$ into non-dissipative and dissipative parts:

$$\boldsymbol{\sigma} = \boldsymbol{\sigma}^{non\text{-}dissipative} + \boldsymbol{\sigma}^{dissipative} \tag{6.16}$$

One can think of $\boldsymbol{\sigma}^{non\text{-}dissipative}$ as being equivalent to the elastic stress and $\boldsymbol{\sigma}^{dissipative}$ as being equivalent to a viscous or plastic stress. Then if we substitute (6.16) into (6.11) we obtain:

$$\left[\boldsymbol{\sigma}^{non\text{-}dissipative} \mathbf{F}^{-T} - \rho \frac{\partial e}{\partial \mathbf{F}} \right] : \dot{\mathbf{F}} + \boldsymbol{\sigma}^{dissipative} : \boldsymbol{\ell} - \frac{1}{T} \mathbf{q} \cdot \nabla T \geq 0 \tag{6.17}$$

If we assume that $\boldsymbol{\sigma}^{dissipative}$ is symmetric, so that we can replace $\boldsymbol{\ell}$ by $\mathbf{d}$, then because the dissipation arising from the action of $\boldsymbol{\sigma}^{dissipative}$ is always positive,

$$\boldsymbol{\sigma}^{dissipative} : \mathbf{d} \geq 0 \tag{6.18}$$

and so, since $-\frac{1}{T} \mathbf{q} \cdot \nabla T \geq 0$, (6.17) can only be satisfied if

$$\boldsymbol{\sigma}^{non\text{-}dissipative} = \boldsymbol{\sigma}^{non\text{-}dissipative}(s, \mathbf{F}) \equiv \rho \frac{\partial e}{\partial \mathbf{F}} \mathbf{F}^{T} \tag{6.19}$$

which is the constitutive relation for a non-dissipative stress. Since (6.18) must be true for all processes the constitutive relation for a dissipative stress must be of the form

$$\boldsymbol{\sigma}^{dissipative} = \boldsymbol{\sigma}^{dissipative}(s, \mathbf{F}, \mathbf{d}) \tag{6.20}$$

With the separation of the total stress into non-dissipative and dissipative parts, the Clausius-Duhem relation becomes

$$\rho T s^{internal} = \boldsymbol{\sigma}^{dissipative} : \mathbf{d} - \frac{1}{T} \mathbf{q} \cdot \nabla T \geq 0 \tag{6.21}$$

The above examples illustrate the use of the Coleman-Noll procedure for examples constrained by some form of steady state. The arguments can be extended to any proposed constitutive relation the issue being that for any such relation, (6.8) must be satisfied for thermodynamic admissibility.

6.5 ELASTIC BEHAVIOUR

The concept of an elastic solid derives from everyday experience with coiled springs where the response to application of a load is more or less instantaneous, the displacement is proportional to the applied load and the spring returns to its initial configuration when the load is removed so that the work done in deforming the spring is recovered on unloading. This latter observation is idealised as a lack of dissipation of energy during the loading–unloading cycle. All of these effects are independent of how fast the spring is loaded. Of course all of these observations are oversimplifications. For instance, the spring may heat up with repeated loading–unloading cycles. Such behaviour leads to problems in a meaningful definition of elasticity when one attempts to reconcile experimental observations with mathematical descriptions and thermodynamic considerations. The idealisation of elastic behaviour thus involves the following concepts. First an undeformed reference configuration exists from which deformations can be measured. Second, the stress in the spring is a function solely of the deformation measured from that configuration and is rate independent. Third, the deformation is reversible and involves no dissipation of energy.

Classical elasticity is commonly expressed as Hooke's law: *The stress is proportional to the strain* and is written in its most general form for a (symmetrical) Cauchy stress tensor, σ_{ij}, as (Nye, 1957):

$$\sigma_{ij} = E_{ijkl}\varepsilon_{kl} \tag{6.22}$$

where ε_{kl} is the small strain tensor and E_{ijkl} is a fourth order tensor called the *elastic stiffness matrix*. (6.22) represents six equations with 36 components for E_{ijkl}. Nye (1957, his Chapter 8) shows that the number of independent components of E depends on the symmetry of the material and that the introduction of an elastic energy function means that E_{ijkl} is symmetric so that the number of independent components for a material with triclinic symmetry reduces to 21. The number of independent components also decreases with a further increase in material symmetry until for an isotropic material there are just two independent components. Experimentally it is possible to measure four coefficients that describe the elastic response of an isotropic elastic material and these are the Young's modulus, E, the shear modulus, G, the bulk modulus, K, and the Poissons ratio, v. Only two of these are independent. These moduli are related by (Jaeger, 1969, p. 57):

$$E = \frac{9KG}{3K+G}, \quad G = \frac{E}{2(1+v)}, \quad K = \frac{2(1+v)G}{3(1-2v)} = \frac{E}{3(1-2v)}, \quad v = \frac{3K-2G}{2(3K+G)} \tag{6.23}$$

where $-1 < v < 0.5$. In addition it is often convenient to define two parameters, known as Lamé's parameters, λ and γ. These are given by

$$\lambda = G\frac{2v}{1 - 2v} = \frac{Ev}{(1 + v)(1 - 2v)} \quad \text{and} \quad \gamma = G \qquad (6.24)$$

For an isotropic elastic material at small strains the constitutive relation becomes

$$\sigma_{ij} = K\varepsilon_{kk}\delta_{ij} + 2G\varepsilon'_{ij} \qquad (6.25)$$

where ε'_{ij} is the deviatoric small strain tensor. However, (6.22) is not necessarily the most convenient or accurate way to describe elasticity. On the one hand there are generally experimental difficulties in establishing where elastic behaviour stops and some other form of behaviour such as plasticity begins. There may be a region in between where linear elastic behaviour stops (*the proportional limit*) and permanent plastic deformation begins. The difficulty is shown in Figure 6.2(a) where there exists a nonlinear region after the proportional limit and before the gradient in stress decreases substantially. This may be an artefact and arise from lack of sensitivity in the experimental apparatus in this region of the stress strain curve, especially if the loading machine is elastically 'soft'. On the other hand in some instances the effect is real and the elastic behaviour becomes *nonlinear* (see Hamiel et al., 2004; Jaeger and Cook, 1969, Chapter 4). Precise measurements at low strains under dynamic loading also show nonlinear behaviour (Guyer and Johnson, 2009; Winkler et al., 1979). Typical stress strain curves are shown in Figure 6.2. If we neglect the nonlinear effects described by Guyer and Johnson (2009), the early parts of the stress strain curve are commonly linear, although even this region may be quite limited in high porosity rocks. Acoustic emissions begin as the material response becomes nonlinear (A in Figure 6.2(b)). If the material is unloaded before the stress reaches A the strain returns to zero. At stresses greater than that corresponding to A in Figure 6.2(b) the unloading curve may not

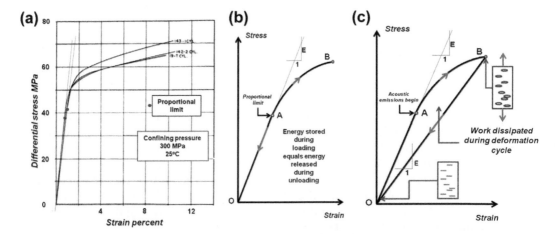

FIGURE 6.2   Linear and nonlinear elastic response. (a) Experimental stress strain curve for Solenhofen limestone at 300 MPa confining pressure and 25 °C; strain rate $2 \times 10^{-4}\,\text{s}^{-1}$ *(From Heard (1960).).* The proportional limit where the experimental stress strain curve ceases to be linear is shown as a red dot. (b) Schematic diagram showing ideal nonlinear elastic behaviour. The loading curve corresponds to the unloading curve and the work stored during loading equals the work used during unloading. No work is dissipated during the deformation cycle. (c) Schematic diagram showing transition from linear to nonlinear response on loading and a decrease in elastic modulus from E (linear loading) to $\bar{E}$ on unloading. The energy dissipated during the loading cycle corresponds to the area OABO. The only reason that such a response might be called elastic is that there is no permanent deformation.

correspond to the loading curve (Figure 6.2(c)) and the elastic modulus has now degraded from E in the linear elastic range to a lower value, $\overline{E}$, in the nonlinear range. These effects are attributed to the opening of penny-shaped cracks in extension and reversible closing of these cracks during unloading. We examine these effects in greater detail in Chapter 10. The degradation in elastic modulus is sometimes written (Whiting and Hunt, 1997):

$$\overline{E} = \frac{E}{n} \sinh^{-1}(n\varepsilon) \approx E\left(\varepsilon - \frac{1}{6}n^2\varepsilon^3 + higher\ order\ terms\right) \tag{6.26}$$

This short discussion of elastic behaviour in the laboratory indicates that real materials behave in quite complicated manners and we examine this in a little more detail in Chapter 10. One prescription of elastic behaviour involves defining a constitutive behaviour where the stress is solely a function of the strain (or for large deformations, of the deformation gradient). This is a *Cauchy elastic* material where the constitutive relation is given by

$$\sigma_{ij} = f_{ij}(\varepsilon_{ij}) \tag{6.27}$$

$f_{ij}(\varepsilon_{ij})$ is a second-order tensor valued function of the strains, ε_{ij}, and may be linear or nonlinear. Here the strain (or deformation gradient) is measured from an undeformed reference state. If $f_{ij}(\varepsilon_{ij})$ is linear then (6.27) can be written as

$$\sigma_{ij} = f_{ij}(\varepsilon_{ij}) = E_{ijkl}\varepsilon_{kl} \tag{6.28}$$

which is the same as (6.22). The incremental form of (6.28) is

$$\dot{\sigma}_{ij} = \frac{\partial f_{ij}(\varepsilon_{ij})}{\partial \varepsilon_{kl}}\dot{\varepsilon}_{kl} \tag{6.29}$$

Although such a definition may sound reasonable and consistent with experience a problem arises that for some closed cycles of stress (or strain) materials described by (6.28) can create or destroy energy which is counter to the first law of thermodynamics. The issue here is that in the definition of Cauchy elasticity no constraints are placed on the behaviour arising from the stored elastic energy so that there are no constraints on the symmetry of E. The number of independent elastic coefficients is 36 for a triclinic material and not 21 as derived by Nye (1957). Consider the loading and unloading histories of an anisotropic linear Cauchy elastic material in two dimensional principal stress space (σ_1, σ_2) as shown in Figure 6.3. The discussion is taken from Chen and Saleeb (1982, pp. 161–163). The principal strains are ε_1 and ε_2. We describe the stress strain relation as

$$\varepsilon_1 = a_{11}\sigma_1 + a_{12}\sigma_2 \quad and \quad \varepsilon_2 = a_{21}\sigma_1 + a_{22}\sigma_2$$

where a_{11}, a_{12}, a_{21} and a_{22} are elastic constants. We distinguish a loading path from $(0, 0)$ to $(\sigma_1^*, 0)$ and from $(\sigma_1^*, 0)$ to (σ_1^*, σ_2^*) and an unloading path from (σ_1^*, σ_2^*) to $(0, \sigma_2^*)$ and then from $(0, \sigma_2^*)$ to $(0, 0)$ so that a closed cycle of deformation is achieved. The Gibbs energy, $G^{loading}$, associated with the loading path is

$$G^{loading} = \int_{(0,0)}^{(\sigma_1^*,0)}(\varepsilon_1 d\sigma_1 + \varepsilon_2 d\sigma_2) + \int_{(\sigma_1^*,0)}^{(\sigma_1^*,\sigma_2^*)}(\varepsilon_1 d\sigma_1 + \varepsilon_2 d\sigma_2)$$

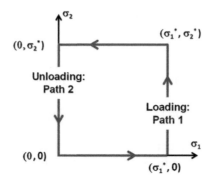

FIGURE 6.3   Loading and unloading paths for a Cauchy elastic solid.

Since $d\sigma_2 = 0$ for this path this integral reduces to

$$G^{loading} = \frac{1}{2}a_{11}\sigma_1^{*2} + a_{21}\sigma_1^*\sigma_2^* + \frac{1}{2}a_{22}\sigma_2^{*2}$$

Similarly the Gibbs energy for the unloading path (where $d\sigma_1 = 0$) is

$$G^{unloading} = \int_{\left(\sigma_1^*,\sigma_2^*\right)}^{\left(0,\sigma_2^*\right)} (\varepsilon_1 d\sigma_1 + \varepsilon_2 d\sigma_2) + \int_{\left(0,\sigma_2^*\right)}^{(0,0)} (\varepsilon_1 d\sigma_1 + \varepsilon_2 d\sigma_2)$$

This gives

$$G^{unloading} = -\left[\frac{1}{2}a_{11}\sigma_1^{*2} + a_{12}\sigma_1^*\sigma_2^* + \frac{1}{2}a_{22}\sigma_2^{*2}\right]$$

Thus the total energy for the cycle is

$$G^{cycle} = G^{loading} + G^{unloading} = (a_{21} - a_{12})\sigma_1^*\sigma_2^*$$

Hence, during a closed cycle, the system can dissipate energy for $a_{21} > a_{12}$ or create negative energy for $a_{21} < a_{12}$. The former case is counter to the concept of an elastic solid and the latter case also counters the first law of thermodynamics. Only for $a_{21} = a_{12}$ does the system conform to the concept of an elastic solid; this is the situation for hyperelasticity (or Green elasticity) when the matrix for the elastic moduli is always symmetrical.

An alternative form of constitutive relation that does not suffer from the disadvantages of (6.27) is (6.30) where the stresses are expressed in terms of a *strain energy potential*, Ψ, which is equal to the Helmholtz energy:

$$\sigma_{ij} = \frac{\partial \Psi\left(\varepsilon_{ij}\right)}{\partial \varepsilon_{ij}} \tag{6.30}$$

If $\Psi(\varepsilon_{ij})$ is a quadratic function of the ε_{ij} then the material is linearly elastic and $\frac{\partial^2 \Psi\left(\varepsilon_{ij}\right)}{\partial \varepsilon_{ij}\partial \varepsilon_{kl}} = E_{ijkl} = E_{klij}$ so that $\mathbf{E}$ is symmetrical. A material with a constitutive relation given by (6.30) is called a *hyperelastic* material and the response is referred to as *Green elasticity*; this is the preferred thermodynamically admissible constitutive relation for elasticity. The incremental form is

$$\dot{\sigma}_{ij} = \frac{\partial^2 \Psi\left(\varepsilon_{ij}\right)}{\partial \varepsilon_{ij} \partial \varepsilon_{kl}} \dot{\varepsilon}_{kl} \qquad (6.31)$$

(6.30) may be extended to finite strains when the stress becomes the second Piola-Kirchhoff stress and Ψ is a function of $\mathbf{F}$ (see Gurtin et al., 2010).

In addition, another form of elasticity is *hypoelasticity* (or *Truesdell elasticity*) defined by an incremental constitutive relation:

$$\dot{\sigma}_{ij} = f_{ijkl}\left(\sigma_{ij}, \varepsilon_{ij}\right) \dot{\varepsilon}_{kl}$$

where f_{ijkl} is a fourth order tensor function of the stresses or strains. If $f_{ijkl}(\sigma_{ij}, \varepsilon_{ij})$ is a constant then the material is linearly elastic and $f_{ijkl}(\sigma_{ij}, \varepsilon_{ij}) = E_{ijkl}$. Hyperelasticity is a subset of Cauchy elastic behaviour which in turn is a subset of hypoelasticity (Figure 6.4). Again, under some conditions a hypoelastic material can create or destroy energy during a closed cycle and under some conditions may not even return to the initial unstrained state upon unloading so that hypoelasticity includes non-elastic behaviour as a subset (Olsen and Bernstein, 1984).

We complete this survey of elasticity with some thermodynamic considerations. If we take (6.28) as the constitutive relation for linear elasticity at small strains, then for situations with no thermal effects, the Helmholtz energy, Ψ, becomes equal to the internal energy, e, so that

$$\Psi = e = \sigma_{ij}\varepsilon_{ij} = K\frac{\varepsilon_{ii}\varepsilon_{jj}}{2} + G\varepsilon'_{ij}\varepsilon'_{ij} \qquad (6.32)$$

where ε'_{ij} is the deviatoric small strain tensor. If there are no thermal effects the Gibbs energy, G, is equal to the enthalpy, H, so that from Table 5.4,

$$G = H = e - \sigma_{ij}\varepsilon_{ij} = \frac{\sigma_{ii}\sigma_{jj}}{18K} - \frac{1}{4G}\sigma'_{ij}\sigma'_{ij} = -\Psi \qquad (6.33)$$

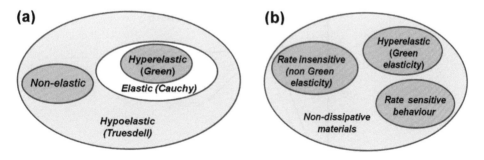

FIGURE 6.4 Classes of elasticity and of non-dissipative materials. (a) Elastic (Cauchy) and hyperelastic (Green) behaviour as a subset of hypoelasticity (Truesdell). The preferred, thermodynamically admissible, constitutive behaviour for elasticity is that of hyperelasticity. Some forms of non-elastic behaviour are also a subset of hypoelasticity. (b) Non-dissipative materials. Hyperelasticity is but one subset of non-dissipative materials. Other classes of rate insensitive and rate sensitive constitutive relations exist (Rajagopal and Srinivasa, 2009).

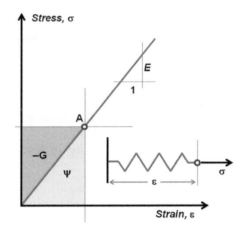

FIGURE 6.5   One dimensional elasticity. Ψ is the Helmholtz energy and G is the Gibbs energy.

where σ'_{ij} is the deviatoric Cauchy stress tensor. These relations are shown in Figure 6.5 for a one dimensional situation. In elasticity theory the Gibbs energy is commonly referred to as the *complementary energy* because of the relation shown in Figure 6.5.

6.6 PLASTIC BEHAVIOUR

6.6.1 Plastic Response

Plasticity is motivated by experiments such as those shown in Figure 6.6 where after a regime of elastic loading some form of permanent deformation occurs. This means that when the specimen is unloaded the strain no longer returns to zero (Figure 6.6) so that the material is permanently deformed and energy has been dissipated in the process. The point

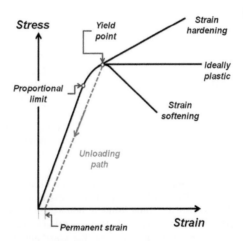

FIGURE 6.6   Various types of plastic behaviour.

on the stress strain curve where permanent deformation sets in is called the *plastic yield point*; this point may be difficult to determine experimentally. After the yield point the stress may continue to rise as in Figure 6.6; this is called *strain hardening*. In some instances the stress may remain constant; this is called *ideal plasticity*. Or the stress may decrease and this is called *strain softening* (Figure 6.6). Independently of the behaviour after yield, for linear elasticity, the following equation remains true:

$$\varepsilon^{total} = \varepsilon^{elastic} + \varepsilon^{plastic} \qquad (6.34)$$

This is one of the fundamental equations of classical plasticity.

The next step is to describe how the yield stress changes as we change the stress state responsible for yield. This means we want to describe how the yield stress varies as we vary the principal stresses, σ_1, σ_2, and σ_3. Experimentally we observe that rocks do not undergo plastic deformation by mechanical mechanisms if loaded solely by hydrostatic stresses. This is true for hydrostatic pressures met within the lithosphere but may not be true for greater depths. Any permanent deformation that arises from hydrostatic pressure arises from mineralogical phase changes and we do not consider such deformations now but leave them until Chapter 14 and Volume II. The above statements do not consider deformation of porous materials due to compaction and grain crushing under hydrostatic stress (Borja and Aydin, 2004). We will consider this in Chapter 8.

The absence of permanent deformation by hydrostatic pressure means that any dependence of the yield stress must be expressed in terms of the deviatoric stresses. Since we require that the conditions for yield be independent of any set of material coordinate axes the expression we seek must be in terms of the *invariants of the deviatoric stress*. The first three of these are (Jaeger, 1969)

$$J_1 = \sigma_1' + \sigma_2' + \sigma_3' = 0$$

$$J_2 = \frac{1}{2}(\sigma_1'^2 + \sigma_2'^2 + \sigma_3'^2) = \frac{1}{6}\left[(\sigma_2 - \sigma_3)^2 + (\sigma_1 - \sigma_2)^2 + (\sigma_3 - \sigma_1)^2\right] \qquad (6.35)$$

$$J_3 = \frac{1}{3}\sigma_{ij}'\sigma_{jk}'\sigma_{ki}'$$

where σ_i' are the deviatoric principal stresses.

Since $J_1 = 0$, the simplest expression to propose is that, at yield, J_2 is a constant and so we write

$$2J_2 = (\sigma_1'^2 + \sigma_2'^2 + \sigma_3'^2) = 2k^2/3$$

or $\qquad (6.36)$

$$\left[(\sigma_2 - \sigma_3)^2 + (\sigma_1 - \sigma_2)^2 + (\sigma_3 - \sigma_1)^2\right] = 2k^2$$

where k is a constant with the dimensions of stress; the division by three in $(6.36)_1$ is included so that, as we will see below, k is the yield stress in simple shearing. For pure shearing, $\sigma_1 = -\sigma_3$ and $\sigma_2 = 0$ so that $k = \sqrt{3}\sigma_1$ at yield. This derivation can be given a physical significance by returning to the expressions for the Helmholtz energy in a general loading situation, (6.32 and 6.33). These expressions consist of a part arising from a volume change and a part, $\frac{1}{4G}\sigma_{ij}'\sigma_{ij}' = \frac{J_2}{2G}$, arising from distortion. So the proposal that yield occurs when J_2 reaches a critical value is equivalent to proposing that *yield occurs when the elastic energy associated with distortion reaches a critical value*. This very simple criterion for yield has widespread experimental support in metals (see Taylor and Quinney (1934) for the classical study). Unfortunately there is very little experimental work on rocks at high pressures and temperatures but we assume this criterion is also relevant. The criterion is commonly known as the *von Mises criterion*. Nadai (1950) has also expressed this yield condition in terms of the octahedral stress, (4.17):

$$\tau^{octahedral} = \frac{1}{3}\left[(\sigma_2 - \sigma_3)^2 + (\sigma_1 - \sigma_2)^2 + (\sigma_3 - \sigma_1)^2\right]^{\frac{1}{2}} = \left(\frac{2J_2}{3}\right)^{\frac{1}{2}} \tag{6.37}$$

so that the von Mises condition may be restated as *yield occurs when the octahedral shear stress is equal to $\sqrt{2/3}$ times the yield stress in pure shearing*.

6.6.2 The Yield Surface

If one accepts the von Mises criterion for yield then $(6.36)_2$ defines a cylinder in stress space with an axis parallel to the line $\sigma_1 = \sigma_2 = \sigma_3$ as shown in Figure 6.7(a). If we write

$$f(\sigma_{ij}) = \left[(\sigma_2 - \sigma_3)^2 + (\sigma_1 - \sigma_2)^2 + (\sigma_3 - \sigma_1)^2\right] - 2k^2 \tag{6.38}$$

then the yield surface is defined by $f(\sigma_{ij}) = 0$. If $f(\sigma_{ij}) < 0$, (that is, for stress states within the yield surface) the material is in an elastic state. Stress states corresponding to $f(\sigma_{ij}) > 0$ (that is, stress states outside the yield surface) cannot exist. In plane stress (for $\sigma_2 = 0$) the intersection of the yield surface with the $\sigma_1 - \sigma_3$ plane is shown by the ellipse in Figure 6.7(b). In pure tension ($\sigma_2 = \sigma_3 = 0$) the yield stress is $\sqrt{3}k$. In simple shearing when $\sigma_2 = 0$ and $\sigma_1 = -\sigma_3$, the yield stress is k.

An alternative, but much earlier, proposal for a yield criterion is the *Tresca criterion* that states that yield occurs when the shear stress on a particular plane reaches a critical value. This is written in terms of the stress deviators as

$$\sigma_1' - \sigma_3' = \sigma_1 - \sigma_3 = k \tag{6.39}$$

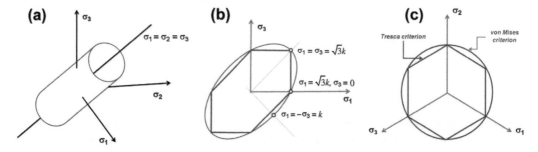

FIGURE 6.7   Von Mises and Tresca yield criteria. (a) von Mises criterion: a cylinder parallel to the hydrostatic axis, $\sigma_1 = \sigma_2 = \sigma_3$. (b) The intersection of the von Mises criterion (ellipse) and the Tresca criterion (elongate polygon) with the $\sigma_1 - \sigma_3$ plane. (c) The intersection of the von Mises and Tresca criteria with the π-plane.

Since the principal axes of stress are independent of the coordinate frame the difference $\sigma_1 - \sigma_3$ is also independent of the coordinate frame. The yield surface now is a prism parallel to the line $\sigma_1 = \sigma_2 = \sigma_3$ but with a regular hexagon (see Jaeger, 1969, p. 96) as cross section (Figure 6.7(c)).

For many materials, it is experimentally observed that at yield the ratios between the components of the incremental plastic strain are constant and independent of the stress increments at yield. A convenient way of expressing this is to define a *plastic potential*, $g(\sigma_{ij})$, so that the plastic increment is normal to the surface defined by this potential. This means that

$$\dot{\varepsilon}_{ij}^{plastic} = \lambda \frac{\partial g(\sigma_{ij})}{\partial \sigma_{ij}} \tag{6.40}$$

where λ is called a *plastic multiplier*. (6.40) is known as a *flow rule* and says that the incremental plastic strain is normal to the potential surface and has a magnitude λ times the potential gradient. In metal plasticity, and presumably for rocks at high temperatures and slow deformation rates, the surface corresponding to the plastic potential, the *potential surface*, coincides with the yield surface so that the plastic incremental strain is normal to the yield surface. This is known as *associative plasticity*. For rocks at low temperatures and confining pressures and/or fast strain rates the potential surface does not coincide with the yield surface (see Lade, 1993 and Paterson and Wong, 2005, for reviews); this is known as *non-associative plasticity* (Figure 6.8).

6.6.3 Strain Hardening

Two ideal end member types of strain hardening are recognised. One is *isotropic hardening* (Figure 6.9(a)) where the yield surface increases in size during loading but maintains its position in stress space. The other is *kinematic hardening* where the yield surface maintains its size during deformation but shifts its position in stress space (Figure 6.9(c)). Experimentally

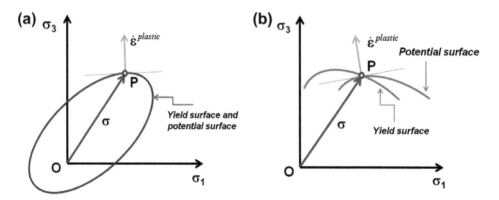

FIGURE 6.8   The flow rule. (a) *Associative plasticity* where the yield and potential surfaces coincide. The plastic increment is normal to the potential surface (and the yield surface) at P where the stress vector touches the surfaces. (b) A sketch illustrating *non-associative plasticity* where the yield and potential surfaces coincide only at P. The plastic increment is normal to the potential surface at P where the stress vector touches both surfaces.

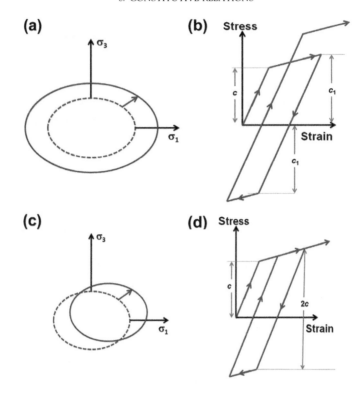

FIGURE 6.9 The two end member forms of strain hardening: (a) and (b) *Isotropic hardening*. The yield surface (a) expands to remain the same shape during deformation. (b) A typical loading–unloading–reloading path for ideal isotropic hardening. (c) and (d) *Kinematic hardening*. (c) The yield surface remains the same size during loading but translates in stress space. (d) A typical loading–unloading–reloading path for ideal kinematic hardening.

these two types of hardening behaviour can only be distinguished during loading–unloading–reloading cycles (Figure 6.8(b and d)). For detailed discussion of hardening behaviour see Fung (1965) and Houlsby and Puzrin (2006a).

6.6.4 Pressure Sensitive Plasticity

Most rocks are sensitive to changes in confining pressure when deformed at low temperatures, low confining pressures and fast strain rates (see Paterson and Wong, 2005). A common constitutive relation to describe such behaviour is the *Mohr-Coulomb material*. A Mohr-Coulomb frictional material with cohesion, c, subjected to a normal stress, σ, and shear stress, τ, on a particular plane in the material, with corresponding strains, ε and γ, fails when the shear stress on that plane equals $(c + \mu^*\sigma)$ where μ^* is the apparent coefficient of internal friction. Here σ is taken as negative in compression. Hence we define a yield surface (Figure 6.10(a)), f, by

$$f = |\tau| + \mu^*\sigma - c = 0 \qquad (6.41)$$

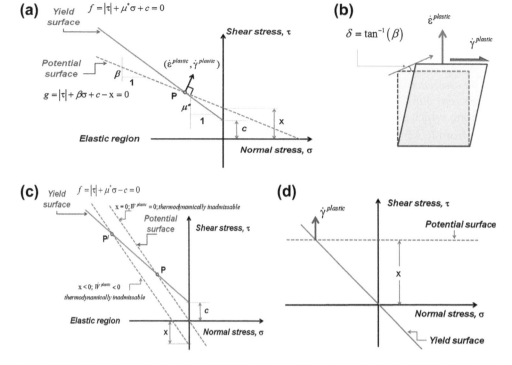

FIGURE 6.10 Mohr-Coulomb plasticity. (a) Yield and potential surfaces for frictional Mohr-Coulomb yield. (b) The significance of β as a measure of dilation; δ is known as the *dilatancy angle*. (c) Potential surfaces for which x in (a) is zero (corresponding to the stress state P) or negative (corresponding to the stress state P'). The corresponding value of β is a thermodynamically inadmissible constitutive parameter. (d) The Byerlee Law as a Mohr-Coulomb constitutive relation.

Within the yield surface the behaviour is elastic and given by

$$\varepsilon = \frac{\sigma}{K}, \quad \gamma = \frac{\tau}{G}$$

During plastic flow the stress strain behaviour is described by the flow rule:

$$\dot{\varepsilon}^{plastic} = \lambda \frac{\partial g}{\partial \sigma}, \quad \dot{\gamma}^{plastic} = \lambda \frac{\partial g}{\partial \tau} \tag{6.42}$$

where g is the plastic potential defined by

$$g = |\tau| + \beta \sigma - x = 0 \tag{6.43}$$

Here β is a measure of the dilation and x is a quantity that ensures that f and g coincide at the current stress state (see Figure 6.10(a)). If $\beta = \mu^*$ then $f = g$ and the flow rule is associated and (6.42) gives

$$\dot{\varepsilon}^{plastic} = \lambda \beta = \lambda \mu^* \quad \text{and} \quad \dot{\gamma}^{plastic} = \mathbb{S}(\tau) \lambda$$

where $\mathbb{S}(\tau)$ is the signum function such that $\mathbb{S}(\tau) = -1$, if $\tau < 0$, $\mathbb{S}(\tau) = +1$, if $\tau > 0$ and $\mathbb{S}(\tau)$ has some value between -1 and $+1$ if $\tau = 0$. The rate of plastic work is

$$\dot{W}^{plastic} = \sigma\dot{\varepsilon}^{plastic} + \tau\dot{\gamma}^{plastic} = \lambda\mu^*\sigma + \mathbb{S}(\tau)\lambda\tau = \lambda(\mu^*\sigma + |\tau|) = \lambda x = \lambda c$$

Thus *an associated Mohr-Coulomb material ($\beta = \mu^*$ and hence $x = c$) does plastic work during deformation controlled solely by the value of the cohesion.* If the material is cohesionless then no work is done which is counter to the first law of thermodynamics.

We see from (6.42) that $\dot{\varepsilon}^{plastic} = \beta|\dot{\gamma}^{plastic}|$ and so one can define a *dilatancy angle, δ,* such that $\delta = \tan^{-1}\beta$ (Figure 6.10(b)). Examination of Figure 6.10(a) shows that if $\beta < \mu^*$ then $\dot{W}^{plastic} = \lambda x$ which is always positive, whereas if $\beta > \mu^*$ then $\dot{W}^{plastic}$ is negative for a range of β values dependent on the imposed stress as shown in Figure 6.10(c). Thus *Mohr-Coulomb materials with non-associative flow, where the dilatancy angle is greater than the frictional angle, are not thermodynamically admissible materials for a range of conditions.* Moreover, the counter-intuitive situation exists that as the dilatancy is increased with $\beta \leq \mu^*$, and for a given stress state, the plastic work decreases.

The *Byerlee Law* (Byerlee (1978)) which is widely used as an upper bound to the stress level in the upper crust is a form of the Mohr-Coulomb model where $c = 0$ and $\beta = 0$, and hence $x = \tau$, so that $\dot{\varepsilon}^{plastic} = 0$ and $\dot{W}^{plastic} = \mathbb{S}(\tau)\lambda\tau$ (Figure 6.10(d)).

Although the Mohr-Coulomb constitutive relation is widely used in the geosciences, it does not have a lot of experimental support. This relation, as expressed in (6.41), is defined by a cone in stress space with an irregular hexagon (Jaeger, 1969) as cross section and straight edges on the cone as shown in Figure 6.11. This figure also shows various other relations, such as Drucker-Prager, that have been proposed that are similar to Mohr-Coulomb relation. In their simple representations these models have linear cross sections in any plane parallel to the hydrostatic line $\sigma_1 = \sigma_2 = \sigma_3$. Note that the plane normal to the hydrostatic axis, in stress space, is commonly referred to as the π-plane.

Many proposals for yield surfaces applicable to rocks at low temperatures and pressures have been put forward by Hill (1950), Drucker and Prager (1952), Murrell (1963), Bishop (1966), Matsuoka and Nakai (1974), Lade and Duncan (1975), Lade (1977), Hoek and Brown (1980a,b) and Kim and Lade (1984). Experimental data on the shapes of yield surfaces for rocks are reviewed by Lade (1993). He points out that the data indicate that, contrary to the surfaces shown in Figure 6.11(a), the yield surface is bullet shaped with curved intersections with any plane parallel to the hydrostatic axis and a cross section with rounded corners that is closer to a triangle than a hexagon as shown in Figure 6.12. There are strong suggestions that the yield surface may be close to triangular shaped in cross section at low confining pressures and evolve to close to circular at high confining pressures. Unfortunately there are very few experiments on rocks at high confining pressures at other than room temperature; experiments by Byerlee (1967) exploring the influence of confining pressure (up to 1 GPa at room temperature) are reviewed by Lade (1993). We have no indications on how the yield surface may evolve with increasing temperature at high confining pressures and as a function of strain rate. We add, however (as a hint to budding experimentalists), that such data are, in principle, relatively easy to

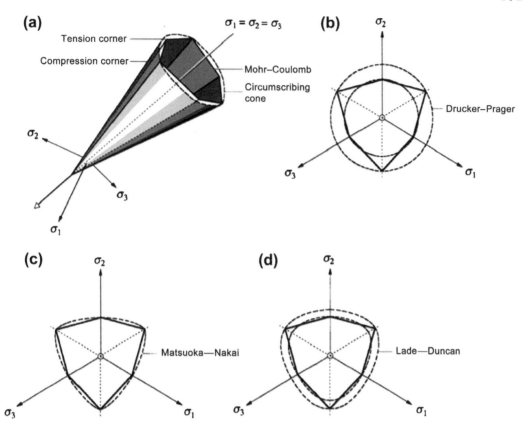

FIGURE 6.11 Various forms of constitutive relations proposed for pressure-sensitive materials. (a) The Mohr-Coulomb yield surface in principal stress space. A circular cone is circumscribed around the yield surface. This corresponds to a Drucker-Prager yield surface. (b) View on the π-plane showing a Mohr-Coulomb cross section and two Drucker-Prager sections (inscribed and circumscribed). (c) The Matsuoka yield surface cross section on the π-plane. (d) Cross section of the Lade-Duncan yield surface on the π-plane. *From Borja and Aydin (2004).*

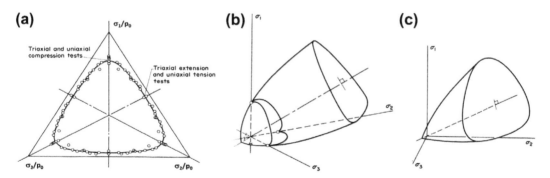

FIGURE 6.12 Summary of experimental constraints on the shapes of yield surfaces for frictional materials. *(From Lade (1993) .)* (a) Experimental results plotted in the π-plane from Akai and Mori (1967). (b) and (c) Two views of a general yield surface as revealed by a synthesis of experimental data. This is a 'best guess' as to a yield surface for rocks that is compatible with the experimental data.

obtain from modern high pressure–high temperature apparatus with combined torsion and compression facilities.

The various models of yield surfaces for frictional, plastic materials can be unified under a proposal by Borja and Aydin (2004) for a family of failure criteria for rocks that involves the three stress invariants (see also Lade, 1993) and is given by

$$f = -\alpha^n \bar{I}_1 + \kappa; \quad \alpha = -c_0 + c_1\alpha_1 + c_2\alpha_2 + c_3\alpha_3 > 0 \qquad (6.44)$$

where the three stress invariants used, $\bar{I}_1, \bar{I}_2, \bar{I}_3$, are defined as

$$\bar{I}_1 = \bar{\sigma}_1 + \bar{\sigma}_2 + \bar{\sigma}_3; \quad \bar{I}_2 = \bar{\sigma}_1\bar{\sigma}_2 + \bar{\sigma}_2\bar{\sigma}_3 + \bar{\sigma}_3\bar{\sigma}_1 \text{ and } \bar{I}_3 = \bar{\sigma}_1\bar{\sigma}_2\bar{\sigma}_3$$

with

$$\bar{\sigma}_1 = \sigma_1 - a; \quad \bar{\sigma}_2 = \sigma_2 - a \text{ and } \bar{\sigma}_3 = \sigma_3 - a$$

$a > 0$ is a parameter that shifts the stress according to the value of the cohesion. Also the functions α_1, α_2 and α_3 are given by

$$\alpha_1 = \frac{\bar{I}_1^2}{\bar{I}_2}; \quad \alpha_2 = \frac{\bar{I}_1\bar{I}_2}{\bar{I}_3} \text{ and } \alpha_3 = \frac{\bar{I}_1^3}{\bar{I}_3}$$

In (6.44) $n > 0$ is a material parameter that defines the shape of the yield surface on planes parallel to the hydrostatic axis and κ is an internal parameter given by $\kappa = -\frac{\partial\Psi(\varepsilon_{ij}^{elastic}, V^{plastic})}{\partial V^{plastic}}$ where $V^{plastic}$ is the compactive volumetric strain. The various yield surfaces proposed in the literature are then defined by various values of c_0, c_1, c_2 and c_3 in (6.44). We are interested in models where $c_0 = 3c_1 + 9c_2 + 27c_3$. These intersect the hydrostatic axis at $\bar{I}_1 = 0$ and open outwards along this axis. In particular (see Borja and Aydin, 2004, p. 2687), for $n \to \infty$ the following models (Figure 6.13) are produced:

- for $c_1 \neq 0, c_2 = c_3 = 0$ we obtain the Drucker–Prager model
- for $c_2 \neq 0, c_1 = c_3 = 0$ we obtain the Matsuoka–Nakai model
- for $c_3 \neq 0, c_1 = c_2 = 0$ we obtain the Lade–Duncan model.

If $c_0 < 3c_1 + 9c_2 + 27c_3$, the yield surface closes along the hydrostatic axis to form a *cap*. We return to these types of models in Chapter 8.

6.6.5 Hyperplasticity

Classical plasticity developed as an empirical or experimental subject and it was natural to arrive at the concept of a yield surface as the foundation for the subject (Tresca, 1864). However, this has lead to a vast range of empirically developed yield criteria, and hardening and flow rules to suit specific applications. A *hyperplastic* approach brings considerable order to and insight into the subject and arrives at the concept of a *dissipation surface* as the foundation for the subject. The yield function then arises as the Legendre transform of the dissipation function. The approach places the development of hardening and flow rules within a framework that ensures that such rules are thermodynamically admissible.

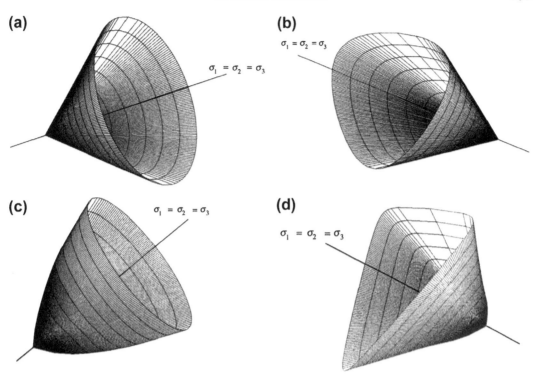

FIGURE 6.13 Various yield surfaces given by (6.44). (a) and (b) correspond to $n \to \infty$ in (6.44); (c) and (d) correspond to a finite positive value of n. (a) The classical Drucker-Prager yield surface with linear intersections on planes parallel to the hydrostatic axis. (b) The Matsuoka yield surface with linear intersections on planes parallel to the hydrostatic axis. (c) The 'enhanced' Drucker-Prager yield surface with curved intersections on planes parallel to the hydrostatic axis. (d) The 'enhanced' Matsuoka yield surface with curved intersections on planes parallel to the hydrostatic axis. *From Borja and Aydin (2004).*

Hyperplasticity has a long history starting perhaps with Biot (1958) and Ziegler (1963) and developing through works such as Rice (1970, 1971, 1975), Maugin (1999), Collins and Houlsby (1997), Houlsby and Puzrin (2006a) and Haslach (2011). The framework proposes that all aspects of the constitutive behaviour may be derived from just two potentials. One potential is any one of the specific internal energy, e, the specific Helmholtz energy, Ψ, the specific enthalpy, H, or the specific Gibbs energy, G, the choice depending on the problem to be investigated. A strain-based approach would use either e or Ψ, whereas a stress-based approach would use either H or G. If we select e as the first potential then a convenient and simple way of expressing e in a small strain formulation is

$$e = e\left(\varepsilon_{ij}, \alpha_{ij}, s\right) \tag{6.45}$$

where α_{ij} is an internal variable that is kinematic in nature (such as the plastic strain). We follow the small strain approach because the mathematical structure is straight forward. Purists interested in a large deformation approach should see Gurtin et al. (2010). One could add many more such variables of the same nature as α_{ij} but one is enough to make the argument. We could also write $\Psi = \Psi(\varepsilon_{ij}, \alpha_{ij}, T)$, $H = H(\varepsilon_{ij}, \alpha_{ij}, s)$ and $G = G(\varepsilon_{ij}, \alpha_{ij}, T)$. Corresponding to the kinematic variable α_{ij} we define a *generalised stress*, $\overline{\chi}_{ij} = -\frac{\partial e}{\partial \alpha_{ij}}$. Then we have also that

$$\overline{\chi}_{ij} = -\frac{\partial e}{\partial \alpha_{ij}} = -\frac{\partial \Psi}{\partial \alpha_{ij}} = -\frac{\partial H}{\partial \alpha_{ij}} = -\frac{\partial G}{\partial \alpha_{ij}} \tag{6.46}$$

The first law of thermodynamics gives us

$$\rho \dot{e} = \sigma_{ij} \dot{\varepsilon}_{ij} - \frac{\partial q_k}{\partial x_k} \tag{6.47}$$

The second potential is the specific mechanical dissipation, $\Phi^{mechanical}$, which is a function of the strain, ε_{ij}, the kinematic variables, α_{ij}, $\dot{\alpha}_{ij}$, and the specific entropy, s:

$$\Phi^{mechanical} = \Phi^{mechanical}\left(\varepsilon_{ij}, \alpha_{ij}, \dot{\alpha}_{ij}, s\right) \tag{6.48}$$

As a guide to the physical significance of the dissipation function, the stress strain curve for a one dimensional elastic-plastic material is shown in Figure 6.14 for isotropic hardening. The area CBDC corresponds to the Helmholtz energy, the area OEBCO corresponds to the Gibbs energy and the area OABCO is a measure of the dissipation.

$\Phi^{mechanical}$ satisfies the second law of thermodynamics so that

$$T\dot{s} + \frac{1}{\rho}\frac{\partial q_k}{\partial x_k} = \Phi^{mechanical} \geq 0 \tag{6.49}$$

If we add (6.47) and (6.49) we obtain

$$\dot{e} + \Phi^{mechanical} = \frac{1}{\rho}\sigma_{ij}\dot{\varepsilon}_{ij} + T\dot{s} \tag{6.50}$$

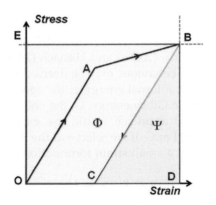

FIGURE 6.14 Cycle of loading and unloading (OABC) for a one dimensional elastic-plastic material with isotropic hardening. Dissipation, Φ (OABCO), Helmholtz, Ψ (BDCB-grey), and Gibbs, G (OEBCO-cream) functions. *After Collins and Houlsby (1997).*

(6.45) and (6.48) give us

$$\dot{e} = \frac{\partial e}{\partial \varepsilon_{ij}}\dot{\varepsilon}_{ij} + \frac{\partial e}{\partial \alpha_{ij}}\dot{\alpha}_{ij} + \frac{\partial e}{\partial s}\dot{s} \tag{6.51}$$

and

$$\Phi^{mechanical} = \frac{\partial \Phi^{mechanical}}{\partial \dot{\alpha}_{ij}}\dot{\alpha}_{ij} \tag{6.52}$$

If we substitute (6.51) and (6.52) into (6.50) we obtain

$$\left(\frac{\partial e}{\partial \varepsilon_{ij}} - \sigma_{ij}\right)\dot{\varepsilon}_{ij} + \left(\frac{\partial e}{\partial s} - T\right)\dot{s} + \left(\frac{\partial e}{\partial \alpha_{ij}} + \frac{\partial \Phi^{mechanical}}{\partial \dot{\alpha}_{ij}}\right)\dot{\alpha}_{ij} = 0 \tag{6.53}$$

If we only consider deformations where the *processes* of producing entropy during straining and the production of entropy by thermal mechanisms are considered to operate independently we have

$$\sigma_{ij} = \rho\frac{\partial e}{\partial \varepsilon_{ij}} \quad \text{and} \quad T = \frac{\partial e}{\partial s} \tag{6.54}$$

We now define a *generalised dissipative stress*, $\chi_{ij} = \frac{\partial e}{\partial \alpha_{ij}}$. Then the remaining term in (6.53) becomes

$$\left(\chi_{ij} - \overline{\chi}_{ij}\right)\dot{\alpha}_{ij} = 0 \tag{6.55}$$

This seems to imply that $(\chi_{ij} - \overline{\chi}_{ij}) = 0$ since the $\dot{\alpha}_{ij}$ are arbitrary. However, χ_{ij} may be a function of the $\dot{\alpha}_{ij}$ and so the most one can derive from (6.55) is that $(\chi_{ij} - \overline{\chi}_{ij})$ is orthogonal to $\dot{\alpha}_{ij}$. This is known as *Ziegler's orthogonality condition*. This condition is a statement that for some processes the stress is normal to the dissipation surface which in turn implies a condition corresponding to maximum entropy production (see Chapter 5). Ziegler (1983) proposes also that $(\chi_{ij} - \overline{\chi}_{ij}) = 0$ is true. Notice that this same condition written from (6.53) as

$$\left(\frac{\partial e}{\partial \alpha_{ij}} + \frac{\partial \Phi^{mechanical}}{\partial \dot{\alpha}_{ij}}\right) = 0 \tag{6.56}$$

is the equation derived by Biot (1958) and used extensively by him throughout his lifetime. (6.56) is commonly known as *Biot's equation*.

The development above supplies the tools to develop a thermodynamically admissible constitutive relation from any one of the potentials e, Ψ, H and G together with the dissipation function, Φ. We illustrate with the example of a one dimensional frictional model (Figure 6.15) consisting of an elastic spring with modulus E, in series with a frictional sliding block that slides when the shearing stress reaches a value, k. $\varepsilon^{elastic}$ is the strain in the spring, α is the strain arising from sliding of the block and ε is the total strain in the system. We take the Helmholtz energy to be given by

$$\Psi = \frac{E}{2}(\varepsilon - \alpha)^2$$

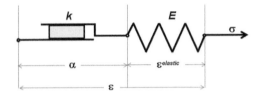

This means that $\sigma = \frac{\partial \Psi}{\partial \varepsilon} = E(\varepsilon - \alpha)$. The Legendre transform of Ψ gives the Gibbs energy:

$$G = \Psi - \sigma\varepsilon = \frac{E}{2}(\varepsilon - \alpha)^2 - E(\varepsilon - \alpha)\varepsilon$$

$$= -\frac{\sigma^2}{2E} - \sigma\alpha$$

By differentiating G we find as expected that

$$\varepsilon = -\frac{\partial G}{\partial \sigma} = \frac{\sigma}{E} + \alpha$$

so that the total strain consists of the elastic strain, $\frac{\sigma}{E}$, and the plastic strain, α. It is important to note that it is the $-\sigma\alpha$ term in the Gibbs energy that establishes α as the plastic strain. Also note that $\overline{\chi} = -\frac{\partial G}{\partial \alpha} = \sigma$ and that $\overline{\chi} = -\frac{\partial \Psi}{\partial \alpha} = -E(\varepsilon - \alpha) = \sigma$.

The dissipation is given by

$$\Phi = k|\dot{\alpha}|$$

Differentiation gives

$$\chi = \frac{\partial \Phi}{\partial \dot{\alpha}} = k\mathbb{S}(\dot{\alpha})$$

Thus when $\dot{\alpha} \neq 0$, $|\chi| = k$ and when $\dot{\alpha} = 0$, $-k < \chi < k$. Thus the magnitude of the generalised stress, $|\chi|$, is below k when there is no plastic deformation and equals k when plastic deformation occurs. Thus the yield surface is defined by $|\chi| - k = 0$.

Although this is a very simple example it illustrates the steps that need to be taken to define the yield criterion for quite complicated materials. The complete behaviour of the material is derived from the two potentials, the Helmholtz (or the Gibbs) energy and the dissipation function. Extensions to a number of other materials are discussed by Houlsby and Puzrin (2006a).

6.7 VISCOUS–PLASTIC–ELASTIC BEHAVIOUR

6.7.1 Rate Dependent Behaviour

The previous section was concerned with rate independent, dissipative deformations defined by a yield surface that is independent of temperature and deformation rate. Such constitutive behaviour is an idealisation but quite a good approximation to the behaviour of many materials at low temperatures (for metals below 0.35 of the melting temperature) and fast deformation rates (see Kocks et al., 1975 for a detailed discussion). However, most processes operating in solids during deformation are thermally activated. Even frictional sliding is thermally activated, although the activation energy is relatively low; Stesky (1978) reports an activation energy of 130 kJ mol^{-1} for temperatures less than 500 °C and 360 kJ mol^{-1} for temperatures greater than 500 °C for faulted Westerly granite at 250 MPa confining pressure and sliding rates of 10^{-3} to 1 ms^{-1}. This means that in reality the pressure dependent yield surfaces described in Section 6.6.4 are all weakly temperature and deformation rate dependent. At high temperatures and confining pressures and slow deformation rates all minerals deform by thermally activated processes such as mass transport, involving intra-grain diffusion, dissolution and precipitation and chemical reactions, dislocation and disclination motion and/or grain boundary sliding. The activation energies for these latter processes are at least an order of magnitude greater than that for frictional sliding so these processes are highly temperature and deformation rate dependent. This means that the yield surface changes its shape and size as the temperature and deformation rate change. Moreover, some of these processes, such as dislocation motion, induce a crystallographic preferred orientation (CPO) in the material which in turn leads to anisotropy of the yield stress and a consequent evolution of yield surface shape during deformation. Such anisotropy is important for the creation of anisotropy in the elastic properties of deformed rocks, for seismic wave propagation in the Earth (Mainprice et al., 2011), for an enhanced tendency for localisation of deformation and for the evolution of anisotropy of strength (Backofen, 1972; Kocks et al., 1975).

The rate dependency of deformation is expressed as *viscosity* which is a rate and temperature dependent material property that characterises the resistance of the material to flow; viscosity is commonly expressed as the ratio of the effective stress to the stretching rate or in the small deformation formulation as the ratio of the effective stress to the strain rate. If the response to strain rate variations is linear then one refers to *Newtonian viscosity*. If the response is nonlinear then two different viscosities, η_T and η_S, can be defined known as the *tangent* and *secant* viscosities (Figure 6.16(a)) and are given by

$$\eta_T = \frac{d\tau}{d\dot{\gamma}} \quad \text{and} \quad \eta_S = \frac{\tau}{\dot{\gamma}} \tag{6.57}$$

where the shear strain rate, $\dot{\gamma}$, and the shear stress, τ, are given by

$$\dot{\gamma} = \sqrt{2d_{ij}d_{ij}} \quad \text{and} \quad \tau = \sqrt{\frac{1}{2}\sigma'_{ij}\sigma'_{ij}} \tag{6.58}$$

Some authors (Fletcher, 1974; Johnson and Fletcher, 1994; Smith, 1977) refer to η_T and η_S as the normal and shear viscosities, respectively. The common constitutive relation for a purely viscous material is

$$\sigma'_{ij} = 2\eta_S d_{ij}, \quad d_{kk} = 0 \tag{6.59}$$

$$\eta_S = \eta_{0S}\left(\frac{\dot{\gamma}}{\dot{\gamma}_0}\right)^{N-1} \tag{6.60}$$

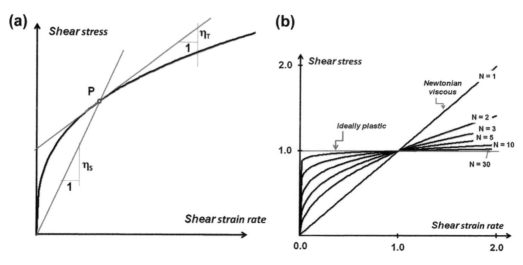

FIGURE 6.16   Power law viscous behaviour. (a) Plot of shear stress versus shear strain rate for a power law viscous material with $N > 1$. The tangential and secant viscosities, η_T and η_S, corresponding to the point P are shown. (b) Plot of shear stress versus shear strain rate for a power law viscous material for various values of N. $N = 1$ corresponds to Newtonian viscous behaviour. As N increases the behaviour becomes more rate insensitive until at large values of N the behaviour approaches ideal rate insensitive plasticity.

N is a material constant. If $N = 1$ then the viscosity is Newtonian; if $N > 1$ as it is mostly for rocks, the material is known as a *power law viscous material*. Rocks with $N < 1$ have not been described. Although power law viscous behaviour is routinely assumed for visco-plastic rock constitutive behaviour it is wise to realise that for the most part such expressions are purely empirical and based on conditions accessible in the laboratory. Only for $N = 1$ are there model-based derivations of constitutive relations (Paterson, 1995). In Section 6.7.3 we explore other expressions for rock constitutive behaviour in the viscous regime.

Note that

$$\frac{d\eta_S}{d\dot{\gamma}} = \frac{N-1}{\dot{\gamma}_0} \eta_{0S} \left(\frac{\dot{\gamma}}{\dot{\gamma}_0} \right)^{N-2} = (N-1)\eta_S \dot{\gamma}^{-1} \quad \text{and} \quad \eta_T = N\eta_S \qquad (6.61)$$

so that for $N > 1$, power law viscous materials are *viscosity strain rate softening*; an increase in strain rate results in a decrease in secant viscosity and hence also in the tangential viscosity. However, $\frac{d\tau}{d\dot{\gamma}} = \eta_T$, so that an increase in strain rate leads to an increase in τ. Thus a power law viscous material with $N > 1$ is *strain rate hardening* and *viscosity strain rate softening*.

An important characteristic of power law viscous materials is that they are intrinsically anisotropic from a constitutive point of view even though the material itself is isotropic. This can be seen by considering a body undergoing deformation at an instant where the instantaneous strain rate is $\dot{\varepsilon}_{ij}$. We see from $(6.58)_1$ that the shear strain rate is given in terms of the second invariant of the stretching tensor as $\dot{\gamma} = \sqrt{2d_{ij}d_{ij}}$. We follow Smith (1977) and consider only plane straining in two dimensions. The second invariant of the instantaneous strain rate is $II_{\dot{\varepsilon}_{ij}} = \dot{\varepsilon}_{11}^2 + 2\dot{\varepsilon}_{12}^2 + \dot{\varepsilon}_{22}^2$. If we perturb the strain rate by $\delta\dot{\varepsilon}_{ij}$ and neglect terms such as $(\delta\dot{\varepsilon}_{ij})^2$ we have the perturbed invariant as

$$
\begin{aligned}
\widetilde{II}_{\dot{\varepsilon}_{ij}} &= (\dot{\varepsilon}_{11} + \delta\dot{\varepsilon}_{11})^2 + 2(\dot{\varepsilon}_{12} + \delta\dot{\varepsilon}_{12})^2 + (\dot{\varepsilon}_{22} + \delta\dot{\varepsilon}_{22})^2 \\
&= II_{\dot{\varepsilon}_{ij}} + 2(\dot{\varepsilon}_{11}\delta\dot{\varepsilon}_{11} + \dot{\varepsilon}_{12}\delta\dot{\varepsilon}_{12} + \dot{\varepsilon}_{22}\delta\dot{\varepsilon}_{22})
\end{aligned}
\tag{6.62}
$$

For pure shearing $\dot{\varepsilon}_{12} = 0$ and hence $\widetilde{II}_{\dot{\varepsilon}_{ij}} = II_{\dot{\varepsilon}_{ij}} + 2(\dot{\varepsilon}_{11}\delta\dot{\varepsilon}_{11} + \dot{\varepsilon}_{22}\delta\dot{\varepsilon}_{22})$. Thus in this case, perturbations in the normal strain rates influence the invariant but perturbations in the shear strain rate have no effect.

For simple shearing $\dot{\varepsilon}_{11} = \dot{\varepsilon}_{22} = 0$ and hence $\widetilde{II}_{\dot{\varepsilon}_{ij}} = II_{\dot{\varepsilon}_{ij}} + 2\dot{\varepsilon}_{12}\delta\dot{\varepsilon}_{12}$. Thus in this case, perturbations in the shear strain rate influence the invariant but perturbations in the normal strain rates have no effect. Thus a power law viscous material with $N > 1$ behaves differently to perturbations in strain rate depending on whether the deformation involves pure or simple shearing. The situation is discussed by Biot (1965, p. 390) who points out that the viscosity for the bulk deformation is controlled by the secant viscosity, whereas the incremental stresses are controlled by the tangential viscosity.

6.7.2 Common Constitutive Models

It is useful to think of many constitutive relations in terms of analogue models comprised of combinations of elastic elements (springs), viscous elements (viscous dashpots) and yielding elements (sliding blocks that do not move until the stress reaches a critical stress, σ^{yield}). Some common combinations of these elements are shown in Table 6.3. Of these models the Bingham material, comprising the three elements in series, comes closest to a simple model of elasto-visco-plastic behaviour. Under an increasing imposed stress, this model first loads elastically and when the stress reaches the critical value, σ^{yield}, the block begins to slide, the model behaving viscously at the same time. Notice that this model is also used to describe the behaviour of some elasto-viscous fluids (Barnes, 1999).

6.7.3 Models Based on Dislocation Glide and Extrapolation to Conditions

It is commonly assumed that the constitutive behaviour of rocks can be extrapolated from those established under laboratory conditions to the strain rates and stresses operating under geological deformation conditions, although the pitfalls in doing so have been discussed by Paterson (1987, 2001) and Poirier (1985). This means that the power law viscous behaviour of many geological materials established in the laboratory at strain rates of $10^{-4}\,\mathrm{s}^{-1}$ to perhaps $10^{-8}\,\mathrm{s}^{-1}$ is routinely extrapolated and used to model mechanical behaviour at strain rates of $10^{-9}–10^{-15}\,\mathrm{s}^{-1}$.

TABLE 6.3 Common Constitutive Models for Elastic, Plastic, Viscous, Visco-Elastic and Visco-Elastic-Plastic Materials

Name	One Dimensional Model	One Dimensional Constitutive Relation
Perfectly elastic – Hookean material		$\sigma = E\varepsilon$
Perfectly plastic material		$\sigma \leq \sigma^{yield}$
Elasto-plastic material		$\sigma = E\varepsilon; \; \sigma \leq \sigma^{yield}$
Newtonian viscous material		$\sigma = \eta\dot{\varepsilon}$
Power law viscous material		$\sigma = \eta^{effective}\dot{\varepsilon}$ $\eta^{effective} = \eta_0\left(\dfrac{\dot{\varepsilon}}{\dot{\varepsilon}_0}\right)^{N-1}$
Kelvin–Voigt elasto-viscous material		$\sigma = \eta\dot{\varepsilon} + E\varepsilon$
Maxwell material		$\dot{\varepsilon} = \left(\dfrac{\dot{\sigma}}{E}\right) + \left(\dfrac{\sigma}{\eta}\right)$
General linear material		$\sigma + \tau_0\dot{\sigma} = E_2(\varepsilon + \tau_1\dot{\varepsilon})$ $\tau_0 = \dfrac{\eta}{E_1}; \quad \tau_1 = \dfrac{\eta(E_1 + E_2)}{E_1 E_2}$
Bingham material		For constant stress:$\varepsilon = \left(\dfrac{\sigma}{E}\right); \quad \sigma < \sigma^{yield}$ $\varepsilon = (\sigma - \sigma^{yield})t + \left(\dfrac{\sigma}{E}\right);$ $\sigma > \sigma^{yield}$

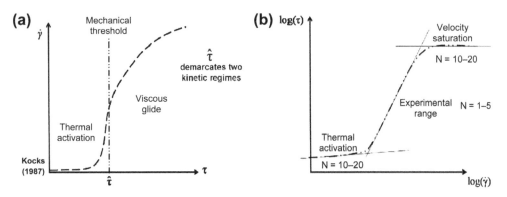

FIGURE 6.17   Dependence of the power law exponent, N, on stress. (a) Proposed relationship by Kocks (1987) between shear strain rate and shear stress. $\hat{\tau}$ defines the boundary between a region where dislocation motion is controlled by thermally activated kink migration and a region where dislocation motion is dominated by viscous glide. (b) Summary of relationships. *(Motivated by Cordier et al. (2012) .)* At low stresses the behaviour is relatively rate insensitive with large values of N. Within the experimentally accessible region values of N for minerals are in the range of 1 to about 5. At high stresses the dislocation velocity saturates near the velocity of sound and the behaviour is again characterised by large values of N.

There are two aspects to this dogma that need further investigation. First, Kocks (1987) and Kocks et al. (1975) point out that in metals the relationship between stress and strain rate can be divided into two regimes (Figure 6.17(a)). There exists a high stress regime where the stress is larger than a critical value, $\hat{\tau}$, which is temperature dependent. This regime Kocks calls the *viscous glide regime*. If $\tau < \hat{\tau}$, dislocation motion is controlled by thermally activated drift. This is typical of behaviour at low stresses and low strain rates. The overall material behaviour can be described by a power law of the form (6.59 and 6.60) and in the limit $N \to \infty$ so that the material is perfectly plastic although still thermally activated which means that η_{0S} in (6.60) is a function of temperature. This regime is *the thermally activated regime*. Figure 6.17(a) is the same as Figure 6.16(b) except that the axes have been interchanged. In addition Kocks has added an upper cut-off in strain rate representing dislocation velocities limited by the speed of sound in the material. The interpretation by Kocks and co-workers is that although $\hat{\tau}$ can be identified with a yield stress there is still thermally activated deformation below this yield stress, although the deformation rates are small. Larger deformation rates are observed at and above $\hat{\tau}$. In this sense $\hat{\tau}$ is better labelled a *flow stress* rather than a *yield stress*.

Second, an assumption that power law constitutive behaviour established at laboratory accessible strain rates can be used at slow geological strain rates means that one assumes that N is independent of strain rate; recent work by Cordier and co-workers (Amodeo et al., 2011; Carrez et al., 2009; Cordier et al., 2012), and also Monnet et al. (2004), Naamane et al. (2010), Nabarro (2003) and Tang et al. (1998) proposes that N is a function of strain rate and becomes large at low stresses (and hence small strain rates). This proposition is impossible to confirm for rocks in the laboratory since the strain rates required are experimentally inaccessible. However, there is no theoretical justification for (6.59 and 6.60) except for N = 1 so that extrapolation to geological strain rates is more a matter of optimism and convenience than being based on theoretical grounds. Computer simulations reported by the above group of workers have studied the relationship between dislocation velocity and stress for various materials at various temperatures and pressures using molecular and dislocation dynamics and compared the simulated results (as far as is possible) with experimental observations. Both the viscous glide and thermally activated regimes of Kocks (1987) are recognised. The above work proposes that (6.59 and 6.60) should be replaced by

$$\dot{\gamma} = A \frac{L}{w^*(\tau)^2} \sinh\left[\frac{\Delta H^-(\tau) - \Delta H^+(\tau)}{kT}\right]$$
(6.63)

where A is a temperature-dependent material constant that depends on dislocation geometry and kink structure, L is the dislocation length, w^* is the width of dislocation kinks and is stress dependent, ΔH^- and ΔH^+ are the changes in activation enthalpies associated with dislocation−obstacle interactions for backward and forward jumps of the dislocation kinks, T is the absolute temperature and k is the Boltzmann constant. Notice that ΔH is a function of the shear stress. An expression for ΔH, given by Kocks et al. (1975), is

$$\Delta H(\tau) = \Delta H_0\left[1 - \left(\frac{\tau}{\tilde{\tau}}\right)^p\right]^q$$
(6.64)

where ΔH_0 is the activation enthalpy at 0 K, $\tilde{\tau}$ is the Peierls stress at 0 K and p, q are parameters such that $0 \leq p \leq 1$ and $1 \leq q \leq 2$. Values of p and q are obtained from experiments and/or from computer simulations but Kocks et al. (1975) suggest that $p = 0.5$ and $q = 1.5$ are representative values. In experiments on olivine, Mei et al. (2010) use $p = 0.5$ and $q = 1.0$. Cordier et al. (2012) have explored the behaviour of MgO at various pressures, strain rates and temperatures and show that under geological conditions MgO is likely to behave as a material with large values of N to the extent that it may resemble more a perfectly plastic, rate insensitive material than a power law material with relatively small values of N. At the other extreme of behaviour at high stresses and/or high strain rates dislocation velocity is limited by the velocity of sound in the crystal so that the constitutive behaviour at high strain rates is again approximated by a power law with large values of N (Figure 6.17(b)).

6.7.4 An example − Anisotropic Yield in Polycrystalline Aggregates

The constitutive relations that have received considerable attention over the past 150 years since the pioneering work of Tresca (1864, 1865, 1867, 1868, 1869, 1870a, 1870b) are those concerned with the deformation of metals. In particular, since the early studies of single crystal plasticity and its application to CPO development by Taylor (1938), the emphasis has been on establishing the evolution of the shape of the yield surface, as anisotropy develops, associated with CPO evolution. Although the deformation of metals is rate dependent the value of N for a power law constitutive relation is relatively large (in the range 30 to at least 300 at low temperatures, less than 0.35 of the melting temperature, and fast strain rates, $\approx 10^{-3}$

to 10^2 s^{-1}). It is only at higher temperatures that values for N of say five, comparable with many minerals under laboratory conditions (Evans and Kohlstedt, 1995), occur. This means that metals at low temperatures are almost rate insensitive and so approximate the behaviour of ideal plastic materials.

The single crystal, rate insensitive yield surface has been investigated by a large number of workers including Taylor (1938, pp. 218–224), Bishop and Hill (1951a, 1951b), Asaro and Needleman (1985), Lister and Hobbs (1980), Hobbs (1985) and Kocks et al. (1998). The rate

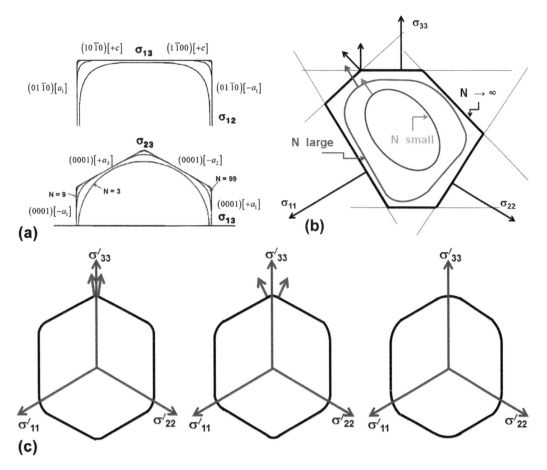

FIGURE 6.18 Yield and flow surfaces. (a) Cross sections through the yield surface for quartz (outer solid lines) showing the slip systems that are active to define that part of the surface for ideal rate insensitive plasticity. Inner lines show flow surfaces for rate dependent flow with N = 3, 9 and 99. *After Hobbs (1985) and Wenk et al. (1989).* (b) Sketch of a cross section through a yield surface (N → ∞) showing sharp corners and ambiguity in the plastic strain increment at the corners. Inner surfaces are flow surfaces for large (red) and small (blue) N. (c) Transition from yield surface with sharp corners to a slightly rate dependent surface with rounded corners. *Adapted from Kocks et al., 1998.*

insensitive yield surface is bounded by planar segments that each represent yield on a particular slip system (see Chapter 13 for details). Since there are five independent strains for a constant volume deformation, this surface is constructed in five dimensions and five independent slip systems are necessary to achieve a given imposed homogeneous strain (Kelly and Groves, 1970; Paterson, 1969). Thus the yield surface in a two dimensional cross section through five dimensional stress space is a planar-faced polygon as shown in Figure 6.18(a) for quartz. The intersections of the planar faces define sharp corners on the yield surface where five independent slip systems are defined in five dimensional space. However, the introduction of strain rate sensitivity (as would be true for values of N of 3 or 4, typical of minerals under experimental conditions) means that these corners are smoothed out. Two effects are in operation here: (i) rate sensitivity means that the stress at a given strain rate is less than the rate insensitive yield stress (see Figure 6.16(b)) and (ii) rate sensitivity means that for each imposed stress there is a unique strain rate corresponding to the operation of the easiest slip system for the imposed stress. This has the effect of smoothing the corners (Figure 6.16(b)) on the yield surface (Asaro and Needleman, 1985; Canova and Kocks, 1984; Rice, 1970). Thus Wenk et al. (1989) show that the polygonal rate insensitive yield surface with sharp corners for quartz becomes almost equivalent to a circular von Mises criterion if realistic experimental values for N are considered (Figure 6.18(a)). The rate sensitive surface is more aptly called a *flow surface* rather than a yield surface.

The absence of sharp corners on a yield or flow surface means that the propensity for shear localisation arising from the corner effect is decreased or eliminated. Thus in Figure 6.18(b and c) ambiguity exists at a corner with respect to the incremental plastic vector which in turn means that a small change in stress results in different deformations and hence the likelihood of localisation in the deformation. The more rounded the apex becomes the less likely is this effect until for smooth apices a unique incremental strain vector is defined.

We return to this general subject in Chapter 13 when we consider CPO development in rocks. An example from Kocks et al. (1998) sets the scene (Figure 6.19). Results of computer modelling of CPO and yield surface evolution are shown for von Mises effective strains of 1 (left column), 2 (middle column) and 3 (right column). A *von Mises strain* is defined as $\varepsilon^{von\ Mises} = \frac{1}{1+v}\left[\frac{1}{2}\{(\varepsilon_1 - \varepsilon_2)^2 + (\varepsilon_2 - \varepsilon_3)^2 + (\varepsilon_3 - \varepsilon_1)^2\}\right]^{\frac{1}{2}}$ where v is Poisson's ratio. The assumption is of Taylor homogeneous strain. The yield surfaces shown in red representing operational slip systems in frames (a) through to (f) are for rate insensitive constitutive parameters (N = 300). The circular surfaces inside these yield surfaces represent the results for rate sensitive materials (N = 5). Frames (a) to (c) are for equiaxed grains so that the situation is meant to represent deformation with accompanying recrystallisation that does not alter the CPO. Frames (d) to (f) represent the situation where grains are deformed and grain shape has an influence through the imposition of *relaxed constraints* where the number of independent active slip systems required to produce the imposed deformation is reduced below five (Canova and Kocks, 1984; Kocks and Canova, 1981; Ord, 1988). We see that the overall shape of the yield and flow surfaces and details of corner geometry on the yield surfaces are

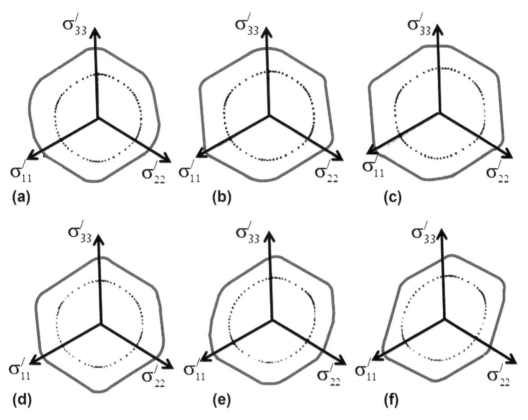

FIGURE 6.19 Yield surfaces for increasing strain and changes in grain shape. Von Mises strain = 1, 2, 3 for columns 1, 2, 3 respectively. Row 1 corresponds to equiaxed grain shape. Row 2 corresponds to grain shape that reflects the strain. *Modified from Kocks et al. (1998).*

quite sensitive to the amount of strain and to the grain shape. We return to this subject in Chapter 13.

6.8 FLUIDS AND SOLIDS

There is evidently a complete spectrum of behaviour between what might commonly be called a *fluid* and what would be called a *solid* (see Rajagopal, 1995; Truesdell, 1966). At the fluid end of the spectrum is the ideal gas which displays no viscosity and can support a hydrostatic pressure as the only stress state. The next level of complexity is the compressible linearly viscous fluid that can support a shear stress that depends on the rate of straining while undergoing deformation; at rest the fluid can only support a hydrostatic pressure. Both of these materials have no knowledge of a preferred or reference stress-free state and the stress in the material depends only on the current state of the material. The mathematical framework for such materials is Eulerian. Both of these materials display elasticity. They do not exhibit localised states of deformation. The mechanism of deformation of these materials comprises the directed flow of simple molecules arising from stress (including pressure) gradients.

At the other end of the spectrum is the rate insensitive elastic-plastic solid which can exhibit two states of deformation depending on the history of loading. One state is an elastic state where the material returns to a preferred stress-free reference state upon unloading. The stress in the material is a function of how far the material has been removed from that reference state. The other type of deformed state is the plastic state where the material is permanently deformed having passed a yield point. The stress in the material is the elastic stress at yield (and hence is a function of how far the material has been deformed from the reference configuration) or a modification of this stress due to hardening or softening mechanisms. The mathematical framework for such materials is Lagrangian. The incremental strain in the material is prescribed by the normal to a potential surface in stress space or the normal to a dissipation surface in dissipation space. Plasticity arises from the rate independent migration of microstructural defects in the solid driven by gradients in deformation. Localisation of deformation is a common mode of deformation and depends on the evolution of the shape of the yield surface. Localised deformation is commonly associated with unloading of the localised zone to an elastic state. Although the elastic strain is always small in such materials, elasticity cannot be neglected since its existence allows localised regions of deformation to unload. The stress distribution in the material is determined by the history of motions the material has experienced.

Between these two end members is a broad range of materials that display viscosity arising from a rate dependence of the deformation mechanisms. Examples of these materials are power law viscous materials. All display elasticity. Many of these materials also display yield phenomena (see Barnes, 1999) but the 'yield point' may separate two contrasting regimes of rate dependent behaviour as opposed to the yield point of an ideal elastic-plastic material which separates an elastic regime from a rate independent deformation regime. Although such yield behaviour is characteristic of many non-Newtonian fluids (Barnes, 1999) it has also been proposed for metals and ceramics (and hence presumably rocks) by Kocks et al. (1975); (see Figure 6.17(a) of this book). They propose a stress level, $\hat{\tau}$, that separates viscous glide from thermally activated dislocation migration. This is proposed to act as a 'yield point' for rate dependent crystal plasticity. The concept carries with it the concept of a *flow surface* in contrast to a *yield surface*. For associated flow the flow surface defines the orientation of the incremental flow vector as discussed in Section 6.7.4. Thus for such materials, although they flow at very small stresses, they still behave as plastic materials especially if the exponent N in (6.60) is greater than about 5. The unresolved issue as far as deformation of rocks is concerned is: *what is the value of N for rocks flowing under geological conditions?* As long as N > 1, these materials behave more like a plastic solid than a viscous fluid.

Another aspect of a power law viscous fluid is that it shows no or weak localisation of deformation at small values of N even if strain softening is present. Thus a material with N = 1 shows no localisation. If N = 3 the localisation is weak and for N = 10 localisation with strain softening is highly probable (see Mancktelow, 2002, for some exploration of this subject). This arises because, as discussed in Section 6.7.1 and (6.61), power law viscous materials with N > 1 are strain rate hardening and viscosity strain rate softening. For large values of N the viscosity strain rate softening effect overrides the strain rate hardening effect and promotes localisation. Thus fluid-like materials show no localisation, whereas rate insensitive plastic solid-like materials are prone to localisation. Localisation is also possible in more complicated fluid-like materials such as the differential grade two fluids studied by Patton and Watkinson (2005, 2010, 2013).

However, the essential difference between fluid- and solid-like behaviour lies in the extra variables that one needs in order to adequately describe solids. The evolution of internal microstructural deformation processes in solids that are not present in many fluids means that internal variables need to be incorporated in the constitutive relations for solids that are not necessary for fluids. This means that the mathematical apparatus needed for fluid mechanics is quite different to that needed for rate dependent plasticity.

Recommended Additional Reading

Fung, Y.C. (1965). *Foundations of Solid Mechanics*. Prentice-Hall.
  This book contains an excellent treatment of constitutive relations for elastic, plastic and viscous materials.

Gurtin, M.E., Fried, E., Anand, L. (2010). *The Mechanics and Thermodynamics of Continua*. Cambridge University Press.
  A rigorous, advanced treatment of constitutive relations and their thermodynamic foundations.

Houlsby, G.T., Puzrin, A.M. (2006). *Principles of Hyperplasticity*. Springer.
  The definitive text on hyperplasticity (especially for soil- and rocklike materials) and the thermodynamic constraints on constitutive relations.

Jaeger, J.C. (1969). *Elasticity, Fracture and Flow*. Methuen.
  A classical, highly readable work with considerable insight into the constitutive relations for various materials.

Kocks, U.F., Tome, C.N., Wenk, H.−R. (1998). *Texture and Anisotropy*. Cambridge University Press.
  An important book that brings many aspects of CPO together with an emphasis on anisotropy and its influence on the yield surface.

Kocks, U.F., Argon, A.S., Ashby, M.F. (1975). *Thermodynamics and kinetics of slip*. Progress in Material Science 19, 1−288.
  A thermodynamic treatment of crystal slip with a discussion of viscous glide and thermally activated slip regimes for crystal plasticity.

Malvern, L.E. (1969). *Introduction to the Mechanics of a Continuous Medium*. Prentice-Hall.
  An excellent treatment of constitutive relations.

Tadmor, E.B., Miller, R.E., Elliot, R.S. (2012). *Continuum Mechanics and Thermodynamics*. Cambridge University Press.
  A modern mathematical treatment of constitutive relations with an emphasis on thermodynamic constraints.

[illegible] of [illegible] [illegible] softening [illegible] the strain [illegible] and [illegible] [illegible] [illegible] [illegible] [illegible] Localisation [illegible] [illegible] [illegible] grade two [illegible] [illegible] [illegible] 2019).

However [illegible] and [illegible]-like behaviour lies in the [illegible] variable [illegible] in order to [illegible] solids. The evolution of internal microstructural deformation processes in solids that are not present in many fluids means that internal variables need to be incorporated in the constitutive relations for solids that are not necessary for fluids. This means that the mathematical apparatus needed for fluid mechanics is quite different to that needed for [illegible] plasticity.

Recommended [illegible] Reading

[illegible]

[illegible]

[illegible]

[illegible]

Hill, R. [illegible] Cambridge University Press.

[illegible] (2019). [illegible] Progress in Materials Science [illegible]

[illegible] Cambridge University Press.

Nonlinear Dynamics

7.1 INTRODUCTION

7.1.1 What is Nonlinear Dynamics?

Dynamics is the study of the *forces* that cause *flows* of physical and chemical quantities in deforming, chemically reacting systems. In a generalised sense these forces are *thermodynamic forces* or *affinities* as discussed in Chapter 5. Thus, in general, the forces are represented by *gradients* in the deformation, the inverse of the temperature, hydraulic potential or chemical potential. Such systems undergo *evolutionary trajectories* and it is the study of such trajectories that occupies much of the current literature and the literature over the past 30 years or so on *dynamical systems*. The observation that many of the processes that operate in such systems are *coupled* means that the behaviour of these systems is *nonlinear*. Thus in keeping with common usage, although driving forces may not be explicitly included, we refer to the study of the *evolution* of nonlinear systems with time as *nonlinear dynamics*.

A linear system is one where the output of the system is proportional to the input. Thus perfectly elastic materials respond to stress so that the strain is proportional to the stress; another example is the stretching of a linear viscous material which is proportional to the imposed stress. The mathematical *law of superposition* holds for linear systems so that if f_1 and f_2 are functions that satisfy the governing equations for a system then $\alpha f_1 + \beta f_2$ also satisfies the governing equations where α and β are arbitrary constants. Thus Fourier methods are useful in solving the governing equations for linear systems. The solutions to the equations for a particular linear system are commonly stable in the sense that small perturbations of the system away from a solution result in the system returning to that solution.

Most systems in the natural world are nonlinear. For nonlinear systems a small change in the input can result in large and non-intuitive outputs. Mathematical solutions to nonlinear problems are rare and we will see that for some nonlinear systems there is in principle no unique solution (Champneys and Toland, 1993; Knobloch, 2008). Even if a solution can be found it is commonly unstable so that a small perturbation of the system away from that solution results in evolution to another (possibly unstable) solution. The intent of this chapter is to provide an introduction to the study of the evolution of nonlinear systems. The chapter is meant to give the reader examples and a vocabulary that enables more advanced treatments of the subject to be accessed.

Many nonlinear systems can be understood by first considering their behaviour close to some homogeneous ground state where the behaviour is essentially homogeneous and linear (Figure 7.1). The procedure is to perturb the system by a small amount and analyse whether this perturbation is stable (decays back to the ground state) or unstable (grows to produce a pattern). This procedure is called a *linear stability analysis*. If the perturbation is unstable the

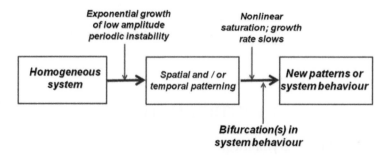

FIGURE 7.1 Behaviour of many nonlinear systems. An initially homogeneous system can become unstable leading to the growth of spatial and/or temporal patterning. The initial growth is commonly exponential and the pattern periodic consisting of spots, stripes or linear features in space or some kind of periodic oscillations in space and/or time. The exponential growth slows once the influence of nonlinear terms becomes large enough. The system may also undergo one or more bifurcations so that the system switches to a qualitatively different kind of behaviour.

initial growth is commonly exponential in time and sinusoidal in space since such a solution is an eigenfunction of the governing equations. This leads to some form of *patterning* in space. However, the exponential growth can only continue until the nonlinearities that were neglected in the linear stability analysis become important. At such stage many systems undergo a slowing in the growth rate. This is called *nonlinear saturation* (Cross and Greenside, 2009). However, the nonlinearities may also lead to a sudden switch in overall system behaviour so that the initially formed patterns switch to new patterns with their own set of growth rates. This is called a *bifurcation* and many systems continue to undergo bifurcations as the system continues to evolve. Such switches can, for instance, be from spatial to temporal instabilities, or from periodic, to localised, to chaotic instabilities. In buckling problems the buckling layer may undergo sequential buckling so that each bifurcation represents the addition of one or more new buckles (Burke and Knobloch, 2007). We explore aspects of the spatio-temporal evolution of nonlinear systems in the remainder of this chapter.

The nonlinearities that lead to the behaviour described above can be of *geometrical* or *constitutive* origins. Thus the nonlinear buckling behaviour of the layers of spheres described in Figure 1.10 arises from the geometrical necessity that defects (holes) in the packing of the spheres need to form as the layers deform (Hunt and Hammond, 2012). Another geometrical nonlinearity is the decrease in resolved shear stress on a plane such as a crystallographic slip plane in a crystal as it rotates in a stress field (Ortiz and Repetto, 1999). Constitutive nonlinearities most commonly arise from nonlinear behaviour such as nonlinear elasticity, power law viscosity, anisotropy or some form of mechanical or chemical softening.

The behaviour of nonlinear systems can be divided into two end members. One involves the temporal evolution of the system which may be steady, oscillatory or chaotic. The other involves the evolution of spatial patterns; again these can be sinusoidal, periodic, quasiperiodic, localised or chaotic. We concentrate on the spatial evolution of nonlinear systems in this chapter because that is what we finally get to look at in the field. However, the temporal evolution of such systems is just as important but is more difficult for a metamorphic geologist to grapple with from an observational viewpoint.

Cross and Greenside (2009) classify nonlinear systems that produce patterns on the basis of three parameters (Figure 7.2(a)). The first is the ratio, $\mathbb{E}$, of the power associated with the driving force for evolution of the system to dissipation within the system. As we have seen in Chapter 5 the driving forces in metamorphic systems are gradients in the deformation, the inverse of the temperature, hydraulic potential and chemical potentials of chemical components. These drive dissipation through the development of fluxes in momentum, heat, fluid and mass. The driving forces tend to move the system away from equilibrium, whereas the dissipation tends to return the system towards equilibrium. Thus the ratio $\mathbb{E}$ is a measure of how far the system is from equilibrium. The second parameter is $\mathbb{L}$ the ratio of the length scale of the system to the length scale of the pattern that forms. For a metamorphic system, this ratio may vary from something like 10^3 for structures considered at the outcrop scale, to 10^6 at the regional scale. For experimental systems in convection or reaction-diffusion systems this ratio is more like 10^2. The third parameter, $\mathbb{N}$, is the number of interacting components comprising the system. For a reacting metamorphic system, a climate system or a biological system this is likely to be in the range 10 to $\gg 100$, whereas in experimental reaction-diffusion systems $\mathbb{N}$ is more likely to be less than 10. Cross and Greenside (2009)

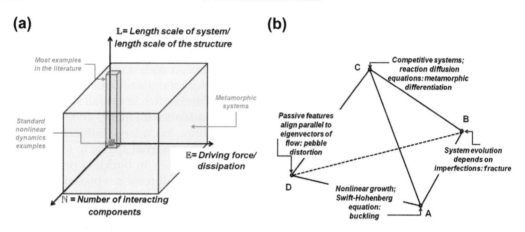

FIGURE 7.2 Classifications of nonlinear systems. (a) Classification in terms of departure from equilibrium, $\mathbb{E}$, ratio of length scale of system to length scale of structure, $\mathbb{L}$, and number of interacting components, $\mathbb{N}$. Classical nonlinear systems tend to be near the origin. Many examples involving convecting fluids and simple chemical reactions occupy a tube elongate parallel to the $\mathbb{L}$-axis. Most metamorphic, climate and biological systems occupy a much larger region of space. *(From Cross and Greenside (2009).)* (b) Classification of nonlinear systems in terms of evolutionary behaviour. Systems involving the buckling of layers tend to be near apex A; systems that depend on the presence of imperfections are near B — the development of fracture patterns is an example; competitive processes, in particular reaction-diffusion systems, are near C; passive features that are amplified by alignment with eigenvectors of the flow (Chapter 3) are near D. In reality many processes lie within the pyramid and involve the interaction of a number of mechanisms.

point out that the classical nonlinear systems that readers may be familiar with (the three variable Lorenz-type 'butterfly' instability and the driven Duffing equation for a pendulum — see Lynch, 2007; Wiggins, 2003) involve relatively small values of $\mathbb{E}$, $\mathbb{L}$ and $\mathbb{N}$ so that in Figure 7.2(a) these systems tend to be in a small box close to the origin. For experimental convection and reaction-diffusion systems (Cross and Hohenberg, 1993) $\mathbb{L}$ may be relatively large but $\mathbb{E}$ and $\mathbb{N}$ are relatively small so these systems occupy a long parallelepiped parallel to the $\mathbb{L}$ axis. Metamorphic systems, in common with biological and climate systems, on the other hand involve large ranges in all of $\mathbb{E}$, $\mathbb{L}$ and $\mathbb{N}$ and so occupy a relatively large box in Figure 7.2(a). This comparison is important because even for the 'classical' nonlinear systems few analytical solutions exist and the phase space is small. For the experimental systems even fewer analytical solutions exist and the phase space may be very large. For metamorphic systems the problem is intractable with present knowledge. It is, however, possible to explore relatively simple systems that give us some insight into how the whole system, or parts of it, may operate. This is the approach adopted here and is that adopted so far in the study of climate and biological systems. Just as important is the concept of a *linear instability analysis* which we will consider in Section 7.4. This involves making the governing equations linear, without (hopefully) throwing away too much of the physics, and investigating the conditions under which such a system becomes unstable together with the embryonic growth rates for such instabilities. Of course the results of such a procedure only hold for as long as the assumptions involved in making the governing equations linear hold but the approach can supply considerable insight into the initial behaviour of the system.

A second way of classifying nonlinear systems is shown in Figure 7.2(b) in the form of four different end members that involve different modes of evolution of the system. These are as follows: (1) Systems where a change in behaviour early in the growth of the system leads to instability and the nonlinear growth of the instability. The most relevant system here is the buckling of a layer embedded in a nonlinear material such as a 'power law viscous' material. We will consider such systems in detail in Volume II. A well explored equation governing the behaviour of such systems is the *Swift-Hohenberg equation* (Cross and Greenside, 2009; Cross and Hohenberg, 1993; Peletier and Troy, 2001). (2) Systems that rely on the presence of imperfections for their behaviour and evolve in different manners depending upon the nature and distribution of the imperfections. Various forms of *criticality* including *self-organised criticality* (Bak, 1996) lie near this apex. Fracture is the classical example of such behaviour which we look at in Chapter 8 and Volume II. (3) Systems that evolve because of *competition* between processes. Although competition is the hallmark of all systems not at equilibrium, these systems have competition built directly into their governing equations. The *reaction-diffusion equation* is the archetype for these systems and we explore this equation for the development of metamorphic differentiation in Volume II. (4) Systems that evolve such that passive markers align with one or more *eigenvectors* of the flow. We have looked at these in Chapter 3 but will consider them in greater depth in later chapters. We first look at metamorphic systems in a little more detail and some archetype examples of nonlinear systems before examining the concept of linear stability analysis, the types of nonlinear growth that can occur and how we can quantitatively describe the patterns that form.

7.2 PATTERNS IN METAMORPHIC ROCKS

Spatial patterns are ubiquitous in deformed metamorphic rocks (Figure 7.3). Examples include folds, boudins, mullions, metamorphically differentiated layering, spiral garnets, joint and vein systems, zoned mineral assemblages in veins and mineral segregations that constitute lineations and leucosomes. Important questions are: *Is there some form of order in the patterns we see? Are these structures monofractals, multifractals or something else? If the answer is 'something else' what does that mean? If some of these structures have a fractal geometry what does that tell us about the processes involved in the formation of those structures? How can we quantitatively describe and compare these patterns and decide if they are really monofractal or multifractal in character?*

Non-equilibrium systems are characterised by competition between *forcing processes* and *dissipative processes*. As we have seen the forcing processes arise from *gradients* in deformation, the inverse of the temperature, hydraulic potential and chemical potentials that are imposed on the system by external agents. The dissipative processes are responses to this forcing and constitute *flows* of momentum, heat, fluid and chemical components. The forcing processes tend to move the system away from equilibrium, whereas the dissipative processes tend to move the system towards equilibrium. If the forcing is maintained a balance between these processes commonly develops and the system 'locks into' a *non-equilibrium stationary state* where the fluxes are constant in time and where the production of entropy is a maximum if the constraints on the system allow this to happen. Such a state may or may

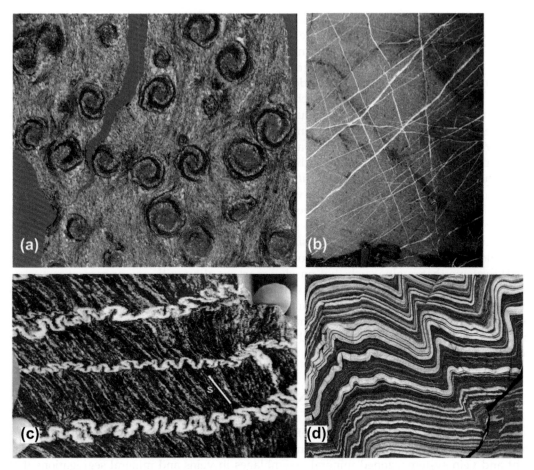

FIGURE 7.3   Patterns in deformed metamorphic rocks. (a) Spiral garnet from Vermont, USA. Thin section in crossed polars with gypsum plate. *(Photo from John Rosenfeld.)* Image about 30 mm across. (b) Veins in deformed dolomite. Ruby Gorge, Central Australia. (c) Ptygmatically folded veins crossed by gneissic layering. *(Photo: Haakon Fossen.)* (d) Folded quartz-biotite schists. Harvey's Retreat, Kangaroo Island, Australia. Outcrop about 1 m across.

not be stable and the system may evolve or switch to a new stationary state with small changes in the forcing or as the system evolves further. We have discussed some aspects of this behaviour in Chapter 5 and will consider other systems in detail in later chapters.

The important point to grasp at this stage is that, although the system may be constrained to settle into a stationary state it may not be possible from a geometrical or physical point of view for this state to be *homogeneous*. Thus in a fluid system uniformly heated at its base and once the temperature gradient through the fluid reaches a critical value, contrasts in buoyancy mean that there is a tendency for the cold layer at the top to sink to the base of the system and for the hot layer at the bottom to rise to the top. However, because elements of the fluid cannot interpenetrate or overlap this cannot happen simultaneously everywhere and a compromise develops whereby some cold parts sink and localised hot plumes rise. This compromise results in a *convection pattern* that is a stationary state so long as the temperature gradient is maintained but may be unstable to small perturbations in the temperature gradient (Cross and Greenside, 2009, their Figure 1.12). In such a system the production of hot, less dense fluid competes with the production of denser, cold fluid and this density production system is coupled by the advection of heat. Such a system is also competitive in that it is driven away from equilibrium by the gradient in temperature and tends to be returned towards equilibrium by advection of both mass and heat. A *convection pattern* is the response to such competition.

In the same manner, production and consumption of chemical components can compete with each other in a chemical system and are coupled through the diffusion of chemical components. The driving forces are gradients in deformation and chemical potentials which force the system away from equilibrium. Chemical reactions and diffusion dissipate energy and tend to move the system towards equilibrium. These systems also produce patterns − Turing patterns (Turing, 1952) − which are expressed as metamorphically differentiated layering or mineral lineations in deforming/chemically reacting rocks.

We will see that the stationary states within a system not at equilibrium correspond to both stable and unstable energy states. This arises because the Helmholtz energy of such systems is *non-convex* − in exactly the same way that the Gibbs energy is non-convex at a phase transition in a chemical system. This gives rise to another, but equivalent, way of considering pattern formation. For a non-convex Helmholtz energy the system can minimise the energy by dividing into two subsystems which then constitute a pattern. Such a division into two subsystems corresponds to critical behaviour analogous to the phase H_2O dividing into water and steam at the boiling point of the system. In many nonlinear systems the energy function comprises many local bumps and basins so that as the system evolves it switches from one local minimum to another and hence from one pattern to another. We explore some of this behaviour in the remainder of this chapter and also in Chapter 14 and Volume II.

7.3 SOME ARCHETYPE EXAMPLES

There are a number of pattern forming nonlinear systems that have been explored in great detail over the past 50 years so that their general behaviour is well known. Except for the important contributions of Ortoleva (1994) and workers such as Merino (Merino and Canals, 2011; Wang and Merino, 1992) very little work in this regard has been carried out in the geosciences so the general area of study is quite open. Some of the systems that have been studied in detail have direct relevance to metamorphic geology, namely, the Swift-Hohenberg equation and the reaction-diffusion equation. Both of these equations display an enormous range of diversity in their behaviour involving the formation and evolution of spatio-temporal patterning. For detailed analysis one should consult Peletier and Troy (2001) and Murray (1989). We describe these two equations below.

7.3.1 The Swift-Hohenberg Equation

If x is a spatial coordinate, one example of the one-dimensional Swift-Hohenberg equation for some field, $w(x, t)$ is written:

$$\frac{\partial w}{\partial t} = -\frac{\partial^4 w}{\partial x^4} - 2\frac{\partial^2 w}{\partial x^2} + (r - 1)w - w^3 \qquad (7.1)$$

This looks somewhat complicated but it becomes simpler once one understands what the individual terms mean. For a buckling layer (Turcotte and Schubert, 1982, Section 3–9; also Volume II of this book) the term $\frac{\partial^4 w}{\partial x^4}$ arises from the moment produced in the layer by the buckling force initially parallel to the layer, the term $\frac{\partial^2 w}{\partial x^2}$ arises from the curvature due to buckling and the terms involving w and w^3 arise from the reaction forces exerted on the buckling layer by the matrix material. This is one of the simplest of pattern forming equations and is widely used in studies of Rayleigh-Benard convection (Cross and Greenside, 2009; Cross and Hohenberg, 1993). Its importance in metamorphic geology is that it is a commonly used equation for studying the buckling of layers embedded in a weaker medium (Biot, 1965; Smith, 1977; Johnson and Fletcher, 1994) where it is referred to as the *biharmonic equation* (Ramsay, 1967). For the buckling problem a stationary state is often considered where $\frac{\partial w}{\partial t} = 0$. w is the layer deflection at a position x measured from some reference point along the layer (Figure 7.4); as indicated above, the function $[(r - 1) w - w^3]$ is related to the reaction force exerted by the embedding medium on the deflecting layer. It can be replaced by other functions depending on the constitutive behaviour of the embedding medium. It is straightforward to convince one's self that w = 0 is a solution to (7.1); this means that a homogeneously shortening layer is one solution to (7.1). We are interested in the behaviour of (7.1) as the parameter r is varied. We want to establish the conditions under which the undeflected configuration becomes unstable so that small perturbations from the un-deflected state begin to grow. We look at this problem in Section 7.4 below.

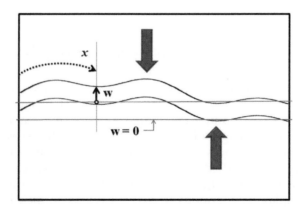

FIGURE 7.4　The deflection, w, of a layer at distance x measured along the layer from a reference point. The blue layer represents the un-deflected state. The red arrows represent the reaction forces exerted on the deflecting layer by the embedding matrix.

7.3.2 Reaction-Diffusion Equations

Reaction-diffusion equations describe the behaviour of a large range of chemical systems where diffusion of material competes with the production of that material by some form of chemical reaction. Many other kinds of systems are described by the same type of relation. Thus systems where heat (or fluid) is produced and diffuses away from the heat (or fluid) production site are described by the same form of equation. For chemical systems the rate of change of the concentration, c, of a chemical component, C, consists of a diffusive term and a production/consumption term:

$$\frac{\partial c}{\partial t} = D_c\frac{\partial^2 c}{\partial x^2} + R_c \tag{7.2}$$

$$\begin{bmatrix} \text{Time rate of} \\ \text{change} \\ \text{of concentration} \\ \text{of chemical} \\ \text{component} \end{bmatrix} = \begin{bmatrix} \text{Change in} \\ \text{component} \\ \text{due to} \\ \text{diffusion} \end{bmatrix} + \begin{bmatrix} \text{Rate of formation} \\ \text{of component} \end{bmatrix} - \begin{bmatrix} \text{Rate of} \\ \text{consumption} \\ \text{of} \\ \text{component} \end{bmatrix} \tag{7.3}$$

where D_c is the diffusivity of C and R_c is the net rate at which C is produced in a chemical reaction. If there are $\aleph$ coupled chemical reactions involving chemical components, C_i, then the array of reaction-diffusion equations is

$$\frac{\partial c_i}{\partial t} = D_i\frac{\partial^2 c_i}{\partial x^2} + R_i(c_i), \quad i = 1, 2, \dots \aleph \tag{7.4}$$

where c_i is the concentration of the ith component. These equations describe situations where chemical components can diffuse within the system and can be produced or consumed in a competitive manner by simultaneous coupled chemical reactions. In these systems the competition between chemical reaction rates tends to produce heterogeneities while diffusion tends to homogenise the system. The final result is some form of spatio-temporal patterning. Notice that we use arial font to denote chemical components (in non-italic upper case font) and their concentrations (in italic lower case font) throughout.

7.4 LINEAR STABILITY ANALYSIS

Having obtained the equations that govern the evolution of a system, one way of studying the stability and the early evolution of the system is to discover (by trial and error or intuition) a time-independent state, w_B, that is also uniform with respect to spatial coordinates and is a solution to the evolutionary equations. This state is called the *base state*. The nonlinear evolution equation is then linearised about the base state by considering a small perturbation, δw_B, from the base state, substituting this perturbation into the evolutionary equations and neglecting terms that are of second order or higher in the perturbation and in derivatives of the perturbation. One then searches for solutions to what are now evolutionary equations linear in δw_B by proposing solutions of the form

$$\delta w_B = A_k \exp(\omega_k t) \exp(ikx) \tag{7.5}$$

where $i \equiv \sqrt{-1}$. Solutions of this form are eigenfunctions (Boyce and DiPrima, 2005) with respect to the Laplace operator: $\left[\frac{\partial^2}{\partial x^2}\right]$ since they are not altered, except for a constant multiplier, by the operator. In this case $\frac{\partial^2 \delta w_B}{\partial x^2} = -A_k k^2 \exp{(\omega_k t)} \exp{(ikx)}$. Of course the most difficult step in this process is the initial step of finding the evolutionary equations for the system. The remaining steps are reasonably straightforward using codes such as *Mathematica*® or *Matlab*®. We give examples of linear stability analyses below for the Swift-Hohenberg equation, for a system of two coupled chemical reactions and for a reaction-diffusion equation.

7.4.1 The Swift-Hohenberg Equation

We denote the base state, $w = 0$ as $w_B(x,t)$ and perturb this state by a small displacement, w_P. The question is: *Does the new perturbation field, $w_P(x,t)$, grow or relax back to the base field?* This perturbation field is

$$w_P(x, t) = w(x, t) - w_B \tag{7.6}$$

If we substitute (7.6) into (7.1) then w_P evolves according to

$$\frac{\partial w_P}{\partial t} = (r - 1)[w_B + w_P] - 2\frac{\partial^2 [w_B + w_P]}{\partial x^2} - \frac{\partial^4 [w_B + w_P]}{\partial x^4} - [w_B + w_P]^3 - (r - 1)[w_B]$$

$$+ 2\frac{\partial^2 [w_B]}{\partial x^2} + \frac{\partial^4 [w_B]}{\partial x^4} + [w_B]^3 \tag{7.7}$$

If we assume w_P is small enough that the square of w_P and higher powers can be neglected along with the spatial derivatives of these terms, we arrive at

$$\frac{\partial w_P}{\partial t} = (r - 1)w_P - 2\frac{\partial^2 w_P}{\partial x^2} - \frac{\partial^4 w_P}{\partial x^4} - [3w_B^2]w_P \tag{7.8}$$

Referring (7.8) to the state $w_B = 0$ we obtain

$$\frac{\partial w_P}{\partial t} = (r - 1)w_P - 2\frac{\partial^2 w_P}{\partial x^2} - \frac{\partial^4 w_P}{\partial x^4} \tag{7.9}$$

A well-established solution to this type of equation is (Boyce and DiPrima, 2005):

$$w_P = A \exp{(\omega t)} \exp{(\alpha x)} \tag{7.10}$$

where ω is the growth rate of the perturbation and α is a constant that can be real or complex. If we substitute (7.10) into (7.9) we obtain

$$\omega = r - [\alpha^2 + 1]^2 \tag{7.11}$$

Cross and Greenside (2009, p. 65) show that α depends on the boundary conditions but for both infinite and periodic boundary conditions the general solution to (7.9) is

$$w_P = A \exp{(\omega t)} \exp{(ikx)} \tag{7.12}$$

where $i \equiv \sqrt{-1}$ and k is a wave number for the deflection; for infinite domains k can be any real number while for finite periodic domains of length, L,

$$k = m\frac{2\pi}{L}, \quad m = 0, \pm1, \pm2, \ldots \tag{7.13}$$

Thus, substituting $\alpha = ik$ into (7.11) the growth of a wave-number, k, is given by

$$\omega_k = r - [k^2 - 1]^2 \tag{7.14}$$

If ω_k is negative then perturbations do not grow and relax back to the un-deflected state, but if ω_k is positive the perturbations grow at an exponential rate as given by (7.12). We look at this condition further in Section 7.5. However, this exponential growth continues only for as long as the simplifying assumptions involved in deriving (7.8) are true. Once these assumptions fail other effects emerge to control the growth rate and in nonlinear systems one cannot assume anything about this subsequent growth given only a linear stability analysis. In general a phenomenon called *nonlinear saturation* emerges (Cross and Greenside, 2009, their Section 4.1).

7.4.2 Coupled Chemical Reactions

Many mineral reactions occurring during metamorphism are *coupled* in the sense that the product or products of one mineral reaction are involved as reactants in other mineral reactions that are occurring at the same time. This was pointed out in a classical paper by Carmichael (1969) but many other examples have been discussed since then (see Chapter 14). These reactions are also referred to as *networked* (Epstein and Pojman, 1998) and *cyclic* (Vernon, 2004) reactions. The definitive treatment for coupled chemical systems, both with and without coupled diffusion, is by Murray (1989) and see also Epstein and Pojman (1998). The systematics behind the behaviour of such systems are important for a variety of processes other than chemical reactions and so a brief review is given below. One should consult the above two books for details.

As an example we take the system of coupled reactions described by Whitmeyer and Wintsch (2005) and illustrated in Figure 7.5. The reactions involved here are

Sillimanite → Muscovite

$3Al_2SiO_5 + 3SiO_2 + 3H_2O + 2K^+ \rightarrow 2KAl_3Si_3O_{10}(OH)_2 + 2H^+$

Biotite → Chlorite

$2K(Mg, Fe)_3\,AlSi_3O_{10}(OH)_2 + 4H^+ \rightarrow (Mg, Fe)_5Al_2Si_3O_{10}(OH)_8 + 3SiO_2 + 2K^+ + (Mg, Fe)^{++}$

Plagioclase → Muscovite

$15Na_{0.6}Ca_{0.4}Al_{1.4}Si_{2.6}O_8 + 7K^+ + 14H^+ \rightarrow 7KAl_3Si_3O_{10}(OH)_2 + 18SiO_2 + 9Na^+ + 6Ca^{++}$

Biotite → Muscovite

$3K(Mg, Fe)_3\,AlSi_3O_{10}(OH)_2 + 20H^+ \rightarrow KAlSi_3O_{10}(OH)_2 + 6SiO_2 + 12H_2O + 2K^+ + 9(Mg, Fe)^{++}$

$$\tag{7.15}$$

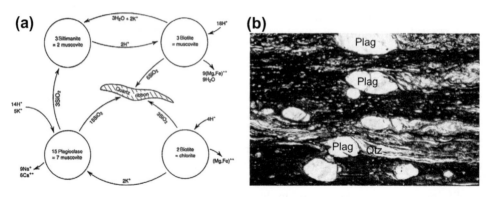

FIGURE 7.5   Cyclic reactions. *(From Whitmeyer and Wintsch (2005).)* (a) The system of coupled or networked reactions. (b) Quartz ribbons produced by networked reactions in (a). Details of specimen are in Whitmeyer and Wintsch (2005). *Photomicrograph supplied by Bob Wintsch.*

This set of coupled reactions consumes 3 moles of SiO_2 and produces 27 moles of SiO_2 and hence is strongly *autocatalytic* in SiO_2. The term *autocatalytic* refers to a situation where the addition of a chemical component to a reacting system produces even more of that component. The result is the microstructure illustrated in Figure 7.5 where ribbons of quartz form.

We consider such reactions in greater detail in Chapter 14 and Volume II but for now let us consider a set of two reactions involving two components A and B with independent concentrations a and b. We write for the rates at which a and b change:

$$\frac{da}{dt} = f(a, b)$$
$$\frac{db}{dt} = g(a, b)$$

(7.16)

where f and g are usually nonlinear functions of a and b. Such reactions commonly move to a stationary state where the production of a by reaction $(7.16)_1$ is balanced by the consumption of a in $(7.16)_2$ with a similar statement for the production and consumption of b (Epstein and Pojman, 1998). This stationary state corresponds to $\frac{da}{dt} = \frac{db}{dt} = 0$ or $f(a,b) = g(a,b) = 0$. We label the stationary state concentrations of a and b, a_{ss} and b_{ss}. To undertake a linear stability analysis we perturb these stationary state concentrations by small amounts, δa and δb:

$$a = a_{ss} + \delta a$$
$$b = b_{ss} + \delta b$$

(7.17)

Substituting (7.17) into (7.16) and expanding the functions f and g in a Taylor series about the stationary state (a_{ss}, b_{ss}) where $f = g = 0$ and assuming that the perturbations are small enough that second and higher order terms may be neglected we arrive at

$$\frac{d\delta a}{dt} = \left(\frac{\partial f}{\partial a}\right)_{ss} \delta a + \left(\frac{\partial f}{\partial b}\right)_{ss} \delta b$$
$$\frac{d\delta b}{dt} = \left(\frac{\partial g}{\partial a}\right)_{ss} \delta a + \left(\frac{\partial g}{\partial b}\right)_{ss} \delta b$$

(7.18)

where the subscript ss means that the relevant quantity is evaluated at the stationary state. Again, as in Section 7.4.1 these equations have solutions of the form (Boyce and DiPrima, 2005):

$$\delta a(t) = A_1 \exp(\lambda t) \quad \text{and} \quad \delta b(t) = A_2 \exp(\lambda t) \tag{7.19}$$

where A_1 and A_2 are constants and λ is an eigenvalue of the characteristic equation given in (7.21) below. We define the Jacobian matrix, $\mathbf{J}$ at the stationary state as

$$\mathbf{J} = \begin{bmatrix} \frac{\partial f}{\partial a} & \frac{\partial f}{\partial b} \\ \frac{\partial g}{\partial a} & \frac{\partial g}{\partial b} \end{bmatrix}_{ss} \tag{7.20}$$

Then the eigenvalues of (7.20) are given by the roots of the characteristic equation:

$$\lambda^2 - Tr\mathbf{J} + det\,\mathbf{J} = 0 \tag{7.21}$$

The stability of the set of coupled equations is now determined by the values that $Tr\mathbf{J}$ and $det\mathbf{J}$ take along with another quantity we will call $\Gamma = (Tr(\mathbf{J}))^2 - 4det\mathbf{J}$. Details of the calculations involved are given by Murray (1989) and Epstein and Pojman (1998) and the results are summarised in Table 7.1 after Hobbs and Ord (2011). Notice the resemblance of these arguments to those in Chapters 2 and 3 and of the phase portraits to those of Figure 3.3 except that

TABLE 7.1 Stability Criteria for Two Component Coupled Chemical Reactions

$Tr\mathbf{J}$	$det\mathbf{J}$	Γ	Behaviour	Phase Portrait
<0	>0	>0	Stable node	
<0	>0	<0	Stable focus	

(Continued)

TABLE 7.1 Stability Criteria for Two Component Coupled Chemical Reactions—cont'd

TrJ	detJ	Γ	Behaviour	Phase Portrait
>0	>0	<0	Unstable focus	
>0	>0	>0	Unstable node	
	<0		Saddle point	
=0	>0		Hopf bifurcation	

After Hobbs and Ord (2011).

A. THE MECHANICS OF DEFORMED ROCKS

the axes of the phase portraits are now chemical concentrations (or activities) rather than velocities. There is, however, one other important difference to Figure 3.3, namely, the presence of a so-called Hopf bifurcation which represents oscillatory (in time) behaviour of the system.

As an example, let us consider one of the simplest expressions for homogeneous coupled reactions that can exhibit instability. The rate equations are linear and given by

$$f = \frac{da}{dt} = \alpha a - \beta b \quad \text{and} \quad g = \frac{db}{dt} = \gamma a - \delta b \tag{7.22}$$

where α, β, γ and δ are (positive) rate constants. We then have $J_{11} = \alpha$; $J_{12} = -\beta$; $J_{21} = \gamma$; $J_{22} = -\delta$ so that $Tr\mathbf{J} = \alpha - \delta$; $det\mathbf{J} = \beta\gamma - \alpha\delta$ and $\Gamma = (\alpha + \delta)^2 - 4\beta\gamma$. We see from Table 7.1 that the following stability states arise if $\beta\gamma > \alpha\delta$; that is, if $det\mathbf{J} > 0$. The system is stable if $\delta > \alpha$, but in the stable mode can exhibit a stable node if $4\beta\gamma < (\alpha + \delta)^2$ or a stable focus if $4\beta\gamma > (\alpha + \delta)^2$. If $\delta < \alpha$ then the system is unstable and exhibits an unstable focus if $4\beta\gamma > (\alpha + \delta)^2$ or an unstable node if $4\beta\gamma < (\alpha + \delta)^2$. On the other hand, if $\beta\gamma < \alpha\delta$ then the system exhibits a saddle point. If $\alpha = \delta$ and $\beta\gamma > \alpha\delta$ a Hopf bifurcation arises. This very simple system is capable of exhibiting all of the simple temporal instabilities that are possible in such coupled systems. It has an additional point of interest in that if diffusion is added to the processes involved then the system is capable of exhibiting spatial instabilities (Turing patterns). An important characteristic of a system with respect to the development of spatial instabilities is that J_{11} must have the opposite sign to J_{22} (Epstein and Pojman, 1998).

In simple systems these kinds of behaviour are relatively easy to explore (Cross and Hohenberg, 1993; Epstein and Pojman, 1998; Murray, 1989) but in highly nonlinear reactions, such as are common in metamorphic rocks, the details of such instabilities may be impossible to establish. By highly nonlinear here we mean that f and g in (7.16) are highly nonlinear functions. An important theorem in this regard is the *Poincare–Bendixson theorem* (Andronov et al., 1966; Epstein and Pojman, 1998; Lynch, 2007; Strogatz, 1994) which states that if a two-component system is confined to a finite region of concentration space then it must ultimately reach either a steady state or oscillate periodically. Thus if one can demonstrate instability then periodic oscillations of the system must exist, although it may prove impossible to define these explicitly. Clearly compositional zoning can form by these oscillatory processes as suggested by Ortoleva (1994) and Wang and Merino (1992) and we explore this topic in Volume II.

Carmichael (1969) presents highly nonlinear coupled reactions where one such reaction is

$$43\text{albite} + 7\text{K}^+ + 18(\text{Mg}, \text{Fe})^{++} + 7\text{H}_2\text{O} \rightleftharpoons 17\text{sillimanite} + 6\text{biotite} + \text{muscovite}$$
$$+ 91\text{quartz} + 43\text{Na}^+$$

In these reactions it will probably prove quite difficult to define the conditions for stability or instability due to the algebraic opacity of the relations involved and for such highly nonlinear systems a number of theoretical and graphical approaches have been developed. An excellent summary of these methods is given in Epstein and Pojman (1998, Chapter 5). In particular, the network methods developed by Clarke (1976, 1980) deserve special consideration. Other developments are discussed by Schreiber and Ross (2003). For open systems a graphical representation presented by Gray and Scott (1994), in the form of *flow diagrams* is particularly powerful.

The behaviour of two-component systems such as given in (7.16) may thus be of the following types:

(1) *Stable behaviour* expressed as either a stable node or a stable focus (Table 7.1). If these systems also involve diffusion of A and B then the system is defined in terms of reaction-*diffusion* equations (such as those given in (7.4)). Then, subject to other conditions considered above, stationary spatial patterns known as Turing instabilities can form spontaneously in an otherwise homogeneous material.
(2) *Unstable behaviour* expressed as an unstable focus, an unstable node, a saddle point or a Hopf bifurcation (Table 7.1). The closed ellipse in the phase portrait of a Hopf bifurcation in Table 7.1 is known as a *limit cycle*.

7.4.3 Reaction-Diffusion Equations

Chemically reacting systems can be open or closed as discussed in Chapter 5; we consider closed systems to begin with and extend the discussion to open systems in Volume II. Most metamorphic systems are postulated to be closed so our example is meant to pertain to such systems. As a specific example we choose two autocatalytic reactions including the Brusselator model which has been widely studied (Epstein and Pojman, 1998). Fisher and Lasaga (1981) present a detailed discussion of the reaction. First we look at a networked series of reactions which result in the net reaction A $\rightarrow$ E with intermediaries Fe^{2+}, Fe^{3+}, B and D and in which no diffusion terms are included. We write this in terms of an autocatalytic redox reaction as

$$
\begin{aligned}
&\textit{Reaction of A to produce } Fe^{2+}: &&A \xrightarrow{k_1} Fe^{2+} \\
&\textit{Non-catalytic step involving B and D}: &&B + Fe^{2+} \xrightarrow{k_2} Fe^{3+} + D \\
&\textit{Autocatalytic step}: &&Fe^{2+} + 2Fe^{3+} \xrightarrow{k_3} 3Fe^{3+} \\
&\textit{Production of E from } Fe^{3+}: &&Fe^{3+} \xrightarrow{k_4} E
\end{aligned}
\tag{7.23}
$$

where the k_i are rate constants. The equations describing the evolution of the system are

$$
\begin{aligned}
\frac{da}{dt} &= -\alpha a \\
\frac{d\left[Fe^{2+}\right]}{dt} &= \alpha a - \left[Fe^{3+}\right]^2\left[Fe^{2+}\right] - \beta\left[Fe^{2+}\right] \\
\frac{d\left[Fe^{3+}\right]}{dt} &= \beta\left[Fe^{2+}\right] + \left[Fe^{3+}\right]^2\left[Fe^{2+}\right] - \left[Fe^{3+}\right] \\
\frac{de}{dt} &= \left[Fe^{3+}\right]
\end{aligned}
\tag{7.24}
$$

where α and β are functions of the rate constants and of b, and we have taken $k_3 = k_4 = 1$. Here the $\left[Fe^{3+}\right]^2\left[Fe^{2+}\right]$ term arises from autocatalytic reactions of the form $(7.23)_3$ and so could be relevant in many networked metamorphic reactions. A solution to (7.24) is shown in

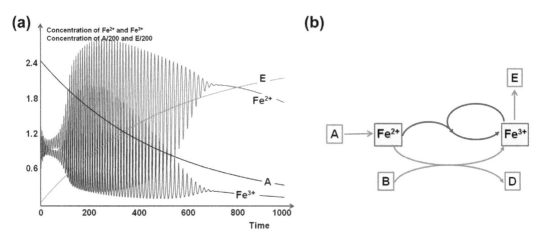

FIGURE 7.6   (a) Autocatalytic reaction (7.24) with no diffusion. $\alpha = 0.002$, $\beta = 0.06$. Solution produced using software in Boyce and DiPrima (2005). (b) Network diagram for reaction (7.23).

Figure 7.6(a). Since this system is closed the concentration of the reactant A proceeds exponentially to zero at equilibrium, whereas the product E increases towards equilibrium. The intermediaries, Fe^{2+} and Fe^{3+} oscillate for a period of time. A schematic diagram of the networked reaction system (7.23) showing the autocatalytic feedback loop is shown in Figure 7.6(b).

A form of the Brusselator reaction (Epstein and Pojman, 1998, Chapter 6) is

$$
\begin{aligned}
\text{\textit{Reaction of} A \textit{to produce} } Fe^{3+}: \qquad & A \xrightarrow{k_1} Fe^{3+} \\
\text{\textit{Non-catalytic step involving} B \textit{and} D: } \qquad & B + Fe^{3+} \xrightarrow{k_2} Fe^{2+} + D \\
\text{\textit{Autocatalytic step}: } \qquad & Fe^{2+} + 2Fe^{3+} \xrightarrow{k_3} 3Fe^{3+} \\
\text{\textit{Production of} E \textit{from} } Fe^{3+}: \qquad & Fe^{3+} \xrightarrow{k_4} E
\end{aligned}
\tag{7.25}
$$

This is a particularly important archetype reaction because, despite its relative simplicity, it displays an enormous range of spatio-temporal patterns. A review is presented by De Wit (1999). We write the rate equations for the Brusselator reaction as

$$
\frac{\partial c_1}{\partial t} = \alpha - (\beta + 1)c_1 + c_1^2 c_2 + D_1 \frac{\partial^2 c_1}{\partial x^2}
$$

$$
\frac{\partial c_2}{\partial t} = \beta c_1 - c_1^2 c_2 + D_2 \frac{\partial^2 c_2}{\partial x^2}
\tag{7.26}
$$

where $c_1 \equiv [Fe^{3+}]$, $c_2 \equiv [Fe^{2+}]$, $\alpha = c_1 \frac{k_1}{k_3}$, $\beta = c_2 \frac{k_2}{k_3}$, $D_1 = D_{c_1}/k_3$; $D_2 = D_{c_2}/k_3$ and D_{c_1} and D_{c_2} are the diffusion coefficients for C_1 and C_2.

The network diagram for the Brusselator is given in Figure 7.7(a) for comparison with reaction (7.23) shown in Figure 7.6(b).

(a)

(b)

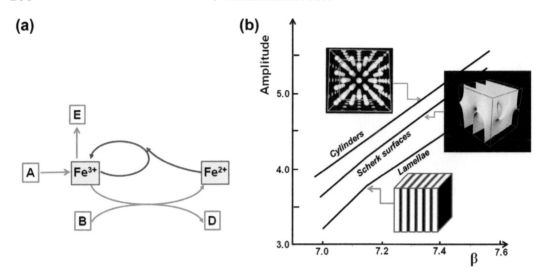

FIGURE 7.7   The Brusselator reaction. (a) Network diagram for reaction (7.25). (b) Bifurcation diagram for the Brusselator reaction in three dimensions. *Adapted from De Wit (1999).*

If one puts $\frac{\partial c_1}{\partial t} = \frac{\partial c_2}{\partial t} = 0$ and $D_1 = D_2 = 0$ one can find the stationary state: $c_1^{ss} = \alpha$ and $c_2^{ss} = \frac{\beta}{\alpha}$. De Wit (1999) shows that a Turing instability occurs for $\beta > \beta^T = \left[1 + \alpha\sqrt{\frac{D_1}{D_2}}\right]^2$ with critical wave number, k_c, given by $k_c^2 = \frac{\alpha}{\sqrt{D_1 D_2}}$. A Hopf bifurcation occurs for $\beta > \beta^H = 1 + \alpha^2$. The threshold of these two instabilities occurs at what is called the co-dimension-two Turing-Hopf point given by $\beta^H = \beta^T$. This occurs for

$$\frac{D_1}{D_2} = \left[\frac{\sqrt{1 + \alpha^2} - 1}{\alpha}\right]^2 \tag{7.27}$$

Near this point a vast range of spatio-temporal instabilities are possible comprising various combinations of Turing-Hopf patterns (De Wit, 1999).

In three dimensions new patterns emerge for the Brusselator. An analysis by De Wit et al. (1997) and De Wit (1999) reveals a number of differently stacked spotlike patterns, sometimes in the form of cylinders, and minimal surfaces including planar lamella structures (Figure 7.7(b)). A bifurcation diagram for the development of these structures is shown in Figure 7.7(b).

In the examples considered above the instability arises because of an autocatalytic chemical reaction. To many this may seem a little exotic, although in Chapter 14 and Volume II we point out that autocatalytic reactions in mineral systems are quite common and the Whitmeyer-Wintsch reaction is representative of many geological examples. By contrast, the case of a single first order exothermic reaction in a closed system would seem to be intrinsically simple. However, it turns out to be quite complicated (Gray and Scott, 1994, Chapter 4) and behaves in much the same manner as the autocatalytic example in Figure 7.6(a). This complexity arises because the reaction rate is temperature dependent and the exothermic nature of the reaction supplies heat that drives the reaction even faster. A balance between heat production and reaction rate ultimately leads to a stationary state that is oscillatory. We consider such thermo-catalytic reactions (Putnis, 2002, 2009) further in Chapter 14. Thus what at first sight would appear to be one of the simplest systems imaginable turns out to have an enormous range of complicated behaviours and becomes even more complicated in open flow systems (Gray and Scott, 1994, Chapter 7; see also Volume II).

The situation becomes more complex if more than two coupled processes operate simultaneously. An example is two independent exothermic reactions in an open flow system considered by Lynch et al. (1982). The system is shown diagrammatically in Figure 7.8 and

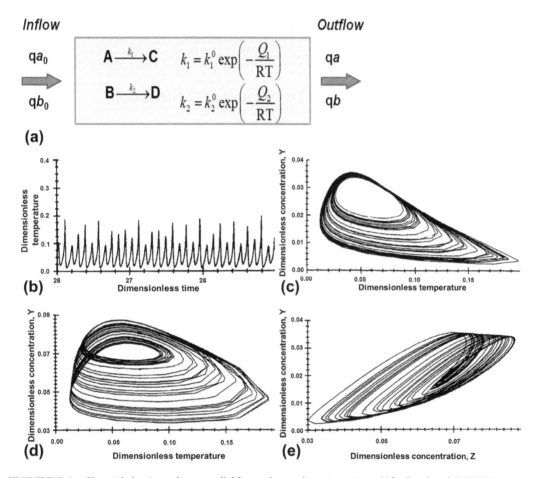

FIGURE 7.8   Chaotic behaviour of two parallel first order exothermic reactions (*After Lynch et al. (1982).*) in an open system. (a) An open system in which the two uncoupled exothermic reactions A → C and B → D proceed with input concentrations a_0 and b_0. k_1 and k_2 are temperature dependent rate constants with activation energies, Q_1 and Q_2. (b) Dimensionless temperature plotted against time. Notice that these fluctuations are relatively large and are about 10–20% of the ambient temperature. (c) Dimensionless concentration Y = a/a_0 plotted against dimensionless temperature. (d) Dimensionless concentration Z = b/b_0 plotted against dimensionless temperature. (e) Dimensionless concentration Y plotted against dimensionless concentration Z.

consists of an open flow reactor fed by a volumetric flow rate, q, with concentrations of re-actants a_0 and b_0. The reactions involved are A → C and B → D for which we define the dimensionless concentrations: $Y = a/a_0$ and $Z = b/b_0$. The dimensionless temperature is defined as $\tilde{T} = (T - T_0)/T_0$ where T is the temperature within the reactor and T_0 is the (constant) temperature of the walls of the reactor. The two reactions are chemically uncoupled but there is thermal feedback between the two. Since there are now more than three processes operating we expect chaotic behaviour and this indeed develops as shown in Figure 7.8. Chemical concentrations oscillate chaotically with time together with the temperature. In principle open systems such as this can be held far from equilibrium indefinitely (Volume II). This is the kind of behaviour that one would expect in any hydrothermal system and in many retrograde metamorphic systems while the exothermic reactions responsible for the development of alteration and retrograde mineral assemblages are in progress (Volume II).

7.5 CLASSIFICATION OF INSTABILITIES

Cross and Greenside (2009) divide nonlinear systems into three types on the basis of the *dispersion relations: the plot of growth rate versus wave number*. Each one of these can be further subdivided depending on whether the system is temporally stable (subscript-s) or oscillatory (subscript-o). Type I systems are shown in Figure 7.9(a and b). The dispersion relations for type I systems are given by

$$\omega_k \approx \frac{1}{\tau_0}\left[\varepsilon - \xi_0^2(k - k_{crit})^2\right] \tag{7.28}$$

where ε is a *bifurcation parameter* that measures how far the perturbed state is from that state where linear instability sets in:

$$\varepsilon = \frac{p - p_{crit}}{p_{crit}} \tag{7.29}$$

In (7.29) p is a parameter that measures the evolution of the system and p_{crit} is a critical value of that parameter which marks the development of a bifurcation. ω_k is the *growth rate* of the wave number, k, ξ_0 is the *coherence length* which measures the spatial scale over which some local imperfection perturbs the pattern and k_{crit} is the wave number that grows at $\varepsilon = 0$. τ_0 is a characteristic time scale for the growth of the instability. In Figure 7.9 we take $\tau_0 = 1$. When $\xi_0^2/\tau_0 < 1$ the plot of ω_k against k is broadened relative to that at $\xi_0^2/\tau_0 = 1$ so that the effect of decreasing the scale of the coherence length (for constant τ_0) is to broaden the range of wavelengths that are likely to grow (Figure 7.9(b)). Type I instabilities are characteristic of the Swift-Hohenberg equation and hence feature in buckling problems. They also feature in reaction-diffusion equations (De Wit, 1999). Type I instabilities can be I_s or I_o.

The behaviour of type II instabilities is given by

$$Re\ \omega_k \approx D\left[\varepsilon k^2 - \frac{1}{2}\xi_0^2 k^4\right] \tag{7.30}$$

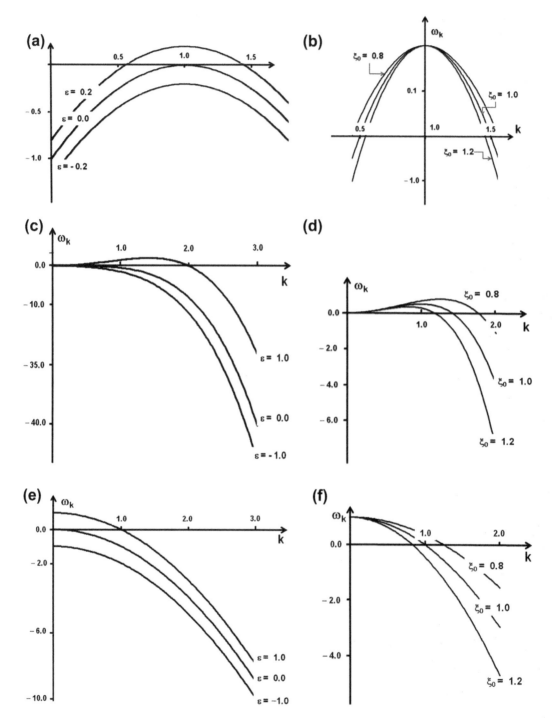

FIGURE 7.9 Classification of instabilities according to Cross and Greenside (2009). (a) Type I instability. Growth rate given by (7.28). (b) Type I instability for various values of the coherence length ξ_0. $\tau_0 = 1$, $\varepsilon = 0.2$ in (7.28). (c) Type II instability. Growth rate given by (7.30). (d) Type II instability for various values of the coherence length ξ_0. $D = 1$, $\varepsilon = 1.0$ in (7.30). (e) Type III instability. Growth rate given by (7.31). (f) Type III instability for various values of the coherence length ξ_0. $\tau_0 = 1$, $\varepsilon = 1.0$ in (7.31).

and is shown in Figure 7.9(c and d). The growth of instabilities is always zero at $k = 0$ and for $\varepsilon = 0$ the growth rate is less than zero for all values of $k > 0$. The maximum growth rate occurs for $\varepsilon > 0$ for $k = \sqrt{\varepsilon}/\xi_0$. Type II instabilities can be II_s and II_o. Figure 7.9(d) shows the broadening of the dispersion curve arising from decreases in the coherence length for $\varepsilon > 0$.

Type III instabilities are given by

$$\mathrm{Re}\,\omega_k \approx \frac{1}{\tau_0}\left[\varepsilon - \xi_0^2 k^2\right] \tag{7.31}$$

and shown in Figure 7.9(e and f). The growth of instabilities is always zero at $k = 0$ for all values of ε. For $\varepsilon > 0$ there is a range of positive growth rates for $0 \leq k \leq \sqrt{\varepsilon}/\xi_0$. Type III instabilities are normally III_o. Figure 7.9(f) again shows the broadening of the dispersion curve arising from decreases in the coherence length for $\varepsilon > 0$.

7.6 BIFURCATIONS

In the previous discussion we have referred to bifurcations in evolving systems and to various forms of bifurcations. In this section we formalise the discussion and present some definitions and examples of such behaviour. For many nonlinear systems the behaviour can be described by the variation of a parameter such as temperature, strain, or chemical concentration. If the overall character of the behaviour changes suddenly for some value of this parameter, the system is said to *bifurcate*. At a point of bifurcation stability may be gained or lost. An example of an unstable bifurcation is the failure by localisation of a specimen once some critical condition is reached (Rudnicki and Rice, 1975); at the bifurcation point the behaviour of the specimen changes from homogeneous deformation to localised (Figure 7.10(a)). For other systems multiple stationary states may exist and the system undergoes successive bifurcations as it evolves from one stationary state to another. Many chemical (particularly open chemical systems) and mechanical systems show such behaviour. We reiterate here that stable means that a small perturbation away from a stationary state results in the system returning to that state. An unstable state represents a situation where a small perturbation grows and the system moves to another stationary state. For classifications of bifurcations, see Thompson (1982), Thompson and Stewart (2002) and Thompson and Sieber (2010). Four common classes of bifurcations that are controlled by the variation of a single parameter are described below; detailed discussions of these examples may be found in Lynch (2007) and in Wiggins (2003).

In order to succinctly describe bifurcation behaviour it is useful to define a number of terms, some of which we have already used.

Bifurcation: A bifurcation is a point in the evolution of a system where a qualitative (including discontinuous) change in system behaviour occurs (Figure 7.10(a)). A bifurcation normally occurs when a parameter that describes the evolution of the system passes through some critical value. Some systems may be controlled by more than one parameter. An example is the buckling of a layer controlled by first the value of the axial load and second by the value of the reaction forces normal to the layer (Thompson, 1982). Stability may be lost or gained at a bifurcation point. In the localisation problem illustrated in

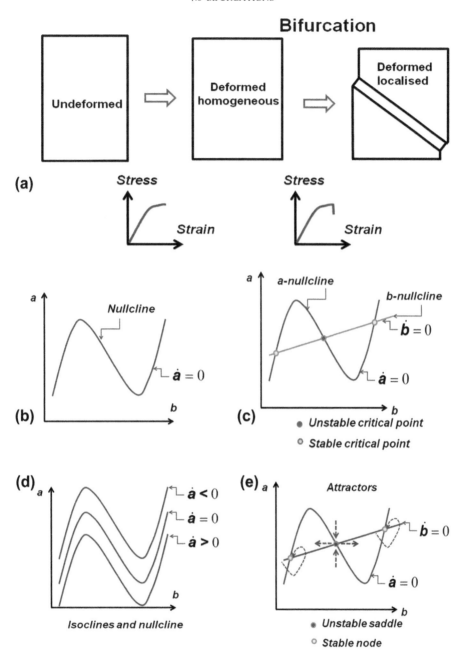

FIGURE 7.10 Some concepts involved in the dynamics of nonlinear systems. (a) A bifurcation. (b) The a-nullcline which is the locus of all stationary states for the evolution of the concentration of A. (c) Critical points in the A–B system. (d) Isoclines and the nullcline for the concentration of A. (e) Attractors in the A–B system.

Figure 7.10(a) bifurcation may occur when the slope of the stress–strain curve reaches a critical value (see Rudnicki and Rice, 1975).

Stationary state, steady state: A stationary state is one where the rate of change of one or more quantities is zero (Figure 7.10(b and c)). Equilibrium is one stationary state for a system (see Chapter 5) but there may also be one or more non-equilibrium stationary states. The terms *steady state* and *stationary state* are commonly used synonymously. However, a system at a stationary state can oscillate about that state and so we prefer the term *stationary* rather than *steady*. The *mean value* of a quantity at a stationary state remains constant with time.

Critical point: A critical point for a system refers to a point in a phase portrait which corresponds to a stationary state for the system. This means that if the system evolution is described by $\dot{x} = f(x, y)$ and $\dot{y} = g(x, y)$ then the critical points correspond to $\dot{x} = \dot{y} = 0$. A critical point can be stable or unstable so that it may act as an attractor or a repeller for the system. If we write this evolution equation in the form

$$\dot{x}_1 = f(x_1, x_2), \quad \dot{x}_2 = g(x_1, x_2)$$

and define the Jacobian matrix, J, as

$$J = \begin{bmatrix} \dfrac{\partial f}{\partial x_1} & \dfrac{\partial f}{\partial x_2} \\ \dfrac{\partial g}{\partial x_1} & \dfrac{\partial g}{\partial x_2} \end{bmatrix}$$

then the critical point is said to be *hyperbolic* if the real parts of the eigenvalues of J are non-zero. If the real parts of the eigenvalues are equal to zero the critical point is *non-hyperbolic*. Critical points are also known as *fixed points* or *stationary points*. A plot of the behaviour of critical points as a control parameter is varied is known as a *bifurcation diagram*. Examples are given in Figures 7.12, 7.15, and 7.16.

Isocline, nullcline: The line on a phase portrait that marks a constant rate of change of a quantity is an *isocline*. If the rate of change is zero this is a *nullcline* and marks the locus of all non-equilibrium stationary states (Figures 7.10(b,c and d)).

Manifold: A manifold in the context of a two dimensional phase portrait is a line parallel to (or tangent to) an eigenvector of the flow and that passes through a critical point. Manifolds need not be straight lines. If the eigenvalue at the critical point is positive the *manifold is stable*; if the eigenvalue is negative the *manifold is unstable*. In three dimensions the manifold can be a convoluted surface (see Lynch, 2007, Chapter 7). The *fabric attractors* referred to by Passchier (1997) are manifolds.

Attractor: A stable stationary state of the system (Figure 7.10(e)). The attractor can be a critical point, a limit cycle, a torus or a strange attractor. The term *strange* means that the attractor has a fractal geometry and its geometry is sensitive to initial conditions.

7.6.1 Saddle-Node Bifurcation

An example of a saddle-node bifurcation is the vector field given by

$$\dot{x}_1 = \mu - x_1^2 \quad \dot{x}_2 = -x_2$$

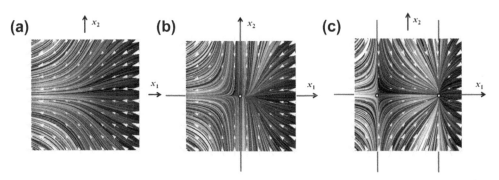

FIGURE 7.11   A saddle-node bifurcation. (a) $\mu < 0$. (b) $\mu = 0$. (c) $\mu > 0$. The traces of manifolds are marked in red. Critical points are white dots. There are no manifolds or critical points for $\mu < 0$. Just as μ passes through zero the phase portrait shows a bifurcation into a saddle and a node.

This vector field changes character depending on the value of μ as shown in Figure 7.11. The critical points are defined by $\dot{x}_1 = \dot{x}_2 = 0$. When $\mu < 0$ there are no critical points, the flow is continuous from right to left and flow on the x_1-axis is invariant. When $\mu = 0$ there is one non-hyperbolic critical point at the origin. The vector field is given in Figure 7.11(b). The flow is invariant along both the x_1- and x_2-axes. When $\mu > 0$ there are two critical points at $(\sqrt{\mu}, 0)$, which is stable, and $(-\sqrt{\mu}, 0)$ which is unstable. The manifolds (where they exist) are also shown in Figure 7.11. Thus the qualitative behaviour of the system changes as μ passes through zero as shown in Figure 7.12(a). For $\mu < 0$ there are no critical points. At $\mu = 0$ one critical point appears and as μ increases past zero two critical points appear (bifurcate) and move further apart as μ increases. One critical point is a saddle and the other a node. This is a particularly common form of bifurcation in both mechanical and chemical systems. Such bifurcations can be stacked in phase space as shown in Figure 7.22(a) when they form part of a *snakes and ladders system* (see Section 7.7).

7.6.2 Transcritical Bifurcation

An example here is the velocity field given by

$$\dot{x}_1 = \mu x_1 - x_1^2, \quad \dot{x}_2 = -x_2$$

The bifurcation diagram is shown in Figure 7.12(b) with the bifurcation point given by $\mu = 0$. There is always at least one critical point for this system. For $\mu < 0$ there are two critical points at $(0, 0)$, a stable node, and at $(\mu, 0)$, a saddle point. There are two manifolds as shown in Figure 7.13(a). For $\mu = 0$ there is one critical point at $(0, 0)$ and two manifolds as shown in Figure 7.13(b). For $\mu > 0$ there are again two critical points at $(0, 0)$, a saddle point, and at $(\mu, 0)$, a stable node. The manifolds are shown in Figure 7.13(c).

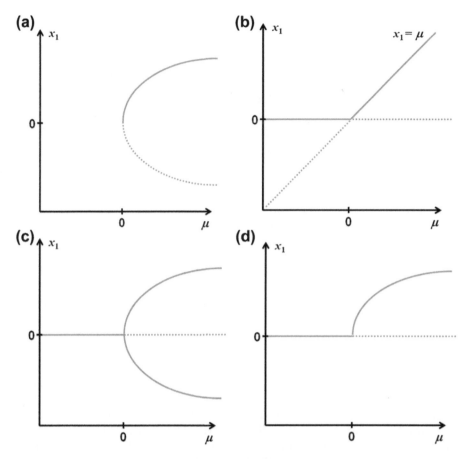

FIGURE 7.12   Bifurcation diagrams for the four examples given in the text. (a) Saddle-node bifurcation. (b) Transcritical bifurcation. (c) Pitchfork bifurcation. (d) Supercritical Hopf bifurcation. These diagrams are obtained by taking $\dot{x}_1 = 0$ in the relevant equation describing the evolution of the system. Full lines are stable states, dotted lines are unstable states.

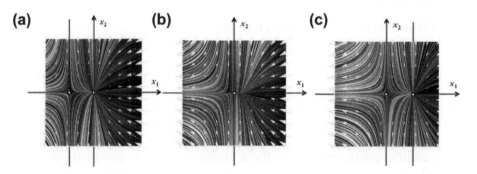

FIGURE 7.13   A transcritical bifurcation. (a) $\mu < 0$. (b) $\mu = 0$. (c) $\mu > 0$. Critical points are white dots. Traces of the manifolds are shown in red.

7.6.3 Pitchfork Bifurcation

An example here is the velocity field given by

$$\dot{x}_1 = \mu x_1 - x_1^3, \quad \dot{x}_2 = -x_2$$

The bifurcation diagram is shown in Figure 7.12(c) where again the bifurcation occurs at $\mu = 0$. Now there are either one or three critical points as shown in Figure 7.14. When $\mu < 0$ there is one critical point at $(0, 0)$ which is a stable node. There are two manifolds as shown in Figure 7.14(a). When $\mu = 0$ there is one non-hyperbolic critical point at $(0, 0)$ as shown in Figure 7.14(b). For $\mu > 0$ there are three critical points at $(0, 0)$, a saddle point, at $(\sqrt{\mu}, 0)$, a stable node, and at $(-\sqrt{\mu}, 0)$ there is another stable node. These critical points and the manifolds are shown in Figure 7.14(c).

7.6.4 Hopf Bifurcation

A Hopf bifurcation is different in character to the previous three bifurcations and represents a situation where a system that is steady with time suddenly begins to oscillate as a parameter is varied. The behaviour of a spring-slider system with velocity weakening friction (Gu et al., 1984) is an example. The system slides in a stable manner for values of the spring constant larger than a critical value but the system undergoes stick-slip behaviour once the spring constant drops below this critical value. Another example is given by

$$\dot{r} = r(\mu - r^2), \quad \dot{\theta} = -1$$

The bifurcation diagram is shown in Figure 7.12(d). Bifurcation occurs at $\mu = 0$. For $\mu \leq 0$ a stable focus occurs at $(0, 0)$ and there is no limit cycle. For $\mu > 0$ a limit cycle develops and grows in diameter, at $r = \sqrt{\mu}$, as μ increases.

There are two types of Hopf bifurcation. The one shown in Figures 7.12(d) and 7.15(a and b) is a *supercritical Hopf bifurcation*. In addition there exist *subcritical Hopf bifurcations* shown in Figure 7.15(c and d) and in Table 7.1.

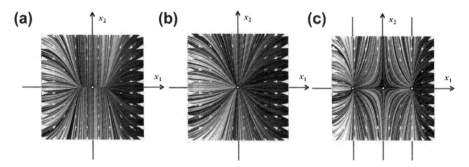

FIGURE 7.14   A pitchfork bifurcation. (a) $\mu < 0$. (b) $\mu = 0$. (c) $\mu > 0$. Critical points are white dots. Traces of the manifolds are shown in red.

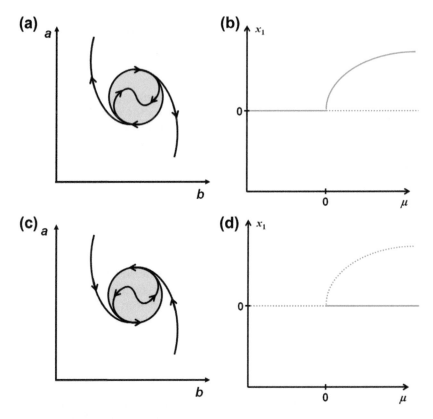

FIGURE 7.15 Supercritical (a, b) and subcritical (c, d) Hopf bifurcations. (a, c) Phase portraits for chemical systems with concentrations *a* and *b*. The circle is a *limit cycle*. (b, d) Bifurcation diagrams. Full lines are stable; dotted lines are unstable.

7.6.5 Bistability and Multistability

A system is said to be *multistable* if for a given value of a parameter, μ, there exists more than one stationary state. If there are two such states the system is *bistable*. Such systems are common in nonlinear mechanical and chemical systems (especially open flow chemical systems) as we will see particularly when we consider hydrothermal systems in Volume II. An example has already been given in Figure 1.9 where for a given value of the load there is more than one value of the deflection. An instructive example is the system described by Lynch (2007):

$$\dot{r} = r\left(\mu - 0.28r^6 + r^4 - r^2\right), \quad \dot{\theta} = -1 \tag{7.32}$$

The bifurcation diagram is shown in Figure 7.16 and consists of a supercritical Hopf bifurcation at O followed by a saddle-node bifurcation at B as the parameter μ is increased. As μ is decreased from some high value the system undergoes a saddle-node bifurcation at A followed by a Hopf bifurcation back to a stable state at O.

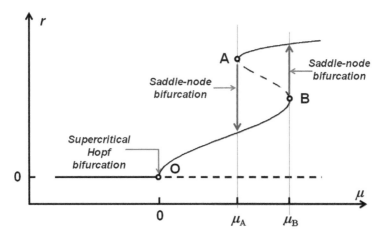

FIGURE 7.16 Bifurcation diagram for the system described by (7.32). Dotted lines are unstable. The system is bistable and typical of many nonlinear chemical systems. As μ is increased there is a supercritical Hopf bifurcation at O with $\mu = 0$ and a saddle node bifurcation at B with $\mu = \mu_B$. As μ is decreased from μ_B there is another saddle node bifurcation at $\mu = \mu_A$. See also Figure 1.9 for another example of this kind of behaviour.

7.7 ENERGY MINIMISATION AND THE GROWTH OF FRACTAL STRUCTURES

In Chapter 5, following the classical work of Rice (1976) we pointed out that the stress–strain curve for a material may be derived as the derivative of the Helmholtz energy with respect to strain. In this section we explore the various forms that the Helmholtz energy may take and how these forms are expressed as the stress–strain relations. In particular the Helmholtz energy can be non-convex and minimisation of the Helmholtz energy then results in the development of (at least) two new 'phases' (Ball and James, 1987; Bhattacharya, 2003).

The various forms of the relationship between the stored energy and the deformation gradient are shown in Figure 7.17. It is common in mechanics (Houlsby and Puzrin, 2006a) to assume that Ψ is a *convex* function of the deformation gradient, **F**. The relationship between Ψ and the strain is discussed by Pipkin (1993). The mathematical treatment of this subject rapidly becomes complicated and depends heavily upon the theory of convex and non-convex functionals (a functional is a quantity that is a function of a function). We do not go down the route of mathematical rigour; the interested reader is referred to papers by Ball (Ball, 1977, 2004; Ball and James, 1987) and the book by Silhavy (1997). Ψ as a *convex* function of **F** is shown in Figure 7.17(a). By convex here we mean the type of relation illustrated in Figure 7.17(a) where the stored energy increases with the deformation gradient so as to always be convex towards the deformation gradient axis. If we identify Ψ with the specific Helmholtz energy defined in the usual way for a deforming material with a density, ρ (Houlsby and Puzrin, 2006a), by

$$\Psi = e - Ts = \Psi\left(F_{ij}, T\right) \tag{7.33}$$

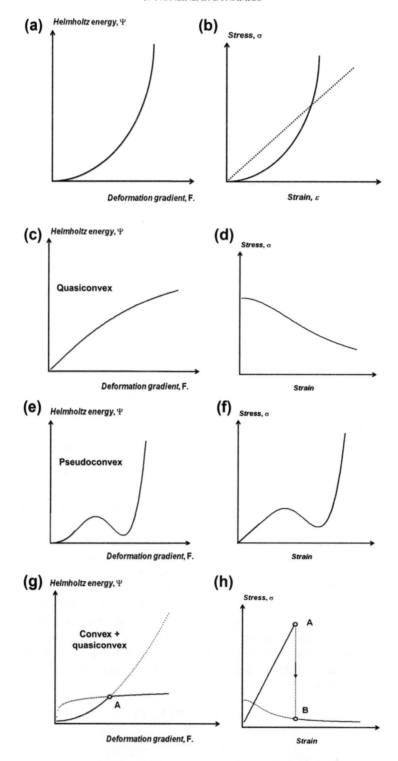

where e is the specific internal energy, T is the absolute temperature and s is the specific entropy then the Cauchy stress, σ_{ij}, is related to the Helmholtz energy by $\sigma_{ij} = \rho \frac{\partial \Psi}{\partial F_{ij}}$ (see Chapter 5) or by $\sigma_{ij} = \rho \frac{\partial \Psi}{\partial \varepsilon_{ij}}$ in the small strain, small rotation case (Houlsby and Puzrin, 2006a). Hence, the stress–strain curve can be obtained from the Ψ–F relation by differentiation. Thus the stress–strain curve corresponding to Figure 7.17(a) is given in Figure 7.17(b). If, for instance, the Ψ–F relation is a quadratic as it commonly is in classical elasticity then the stress–strain curve is a straight line (dotted in Figure 7.17(b)). If the Ψ–F relation is of higher order than quadratic then the stress–strain curve shows increasing hardening with strain (full curve in Figure 7.17(b)). A convex form of the Ψ–F relation is favoured in classical continuum mechanics because it guarantees stability of the deformation so that the deformation remains homogeneous. However, we know that deformation in most materials is never stable and heterogeneities in the deformation are ubiquitous. These heterogeneities are expressed as subgrains defined by dislocation and disclination walls, deformation bands, shear zones, folds and various forms of foliations and lineations. Thus, 'interesting' deformation behaviour does not arise from convex Ψ–F relations, which lead to homogeneous deformations, and our emphasis switches to various non-convex Ψ–F relations.

In order for unstable behaviour to develop, leading to microstructure formation, the Ψ–F relation needs to be non-convex and the two important classes of non-convexity are *quasi-convexity* (Figure 7.17(c)) and *pseudo-convexity* (Figure 7.17(e)). For rigorous definitions of these terms see Mangasarian (1994) and Silhavy (1997). The stress–strain curves that result from these two different Ψ–F relations are shown in Figure 7.17(d and f), respectively. These figures mean that for instabilities and hence microstructure to develop in deforming solids the response of the material to deformation must be nonlinear in the ways shown in Figures 7.17(d and f). Many experimentally deformed geological materials show stress–strain curves of this nature (for example, Heilbronner and Tullis, 2006 and Chapter 6). For some materials, especially those that exhibit fracture, the Ψ–F relation is better represented by a diagram such as Figure 7.17(g) with the resultant stress–strain curve shown in Figure 7.17(h) (Del Piero and Truskinovsky, 2001).

In fluids and gases the convex relation shown in Figure 7.17(a) has its analogue in the plot of Gibbs energy against the specific volume for a single phase and the stress–strain plot of Figure 7.17(b) has its analogue in the pressure-specific volume plot for a single phase (Kondepudi and Prigogine, 1998, p. 190). These kinds of relationships have been well known since the work of Gibbs (1906) and are the basis for Equilibrium

FIGURE 7.17   Various relations between the Helmholtz energy function, Ψ, and the deformation gradient, F. Also shown are the stress–strain curves corresponding to each energy function. (a) Convex Ψ–F relation. (b) Stress–strain curves corresponding to (a). If the Ψ–F relation in (a) is a quadratic then the stress–strain curve is linear as shown by the dotted line. If the Ψ–F relation is of higher order than quadratic the stress–strain curve shows increasing hardening with strain as shown by the full curve. (c) Quasi-convex Ψ–F relation. (d) Stress–strain curve resulting from the quasi-convex Ψ–F relation in (c). (e) Pseudo-convex Ψ–F relation. (f) Stress–strain curve resulting from the pseudo-convex Ψ–F relation in (e). (g) Combined convex–quasi-convex Ψ–F relation with a discontinuity. (h) Stress–strain curve resulting from the Ψ–F relation in (g). This resembles stress–strain curves that arise from brittle behaviour.

Chemical Thermodynamics where the minimum in the Gibbs energy corresponds to the specific volume of the stable phase. The pseudo-convexity of Figure 7.17(e) has a direct analogue in Equilibrium Chemical Thermodynamics for two phase materials where the Gibbs energy becomes non-convex (Kondepudi and Prigogine, 1998. p. 193) and a tangent construction (Cahn and Larche, 1984) gives the specific volumes of the two stable coexisting phases; the pressure-specific volume plot shows a region where the pressure increases with the specific volume corresponding to a spinodal region. In such a plot the specific volumes of the two coexisting stable phases are obtained from the *Maxwell construction* by equating the chemical potentials of the two phases (Kondepudi and Prigogine, 1998, p. 190).

In exactly the same way the *Maxwell construction* can be used for deforming systems (Hunt et al., 2000). The tangent construction shown in Figure 7.18(a) shows the way in which the Helmholtz energy of the system can be minimised for an imposed deformation

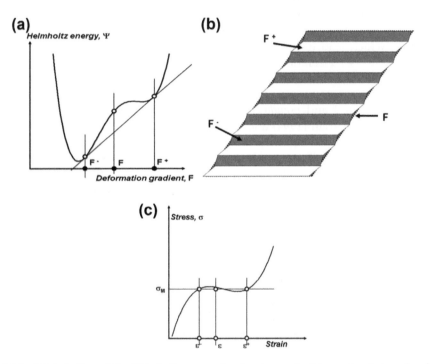

FIGURE 7.18   Minimisation of the Helmholtz energy by development of two sets of shear bands. (a) The Helmholtz energy is pseudo-convex and an imposed deformation gradient is represented by F. The energy can be minimised by dividing the deformation into two sets of shears, F^- and F^+ which correspond to points where the tangent line touches the Helmholtz energy function. (b) The resulting microstructure. The imposed homogeneous deformation, F, is represented by the dotted outline; the structures that minimise the energy are F^- (grey) and F^+ (white). (c) The stress–strain curve corresponding to the Helmholtz energy in (a). σ_M is the Maxwell stress and corresponds to that stress where the areas between the stress–strain curve and the horizontal line through σ_M above and below the stress–strain curve are equal. ε corresponds to the imposed strain and ε^+, ε^- correspond to the strains in the two domains.

gradient **F**. Two stable deformations **F**$^+$ and **F**$^-$ exist and the system divides into domains corresponding to these two deformation gradients which are defined by the two points where the common tangent touches the Ψ–**F** curve. The resultant microstructure is shown in Figure 7.18(b). The stress–strain curve corresponding to Figure 7.18(a) is shown in Figure 7.18(c) and σ$_M$ is the *Maxwell stress* which represents the normal component of the Eshelby energy–momentum tensor (Eshelby, 1975; Silhavy, 1997) such that the area between the horizontal line through σ$_M$ and the stress–strain curve above is equal to the area between the line and the curve below. The significance of σ$_M$ is that it represents the stress where the stored elastic energy is sufficient to supply the energy to drive the formation of the microstructure. The stress σ$_M$ plays the same role (Silhavy, 1997) as the chemical potential in chemical systems (Kondepudi and Prigogine, 1998, p. 190). Similar arguments have been followed by Hunt and co-workers for the development of kink and chevron folds (Hunt et al., 2000).

However, the development of these two deformation gradients cannot fully match the imposed deformation and gaps always remain as shown in Figure 7.19 with the implication that long-range stresses exist so that the stored energy is not fully minimised. One way of overcoming this situation is to produce the two deformations on a finer and finer scale as shown in Figure 7.19. Gaps still remain and the next stage in minimising the stored energy is to produce fine-scale structure within the gaps (Ball and James, 1987, their Figure 6). An example is shown in Figure 7.20 where self-similar refinement of the broad-scale kinking comprising simply **F**$^+$ and **F**$^-$ is illustrated. This refinement process is referred to as *sequential lamination* by various authors including Kohn (1991) and Ortiz and Repetto (1999). Examples of such self-similar refinement are shown for a propagating fracture in Figure 7.21(a) and for twinning in microcline matching an interface with albite in Figure 7.21(b).

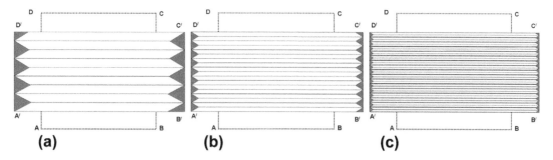

(a)   **(b)**   **(c)**

FIGURE 7.19   The square ABCD is deformed to become the rectangle A′B′C′D′. The series a, b, c shows progressive refinement of the microstructure in order to minimise the energy of the system. We discuss this in greater detail in Chapter 8. The deformation is approximated by alternating sheared zones with equal thicknesses but opposite senses of shear. As the thickness of the sheared zones is decreased the inhomogeneous deformation field comes closer to approximating the imposed homogeneous deformation. The error in matching (denoted by the grey areas) becomes smaller with the decrease in thickness and in this case is proportional to the thickness of the individual sheared zones.

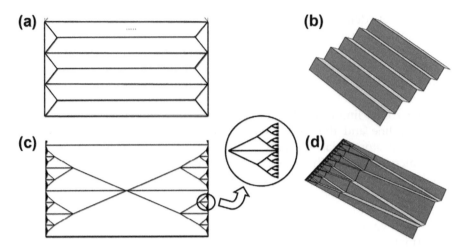

FIGURE 7.20   Self-similar refinement of a microstructure to attempt to match an imposed deformation. (a) Coarse microstructure with gaps represented in three dimensions in (b). (c) Self-similar refinement of the microstructure to fit the imposed deformation. (d) Three dimensional arrangement of the self-similar microstructure. *Adapted from Ortiz (2008). Presentation at the International Congress of Theoretical and Applied Mechanics, Adelaide, Australia.*

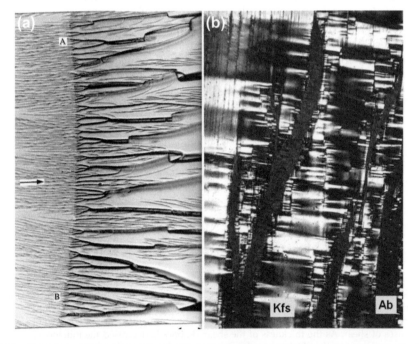

FIGURE 7.21   Fractal structures arising from self-similar branching. (a) Self-similar steps in a fracture surface resulting from mixed mode I/III loading. The arrow indicates the direction of propagation of the fracture tip, A—B. *(From Hull (1999).)* Photo is about 0.8 mm across. (b) Self-similar refinement of twin structure in K-feldspar as an interface with albite is approached. *(Photo from Ron Vernon.)* Photo is about 1.3 mm in height.

7.7.1 What Causes the Stored Energy to be Non-Convex?

We discuss many aspects of non-convex energy functions in this book. As with the origins of nonlinear behaviour, non-convexity can originate for geometrical or constitutive reasons. Two processes associated with crystal plasticity, for instance, have been identified as producing non-convex stored energy functions (Ortiz and Repetto, 1999). These processes are geometrical softening arising from single slip and constitutive latent hardening. Non-convex behaviour, as can be seen from Figures 7.17 and 7.18 arises from some form of softening behaviour in the stress—strain curve. However, a softening in the sense of a decrease in the load bearing capacity is not the only cause of non-convexity. A decrease, with increasing deformation or rate of deformation, in the magnitude of material properties such as elastic modulus or viscosity is just as effective. In addition, in deforming metamorphic rocks, softening in both load-bearing capacity and in material properties arising from mineral reactions is likely to be important.

7.7.2 Snakes and Ladders Behaviour

In many systems the bifurcations are arranged as an intertwined cascade of saddle-node bifurcations (Figure 7.22(a)). The system evolves by 'snaking' its way from one bifurcation to another along intertwined paths symmetrically disposed about a Maxwell stress labelled r_M. This behaviour is characteristic of nonlinear buckling systems (Burke and Knobloch, 2007; Hunt and Hammond, 2012; Hunt et al., 2000a; Knobloch, 2008; Peletier and Troy, 2001) where the buckles form sequentially, with one or more extra buckles being added at each bifurcation. Part of such sequential buckling evolution is shown in Figure 7.22(a and b) where the frames shown in Figure 7.22(b) in the sequence (a) → (b) → (c) correspond to sequential buckles added on one bifurcation path and the sequence (d) → (e) → (f) to the other path. There are other ways in which sequential buckles can be added and these correspond to unstable (and asymmetric) points on 'cross-rungs' between the two bifurcation paths in Figure 7.22(a). For details see Burke and Knobloch (2007). Here the intertwined bifurcation paths are called 'snakes' and the 'cross-rungs' are 'ladders'.

Another example of snakes and ladders behaviour is shown in Figure 7.22(c,d, and e). Here the system is that described in Figure 1.10 where three layers of close packed balls are shortened. Plots of axial load against displacement are shown with two intertwined snakes. The deformation consists of sequential thickening of a kink band by a mechanism that adds one or more balls to the band in a sequence of bifurcations. There are three mechanisms for kink band thickening: (1) The system follows the bifurcation path $2_{min} \rightarrow 2_{max} \rightarrow 4_{min} \rightarrow 4_{max} \rightarrow \ldots$ with two balls being added to the kink band just after each load maximum so that the number of balls in a kink band remains an even number. (2) The system follows the bifurcation path $3_{min} \rightarrow 3_{max} \rightarrow 5_{min} \rightarrow 5_{max} \rightarrow \ldots$ again with two balls being added to the kink band just after each maximum so that the number of balls in a kink band remains an odd number. (3) The system on one or the other of these two paths can jump to the other down a 'ladder' to a lower energy configuration by the addition of one ball so that an even sequence of kink widening is converted to an odd sequence.

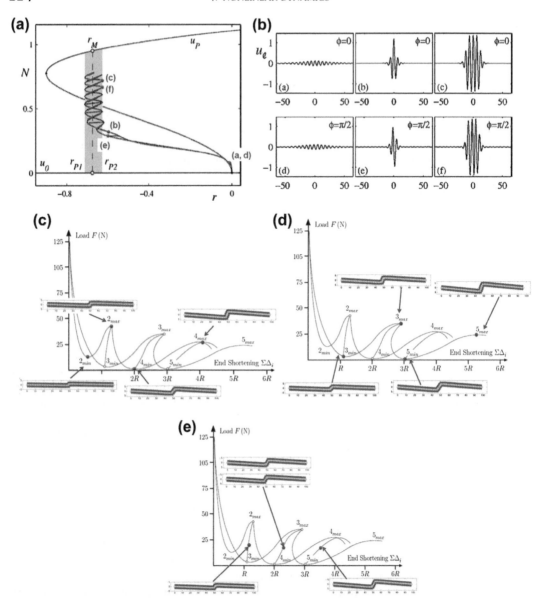

FIGURE 7.22 Snakes and ladders behaviour. (a) Plot showing snakes and ladders behaviour. r is a control parameter and for a buckling layer can be thought of as the load or stress in the layer. N can be thought of as the end displacement of the layer. Two bifurcation paths are shown. The layer begins to buckle at (a) and (d) and follows a snaking series of sequential buckles once the load drops to a Maxwell stress, r_M. The sequence of buckles is shown in frame (b). Other buckles are added on the cross-rungs connecting the two snakes. (*After Burke and Knobloch (2007).*) (b) Sequential buckles developed along intertwining 'snakes'. (*After Burke and Knobloch (2007).*) (c), (d) and (e) Snakes and ladders behaviour for the widening of a kink band. (c) The $2 \to 4$ ball transitions in kink band widening. (d) The $3 \to 5$ ball transitions in kink band widening. (e) The single ball transitions in kink band widening on cross-rungs. (*Modified from Hunt and Hammond (2012).*) The end shortening is measured in units of the ball radius, R.

7.8 FRACTALS AND MULTIFRACTALS

Within a metamorphic system the deformation and metamorphism of an element of the lithosphere is controlled by the velocity boundary conditions imposed by the local plate tectonic regime together with temperature changes arising from the impingement of heat sources and/or advection of the element through the geothermal gradient of the Earth. Such advection also contributes to changes in the pressure. Thus at the coarse scale a deformation gradient drives momentum transfer while changes in temperature and pressure induce changes in chemical potentials that drive mineral reactions. In this process, forcing at lithospheric scales drives a cascade of dissipation at finer and finer scales within the lithosphere until at a very fine scale the dissipation is negligible. The development of lithospheric scale buckles and fault systems is part of a dynamical cascading process whereby structure is developed at finer and finer scales. At the coarse scale dissipation is dominated by deformation coupled with heat diffusion; at finer scales heat diffusion ceases to be dominant as a dissipative mechanism and the dissipative processes are dominated by local deformations, metamorphic mineral reactions and microstructural rearrangements such as recrystallisation, crystallographic preferred orientation development and microfracturing. Such a cascading process is similar to that proposed for fluid turbulence where large eddies break down to shed eddies at smaller and smaller scales (Drazin and Reid, 1981; Frisch, 1995; Kestener and Arneodo, 2003; Richardson, 1922). These cascading processes lead to multifractal geometries and so it is of importance to develop efficient ways of measuring and characterising the scaling properties of structures in deformed rocks since there is the potential that such properties reflect the details of the cascading process. We put forward the wavelet transform method as a means of achieving this characterisation.

In Section 7.7 we saw examples of how fractal structures develop in a system as a means of minimising the energy of the system. In general such processes involve an iterative procedure of some kind whereby the system continuously branches in a self-similar manner to produce replicas on a finer and finer scale. Examples are the fine structure at the edge of an advancing crack (Figure 7.21(a)) or the twin structure at the boundaries of K-feldspar phases (Figure 7.21(b)). Such iterative processes are a mechanism of producing self-similar structures at a finer and finer scale and the *Iterative Function System* (IFS) of Barnsley (1988) is one way of expressing such a process. Other ways of producing fractal structures involve escape functions and random walk processes (Feder, 1988; Schroeder, 1991). We will see an example of an escape function that leads to fractal buckling structures in Volume II. The development of percolation networks during flow in porous media is an example of a random walk process (Arneodo et al., 1995).

Formally a *fractal is an object that has structure at all length scales and some measure of this structure is invariant with respect to an affine transformation* (Mandelbrot, 1982). This means that the object has structure within structure and no matter what scale one views the object one sees much the same kind of structure. From a mathematical point of view the way in which the structure scales with length is a power law:

$$\mu \sim \varepsilon^D \tag{7.34}$$

where μ is some measure of the structure, ε is a length scale and D is commonly known as the *fractal dimension*. It should be appreciated that not all fractals are self-similar. *Self-similar fractals* remain the same under the influence of any transformation that is a *dilation* (a contraction or expansion of equal magnitude in all directions). Fractals also exist that are self-affine; these are objects that remain the same under other *affine transformations*. Thus a surface such as a fault plane appears smooth at a large length scale but is rough at a fine length scale. These surfaces are *self-affine fractals*. Such scale invariance cannot continue indefinitely and so fractal geometries are practically restricted in metamorphic systems to three to perhaps five orders of magnitude in length scales (corresponding say to scaling from 1 mm to 100 m).

Fractals are *singular functions* from a mathematical point of view since they cannot be differentiated and so D is also known as a *singularity measure*. The interesting feature of (7.34) is that D is commonly a non-integer. The reason that self similar geometries follow a power law of the form (7.34) is that it is the only form of relation where changing the size of ε leaves the form of the equation unaltered. Thus if we double the size of ε we get

$$\mu \sim (2\varepsilon)^D = 2^D \varepsilon^D$$

which is of the same form as (7.34).

The method for determining D in (7.34) consists of placing a box (or ball) of size ε on the space occupied by the object and observing if part of the object lies in the box. This is repeated until all the space occupied by the object is covered and the proportion of boxes that cover the object are established. The size of the box is then altered and the process repeated. After a large number of iterations of this process one plots the logarithm of the proportion of boxes for a given ε that contained parts of the object against the logarithm of ε. If the object is a (mono) fractal then this plot is a straight line with slope $-D$ (Feder, 1988; Schroeder, 1991). That is the relation is of the form

$$N(\varepsilon) = \varepsilon^{-D} \tag{7.35}$$

where $N(\varepsilon)$ is the number of boxes of size ε that contain part of the object. This process is called *box counting* and descriptions of the process with applications to structural geology are given in Kruhl (1994). Other examples include applications to sutured grain boundaries associated with recrystallisation (Kruhl and Nega, 1996), sub-grain size (Hahner et al., 1998; Streitenberger et al., 1995) and grain shape (Mamtani, 2010). The fractal characteristics of these microstructures have been used to indicate temperature and strain rate (Kruhl and Nega, 1996; Mamtani, 2010; Mamtani and Greiling, 2010; Takahashi, 1998). Applications to mineral systems are given by Arias et al. (2011), Bastrakov et al. (2007), Ford and Blenkinsop (2009), Hunt et al. (2007), Li et al. (2009), Oreskes and Einaudi (1990), Qingfei et al., (2008), Riedi (1998) and Sanderson et al. (2008). Fractal (or at least, power law) distributions of mineralisation are reported by Carlson (1991) and Schodde and Hronsky (2006). D is commonly known as the *box-counting dimension*.

Other dimensions of use in describing fractal geometries are as follows.

The information dimension which is related to the sum of the probabilities, p_k, of finding a part of the object in the kth box:

$$D^{information} = \lim_{r \to 0} \frac{-\sum_k p_k \log p_k}{\log r} \tag{7.36}$$

and *the correlation dimension* which is a measure of the number of pairs of points whose distance apart is less than r:

$$D^{correlation} = \lim_{r \to 0} \frac{-\sum_k \log C(r)}{\log r} \quad \text{where} \quad C(r) = \frac{1}{N^2} \sum_{i \neq j} H\left[r - |x_i - x_j|\right] \tag{7.37}$$

where H is the Heaviside function: $H(x) = 0$ for $x \leq 0$ and $H(x) = 1$ for $x > 0$. We will see that both these dimensions have special meanings within a multifractal spectrum.

The geometry of metamorphic fabrics is, however, more complicated than being characterised by a single value of D as we will see below. If the geometry of an object is characterised by a single value of D then it is called a *monofractal*. Some geometries are characterised by two values of D and these are called *bifractals*. However, if the geometry is characterised by different values of D at different points then the geometry is clearly more complicated and is called a *multifractal* (Mandelbrot, 1974); a spectrum of D values arises. A multifractal therefore consists of a set of interwoven fractals and is characterised by a spectrum of singularity measures. An example consists of a precipitation process, for say gold, that is itself fractal but is controlled spatially by a fractal distribution of the pore or fracture space within which precipitation occurs. The spatial distribution of gold is then multifractal (Volume II). For reviews of the development of the multifractal concept see Feder (1988), Bohr and Tel (1988), Arneodo et al. (1995), Schroeder (1991).

An example of a complex fractal geometry is given by Ord (1994). A computer simulation of shear band development in a non-associative plastic material was analysed using the 'delay' methods developed by Packard et al. (1980), Takens (1981) and Crutchfield et al. (1986) for temporal data sets. The velocity of growth of shear bands at a particular value of the shear strain was analysed by recording values of the x-component (Figure 7.23(a)) of the velocity, $(v_x)_{ij}$ at a spatial point i, j (after the mean velocity for the deformation is removed) at fixed spatial delays, S, where the value of S ranges from 1 to some larger number (Figure 7.23(a and b)). Thus a vector, $(v_x)_{ij}^{(p)}$, of p dimensions is formed by the array

$$(v_x)_{ij}^{(p)} = \left[(v_x)_{i,j}, (v_x)_{i+S,j}, \ldots\ldots\ldots, (v_x)_{i+(p-1)S,j} \right] \tag{7.38}$$

where the comma in the subscript is there for clarity, not to denote differentiation. Other vectors are then constructed for different values of S. The final array of vectors defines an attractor in p-space. The attractor in 3-space for this model is shown in Figure 7.23(c). Calculations were performed to establish the *embedding dimension* for this attractor; the embedding dimension is that dimension for which the calculated fractal dimension no longer increases. It was established that the embedding dimension for this attractor is between 4 and 5. Figure 7.23(c) represents a *strange attractor* with a fractal dimension of 2.3. The wavelet transform of this data set is shown in Figure 7.23(d and e) and is analysed further in Figure 7.28.

Although box-counting procedures can be used to explore multifractal geometries, the process is quite cumbersome and can produce erroneous results especially if the singularity measure varies significantly within box sizes greater than a given value (see Arneodo et al., 1995; Arneodo et al., 1987). We require a fast, compact and quantitative characterisation of seemingly complex data sets that is readily applicable to one-, two- and three-dimensional situations and so we turn to a *wavelet* based system. The wavelet approach has many advantages over box-counting procedures, although a wavelet is basically a 'generalised box'. Methods of multifractal analysis based on the wavelet transform are particularly applicable to self-similar, intermittent data sets where the wavelet acts as a 'microscope' that can zoom into the details of the signal and define local structure and singularities. Wavelet-based software now exists that makes fractal and multifractal analysis fast and efficient so that complex data sets can be completely analysed within minutes using a laptop computer. In addition the wavelet approach is reasonably well established within a thermodynamic framework

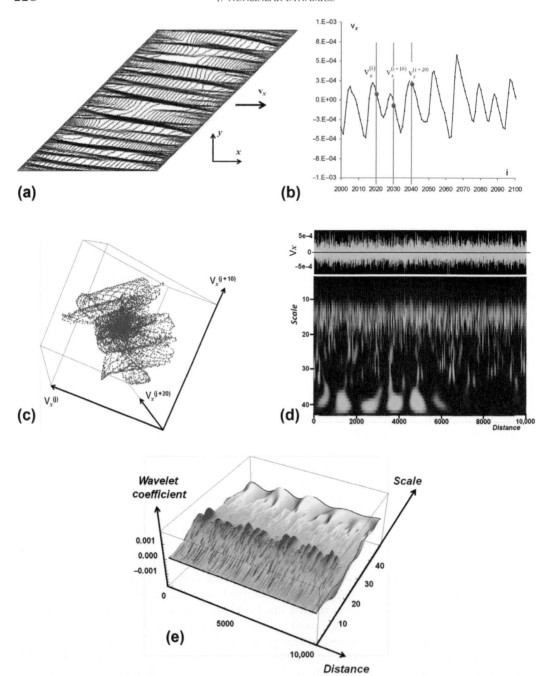

FIGURE 7.23   Strange attractor for shear zone development. (a) Finite difference model of shear zone development after a shear through 45°. The material is initially homogeneous. Contours of x-velocity shown. (b) Part of the spatial distribution of the x-velocity signal. The x-velocity at position 2020 is compared with that at position 2030 and 2040. These three values are plotted to generate a point in the attractor (c). (d) x-velocity signal at top, wavelet transform scalogram at bottom. (e) Three dimensional plot of scalogram.

(Arneodo et al., 1995; Bohr and Tel, 1988) so that the procedures and results can be placed within a broader mechanics framework.

First let us explore the multifractal concept in greater detail. As we have indicated a multifractal consists of groups of fractal objects interwoven in space. The resulting geometry may or may not appear self-similar depending on the complexity of the inter-relations between the various fractals. It is common (Feder, 1988) to represent the range of singularity measures for multifractals by the expression

$$N_\alpha(\varepsilon) = \varepsilon^{-f(\alpha)} \tag{7.39}$$

which is a generalisation of (7.35) so that D in (7.35) is replaced by a spectrum, $f(\alpha)$, of singularity measures. α is commonly known as the *Holder exponent* and is also known as the Lipshitz exponent. The function $f(\alpha)$ appears as a hump-shaped curve represented by a plot of $f(\alpha)$ against α as shown in Figure 7.24. As complex geometries continue to be explored other representations and methods of analysis of multifractals appear (Arneodo et al., 1995; Venugopal et al., 2006) with extensions to two and three dimensions (Arneodo et al., 2003). The future development of these procedures represents an enormous opportunity to explore and quantify the geometry of fabrics in metamorphic rocks and so add to our knowledge of complexity, and the associated mechanisms, in metamorphic fabrics and processes.

In general the spectrum of singularity measures needs to be established by direct measurement but it is important in the interpretation of these spectra to be able to make comparisons with spectra generated by specific models. Some examples are shown in Table 7.2 adapted largely from Halsey et al. (1986) and from Arneodo et al. (1995). In the table, D_0 is the value of $[f(\alpha)]^{max}$ and D_∞, $D_{-\infty}$ are the values of α for $f(\alpha) = 0$. The last row in the table is adapted

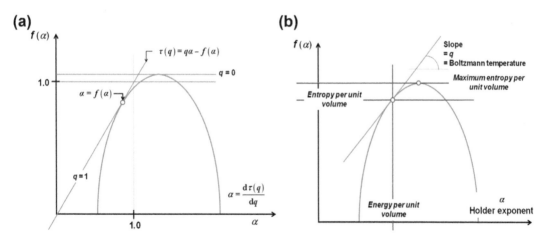

FIGURE 7.24 Some features of the singularity spectrum. (a) τ as the Legendre transform of $f(\alpha)$: $\tau(q) = \frac{df(\alpha)}{d\alpha}\alpha - f(\alpha)$, $q = \frac{df(\alpha)}{d\alpha} = 0$ corresponds to the tangent at $[f(\alpha)]^{max}$. (b) The thermodynamic interpretation of the singularity spectrum. $f(\alpha)$ corresponds to the entropy per unit volume while α corresponds to the energy per unit volume. $[f(\alpha)]^{max}$ corresponds to the maximum entropy per unit volume while the slope of the tangent to the singularity spectrum corresponds to the Boltzmann temperature $\mathcal{B}$ (see Arneodo et al., 1995; Bohr & Tel, 1988).

TABLE 7.2 Fractal Distributions Resulting from Various Recursive Models

Recursive Model	Result	Singularity Spectrum
Power law probability distribution $p(x) = \tilde{\alpha}x^{\tilde{\alpha}-1}$	Bifractal Fractal states at (0,0) and (1,1)	
Uniform Cantor set $p_1 = p_2 = p_3 = \frac{1}{2}$ $l_1 = l_2 = l_3 = \frac{1}{3}$	Monofractal Fractal state at $\left[\frac{\ln 2}{\ln 3}, \frac{\ln 2}{\ln 3}\right]$	
Two-scale Cantor set $p_1 = \frac{3}{5};\quad p_2 = \frac{2}{5}$ $l_1 = \frac{1}{4};\quad l_2 = \frac{2}{5}$	Multifractal	

(Continued)

TABLE 7.2   Fractal Distributions Resulting from Various Recursive Models—cont'd

Recursive Model	Result	Singularity Spectrum
Generalised Cantor set-I $p_1 + 2p_2 = 1$ $l_1 + 2l_2 = 1$ $p_2/l_2 > p_1/l_1$ $l_2 > l_1$	Incomplete multifractal	
Generalised Cantor set-II $p_1 = p_3 \neq p_2$ $l_1 = l_3 \neq l_2$	Bifractal $\tilde{\alpha} = \frac{\ln p_2}{\ln l_2}; \quad \tilde{f} = \frac{\ln\left(\frac{1}{2}\right)}{\ln l_2}$	
Period doubling cascade $x' = \lambda(1 - 2x^2)$	Multifractal	

(*Continued*)

A. THE MECHANICS OF DEFORMED ROCKS

TABLE 7.2   Fractal Distributions Resulting from Various Recursive Models—cont'd

Recursive Model	Result	Singularity Spectrum
Mode locking cascade	Multifractal	
Quasi-periodic circle map	Multifractal	
Multifractal devil's staircase with superimposed sine wave	Multifractal with phase change: non-singular to singular. (See Arneodo et al., 1995)	

TABLE 7.3 Significance of Various Values of α

Holder Exponent, α	Singularity Type	Example
$\alpha > 1$	Continuous and differentiable	Smooth curve, $y = \sin(x)$
$\alpha = 1$	Continuous, differentiable almost everywhere	Brownian motion
$0 < \alpha < 1$	Continuous, non-differentiable	Heaviside function
$-1 < \alpha \leq 1$	Discontinuous, non-differentiable	Gaussian noise
$-1/2 < \alpha < 0$	Data are positively correlated	$\alpha = -1/2$ implies white noise
$-1 < \alpha < -1/2$	Data are negatively correlated	
$\alpha \leq -1$	Non-locally integrable	Dirac pulse

From Holdsworth et al. (2012) see also Mallat (1991).

from Arneodo et al. (1995). In addition, one should be aware of the significance of various values of α as indicated in Table 7.3.

In exploring multifractal geometries it is convenient to define the generalised fractal dimension, D_q, which is related to the scaling exponent for the qth moment of the measure μ. If we have a set of measures, say the concentration of a chemical component such as Na distributed over a fabric, we define the qth moment (or partition function) $\mathbb{Z}(q, \varepsilon)$ for the box size ε as (Lynch, 2007):

$$\mathbb{Z}(q, \varepsilon) = \sum_{i=1}^{N(\varepsilon)} \mu_i^q(\varepsilon) \tag{7.40}$$

Then we can also define $\tau(q)$ as

$$\tau(q) = \lim_{\varepsilon \to 0} \frac{\ln \mathbb{Z}(q, \varepsilon)}{-\ln \varepsilon} \tag{7.41}$$

The generalised fractal dimensions D_q are given by

$$\tau(q) = D_q(1 - q) \tag{7.42}$$

In the limit $\varepsilon \to 0^+$, $\mathbb{Z}(q, \varepsilon)$ behaves as a power law:

$$\mathbb{Z}(q, \varepsilon) \sim \varepsilon^{-\tau(q)} \tag{7.43}$$

The relations between the $f(\alpha)$ singularity spectrum, q, and the $\tau(q) = (1 - q)D_q$ spectrum are given (Arneodo et al., 1995) by

$$q = \frac{df(\alpha)}{d\alpha}$$
$$\frac{d^2 f(\alpha)}{d\alpha^2} < 0 \tag{7.44}$$
$$\tau(q) = q\alpha - f(\alpha)$$

These relations are illustrated in Figure 7.24(a).

Thus there are strong analogies with thermodynamic functions and there is a basis in statistical mechanics for interpreting these functions from a thermodynamic point of view (Arneodo et al., 1995; Bohr and Tel, 1988). The function $f(\alpha)$ is identified with the entropy per unit volume, α_i, and with the energy, E_i, per unit volume of a microstate, i, q with the Boltzmann temperature, $\mathbb{B} = \frac{1}{KT}$, and the $\tau(q)$ spectrum is the Legendre transform (see Chapter 5) of the $f(\alpha)$ spectrum. $\mathbb{Z}$ can be written

$$\mathbb{Z}(\mathbb{B}) = \sum_i \exp\left(-\mathbb{B}E_i\right) \tag{7.45}$$

Some of these interpretations are illustrated in Figure 7.24(b). This thermodynamic interpretation of fractals and particularly of multifractals is explored by Bohr and Tel (1988) and by Arneodo et al. (1995, 2003) who point out that such thermodynamic properties of multifractals place limitations on the characterisation of multifractals by the box-counting procedure and why a resort to wavelet analysis is a better approach. Notice that for $q = 0, 1, 2, D_q$ corresponds to well-known functions. Thus D_0 is the *box dimension* (7.35), D_1 is the *information dimension* (7.36) and D_2 is the *correlation dimension* (7.37).

7.8.1 Wavelet Analysis. What is a Wavelet?

We introduced the concept of *wavelets* in Chapter 1. A *wavelet* can be thought of as a *generalised box*. In fact some wavelets (for instance, the Haar wavelet, Figure 7.25(a)) have a boxlike shape. For the most part a wavelet has a localised wave-like shape and is designed to emphasise particular aspects of the signal to be analysed. Libraries of wavelets can be found in *Mathematica*® and some examples are shown in Figure 7.25.

The wavelet transform (WT) of a signal, $\hat{s}$, consists of decomposing $\hat{s}$ into space-scale contributions that are defined by the *analysing wavelet*, ψ, which is chosen to be localised in space and commonly of zero mean, although $g^{(0)}$ (see (7.46)) is sometimes used where $g^{(0)}$ is the Gaussian function (Arneodo et al., 1995). A class of commonly used wavelets is defined by the successive derivatives of the Gaussian function:

$$\psi(x) = g^{(N)}(x) = (-1)^{N+1} \frac{d^N}{dx^N}\left[\exp\left(-x^2/2\right)\right] \tag{7.46}$$

$g^{(2)}$ and $g^{(4)}$ are shown in Figure 7.25(b and c). Note that various conventions are adopted in the literature with respect to the form of (7.46) and in some instances the $(-1)^{N+1}$ term is omitted. The WT of the function $\hat{s}$ is defined as the convolution of $\overline{\psi}$ with $\hat{s}$:

$$W_\psi[\hat{s}](b,a) = \frac{1}{a} \int\limits_{-\infty}^{+\infty} \overline{\psi}\left(\frac{x-b}{a}\right)\hat{s}(x)dx \tag{7.47}$$

where b is the space parameter, $a > 0$ is the scale parameter and $\overline{\psi}$ is the complex conjugate of ψ. The quantity $W_\psi[\hat{s}](b,a)$ is known as the *wavelet coefficient* at the scale a and around the point $x = b$. The procedure involved in a wavelet analysis is to select a *mother wavelet*, ψ, and contract or extend ψ by successive scales a. For each scale the wavelet is scanned

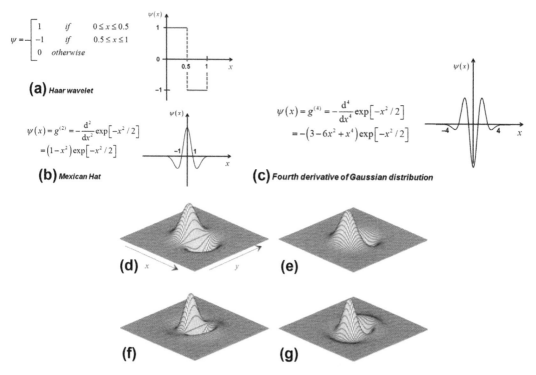

$$\psi = \begin{bmatrix} 1 & if & 0 \le x \le 0.5 \\ -1 & if & 0.5 \le x \le 1 \\ 0 & otherwise \end{bmatrix}$$

(a) Haar wavelet

$$\psi(x) = g^{(2)} = -\frac{d^2}{dx^2}\exp\left[-x^2/2\right]$$
$$= \left(1 - x^2\right)\exp\left[-x^2/2\right]$$

(b) Mexican Hat

$$\psi(x) = g^{(4)} = -\frac{d^4}{dx^4}\exp\left[-x^2/2\right]$$
$$= -\left(3 - 6x^2 + x^4\right)\exp\left[-x^2/2\right]$$

(c) Fourth derivative of Gaussian distribution

(d) x y

(e)

(f)

(g)

FIGURE 7.25 Examples of one- and two-dimensional wavelets. (a) to (c) One-dimensional wavelets. (a) The Haar wavelet. (b) The Mexican hat wavelet; this is $g^{(2)}$. (c) The fourth derivative of a Gaussian distribution: $g^{(4)}$. (d) to (g) Two-dimensional wavelets after Arneodo et al. (2003). (d), (e) First-order derivatives with respect to x and y for a Gaussian function. (f), (g) First-order derivatives with respect to x and y for a Mexican hat function.

across the image with the same procedure as for box counting so that $W_\psi[\hat{s}](b,a)$ is evaluated at each point, b, and for each scale a. The local behaviour of $\hat{s}$ is reflected in the WT which behaves as

$$W_\psi\left[\hat{s}\right](x_0, a) \sim a^{\alpha(x_0)} \tag{7.48}$$

where x_0 is a selected point and α is the *Holder exponent*. Some examples of WTs are given in Figures 1.11 and 7.23(d and e). The WT contains all the information needed to establish the fractal geometry of an object (Arneodo et al., 1995).

7.8.2 The WTMM Method

The wavelet transform modulus maxima (WTMM) method (Mallat and Hwang, 1992) consists of evaluating the *partition function*:

$$\mathbb{Z}(q,a) = \int \left|W_\psi\left[\hat{s}\right](x,a)\right|^q dx \tag{7.49}$$

at each point on the signal $\hat{s}$. This is straightforward for $q \geq 0$ but fails for $q < 0$ whenever $W_\psi[\hat{s}](x_0, a) = 0$ since $\mathbb{Z}$ becomes infinite. For this reason, $\mathbb{Z}(q, a)$ is evaluated for each value of a by noting the maximum in $W_\psi[\hat{s}](x_0, a)$ for all scales a' where $a' \leq a$. This means that (7.49) is replaced by:

$$\mathbb{Z}(q, a) = \sum_l \left(\sup_{(x, a')} |W_\psi[\hat{s}](x, a')|^q \right)$$

(7.50)

This procedure allows one to establish the exponents $\tau(q)$ as

$$\mathbb{Z}(q, a) \sim a^{\tau(q)}$$

(7.51)

Then by taking the Legendre transform of $\tau(q)$ we obtain

$$f(\alpha) = \min_q(q\alpha - \tau(q))$$

(7.52)

This is best done (Arneodo et al., 1995) by first calculating

$$\widehat{W}_\psi[\hat{s}](q, l, a) = \frac{\left| \sup_{x, a'} W_\psi[\hat{s}](x, a') \right|^q}{\mathbb{Z}(q, a)}$$

(7.53)

Then $\alpha(q, a)$ follows as

$$\alpha(q, a) = \sum_l \ln \left| \sup_{x, a'} W_\psi[\hat{s}](x, a') \right| \widehat{W}_\psi[\hat{s}](q, l, a)$$

(7.54)

and

$$f(q, a) = \sum_l \widehat{W}_\psi[\hat{s}](q, l, a) \ln \left[\widehat{W}_\psi[\hat{s}](q, l, a) \right]$$

(7.55)

These calculations can be performed in a relatively painless and efficient manner using software such as *LastWave* (Arneodo, A., Audit, B., Kestener, P., and Roux, S. (2008). Wavelet-based multifractal analysis. *Scholarpedia*, *3*, 4103. http://dx.doi.org/10.4249/scholarpedia.4103. http://www.cmap.polytechnique.fr/~bacry/). Alternative software is *WaveLab* (Buckheti et al. (December, 2005). About WaveLab. Version .850. http://www-stat.stanford.edu/~wavelab/). The method has been extended to two dimensions by Roux et al. (2000), Decoster et al. (2000) and Arneodo et al. (2003) and to three dimensions by Kestener and Arneodo (2003, 2004, 2008) and Arneodo et al. (2003). Some comparisons between different methods for analysing singularity spectra and results are given by Turiel et al. (2006).

We give an example below of the WTMM method applied to a non-uniform Cantor set (Figure 7.26(a)). The example is taken from a demonstration included in the *LastWave* software. We also show the uniform triadic Cantor set in Figure 7.26(b). This is formed by uniformly removing the central one-third of the signal and then repeating the procedure as shown in the figure. The WTMM method is applied to this fractal by Arneodo et al. (1995), Section 3.3.2, Figure 3 and the result is that for a monofractal with a fractal dimension of $(\ln 2 / \ln 3)$.

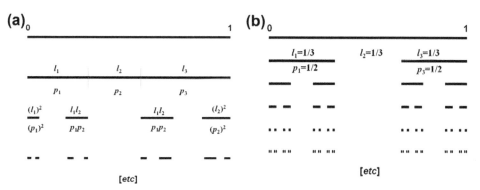

FIGURE 7.26   Cantor sets. (a) A generalised Cantor set divided initially into lengths, l_1, l_2 and l_3 with an initial probability of selection of p_1, p_2 and p_3. (b) A uniform Cantor set where $l_1 = l_2 = l_3$ and $p_1 = p_2 = p_3$.

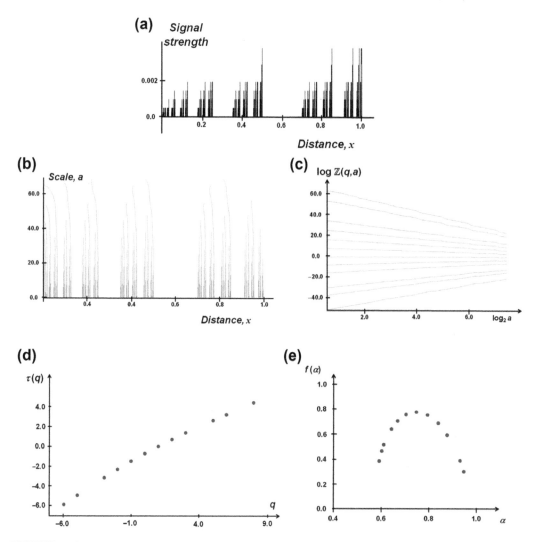

FIGURE 7.27   WTMM analysis of nonuniform Cantor set. (a) The signal to be analysed. (b) Map of the maxima lines from the wavelet scalogram. (c) The $\mathbb{Z}(q,a)$ spectrum. (d) The $\tau(q)$ spectrum. (e) The $f(\alpha)$ spectrum.

The production of monofractals such as this requires an ordered and coordinated process whereby each level of subdivision follows a precisely defined law. It is amazing that some DNA sequences are monofractals (Arneodo et al., 1995). For the most part in natural examples such order is not followed and multifractals develop. An example is the generalised non-uniform Cantor set shown in Figure 7.27(a). The maxima lines from the WT of the signal are shown in Figure 7.27(b). Plots of $\mathbb{Z}(q, a)$ against $\log_2 a$ are shown in Figure 7.27(c) from which the $\tau(q)$ spectrum may be derived in Figure 7.27(d). The observation that the $\tau(q)$ versus q curve is nonlinear is the hallmark of a mutifractal distribution; for the uniform Cantor set such a curve is linear. Finally the $f(\alpha)$ spectrum is given in Figure 7.27(e) showing a well-defined multifractal spectrum with a support given by $f(\alpha) \approx 0.8$ at $\alpha \approx 0.75$.

A second example is taken from the model presented in Figure 7.23 where the WT is shown (using $g^{(2)}$). In Figure 7.28, the maxima plot, the plot of $q(a)$ against $\log_2 a$, the $\tau(q)$ spectrum and the singularity spectrum are shown. As with many data sets the $q(a)$ plots are irregular for negative values of q and become irregular for high positive values of q. The

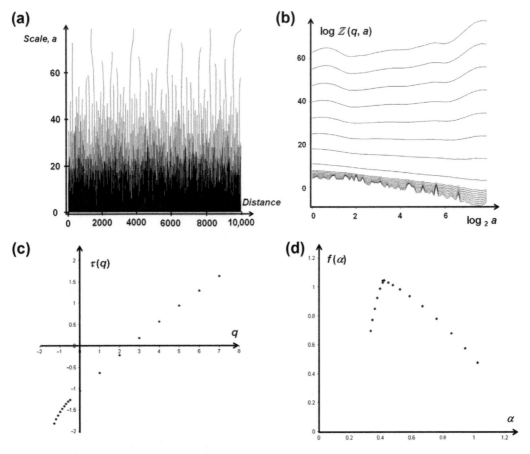

FIGURE 7.28   WTMM analysis of the structure shown in Figure 7.23. (a) Map of maximum lines on the scalogram. (b) Plot of log $\mathbb{Z}$ against $\log_2 a$. (b) Plot of τ against q. (d) Multifractal spectrum.

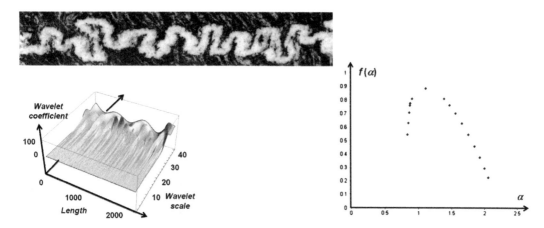

FIGURE 7.29   Wavelet transform of the ptygmatic fold structure illustrated (see Figure 7.3(c)). The wavelet transform is shown on the left and the multifractal spectrum on the right.

singularity spectrum is quite well defined; it is skewed to low values of α and has a support value of $f(\alpha)$ close to unity corresponding to $\alpha \approx 0.4$. The overall shape of this distribution (asymmetrical and skewed to lower values of α) resembles many other geological examples we present in this book. This example demonstrates the power of the WTMM method as a rapid means of quantifying the geometry of apparently structure-less data sets and deciding whether the geometry is multifractal or not. In particular in this case the details of the *multi-fractal* nature of the data set are established.

As another example, in Figure 7.29 we present a WT of one of the structures shown in Figure 7.3, namely, one of the ptygmatic folds in Figure 7.3(c). The data set is taken from a horizontal line through the centre of the image in Figure 7.29(c). The mother wavelet is $g^{(4)}$.

This completes Section A of this book. The intent has been to present an introduction to most of the tools required to examine the structures in metamorphic rocks in greater detail. We will use much of this material in the remainder of the book and in Volume II where we expand on some of the more useful concepts with specific examples. The hope is that we have supplied a vocabulary and a context that enable more advanced treatments of mechanics to be accessed by the reader.

Recommended Additional Reading

Arneodo, A., Bacry, E., Muzy, J.F. (1995). *The thermodynamics of fractals revisited with wavelets. Physica A* 213, 232–275.
   An important paper that summarises the thermodynamic properties of multifractals and the link to wavelets. It presents the wavelet transform maximum modulus (WTMM) method of deriving multifractal spectra from wavelet transforms.

Bak, P. (1996). *How Nature Works.* Copernicus, Springer-Verlag.
   This book presents the concept of self-organised criticality in a nonmathematical manner.

Boyce, W.E., DiPrima, R.C. (2005). *Elementary Differential Equations and Boundary Value Problems*, eighth ed. Wiley.
  An introduction to the elements of differential equations with many examples of systems relevant to metamorphic geology.

Cross, M., Greenside, H. (2009). *Pattern Formation Dynamics in Nonequilibrium Systems*. Cambridge University Press.
  This book presents the theory behind pattern formation with details of pattern selection, growth and stability.

Cross, M.C., Hohenberg, P.C. (1993). *Pattern formation outside of equilibrium*. Reviews of Modern Physics 65, 851–1112.
  This very long paper reviews developments in solutions to the Swift-Hohenberg and reaction-diffusion equations and gives a large number of examples.

Epstein, I.R., Pojman, J.A. (1998). *An Introduction to Nonlinear Chemical Dynamics*. Oxford University Press.
  One of the definitive text books devoted to nonlinear chemical systems. It has clearly written chapters on the basic principles, linear stability analyses and the spatiotemporal behaviour of chemical systems.

Lynch, S. (2007). *Dynamical Systems with Applications Using Mathematica®*. Birkhauser.
  A clear presentation of nonlinear dynamics and wavelets with many examples using Mathematica®.

Murray, J.D. (1989). *Mathematical Biology*. Springer-Verlag, Berlin.
  An important book on nonlinear systems and nonlinear chemical systems in particular. The basic theory is clearly written with many examples.

Ortoleva, P. (1994). *Geochemical Self-Organisation*. Oxford University Press.
  One of the few explorations of nonlinear chemical systems applied to geology. The text is difficult to read but worth pursuing.

Peletier, L.A., Troy, W.C. (2001). *Spatial Patterns: Higher Order Models in Physics and Mechanics*. In: *Progress in Nonlinear Differential Equations and Their Applications*, vol. 45. Birkhauser, USA, 320 pp.
  This book is devoted solely to solutions of the Swift-Hohenberg equation.

Thompson, J.M.T. (1982). *Instabilities and Catastrophes in Science and Engineering*. Wiley.
  An excellent introduction to nonlinear dynamics with many informative diagrams.

Thompson, J.M.T., Stewart, H.B. (2002). *Nonlinear Dynamics and Chaos*. John Wiley and Sons.
  A more advanced but geometrical treatment of nonlinear dynamics with many applications. Also included is an excellent treatment of the classification of bifurcations and an illustrated glossary.

Walgraef, D. (1996). *Spatio-Temporal Pattern Formation*. Springer-Verlag.
  An in-depth analysis of pattern formation.

Wiggins, S. (2003). *Introduction to Applied Nonlinear Dynamical Systems and Chaos*, second ed. Springer-Verlag.
  An advanced mathematical text on nonlinear systems, bifurcation behaviour and chaotic systems.

Processes Involved in the Development of Geological Structures: Overview of Section B

In Section A of this volume we assembled the various conceptual and mathematical tools necessary in order to understand and explore the mechanics of metamorphic systems. We defined a metamorphic system as a large body of rock of lithospheric scale, whose behaviour is influenced by changes in the velocity field and perhaps also in the temperature field at its boundaries resulting from the impingement of rising molten material or underplated material from deeper in the earth, or some form of delamination of the base of the lithosphere. These changes in boundary conditions induce new behaviour in the lithosphere driven by new gradients in deformation, temperature, hydraulic potential and chemical potentials. The processes that now operate are flows in momentum, heat, fluid and mass and the observation is that these processes occur in a cascade of spatial scales ranging from the lithosphere to the nanometre. Section B of this volume is concerned with the processess that underlie the development of structures and metamorphic assemblages across this vast range of length scales.

Chapters 8, 9 and 10 consider momentum transfer which, depending on the physical environment, results in brittle or visco-plastic behaviour and is expressed as damage and other microstructural arrangements. Of particular importance here is the concept of critical behaviour and a class of this behaviour is commonly referred to as self-organised criticality or SOC. We also emphasise in these chapters the importance of defects as the dominant mode of deformation. Thus we consider the development of fracture systems not from the classical fracture mechanics viewpoint but the fractures themselves as deformation mechanisms which are defects in an elastic continuum. Classical dislocations are considered in Chapter 9 but the emphasis is on disclinations and disconnections. In particular we explore the concept of coupled grain boundary migration, which exerts an overwhelming control on the development of metamorphic fabrics. In particular we propose coupled grain boundary migration as a mechanism for the development of curved inclusion trails in porphyroblasts.

Chapter 11 discusses conductive heat transfer driven by temperature gradients in the lithosphere with an emphasis on the time scales involved in metamorphic processes. Chapter 12 is concerned with fluid flow during metamorphism driven by gradients in hydraulic head and inevitably, the coupling with heat transfer by some advective process. The interest lies in the influence of heterogeneities such as fractures, faults and shear zones on the rates and patterns of fluid and heat flow, particularly in three dimensions, and the controls on convection within the crust of the Earth.

In Chapter 13 we examine many of the issues to do with microstructure development and its influence on the development of both elastic and plastic anisotropy. Elastic anisotropy is a property that is usually thought of as important for seismic wave propagation. However it has equal importance in controlling the conditions for localisation of deformation and for controlling the orientations of shear bands.

Chapters 14 and 15 are concerned with mineral reactions in metamorphic systems driven by gradients in chemical potential but not from the conventional view of the mineral assemblages at equilibrium states. We take such a view as established background material and consider the processes that operate during the route to equilibrium, the influences these processes have on the mechanical behaviour of deforming rocks and on the resulting metamorphic fabrics.

CHAPTER

8

Brittle Flow

8.1 INTRODUCTION: WHAT ARE THE ISSUES AND PROBLEMS? WHAT DO WE MEAN BY BRITTLE AND DUCTILE?

What do we mean by brittle? We use the term *brittle* to mean those *processes* whereby rock masses undergo macroscopic deformation by the breaking of atomic bonds within individual crystal structures comprising the minerals in the rock or at grain boundaries (Figure 8.1). The

Structural Geology
http://dx.doi.org/10.1016/B978-0-12-407820-8.00008-4

243

(a)   **(b)**   **(c)**

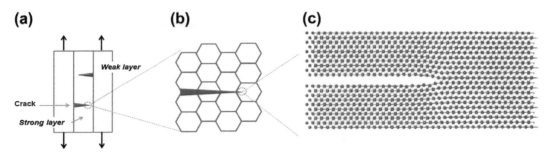

FIGURE 8.1   The spatial scales involved in considering brittle deformation. (a) A brittle layer bonded to two weak deforming layers. Wedge-shaped cracks form in the extending brittle layer. These wedge-shaped cracks are *wedge disclinations* and must form in opposing pairs to maintain deformation compatibility. (b) The tip of the crack at the grain scale. (c) The tip of the crack at the atomic scale. *From Marder and Fineberg (1996).*

macroscopic deformation can be discontinuous as in the generation of a large earthquake on a single slip fracture in a wide fault zone but we concentrate on situations where the macroscopic deformation is continuous. The mechanisms involved in the brittle process are the generation and propagation of *fractures* (surfaces where the cohesion of the rock mass is broken), and subsequent sliding on, or opening of, these fractures, at all scales from the nanometre to the kilometre. There is a transition to deformation mechanisms where crystal defects, such as point and line defects, dominate the deformation process. That transition is characterised by otherwise sharp fracture tips being blunted by the generation of dislocations at the tip (Abraham, 2003; see Figure 1.5). This zone of crystal plastic deformation reduces the stress concentration at the tip and inhibits fracture propagation.

The terminology to do with brittle processes is quite confusing. This confusion arises from historical reasons to do with usages that mix up terms that describe processes, constitutive behaviour and macroscopic behaviour. Confusion also arises because there is a complete gradation, rather than a distinct transition, between deformation modes characterised by loss of cohesion at the crystal structure scale and deformation modes characterised by the rate- and temperature-dependent motion of crystal defects such as vacancies, dislocations and disclinations (Chapter 9).

First, the situation is made complicated by the fact that *fracture* can be both *brittle* and *ductile* (Ashby et al., 1979; Budiansky et al., 1982; Needleman et al., 1995). There have been suggestions that ductile fracture is important in melt segregation and migration (Brown, 2010; Hobbs and Ord, 2010 and references therein); we do not consider the topic of ductile fracture any further here. Second, the term *brittle* is commonly associated with loss of cohesion of an experimental specimen at failure and the development of a localised, through-going zone of deformation. Such behaviour is common for brittle materials with soft (force controlled, see Figure 4.1) loading conditions but not necessarily developed for brittle materials with hard (displacement-controlled) loading conditions. We do not use this geometrical definition of brittle and restrict our attention to displacement-controlled situations where the deformation within the specimen must be such as to be compatible with the imposed deformation and hence the boundaries of the specimen remain smooth.

Third, the term *brittle* is often used as a contrast to so-called *ductile* behaviour and one talks of the *brittle–ductile transition*. We do not adhere to this usage. To us the term *ductile* means *capable of large permanent deformation without loss of macroscopic continuity.* Thus in a displacement controlled situation where all the boundaries are prescribed during the deformation, brittle materials can show large ductility through mechanisms involving fracture and cataclasis (see Wong and Baud, 2012).

Fourth, the term *brittle* is sometimes contrasted with the term *plastic*. *Plastic* is a term we use for any material whose deformation is characterised by a yield (or flow) surface (Chapter 6). Thus a brittle material might be described by a plastic Mohr—Coulomb or Drucker—Prager constitutive relation, whereas a material deforming by crystal slip or diffusion might be described by a plastic Tresca or von Mises constitutive relation. Both classes of behaviour are characterised by yield surfaces and both are plastic materials.

We believe the best way of thinking of the differences in these various types of behaviour is solely in terms of the processes operating and there is a complete spectrum in behaviours between relatively rate- (and temperature-) insensitive process characterising *brittle fracture* and rate- (and temperature-) sensitive processes, such as the motion of dislocations, disclinations and point defects in crystal structures and at grain boundaries, characterising *viscoplastic processes*. At the brittle end of the spectrum the constitutive behaviour tends to be sensitive to changes in hydrostatic pressure, whereas at the visco-plastic end of the spectrum the constitutive behaviour is relatively insensitive to changes in pressure. Thus instead of using the term *brittle—ductile transition* one should use a term such as *brittle—visco-plastic transition*.

Two different classes of theory are developed for brittle behaviour. One involves *dynamic crack growth* where the inertia of the system plays an important role in driving crack propagation; such theories are important for earthquake dynamics. The other class involves quasi-static crack growth which means that the kinetic energy of the system can be neglected. Motivated by the discussion of Nguyen and Einav (2009), we distinguish five different applications of *Fracture Mechanics* (Figure 8.2). One, *Rupture Mechanics*, relevant to the propagation of faults, is a dynamic process where the kinetic energy of the system is important during initiation and growth of the fracture. A second is *Damage Mechanics*, applicable to many different loading conditions and deformation processes and concentrating on distributed

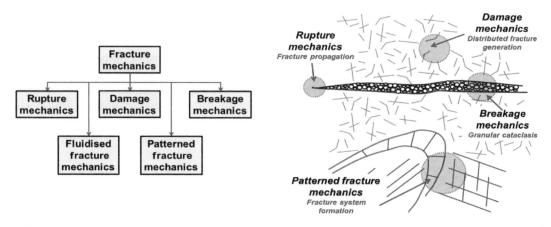

FIGURE 8.2   The different approaches to Fracture Mechanics. *Inspired by Nguyen and Einav (2009).*

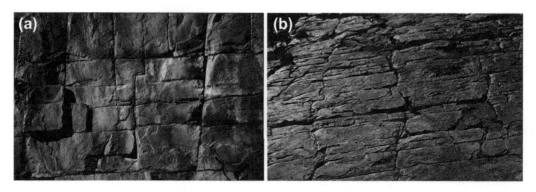

FIGURE 8.3   Fracture patterns in quartzite. These two images (a) and (b) are from adjacent outcrops of Heavitree Quartzite in Ormiston Gorge, Northern Territory, Australia. Although near-orthogonal patterns of joints are developed in both localities, the nature of the fracture surfaces is quite different. A nearby outcrop is shown in Figure 8.4(c). In (a) which is a photograph of a near-horizontal surface, plumose markings can be observed on this surface indicating that it also is a fracture surface. Thus, although bedding is also approximately coincident with this surface, the rock mass is broken into three joint sets that are close to being mutually orthogonal as shown in Figure 8.4(c). In (b) one of the joint sets has become irregular.

fracture commonly in the vicinity of strongly deformed bodies of rock. We consider the subject in some detail in Chapter 10.

A third application, *Breakage Mechanics*, concerns the fragmentation and particle size reduction processes within fault and shear zones and we look at this aspect in Section 8.6. We believe it is important to emphasise another class of behaviour, *Patterned Fracture Mechanics*, characterised by the patterned development of fractures (Figure 8.3). This fourth kind of behaviour is of main concern to structural geologists and the reason we distinguish this as a distinct class is that it has received relatively little attention and emphasis in the literature. Attempts to analyse the development of fracture patterns (Pollard and Aydin, 1988; Mandl, 2005) have mainly relied on linear elastic fracture mechanics and have considerable trouble in discussing the origin of fracture patterns, particularly multiple fracture systems, and the length scales involved in their development. We do not pretend that significant progress has been made in this area since the work of Hobbs (1967), Pollard and Aydin (1988) and Mandl (2005) but recent work that treats the development of structural patterns as energy minimising mechanisms forced by the necessity for deformation compatibility offers considerable promise and insight. Such approaches focus on energy and compatibility issues rather than on stress distributions as controls on fracture development as is the traditional fracture mechanics approach. It is unfortunate that the majority of experimental observations have been obtained for dynamic situations where the kinetic energy of the loading frame is an important consideration (soft loading conditions) rather than for fully displacement constrained (hard) loading conditions.

Another, fifth, application of fracture mechanics is that pertinent to the development of breccias as illustrated in Figure 8.4(d). This area presumably overlaps with Breakage Mechanics but is distinguished by the presence of fluids which facilitate fragment formation, rotation, transport and multiple cycles of these processes. We call this application *Fluidised Fracture Mechanics* but leave detailed discussion of the processes involved to Volume II.

FIGURE 8.4   Patterns of rock fracture. (a) Complicated fracture systems. Hilton Mine, Mt. Isa, Australia. Scale: Base is 1 m across. (b) Quartz vein systems, Fountain Springs, Queensland, Australia. Scale: Outcrop is 2 m across. (c) Orthogonal joint patterns in Heavitree Quartzite, Ormiston Gorge, Central Australia. (d) Breccia, Fountain Springs, Queensland, Australia. Scale: Outcrop is 1 m across. (e) Cataclasite from Namibia. *From Trouw et al. (2010).* Scale: Base is 16 mm across. (f) 'Domino' structure resulting from brittle deformation of K-feldspar with recrystallised quartz filling gaps between feldspar fragments. West-central Arizona. *From Singleton and Mosher (2012).*

This chapter is mainly concerned with the *fracture patterns* that are developed in deformed metamorphic rocks. The subject is important because it supplies examples of one end member of a spectrum of structure development that spans the complete range from brittle to visco-plastic behaviour. All of the structures that develop in this spectrum, whether they be joints, faults, veins, shear zones, folds, boudins or foliations are characterised by more or less periodic spatial patterns, although the patterns at the brittle end of the spectrum tend to be by far the most complex. The question is: *why does this trend towards periodicity exist and what controls the length scales involved*? We also include here the controls on the evolution of grain size distribution that arise within zones of cataclasis. An understanding of such evolution is important since it is one of the few measurable features of strongly deformed brittle rocks that tell us something about process.

Of the various modes of rock deformation, brittle deformation is the least understood. This is not to say that some of the physics is not well understood (see Paterson and Wong, 2005) the issue is that the *patterns of deformation* that form during brittle deformation appear to have little regularity either in space or time so that in many instances a large number of different orientations of fractures develop and overprint each other in bewildering complexity (Figure 8.4(a) and (b)). The essential problem is to understand this complexity at least to the extent that it supplies information on the conditions and the kinematics of deformation.

The study of brittle deformation is in its infancy compared to developments in crystal plasticity; a number of different but overlapping approaches exist. One is the experimental study of rock fracture typified by work reported in Paterson and Wong (2005). A second approach is to consider the development of rock fractures essentially in terms of the stress fields that are associated with their development; excellent treatments in this regard are Mandl (2005) and a review by Pollard and Aydin (1988). A third approach is to develop constitutive relations for the deformation of rock by brittle flow; a review is given by Brantut et al. (2013). A fourth approach is to try to understand the role that rock fractures play in the kinematics of rock deformation and that is the route we concentrate on here.

The essential questions we ask are: *What is meant by brittle behaviour and how does it differ from plastic or ductile behaviours? What controls the orientations of fracture systems in deformed rocks (Figure 8.3)? What controls their spacings? Why do complicated overprinting relations such as illustrated in Figure 8.4(a) and(b) develop and what do such fabrics mean? What controls the development of breccias such as that in Figure 8.4(d)? How do cataclasites (Figure 8.4(e)) contribute to rock deformation? What are the relations between rock fracture and other processes such as dissolution, compaction, shear heating, mineral reactions and fluid transport and how are such relations expressed in the rock fabric*?

We begin this chapter (Section 8.2) by examining the general principles that govern the development of patterned structures during deformation no matter if the constitutive behaviour is brittle or visco-plastic, and then concentrate in the remainder of the chapter on those patterns that arise from brittle processes. This part of the chapter involves theories to do with minimisation of energy and the necessity for continuous refinement of structure in order to achieve this minimisation and simultaneously match an imposed deformation (see Chapter 7).

In Section 8.3 we consider some aspects of the mechanics of fracture. The traditional way of treating brittle behaviour is to try and extend the studies involving simple mechanical tests where specimens are subjected to point loads or uniaxial soft loading conditions; for the most part in such experiments there is little or no constraint on the behaviour of the specimen once failure occurs so that brittle failure is commonly associated with disintegration of the specimen. Such disintegration and/or catastrophic failure is not what we are concerned with in the development of fracture systems in rock masses where macroscopic continuity is maintained. The emphasis in the traditional approach is on the *stress field* associated with a single fracture. We briefly summarise some of this work in Section 8.3. The classical approaches of Griffith and Barenblatt are included here with the associated concepts of stress fields at crack tips, stress intensity factors and fracture toughness. These approaches have been very successful in explaining many aspects of brittle behaviour but they have considerable difficulty in treating multiple, overprinting fracture systems that develop in well-defined spatial patterns and particularly in arriving at length scales that define fracture spacing in these patterns.

Since about 1975 (de Giorgi, 1975), an approach to brittle fracture based on the two concepts of energy minimisation and global compatibility of the kinematics of the fracture system with the imposed deformation has had some success in addressing the issue of fracture patterning. We consider these concepts in Section 8.3 where we build on the material in Section 8.2. In particular we emphasise that many of the structures that develop in deformed metamorphic rocks develop by the motion of *defects*. Instead of the traditional classification of fractures into Mode I, II and III we find it more instructive to view these features as special cases of *Volterra defects* in elastic materials and refer to *dislocations* and *disclinations*. Although such concepts are widely attributed to processes that occur in the plastic deformation of single crystals, the generalised concepts are just as applicable to any elastic medium and their usage makes it clear why various kinds of accommodation processes such as compaction, dissolution and chemical reactions may be associated with fracture networks. We also find it useful to introduce another defect common in fractured rocks, namely, *dilclinations*, which are linear *dilational* defects expressed as veins, stylolites and/or compaction bands. We also consider kinematic controls on the number of fracture systems that develop, on fracture orientation and the compatibility requirements that enable complex multiple fracture systems to develop.

In Section 8.4 we examine the kinematics of fracture formation and the relation between the eigenvectors of the flow field and the orientation of fractures. We integrate this geometrical relation with a general energy-kinematic approach to the formation of patterned fracture systems. Compaction appears to be an important deformation mechanism that commonly acts as an accommodation process to enable fractures to form in a macroscopic isochoric deformation and we consider the extreme end member in this process, *compaction banding*, in Section 8.5. In Section 8.6 we examine a model for a general deformation of a rock mass by brittle processes using the breakage mechanics approach of Einav (Einav, 2007a, b, c; Einav et al., 2007).

We first address some semantic issues: A distinction is sometimes made between a *joint* and a *fault*, both considered to have formed by fracture of the rock mass, on the basis of *detectable* displacements parallel to the plane of fracture (Mandl, 2005); a joint is said to have no *observable* displacement parallel to the joint surface (although it can have displacements normal to the joint surface), whereas a fault is characterised by displacements parallel to the fault. We consider such a distinction to be somewhat arbitrary and opportunistic and also subjective. In what follows we use the term *joint* to mean *any planar fracture surface in a rock mass arising from deformation, independently of the displacement field involved in its formation*. The term *fault* is used when one wants to emphasise the displacement parallel to the fracture. In many instances fracture surfaces in deformed rocks are now delineated by quartz, carbonate or other minerals so that they are identified as *veins*. We make no distinction here. All such surfaces are fractures, although the possibility of displacive veins (Merino and Canals, 2011) is recognised. If we want to emphasise the fact that the features are veins then we use that term. We are not concerned here with joints that develop during cooling or unloading of rock masses or during compaction arising from a superimposed sedimentary load.

8.2 THE MECHANICS OF PATTERN FORMATION DURING DEFORMATION

In this section we discuss approaches developed over the past 40 years or so that concentrate on the concept that structures formed during deformation, no matter the scale or the process involved, develop into patterns that minimise the energy of the system. Importantly, these patterns also enable the available deformation mechanisms (fracture, twinning, dislocation glide, subgrain rotation and so on) to accommodate the imposed deformation so that the deforming material remains compatible with the macroscopically imposed displacement boundary conditions.

The classical theory of the plasticity of solids, developed by writers such as Hill (1958) and Drucker (1959) emphasises stability of the loading problem and is expressed in *Drucker's postulate* that the energy functional is convex (a *functional* is a function that is a function of a function, although we will mainly use the term *function* unless the distinction is important). This ensures *stability* of the deformation. However, a ubiquitous feature of deformed solids, especially deformed metamorphic rocks, is the development of *instabilities* in the form of localised fracturing, shearing, necking, kinking, subgrain formation and dislocation patterning that result in some form of *structure*, commonly characterised by various fractal scaling laws. Understanding the formation of structure is essential to understanding the strength of materials especially at the microscale since materials with microstructure can be stronger or weaker than materials without microstructure and develop anisotropic physical properties. At length scales larger than the grain scale, other patterned structures, such as folding and delamination, develop in anisotropic materials and are also fundamental to the performance of the materials. In geological materials an understanding of such features associated with the development of structure at all scales is critical to deciphering Earth history.

Departures from the postulate of convex energy functionals that characterise classical plasticity and the Drucker postulate were first studied in the field of nonlinear finite elasticity with particular application to martensitic transformations and the development of materials with memory of their initial geometry. Such developments are based on minimisation of non-convex energy functionals and for non-dissipative materials employ theorems in the Calculus of Variations by Tonelli (1921, 1923) and Morrey (1952) that involve conditions for convergence and stability for sequences that minimise the energy functional. de Giorgi (1975) was responsible for drawing the analogy between such minimisation sequences and the development of microstructure. The overall development of the theory of microstructures in nonlinear elastic materials has been heavily influenced by Ericksen (1975), Truskinovsky (1996), Truskinovsky and Zanzotto (1995, 1996), and Ball and James (Ball, 1977, 2004; Ball and James, 1987; Ball et al., 1991) who also emphasise that the minimising sequences can result in fractal geometries. The subject is now fairly mature except for two aspects: The microstructures described by these energy minimisation procedures are based on observations of microstructures actually developed in martensitic materials. First, there is no theory for determining the detailed geometry of the microstructure or whether a given microstructure that minimises the energy is unique, except for the geometrical constraint that the deformation from one part of the microstructure to another must satisfy certain jump conditions (see Section 2.10). Second, the existing theory is essentially 'static' with no overarching principle that describes how a given microstructure might evolve for arbitrary finite deformations.

The theory involving non-convex energy functionals has been extended to the development of fracture systems by Truskinovsky (1996), Francfort and Marigo (1998), Del Piero and Owen (1993), Del Piero and Truskinovsky (2001) and Choksi et al. (1999). Again, as with the theoretical development for nonlinear elastic materials, there is as yet no theorem to show whether a given pattern of fractures is unique in minimising the energy functional.

8.2.1 Energy Minimisation and Structural Refinement

We follow proposals that have a basis in the development of microstructures in systems characterised by nonlinear elasticity and that are particularly well developed for martensitic microstructures (Ball and James, 1987; Bhattacharya, 2003). The approach treats the development of microstructure, and in our case, the development of fracture systems as a phase transition (Truskinovsky, 1996) where the microstructure develops in order to minimise a non-convex energy function in exactly the same manner as two chemical phases develop at a chemical phase transition (Powell, his Chapter 2, 1978). The non-convex energy in mechanical systems arises from geometrical and/or constitutive softening or from contributions to the total energy from a number of subsystems as we explore below. The basic principle is outlined thus: Consider a system deforming under the influence of some imposed boundary conditions and where the energy of the system, as a function of the imposed deformation measured by the deformation gradient, F, consists of a number of wells as shown in Figure 8.5(a). The system tends to deform so that the total energy of the system is minimised and in the case of the energy function shown in Figure 8.5(a) this would ideally correspond to the bottom of one of the wells. If the imposed deformation, F, corresponds to the base of one of the wells then the deformation can be achieved in a homogeneous manner. If, however, the deformation does not correspond to one of the wells but to the average of two or more wells, the deformation can be achieved by combining some mixture of the deformations corresponding to each well. However, such a mixture cannot be arbitrary since there is a kinematic constraint that the average deformation be compatible with the imposed deformation. We will see below that this compatibility requirement can only be met by developing microstructure at a number of spatial scales. If more than one orientation of microstructural elements is required for compatibility, quite complex patterns can arise with a number of length scales involved. We first look at a one-dimensional example below taken from Bhattacharya (2003).

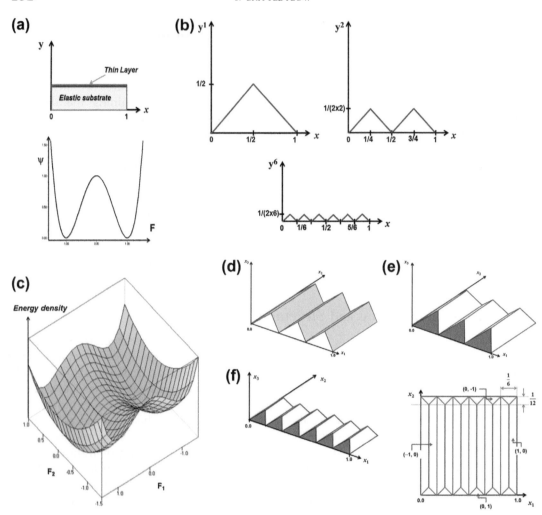

FIGURE 8.5 Examples of energy minimising microstructural arrangements. (a) A one-dimensional layer bonded to an elastic substrate (above) and the elastic energy plotted against the deformation gradient, **F** (below). (b) The succession of deformations that comprises a sequence that gets closer and closer to minimising the energy in (a) as the deformations become finer and finer. (c) The elastic energy for an elastic two-dimensional layer constrained by two deformations with gradients, **F**$_1$ and **F**$_2$. (d), (e) and (f) are adapted from Bhattacharya (2003). The succession of deformations that comprises a sequence becomes closer and closer to minimising the energy in (c) as the deformations become finer and finer. In (e) and (f) another set of deformations is added at the ends of the chevrons to maintain compatibility with the imposed deformation, **F**$_2$.

Consider a bar of unit length bonded to an elastic substrate and undergoing deformation parallel to its length. The coordinate system is shown in Figure 8.5(a). The deformation is $y = y(x)$, so that the deformation gradient, **F**, is the scalar:

$$\mathbf{F} = \frac{dy}{dx}$$

We assume that the energy density, $\psi(x)$, is given by

$$\psi(x) = \left(F^2 - 1\right)^2$$

This function is shown in Figure 8.5(a) and has two minima at $F = +1$ and $F = -1$. Such an energy function arises from geometrical reorientation of slip systems or from any form of nonlinear constitutive behaviour that involves softening of a material parameter such as the elastic modulus or softening of the load bearing capacity (see Chapter 7).

The total energy of the system is given by

$$E[y] = \int_0^1 \left\{ \left[\psi(F(x)) + (y(x))^2 \right] \right\} dx = \int_0^1 \left\{ \left[(F(x))^2 - 1 \right]^2 + (y(x))^2 \right\} dx \qquad (8.1)$$

Note that the term $(y(x))^2$ is the energy of the elastic substrate. We now attempt to minimise the total energy with respect to the deformation, y. The absolute minimum for the total energy is zero. This requires that both terms in the integrand of (8.1) are zero. Hence we need to find a deformation that satisfies both

$$F(x) = \frac{dy}{dx} = \pm 1 \quad \text{and} \quad y(x) = 0 \qquad (8.2)$$

However, if $y(x) = 0$ everywhere then $F(x) = \frac{dy}{dx} = 0$ and this does not satisfy the first requirement in (8.2). Hence it is evident that *there is no single deformation, y(x), that minimises the total energy of the system.*

However, it turns out that if we allow more than one deformation we can get as close as we wish to minimising the energy. In fact we can find a series of deformations, $y^{(n)}$ with $n = 1, 2, 3\ldots$, which has progressively smaller and smaller energy as $n \to \infty$. Let the first term in $(8.1)_2$ be zero and so consider the first deformation shown in Figure 8.5(b):

$$y^{(1)}(x) = x \quad \text{if } 0 < x < \frac{1}{2}$$

$$= (1 - x) \quad \text{if } \frac{1}{2} \le x < 1$$

This is a sawtooth or kink-like deformation with slopes $F^{(1)} = 1$ on the left half and $F^{(1)} = -1$ on the right half. The total energy is now given by

$$E\left[y^{(1)}\right] = \int_0^{1/2} x^2 dx + \int_{1/2}^1 (1 - x)^2 dx = \frac{1}{12}$$

If there are n deformations given by

$$y^{(n)}(x) = x \quad \text{if } 0 < x < \frac{1}{2n}$$

$$= \left(\frac{1}{n} - x\right) \quad \text{if } \frac{1}{2n} \le x \le \frac{2}{2n}$$

then

$$E\left[y^{(n)}\right] = n \int_{0}^{1/2n} x^2 dx + n \int_{1/2n}^{2/2n} \left(\frac{1}{n} - x\right)^2 dx = \frac{1}{12n^2}$$

Examples for n = 2 and 6 are shown in Figure 8.5(b). Thus we cannot minimise the total energy of the system with just one homogeneous deformation but we can get *arbitrarily close to minimising the energy with combinations of more than one deformation at finer and finer scales. This is the fundamental principle behind the development of structure in deformed rocks at all scales.*

Bhattacharya (2003) continues and explores a two dimensional deformation where the deformation gradient, **F**, is now a vector:

$$\mathbf{F} = \nabla y = \left\{\frac{\partial y}{\partial x_1}, \frac{\partial y}{\partial x_2}\right\} = \{F_1, F_2\}$$

The plate to be deformed is square with edges of length, L. Now there is no need to introduce an elastic substrate to introduce problems in energy minimisation into the problem and the energy is given (Figure 8.5(c)) by

$$\psi(\mathbf{F}) = \left(F_1^2 - 1\right)^2 + F_2^2$$

where problems in energy minimisation arise through the deformation being required to be compatible with both F_1 and F_2. With exactly the same argument as in the one-dimensional example one can show that continuously refining the structure in the form of kink-like deformations can bring the energy arbitrarily close to zero; however, the resulting structure (Figure 8.5(d)) is not compatible with the boundary conditions and the kink structures need to be cut off in the x_2 direction (Figure 8.5(e) and (f)) in order to achieve compatibility with the imposed deformation. The resulting fine-scale structure consists of a kink-like structure with the deformation gradient alternating between $\nabla y^{(n)} = \{1,0\}$ and $\nabla y^{(n)} = \{-1,0\}$ capped by deformations with deformation gradients $\nabla y^{(n)} = \{0,1\}$ and $\nabla y^{(n)} = \{0,-1\}$ (Figure 8.5(f)). The energy of the system can be shown to equal (L^2/n) and so approaches zero as $n \to \infty$. Ortiz and Repetto (1999) have shown that an alternative way of maintaining compatibility with the boundary conditions consists of continuous refinement (Figure 7.20) of the structure in the x_2 direction instead of the single cut-offs shown in Figure 8.5(e) and (f). Such refinement structures are common in deformed rocks and two examples are given in Figure 7.21. Another example is the core and mantle structure common in partially recrystallised aggregates.

These arguments also give some insight into a length scale that may be associated with the structure. The discussion indicates that we can reduce the energy of the system, (L^2/n), and maintain compatibility with the imposed deformation by increasing n and so by making the structure finer and finer. However, there is a lower limit to the fineness of the structure set by the interfacial energy required to produce the interfaces between the two deformations. This might be the interfacial energy of a twin, the energy required to produce a fracture surface or the interfacial energy associated with the formation of a subgrain. We call this interfacial energy ψ_0 per unit length of boundary. Then the total energy of the system is

$$\underbrace{E}_{\text{Total energy of the system}} = \underbrace{\frac{L^2}{n}}_{\text{Bulk energy}} + \underbrace{(\psi_0 Ln)}_{\text{Surface energy} \times \text{length} \times \text{number of interfaces}}$$

Notice that this is not minimised as $n \to \infty$. E is minimised for $n = \left(\frac{2}{\psi_0}\right)^{1/2} \sqrt{L}$ so that the density of the minimising structure that minimises the total energy is related to a length scale set by the specimen. This is in general accord with what we see where the length scale of the microstructure is set by layer thickness, grain size or some other length scale characteristic of the specimen.

We repeat, the essential and fundamental principle that arises from this discussion is that a *homogeneous deformation cannot minimise the total energy of a deforming system (with a non-convex energy function) and at the same time satisfy the constraints of the imposed deformation; the development of structure at all scales in deforming rocks is the response to satisfying these two conditions.*

8.3 THE GEOMETRY AND PHYSICS OF FRACTURE

8.3.1 Stress-Based Approaches to Fracture: Griffith and Barenblatt

Classical fracture mechanics is a well-developed field with an enormous literature. We can only hope to cover a few aspects of the subject here. The interested reader should consult Orowan (1949), Irwin (1957), Liebowitz (1968), Paterson (1978) and Paterson and Wong (2005) for detailed discussion and further references. The theory begins by noting that the strength of a rock that fails by brittle mechanisms is far below the theoretical estimates for a homogeneous, defect-free material. The argument is as follows: The strain at failure by fracture is commonly of the order of 1% (shortening or elongation). Assuming linear elasticity with a Young's modulus, E, the stress at failure should be of the order of $\sigma^{failure} = E/100$. Since a reasonable value for E for rocks is of the order of 5×10^{10} Pa this means the theoretical value of $\sigma^{failure}$ should be about 0.5 GPa which is at least an order of magnitude higher than what is observed. Thus the conclusion is that defects are initially present in the material in the form of microcracks or pores and fracture initiates due to local stress concentrations induced by these defects.

The enduring contribution to fracture mechanics was made by Griffith (1921) who proposed that initial imperfections in an elastic material loaded with a tensile stress σ_T, provided sites for fractures to grow. He calculated that the decrease in elastic energy, $W^{elastic}$, of the material due to the development of an elliptical crack of length $2l$ is

$$W^{elastic} = \frac{\pi l^2 \sigma_T^2}{E}$$

Then, if γ is the surface energy density of the newly formed crack,

$$W^{surface} = 4l\gamma$$

Thus the energy, W, of the system is expressed by

$$W = W^{surface} - W^{elastic} = \frac{\pi l^2 \sigma_T^2}{E} - 4l\gamma \tag{8.3}$$

Griffith proposed that an elliptical crack grows if W is minimised. That is, $\frac{\partial W}{\partial l} = 0$, which gives the stress at fracture, $\sigma^{fracture}$ as

$$\sigma^{fracture} = \sqrt{\frac{2E\gamma}{\pi l^{critical}}}$$

where $l^{critical}$ is a critical length for the crack; beyond that length the energy of the system drives crack growth. The critical length corresponds to a minimum energy, $W_{min} = -2\gamma l^{critical}$; the quantity 2γ is variously known as the *critical energy release rate* or the *modulus of cohesion* and is clearly a material-dependent parameter. Sometimes this quantity is called the *fracture toughness.* Also used in fracture mechanics are quantities called *stress intensity factors* (Paterson and Wong, 2005) that describe how the stress at a crack tip scales with crack length; different stress intensity factors are defined for Mode I, II and III cracks.

If we identify W with the Helmholtz energy of the system, then we see that the energy of the system is a convex function of l as shown in Figure 8.6(a). Notice that the Griffith energy of the crack is proportional to a surface area whilst the bulk energy of the material is proportional to a volume so that the ratio of bulk to surface energy is geometry dependent. Hence the Griffith theory says that the breakage pattern should be length dependent and not just a function of a breakage criterion. This is sometimes put forward as one of the many objections to the Griffith theory. The Griffith theory has been modified by many subsequent workers. One of the important modifications is due to Barenblatt (1959, 1962) who proposed that the breakage process is not simply an on or off process where atomic bonds are either broken or are intact, as in the Griffith theory, but that atomic bonds stretch before breaking giving a concave, nonlinear energy as shown in Figure 8.6(b).

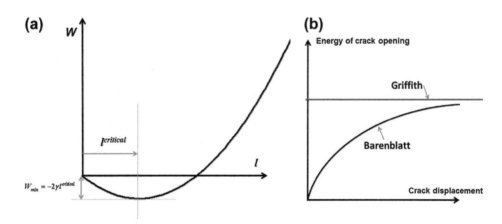

FIGURE 8.6   Energy functions for fracture. (a) The total energy of a fracturing elastic system according to the Griffith theory, (8.3). (b) The cohesive term in (8.3) according to the Griffith and Barenblatt models.

8.3.2 Energy-Based Approaches with Deformation Compatibility

Although the Griffith theory (with subsequent modifications) has been amazingly successful in engineering design problems, it suffers from four major drawbacks (Bourdin et al., 2008): (1) There is no criterion for the initiation of a fracture; (2) irreversibility, or the existence of a threshold beyond which the advance of a crack cannot be reversed is not treated; (3) the path of the crack as it grows is not explained and (4) crack patterns including branching and the development of multiple cracks with a more or less uniform spacing are not included in the theory (in fact one can show (Del Piero and Truskinovsky, 2001) that given Griffith-type arguments for the energy of the system the energy of a single bar with a given elongation is minimised by the development of a single fracture). The last drawback is of course of fundamental importance for structural geology.

The Griffith theory of fracture assumes that the surface energy density is a constant that is independent of fracture opening (Figure 8.6(b)); the assumption is that atomic bonds simply break without prior stretching. The Barenblatt theory assumes some bond distortion prior to breakage so that the surface energy density is an increasing, concave function of the displacement discontinuity (Figure 8.6(b)). The two models are consistent if one assumes that the Barenblatt energy approaches the Griffith energy asymptotically. If this is the case the two models are formally equivalent (Willis, 1967; Rice, 1968; Marigo and Truskinovsky, 2004). The classical theories of fracture therefore involve the assumption of a convex energy function combined with a concave function. Other energy functions have been explored by Francfort, Truskinovsky and others (Truskinovsky and Zanzotto, 1995; Francfort and Marigo, 1998; Del Piero and Truskinovsky, 2001; Marigo and Truskinovsky, 2004; Francfort, 2006; Bourdin et al., 2008). These authors have selected energy functions motivated by the force–displacement curves that are observed during fracturing experiments. To see some of this variety we present the classical energy density of Griffith in Figure 8.7(a) as a plot of the energy density, ψ, against some measure of the deformation gradient, $\mathbf{F}$. Since the force is obtained from the derivative of the ψ–$\mathbf{F}$ curve, the corresponding force–displacement curve is shown in Figure 8.7(b) which corresponds to linear elasticity. Increasing levels of detail in the energy curves are shown in Figure 8.7(c), (e) and (g) with the corresponding force–displacement curves shown in Figure 8.7(d), (f) and (h). All of these energy functions result in the development of fracture systems as minimising sequences rather than just a single fracture (Truskinovsky and Zanzotto, 1995; Francfort and Marigo, 1998). An example is shown in Figure 8.8 where a *sequential* development of fractures is shown as a bar is extended. Notice that the cracks are not spaced in a strictly periodic manner.

8.3.3 Modes of Fracture

In Fracture Mechanics, fractures are classified (see Paterson and Wong, 2005) as Mode I (opening cracks), Mode II (shear cracks) and Mode III (rotational cracks). Although such a classification is clearly insightful and useful in designing experiments and calculating stress fields associated with crack tips, it does not provide insight into the long-range stresses and deformation fields associated with the cracks. A more informative way of considering fractures is to consider them as defects in an elastic medium and treat the deformation as arising

FIGURE 8.7 Various energy functions and the corresponding force–displacement curves. All of the force–displacement curves are observed in experiments on brittle materials. *See Jaeger and Cook, 1969, Brady and Brown, 1993.*

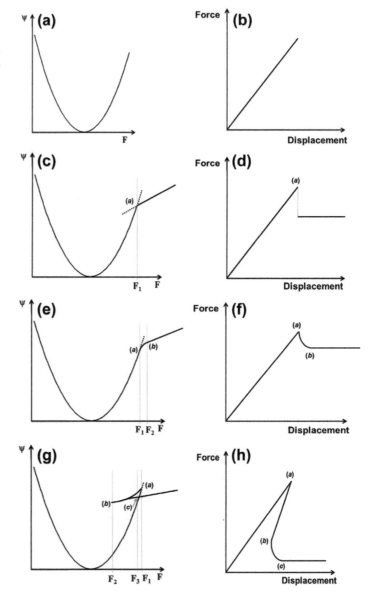

from the motion of these defects just as one would consider dislocations and disclinations in a plastically deforming polycrystal.

An important way of considering such defects was introduced by Volterra (1907) by portraying defects as modifications to a cut made in a hollow elastic cylinder (Figure 8.9(a)) whose axis is **L**. The surfaces of the cut are displaced relative to each other by some amount. If the displacement is a simple translation the defect is known as a *dislocation*; in crystals the

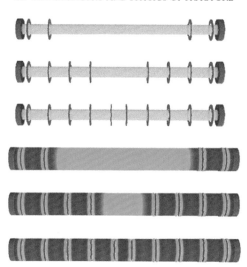

FIGURE 8.8 Sequential development of a multifracture system in a rod undergoing extension. The energy function is non-convex. *From Bourdin et al. (2008).* Note that the fractures develop sequentially and not all at once. Also the spacing is not strictly periodic.

displacement is known as the *Burgers vector*, **b**, and its magnitude is related to the crystal structure and both **L** and **b** are commonly rational crystallographic directions. In non-crystalline materials the displacement, **s**, is controlled by the imposed deformation. If the displacement is a rotation then the defect is known as a disclination. The displacement is described by the *Frank vector*, **ω**, which is a vector along the axis of rotation for the defect with a magnitude equal to the magnitude of the rotation. The magnitude of **ω** is known as the *strength* of the disclination. Again, in crystalline materials both **L** and **ω** are commonly rational crystallographic directions and the magnitude of **ω** is controlled by the crystal structure. For many years it was thought that disclinations involved too much energy of formation to exist in crystalline materials but if the disclinations occur in suitable configurations where the distortion of one disclination is offset by the distortion of another they are stable and have now been widely recognised in deformed crystals (Taupin et al., 2013; Cordier et al., 2014; see Chapter 9). Disclinations are common in fractured materials as we will see.

Thus the deformation of materials can be viewed as the motion of linear defects. In what we have considered so far these defects are of two kinds: those that separate slipped from non-slipped regions of the material (*dislocations*) and those that separate rotated from non-rotated regions (*disclinations*). We will see (see Figure 8.9) that two other forms of linear defects can be identified that are important deformation mechanisms: those that separate dilated from non-dilated regions (*dilclinations*) and those that separate regions of phase change from regions of no phase change (*disconnections*).

This way of considering defects makes it clear what the long-range distortions of the cylinder must be in order to accommodate the defect and distinguishes between linear defects that are associated solely with translation (dislocations or Mode II and III cracks) and those associated with rotation (*disclinations* as Mode I cracks are one example). Some forms of these defects are shown in Figure 8.9 and many others are considered by Volterra (1907).

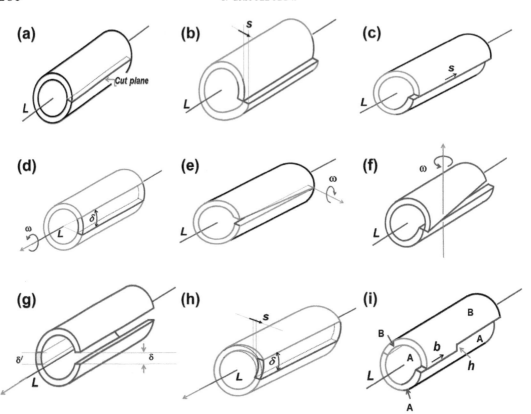

FIGURE 8.9   The geometry of some defects in an elastic cylinder. (a) Original undeformed cylinder with axis, **L**. A cut is made along the surface indicated. (b) Edge dislocation. (c) Screw dislocation. (d) Wedge disclination. (e) Wedge disclination. (f) Twist disclination. (g) Distorted cylinder with additional material added resulting in a displacement δ. This is a positive dilclination and corresponds to a Mode I crack filled with vein material or, in the case of crystal plasticity, the insertion of a plane of new atoms. If material is removed, as in the case of a stylolite, dissolution seam or a plane of atoms during climb of a dislocation then the defect is a negative dilclination. (h) A positive dilclination coupled to an edge dislocation or Mode II crack with slip vector, **s**. (i) A disconnection relating one mineral phase, A, to another, B. This defect is characterised by a Burgers vector, **b**, and a step height, h.

The defects shown in Figure 8.9 are (b) an edge dislocation with $\mathbf{s} \perp \mathbf{L}$, (c) a screw dislocation with $\mathbf{s} \| \mathbf{L}$, (d) a wedge disclination with $\boldsymbol{\omega} \| \mathbf{L}$, (e) a wedge disclination with $\boldsymbol{\omega} \perp \mathbf{L}$, (f) a twist disclination with $\boldsymbol{\omega} \perp \mathbf{L}$, (g) a dilclination where a dilation measured by the opening, δ, distorts an initially circular cylinder into an elliptical cross-section cylinder with elastic elongation measure by δ' and (h) a combined dilclination with an edge dislocation. For completeness in Figure 8.9(i) we add one form of disconnection where one mineral phase, A, abuts another phase, B. The defect with Burgers vector, **b**, and height, h, is a disconnection. A grows into B (or vice versa) by the motion of the disconnection.

    The detailed geometry of dislocations and disclinations is considered by Kroner and Anthony (1975) and Romanov and Kolesnikova (2009). An analysis of disconnections is given by Howe et al. (2009). We revisit these concepts with respect to crystal plasticity in Chapter 9.

8.3.4 How Many Independent Fracture Systems are Required for a General Brittle Deformation?

This is the familiar question asked in crystal plasticity and one common answer in that case is known as the *von Mises criterion*: five independent slip systems are required in crystal plasticity to accommodate a general imposed strain. Clearly if there is no opening normal to the fracture, so that all fractures are dislocations, then the von Mises condition holds for deformation by slip on fractures just as it does for crystal slip, although the scale is different. The question addressed in this chapter is: *what is the equivalent of the von Mises criterion if veins (dilclinations or disclinations) develop also?*

8.3.4.1 The Deformation Gradient Associated with Fracture

We consider a situation where a vein forms as a fracture slips and the vein opening is normal to the fracture. More complicated patterns of opening (see Bons et al., 2012) could be considered using the same arguments. Consider a point, P, lying on the fracture plane (Figure 8.10), with position vector, $\mathbf{r}_0$, unit normal, $\mathbf{m}$, and unit slip direction, $\mathbf{s}$. During slip parallel to $\mathbf{s}$ and an opening, δ, parallel to $\mathbf{m}$, P is displaced to P' with position vector, $\mathbf{r}$. Then, if γ is the shear strain associated with the slip and ε is the stretch associated with displacements normal to the fracture (in the form of vein opening), the displacement, $\mathbf{u}$, is given by

$$\mathbf{PP'} = \mathbf{u} = \mathbf{r} - \mathbf{r}_0 = \gamma(\mathbf{r}_0 \cdot \mathbf{m})\mathbf{s} + \varepsilon(\mathbf{r}_0 \cdot \mathbf{m})\mathbf{m}$$
$$= (\mathbf{r}_0 \cdot \mathbf{m})(\gamma \mathbf{s} + \varepsilon \mathbf{m})$$

so that the deformation, $\mathbf{r}(x)$, is given by

$$\mathbf{r} = \mathbf{r}_0 + (\mathbf{r}_0 \cdot \mathbf{m})(\gamma \mathbf{s} + \varepsilon \mathbf{m})$$

Now $\mathbf{r}_0$, $\mathbf{m}$ and $\mathbf{s}$ are given in terms of the unit vectors, $\mathbf{i}$, $\mathbf{j}$ and $\mathbf{k}$ (Figure 8.10(a)) as

$$\mathbf{r}_0 = x_1 \mathbf{i} + x_2 \mathbf{j} + x_3 \mathbf{k}$$

and

$$\mathbf{m} = m_1 \mathbf{i} + m_2 \mathbf{j} + m_3 \mathbf{k} \quad , \quad \mathbf{s} = s_1 \mathbf{i} + s_2 \mathbf{j} + s_3 \mathbf{k}$$

Then if $\mathbf{F}$ is the deformation gradient for the deformation arising from slip and opening,

$$F_{11} = \frac{\partial r_1}{\partial x_1} = \frac{\partial}{\partial x_1}\left[x_1 + (\mathbf{r}_0 \cdot \mathbf{m})(\gamma s_1 + \varepsilon m_1)\right]$$

$$= \frac{\partial}{\partial x_1}\left[x_1 + (x_1 \mathbf{i} + x_2 \mathbf{j} + x_3 \mathbf{k}) \cdot (m_1 \mathbf{i} + m_2 \mathbf{j} + m_3 \mathbf{k})(\gamma s_1 + \varepsilon m_1)\right]$$

$$= \frac{\partial}{\partial x_1}\left[x_1 + (x_1 m_1 + x_2 m_2 + x_3 m_3)(\gamma s_1 + \varepsilon m_1)\right]$$

$$= 1 + m_1(\gamma s_1 + \varepsilon m_1)$$

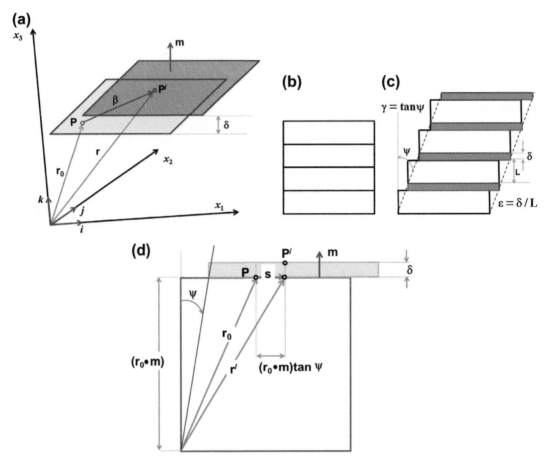

FIGURE 8.10   Geometry of fracture/vein formation. A point P (with position vector, r_0) is displaced an amount β to a new point P$'$ (with position vector, r) involving a component of slip on the fracture parallel to the unit slip vector, **s**, within the fracture plane with unit normal, **m**. **i**, **j**, **k** are unit vectors parallel to the coordinate axes. δ is the displacement arising from vein opening.

Similarly,

$$F_{23} = \frac{\partial r_2}{\partial x_3} = \frac{\partial}{\partial x_3}[x_2 + (r_0 \cdot m)(\gamma s_2 + \varepsilon m_2)]$$

$$= \frac{\partial}{\partial x_3}[x_2 + (x_1 m_1 + x_2 m_2 + x_3 m_3)(\gamma s_2 + \varepsilon m_2)]$$

$$= m_3(\gamma s_2 + \varepsilon m_2)$$

Similar arguments enable the complete deformation gradient to be written:

$$\mathbf{F} = \begin{bmatrix} 1 + m_1(\gamma s_1 + \varepsilon m_1) & m_2(\gamma s_1 + \varepsilon m_1) & m_3(\gamma s_1 + \varepsilon m_1) \\ m_1(\gamma s_2 + \varepsilon m_2) & 1 + m_2(\gamma s_2 + \varepsilon m_2) & m_3(\gamma s_2 + \varepsilon m_2) \\ m_1(\gamma s_3 + \varepsilon m_3) & m_2(\gamma s_3 + \varepsilon m_3) & 1 + m_3(\gamma s_3 + \varepsilon m_3) \end{bmatrix} = \mathbf{I} + \gamma \mathbf{m} \otimes \mathbf{s} + \varepsilon \mathbf{m} \otimes \mathbf{m}$$

The trace of the deformation gradient is $3 + \gamma(m_1 s_1 + m_2 s_2 + m_3 s_3) + \varepsilon(m_1^2 + m_2^2 + m_3^2)$. Since $\mathbf{m}$ and $\mathbf{s}$ are not orthogonal, the trace is not zero and the deformation is not isochoric.

The small strain tensor, $\varepsilon_{ij} = \frac{1}{2}\left(\frac{\partial u_i}{\partial x_j} + \frac{\partial u_j}{\partial x_i}\right)$ is

$$\varepsilon_{ij} = \begin{bmatrix} m_1(\gamma s_1 + \varepsilon m_1) & \frac{\gamma}{2}(m_1 s_2 + m_2 s_1) + \varepsilon m_1 m_2 & \frac{\gamma}{2}(m_1 s_3 + m_3 s_1) + \varepsilon m_1 m_3 \\ \frac{\gamma}{2}(m_1 s_2 + m_2 s_1) + \varepsilon m_2 m_1 & m_2(\gamma s_2 + \varepsilon m_2) & \frac{\gamma}{2}(m_3 s_2 + m_2 s_3) + \varepsilon m_2 m_3 \\ \frac{\gamma}{2}(m_1 s_3 + m_3 s_1) + \varepsilon m_3 m_1 & \frac{\gamma}{2}(m_3 s_2 + m_2 s_3) + \varepsilon m_3 m_2 & m_3(\gamma s_3 + \varepsilon m_3) \end{bmatrix}$$

(8.4)

and the rotation is

$$\omega_{ij} = \gamma \begin{bmatrix} 0 & \frac{1}{2}(m_2 s_1 - m_1 s_2) & \frac{1}{2}(m_3 s_1 - m_1 s_3) \\ -\frac{1}{2}(m_2 s_1 - m_1 s_2) & 0 & \frac{1}{2}(m_3 s_2 - m_2 s_3) \\ -\frac{1}{2}(m_3 s_1 - m_1 s_3) & -\frac{1}{2}(m_3 s_2 - m_2 s_3) & 0 \end{bmatrix}$$

(8.5)

which is independent of the stretch, ε.

In order to see how many fracture systems are required to accommodate a given imposed strain consider a single fracture system similar to that in Figure 8.10(a) with $\mathbf{m}$ parallel to the x_1-axis (Figure 8.11) and the slip direction, $\mathbf{s}$, inclined at θ to the x_2-axis.

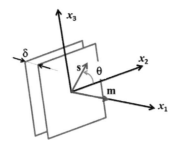

FIGURE 8.11   Coordinate system for a single vein set with normal parallel to the x_1 axis.

Then, $m_1 = 1, m_2 = 0, m_3 = 0$ and $s_1 = 0, s_2 = \cos\theta, s_3 = \sin\theta$.
Thus the deformation gradient becomes

$$
\mathbf{F} = \begin{bmatrix} 1+\varepsilon & 0 & 0 \\ \gamma\cos\theta & 1 & 0 \\ \gamma\sin\theta & 0 & 1 \end{bmatrix}
$$

and the volume change given by the Jacobian, J, of the deformation gradient, is $J = 1 + \varepsilon$. The small strain tensor is

$$
\varepsilon_{ij} = \begin{bmatrix} \varepsilon & \dfrac{1}{2}\gamma\cos\theta & \dfrac{1}{2}\gamma\sin\theta \\[2ex] \dfrac{1}{2}\gamma\cos\theta & 0 & 0 \\[2ex] \dfrac{1}{2}\gamma\sin\theta & 0 & 0 \end{bmatrix}
$$

Thus the vein system can contribute to three of the strain components, namely, ε_{11}, ε_{12} and ε_{13}. Similarly a vein set with $\mathbf{m}$ parallel to the x_2 axis can contribute to ε_{22}, ε_{21} ($= \varepsilon_{12}$) and ε_{23}. The missing component, ε_{33}, can be supplied by any vein system which has an opening direction that can be resolved in the x_3 direction. Thus three vein systems with normals not coplanar are all that is necessary to accommodate a general imposed strain. This result is similar to that reached by Groves and Kelly (1969) for the number of independent slip systems accompanied by climb in crystal plasticity. The additional arguments by Groves and Kelly for combined climb and slip are relevant to vein systems as well.

The strain produced by the operation of a number of fracture systems is

$$
\varepsilon_{ij} = \begin{bmatrix} \sum\limits_{\alpha=1}^{3}\left[m_1^{(\alpha)}\left(\gamma^{(\alpha)}s_1^{(\alpha)}+\varepsilon^{(\alpha)}m_1^{(\alpha)}\right)\right] & \sum\limits_{\alpha=1}^{3}\left[\dfrac{\gamma^{(\alpha)}}{2}\left(m_1^{(\alpha)}s_2^{(\alpha)}+m_2^{(\alpha)}s_1^{(\alpha)}\right)+\varepsilon^{(\alpha)}m_1^{(\alpha)}m_2^{(\alpha)}\right] & \sum\limits_{\alpha=1}^{3}\left[\dfrac{\gamma^{(\alpha)}}{2}\left(m_1^{(\alpha)}s_3^{(\alpha)}+m_3^{(\alpha)}s_1^{(\alpha)}\right)+\varepsilon^{(\alpha)}m_1^{(\alpha)}m_3^{(\alpha)}\right] \\[3ex] * & \sum\limits_{\alpha=1}^{3}\left[m_2^{(\alpha)}\left(\gamma^{(\alpha)}s_2^{(\alpha)}+\varepsilon^{(\alpha)}m_2^{(\alpha)}\right)\right] & \sum\limits_{\alpha=1}^{3}\left[\dfrac{\gamma^{(\alpha)}}{2}\left(m_3^{(\alpha)}s_2^{(\alpha)}+m_2^{(\alpha)}s_3^{(\alpha)}\right)+\varepsilon^{(\alpha)}m_2^{(\alpha)}m_3^{(\alpha)}\right] \\[3ex] * & * & \sum\limits_{\alpha=1}^{3}\left[m_3^{(\alpha)}\left(\gamma^{(\alpha)}s_3^{(\alpha)}+\varepsilon^{(\alpha)}m_3^{(\alpha)}\right)\right] \end{bmatrix}
$$

(8.6)

where the asterisks represent the symmetrically equivalent components. The rotation accomplished by these mechanisms is

$$
\omega_{ij} =
\begin{bmatrix}
0 & \sum\limits_{\alpha=1}^{3}\left[\dfrac{\gamma^{(\alpha)}}{2}\left(m_1^{(\alpha)}s_2^{(\alpha)}-m_2^{(\alpha)}s_1^{(\alpha)}\right)\right] & \sum\limits_{\alpha=1}^{3}\left[\dfrac{\gamma^{(\alpha)}}{2}\left(m_1^{(\alpha)}s_3^{(\alpha)}-m_3^{(\alpha)}s_1^{(\alpha)}\right)\right] \\[3ex]
-\sum\limits_{\alpha=1}^{3}\left[\dfrac{\gamma^{(\alpha)}}{2}\left(m_1^{(\alpha)}s_2^{(\alpha)}-m_2^{(\alpha)}s_1^{(\alpha)}\right)\right] & 0 & \sum\limits_{\alpha=1}^{3}\left[\dfrac{\gamma^{(\alpha)}}{2}\left(m_3^{(\alpha)}s_2^{(\alpha)}-m_2^{(\alpha)}s_3^{(\alpha)}\right)\right] \\[3ex]
-\sum\limits_{\alpha=1}^{3}\left[\dfrac{\gamma^{(\alpha)}}{2}\left(m_1^{(\alpha)}s_3^{(\alpha)}-m_3^{(\alpha)}s_1^{(\alpha)}\right)\right] & -\sum\limits_{\alpha=1}^{3}\left[\dfrac{\gamma^{(\alpha)}}{2}\left(m_3^{(\alpha)}s_2^{(\alpha)}-m_2^{(\alpha)}s_3^{(\alpha)}\right)\right] & 0
\end{bmatrix}
$$

$$(8.7)$$

This discussion has involved the number of fracture systems needed to accommodate a given imposed *strain* which has *five* independent components if the volume change is prescribed. The situation becomes a little more involved if one seeks to find the brittle mechanisms necessary to accommodate a given imposed isochoric *deformation* which has *eight* independent components and we consider this aspect in Section 8.4.1.

8.4 A KINEMATIC VIEW OF FRACTURE DEVELOPMENT

It is commonly observed that the brittle deformation of rocks is inhomogeneous at one or more spatial scales so that patterns of fracture systems develop as shown in Figures 8.3, 8.4 and 8.12. The study of rock fractures is difficult because it is not immediately clear what controls the complex array of fractures we commonly see and if there is any order in the patterns that develop. The traditional view is to interpret these structures in terms of stress patterns or trajectories (Mandl, 2005; Pollard and Aydin, 1988; Bai and Pollard, 2000) and such approaches are quite successful for simple fracture systems such as extension joints or veins

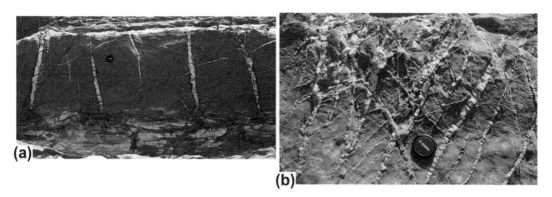

FIGURE 8.12   Vein systems in greenschist facies turbidites, Bermagui, Australia. (a) Approximately equally spaced veins normal to bedding and confined only to the competent beds. Even here there is complexity with the presence of other oblique veins and *en echelon* arrays. (b) More complex arrays in outcrop close to (a) with the same geographical orientation as (a).

that occur as a single set normal to competent beds in deformed rocks (Mandl, 2005, his Figure 4.3). Attempts to consider the development of fracture patterns have met with limited success. Two papers that provide some insight are Muhlhaus et al. (1996) and Chau et al. (1998). Part of the problem in applying classical fracture mechanics to fracture pattern development is that no length scale is incorporated in the constitutive relations. This presumably can be remedied using a non-local approach.

A different approach is to interpret the fracture patterns in terms of the *kinematics* of the deformation that produced the patterns. In such an approach the stress field is not assumed to be homogeneous but to evolve according to the displacement field that constitutes the deformation and the resultant interaction with the constitutive relations for the material in question. This is particularly important because the displacement field is commonly strongly localised in the form of distinct fractures, narrow shear zones and diffuse shear zones that commonly host *en echelon* arrays of fractures and/or veins.

In adopting a kinematic approach we ask the following questions:

- *What controls the orientations of fractures and of other localised planes of deformation?* These other zones of localisation include narrow and diffuse shear zones and diffuse compaction zones.
- *How many fracture systems and other zones of localisation are required to accommodate the imposed deformation? How are these geometrically necessary deformation systems arranged in order to accommodate the imposed deformation?* Thus the fractures and associated structures are viewed more as deformation mechanisms that operate to produce the observed deformation rather than as a group of structures that develop as a result of the imposed stress field.
- *If the overall imposed deformation is constrained to be isochoric, what are the accommodation mechanisms that allow dilatant behaviour at one place (such as opening of vein arrays) to be compensated at other places in the rock mass?*

8.4.1 Kinematic Controls on Fracture Formation

From a kinematic point of view, compatibility of the velocity field needs to be maintained along the length of the fracture during its formation so that no overlaps of material occur. A long fracture can only form if the displacement field associated with the fracture is constant along its length and maintains compatibility with deformations on either side of the fracture. Thus some displacement fields are not permissible (those on the left of Figure 8.13), whereas others are permissible (those on the right). In order for a fracture to grow and not meet problems of incompatibility it must form, at each point, parallel to an eigenvector of the velocity field. We remind the reader that an *eigenvector* is a vector in the velocity field along which lines may shorten or extend or remain constant in length but whose direction is not changed by the deformation. Tangents to the eigenvectors are surfaces called *manifolds*. If particles in the flow are attracted to the manifold it is called a *stable manifold*; if particles are repelled from the manifold it is called *unstable* (see Chapter 3 for further discussion).

In a general two-dimensional affine deformation there are two manifolds, one stable and one unstable (Chapter 3). The exception is a simple shearing deformation where only one

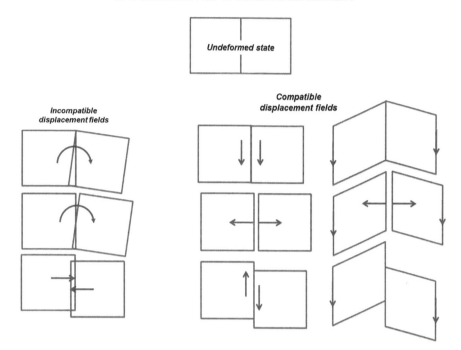

FIGURE 8.13   Incompatible (left) and compatible (right) two-dimensional displacement fields associated with fracture formation. Displacement discontinuities are allowed parallel to the fracture as long as lines either side of the fracture are extended or shortened by equal amounts.

manifold exists, parallel to the shearing plane. In Figure 8.14(a) we represent the displacement field for the deformation defined for Figure 2.3(a):

$$x_1 = 1.5X_1 + 0.5X_2 + 1.25$$
$$x_2 = X_1 + X_2 + 1.0$$

Two manifolds are defined by the lines A-B and C-D. We expect fracture systems similar to those sketched in Figure 8.14(b) to develop from this deformation. One of these, parallel to A-B, we expect to be essentially a shear fracture whilst the other, parallel to C-D, we expect to be an extension fracture. Both, however, can have shear and normal displacements if the imposed deformation prescribes such displacements as is the case for this deformation.

In Chapter 13 we consider deformations similar to that depicted in Figure 8.14 but also derive the eigenvectors of the stretching tensor. These eigenvectors must always be orthogonal and hence cannot coincide with the eigenvectors of a general two dimensional affine deformation where the eigenvectors are never orthogonal except for a pure shear. In general, the stable manifold for the deformation is approximately parallel to a shortening principal axis of the stretching tensor so that it is essentially a plane which is undergoing shortening or elongation with extension normal to the manifold. The unstable manifold is essentially

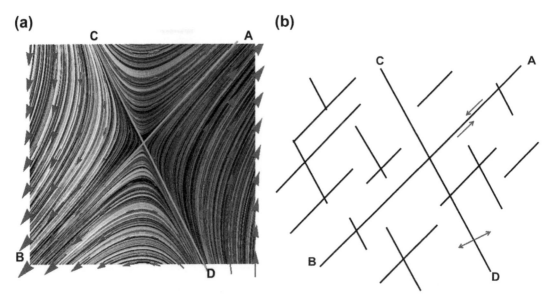

(a)   **(b)**

FIGURE 8.14   Fracture systems developed parallel to the manifolds of the deformation. (a) Phase portrait for the deformation given in Figure 2.3(a). Two manifolds are marked. One, A-B, is unstable whilst the other, C-D, is stable. (b) The fracture system expected for this deformation.

a shear plane. For many materials the constitutive relation connecting the stress, $\boldsymbol{\sigma}$, and the stretching, $\mathbf{d}$, tensors is of the form

$$\boldsymbol{\sigma} = \alpha \mathbf{d}$$

so that the principal axes of stress are parallel to the principal axes of stretching. Such a constitutive relation is said to be *coaxial*; such usage is not to be confused with the common usage of the word *coaxial* to mean coincidence of the principal axes of the stretching tensor throughout a deformation history. Thus a fracture forming parallel to the stable manifold would be interpreted as a 'tension fracture' if one wanted to interpret the fracturing process in terms of stress, whereas a fracture forming parallel to the unstable manifold would be interpreted as a shear fracture. Notice, however, that both can open during deformation and both can have shear displacements parallel to the fracture.

8.4.2 A General Deformation by Brittle Processes

We consider only the two-dimensional case. The kinematic condition for the formation of a fracture is that it is a surface across which the deformations on either side of the surface are continuous, although the deformation gradients can be discontinuous at least with respect to some components (see Figure 2.20(b) and Figure 8.13). That is, the fracture is an invariant surface in the deformation field. This is simply a statement of the characteristic that all lines within the plane of the fracture are distorted in the same manner and in a manner that is compatible with deformations immediately adjacent to the fracture. An invariant surface in the deformation field is a manifold of the velocity gradient and hence we expect fractures to form parallel to the two manifolds of the deformation. Thus patterns of fractures that might arise are shown in Figure 8.15.

If incompatibilities in the velocity gradient field do arise then, as in crystal plasticity (Chapters 9 and 13), the incompatibility is accommodated by the development of defects: dislocations and disclinations. In particular, wedge disclinations are common. Examples are the ends of some en echelon vein arrays and contractions and swellings of veins that develop in the axial planes of folds (Figure 8.16). This raises the question: *how many brittle fracture processes are required to accommodate a general three-dimensional affine deformation?*

To make clear what the above question means we need to remember (Chapter 2) that a general three-dimensional affine deformation with coordinates, X, in the undeformed state and components, x, in the deformed state is described by the deformation gradient, F, that has *nine* independent components:

$$F = \begin{bmatrix} \dfrac{\partial X_1}{\partial x_1} & \dfrac{\partial X_2}{\partial x_1} & \dfrac{\partial X_3}{\partial x_1} \\[2mm] \dfrac{\partial X_1}{\partial x_2} & \dfrac{\partial X_2}{\partial x_2} & \dfrac{\partial X_3}{\partial x_2} \\[2mm] \dfrac{\partial X_1}{\partial x_3} & \dfrac{\partial X_2}{\partial x_3} & \dfrac{\partial X_3}{\partial x_3} \end{bmatrix}$$

and hence requires *nine* independent deformation processes to be accommodated. We have seen that the operation of three fracture systems involving slip and opening and with non-coplanar normals can provide *five* components of strain and *three* components of rotation. The additional constraint is supplied by the volume change which is determined by the Jacobian, J, of F; if the deformation is isochoric then $J = 1$ (Chapter 2). However, the rotation described by (8.7) in general will not be the same as the *imposed rotation* as illustrated in Figure 8.16(a). In general other rotation mechanisms will be required to match the imposed rotations. These additional rotational mechanisms are supplied by wedge or twist disclinations. Some examples are shown in Figure 8.16. In Figure 8.16(a) an asymmetric shear zone is shown where the deformation is isochoric every-where, the regions either side of the shear zone undergo a pure shearing given by

$$x_1 = 1.118X_1, \quad x_2 = 0.894X_2$$

whilst the shear zone undergoes simultaneous shearing and shortening according to the deformation

$$x_1 = 1.1X_1 - 0.2X_2$$
$$x_2 = -0.2X_1 + 0.946X_2$$

The deformations are constrained to meet along a common plane where there is complete compatibility of the deformation. Part of the deformation shown in Figure 8.16(a) must un-dergo a rotation to meet the compatibility constraints and this can be met by introducing wedge disclinations in the form of wedge-shaped veins as shown in Figure 8.16(c). A possible example of this kind of geometry is shown in Figure 8.16(d).

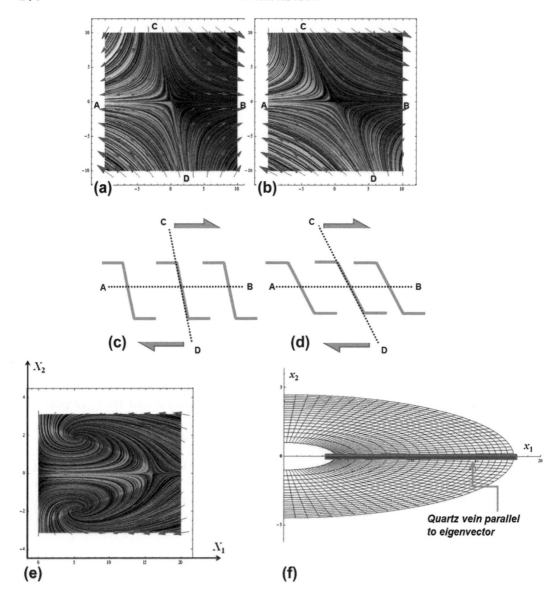

FIGURE 8.15 Eigenvectors and veins. (a) and (b) Eigenvectors for velocity gradient corresponding to the deformation (3.5). Both represent deformations that involve dextral shearing parallel to A-B and shortening normal to A-B. The shortening is larger compared to the shearing in (b) than in (a). A-B and C-D are the manifolds. (c) and (d) En echelon vein systems corresponding to the manifolds in (a) and (b), respectively. (e) Eigenvectors for the displacement gradient corresponding to the deformation that produces Type IC folds described in Figure 2.14(f) The resulting fold and a quartz vein parallel to the only planar manifold in (e).

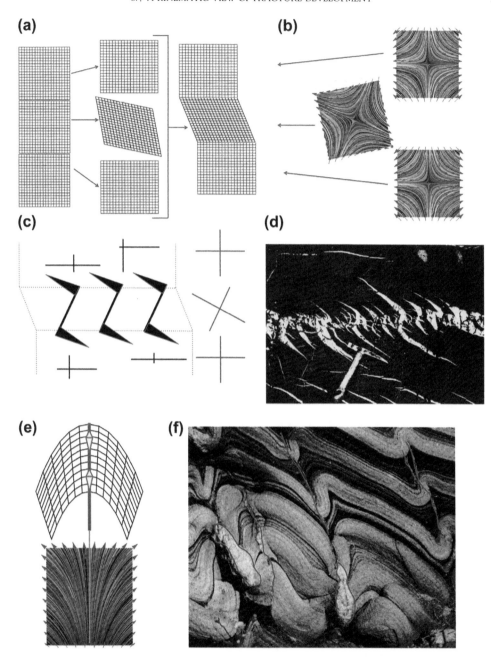

FIGURE 8.16 Fracture systems and eigenvectors. (a) A deformation to produce a shear zone with shortening normal to the boundaries of the shear zone. Left shows the undeformed state. The middle column shows the deformation with no rotation to produce compatibility. A 12° anticlockwise rotation is required to produce compatibility. The right-hand side shows the final deformation after rotation and translations. (b) Eigenvectors for the deformations in the final rotated state. (c) A cartoon showing manifolds in red and vein systems that might form parallel to the manifolds. Wedge-shaped veins (wedge disclinations) accommodate the rotation required for compatibility. (d) An en echelon vein system from Ramsay (1980b). Possible manifolds are shown in red to left of (d). (e) Stable manifold for a fold formed with shearing parallel to the folded surfaces (see Hobbs, 1971, Figure 8). A vein is shown parallel to this manifold with local disclinations. (f) A fold system from Kangaroo Island, Australia, with quartz veins parallel to the axial planes and bulbous expansions of these veins formed to accommodate incompatibilities in the deformation.

Similarly, if the deformations in abutting limbs of a fold are different any incompatibility or rotations can be accommodated by wedge disclinations in the form of bulbous veins in the axial plane (Figure 8.16(f)) which is always an unstable manifold of the non-affine deformation (Figure 5.15(e)).

As we have indicated, fracture systems in deformed metamorphic rocks are commonly far more complicated geometrically than conforming to the orientations of two or three manifolds of a general three-dimensional affine deformation. To some extent this may arise because the total energy of the system is expressed by more complicated functions than the relatively simple functions illustrated in Figure 8.7. Some of these energy functions have been explored by Del Piero and Truskinovsky (2001) and are shown in Figure 8.17. These functions are motivated by experimental force displacement curves observed under hard loading conditions and each develop multiple sets of fractures. If we combine such complications with the requirement for deformation compatibility and hence continuous refinement of fracture systems then *fracture systems within fracture systems* need to form just as *twins within twins* form in martensitic transformations (Bhattacharya, 2003).

FIGURE 8.17 Complicated energy forms (left) that result in realistic force-displacement behaviour (right). *From Del Piero and Truskinovsky (2001).*

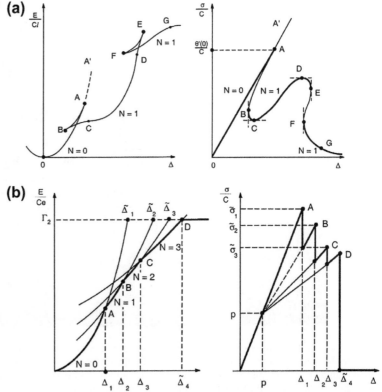

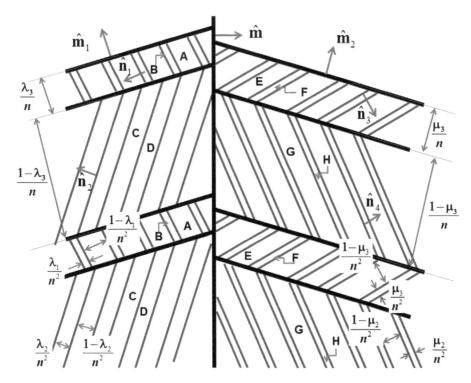

FIGURE 8.18   Complex arrays of twins. *Twins within twins structure. Adapted from Bhattacharya (2003).*

Figure 8.18 shows an example of *twins within twins structure* comprising a 'first order' twin with a twin plane normal, $\hat{\mathbf{m}}$. Within this twin other twins form as pairs labelled A and B (with twin plane normal, $\hat{\mathbf{n}}_1$), C and D (twin plane normal, $\hat{\mathbf{n}}_2$), E and F (twin plane normal, $\hat{\mathbf{n}}_3$) and G and H (twin plane normal, $\hat{\mathbf{n}}_4$). These pairs are arranged within larger twinned domains with normals $\hat{\mathbf{m}}_1$ and $\hat{\mathbf{m}}_2$. The rules that govern the formation of this geometry are

- The geometry forms in order to minimise the energy of the system. Since this energy function is non-convex a single homogeneous deformation cannot minimise the energy and a number of twins must form with continuous refinement in the structural pattern.
- Deformation compatibility must exist across all twin boundaries. This is expressed by the compatibility requirement, (2.17), so that the condition for compatibility across the A–B twin system is $\mathbf{A} - \mathbf{B} = \mathbf{a}_1 \otimes \hat{\mathbf{n}}_1$ where $\mathbf{A}$ and $\mathbf{B}$ represent the deformation gradients in A and B, respectively, and $\mathbf{a}_1$ is the vector defined with (2.17).
- The proportion of twins (and hence their spacing) is governed by the requirement that the average of the local deformation gradient is the weighted mean of the deformation gradients of individual twin sets making up that local region (see Figure 7.18(a) and (b)). Thus for A and B the mean deformation gradient is $\mathbf{F} = \lambda_1 \mathbf{A} + (1 - \lambda_1)\mathbf{B}$ where the spacing of A and B is in the ratio $\frac{\lambda_1}{1-\lambda_1}$.

With these three rules in mind the geometry of Figure 8.18 is expressed by

$$\mathbf{A} - \mathbf{B} = \mathbf{a}_1 \otimes \hat{\mathbf{n}}_1$$

$$\mathbf{C} - \mathbf{D} = \mathbf{a}_2 \otimes \hat{\mathbf{n}}_2$$

$$\mathbf{E} - \mathbf{F} = \mathbf{a}_3 \otimes \hat{\mathbf{n}}_3$$

$$\mathbf{G} - \mathbf{H} = \mathbf{a}_4 \otimes \hat{\mathbf{n}}_4$$

$$(\lambda_1 \mathbf{A} + (1 - \lambda_1)\mathbf{B}) - (\lambda_2 \mathbf{C} + (1 - \lambda_2)\mathbf{D}) = \mathbf{b}_1 \otimes \hat{\mathbf{m}}_1$$

$$(\mu_1 \mathbf{E} + (1 - \mu_1)\mathbf{F}) - (\mu_2 \mathbf{G} + (1 - \mu_2)\mathbf{H}) = \mathbf{b}_2 \otimes \hat{\mathbf{m}}_2$$

$$(\lambda_3(\lambda_1 \mathbf{A} + (1 - \lambda_1)\mathbf{B}) + (1 - \lambda_3)(\lambda_2 \mathbf{C} + (1 - \lambda_2)\mathbf{D}))$$

$$- (\mu_3(\mu_1 \mathbf{E} + (1 - \mu_1)\mathbf{F}) + (1 - \mu_3)(\mu_2 \mathbf{G} + (1 - \mu_2)\mathbf{H}) = \mathbf{b} \otimes \hat{\mathbf{m}}$$

Similar rules must hold for complicated geometries developed as *fracture systems within fracture systems* in deformed rocks. The only difference is that twin orientations are controlled by the crystallography of the host material, whereas the orientations of fractures are controlled by the kinematics of the deformation.

8.5 ACCOMMODATION MECHANISMS

8.5.1 Volume Change Compensating Mechanisms

When a vein system forms there is a displacement with a component normal to the walls of the vein that results in a local change in volume of the system. The volume change is usually small since the shortening of the material between veins is commonly a few percent. Nevertheless if the macroscopic deformation is isochoric the expansion due to vein formation must be accommodated in the system by some form of contraction. Such a contraction can take on a number of forms. The contraction may arise from diffuse compaction spread over a large volume or the compaction can be strongly localised in the form of *compaction bands* (see Figures 8.20 and 8.21). Other modes of compensation involve the development of solution seams such as stylolites. Commonly a slaty cleavage or a differentiated foliation accompanies the vein formation; this represents another form of dissolution in the system that compensates for any volume change associated with veining and in fact may supply the material that constitutes the vein. In any study of veins attention should be paid to determining these volume compensating mechanisms. We concentrate here only on compaction.

Most studies of compaction involve rocks with initially high porosity, perhaps as large as 20%. This is not the case for deformed metamorphic rocks where the porosity is a few percent at most. Compaction bands in rocks with initially high porosity are tabular zones of deformation localisation where the deformation consists of compaction by grain crushing (Holcomb et al., 2007; Wong and Baud, 2012, and references therein). The grains within the band of localisation are fractured and the porosity is noticeably reduced (Figure 8.20(d)). For pure compaction bands there is no shear displacement parallel to the localisation zone.

The condition for localisation in the form of compaction bands in porous materials has been studied by a number of workers and reviews are given by Holcomb et al. (2007) and

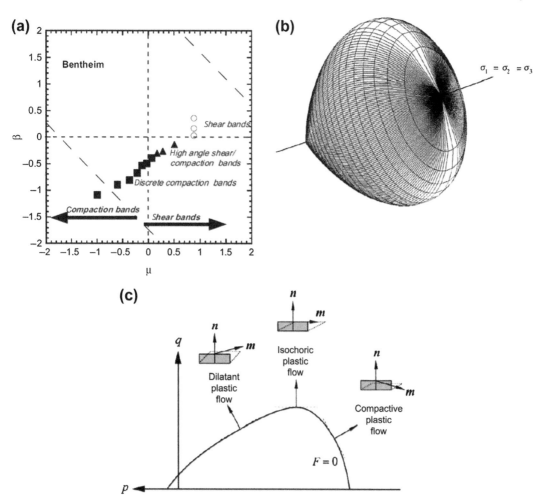

FIGURE 8.19 Characteristics of compaction band formation. (a) A plot of dilatancy factor, β, against friction, μ, showing conditions for compaction and shear bands to form for Bentheim sandstone. *From Wong and Baud (2012)*. (b) A yield surface with a cap. *From Borja and Aydin (2004)*. (c) A plot of q versus p showing conditions for dilatant plastic flow, isochoric plastic flow and compactive plastic flow. *From Borja and Aydin (2004)*. The yield surface is expressed by $F = 0$.

Wong and Baud (2012). For the most part the formation of a compaction band is viewed as a constitutive rather than a geometrical instability arising from some critical value of a bifurcation parameter, which involves the dilation of the material, being reached. The theoretical development for compaction bands follows the same path as outlined for shear localisation in non-associated materials by Rudnicki and Rice (1975). The material is regarded as deforming homogeneously and conditions are defined for the onset of localised deformation by compaction (Issen and Rudnicki, 2000, 2001; Bésuelle and Rudnicki, 2004; Borja and Aydin, 2004). Pure compaction and shear compaction bands are recognised (Figure 8.19(a)), the conditions

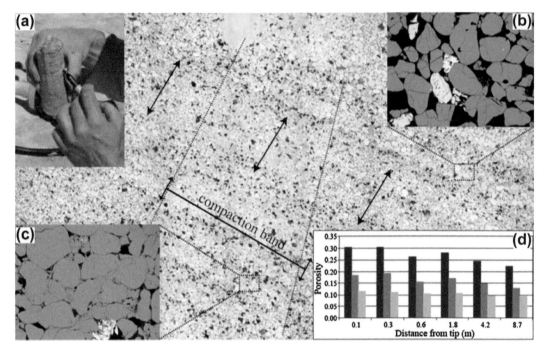

FIGURE 8.20   Microstructure of compaction band. A compaction band (delineated by red dotted lines) 10 mm wide with no shear displacement of bedding crossing at right angles. (a) View of hand specimen. (b) Zoom into uncompacted sandstone. (c) Zoom into microstructure in compaction band. (d) Porosity decrease within compaction band. *From Sternlof (2006). See also Holcomb et al. (2007) for discussion.*

for formation depending on the ratios of a dilatancy factor, β, and a friction coefficient, μ (see Wong and Baud, 2012). The analysis depends on the introduction of a cap on the yield surface so that the yield surface appears as in Figures 8.19(b) and (c) rather than as in Figure 6.13 and the yield surfaces are closed along the hydrostatic axis. If yield is expressed as a surface in $p-q$ space (Figure 8.19(c)) then three regions of localised failure are recognised, namely, dilatant plastic flow, isochoric plastic flow and compactive plastic flow. Here p is the effective mean stress (the difference between the total mean stress and the pore pressure) and q is the square root of the second invariant of the deviatoric stress. Some examples of compaction bands are given in Figures 8.20 and 8.21.

Other modes of compaction that have been extensively studied involve coupling to fluid pressure (Vardoulakis and Sulem, 1995; Veveakis et al., 2013) or to chemical reactions (Stefanou and Sulem, 2014). These coupled studies are likely to be more relevant to metamorphic rocks than the ones that involve high initial porosities, although the high-porosity studies indicate the general principles involved. Patterns of alternating compaction and dilation that develop in deforming fluid-saturated rocks are shown in Figure 8.22.

Stefanou and Sulem (2014) propose a model for coupled deformation fluid chemical reactions that results in localised compaction. Their model suggests a more general model for metamorphic rocks: The nonlinearity results from feedback between deformation-induced

FIGURE 8.21 Compaction bands in Aztec Sandstone, Nevada. *From Sternlof (2006).*

damage that increases the specific area of reactants and/or the stored energy of deformation which in turn accelerates chemical reactions including dissolution. This results in chemical softening that, in turn, can trigger the localisation or compaction instability (Figure 8.23(a)). For the model adopted by Stefanou and Sulem (2014) the spacing of compaction bands is influenced by the hydraulic diffusivity (Chapter 12) as shown in Figure 8.23(b). The larger the hydraulic diffusivity, the larger the wavelength for localisation bands. The dominant wavelength is shown by the red dots in Figure 8.23(b).

8.6 BREAKAGE MECHANICS

The formation of cataclasites and ultra-mylonites in brittle shear zones is especially widespread in the upper parts of the crust. Understanding the constitutive behaviour of these rocks and the evolution of the grain size distribution is important for models of fault

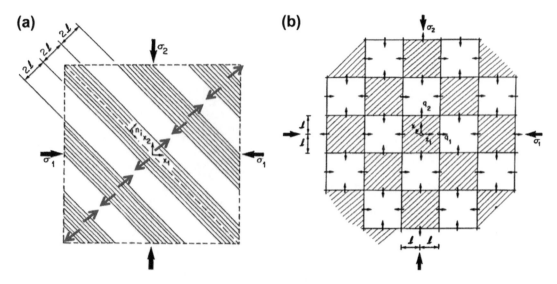

FIGURE 8.22   Compaction and dilating patterns in deforming fluid-saturated media. *From Vardoulakis and Sulem (1995).* The red arrows (a) and small black arrows (b) indicate the direction of fluid flow during deformation. See Vardoulakis and Sulem (1995) for discussion of the length scale, l.

instability and the evolution of fluid permeability, important for earthquake modelling and the development of some hydrothermal systems. The development of microstructures within these zones has been treated under the heading of *Breakage Mechanics* by Einav (2007,a,b,c). A link between *Breakage Mechanics* and *Fracture Mechanics* is explored by Einav (2007c). The

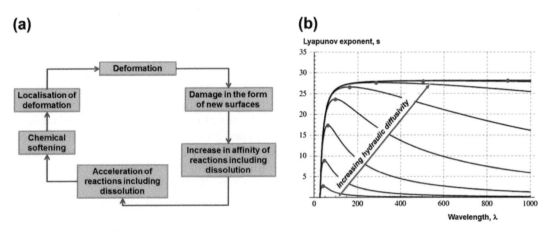

FIGURE 8.23   Compaction arising from coupled grain crushing, fluid flow and dissolution. (a) Feedback loop linking increased chemical reaction rate arising from grain crushing and localisation of deformation. (b) Growth rates for chemically induced compaction bands for changes in hydraulic diffusivity. The red dots indicate the maximum growth rate of a particular wavelength for a given hydraulic diffusivity. Small hydraulic diffusivities result in the most localised compaction bands but the growth rate is also the smallest. *From Stefanou and Sulem (2014); see that paper for discussion.*

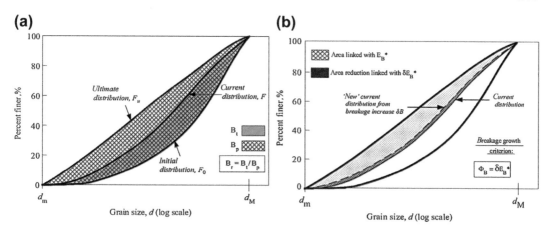

FIGURE 8.24 Definition of breakage terms. (a) The relative breakage, B_r, is defined as the ratio of the two areas, B_t and B_p. *From Einav (2007a).* (b) Breakage propagation criterion. Φ_B is the breakage dissipation and is the energy consumption arising from an increment in breakage, δB. δE_B^* is the incremental reduction in the residual breakage energy. d_m and d_M are the minimum and maximum grain diameters.

theory developed by Einav begins by proposing that a granular material (which may itself be a 'healed' granular material formed in a previous cycle of deformation) with an initial grain size distribution, $p_0(d)$, evolves through breakage in a shear zone towards an ultimate grain size distribution, $p_u(d)$, that is probably fractal (Figure 8.24). Here d is grain diameter. The fractal nature of the ultimate grain size distribution is suggested by observations of natural fault gauges (Sammis et al., 1987; Blenkinsop, 1991; Storti et al., 2003) where power law grain size distributions have been reported. These power law distributions are expected from a two-dimensional section through the Apollonian packing (see Figure 8.27(c)) of different sized spheres (Baram and Herrmann, 2004, 2005; Baram et al., 2004, 2010). If $p(d)$ is the current grain size distribution then the current *cumulative* grain size distribution by mass is given by

$$F(d) = F(\Delta < d) = \int_{d_m}^{d} p(\Delta)d\Delta$$

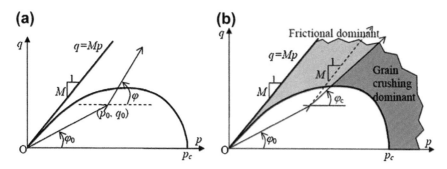

FIGURE 8.25 The physical meaning of the breakage growth criterion. *From Nguyen and Einav (2009).*

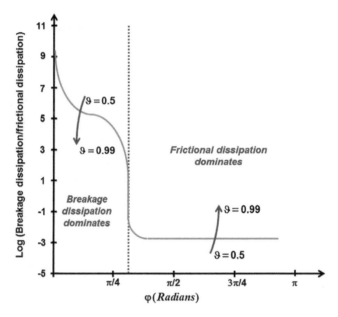

FIGURE 8.26   Dissipation as a function of φ and ϑ. *After Nguyen and Einav (2009).*

where Δ denotes the number of particles less than d in diameter, $p(\Delta)$ is the current grain size distribution, $p(\Delta)d\Delta$ is the current probability of a particle to be in the fraction $d\Delta$, and d_m is the minimum particle size. Similarly we can define the initial, F_0, and ultimate, F_u, cumulative grain size distributions by mass as

$$F_0(d) = F_0(\Delta < d) = \int_{d_m}^{d} p_0(\Delta)d\Delta$$

$$F_u(d) = F_u(\Delta < d) = \int_{d_m}^{d} p_u(\Delta)d\Delta$$

where $p_0(\Delta)$ and $p_u(\Delta)$ are the initial and ultimate grain size distributions. A quantity B_r, the relative breakage, is defined as the ratio of the areas between the cumulative grain size distribution curve and the initial cumulative grain size distribution curve at time t and in the ultimate state (Figure 8.24(a)), $B_r = \frac{B_t}{B_p}$. In Figure 8.24, d_M is the maximum grain size.

A quantity called the *fractional breakage*, B is defined as

$$B = \frac{p_0(d) - p(d)}{p_0(d) - p_u(d)}$$

and Einav (2007a) shows that $B = B_r$ so that B may be experimentally determined by measuring B_r. Hence,

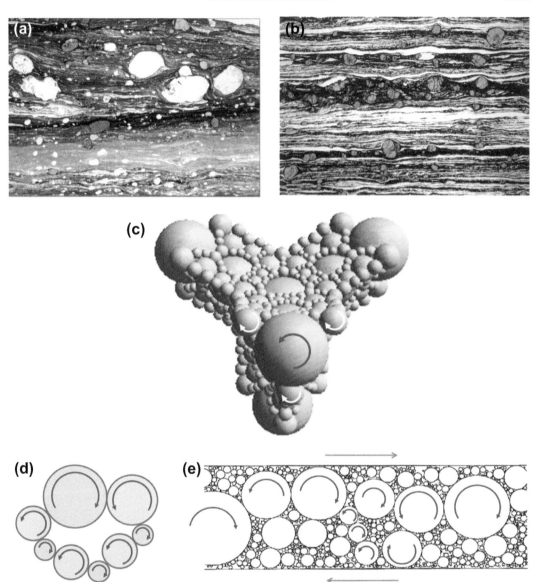

FIGURE 8.27 Cataclasites and ultra-mylonites as roller bearing assemblies. (a) Mylonite. *From Trouw et al. (2010).* Scale: 16 mm across base. (b) Mylonite. Scale: 13 mm across base. *From Passchier and Trouw (2005).* (c) Apollonian packing of spheres. The fractal dimension of a two-dimensional section through this assembly is 1.3. Constructed using Mathematica (Weisstein, Eric W. 'Apollonian Gasket'. From MathWorld A Wolfram Web Resource. http://mathworld.wolfram.com/ApollonianGasket.html). (d) If the grains in an ultra-mylonite are to act as ball bearings then the rotation sense in one grain must be the opposite to that in immediately adjacent grains. This means that in any loop through the aggregate the rotations should match throughout. Here all grey balls rotate clockwise, whereas all orange balls (that alternate with grey balls) rotate anticlockwise. (e) A ball bearing assembly in a simple shearing deformation. The rotation of balls next to the moving interface must match the shearing sense of the walls. *Modified from Herrmann et al. (1990).*

$$p(d) = p_0(d)(1 - B) + p_u(d)B$$

The grain size cumulative distribution by mass is

$$F(d) = F_0(d)(1 - B) + F_u(d)B$$

We assume that the ultimate cumulative size distribution is fractal with fractal dimension, α:

$$F_u(d) = \left(\frac{d}{d_{\mathrm{m}}}\right)^{3-\alpha}$$

In the spirit of Chapter 5, Einav explores the development of a constitutive relation for breakage and an evolutionary law for grain size distributions that are thermodynamically admissible and defines the Helmholtz energy of the system for isothermal breakage in terms of the strain, ε, and the breakage, B, which is assumed to be a macroscopic internal variable:

$$\Psi = \Psi(\varepsilon, B)$$

Then if we define a quantity, ϑ, as a measure of the distance between the initial and ultimate grain size distributions,

$$\vartheta = 1 - \frac{\langle d^2 \rangle_u}{\langle d^2 \rangle_0}$$

where $\langle d^2 \rangle_u$ and $\langle d^2 \rangle_0$ are the second-order moments by mass of the ultimate and initial grain size distributions,

$$\Psi = (1 - \vartheta B)\Psi_{reference}(\varepsilon)$$

Here, $\Psi_{reference}(\varepsilon)$ is the elastic strain energy expressed in terms of the total strain and a reference grain size defined as

$$d_{reference} = \sqrt{\int_{d_m}^{d_M} d^2 p_0(d)\mathrm{d}d} = \langle d^2 \rangle_0^{1/2}$$

The energy conjugate to Ψ is

$$E_B = -\frac{\partial \Psi}{\partial B} = -\vartheta\Psi_{reference}(\varepsilon) \tag{8.8}$$

and the yield criterion is

$$y_B = (1 - B)^2 E_B - E_C = 0$$

where E_C is a strain energy material constant. Einav (2007a) shows that

$$E_B = \frac{1}{B}(\Psi_0 - \Psi)$$

so that

$$B = \frac{(\Psi_0 - \Psi)}{(\Psi_0 - \Psi_u)}$$

Ψ_0 and Ψ_u are the elastic strain energies in the initial and ultimate states. Thus the breakage can be thought of as the relative stored energy that has been released during crushing. Initially the energy is all stored in unbroken particles so that $\Psi = \Psi_0$ and $B = 0$. As the particles are crushed the stored energy approaches Ψ_u and B approaches 1.

In addition to the energy, E_B, defined in (8.8) one can define a dissipation, Φ_B, arising from the creation of new surface area during breakage:

$$\Phi_B = \delta E_B^* \tag{8.9}$$

where

$$E_B^* = (1 - B)E_B$$

so that E_B^* is the available energy in the system for the breakage process and is defined in Figure 8.24(b).

The model developed by Nguyen and Einav (2009) couples the release of surface energy with other dissipative processes such as dissipation arising from frictional sliding and the reorganisation of fragments following crushing so that the Helmholtz energy is written in terms of the total volumetric strain, $\varepsilon_{volumetric}$, the elastic volumetric strain, $\varepsilon_{volumetric}^{elastic}$, the total shear strain, ε_{shear}, the elastic shear strain, $\varepsilon_{shear}^{elastic}$, the mean effective stress, p, and the shear stress, q, as

$$\Psi = (1 - \vartheta B)\left[\psi_{volumetric}\left(\varepsilon_{volumetric}^{elastic}\right) + \psi_{shear}\left(\varepsilon_{voluemetric}^{elastic}, \varepsilon_{shear}^{elastic}\right)\right]$$

$$\Phi_{total} = \sqrt{\Phi_B^2 + \left(\Phi_{plastic}^{volumetric}\right)^2 + \left(\Phi_{plastic}^{shear}\right)^2}$$

The energy functions $\psi_{volumetric}$ and ψ_{shear} govern the elastic volumetric and shear behaviours of the model and represent the unbroken stored energy. The dissipation potential is expressed as a function of the three parts, the breakage dissipation, Φ_B, the plastic volumetric (or particle reorganisational) dissipation, $\Phi_{plastic}^{volumetric}$, and the plastic shear (or frictional) dissipation, $\Phi_{plastic}^{shear}$ where

$$\Phi_B = \frac{\sqrt{2E_B E_C}}{(1 - B)}\delta B$$

$$\Phi_{plastic}^{volumetric} = \frac{p}{(1 - B)}\sqrt{\frac{2E_C}{E_B}}\delta\varepsilon_{plastic}^{volumetric}$$

$$\Phi_{plastic}^{shear} = Mp\left|\delta\varepsilon_{plastic}^{shear}\right|$$

Nguyen and Einav (2009) finally arrive at a constitutive relation:

$$y = \frac{E_B(1 - B)^2}{E_C} + \left(\frac{q}{Mp}\right)^2 - 1 \leq 0 \tag{8.10}$$

where $M = \frac{q_u}{p_u}$, which is the ratio between the ultimate shear stress and ultimate volumetric stress at failure. This constitutive relation for mixed breakage and plastic yield is non-associative and expresses hardening behaviour during breakage as the balance of competition between the term E_B and the term $(1-B)^2$. As B grows towards ultimate breakage at $B = 1$, E_B must increase to balance the inequality in (8.10). Other embellishments to this constitutive law are explored by Nguyen and Einav (2009) including the incorporation of linear elasticity and pressure-dependent elasticity. The final outcome is that the constitutive behaviour can be represented by the elastic parameters, the shear and bulk moduli, the index, ϑ, which describes how far one is from the ultimate grain size distribution, and two parameters, E_C and M. Typically the predicted stress–strain behaviour shows hardening in agreement with experimental results and the energy balance is governed by four factors:

1. The initial stress state represented as the angle, $\varphi_0 = \frac{p_0}{q_0}$ (Figure 8.25).
2. The index, ϑ, which measures how far the initial grain size distribution is from the ultimate grain size distribution.
3. The loading path, measured by the angle, $\varphi = \tan^{-1}\frac{\delta p}{\delta q}$ in Figure 8.25(a).
4. The shear strain, ε^{shear}.

The style of dissipation also depends on φ and on ϑ as shown in Figure 8.26. Below a critical value of φ, given by $\varphi^{critical} = \tan^{-1}(M)$, the dissipation is dominated by breakage; above this critical value the dissipation is dominated by frictional behaviour. For $\varphi < \varphi^{critical}$ the ratio of breakage to frictional dissipation tends to decrease as ϑ increases. For $\varphi > \varphi^{critical}$ this ratio increases as ϑ increases (Figure 8.26). These conclusions are important in considering the energy budget of earthquakes.

The grain shape evolution and fractal dimension changes during cataclasis have been explored by Heilbronner and Keulen (2006), Storti et al. (2007) and Stunitz et al. (2010). Perhaps, as suggested by Baram et al. (2004a,b, 2005, 2010), Baram and Herrmann (2004, 2005), Baram et al. (2004, 2010) and Lind et al. (2008), the mechanism of deformation within cataclasites (and perhaps ultra-mylonites) approaches that of a ball bearing roller system where more or less rounded grains rotate as shown in Figure 8.27(d) and (e). The important principles here are (1) in order to maintain continuity, adjacent grains must rotate in opposite directions which means that within any loop of contacting particles (Figure 8.27(e)) adjacent rotations should match throughout and (2) the tangential velocity of adjacent grains is equal and therefore constant (in magnitude) for all grains. Thus large grains have a smaller angular velocity than small grains. An important outcome of such a model is that a sense of shear may be impossible to detect since adjacent grains rotate in opposite senses and the absolute rotation of large grains may be very small. This may be the explanation for why it is commonly difficult to determine a shear sense within cataclasites (see figures and comments in Chapter 2 of Trouw et al., 2010) and some ultra-mylonites as shown in Figure 8.27(a) and (b); even though these rocks have undergone large shear strains, there is little if any indication of rotation of grains relative to the finer grained matrix.

Recommended Additional Reading

Atkinson, B.K. (1987) (Ed.), *Fracture Mechanics of Rock*. Academic Press.
This book, with contributions from many authors, is a comprehensive discussion of fracture mechanics applied to geological situations

Bhattacharya, K. (2003). *Microstructure of Martensite: Why it Forms and How it Gives Rise to the Shape-memory Effect*. Oxford Series in Materials Modelling.
Although this book is essentially concerned with martensitic transformations it has an easily readable discussion of compatibility, energy minimisation principles and the development of microstructure that is applicable to many geological examples

Mandl, G. (2005). *Rock Joints. The Mechanical Genesis*. Springer.
A comprehensive treatment of rock joints with extensive use of the Mohr circle to illustrate stress distributions

Paterson, M.S., Wong, T.-f. (2005). *Experimental Rock Deformation. The Brittle Field*. Springer.
This is the definitive book on experimental brittle deformation of rocks with an appendix on fracture mechanics

9

Visco-Plastic Flow

9.1 INTRODUCTION

We use the term visco-plastic flow in the following way: a material deforms in a visco-plastic manner if it is able to undergo very large strains by processes that are significantly temperature and rate dependent; this commonly (but not always) means that the deformation occurs without fracturing in some manner such as opening or sliding on joints or undergoing cataclastic flow. A visco-plastic material may develop localised zones of deformation but so long as the mechanisms of deformation are not dominated by fracture then the deformation is said to be visco-plastic. The mechanisms of deformation involved in visco-plastic flow are numerous in detail but at the atomic level involve the motion of crystal defects or defects in the packing of grains or of subgrains. A fundamental characteristic of visco-plastic flow is the existence of a yield or flow surface which marks the change over from elastic behaviour for stress states inside the surface to visco-plastic behaviour for stress states on the yield surface. The shape and size of the yield surface depends on both the temperature and the deformation rate, although we will see below (Section 9.7) that there is a discussion yet to be had concerning the rate dependency at geological strain rates.

The migration of point defects constitutes the process known as diffusion which we will not dwell upon. A full treatment is given by Paterson (2013 and references therein). We concentrate in this chapter on line defects with an emphasis on two types of line defects that have been recognised over the past few years as being of fundamental importance; these are disclinations and disconnections. Just as a dislocation is a line defect that marks the boundary between slipped and unslipped parts of a crystal, a disclination is a line defect that marks the boundary between rotated and unrotated parts of a crystal. A disconnection is a linear defect that marks the boundary between where a phase change has or has not occurred. The term phase change is used here in its widest sense to mean changes with and without changes in chemical composition but always involves a change in crystallographic orientation. The operation of disclinations and disconnections, facilitated by dislocation motion and/or diffusion, as mechanisms of deformation, has profound implications for a variety of processes in deformation and metamorphism including subgrain rotation, the development of curved inclusion trails in porphyroblasts, mineral phase growth and the development of foliations and lineations defined by the crystallographically controlled shapes of grains.

Since dislocations were first observed using transmission electron microscopy in naturally occurring silicates by McLaren and Phakey (1965) the structural geology community has emphasised these defects as the dominant mechanism of deformation at the grain scale except at high temperatures and/or small grain sizes where diffusion is proposed as the dominant mechanism (Paterson, 2013). Again, we assume that the reader is well versed in such topics and restrict ourselves to more recent developments. A discussion of dislocation motion is in Paterson (2013, and references therein).

In Figure 9.1 we illustrate many of the visco-plastic mechanisms involved in the deformation and growth/reduction of grain size in a polycrystalline aggregate. Dislocations, characterised by a slip plane and a Burgers vector, $\mathbf{b}$, operate within the grain. A dislocation in this chapter is a linear crystal defect that enables translational deformation of a crystal; the motion of a dislocation may be facilitated by diffusion resulting in the climb of the dislocation.

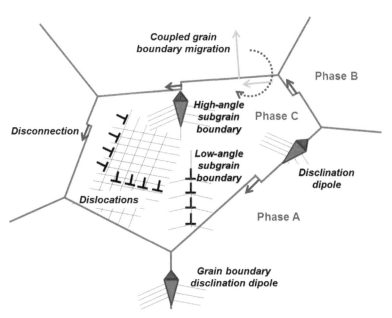

FIGURE 9.1   Various modes of visco-plastic deformation at the grain scale. These modes constitute both deformation and grain size growth/reduction mechanisms for a grain within a polycrystalline aggregate.

Interacting dislocation systems spontaneously develop into subgrain boundaries within the deforming grain. As the misorientation of a subgrain increases, and the cores of dislocations begin to overlap, the misorientation is accommodated by disclination arrays. A disclination in this chapter is a linear crystal defect characterised by a line and a rotation vector known as the Frank vector, ω. The sign of ω distinguishes a positive and negative disclination. The migration of both small and large misorientation subgrain boundaries is an important deformation mechanism; the migration of arrays of disclinations can result in large shear deformations. Disclinations are also important at grain boundaries where again much of the misorientation is accommodated. The growth of one grain into another is accomplished by the motion of disconnections, characterised by a Burgers vector, $\mathbf{b}$, and a step height, h. During the growth of one grain into another the movement of disconnections involves the local shuffling of atoms if both grains have the same chemical composition. Otherwise if the two grains have different chemical compositions (say garnet growing into biotite) then long-range diffusion is important for disconnection motion.

An additional mechanism of deformation at the grain scale is coupled grain boundary migration. This is a phenomenon whereby the motion of a grain (or subgrain) boundary is coupled to shear deformation within the grain. Thus the motion of a grain boundary normal to itself induces a shear deformation in the grain parallel to the grain boundary. Conversely, a shear parallel to a grain boundary induces a motion of the grain boundary normal to itself. We will see that this coupled motion is facilitated by diffusion, and the motion of dislocations, disclinations and disconnections. The process is characterised by a coupling factor, β, which is the ratio of the velocities parallel and normal to the grain boundary. The value and sign of β is material and orientation dependent and also temperature dependent so that $\beta \rightarrow 0$ at high homologous temperatures in most materials where coupled grain boundary motion is replaced by grain boundary sliding.

This chapter concerns the above topics. In Section 9.2 we briefly visit the basic mechanics of visco-plastic flow to define the essential principles involved in the description of visco-plastic deformations in crystalline materials. Section 9.3 defines the basic geometries of disclinations and disconnections and includes the experimental observations on these defects and ways of calculating their densities. Section 9.4 revisits the question of how defects are arranged to accommodate a general affine deformation where nine independent components exist. Section 9.5 examines the processes involved in the formation of subgrain boundaries and mechanisms of rotation within grains with an emphasis on coupled grain boundary migration. The discussion is extended in Section 9.6 to include a range of processes that occur during deformation and metamorphism including subgrain rotation, bulge recrystallisation, grain growth during metamorphism, the development of curved inclusion trails within porphyroblasts and the development of mineral lineations and foliations. Section 9.7 comments upon constitutive relations used in the structural geology community and explores extrapolation of experimentally determined relations to geological conditions.

9.2 THE MECHANICS OF CRYSTAL PLASTIC FLOW

9.2.1 The Deformation Gradient

We consider below the deformation of a crystal by slip on a crystallographically defined plane in a crystallographically defined direction. The following is a simplified version of a more general argument developed for fracture and vein systems in Chapter 8 and follows arguments given by Kelly and Groves (1970). We consider a point, P, with position vector, $\mathbf{r}_0$, lying on the slip plane (Figure 9.2) with unit normal, $\mathbf{m}$, and unit slip direction, $\mathbf{s}$. During slip parallel to $\mathbf{s}$, P is displaced to P' with position vector, $\mathbf{r}$. Then, if γ is the shear strain associated with the slip,

$$\mathbf{PP'} = \mathbf{r} - \mathbf{r}_0 = \gamma(\mathbf{r}_0 \cdot \mathbf{m})\mathbf{s}$$

so that

$$\mathbf{r} = \mathbf{r}_0 + \gamma(\mathbf{r}_0 \cdot \mathbf{m})\mathbf{s}$$

Now $\mathbf{r}_0$, $\mathbf{m}$ and $\mathbf{s}$ are given in terms of the unit vectors, i, j and k (Figure 9.2) as

$$\mathbf{r}_0 = x_1 i + x_2 j + x_3 k$$

and

$$\mathbf{m} = m_1 i + m_2 j + m_3 k, \quad \mathbf{s} = s_1 i + s_2 j + s_3 k$$

Then if $\mathbf{F}$ is the deformation gradient for the deformation arising from slip and $\mathbf{u}$ is the displacement, $\mathbf{u} = (\mathbf{r} - \mathbf{r}_0)$.

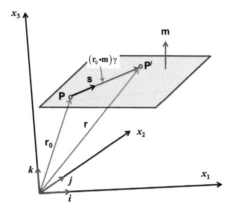

FIGURE 9.2   The geometry of crystal slip. A point P (with position vector, $\mathbf{r}_0$) is displaced to a new point P′ (with position vector, $\mathbf{r}$) during crystal slip parallel to the unit slip vector, $\mathbf{s}$, within the slip plane with unit normal, $\mathbf{m}$. $\mathbf{i}, \mathbf{j}, \mathbf{k}$ are unit vectors parallel to the coordinate axes.

$$F_{11} = \frac{\partial r_1}{\partial x_1} = \frac{\partial}{\partial x_1}[x_1 + \gamma(\mathbf{r}_0 \cdot \mathbf{m})s_1]$$

$$= \frac{\partial}{\partial x_1}\{x_1 + \gamma s_1[(x_1\mathbf{i} + x_2\mathbf{j} + x_3\mathbf{k}) \cdot (m_1\mathbf{i} + m_2\mathbf{j} + m_3\mathbf{k})]\}$$

$$= \frac{\partial}{\partial x_1}[x_1 + \gamma s_1(x_1 m_1 + x_2 m_2 + x_3 m_3)]$$

$$= 1 + \gamma s_1 m_1$$

Similar arguments enable the complete deformation gradient to be written:

$$\mathbf{F} = \begin{bmatrix} 1 + \gamma m_1 s_1 & \gamma m_2 s_1 & \gamma m_3 s_1 \\ \gamma m_1 s_2 & 1 + \gamma m_2 s_2 & \gamma m_3 s_2 \\ \gamma m_1 s_3 & \gamma m_2 s_3 & 1 + \gamma m_3 s_3 \end{bmatrix} = \mathbf{I} + \gamma \mathbf{m} \otimes \mathbf{s} \tag{9.1}$$

where $\mathbf{I}$ is the identity matrix and $\otimes$ denotes the dyadic product between two vectors. Notice that the deformation is isochoric since $\mathbf{s}$ and $\mathbf{m}$ are orthogonal. The isochoric nature can be seen from an expansion of the determinant of $\mathbf{F}$, J, remembering that $\gamma(m_1 s_1 + m_2 s_2 + m_3 s_3) = \gamma \mathbf{m} \cdot \mathbf{s} = 0$. One finds that $J = 1$. The deformation can be decomposed into a small strain tensor, $\varepsilon_{ij} = \frac{1}{2}\left(\frac{\partial u_i}{\partial x_j} + \frac{\partial u_j}{\partial x_i}\right)$:

$$\varepsilon_{ij} = \gamma \begin{bmatrix} m_1 s_1 & \frac{1}{2}(m_1 s_2 + m_2 s_1) & \frac{1}{2}(m_1 s_3 + m_3 s_1) \\ \frac{1}{2}(m_1 s_2 + m_2 s_1) & m_2 s_2 & \frac{1}{2}(m_3 s_2 + m_2 s_3) \\ \frac{1}{2}(m_1 s_3 + m_3 s_1) & \frac{1}{2}(m_3 s_2 + m_2 s_3) & m_3 s_3 \end{bmatrix} \tag{9.2}$$

and a rotation:

$$
\omega_{ij} = \gamma
\begin{bmatrix}
0 & \dfrac{1}{2}(m_2 s_1 - m_1 s_2) & \dfrac{1}{2}(m_3 s_1 - m_1 s_3) \\[2ex]
-\dfrac{1}{2}(m_2 s_1 - m_1 s_2) & 0 & \dfrac{1}{2}(m_3 s_2 - m_2 s_3) \\[2ex]
-\dfrac{1}{2}(m_3 s_1 - m_1 s_3) & -\dfrac{1}{2}(m_3 s_2 - m_2 s_3) & 0
\end{bmatrix}
\tag{9.3}
$$

9.2.2 Energy and Softening Associated with Single Slip

When a single crystal is deformed in extension with a single slip system operating, the slip plane tends to rotate, as shown in Figure 9.3, depending on the constraints on the crystal, so that ultimately the slip plane and the slip direction align with the extension direction. Similarly, in compression, a single slip system rotates so that the normal to the slip plane tends to align with the compression axis. In order to demonstrate that such rotation leads to geometrical softening and, more importantly, non-convex relations between the Helmholtz energy, Ψ, and the deformation gradient, $\mathbf{F}$, we begin with a slip plane whose normal is $\mathbf{m}$ and slip direction $\mathbf{s}$ as shown in Figure 9.3(a). The discussion follows Ortiz and Repetto (1999). The initial angle between the slip plane and the extensional stress, $\boldsymbol{\sigma}$, is θ (see also Figure 9.4). After deformation, if the ends are constrained, the rotation of the slip plane is α. If $\mathbf{R}$ is the rotation associated with the deformation then the vectors $\mathbf{m}$ and $\mathbf{s}$ become $\mathbf{Rm}$ and $\mathbf{Rs}$ after the deformation where

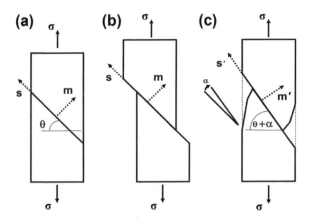

FIGURE 9.3 Rotation of slip plane during deformation. (a) Initial geometry. $\mathbf{m}$ is the normal to the slip plane and $\mathbf{s}$ is the slip vector. $\boldsymbol{\sigma}$ is the extensional stress. (b) After deformation with the ends free to move. (c) After deformation with the ends constrained. $\mathbf{m}'$ is the new orientation of the slip plane normal and $\mathbf{s}'$ is the new orientation of the slip vector. The slip plane rotates through the angle α towards the extension direction.

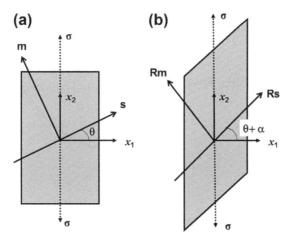

FIGURE 9.4   Geometry of single slip deformation during extension. Vectors **m** and **s** are the slip plane normal and slip direction in the undeformed state and these become the vectors **Rm** and **Rs** in the deformed state through a rotation **R**.

$$\mathbf{R} = \begin{bmatrix} \cos\alpha & -\sin\alpha & 0 \\ \sin\alpha & \cos\alpha & 0 \\ 0 & 0 & 1 \end{bmatrix} \text{ and } \mathbf{Rm} = \begin{bmatrix} -\sin(\theta + \alpha) \\ \cos(\theta + \alpha) \\ 0 \end{bmatrix}, \quad \mathbf{Rs} = \begin{bmatrix} \cos(\theta + \alpha) \\ \sin(\theta + \alpha) \\ 0 \end{bmatrix} \quad (9.4)$$

The slip system is activated if the shear stress on the slip plane in the direction of slip is

$$\tau = \sigma \sin(\theta + \alpha)\cos(\theta + \alpha) = g \quad (9.5)$$

where g is the critical resolved shear stress for the slip system. The deformation gradient for crystal plasticity involving one slip system is given in (9.1) (Asaro, 1983; Rice, 1971).

$$\mathbf{F} = \mathbf{R}[\mathbf{I} + \gamma\mathbf{s} \otimes \mathbf{m}] \quad (9.6)$$

We take $\mathbf{F} = \begin{bmatrix} 0 \\ \lambda \\ 0 \end{bmatrix}$, representing an extension along x_2 where γ is the shear strain due to slip and λ is the stretch parallel to $\boldsymbol{\sigma}$. If we expand (9.6) we eventually arrive at the following expressions for σ, γ and λ in terms of g, θ and α (Ortiz and Repetto, 1999):

$$\sigma = \frac{g}{\sin(\theta + \alpha)\cos(\theta + \alpha)}; \quad \gamma = \frac{\sin\alpha}{\cos(\theta + \alpha)\cos\theta}; \quad \lambda = \cos\alpha + \frac{\sin\alpha \sin(\theta + \alpha)}{\cos(\theta + \alpha)} \quad (9.7)$$

This enables us to plot the stored energy, $\Psi = g\gamma$, against $\log\lambda$, as shown in Figure 9.5 for $\theta = \pi/10$. Also shown is the resulting stress–strain curve.

Thus, geometrical softening arises from constrained single slip and results in a non-convex energy function. Following arguments in Section 8.2 this means that for most deformations the energy cannot be minimised by a single homogeneous deformation and at the same

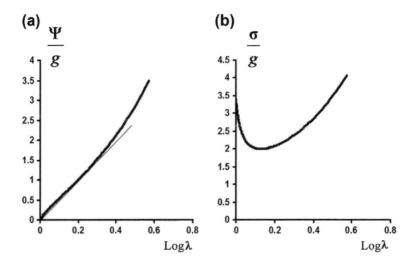

FIGURE 9.5 Geometrical softening arising from single slip with the geometry shown in Figures 9.3 and 9.4 and with $\theta = \pi/10$. (a) Stored energy plotted against the logarithm of the stretch. The energy is weakly non-convex. (b) Plastic part of the stress–strain curve derived from (a). g is the critical resolved shear stress.

time maintain compatibility with the imposed deformation. The system can minimise the energy to within an arbitrarily small amount and simultaneously satisfy the compatibility constraints by developing an inhomogeneous deformation variously expressed as arrays of kinks or subgrains. This is one reason for the development of subgrains during deformation; any softening mechanism has a similar outcome.

9.2.3 Mechanics of Plastic Deformation

As we have seen the plastic deformation gradient for single slip can be written:

$$\mathbf{F}^{plastic} = \mathbf{I} + \gamma \mathbf{s} \otimes \mathbf{m}$$

Figure 9.6 makes it clear that in general there is also a rotation, prescribed by $\mathbf{F}^{rotation}$, associated with a deformation. Also there is an elastic deformation, $\mathbf{F}^{elastic}$, associated with each deformation. The total deformation gradient is (Rice, 1971; Asaro, 1983):

$$\mathbf{F} = \mathbf{F}^{elastic}\mathbf{F}^{rotation}\mathbf{F}^{plastic}$$

With reference to Figure 9.6 a general affine deformation can be written:

$$\mathbf{F} = \mathbf{F}^{*}\mathbf{F}^{plastic}$$

where $\mathbf{F}^{*}$ is a deformation gradient that represents (elastic) stretching and rotation of the crystal structure.

As the crystal deforms, structural bonds are stretched and rotated according to $\mathbf{F}^{*}$. In the elastically deformed configuration the slip direction for the system, α, is given by

$$\mathbf{s}^{*(\alpha)} = \mathbf{F}^{*} \cdot \mathbf{s}^{(\alpha)}$$

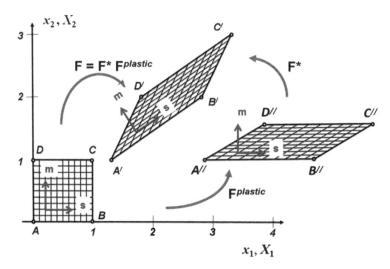

FIGURE 9.6   The decomposition of a general affine deformation of a crystal into a plastic deformation, $\mathbf{F}^{plastic}$, and another deformation, $\mathbf{F}^*$, which comprises a rigid body rotation, $\mathbf{F}^{rotation}$, and an elastic deformation, $\mathbf{F}^{elastic}$. The orientation of one of the slip systems responsible for this deformation is shown with slip direction, $\mathbf{s}$, and slip plane normal, $\mathbf{m}$. During this deformation, which consists of a shear parallel to A-B and a shortening normal to A-B, the initial square ABCD is deformed to $A'B'C'D'$. At least two slip systems are required to achieve this deformation. A rigid body rotation plus an elastic strain brings this configuration to $A''B''C''D''$ which represents the plastic deformation.

The normal to the slip plane in the elastically deformed state is

$$\mathbf{m}^{*(\alpha)} = \mathbf{m}^{(\alpha)} \cdot \mathbf{F}^{*-1}$$

$\mathbf{m}^*(a)$ and $\mathbf{s}^*(a)$ are in general not unit vectors but remain orthogonal in the elastically deformed state since the crystal structure is preserved during deformation so that $\mathbf{s}^{*(\alpha)} \cdot \mathbf{m}^{*(\alpha)} = \mathbf{s}^{(\alpha)} \cdot \mathbf{m}^{(\alpha)} = 0$.

Asaro (1983) goes on to show that for n slip systems the plastic stretching, $\mathbf{D}^{plastic}$, and spin, $\Omega^{plastic}$, are related by

$$\mathbf{D}^{plastic} + \Omega^{plastic} = \sum_{\alpha=1}^{n} \dot{\gamma}^{(\alpha)} \mathbf{s}^{*(\alpha)} \mathbf{m}^{*(\alpha)}$$

so that in the reference configuration,

$$\dot{\mathbf{F}}^{plastic} \mathbf{F}^{plastic-1} + \Omega^{plastic} = \sum_{\alpha=1}^{n} \dot{\gamma}^{(\alpha)} \mathbf{s}^{(\alpha)} \mathbf{m}^{(\alpha)}$$

In general $\mathbf{F}^*$ is accommodated by the insertion of dislocations and disclinations as we shall see in Section 9.3.

An important additional element of crystal plasticity is that the Helmholtz energy, Ψ, of the crystal acts as a potential for the stress so that

$$\sigma = \frac{\partial \Psi}{\partial \mathbf{F}}$$

(Rice, 1975). This is a fundamental result that enables figures such as Figure 9.5(b) to be constructed from Figure 9.5(a).

9.3 DEFORMATION MECHANISMS IN POLYCRYSTALLINE AGGREGATES

The basic building blocks that enable the plastic deformation of polycrystals are defects in the crystal structures of the individual grains, namely, point defects and line defects such as edge and screw dislocations. In addition, there are a number of other processes that involve the motion of surfaces such as twin, subgrain and grain boundaries, including boundaries between different mineral phases. The processes that involve the movement of surfaces invariably involve the motion of disclinations (Romanov and Kolesnikova, 2009; Clayton, 2011) and disconnections (Howe et al., 2009); the motion of disconnections is especially important if the surface marks the boundary between two different mineral phases. The hallmark of the motion of dislocations is deformation dominated by translation, whereas the hallmark of the motion of disclinations is deformation dominated by rotation. The motion of point defects (diffusion) and of dislocations is widely treated in a number of texts and reference should be made to Paterson (2013) and references therein. Here we concentrate on the motion of surfaces as important contributions to the deformation of polycrystalline aggregates including multiphase aggregates. Clearly, as the grain size decreases, the motion of surfaces such as grain or subgrain boundaries increases in importance until at the nano-scale such motion dominates the deformation process since grain boundaries then dominate the volume of the material. However, the motion of boundaries as deformation mechanisms is important in coarse-grained polycrystalline aggregates as well. Another class of deformation involves the motion of larger building blocks such as subgrains or grains. Again the deformation is achieved by the motion of defects, in this instance, defects in the packing of grains or subgrains. These defects are termed cellular dislocations or cellular disclinations. We do not spend any time on these; reference should be made to Sato et al. (1990) and Sherwood and Hamilton (1991, 1992, 1994). We consider these defects again in Chapter 13.

Below we give a brief survey of the geometry of crystal *disclinations* and *disconnections* including the experimental procedures involved in measuring the density of disclinations and then proceed in Section 9.4 to consider their roles in accommodating an imposed deformation.

9.3.1 The Geometry of Disclinations

Just as dislocations are linear defects in crystals that separate slipped regions of the crystal from un-slipped regions, disclinations are linear defects that separate rotated from unrotated parts of the crystal. The basic geometry of disclinations is shown in Figure 8.9 in terms of Volterra distortions of an elastic cylinder. A disclination is characterised by a line, L, and a rotation vector, ω, whose magnitude measures the amount of rotation associated with the defect and whose direction is the axis of rotation. The vector, ω, is known as the Frank vector and its magnitude is the strength of the disclination. The Frank vector for a disclination is the analogue of the Burgers vector for a dislocation. The geometrical relation between the Frank

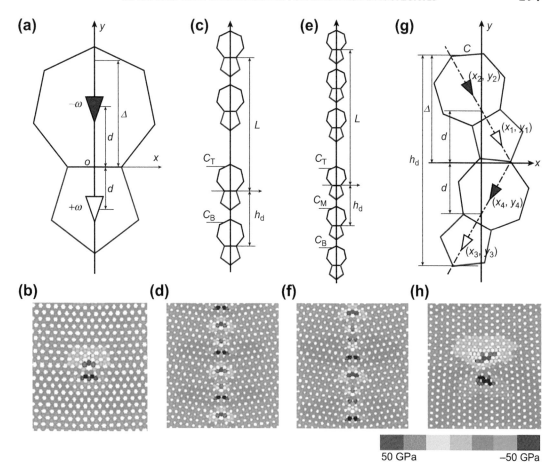

FIGURE 9.7 Disclination arrays in graphene. *From Wei et al. (2012).* (a) An example of a positive and negative disclination pair. This low-energy configuration is known as a disclination dipole. (b) A representation of the disclination dipole in (a) within a lattice of hexagons. The stress field generated by the dipole is shown with the legend to the lower right. The pairs of images, (c) and (d), and (e) and (f) are similar images for larger misorientations of the lattice. (g) and (h) represent another low-energy configuration of dipoles. *For details of these images see Wei et al. (2012).*

vector and L defines three types of disclination; one is a wedge disclination with $\omega \perp L$, a second is also a wedge disclination with $\omega \| L$. The third is a twist disclination with $\omega \perp L$. These defects are shown in Figure 8.9 (d, e, f), respectively. They exist in what are known as positive and negative forms (Figure 9.7) distinguished by the sign of ω. Although many disclinations can be modelled as arrays of dislocations, it is energetically more favourable for disclinations to form as the dislocation density becomes large and dislocation cores overlap (Li, 1972).

Although disclinations were originally postulated for crystal plastic deformation by Nabarro (1967), calculations showed that the energy of an isolated disclination is far too large for such defects to exist within crystals (Friedel, 1967) and for many years these defects have been classed as unimportant in crystal plasticity. However, certain configurations of disclinations cancel out the elastic distortions associated with a single disclination, resulting in decreased elastic energy; such configurations are now recognised as fundamentally important in the deformation of crystals and especially polycrystals (Gutkin and Ovid'ko, 2004; Romanov and Kolesnikova, 2009; Clayton, 2011; Fressengeas et al., 2011). Examples of disclination arrays defining grain boundaries of various misorientations are shown in Figure 9.8.

Disclination arrays can act as powerful mechanisms of plastic shear in crystals. An example is the disclination array identified in olivine by Cordier et al. (2014) and illustrated in Figure 9.9. Here an array of wedge disclinations defines a subgrain boundary. The migration of the boundary normal to itself induces a shear strain in the material through which it moves. This type of deformation can supply an extra mode of shear deformation other than that induced by dislocation motion. Cordier et al. (2014) propose this mechanism as supplying the extra deformation mechanisms that allow a general deformation of olivine which lacks five independent slip systems.

9.3.2 Experimental Determination of Dislocation and Disclination Densities Using Electron Back Scattered Diffraction

Physically, incompatibility of the plastic deformation gradient is equivalent to an array of dislocations and/or disclinations (Figures 9.7, 9.8 and 9.9)). It is convenient to introduce the *Nye dislocation density tensor*, A_{ij}, (Nye, 1953) which is a way of quantifying, in a continuum manner, the dislocation distribution in a deformed crystal. If b is a Burgers vector and t is a unit vector tangent to the dislocation line, L, then the Nye dislocation density tensor is

$$A_{ij} = \frac{1}{V} \int_L b_i t_j ds$$

so that A_{ij} is the integral along all the dislocation lines of the scalar product of the Burgers vector and the tangent to the dislocation line at each point in the crystal (Arsenlis and Parks, 1999). The components of A_{ij} can be established using high-resolution electron back scattered diffraction (EBSD) (Pantleon, 2008; Hardin et al., 2011). One establishes the derivatives of the infinitesimal elastic deformation gradient tensor, β_{ij}, which measures not only the local symmetric part of the lattice strain but also local rotations of the lattice. Thus if $\mathbf{u}$ is the local (elastic) lattice displacement,

$$\beta_{ij} = \frac{\partial u_i}{\partial x_j}$$

and the Nye tensor is given by

$$A = curl(\beta)$$

The use of EBSD to determine the densities of dislocations and disclinations is described by Beausir and Fressengeas (2013) and Cordier et al. (2014). A square grid is superposed on the EBSD image with a coordinate system x_1 and x_2 in the plane of the specimen and x_3 normal to this plane and unit vectors parallel to these coordinate axes, $\mathbf{e}_1, \mathbf{e}_2, \mathbf{e}_3$. The disorientation vector between two neighbouring points, A and B, is $\Delta\theta \cdot \mathbf{r} = \Delta\theta_i \mathbf{e}_i$ where $\Delta\theta$ is the

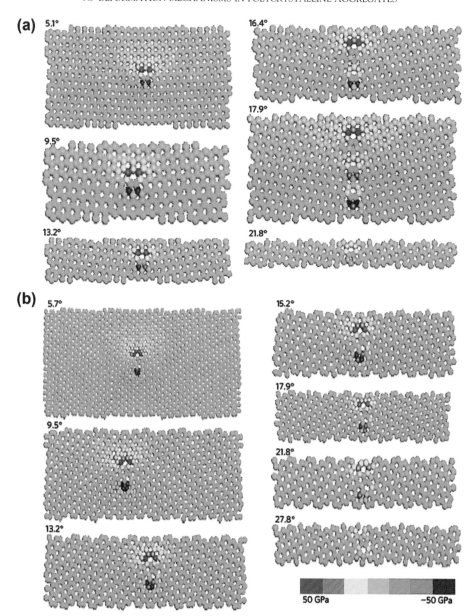

FIGURE 9.8 Examples of disclination configurations defining boundaries of different misorientations in graphene. The misorientation is shown on each figure. *From Wei et al. (2012).* (a) shows what are called 'arm chair' arrays whilst (b) shows 'zigzag' arrays. Notice that for both models misorientations of 21.8° (armchair) and 27.8° (zigzag) show very small stresses.

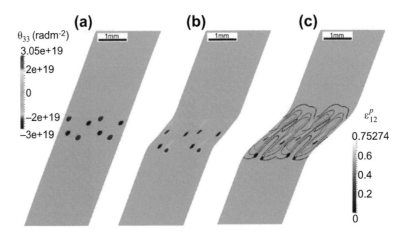

FIGURE 9.9   Coupled shearing and grain boundary migration of a (011)/[100] tilt boundary in olivine. The wedge disclination density (θ_{33} in rad m^{-2}, see Section 9.3.2 below) is colour coded from blue to red. The figure shows the progressive downward migration of the disclinations when a positive shear strain is applied. In C black/white contours show the positive shear strain, $\varepsilon_{12}^{plastic}$, produced by the downward grain boundary migration. *From Cordier et al. (2014).*

disorientation angle and $\mathbf{r}$ is the rotation axis. From the disorientations, $\Delta\theta_i$, between neighbouring points separated by Δx_i only six of the nine components of the elastic curvature tensor, $\kappa_{il}^{elastic}$ can be determined:

$$\kappa_{il}^{elastic} = \frac{\Delta\theta_i}{\Delta x_l} \quad i = 1,2,3; \ l = 1,2$$

The other three components are not accessible because disorientations along the x_3 axis cannot be determined from a plane section. If access to this information is available from detailed micro-sectioning then all nine components can in principle be determined.

One can also define disorientation tensors, $\mathbf{g}_A$ and $\mathbf{g}_B$, which specify the rotation of the lattice at A and B. Then $\Delta\mathbf{g} = \mathbf{g}_A^{-1} \cdot \mathbf{g}_B$. If there are no disclinations present then a quantity, α, can be defined as

$$\alpha = \mathbf{curl} \ \mathbf{g}$$

Five components of the Nye dislocation density tensor, A, can then be obtained from

$$A = \frac{1}{2}\mathrm{tr}(\alpha)\mathbf{I} - \alpha^T$$

These components are α_{12}, α_{13}, α_{21}, α_{23}, and α_{33} with units per metre.

If disclinations are present then the disclination density tensor, θ, is given by

$$\theta_{ij} = e_{jkl}\frac{\partial\kappa_{il}^{elastic}}{\partial x_k}$$

where e_{jkl} are the components of the third-order Levi-Civita tensor which in three dimensions is defined by

$$e_{jkl} = +1 \text{ if } (j,k,l) \text{ is } (1,2,3), (2,3,1) \text{ or } (3,1,2)$$
$$= -1 \text{ if } (j,k,l) \text{ is } (3,2,1), (1,3,2) \text{ or } (2,1,3)$$
$$= 0 \text{ if } j = k, k = l, \text{ or } l = j.$$

This enables the three components, θ_{13}, θ_{23}, θ_{33} of the disclination density tensor to be determined. The units here are radians per square metre.

Examples of some results are shown in Figures 9.10, 9.11 and 9.12.

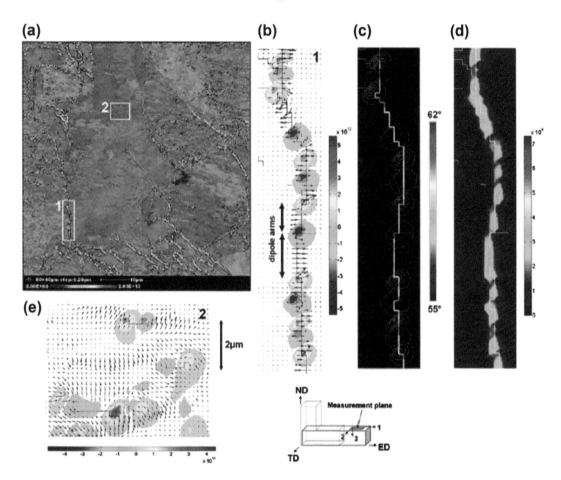

FIGURE 9.10 Determination of disclination densities in deformed copper using electron back scattered diffraction (EBSD). *From Beausir and Fressengeas (2013).* (a) Disclination scalar measure, $\theta = \sqrt{\theta_{13}^2 + \theta_{23}^2 + \theta_{33}^2}$, in radians per square metre. (b), (c) and (d) are zooms into the region 1 and show the density of wedge disclinations in θ_{33} radians per square metre. The scalar dislocation measure is $\sqrt{\alpha_{13}^2 + \alpha_{23}^2}$ in per metre. This is the length of the local Burgers vector per unit surface area resulting from the dislocation densities (α_{13} and α_{23}) along the boundary. (e) shows the θ_{33} component in region 2 together with the local density of dislocations. The arrows in (b) and (e) represent the local Burgers vector with a maximum length corresponding to $3.85 \times 10^6 \text{ m}^{-1}$.

(a)   **(b)**

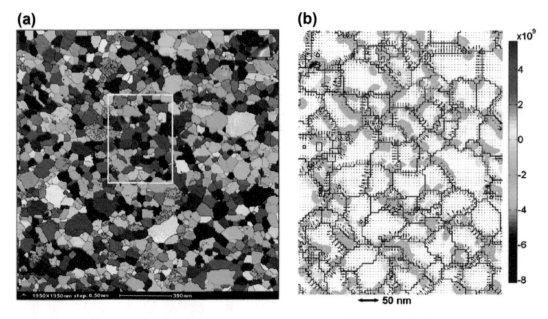

FIGURE 9.11   Disclination densities in aluminium. *From Beausir and Fressengeas (2013).* (a) is a map of grain orientations with inverse pole figure in the top right-hand corner. (b) Zoom into area marked in (a) showing density of wedge disclinations, θ_{33}. The arrows are local Burgers vectors resulting from the dislocation densities (α_{13}, α_{23}); their horizontal components are α_{13} and α_{23} (in per metre). Maximum Burgers vector length, 1.2×10^{8} m^{-1}.

9.3.3 The Geometry of Disconnections

A *disconnection* is a linear defect that marks the boundary between material that has undergone a phase change and material that has not changed. The term *phase change* is used here in its widest sense to include situations where two adjacent grains are of identical crystallography and chemical composition but different orientations, an example being a subgrain of quartz growing into quartz, through to situations where adjacent grains are of completely different crystallography and chemical composition, an example being garnet growing into a biotite grain. A disconnection is commonly seen as a step in a grain boundary and is characterised by a vector, called the Burgers vector for the disconnection, **b**, and a step height, h. The step can be regarded as a site where a dislocation in one grain displaces the grain boundary. For some situations, h = 0. In Figure 9.13(a) we consider two grains, labelled α and γ. The Burgers vector for the dislocation in α that displaces the grain boundary is labelled $\mathbf{t}_{\alpha}$, whilst the dislocation in γ that displaces the grain boundary has a Burgers vector, $\mathbf{t}_{\gamma}$. The Burgers vector for the disconnection is $\mathbf{b} = \mathbf{t}_{\alpha} - \mathbf{t}_{\gamma}$ (Figure 9.13(a)). The array of steps in the grain boundary is called the terrace plane and some mean orientation of the terrace defines the habit plane for the interface between the two phases. In general the strains that exist at the interface cannot be accommodated by the steps and the associated dislocation ξ^{D} and other defects are necessary such as that labelled ξ^{L} in Figure 9.13(b). This dislocation, ξ^{L}, is associated with its own Burgers vector but has zero step height. As the terrace plane moves it produces a deformation in the surrounding material governed by the nature of ξ^{D} and ξ^{L}.

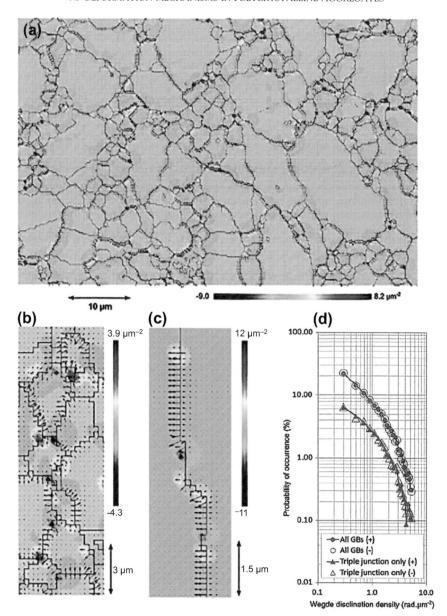

FIGURE 9.12 Densities of wedge disclinations in olivine, θ_{33} (in radians per square metre). Local Burgers vectors arising from edge dislocation are the blue arrows; their horizontal and vertical components are α_{13} and α_{23} (in μm^{-1}). (d) is the probability of occurrence of positive and negative wedge disclination densities (θ_{33}). *From Cordier et al. (2014). See that paper for discussion.*

(a)

(b)

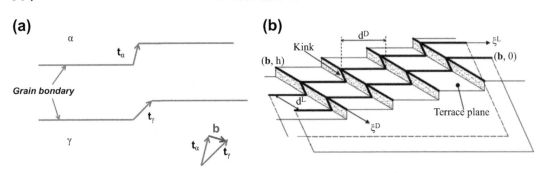

FIGURE 9.13　The geometry of disconnections. (a) A grain boundary between two phases, α and γ. $t_α$ and $t_γ$ are the Burgers vectors for two dislocations, one in each phase, that intersect the grain boundary producing steps. The difference between these two vectors is the Burgers vector, **b**, for the disconnection. (b) Geometry of the terrace and habit planes in a two phase material. *From Howe et al. (2009).*

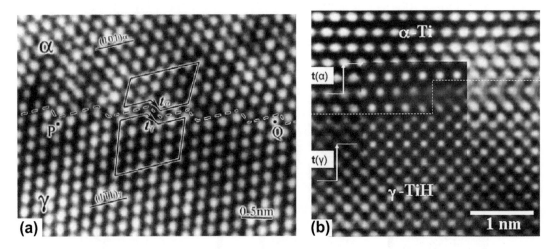

FIGURE 9.14　Transmission electron microscope images of disconnections. (a) Boundary between two phases, α and γ. The Burgers circuits in α and γ define the two Burgers vectors, $t_α$ and $t_γ$. (b) Boundary between two phases of different chemistry, α-Ti and γ-TiH. The Burgers vectors, $t_α$ and $t_γ$ are shown. *From Howe et al. (2009).*

　　Examples of two disconnection systems are shown in Figure 9.14. In Figure 9.14(a) the stepped interface between two phases is clearly shown with Burgers vectors in the two phases, $t_α$ and $t_γ$. In Figure 9.14(b) the boundary is between two phases with different chemistry. $t_α$ and $t_γ$ are shown. In this case the Burgers vector for the disconnection is parallel to the trace of the step face.

　　The general case of the form of defects that constitute a general grain boundary between two phases of completely different orientation and chemistry is yet to be established. In metals, a boundary between phases of identical chemistry but different crystallography is recognised and can be different to a simple array of disconnections. Such a phase transition is called a massive transformation (Aaronson, 2002). Other boundaries that involve diffusional transfer have been considered by Howe et al. (2000) and Massalski et al. (2006). These grain boundaries sometimes consist of single planes of transformed material that move across the interface as a plane but move in a jerky manner with distinct wavelengths for the motion (Raffler and Howe, 2006). There is clearly a rich field of future endeavour in understanding the motion of grain boundaries in transformations typical of metamorphic reactions where gross changes in chemistry occur as well as crystallographic orientation. However, we propose that whatever the mechanisms turn out to be, they act as powerful deformation mechanisms as well as phase change mechanisms.

9.4 THE GEOMETRY OF CRYSTAL DEFORMATION

In order to see how many slip systems are required to accommodate a given imposed strain consider a single slip system shown in Figure 9.15 with $\mathbf{m}$ parallel to the x_1-axis and the slip direction, $\mathbf{s}$, inclined at θ to the x_2-axis. The argument is similar to that presented in Section 8.3.4. Then, $m_1 = 1$, $m_2 = 0$, $m_3 = 0$ and $s_1 = 0$, $s_2 = \cos\theta$, $s_3 = \sin\theta$. Thus the deformation gradient becomes

$$\mathbf{F} = \begin{bmatrix} 1 & 0 & 0 \\ \gamma\cos\theta & 1 & 0 \\ \gamma\sin\theta & 0 & 1 \end{bmatrix}$$

and the dilation given by the Jacobian, J, of the deformation gradient, is $J = 1$. The small strain tensor is

$$\varepsilon_{ij} = \begin{bmatrix} 0 & \frac{1}{2}\gamma\cos\theta & \frac{1}{2}\gamma\sin\theta \\ \frac{1}{2}\gamma\cos\theta & 0 & 0 \\ \frac{1}{2}\gamma\sin\theta & 0 & 0 \end{bmatrix}$$

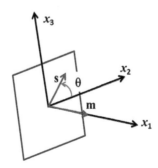

FIGURE 9.15 Strain from single slip.

To satisfy a general affine, isochoric strain, for which there are five independent components, five independent deformation mechanisms are required. This requirement is known as the von Mises criterion for compatibility of strain in a polycrystalline aggregate. For a detailed discussion see Kelly and Groves (1970). Some minerals, such as quartz, possess the required number of slip systems necessary to satisfy an imposed general strain (Paterson, 1969) but some, such as olivine, do not. Cordier et al. (2014) propose that the motion of disclination arrays can supply extra shear components so that five independent mechanisms operate. Another interpretation is that the slip systems available in olivine operate to achieve some components of the imposed strain but incompatibilities develop that are accommodated at grain boundaries by disclinations. The two views are not incompatible.

For the general case, if we restrict ourselves to small deformations, we can write

$$\mathbf{F} = \varepsilon + \omega$$

where $\mathbf{F}$ is the deformation gradient, ε is the small strain tensor and ω is the rotation matrix. We see that five independent slip systems can produce the same ε in each grain but in general the operation of different sets of five slip systems means that ω will be different in each grain. In order to maintain deformation compatibility an extra rotation is required and this may involve elastic distortions that are supplied by disclinations.

One way of considering this issue is to express a general isochoric (not necessarily affine) deformation by the deformation gradient, which has eight independent components and ask: how do we find eight independent deformation mechanisms to accommodate the deformation? The simplest way is to propose that five independent slip systems exist (supplied by dislocations) plus three independent rotational mechanisms (supplied by disclinations). Other possibilities exist and an example is supplied by Basinski and Basinski (2004) who identify eight independent mechanisms in deformed copper single crystals. We will see below (Section 9.5.2) that other geometrical effects arising from coupled grain boundary migration can also introduce systematic, local rotations into a deforming polycrystal.

9.5 THE FORMATION OF SUBGRAINS AND DEFORMATION BY GRAIN BOUNDARY MIGRATION

9.5.1 Dislocation Patterns Arising from Reaction–Diffusion Equations

Walgraef and Aifantis (1985a, b, c) and Aifantis (1986) have applied the concepts of pattern formation arising from reaction–diffusion equations (see Chapter 7) to the development of localised deformation patterns in metals. The results of these latter analyses indicate that such localisation should develop both as temporal oscillations in deformation intensity (a Hopf instability) and spatial fluctuations in deformation intensity (a Turing instability). Both of these kinds of instabilities are observed in deforming metals (Aifantis, 1987). The approach is motivated by work on dislocation dynamics (Walgraef and Aifantis 1985a, b, c; Aifantis 1986; Pontes et al. 2006; Zbib et al. 1996; Shizawa and Zbib 1999; Shizawa et al. 2001). Such an approach suggests that dislocation-pattern formation results from interactions during deformation between different dislocation populations, a process that tends to localise dislocation densities, and diffusion which tends to homogenise dislocation densities.

Patterns develop when a critical stress is reached corresponding to a balance between these two competing processes. In the original version of this model (Walgraef and Aifantis, 1985a) two populations of dislocations were considered: relatively mobile and immobile populations with instantaneous densities ρ_m and ρ_{im}. The four processes that contribute to the total dislocation density in a one dimensional model are proposed as follows (Schiller and Walgraef, 1988).

Dislocation diffusion which contributes terms such as $D_k\frac{\partial^2 \rho_k}{\partial x^2}$ to the total rate of change of the density of the kth population where D_k is the diffusion coefficient for the kth population and x is a spatial coordinate. As in the models for the development of classical Turing instabilities it is important that the diffusion coefficient of mobile dislocations is much larger than that of immobile dislocations.

Dislocation interaction and pinning leading to a nonlinear source term in the rate of change of immobile dislocations and a corresponding sink term in that for mobile dislocations. In the original model the source term is a quadratic, $+\gamma \rho_m \rho_{im}^2$, where γ is the rate for this process. However, the important point is that this term is nonlinear so more complicated interaction/pinning processes leading to higher order terms lead to the same kinds of results.

Dislocation generation by some mechanism such as Frank–Read sources or the like that adds a term $g(\rho_{im})$ into the balance equation for immobile dislocations, the assumption being that all newly generated dislocations are immediately pinned.

Dislocation liberation which adds a source term, $b\rho_{im}$, to the balance equation for mobile dislocations and a corresponding sink term for immobile dislocations.

The above discussion indicates that the overall evolution of dislocation arrays is described by two coupled reaction–diffusion equations which express the relationships and coupling between diffusion, generation, annihilation and pinning of dislocations:

$$\frac{\partial \rho_{im}}{\partial t} = D_{im}\frac{\partial^2 \rho_{im}}{\partial x^2} + g(\rho_{im}) - b\rho_{im} + \gamma \rho_m \rho_{im}^2 \tag{9.8}$$

$$\frac{\partial \rho_m}{\partial t} = D_m\frac{\partial^2 \rho_m}{\partial x^2} + b\rho_{im} - \gamma \rho_m \rho_{im}^2 \tag{9.9}$$

This is the set of reaction–diffusion equations introduced by Walgraef and Aifantis (1985a) and has been intensively studied over the past three decades. In particular one can confirm that this coupled set of equations has the following properties (see Chapter 7) which are the conditions for a Turing instability to develop:

- The homogeneous, steady state of the system is defined by

$$g(\rho_{im}^o) = 0 \quad \text{and} \quad \rho_m^o = \frac{b}{\gamma \rho_{im}^o} \tag{9.10}$$

- Instability represented by temporal oscillations (a Hopf instability) of the dislocation densities occurs when

$$b = b_{Hopf} = a + \gamma (\rho_{im}^o)^2 \tag{9.11}$$

where a is the rate of change of the initial density of immobile dislocations, ρ_{im}^o.

- An instability represented by spatial patterning (a Turing instability) occurs when the stress becomes high enough that

$$b = b_{Turing} = \left(a^{1/2} + \sqrt{\gamma \rho_{im}^o D_{im}/D_m}\right)^2 \tag{9.12}$$

- The wave vector for such spatial patterning is (Chapter 7)

$$q_{Turing} = \frac{2\pi}{\lambda_{Turing}} = \left(\frac{a\gamma(\rho_{im}^o)^2}{D_{im}D_m}\right)^{1/4} \tag{9.13}$$

where λ_{Turing} is the wavelength of the patterning.
- The Turing instability is reached before the Hopf instability if

$$\frac{D_{im}}{D_m} < \frac{a}{\gamma(\rho_{im}^o)^2}\left(\sqrt{1 + \frac{\gamma(\rho_{im}^o)^2}{a}} - 1\right)^2 \tag{9.14}$$

Thus the analysis indicates that an initial homogeneous distribution of dislocations will spontaneously develop into a spatial pattern whose wavelength depends on the diffusion coefficients of the two families of dislocations, the rates of generation (a) and of interaction (γ) of dislocations and the initial density of immobile dislocations (ρ_{im}^o).

This discussion has been for a one dimensional model but the discussion is readily extended to three dimensions (Walgraef and Aifantis, 1985c) and to more complicated dislocation processes (Pontes et al., 2006 and references therein).

A result that immediately appears from models of this type is that if one takes the Orowan relation, $\dot{\varepsilon} = \rho_m v_m b$, where v_m is the velocity of the mobile dislocations and b is the Burgers vector, then using (9.13) one can arrive at the relation (Schiller and Walgraef, 1988)

$$\lambda_{Turing} = \left(\frac{l_{im}\dot{\varepsilon}}{b^2}\right)^{\frac{1}{2}} \rho^{-\frac{1}{2}} \tag{9.15}$$

where l_{im} is the mean free path of immobile dislocations and ρ is taken to be the total dislocation density. If one considers the Turing wavelength to be equivalent to subgrain size then (9.15) gives a relation between subgrain size and dislocation density which is a function of strain rate and hence stress depending on the constitutive relation.

9.5.2 Rotation during Deformation. Coupled Grain Boundary Migration

Grain boundary migration is widely accepted as a recrystallisation mechanism but until recently has rarely been considered as a deformation mechanism in its own right. There are many examples now of grain boundary migration contributing to deformation and, most importantly, there is considerable experimental and computational evidence that grain boundary migration can induce rotation in arrays of grains or subgrains (Srinivasan and Cahn, 2002; Taylor and Cahn, 2007; Trautt and Mishin, 2012, 2014). Grain (or subgrain) rotation coupled to grain boundary migration is now recognised as an integral part of microstructural evolution in polycrystalline aggregates both in the static recrystallisation process (Cahn and Mishin, 2009) and during grain boundary migration in deforming materials (Taylor and Cahn, 2007).

The new development (Srinivasan and Cahn, 2002) is the recognition that grain boundary migration can be coupled to a simple shearing parallel to the boundary. This process is distinct from grain boundary sliding for which there is no coupling but both processes can occur simultaneously (Figure 9.16). In metals, pure grain boundary sliding replaces coupled grain boundary migration at high temperatures (see Figure 9.20). Thus if v_N and $v_\parallel$ are the magnitudes of the velocities normal and parallel to the interface then the two are related by

$$v_\parallel = \beta v_N$$

where β is known as the coupling factor. If there is also a sliding velocity, $\boldsymbol{v}_S$, parallel to the boundary then

$$v_\parallel = \beta v_N + v_S$$

For perfect coupling, Cahn and Taylor (2004) show that where θ is the misorientation across the boundary

$$\beta = 2\tan\frac{\theta}{2}$$

β, however, can be positive or negative according to the nature of the grain boundary at the atomic scale. The degree of coincidence of atomic structure is measured by the quantity, Σ, which is the ratio of the total number of atomic sites to the number of coincident sites. Thus $\Sigma = 1$ for complete coincidence (as in some low-angle tilt boundaries) up to large numbers if there is no coincidence. One expects the energy of the boundary to increase as Σ increases.

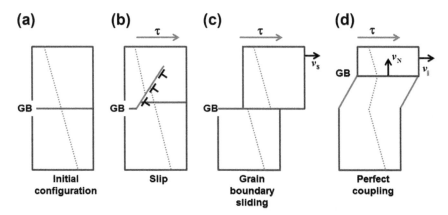

FIGURE 9.16   Tangential and normal motions in a bicrystal parallel and normal to the initial grain boundary, GB. An initial marker is shown as the blue dotted line. (a) Initial bicrystal with a marker line shown in red. (b) Deformation of the grain boundary by dislocation motion. (c) Grain boundary sliding. (d) Coupled grain boundary migration and deformation. $\boldsymbol{v}_N$ and $\boldsymbol{v}_\parallel$ are the velocities normal and parallel to the grain boundary.

Molecular dynamics simulations of grain boundaries in Trautt and Mishin (2012) give values of β shown in Figure 9.18 where β is plotted against θ for a number of coincidence boundaries. β is positive and follows the relation $\beta = 2\tan\frac{\theta}{2}$ up to a misorientation of approximately 34° when it switches to negative values and follows the relation, $\beta = -2\tan\left(\frac{\pi}{4} - \frac{\theta}{2}\right)$. Thus for a given $\boldsymbol{v}_{\parallel}$, the sense of the coupled rotation reverses at about 34°.

9.6 COUPLED GRAIN BOUNDARY MIGRATION DURING DEFORMATION AND METAMORPHISM

Coupled grain boundary migration (Figure 9.17) is a powerful mechanism associated with a number of processes. It can produce lattice rotations, it is associated with grain size reduction and grain growth, it is an important recrystallisation mechanism and is simultaneously an important deformation mechanism. It is also associated with the rotation of material markers within a grain as the grain boundary moves. Systematic and preferred motion of a grain boundary during deformation and grain growth can result in the shape preferred orientation of mineral grains. We address these issues in the following.

9.6.1 Subgrain Rotation

Subgrain rotation is recognised as an important dynamic recrystallisation process in minerals but the mechanism for producing such rotation has never been precisely determined. We revisit this topic in some detail in Chapter 13 but it is now clear in metals that subgrain rotation is a deformation mechanism that arises from a coupled grain boundary migration process. Detailed molecular dynamics simulations such as those carried out by Upmanyu et al. (2006) and Trautt and Mishin (2012) are gradually elucidating the molecular process involved but there is still considerable discussion since the process clearly depends on a number of factors such as the material involved, the current misorientation, the nature of the boundary (that is, the value of Σ), the temperature and the presence or absence of other boundaries. Some examples of these simulations are given in Figures 9.19 and 9.21. The general observation is that a shear stress generates not only a rotation of the subgrain but also a decrease in subgrain size. There are some examples (Cahn et al. 2006) where grain growth can occur during rotation but generally there is a decrease in grain size with continued deformation until finally the subgrain becomes amorphous (glassy) at the nanoscale. Thus subgrain rotation should not simply be viewed as a recrystallisation process. It is an important deformation mechanism and grain size reduction mechanism. It may even be a process whereby some pseudotachylites are generated.

9.6.2 Bulge Recrystallisation

The bulging of grain boundaries at sites where parts of a grain boundary are pinned by small grains of another phase is well documented in deformed rocks (see Vernon, 2004, for examples). The phenomenon is also called *strain-induced grain boundary migration* since the

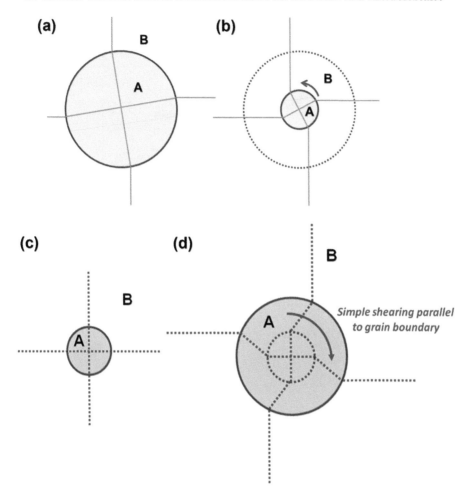

FIGURE 9.17   As a crystal embedded in another crystal decreases in radius, continuity of lattice planes means that the shrinking grain must rotate. (a) Initial crystal A embedded in B. (b) A shrinks and continuity means that A must rotate as shown. One can view the process in the opposite direction and begin with the geometry in (c) as the initial state. The dotted lines are now material lines such as lines of inert inclusions. The lattice of A is not distorted by the plastic shear within the grain as the grain boundary migrates. As A grows to assume the geometry in (d) B must be sheared as it is replaced by A but in the opposite direction to that shown in (b).

process is envisaged to be driven by differences in stored strain energy, ΔG, from one place in an aggregate or grain to another. Cahn and Mishin (2009) point out that the stored strain energy (at least in metals) is too small to drive this process unless the bulging boundary has almost zero curvature. This process cannot depend on classical homogeneous nucleation theory (see Chapter 13) since the energy to do so is insufficient. The process needs to begin with grains of the order of 1–10 μm radius which are relatively strain free. Cahn and Mishin propose that these grains develop by coupled grain boundary migration

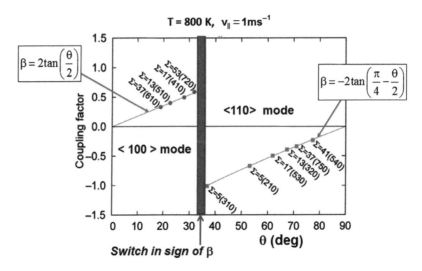

FIGURE 9.18   The switch in sign of the coupling factor, β, as a function of misorientation, θ, and of the coincidence measure, Σ. *After Cahn et al. (2006)*.

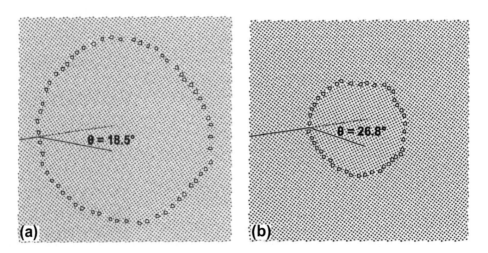

FIGURE 9.19   Molecular dynamics simulation of subgrain rotation arising from coupled grain boundary migration. As the subgrain becomes smaller the rotation increases as in Figures 9.17(a) and (b). *From Trautt and Mishin (2012)*. The small triangles are dislocations.

so that as a small grain (of the order of 1 μm or less) grows it sweeps the dislocations within it away and changes orientation because of the coupling with grain boundary migration. This change in orientation is important because it is the basis for any crystallographic preferred orientation (CPO) that develops during subsequent recrystallisation. Once the radius is large enough it grows into a recrystallised grain driven by ΔG. The process is summarised in Figure 9.22.

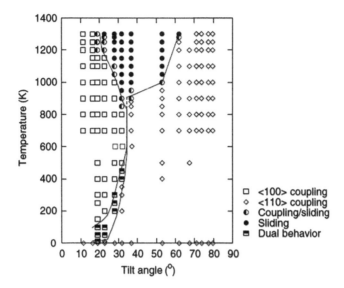

FIGURE 9.20   Diagram showing the behaviour of [001] symmetrical tilt boundaries in copper at various temperatures. Grain boundary sliding replaces coupled grain boundary migration at high temperatures and for a range of tilt angles between about 20° and 60°. *From Cahn et al. (2006).*

9.6.3 Foliations and Lineations

A ubiquitous feature of deformed metamorphic rocks is the preferred orientation of individual mineral phases that are planar, tabular, prismatic or acicular in habit. These preferred orientations constitute foliations in the case of minerals with planar habit such as muscovite and biotite, and mineral lineations in the case of minerals with acicular, bladed or prismatic habit such as sillimanite, kyanite and tourmaline. Note that these preferred orientations of individual minerals constitute only one type of foliation or lineation and it is the type we concentrate on here. The mineral lineations and foliations are commonly paralleled by other fabrics that consist of planar or linear packets of mineral grains.

Three proposals have been put forward for the development of these preferred mineral-shape orientations: (1) the first proposes that these grains form early in a deformation and are rotated during deformation to align with a principal plane or axis of strain, such models have been discussed by Vernon (2004); (2) the second proposes that these grains form to minimise the elastic or stored energy of the grain; an example is the models developed by Kamb (1959), MacDonald (1960) and Brace (1960); (3) the third proposes that internal deformation by crystal slip during deformation aligns the grains with some kinematic framework (see Chapter 13). We explore a different model here whilst admitting that features of all of the above models probably have some relevance.

The model explores two aspects of mineral preferred orientation development comprising *nucleation* and *growth*.

Mineral nucleation: We take as a starting point that during metamorphism new minerals nucleate and begin to grow once the affinity of the relevant mineral reaction becomes large enough. We discuss aspects of this stage of development of a metamorphic fabric in

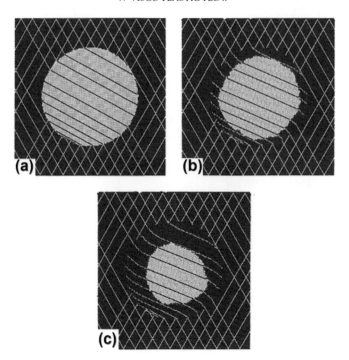

FIGURE 9.21 Subgrain rotation and subgrain grain size reduction. Notice the curvature of the initial lattice in the outer material in (b) and (c). *From Upmanyu et al. (2006).*

Chapter 14. The nucleation stage comprises a competition between the energy of the growing volume of the new grain (proportional to d^3 where d is the grain size) and the increasing energy of the surface energy (proportional to d^2). At a certain critical grain size, increasing the grain size results in a decrease in the Gibbs energy of the grain and grain growth can begin. This theory needs to be modified for solids by the inclusion of the effect of gradients in the chemical potentials of species diffusing to the growing nucleus (Chapter 14) but the basic principle of competition between volume and surface energies remains. The theory, however, makes no comment on the crystallographic orientation of the nucleus and for a non-deforming system there is perhaps no control on the orientation. For a deforming system the situation is different.

As indicated above, Kamb (1959), MacDonald (1960) and Brace (1960) discussed the situation where a nucleus is growing in a stress field and proposed that the ones that grow fastest are those with the minimum stored elastic energy. This effect is undoubtedly always present but it is not clear how important it is relative to other (sometimes competing) effects.

There are of course other influences that may contribute to the growth of a crystal in a solid medium, other than the elastic energy. One of those is anisotropy of the interfacial energies between the growing grain and neighbouring grains. This aspect has been considered by Taylor and Cahn (1988) who addressed the question: *How does the interfacial energy anisotropy cause certain crystallographic orientations to dominate during grain growth?* Taylor and Cahn propose that the anisotropy of both the interfacial energy between the growing crystal and the

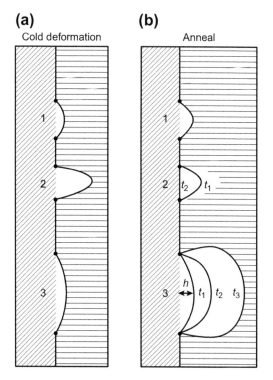

FIGURE 9.22   Bulge recrystallisation. *From Cahn and Mishin (2009).* (a) During deformation coupled grain boundary migration produces parts of original grains that are dislocation free and of various radii. Each one has a different orientation to the initial host because of rotation arising from coupled grain boundary migration during grain growth. (b) Small strain-free grains can only grow if their radii are greater than a critical value. Very-large-curvature grains do nothing. Grains with intermediate curvatures shrink. Grains with small curvatures grow. The time sequence is in the order t_1, t_2, t_3.

surrounding material (the substrate), and the anisotropy of the *difference* between the energy of the substrate and of the crystal/substrate interface, influence the barrier height for nucleation and present some rules for what controls these orientations. These rules are based on the classical Wulff construction (for a freely growing crystal) as modified by Winterbottom (1967) for a crystal growing in a substrate. Further discussion is given by Cahn and Taylor (1988).

There are yet two other influences on the ultimate preferred orientation, namely, anisotropic supply of nutrients to the growing grain and anisotropic growth that selects grains of a specific orientation. We concentrate on anisotropic grain growth below.

9.6.3.1 *Mineral Growth*

The observation of disconnections in many materials suggests that the motion of disconnections may be an important grain growth process. The detailed structure of such disconnections in multiphase metamorphic aggregates is yet to be determined and as we have seen in massive transformations the step height may be a single unit cell but nevertheless

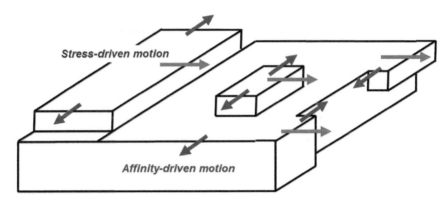

FIGURE 9.23 Driving forces for grain growth during a chemical reaction. The red arrows indicate grain boundary migration driven by the chemical and/or thermally derived affinity of the reaction. The blue arrows indicate the motion of disconnections driven by the applied stress. If the motion arising from affinity is larger than that driven by stress no shape-preferred orientation develops. If the motion driven by stress is larger than that driven by the affinity of the reaction a crystal shape-preferred orientation develops.

the motion of a disconnection type of structure across the surface of a newly growing grain is an attractive model for grain growth and dissolution that needs further investigation. In a stress free environment one might expect the migration of steps to be driven solely by thermal fluctuations and hence to be essentially undirected. In the presence of a stress field we propose that the step motion is strongly correlated and grain boundary migration controlled by the migration of steps (disconnections) constitutes a strong deformation mechanism as well as a preferred growth selection process (Figure 9.23). Thus independently of the processes discussed above involving anisotropy of elastic and interfacial energies, a given distribution of crystal orientations will be further selected by preferred disconnection motion. This is a powerful mechanism for producing the preferred orientations that are seen in foliated mica aggregates (Figure 9.24(b)). In minerals of other habits such as sillimanite or kyanite, a mineral lineation may develop. If the deformation is weak in a particular layer then a decussate structure may form (Figure 9.24(a)) but this does not indicate that the mineral growth is post-tectonic. It simply indicates that the effect of deformation on the preferred migration of disconnection steps is small.

9.6.4 Curved Inclusion Trails in Porphyroblasts

Curved inclusion trails (Figures 9.25 and 9.26) are a common feature within porphyroblasts that have grown at medium grades of metamorphism. They tend to be less common at granulite grades of metamorphism. Discussions of the origins of the curvature are widespread and strongly polarised with fierce defences of various points of view from the sides of the fences involved. A balanced presentation of some of these competing views is in Vernon (2004). The various proponents fit into three camps. The first is the classical camp which claims that the porphyroblast rotates during growth and deformation so that the new porphyroblast overgrows and incorporates the surrounding fabric (Figure 9.25(a)). The rotation is relative to coordinates in the pre-growth state and the process involves a torque generated

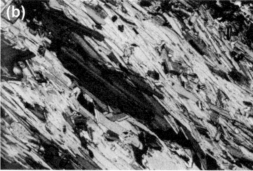

FIGURE 9.24 Microstructures associated with grain shape. (a) Decussate microstructure; weak preferred orientation of blades of kyanite. Western end of South Beach, Fitzgerald National Park, Western Australia. (b) Strong preferred orientation of {001} planes of micas. Stirling Vale, Broken Hill, Australia. Base of photo: 1.75 mm. *From Vernon (2004).*

at the surface of the growing grain exerted by the adjacent deforming matrix. The analogy is commonly drawn with the analysis by Jeffery (1922) regarding the rotation of an ellipsoid in a shearing viscous medium. In modelling these rotations only affine pure or simple shearing deformations are considered.

A second model is due to Ramsay (1962) who proposes that the porphyroblast does not rotate relative to pre-growth reference coordinates and that it is the matrix that rotates relative to the porphyroblast. The non-rotating porphyroblast then simply overgrows and incorporates continuously changing orientations of markers in the deforming matrix (Figure 9.25(b)).

A third model, in some respects quite similar to the Ramsay model, arises from careful microstructural work by Bell and co-workers (Bell, 1985; Bell and Johnson, 1989, 1992; Bell et al., 1992, 1998). This model proposes that the porphyroblast is a composite entity developed during several episodes of growth and dissolution accompanied by alternating periods of foliation development in the matrix. These foliations tend to be orthogonal. No rotation of the composite porphyroblast occurs relative to pre-growth reference axes which are identified with a geographical frame. In the Bell model, although curved trails are admitted, they are inherited from deformation of the matrix and in some instances, the spiral shaped structure is claimed to be apparent and made up of sets of included foliation fabrics.

We do not dwell upon the details of these models here and leave an analysis and discussion to Volume II. It is important to establish the mechanisms involved in this process so that more definitive models can be developed. At present, the classical model (Figure 9.25(a)) gives the opposite sense of shear, for a given inclusion pattern, to that predicted by the Bell model (Figure 9.25(c)). Below we explore the application of the Cahn concepts of grain boundary migration coupled to material rotation to the development of curved inclusion trails in porphyroblasts.

Following the discussion of Section 9.6.3, we propose that porphyroblast growth takes place by the motion of disconnections on the boundary of the porphyroblast. Clearly such motion ultimately arises from the affinity of the chemical reactions involved in the

FIGURE 9.25 Models for the development of curved inclusion trails. (a) The classical model whereby the growing porphyroblast rotates relative to pre-growth coordinates and progressively overgrows the marker as it rotates. (b) The Ramsay (1962) model that proposes the porphyroblast does not rotate relative to pre-growth coordinates and simply overgrows a set of markers in the deforming (and hence rotating) matrix. (c) The model of Bell that proposes the porphyroblast is a composite entity comprised of multiple sets of markers that form in alternating periods of orthogonal shears. The porphyroblast does not rotate relative to geographical coordinates. The sense of shearing interpreted from modes (a) and (c) is opposite to each other. (a) and (c) After Stallard et al. (2002). (b) After Ramsay (1962).

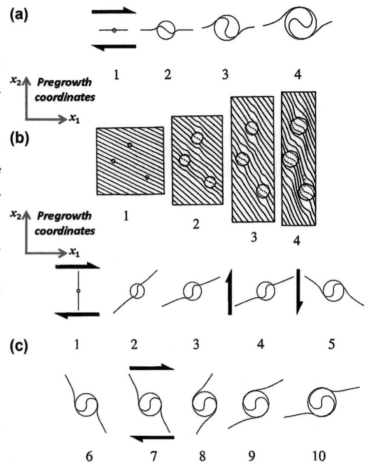

production of the porphyroblast and must depend on diffusion for the supply of nutrients. If there is no deformation, this process is stochastic and the motion of disconnections is not directed (Figure 9.27(a)). In keeping with Section 9.6.3, we propose that under the influence of an imposed deformation, the motion of disconnections is controlled by the deformation and is directed (Figure 9.27(b)).

As considered in Section 9.5.2, the cooperative motion of disconnections at the growing interface induces shear strains parallel to the moving interface. This has the effect of shearing an inert marker such as an inclusion train that is incorporated into the growing porphyroblast. The porphyroblast undergoes plastic deformation by the motion of disconnections so that its crystallographic orientation remains unchanged by the passage of the disconnection. The only evidence of the coupled grain boundary migration-shear deformation is the change in orientation of the inclusion trails (Figure 9.27(c)). The porphyroblast does not rotate relative to pre-growth coordinates during this process.

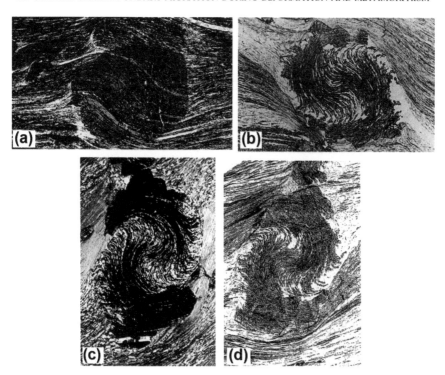

FIGURE 9.26   Examples of curved inclusion trails in garnet. (a) *From Passchier and Trouw (2005).* (b) *From Johnson (1993b).* (c) *Photo: Scott Johnson.* (d) *From Johnson (1993a).*

As Cahn and Taylor (2004) point out, the deformation resulting from the grain boundary migration is a simple shear with shear parallel to the moving boundary. Hence the deformed shape of the inclusion trail reflects the idiomorphic shape of the porphyroblast. If this shape is a rectangle then the deformed shape is a power law curve or a sigmoid depending on the initial orientation of the inclusion trail (Figure 9.27(c)). If the idiomorphic shape is a hexagon, as in garnet, then power-law curves, sigmoids or spirals can develop again depending on the amount of cumulative shear that the inclusion trail experiences (Figure 9.29(c) and (e)).

In Figures 9.27(c), 9.28 and 9.29(b–e), a simplified model is adopted to illustrate the geometrical principles involved. The rules used to construct these figures are:

1. The inclusion trail undergoes a rotation of 20° for each unit of grain boundary migration. This rotation is independent of the pre-rotation orientation of the inclusion trail. The inclusion trail is shortened or stretched in order to preserve continuity of the inclusion trail from one growth increment to the next. The exception is when the inclusion trail becomes parallel to the migrating boundary when no further rotation occurs; the inclusion trail is simply stretched.

2. The surrounding foliation (which is taken to be parallel to what will become the inclusion trial in the porphyroblast) rotates with the deformation so as to always maintain continuity with the inclusion trail in the porphyroblast. This means that these simple models do not consider coupled grain boundary sliding.

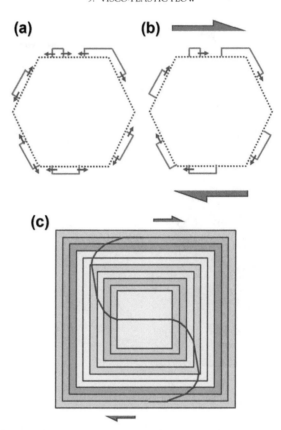

FIGURE 9.27 Growth of porphyroblast by disconnection motion. The red arrows indicate the direction of disconnection motion. (a) Migration of disconnections driven solely by the affinity of the porphyroblast producing mineral reaction with no coupled deformation. (b) Migration of disconnections driven by affinity of the mineral reaction and biased by the imposed deformation. (c) The final microstructure developed in an initially square grain with growth zones outlined by different colours. Shear sense is dextral. The initially horizontal inclusion trail is marked in red. In this simplistic model, the inclusion trail is sheared through 20° and elongated or shortened to maintain continuity for each increment of growth. Once the inclusion trail becomes parallel to the moving grain boundary no further deflection of the inclusion trail occurs.

3. The porphyroblast incrementally overgrows and in doing so distorts the surrounding foliation by coupled grain boundary migration processes.

Notice that this mechanism of coupled grain growth − internal deformation proposed here is the primary mechanism, at the atomic scale, for developing curved inclusion trails in growing, deforming porphyroblasts. The Ramsay mechanism can operate also leading to greater intensity in the reorientation of inclusion trails. It is also possible that alternating periods of growth and dissolution can occur as reflected in the microstructural observations of Bell. It is interesting to speculate that if dissolution is linked to deformation, as in stress-assisted dissolution processes, then shearing within the grain in the opposite direction to that

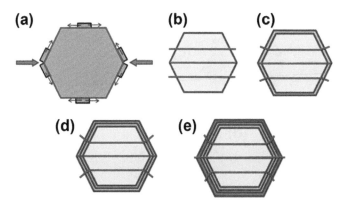

FIGURE 9.28 Development of millipede structure during coaxial shortening and coupled grain boundary migration. (a) Symmetrical migration of disconnections during grain growth under the influence of coaxial deformation history. (b) Initial orientation of hexagonal grain with inclusion trails horizontal. (c) A single growth zone with symmetrical deflection of inclusion trails. (d) and (e) Continued growth with development of millipede structure.

involved in growth could occur. This would require the preferred motion of disconnections during the dissolution process.

If the deformation history is coaxial (Figure 9.28), the shearing coupled to grain boundary migration is symmetrical with respect to the imposed deformation and symmetrical shearing of the inclusion trials evolves resulting in millipede development. If the deformation history is a simple shearing (Figure 9.29), then the shearing coupled to grain boundary migration reflects this asymmetry and successive shearing of the inclusion trails parallel to growing grain boundaries results in spiral shaped trails that cumulatively result in similar folding of the inclusion trails with a curved axial plane (Figure 9.29(d)).

There are interesting geometrical consequences of the inclusion trail assembly being inhomogeneously sheared into nested similar folds. Since adjacent inclusion trails are offset spatially in the initial state, the individual spirals have their centres offset with respect to each other and hence must ultimately merge or intersect. Thus discontinuities in the nested spirals should be common place as a geometrical consequence of the coupled grain boundary migration process. These are indeed common and are called by Bell and Johnson (1989) and Johnson (1993), *foliation truncation zones*. Such zones are commonly attributed to a hiatus in garnet growth; however, they are also a geometrical consequence of the coupled grain boundary migration model. Moreover, a further consequence is that these truncation zones form with a period of approximately $\pi/2$. Thus, in Figure 9.29(d), the truncation zones form at $100°$, $180°$ and $300°$. These relations correspond to the observations of Johnson (1993, his Figure 12).

As the temperature increases, the evidence from metals is that coupled grain boundary migration is replaced by grain boundary sliding. It is then possible that mechanisms proposed by the classical, torque-driven rotation, school may dominate. However, if significant grain boundary sliding occurs simultaneously with coupled grain boundary migration, the development of curved inclusion trails may be accentuated. Despite all these possibilities,

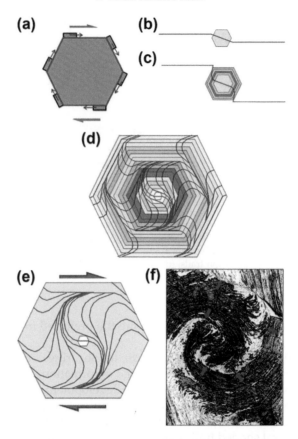

FIGURE 9.29   Shapes of inclusion trails that develop during *coupled grain boundary migration* and with simultaneous rotation of the matrix foliation compatible with the sense of the shearing arrows. (a) Asymmetrical migration of disconnections during grain growth in the sense of the shearing deformation. (b) The early nucleus with deflection of the inclusion trail during early grain boundary migration. (c) Deflection of inclusion trail inside the growing grain due to coupled grain boundary migration. The colours indicate growth zones. Within each growth zone there has been a simple shear of the inclusion trail as the boundary moved through that zone. (d) Late-stage history of grain growth showing the outline of some inclusion trails and with discontinuities in inclusion trails. (e) Part of the image shown in (d) with growth zones removed to show discontinuities more clearly. The initial nucleus is shown in white with the initial orientation of the inclusion trail. (f) Natural garnet porphyroblast with curved inclusion trials and discontinuities in inclusion trails marked. *From Johnson (1993b).*

the fundamental process of producing curved inclusion trails in growing porphyroblasts, in the absence of grain boundary sliding, remains coupled grain boundary migration.

As a final comment, it is important to note that coupled grain boundary migration during porphyroblast growth produces no rotation of the porphyroblast relative to coordinates at the time of nucleation. The grain is internally sheared and this produces a rotation of the inclusion trail within the growing grain relative to coordinates at nucleation but the porphyroblast itself does not rotate relative to these coordinates. The sense of shear given by the shape of the inclusion trail is opposite to the classical interpretation (Figure 9.25(a)).

We leave an in-depth discussion of this process and of other models summarised in Figure 9.25 to Volume II but we emphasise that coupled grain boundary migration is a process grounded firmly in mechanics that offers a quantitative way of addressing the issue of curved inclusion trails and hence is testable with quantitative measurements of geometry.

9.7 CONSTITUTIVE EQUATIONS

The following are some brief comments on the constitutive relations commonly used to describe and model the behaviour of deforming metamorphic rocks. For the most part the relations used assume steady state conditions even though the processes may involve grain size reduction, localisation of deformation, the development of anisotropy and mineral reactions all of which represent non-steady conditions. The experimental 'flow' laws are commonly written for steady state for one dimensional flow in the form

$$\dot{\varepsilon} = A\sigma^{N} \exp\left[\frac{-Q}{RT}\right] \qquad (9.16)$$

where $\dot{\varepsilon}$ is the strain rate, σ is a scalar expression of the Cauchy stress and Q is the activation energy for the rate-controlling process (and sometimes includes an additional pressure-dependent term). N is a material constant that for laboratory-scale experiments lies in the range 1 to about 8 (see compilation by Evans and Kohlstedt, 1995). In order to make this equation dimensionally correct, A has the units $(Pa^{N-1}\ s^{-1})$ so that it is a scalar generalised viscosity.

However, (9.16) cannot be extended to the three dimensional case by writing

$$\dot{\boldsymbol{\varepsilon}} = A\boldsymbol{\sigma}^{N} \exp\left[\frac{-Q}{RT}\right]$$

where $\dot{\boldsymbol{\varepsilon}}$ and $\boldsymbol{\sigma}$ are now second-order tensors. One generalisation of (9.16) to three dimensions is to write

$$\sigma'_{ij} = 2\eta D_{ij}, \quad D_{kk} = 0 \qquad (9.17)$$

where η is the viscosity and given in terms of a reference viscosity η_0 and the shear strain rate, $\dot{\gamma}$, by

$$\eta = \eta_0 \left(\frac{\dot{\gamma}}{\dot{\gamma}_0}\right)^{N-1} \quad \text{where} \quad \eta_0 = \tau_0/\dot{\gamma}_0$$

and

$$\dot{\gamma} = \sqrt{2D_{ij}D_{ij}} \quad , \quad \tau = \sqrt{\frac{1}{2}\sigma'_{ij}\sigma'_{ij}}$$

where primes indicate deviatoric stresses. D_{ij} is the stretching tensor.

The assumption involved in the form of (9.16) and (9.17) is that the principal axes of stretching always parallel the principal axes of stress. Such a constitutive relation is called coaxial where the use of the term is not to be confused with the common use of the term coaxial to describe a deformation history where the principal axes of strain always remain coincident. It is by no means clear that constitutive relations for deforming rocks are coaxial especially as strong elastic and plastic anisotropy evolves during deformations. Strong assumptions are built into (9.16) and (9.17).

A second point concerning (9.16) and (9.17) is that such a material can never localise. It represents the behaviour of a viscosity strain rate weakening viscous fluid with no elasticity and no dependence on the evolution of microstructure. Of course if one constrains the geometry of deformation in a model so that part of the deforming system is pinned in some manner then the deformation must localise to be compatible with the imposed constraints but such behaviour is quite distinct from localisation arising from some form of bifurcation behaviour.

If one adds elasticity to (9.17) then there are many ways to do so (see Chapter 6) but with a complete lack of experimental data for rocks in this regard the choice is made more for convenience than for any regard for reality. Commonly the behaviour of a fluid is again adopted with linear coupling between elasticity and viscosity so that the material is a Maxwell fluid (Chapter 6).

The route towards representing deforming metamorphic rocks by rigorous elastic—plastic—viscous relations is not clear since again we have little in the way of experimental data in this regard. One way is to assume a yield surface (Chapter 6) and then assume that the behaviour of that surface is governed by relation such as (9.17). This is convenient and can readily be implemented in computer codes but nevertheless is still an assumption. Such a procedure, however, does allow localisation and unloading to an elastic state during deformation instabilities if the shape and size of the yield surface is allowed to change with deformation and chemical reactions. This, however, introduces new questions regarding the nature of hardening (or softening). *Is the behaviour isotropic or kinematic* (Chapter 6)? Again there is no experimental work to guide us. Some issues to do with such behaviour are considered in Chapter 13.

Another issue is the incorporation of the influence of chemical species within the constitutive relation. For many silicates a common form of (9.16) is extended to write

$$\dot{\varepsilon} = A f_{H_2O}^M \sigma^N \exp\left[\frac{-Q}{RT}\right] \tag{9.18}$$

where f_{H_2O} is the fugacity of H_2O and M is some exponent. Since the unit of fugacity is Pascals, such an equation cannot be correct unless the units of A are adjusted. Moreover, because of the reaction $2H_2O \rightleftharpoons 2H_2 + O_2$ the fugacity of H_2O needs to be given for a specific value of the fugacity of hydrogen or oxygen. Hence the fugacity of either hydrogen or of oxygen must also appear in equations of the form (9.18). These issues are discussed by Evans and Kohlstedt (1995) and a detailed discussion of constitutive relations is given by Regenauer-Lieb and Yuen (2003).

Of course the criticisms made above are easy to level since the experiments involved in gathering the data required are difficult, expensive and perhaps even impossible to perform because of the difficulty in gathering the required measurements imposed by the required timescales or pressures. One way forward is to model the situations involved using molecular or dislocation dynamical simulations. The results of such modelling needs to be compared with the results of experiments but simulation offers the opportunity to answer some of the

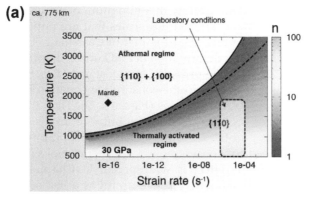

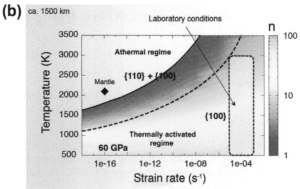

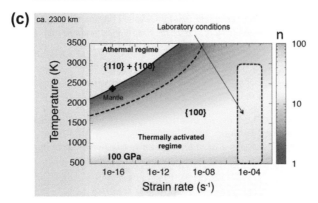

FIGURE 9.30 Molecular dynamic calculations for the variation of the value of the stress exponent, n, for olivine at various temperatures, strain rates and depths within the Earth. (a) 775 km depth; (b) 1500 km depth; (c) 2300 km depth. The dashed curve in each case is the slip system transition from {110} + {100} at high temperatures and low strain rates to {100} slip at low temperatures and fast strain rates. The solid curve marking the boundary between white and red areas is the boundary between the thermally activated dislocation regime (kink migration − Peierls stress controlled) below the curve and the athermal (dislocation interaction) regime, above the curve. Figures from Patrick Cordier. Note that under laboratory conditions the material is always in the thermally activated regime, whereas under mantle conditions the material is in the athermal regime with very large values of n. In the laboratory the material is rate sensitive. Under mantle conditions the material is rate insensitive.

questions posed by the above discussion with a minimum of expense and to perform computer simulations in environments and on timescales that are impossible to access experimentally. An important study in this regard is that of Castelnau et al. (2010) and Amodeo et al. (2011). They have studied the deformation of MgO under mantle conditions and shown that a transition exists with increasing pressure, but most importantly with decreasing strain rate, from relatively small values of N in the constitutive equation under laboratory conditions to very large values at slow strain rates (Figure 9.30). Thus the proposal is that MgO is quite rate sensitive under laboratory conditions but rate insensitive at geological strain rates. This result has wide implications for extrapolation of laboratory data to geological conditions and prompts the need for considerable work on the molecular and dislocation dynamics of silicates to see if the effect is widespread. In Chapter 13 we point out that careful work on CPO measurements may prove useful as a test of such simulations since the transition from rate sensitivity to rate insensitivity is reflected in the shape of the yield or flow surface and so will influence the type of CPO that develops.

Recommended Additional Reading

Clayton, J.D. (2011). *Nonlinear Mechanics of Crystals.* Springer.
    An advanced text including a treatment of the mechanics of crystals in terms of generalised coordinates. Many aspects of elasticity and plasticity are treated together with a thorough treatment of dislocations and disclinations. One might look upon this book as a modern version of Nye (1957).

Gutkin, M., Yu, Ovid'ko, I.A. (2004). *Plastic Deformation in Nanocrystalline Materials.* Springer.
    Although the term nanocrystalline appears in the title of this book, much of the content is just as applicable to coarser grained materials. Topics such as the role of disclinations in deformation and the interactions with dislocations are given extensive treatment.

Kelly, A., Groves, G.W. (1970). *Crystallography and Crystal Defects.* Longman.
    This is an excellent treatment of the geometry of slip, of the theory of dislocations and of crystal interfaces. It includes a discussion of the Wulff construction.

Paterson, M.S. (2013). *Material Science for Structural Geology.* Springer.
    A concise discussion of many aspects of material science with direct application to metamorphic geology. The book contains an excellent treatment of dislocation theory and of diffusion.

Regenauer-Lieb, K., Yuen, D.A. (2003). Modeling shear zones in geological and planetary sciences: solid- and fluid-thermal—mechanical approaches. *Earth-Sci. Rev.* 63, 295—349.
    A critical analysis of constitutive relations used in the geosciences based on a thermo-mechanical framework.

Damage Evolution

10.1 PHENOMENOLOGY — WHAT IS DAMAGE?

Damage is a generic term that refers to any structural or chemical rearrangement within a material that results in degradation of strength or load bearing capacity. The physical parameters that are influenced by damage are the yield strength, the elastic moduli, the rates of plastic hardening and softening and the frictional resistance to sliding. Although degradation in these properties is the hallmark of damage, other processes commonly operate in unison to produce *healing, recovery or annealing* of the damage. The resultant behaviour arises from the competition between the rates of damage accumulation and of damage recovery or healing. The damage (Figure 10.1) may consist of distributed and localised fracturing (Reches and Lockner, 1994), the formation of voids, bubbles or fluid inclusions (Ashby et al., 1979; Billia et al., 2013), chemical effects, such as stress corrosion (Atkinson, 1979; Atkinson, 1987), grain size reduction (Kilian et al., 2011; Rutter and Brodie, 1988) and ductile fracture (Brown, 2004; Eichhubl, 2004; Hobbs and Ord, 2010; Karrech et al., 2011a; Regenauer-Lieb, 1998). Even the generation of dislocations in a deforming crystal is a form of damage (Dimiduk et al., 2010) since single crystals deform plastically by dislocation motion below the theoretical strength.

Ashby et al. (1979) classify damage (Figure 10.2(a)) according to the temperature of deformation, whether the damage is intra- or intergranular and whether the damage comprises microcracks, voids or dynamic recovery and recrystallisation. On the basis of such a classification Ashby et al. (1979) developed *damage mechanism maps* (Figure 10.2(b)), which plot the

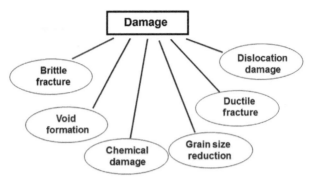

FIGURE 10.1   Classes of damage.

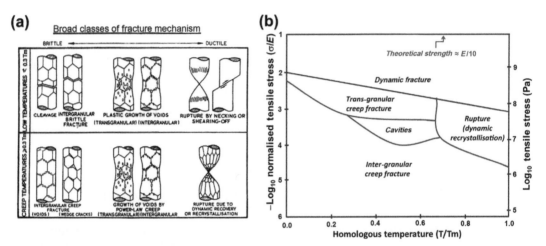

FIGURE 10.2   Classification of damage. (a) *From Ashby et al. (1979).* (b) A sketch of a damage mechanism map motivated by Ashby et al. (1979). All of these mechanisms degrade the theoretical strength of a single crystal given by $\approx E/10$ where E is the Young's modulus. Tm is the melting temperature.

type of damage in stress–temperature space and are useful in interpreting microstructures. Such maps indicate, for instance that cavity development, or void formation, is restricted in temperature–stress space (Figure 10.2(b)). Mancktelow and Pennacchioni (2004) and Fitzgerald et al. (2006) interpret the presence or absence of voids in mylonites solely in terms of 'wet' or 'dry' conditions. Clearly the addition of water adds another layer of complexity to the construction of damage mechanism maps but nevertheless in both 'dry' and 'wet' environments one would expect the presence of cavity development to be restricted in stress-temperature space and not be solely a function of the presence or absence of water. To date there has been little development of this concept for metamorphic rocks although attempts exist (Atkinson, 1982, Atkinson, 1987). Clearly such maps should be drawn for a specific strain rate and grain size and, as indicted above, for 'water content' as well. Two examples of damage in deformed rocks are given in Figure 10.3.

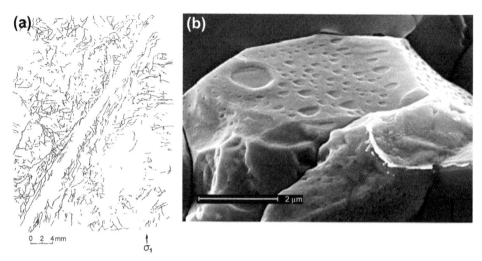

FIGURE 10.3   Examples of damage in deformed rocks. (a) Fractal microfracture development in experimentally deformed granite *From Velde et al. (1991).* (b) SEM image of voids developed on the grain boundaries of quartz grains in a mylonite. *From Mancktelow and Pennacchioni (2004).*

Ashby et al. (1979) distinguish three different modes of failure resulting from damage accumulation: (1) elastic behaviour with little or no dissipation prior to failure: this is commonly referred to as *brittle fracture*; (2) significant rate independent (plastic) dissipation prior to failure: this is *ductile fracture* and (3) significant rate dependent (viscous) dissipation prior to failure: this is *creep* or *viscous fracture*. A review of many of the approaches to failure by damage accumulation is given by Karrech et al. (2011a) and a wider, in depth, review can be found in Voyiadjis (2011). Below we examine some developments specifically devoted to geological examples. The work cited below (Section 10.2) by Lyakhovsky and co-workers is essentially concerned with brittle fracture although such work is extended into creep fracture in Lyakhovsky et al. (2011). The work cited below (Section 10.3) by Karrech and others involves all three modes of failure.

The study of damage over the past decade or so has highlighted a view of deformation that differs from the classical one that deformation, as represented by smooth stress–strain curves, is essentially continuous except where clear localisation or fracturing occurs. Careful experiments have shown that deformation characteristics are scale invariant with discontinuities in deformation occurring at all time and length scales. Even the plastic deformation of single crystals loaded at slow, constant velocities occurs in a jerky manner with avalanches of slip activity separated by periods of relative inactivity (Dimiduk et al., 2006; Zaiser, 2006). Such periods of discontinuous plastic deformation involving avalanches of dislocation slip are associated with acoustic emissions (Zaiser, 2006) in just the same manner as is familiar during the brittle deformation of rocks (Lockner et al., 1991; Reches and Lockner, 1994). The overall behaviour is associated with a loss in convexity of the Helmholtz energy (Ortiz and Repetto, 1999) and the language used in generally describing damage phenomena is to do with *critical phenomena* (Section 10.4), *phase transformations* and *fractal* (or scale invariant) geometries and time series. A review of this approach is given in Girard et al. (2010). Thus the study of damage development is an example of critical behaviour and the resulting scale invariance, subjects that pervade many other areas of metamorphic geology.

10.2 A THERMODYNAMIC THEORY OF DAMAGE

Any useful theory of rock damage should include the observations that the elastic moduli depend on the conditions of deformation (Lockner and Byerlee, 1980) and that subcritical crack growth occurs throughout the loading history accompanied by material degradation as the microcrack density increases (Lockner et al., 1991; Reches and Lockner, 1994). The observed patterns of fractures developed in deformed rocks are commonly fractal in geometry (Velde et al., 1991). In addition the observed time series of acoustic energy release is also scale invariant. Any theory should be able to explain this scale invariance, both spatial and temporal. The theory should provide a criterion for macroscopic failure, including a criterion for damage evolution, and include a description of postfailure evolution including healing of microfractures.

The theoretical developments by Lyakhovsky and co-workers over the past 30 years or so represent important steps in achieving these goals (Lyakhovsky and Myasnikov, 1984; Lyakhovsky et al., 1997, 2011). We present the model in some detail because it provides examples of many of the constitutive and thermodynamic principles discussed in Chapters 5, 6 and 7.

The theory begins with the definition of the Helmholtz energy for a damaging solid. This encompasses the energy arising from the classical linear Hooke's law but is modified by the addition of an energy term that describes the opening and closing of cracks; this results in nonlinear elasticity. The form of this definition needs to take into account that the elastic moduli depend on the nature of the deformation, particularly that the elastic moduli in tension degrade more rapidly than in compression (Lockner and Byerlee, 1980). From this energy the nonlinear constitutive equation for an elastic-viscous material with cracks may be derived. A restriction imposed by the second law of thermodynamics, namely that the entropy production must always be greater or equal to zero, is used to derive a thermodynamically admissible evolution law for subsequent damage. Instability is defined by a loss in convexity of the Helmholtz energy (see Section 7.7); this can coincide with a bifurcation where the governing equations lose ellipticity (See Volume II, and Rudnicki and Rice, 1975) but in general the two criteria for instability do not coincide. Finally the postfailure evolution is modelled using the evolution laws. The result is a geologically realistic, thermodynamically admissible model for damage that can be calibrated using experimental data.

The model proposes that the elastic moduli depend on the microcrack density through a scalar damage parameter, α, which evolves as the deformation proceeds and is spatially inhomogeneous:

$$\alpha = \alpha(x_i, t)$$

$$\lambda = \lambda(\alpha), \quad G = G(\alpha), \quad \varpi = \varpi(\alpha)$$

where x_i are a set of Cartesian coordinates and t is time; λ and G are the Lamé parameters (see (6.24)) and ϖ is another elastic parameter, which appears below in (10.3) and (10.11) and, as can be seen, is also a function of the damage. The evolution of α with time describes the damage evolution and hence the kinetics of degradation (or recovery).

If the material is linearly elastic and we neglect thermal effects we see in Chapter 6 that the specific internal energy can be written, (6.32):

$$e = \frac{1}{\rho}\left(K\frac{\varepsilon_{ii}\varepsilon_{jj}}{2} + G\varepsilon'_{ij}\varepsilon'_{ij}\right) = \frac{1}{\rho}\left(\frac{\lambda}{2}I_1^2 + G\tilde{I}_2\right) \tag{10.1}$$

where I_1 is the first invariant of the small strain tensor, $\tilde{I}_2 = \varepsilon_1^2 + \varepsilon_2^2 + \varepsilon_3^2$, is twice the second invariant of the deviatoric small strain tensor, K and G are the bulk and shear elastic moduli, and λ and G are the Lamé parameters (see Chapter 6). From (10.1) we have Hooke's law (using Table 5.4) for an isothermal deformation:

$$\sigma_{ij} = \rho\frac{\partial e}{\partial \varepsilon_{ij}} = \left(\lambda I_1\delta_{ij} + 2G\varepsilon_{ij}\right) \tag{10.2}$$

The Lyakhovsky model modifies (10.1) by adding another term that involves the energy associated with the opening and closing of cracks:

$$e = \frac{1}{\rho}\left(\frac{\lambda}{2}I_1^2 + G\tilde{I}_2 - \varpi I_1\sqrt{\tilde{I}_2}\right) \tag{10.3}$$

The rationale for adding this extra term is discussed by Lyakhovsky et al. (1997); its form arises from the requirement that the elastic moduli for a solid with microfractures depends on the nature of the deformation (Lockner and Byerlee, 1980). Then (using Table 5.4 again),

$$\sigma_{ij} = \rho\frac{\partial e}{\partial \varepsilon_{ij}} = \left[(\lambda - \varpi\xi^{-1})I_1\delta_{ij} + \left(G - \varpi\xi^{-1}\right)\varepsilon_{ij}\right] \tag{10.4}$$

where

$$\xi = \frac{I_1}{\sqrt{\tilde{I}_2}} = \frac{\varepsilon_1 + \varepsilon_2 + \varepsilon_3}{\sqrt{\varepsilon_1^2 + \varepsilon_2^2 + \varepsilon_3^2}} \tag{10.5}$$

Hence ξ is the ratio of the volumetric to the shear strain and so is a measure of the type of deformation: $\xi = -\sqrt{3}$ corresponds to isotropic compaction ($\varepsilon_1 = \varepsilon_2 = \varepsilon_3 < 0$), $\xi = -1$ to uniaxial compaction ($\varepsilon_1 < 0, \quad \varepsilon_2 = \varepsilon_3 = 0$), $\xi = 0$ to an isochoric deformation, $\xi = +1$ to uniaxial extension ($\varepsilon_1 > 0, \quad \varepsilon_2 = \varepsilon_3 = 0$) and $\xi = +\sqrt{3}$ to isotropic expansion ($\varepsilon_1 = \varepsilon_2 = \varepsilon_3 > 0$).

We can now write the specific Helmholtz energy (Chapter 5) as

$$\Psi = \Psi\left(\varepsilon_{ij}, T, \alpha\right)$$

so that α acts as an internal variable. Then the energy balance, the total entropy balance and the Gibbs equation are given respectively by

$$\frac{de}{dt} = \frac{d}{dt}(\Psi + Ts) = \frac{1}{\rho}\sigma_{ij}\dot{\varepsilon}_{ij} - \nabla_i q_i$$

$$\frac{ds}{dt} = -\nabla_i \left(\frac{q_i}{T}\right) + \Gamma$$

$$d\Psi = -sdT + \frac{\partial \Psi}{\partial \varepsilon_{ij}} d\varepsilon_{ij} + \frac{\partial \Psi}{\partial \alpha} d\alpha$$

where Γ is the *specific internal entropy production* given by

$$\Gamma = -\frac{q_i}{\rho T^2} \nabla_i T + \frac{1}{\rho T} \sigma_{ij}\dot{\varepsilon}_{ij} - \frac{1}{T} \frac{\partial \Psi}{\partial \alpha} \frac{d\alpha}{dt} \geq 0 \qquad (10.6)$$

The first term in (10.6) is the entropy production due to heat conduction. The second term is that arising from deformation and the third is the entropy production due to damage processes. The inequality is an expression of the second law of thermodynamics. We assume that the mechanisms involved in these three processes are independent of each other so that each term in (10.6) must be positive.

Lyakhovsky et al. (1997) expand the entropy production due to damage, Γ_α, as a Taylor series:

$$\Gamma_\alpha = -\frac{1}{T} \frac{\partial \Psi}{\partial \alpha} \frac{d\alpha}{dt} = \Gamma_0(\alpha) + \Gamma_1(\alpha)\frac{d\alpha}{dt} + \Gamma_2(\alpha)\left(\frac{d\alpha}{dt}\right)^2 + \dots \qquad (10.7)$$

where the Γ_i are coefficients for the Taylor expansion. Lyakhovsky et al. (1997) show that the first two terms in the Taylor expansion are zero. Hence to ensure the entropy production is positive,

$$\frac{d\alpha}{dt} = -C\frac{\partial \Psi}{\partial \alpha} \qquad (10.8)$$

where

$$C = \frac{1}{T\Gamma_2}$$

is a positive function of the state variables. Lyakhovsky et al. (1997) assume that $\lambda = \lambda_0 = $ constant, $G = G_0 + \alpha\xi_0\varpi_\Gamma$ and $\varpi = \alpha\varpi_\Gamma$, then the damage evolution takes the form $\frac{d\alpha}{dt} = C_{deg}\tilde{I}_2[\xi - \xi_0]$ for damage that results in degradation.

ξ_0 can be thought of as a generalised internal friction, which separates states associated with degradation and healing (Figure 10.4(a)). In terms of Mohr–Coulomb constitutive behaviour with friction angle φ, one can write

$$\xi_0 = \frac{-\sqrt{3}}{\sqrt{2q^2(\lambda_e/G_e + 2/3)^2 + 1}}$$

where

$$q = \frac{\sin \varphi}{1 - (\sin \varphi)/3} \quad \text{and} \quad \lambda_e = \lambda - \varpi\xi^{-1}, G_e = G - \varpi\xi^{-1}$$

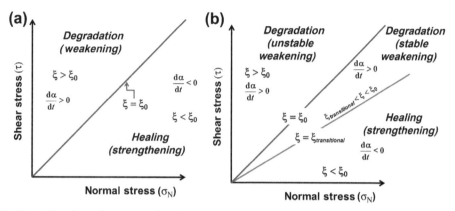

FIGURE 10.4   The physical meaning of ξ_o for a Mohr–Coulomb material. (a) Linear dependence of ϖ on α. The line $\xi = \xi_o$ separates a region in normal stress-shear stress space where degradation is characteristic from one where healing is characteristic. (b) If ϖ is a nonlinear function of α then another field is added to (a). This comprises a region where the damage produces degradation but the deformation remains stable.

Lyakhovsky et al. (1997) give an example for Westerly granite with $\varphi \approx 30°$, resulting in $\xi_o \approx -0.8$.

We now need to understand the conditions for stability of the deformation generated by the distributed damage. In other words, *what are the conditions that the distributed damage should localise?* Two different mathematical approaches to this problem exist. One criterion for such localisation involves the transition from a convex to a nonconvex form of the internal energy, (10.3). If this function is convex then the deformation is stable (Hill, 1998). If the energy is nonconvex it is possible that two solutions exist that can minimise the energy and the system can divide into two different deformations, homogeneous and localised, with each solution occupying different parts of the system (see Section 7.7). The other condition is that instability is marked by the governing equations undergoing a transition from *elliptic* to *hyperbolic* (Rudnicki and Rice, 1975). We discuss this criterion for localisation in Volume II, and the difference between elliptic and hyperbolic partial differential equations in Appendix C. Thus we seek conditions where the system loses either convexity or ellipticity.

The internal energy given by (10.3) is convex if all eigenvalues of the (6 × 6) matrix $\frac{\partial^2 e}{\partial \varepsilon_{ij} \partial \varepsilon_{ij}}$, are positive. Lyakhovsky et al. (1997, Table 1) present this matrix and show that two conditions exist for stability, namely:

$$2G - \varpi\xi \geq 0 \tag{10.9}$$

and

$$(2G - \varpi\xi)^2 + (2G - \varpi\xi)(3\lambda - \varpi\xi) + (\lambda G\xi - \varpi^2)(3 - \xi^2) \geq 0 \tag{10.10}$$

These two conditions emphasise that loss of convexity depends on a critical value of ϖ, and hence of the damage, α, given a particular deformation, ξ. These conditions are plotted in Figure 10.5(a). For (10.9) the condition for loss of convexity is met before loss of ellipticity

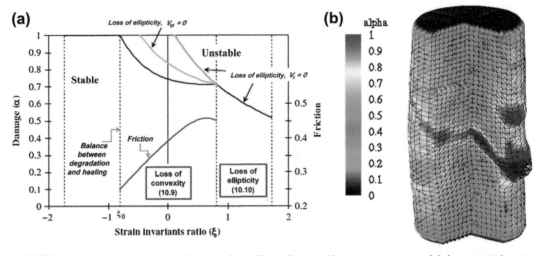

FIGURE 10.5 Behaviour of damaged materials. (a) Phase diagram (damage versus type of deformation) showing conditions for loss of convexity (black curves) and loss of ellipticity (green/black and blue lines). The variation of the coefficient of friction for a damaged surface against type of deformation is shown in red. *From Lyakhovsky et al. (2011).* (b) Simulation of shear localisation arising from accumulated damage. The degree of damage is shown by the colour scale. *From Hamiel et al. (2004).*

which is defined in terms of the velocity of anisotropic S-waves. Condition (10.10) coincides with the loss of ellipticity in terms of the velocity of S-waves.

As an extension of the theory, Hamiel et al. (2004) modify the definition of ϖ to become a nonlinear function of α and so as to better fit experimental data:

$$\varpi = \varpi_1 \frac{\alpha^{1+\beta}}{1+\beta}; \quad 0 < \beta < 1 \tag{10.11}$$

This adds another field to Figure 10.4(a) so that fields of stable and unstable weakening are delineated (Figure 10.4(b)). The result of numerically simulated localisation is shown in Figure 10.5(b).

Another modification to the model is made by Lyakhovsky et al. (2011). Viscous deformation is added formally to the energy equation as a viscous term $\eta \varepsilon_{ij} \dot{\varepsilon}_{ij}$, and the energy is expressed in a non-local manner (see Chapter 6) by the addition of a gradient damage term, $\frac{\kappa}{2} \nabla_i \alpha \cdot \nabla_i \alpha$. η is a viscosity and κ has the units of (stress $\times$ length^2):

$$e = \frac{1}{\rho} \left(\frac{\lambda}{2} I_1^2 + G\tilde{I}_2 - \varpi I_1 \sqrt{\tilde{I}_2} + \eta \varepsilon_{ij} \dot{\varepsilon}_{ij} + \frac{\kappa}{2} \nabla_i \alpha \cdot \nabla_i \alpha \right) \tag{10.12}$$

Using this energy equation, Lyakhovsky et al. (2011) derive an equation for damage evolution, which is a *reaction diffusion equation* (Chapter 7):

$$\frac{\partial \alpha}{\partial t} = C\kappa \nabla^2 \alpha + C\left(G\tilde{I}_2 \varpi I_1 \sqrt{\tilde{I}_2} - \eta \frac{\partial \eta}{\partial \alpha} \varepsilon_{ij} \dot{\varepsilon}_{ij} \right) \tag{10.13}$$

(a) **(b)**

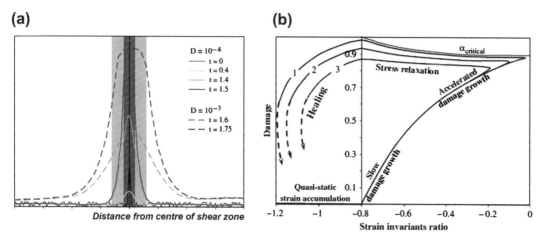

FIGURE 10.6 (a) Localisation of a shear zone with increasing deformation given by t. D is a coefficient for damage diffusion. (b) The earthquake cycle as modelled by a non-local form of the energy involving diffusion and healing of damage. *From Lyakhovsky et al. (2011).*

The first term on the right hand side of (10.13) expresses the diffusion of damage whereas the second term in brackets expresses the nonlinear source of damage.

Lyakhovsky et al. (2011) use this non-local, elasto-viscous formulation to model shear zone development and show that the shear zone localises as deformation proceeds (Figure 10.6(a)), a behaviour seen in many fault zones (Ben-Zion and Sammis, 2003; Chester and Chester, 1998). By incorporating an expression for damage healing,

$$\frac{d\alpha}{dt} = C_{heal\_1} \exp\left[\frac{\alpha}{C_{heal\_2}}\right] \tilde{I}_2[\xi - \xi_0] \tag{10.14}$$

the model also enables details of the earthquake cycle to be reproduced (Figure 10.6(b)).

Thus the Lyakhovsky model reproduces many aspects of damage accumulation commonly observed in deformed rocks namely, differences in elastic moduli depending on the type of deformation, localisation due to accumulation of damage, further localisation as a shear zone evolves and details of the earthquake cycle. We still need to understand why fractal behaviour occurs both spatially in fracture patterns and temporally in patterns of earthquake activity. We look at these issues in Section 10.4.

10.3 DUCTILE DAMAGE

The previous section, derived from the work of Lyakhovsky and others is based specifically on a form of damage comprised of opening and closing and healing of microcracks. Another approach, based on a wider variety of damage mechanisms is that of Karrech et al. (2011b, 2012). Again the approach has a thermodynamic basis but now the evolution of damage is based on maximising the entropy production according to the concepts of Ziegler (1963; see Section 5) rather than relying on the entropy production being non-negative

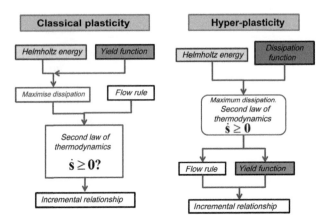

FIGURE 10.7   The different approaches used in classical plasticity and hyperplasticity. *From Karrech et al. (2011b).*

as in the Lyakhovsky model. The approach is summarised in Figure 10.7, which is useful in highlighting differences in using classical approaches to plasticity and using thermodynamic approaches to hyperplasticity (Houlsby and Puzrin, 2006a).

Hyper-plasticity is discussed in Chapter 6. The subject is a formal way of developing thermodynamically admissible theories dealing with the plastic deformation of materials. As shown in Figure 10.7, classical plasticity begins by defining two quantities, the Helmholtz energy and the yield function. Commonly the yield function is an empirical fit to experimental data with no check as to whether the derived yield function obeys the laws of thermodynamics. Both classical (Cauchy) elasticity and Mohr–Coulomb plasticity are of this form and, under some conditions, do not obey the laws of thermodynamics (Chapter 6). Classical plasticity also proposes a flow rule whose precise form is derived by maximising the dissipation. From these procedures an incremental form of the constitutive behaviour can be derived and towards the end of this procedure a check may (or may not) be made to see if the formulation is thermodynamically admissible.

The route taken by Karrech et al. (2011c, 2012) corresponds to hyper-plasticity: two functions are defined initially, namely, the *Helmholtz energy*, Ψ, and the *dissipation function*, Φ. Ψ is the sum of energy contributions from deformation, thermal expansion and damage. An expression for the stress in the material can then be obtained from the Helmholtz energy as $\sigma_{ij} = \rho \frac{\partial \Psi}{\partial \varepsilon_{ij}}$. Φ comprises terms due to dissipation contributions from plastic deformation, volumetric changes, rate sensitive deformation and isotropic damage. Ziegler's principle of maximum dissipation (Section 5.11) is used to establish the *yield functions* for plastic flow and damage together with the *flow rules* from which the *incremental response* follows. Thus one is guaranteed that the resulting response is thermodynamically admissible rather than checking at the end. For details see Karrech et al. (2011c, 2012).

An example of the use of this approach to extensional deformation of the crust is given in Figure 10.8. The results using the damage model with frictional effects are shown in the left column of Figure 10.8 and the results using a damage constitutive behaviour, which resembles the behaviour of a metal (that is, no pressure dependence) on the right. In both cases shear localisation nucleates in the lower crust and propagates upwards with continued

Pressure dependent (t = 1.1 million years)

Von Mises (t = 0.9 million years)

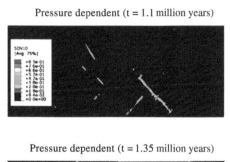

Pressure dependent (t = 1.35 million years)

Von Mises (t = 1.37 million years)

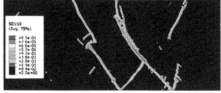
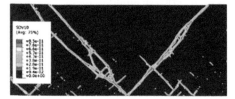

Pressure dependent (t = 2.35 million years)

Von Mises (t = 2.37 million years)

Pressure dependent (t = 4.18 million years)

Von Mises (t = 4.17 million years)

FIGURE 10.8 Damage evolution in a continental cross-section in extension. Left hand column represents pressure dependent response. Right hand column represents a response dominated by nonfrictional processes (von Mises yield surface). The time after the start of extension is labelled for each frame. *From Karrech et al. (2011b)*. Initial model dimensions are 25 × 60 km. Top 5 km is quartz rich and the bottom 20 km feldspar rich. Temperature gradient from top to bottom is 28 °C km^{-1}.

extension. The orientations of the shear zones change with depth in the frictional model and approach a vertical orientation in the uppermost crust whereas they remain constant at approximately 45° to the extension direction in the metal-like model. The behavior of the frictional damage model resembles many natural examples and is not what is expected from Andersonian fault mechanics or from Mohr—Coulomb behaviour. Another example is given by Zhang et al. (2013).

10.4 CRITICALITY AND PHASE TRANSITIONS

The brittle failure of materials has long been recognised to occur discontinuously comprising a series of avalanches of damage formation and propagation (Lockner and Byerlee, 1980) as revealed by acoustic emissions. Moreover, the form of the stress—strain curve is sensitive to the manner in which the material is loaded. If the deformation is controlled by the load applied to the specimen (Figure 4.1(a)) then catastrophic instability is commonly observed (Figure 10.13(c)); such behaviour is referred to as a SNAP instability. If, however, the deformation is displacement controlled (Figure 4.1(b)) then the complete stress—strain behaviour may be followed with a series of small avalanches and with no catastrophic instabilities (Figure 10.13(c)). This type of behaviour may be referred to as a POP instability. These two loading modes correspond to 'soft' and 'stiff' loading frames (Jaeger and Cook, 1969, pp. 167—171).

The behaviours described above are not restricted to brittle deformations; jerky deformation behaviour arising from dislocation slip avalanches is now widely recognised in crystal plasticity (Dimiduk et al., 2010). Figure 10.9(a) shows the results of loading a single crystal of LiF at a constant velocity of 0.2 nm s^{-1}. The displacement versus time plot shows small discontinuous steps associated with bursts or avalanches of localised dislocation slip. Again in Figure 10.9(b) results for loading a single crystal of Ni at a constant velocity of 5 nm s^{-1} are shown; both the stress and displacement records show discontinuities associated with localised slip activity. Such slip avalanches in visco-plastic materials, as is the case for brittle deformation, are associated with acoustic emissions (Imanaka et al., 1973; Weiss and Miguel, 2004; Weiss et al., 2000; Zaiser, 2006).

In both brittle and visco-plastic deformation modes there are power-law relations in the scaling of avalanche events. Figure 10.10 shows the scaling of acoustic emissions in (a) for the visco-plastic deformation of single crystals of ice and (b) seismic events in Southern California. An additional element of complexity is introduced in the earthquake data

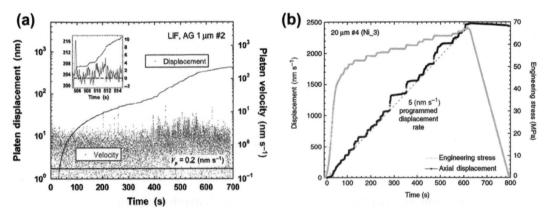

FIGURE 10.9 Intermittent avalanche behaviour during the plastic flow of crystals. (a) Plot of displacement (blue) and velocity (red) during the plastic deformation of a single crystal of LiF against time. The specimen is deformed at a programmed constant velocity of 0.2 nm s^{-1}. Detail in inset. *From Dimiduk et al. (2010).* (b) Plot of engineering stress (grey) and displacement (black) for a single crystal of Ni against time loaded at a programmed constant rate of 5 nm s^{-1}. *From Zaiser (2006).*

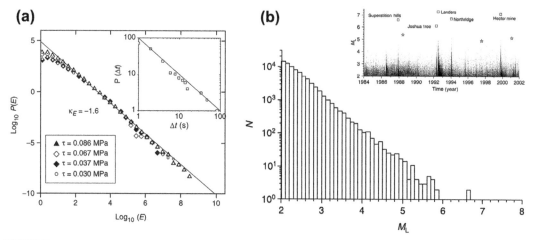

FIGURE 10.10 Power-law statistics for single crystal plasticity and earthquakes in Southern California. (a) Plot of the logarithm of the probability of a dislocation avalanche event of energy E against $\log_{10} E$ for plastic deformation of ice. *From Zaiser (2006).* (b) Logarithm of the number of events of magnitude M_L against $\log_{10} M_L$ for the earthquake time series shown inset for Southern California. *From Ben-Zion (2008).*

in that two modes of power-law behaviour are observed. One is the classical *Gutenberg–Richter* behaviour where the number, N, of seismic events with a magnitude, M_L, bears a power-law relation with M_L up to a cut-off magnitude. The other corresponds to a power-law relation between N and M_L up to a cut-off magnitude but larger earthquakes occur above the cut-off magnitude and are not part of the power-law relation (Figure 10.10(b)). Ben-Zion et al. (1999) refer to this second kind of distribution as *characteristic earthquake statistics* (Figure 10.13(d)). Any particular seismically active region can swap between these two modes of earthquake activity and such behaviour is known as *mode switching* (Ben-Zion, 2008; Ben-Zion et al., 1999).

The types of behaviour described above are explained by viewing the yield behaviour of materials as a *first order phase transition* where the *Helmholtz energy* undergoes a transition from *convex* to *non-convex* (see Chapter 7; Ben-Zion, 2008; Chapter 2 of Powell, 1978). Ortiz (1999) for instance describes the plastic yield of metals in terms of an energy function:

$$f(h_0) = \frac{1}{2}h_0^2 - \kappa \log(\cosh h_0) - hh_0 \qquad (10.15)$$

In (10.15), $h_0 = \tanh^{-1}[2\langle\varepsilon\rangle - 1]$ where $\langle\varepsilon\rangle$ is a measure of the normalised shear strain. The quantity h is given by $h = \frac{\tau}{2} - \frac{1}{2}\log(1 + \exp(\mu/k^B T))$ and so is a measure of the normalised shear stress, τ, at a particular absolute temperature, T, μ is the chemical potential associated with the dislocations, and k^B is Boltzmann's constant. $\kappa = \frac{K}{K_{critical}}$ where $K \approx \frac{Gb^3}{2k^B T}$ is the self-energy of the dislocation divided by $2k^B T$. $K_{critical}$ is the value of K where the energy of the system changes from convex to nonconvex. The critical point also corresponds to a critical temperature, T_{crit}. The energy is convex for $T > T_{crit}$ and nonconvex for $T < T_{crit}$. (10.15) is derived for a very simple deformation mechanism where the glide of a single dislocation is inhibited by forest dislocations that intersect the glide plane. (10.15) is plotted for various

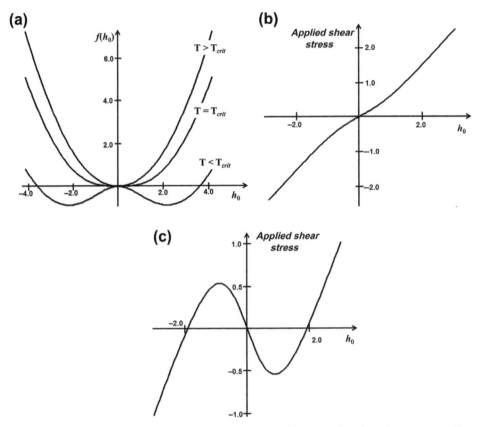

FIGURE 10.11 A critical transition corresponding to plastic yielding as a function of temperature. *From Ortiz (1999).* (a) The transition between a convex energy function for $T > T_{crit}$ to a non-convex function for $T < T_{crit}$. (b) Stress–strain curve for $T > T_{crit}$ where the strain is measured by h_0. (c) Stress–strain curve for $T < T_{crit}$ where the strain is measured by h_0.

values of κ in Figure 10.11(a) and with $h = 0$. The energy passes through a transition from non-convex to convex at the critical temperature, T_{crit}.

Thus below T_{crit} the energy has two minima corresponding to elastic and plastic behaviour (Ortiz, 1999). Above T_{crit} the energy is convex and no yield point exists. Ortiz (1999) shows that T_{crit} is close to the melting point for many materials.

The arguments of Ortiz are extended by Truskinovsky and others (Puglisi and Truskinovsky, 2000, 2002, 2005; Salman and Truskinovsky, 2012). Many of these models are based on the model for a plastic material proposed by Muller and Villaggio (1977). This model comprises a number of *snap-springs* (see Chapter 1) arranged in various arrays. The model proposed by Muller and Villaggio (1977) is shown in Figure 10.12(a). Salman and Truskinovsky (2012) examine various versions of this model and show that a one dimensional model with the snap-springs arranged in series with next to nearest neighbour interactions explains many aspects of rate independent plasticity but fails to reproduce intermittency of flow. A two dimensional arrangement of springs is sufficient to produce avalanche behaviour with power-law statistics and the fractal character of

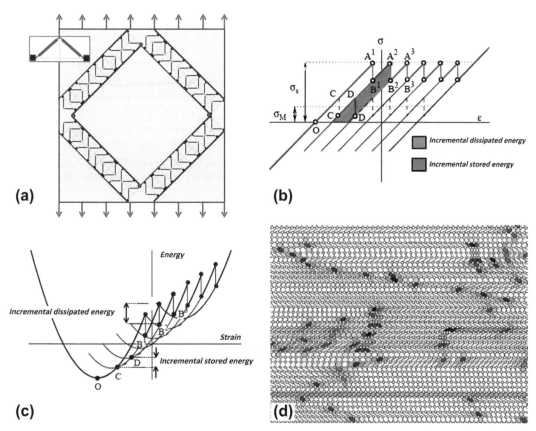

FIGURE 10.12 Avalanche behaviour modelled by a series of snap-springs with next to nearest neighbour interactions. (a) The model proposed by Muller and Villaggio (1977). The unit snap-spring is shown as an insert and is identical to that shown in Figure 1.9. (b) The serrated stress–strain response exhibited by a one dimensional array of snap-springs. (c) The energy function corresponding to the stress–strain behaviour in (b). (d) Spatial distribution of avalanche failure behaviour within a lattice representing a two dimensional array of springs. *From Salman and Truskinovsky (2012).*

microstructure. The stress–strain curve for a one dimensional snap-spring array is shown in Figure 10.12(b) and derives from the nested energy curve shown in Figure 10.12(c). The stress–strain curve is serrated with a sequence of stress drops $A^i \rightarrow B^i$ after yield.

The behaviours discussed above can be summarised as follows. Many aspects of deformation can be described as first-order phase transitions. However instead of the phase changes being sharp as in classical, ideal first-order transitions the ones of interest are extended over a broader range of driving parameters (such as stress or strain) and they show hysteresis, which is manifested as different behaviours under different loading conditions. These behaviours arise because the energy barriers involved (forest dislocations, regions of increased strength or elastic moduli) are large in magnitude compared to thermal fluctuations and are of relatively large spatial extent. The energy landscape consists of numerous local minima and the energy barriers are such that thermal fluctuations play a negligible role. The system can only move from one energy-well to another through the external driving force. This means that thermal fluctuations alone are not enough to drive the phase transition and high levels of stress or of strain are necessary. The high-energy barriers have two origins: (1) initial (quenched) disorder that is inherited from a former state; these determine initial nucleation sites for the new phase; (2) new microstructure, such as fracture surfaces and dislocations, which are spatially localised. This means that the response of the material to a constant driving force is a series of nucleation and collective de-pinning events expressed as discontinuous jumps, as local high strength regions are overcome and neighbouring regions take up the stress, followed by periods of quiescence when propagation of the advancing phase boundary is pinned. The next avalanche is associated with de-pinning of this boundary and this de-pinning event is expressed as an avalanche of acoustic emissions.

Perez-Reche et al. (2008) have proposed that various modes of criticality for a system, loaded with a spring with spring constant k, can be represented on a diagram (Figure 10.13(a)) where the behaviour is influenced by both the value of the spring constant and the degree of imperfection (measured by the standard deviation of the disorder, r) in the system. Figure 10.13(a) is plot of spring stiffness, k, against the standard deviation of disorder, r. The line marked SOC defines the conditions for *self-organised criticality* and separates two fields of behaviour, SNAP (large avalanches of damage) and POP (small scale avalanches). The point r_0 corresponds to a classical order—disorder transition. Figure 10.13(b) is a phase diagram for the system shown in (a). Systems evolve from standard criticality (corresponding to fine tuning of the system) to self-organised criticality (corresponding to slow external forcing) along the diagonal line. Perturbations away from the diagonal line cause the system to evolve either into the SNAP field or the POP field. The inserts show the types of damage patterns that develop in each field with characteristic fractal dimensions (for a two dimensional pattern). Figure 10.13(c) is a stress—strain curve for this nonlinear system with standard deviation of initial damage of 1.5. Soft loading conditions ($k = 0$) correspond to SNAP behaviour with a large avalanche driven by a stress, σ_n, at the nucleation threshold for damage. A spring constant of $k = 0.5$ produces behavior that relaxes to the stress, σ_p, corresponding to the propagation threshold for damage but settles into a damage evolution mode that corresponds to POP behaviour. A hard system, $k = k_\infty$, follows POP behaviour throughout the loading history after yield.

Figure 10.13(d) is a phase diagram showing the field of mode switching in earthquake behaviour reported by Ben-Zion et al. (1999), Ben-Zion (2008) and Dahmen and Ben-Zion (2009). The diagram is a plot of the parameter, ε, which is a measure of the degree of softening, against (1-c) where c is a fraction of the stress drop, associated with an event, that is conserved in the slip region after slip. Below the dotted line the damage statistics correspond to a truncated power law (which is commonly known as Gutenburg—Richter statistics). Above the dotted line the behaviour swaps between Gutenburg—Richter statistics and characteristic earthquake statistics. Perhaps this mode-switching behaviour is an expression of sensitivity to initial conditions in the phase diagram shown in Figure 10.13(b). Slight departures from the diagonal line in the phase diagram will result in a system that evolved into the POP field (Gutenburg—Richter statistics) or result in a system that evolves into the SNAP field (corresponding to characteristic earthquake statistics).

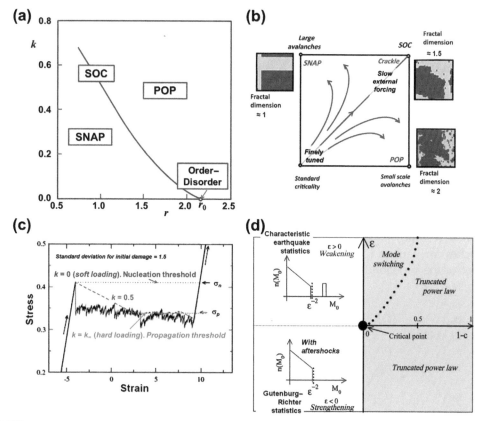

FIGURE 10.13 Dependence of avalanche behaviour on initial disorder and mode of loading. (a) A plot of spring stiffness, k, against the standard deviation of disorder, r. The fields for order–disorder, SOC, SNAP and POP behaviour are outlined. *From Perez-Reche et al. (2008).* (b) A phase diagram for the system shown in (a). *From Perez-Reche et al. (2008).* (c) Stress–strain curve for nonlinear system with standard deviation for initial damage of 1.5. *Modified from Perez-Reche et al. (2008).* (d) Phase diagram showing the field of mode switching in earthquake behaviour. *From Dahmen and Ben-Zion (2009).*

An instructive study that illustrates many aspects of damage evolution and criticality is that of Girard et al. (2010) who model (using finite elements) progressive damage evolution and failure in a Mohr–Coulomb material with initial random distributions of imperfections defined in terms of variations in the values of the cohesion. Two variations are studied, H_1 and H_2; for H_1 the cohesion varies from 5×10^{-4} to 10^{-3} of Y_0 and for H_2 the cohesion varies from 2×10^{-4} to 10^{-3} of Y_0 where Y_0 is the initial value of the Young's modulus. The Young's modulus in the ith finite element is updated after each failure event so that there is a linear degradation proportional to the damage in elements that reach a threshold failure stress. After n failure steps the Young's modulus in the ith element is

$$Y_i(n) = Y_{i,0}d_0^n$$

where d_0 is a preset value for the damage, say $d_0 = 0.9$.

Two control parameters are defined by

$$\Delta_{stress} = \frac{\sigma_{macro-peak} - \sigma_{macro}}{\sigma_{macro-peak}}, \quad \Delta_{strain} = \frac{\varepsilon_{macro-peak} - \varepsilon_{macro}}{\varepsilon_{macro-peak}} \tag{10.16}$$

where Δ_{stress}, Δ_{strain} are the control parameters corresponding to stress controlled and strain controlled loading conditions. $\sigma_{macro-peak}$ and $\varepsilon_{macro-peak}$ are the macroscopic stress and strain at peak load; σ_{macro} and ε_{macro} are the macroscopic stress and strain during deformation. Thus Δ_{stress}, Δ_{strain} vary between 1 at the start of loading to 0 at peak load. For N damage events a spatial correlation integral, C_2, is defined by

$$C_2(r) = 2N_P(r)/[N(N-1)] \tag{10.17}$$

where $N_P(r)$ is the number of pairs of damage events whose separation is less than r. If C_2 scales according to r^{-2} then the spatial pattern of damage is homogeneous and un-clustered.

Figures 10.14(a), (b) and (c) show examples of damage distribution after catastrophic avalanches for various loading conditions and initial damage distributions. Figure 10.14(d) is a plot of the probability of an event of energy E against the energy for various values of

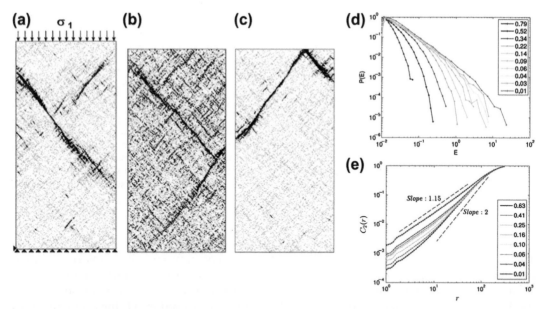

FIGURE 10.14 Brittle failure and critical behaviour. (a) Map of normalised Young's modulus, stress controlled deformation, H_1 initial damage. (b) Map of normalised Young's modulus, stress controlled, H_2 initial damage. (c) Map of normalised Young's modulus, strain controlled, H_1 initial damage. (d) Normalised probability distribution function of avalanche dissipated energy for various values of the control parameter, Δ_{strain}, H_2 initial damage. This diagram illustrates finite size scaling. (e) Spatial correlation integral for damage events for various values of the control parameter, Δ_{strain}, H_1 initial damage. This diagram illustrates the evolution of the spatial patterning of damage from uncorrelated at small strains (large Δ_{strain}) to a fractal distribution close to failure (criticality). *From Girard et al. (2010).*

Δ_{stress} and an initial imperfection distribution, H_2. These probabilities follow a relation known as a *gamma law*:

$$P(E, \Delta) = E^{-\beta} \exp [- E/E^*] \tag{10.18}$$

For small values of E the distribution is fractal with a fractal dimension β; for values of E approaching a cut-off energy E^*, the distribution departs from fractal and decreases rapidly. This is an example of *finite size scaling* and arises because for any real system the energy that can drive a damage event and the size of a damage event are finite. Such behaviour is characteristic of many critical systems and can be seen in the distributions of crustal earthquakes (Figures 10.9(d) and 10.11(b)).

Figure 10.14(E) shows the correlation integrals, (10.18), for strain control, a H_1 distribution of initial imperfections and for various values of Δ_{strain}. For small values of r and large values of Δ_{strain} the distribution is fractal. For large values of r and large values of Δ_{strain} the distribution is homogeneous and unclustered. For systems close to failure (small values of Δ_{strain}) the distribution is always fractal, which means there is no characteristic size for an event. Girard et al. (2010) show that this distribution is multifractal. Again, this kind of scaling for correlation integrals is characteristic of critical systems and demonstrates that at yield (criticality) long wavelength correlations develop that are multifractal in character.

A similar study is conducted by Cowie et al. (1993, 1995) but with a constitutive law that ensures progressive localisation of the fracture system once it nucleates. Again a multifractal behaviour is reported. All elements have identical linear elasticity and heterogeneity is introduced in the local value of the failure stress and in the stress drop after a failure of an element. These two rules result in heterogeneity of failure with a tendency to favour refailure of the largest slip events. This ultimately leads to concentration of failure events near large events and hence localised failure resembling SNAP behaviour. Girard et al. (2010) employ a heterogeneous distribution of cohesion and an element weakens elastically upon failure thus introducing nonlinear elastic behaviour. A failed element has to undergo a larger strain to reach the failure stress than an element that has not failed. Hence the probability that unfailed elements will fail in the next increment of strain is greater than for failed elements. This spreads new failure events rather than localising them as in the Cowie model. Hence the Girard behaviour evolves more to a POP behaviour.

Recommended Additional Reading

Atkinson, B. K. (Ed.). (1987). *Fracture Mechanics of Rock*. Academic Press.
  This is an important and easily read text devoted to various aspects of fracture mechanics of rocks with many aspects of direct application to damage mechanics. It includes (Chapter 5) (by L. S. Costin) on one form of damage mechanics

Dahmen, K. A., & Ben-Zion, Y. (2009). The physics of jerky motion in slow driven magnetic and earthquake fault systems. In R. Meyers (Ed.), *Encyclopedia of Complexity and System Science* (pp. 5021−5037). Berlin: Springer.
  This paper reviews many of the concepts involved in criticality and applies these concepts to earthquake phenomena

Girard, L., Amitrano, D., & Weiss, J. (2010). Failure as a critical phenomenon in a progressive damage model. *Journal of Statistical Mechanics: Theory and Experiment*, 2010, P01013.
  A good review of critical phenomena with detailed application to damage evolution in a Mohr-Coulomb material

Krajcinovic, D. (1989). Damage mechanics. *Mechanics of Materials, 8*, 117−197.
  An important paper devoted to many theoretical developments in damage mechanics

Paterson, M. S., & Wong, T.-f. (2005). *Experimental Rock Deformation—The Brittle Field*. Springer.
  This book contains an appendix devoted to damage mechanics.

Salman, O. U., & Truskinovsky, L. (2012). On the critical nature of plastic flow: one and two dimensional models. *International Journal of Engineering Science, 59*, 219−254.
  A discussion of the application of spring models to critical systems

Voyiadjis, G. Z. (Ed.). (2011). *Micromechanics of Localised Fracture Phenomena in Inelastic Solids*. Verlag: Springer.
  This book contains many applications of damage mechanics with advanced, detailed reviews of many different approaches to the subject

11

Transport of Heat

11.1 FOURIER'S LAW OF HEAT CONDUCTION: DIFFUSION OF HEAT

If a temperature gradient exists in a solid then heat is observed to flow from the hotter to the cooler parts of the body. This is an example of a system not at equilibrium. The thermodynamic force that drives the system away from equilibrium in this case is the gradient of $(1/T)$ where T is the absolute temperature. The thermodynamic flow that dissipates energy and tends to move the system towards equilibrium is heat flow. If the temperature gradient is maintained a non-equilibrium stationary state eventually evolves where the thermal flux adjusts so that it is constant in the body. Such a state is sometimes, erroneously, referred to as thermal equilibrium. Since heat continues to flow, this state, comprising a constant temperature gradient, is a non-equilibrium stationary state and not an equilibrium state.

If the material is isotropic then the flow of heat is proportional to the temperature gradient. This relationship is known as *Fourier's Law* and is expressed as:

$$q = -k(grad T) \tag{11.1}$$

where **q** is the heat flow vector expressed in the units $J\,m^{-2}\,s^{-1}$, and k is a material constant known as the *thermal conductivity* with units $J\,m^{-1}\,s^{-1}\,K^{-1}$; T is the absolute temperature. The minus sign indicates that the heat flows *down* the temperature gradient. If the material is anisotropic the thermal conductivity may also be anisotropic and expressed as a symmetrical second order tensor, k_{ij} (Nye, 1957). Fourier's Law is then written:

$$q_i = -k_{ij}\frac{\partial T}{\partial x_j} \tag{11.2}$$

The flow of heat can be expressed as a diffusion equation by the following argument. Imagine an element of volume at a point P (Figure 11.1) with edges parallel to the coordinate axes (x, y, z) and with lengths 2dx, 2dy, 2dz. For the moment we neglect any internal heat production. The heat flux parallel to the x-axis at P is taken to be q_x. This means that the heat flux entering the element on the face $\mathcal{A}$ normal to the x-axis is:

$$4\left(q_x - \frac{\partial q_x}{\partial x}dx\right)dy\,dz$$

The heat flux leaving the box on the face $\mathcal{A}'$ normal to the x-axis is:

$$4\left(q_x + \frac{\partial q_x}{\partial x}dx\right)dy\,dz$$

Thus the total heat flux in the x-direction is:

$$-8\frac{\partial q_x}{\partial x}dx\,dy\,dz$$

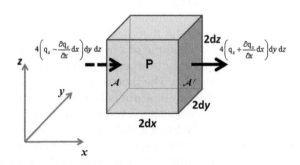

FIGURE 11.1   Heat flow in the x-direction through an element of volume at P.

The same argument applies to the y- and z-axes so that the total heat flux through the element is:

$$-8\left(\frac{\partial q_x}{\partial x} + \frac{\partial q_y}{\partial y} + \frac{\partial q_z}{\partial z}\right) dx\, dy\, dz$$

Now the rate of gain of heat can be also written:

$$8\rho c_P \frac{\partial T}{\partial t} dx\, dy\, dz$$

where ρ is the density and c_P is the specific heat at constant pressure. Therefore,

$$\rho c_P \frac{\partial T}{\partial t} = -\left(\frac{\partial q_x}{\partial x} + \frac{\partial q_y}{\partial y} + \frac{\partial q_z}{\partial z}\right) \tag{11.3}$$

Substitution of (11.1) into (11.3) gives for a homogeneous, isotropic solid with a thermal conductivity independent of temperature:

$$\rho c_P \frac{\partial T}{\partial t} = k\left(\frac{\partial^2 T}{\partial x^2} + \frac{\partial^2 T}{\partial y^2} + \frac{\partial^2 T}{\partial z^2}\right) \equiv k\nabla^2 T \tag{11.4}$$

(11.4) can also be written:

$$\frac{\partial T}{\partial t} = \kappa\left(\frac{\partial^2 T}{\partial x^2} + \frac{\partial^2 T}{\partial y^2} + \frac{\partial^2 T}{\partial z^2}\right) \equiv \kappa\nabla^2 T \tag{11.5}$$

where $\kappa = \frac{k}{\rho c_P}$ is known as the *thermal diffusivity* and has the units $m^2\, s^{-1}$. If temperature changes occur with a characteristic timescale τ then they diffuse a distance $\sqrt{\kappa\tau}$ in that time. On the other hand a time $L^2 \kappa^{-1}$ is required for a temperature change to diffuse over the distance L. The average value of the thermal diffusivity for rocks is $10^{-6}\, m^2\, s^{-1}$. Thus the characteristic timescale for a temperature change imposed at the base of a 30 km thick crust is $(30,000)^2/10^{-6}\, s$ or approximately 28.5 My (taking 1 year $= 3.16 \times 10^7\, s$). Equally a plume arriving at the base of a lithosphere 150 km thick is associated with a characteristic thermal timescale of 712 My. These very long timescales for thermal effects to be felt far from the application of a temperature change need to be taken into account when one postulates the impingement of plumes or delamination events at the base of the lithosphere. A plot of the characteristic time, τ, against the length scale, L, of the system is shown in Figure 11.2 for $\kappa = 10^{-6}\, m^2\, s^{-1}$.

If heat is produced at P at a rate H (expressed as $J\, kg^{-1}\, s^{-1}$) then, for a conductivity independent of temperature, (11.5) becomes

$$\nabla^2 T - \frac{1}{\kappa}\frac{\partial T}{\partial t} = -\frac{H(x,y,z,t)}{\kappa c_P} \tag{11.6}$$

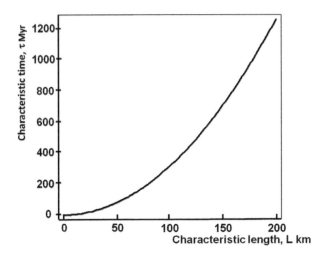

FIGURE 11.2 Plot of characteristic time for heat diffusion over a characteristic length scale. Drawn for $\kappa = 10^{-6}\,\mathrm{m^2\,s^{-1}}$.

11.2 TEMPERATURE AND PRESSURE DEPENDENCE OF CONDUCTIVITY

Some representative values of thermal conductivity are given in Tables 11.1 and 11.2. Although the values in Table 11.1 are from quite old studies they set the scene for what has been further explored in recent studies. One should note that in general the thermal conductivity decreases substantially with an increase in temperature, the exception in the table being gabbro that shows a slight increase with temperature. Anisotropic rocks, such as granite gneiss and slate show higher conductivity parallel to the foliation than normal to the foliation.

The effect of hydrostatic pressure is to increase the thermal conductivity and within the pressure range $10^4\,\mathrm{Pa}{-}120\,\mathrm{MPa}$ the conductivity is given by $k = k_0\,(1 + \alpha P)$ where α is given in Table 11.2 for a hydrostatic pressure measured in Pascals and k_0 is the conductivity at zero P. Although this relation was established at relatively low pressures it seems to be applicable at higher pressures (Kukkonen et al., 1999). At least over this fairly restricted range of pressures the increase in conductivity with increase in pressure is not enough to match the decrease due to increase in temperature.

More recent studies of the influence of temperature and pressure on thermal conductivity include Sass et al. (1992), Kukkonen et al. (1999), Vosteen and Schellschmidt (2003), Abdulagatov et al. (2006) and Mottaghy et al. (2008). Abdulagatov et al. (2006) give a general relationship for the variation of thermal conductivity with temperature for metamorphic rocks:

$$k(\mathrm{T}\,^{\circ}\mathrm{C}) = \frac{k(0)}{0.99 + \mathrm{T}\,^{\circ}\mathrm{C}(a - b/k(0))} \tag{11.7}$$

TABLE 11.1   Representative Values of Thermal Conductivity for Some Rock Types

Rock Type	Thermal Conductivity ($J\,m^{-1}\,s^{-1}\,K^{-1}$)	Temperature (°C)
Granite (Westerly)	2.43	0
	2.34	50
	2.27	100
	2.14	200
Granite gneiss (Pelham Mass.)	3.1 (Parallel to foliation)	0
	2.16 (Normal to foliation)	0
	2.75 (Parallel to foliation)	100
	2.01 (Normal to foliation)	100
Gabbro (Mellen, Wisc.)	1.99	0
	1.99	100
	1.99	200
	2.00	300
	2.01	400
Dunite (Balsam Gap, N. C.)	5.2	0
	4.4	50
	3.9	100
	3.4	200
Marble (Quebec)	1.7	118
	1.5	196
	1.4	245
	1.1	360
Quartzite (Witwatersrand)	4.38 (Feldspathic)	25
	5.94 (Non-feldspathic)	25
	6.54 (Chloritoid bearing)	25
Slate (Wales)	2.8 (Parallel to schistosity)	25
	1.7 (Normal to schistosity)	25

Values taken from Birch et al. (1942). See also Clark (1966).

TABLE 11.2   Influence of Hydrostatic Pressure on Selected Rock Types

Rock Type	Temperature	α
Basalt	30 °C	$+4.7 \times 10^{-10}$
	75 °C	$+2.2 \times 10^{-10}$
Limestone	30 °C	$+1.0 \times 10^{-10}$
	75 °C	$+6.7 \times 10^{-10}$
Talc	30 °C	$+15.7 \times 10^{-10}$
Pipestone	30 °C	$+30 \times 10^{-10}$

Values taken from Birch et al. (1942). See also Clark (1966).

where k(T) and k(0) are the thermal conductivities at temperatures, T °C, and 0 °C and a, b are constants. Mottaghy et al. (2008) give an expression for the thermal diffusivity in terms of thermal conductivity:

$$\kappa(T) = k(T)\frac{1}{mT + n} \cdot 10^{-6} m^2/s \qquad (11.8)$$

together with values for m and n for various rocks. For rocks from Kola in Russia, κ varies from approximately $1 \times 10^{-6} \, m^2 \, s^{-1}$ for $k = 2 \, W \, m^{-2} \, K^{-1}$ to $1.8 \times 10^{-6} \, m^2 \, s^{-1}$ for $k = 4 \, W \, m^{-2} \, K^{-1}$. For these same rocks c_P varies from $2.4 \, MJ \, m^{-3} \, K^{-1}$ at $0 \, °C$ to $3.5 \, MJ \, m^{-3} \, K^{-1}$ at $300 \, °C$.

11.3 EQUATIONS OF HEAT TRANSPORT: TIMESCALES

The use of (11.5) and (11.6) in modelling the thermal evolution of metamorphic rocks is widespread and examples include Thompson and England (1984), Jamieson et al. (1998) and Sandiford and McLaren (2002). In general finite element or finite difference computer codes need to be used to model complicated geometries and histories but a very large number of analytical solutions are available in Carslaw and Jaeger (1959), Turcotte and Schubert (1982) and Boyce and DiPrima (2005). Below we present a solution to (11.5) for a simple geometry and initial conditions to illustrate the principles involved. A detailed discussion of the problem can be found in Turcotte and Schubert (1982, pp. 158–161).

The problem involves the heating of a semi-infinite half-space initially everywhere at a temperature, T_0 (Figure 11.3). The surface temperature of the half-space is increased to T_s at time $t = 0$ and the problem is to calculate the subsequent evolution of temperature in the half-space.

With y measured normal to the surface of the half-space, the initial conditions for this problem are $T = T_0$ at $t = 0$; $T = T_s$ at $y = 0$ and $T = T_0$ for $y \to \infty$. We define

$$\tilde{T} = \frac{T - T_0}{T_s - T_0}$$

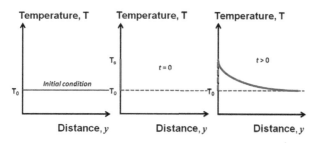

FIGURE 11.3 Heating of a semi-infinite half-space initially at a temperature everywhere of T_0. At time $t = 0$ the surface temperature is increased to T_s. The problem is to calculate the temperature profile through the space at $t > 0$.

Then the equation to be solved is:

$$\frac{\partial \tilde{T}}{\partial t} = \kappa \frac{\partial^2 \tilde{T}}{\partial y^2} \tag{11.9}$$

subject to $\tilde{T}(y, 0) = 0$; $\tilde{T}(0, t) = 1$ and $\tilde{T}(\infty, t) = 0$. We also define a parameter η:

$$\eta = \frac{y}{2\sqrt{\kappa t}} \tag{11.10}$$

Then (see Turcotte and Schubert, 1982, p. 159), (11.9) becomes

$$-\eta \frac{d\tilde{T}}{d\eta} = \frac{1}{2} \frac{d^2 \tilde{T}}{d\eta^2}$$

The solution to this is (Turcotte and Schubert, 1982, p. 160)

$$\tilde{T} = 1 - \frac{2}{\sqrt{\pi}} \int_0^{\eta} \exp(-\eta'^2) d\eta' \tag{11.11}$$

where η' is a dummy variable of integration. The function defined by the integral in (11.11) is common in various branches of mathematics and is known as the *error function*; it is written *erf* (η). Thus the solution (11.11) can be rewritten

$$\tilde{T} = 1 - erf(\eta) \equiv erfc(\eta)$$

where *erfc* (η) is the *complementary error function*. In terms of original variables this is

$$\frac{T - T_0}{T_s - T_0} = erfc\left(\frac{y}{2\sqrt{\kappa t}}\right) \tag{11.12}$$

Values for the error function and the complementary error function are tabulated in Carslaw and Jaeger (1959, pp. 485) and Turcotte and Schubert (1982, p. 161). These two functions are plotted in Figure 11.4.

As an example, (11.12) can be used to calculate the evolution of the temperature profile adjacent to an intrusion. We write

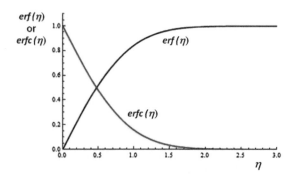

$$T(y, t) \;=\; T_0 + (T_s - T_0)erfc\left(\frac{y}{2\sqrt{\kappa t}}\right) \tag{11.13}$$

We consider a body of rock initially at a temperature $T_0 = 773$ K intruded by an igneous body at 1273 K at $t = 0$ and the igneous body remains at 1273 K. The question is: What is the temperature evolution in the adjacent body of rock? (11.13) becomes:

$$T(y, t) \;=\; 773 + 500 erfc\left(\frac{500 y}{\sqrt{t}}\right) \tag{11.14}$$

where y is measured in metres and t in seconds and κ has been taken to be $10^{-6}\,\mathrm{m^2\,s^{-1}}$. It turns out that because η is dimensionless and because of the value of κ, (11.14) is also true if y is measured in kilometres and t is in years. The evolution of temperature is shown in Figure 11.5. This plot highlights the role that the function $erfc(\eta)$ takes. For a given diffusivity

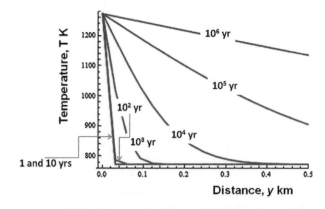

FIGURE 11.5   Thermal history adjacent to an intrusion. Initial temperature of country rock 773 K. Temperature of intrusion is 1273 K and constant. Distance y is measured from the edge of the intrusion. One can see (as an example) that it takes over 10^4 years for the temperature of the country rock 0.4 km from the intrusion to rise above the ambient temperature.

this function simply multiplies the initial temperature by a number between 0 and 1 depending on the value of η (and hence of the ratio $\frac{y}{2\sqrt{t}}$).

11.4 INTERNAL HEAT PRODUCTION

Heat is produced within deforming chemically reacting rocks from the deformation and chemical processes that operate during metamorphism, from the circulation of hot fluids (including magmas) and also from the decay of radioactive isotopes. A large part of the heat that drives many metamorphic processes originates from the radioactive decay of uranium, thorium and potassium mainly within the crust. Here we concentrate solely on this process.

The rate of decay of a radioactive isotope is given by

$$\frac{dN}{dt} = -\lambda^{(1/2)} N \tag{11.15}$$

where N is the number of atoms of the radionuclide present at time t and $\lambda^{(1/2)}$ is known as the *decay constant*. The solution to (11.15) gives the number of atoms present at time t when the number of atoms present at time $t = 0$ is N_0:

$$N = N_0 \exp\left(-\lambda^{(1/2)} t\right) \tag{11.16}$$

The *half-life*, $\tau^{(1/2)}$, of a radionuclide is the time required for half of the atoms present at time $t = 0$ to decay. We can obtain this by writing

$$\frac{N}{N_0} = 0.5 = \exp\left(-\lambda^{(1/2)} \tau^{(1/2)}\right)$$

or,

$$\tau^{(1/2)} = \frac{\ln 2}{\lambda^{(1/2)}} = \frac{0.6932}{\lambda^{(1/2)}}$$

The decay constants and half-lives for the common heat producing radio-nuclides are given in Table 11.3 together with the heat production for each isotope or element.

As an example, one can confirm from Table 11.3 that an Ordovician rock with a current composition (by weight) of 3.5% K_2O, 3.5 ppm uranium and 16.9 ppm thorium has a current heat production rate of 2.4 $\mu W\ m^{-3}$. In the Ordovician this same rock would have had a heat production rate approximately 8% higher.

11.5 TECTONIC MODELS OF THERMAL EVOLUTION

Metamorphic systems are regions of enhanced temperature gradients within the crust of the Earth. The processes that drive such enhanced gradients have been attributed to at least

TABLE 11.3 The Decay Constants, $\lambda^{(1/2)}$, Half-Lives, $\tau^{(1/2)}$ and Rates of Heat Release, H, of the Important Radioactive Isotopes in the Earth

Isotope	Decay Constant, $\lambda^{(1/2)}$ (per year)	Half-life, $\tau^{(1/2)}$ (years)	H (W kg^{-1})
^{238}U	1.55×10^{-10}	4.47×10^9	9.37×10^{-5}
^{235}U	9.85×10^{-10}	7.04×10^8	5.69×10^{-4}
U			9.71×10^{-5}
^{232}Th	4.9×10^{-11}	1.40×10^{10}	2.69×10^{-5}
^{40}K	5.54×10^{-10}	1.25×10^9	2.79×10^{-5}
K			3.58×10^{-9}

Values taken from Turcotte and Schubert, 1982.

four classes of events namely, (1) enhanced thickening of the crust in which heat is produced by radioactive decay; (2) various contributions from regions of enhanced heat production due to radioactive decay that are moved during deformation; (3) plume impingement on the base of the lithosphere and (4) delamination of the lithosphere associated with Rayleigh–Taylor instabilities.

The temperature of a particular elemental volume of rock in a metamorphic system at a point P and its evolution with time is dependent on a number of factors, which can be considered under two headings:

1. *Internal heat production at P* due to internal processes, such as radioactive decay, heat production or absorption arising from chemical reactions, mechanical deformation and fluid flow at P. We examine the influence of radioactive decay below but defer a discussion of the influence of other forms of internal heat production to later chapters.
2. *The relative passage of isotherms past P.* Such movement of isotherms relative to P may be due to end member processes: the first is thermal conduction, which we have seen is a relatively slow process. The second is advection of heat within a deforming body. In this process heat is carried with the deforming body as it moves; this is a relatively fast process.

In the case where chemical equilibrium has been achieved a metamorphic petrologist would sometimes adopt a spatial or Eulerian representation and be only interested in the temperature and pressure at a particular spatial point where the current bulk chemical composition is defined. If the history of metamorphism is considered in the form of PTt diagrams (England and Thompson, 1984) then sometimes a Lagrangian approach is adopted although commonly it is not clear whether a Lagrangian or Eulerian approach is considered. The following examples are meant as illustrations of the distinction between Eulerian and Lagrangian representations of thermal evolution and of the different timescales associated with thermal conduction and thermal advection. We briefly examine four models for the production of heat that drives metamorphic systems as examples of heat flow models.

11.5.1 Thickening of a Radioactive Crust

The temperature of a body of rock within the crust with internal heat production depends on four factors: (1) the internal heat production and its distribution with depth; (2) the flow of heat into the crust at the Moho; (3) the depth of the rock below the surface and (4) if the rock mass is deforming, the thermal Peclet number, $Pe^{thermal}$, which is the ratio of the velocity of the rock to the rate of heat transport by thermal conduction and is given by

$$Pe^{thermal} = \frac{vL}{\kappa} = \frac{vL\rho c_P}{k} \tag{11.17}$$

where v is the velocity and L is a characteristic length scale. These four factors are in general poorly constrained; some discussions are given by Turcotte and Schubert. (1982) and Sandiford and McLaren (2002). The history of temperature evolution is commonly treated as a function of the thickening and thinning of the crust.

As an example of the effect of internal heat production and its distribution with depth we present Figure 11.6 taken from Jamieson et al. (1998). Three different heat distributions of K, U and Th are considered for a crust of thickness 35 km and heat flow from the mantle of 30 mW m^{-2}. These three different assumptions result in three different temperature gradients and three different temperatures at the Moho.

The modelling of England and Thompson (1984) is the classical work for understanding the thermal evolution of a thickening crust. The example taken here is similar to their models

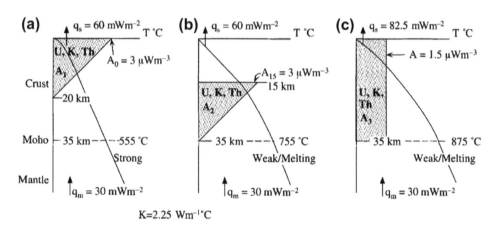

FIGURE 11.6 Different models for incorporation of internal heat production isotopes in the crust. The crust is 35 km thick and the heat flow from the mantle is 30 mW m^{-2} in each model. (a) Heat producing elements are distributed linearly in the upper 20 km of crust with a heat production of 3 µW m^{-3} at the surface. This represents a situation where heat-producing isotopes are concentrated in igneous intrusions high in the crust. Surface heat flow is 60 mW m^{-2} with a temperature at the Moho of 555 °C. (b) Heat producing elements are distributed linearly in the lower 20 km of crust with a heat production of 3 µW m^{-3} at a depth of 15 km. This represents a situation where heat-producing isotopes are concentrated in an old subducted complex. Surface heat flow is 60 mW m^{-2}, with a temperature at the Moho of 755 °C. (c) Heat producing elements are distributed uniformly throughout the crust with a uniform heat production of 1.5 µW m^{-3}. This represents a situation where heat-producing isotopes are distributed within an accreted complex. Surface heat flow is 82.5 mW m^{-2}, with a temperature at the Moho of 875 °C. *From Jamieson et al., 1998.*

in that it involves a one dimensional situation where crustal thickening results from the superposition of thrust sheets on a crust with internal heat producing radio-nuclides followed by erosion of the thrust sheets. We begin with a crust 35 km thick with a uniformly distributed heat production of $1.5\,\mu W\,m^{-3}$ and a heat flux from the mantle of $30\,mW\,m^{-2}$ (Figure 11.7). The situation is identical to the model illustrated by Jamieson et al. (1998) and presented in Figure 11.6(c), The model results in a steady state temperature at the Moho of 875 °C. Five kilometres of material with no radioactive heating and comprising thrust sheets is added instantaneously as on the left of Figure 11.7(a) resulting in the temperature profile shown in the left of Figure 11.7(a) after approximately 15 My; this is not a steady state profile. Erosion of the thrust pile then takes place in five steps at an average erosion rate

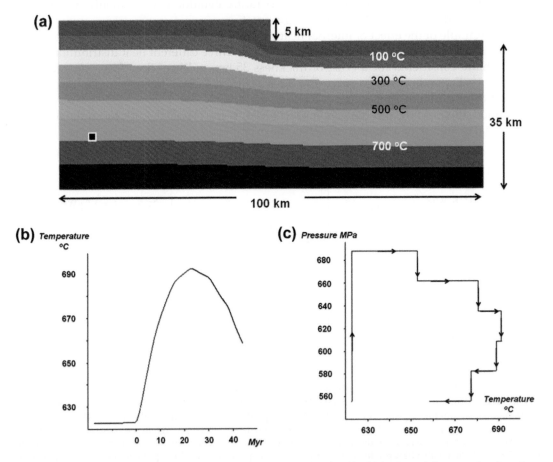

FIGURE 11.7   One dimensional model of PT changes associated with thickening of the crust by thrusting. (a) Temperature distribution before thrusting (right hand side) and approximately 15 My after thrusting (left hand side). PTt path is recorded for the black square. (b) Temperature–time plot for the black square in (a). PTt path for the black square in (a). Heat flow into base of model is $30\,mW\,m^{-2}$. Internal heat production in the crust is uniform at $1.5\,\mu W\,m^{-3}$ as in Figure 11.6(c).

of 0.2 millimetres per year. The resulting temperature evolution is shown in Figure 11.7(b) and the clockwise PTt history for a point (black square) initially (before thrusting) at 20 km depth in Figure 11.7(a). Notice that the time taken for this evolution is approximately 40 My. Notice also that this example, because it is one-dimensional and does not take into account the possibility that the material presently at the black square in Figure 11.7(a) may have come from elsewhere, is essentially a spatial view of the metamorphic evolution. Other examples discussed below adopt a Lagrangian view. The convention is adopted that PTt paths are represented on diagrams where the pressure, P, is the axis of ordinates and the temperature, T, is the axis of abscissae. The path shown in Figure 11.7(c) is, by this convention, clockwise.

11.5.2 Enhanced Heat Production Due to Radioactive Decay

We take as an example here the work of Jamieson et al. (1998) where two dimensional models of a subducting crust are examined and in which various spatial distributions of heat producing concentrations are incorporated. The models are shortened with various rates and characteristic lengths, which lead to thermal Peclet numbers varying from about 50 to 5. We present the results for one of these models in Figure 11.8 where $Pe^{thermal} = 16$. Crustal and lithospheric heat producing regions, (A_3 and A_2 in Figure 11.8(b)) each with $H = 1.5$ µW m^{-3}, correspond respectively to an accreted wedge and a previously subducted wedge of material. The resulting deformation is shown in Figure 11.8(a) with a crustal shear zone developing in

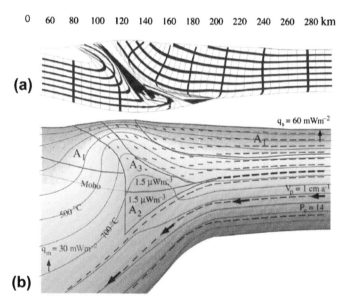

FIGURE 11.8   Deformation of crust and lithosphere during subduction at 10 mm a^{-1}. Initial radioactive decay sites with elevated heat production of 1.5 µW m^{-3} are marked as A_2 and A_3. Mantle heat flow is 30 mW m^{-2}. (a) Deformed grid showing development of shear zone in the crust. This corresponds to the site of elevated temperatures as shown in (b). *From Jamieson et al. (1998).*

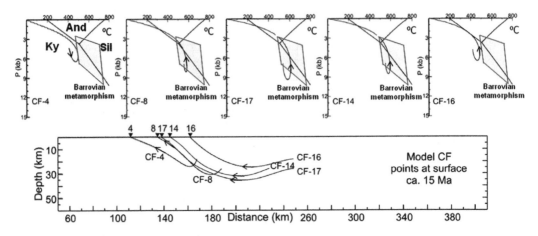

FIGURE 11.9   PT*t* paths corresponding to Figure 11.8. The lower frame shows the trajectories of various material points during 15 Ma of deformation relative to undeformed coordinates. The top frames show the corresponding PT*t* paths in PT space with the field of Barrovian metamorphism marked along with the aluminosilicate stability fields. Notice that according to the convention discussed in the text these are clockwise paths. Because of the high thermal Peclet number temperature changes are quite rapid corresponding to thermal transport dominantly by advection. *From Jamieson et al. (1998).*

the hottest region. The PT*t* histories of various material points are shown in Figure 11.9 together with the Lagrangian trajectories of these points. In some instances material points have been transported over 100 km. Adopting the convention mentioned in Section 11.5.1 these PT*t* paths are clockwise. Notice that this approach is Lagrangian; although the PT*t* paths record the total displacement history of each material point, much of this history is not reflected in the deformation fabrics since much of the displacement history involves rigid body displacements.

11.5.3 Plume Impingement at the Base of the Lithosphere

A common model for introducing heat into the lithosphere is to propose the impingement of a hot (say 1500 °C) plume head on the base of the lithosphere. The models then need to incorporate mechanisms to transport heat through the lithosphere if the lithosphere remains intact. Given lithospheric thicknesses of 100–300 km this involves thermal timescales of 316 My–2.85 Gy, which would seem to be too long to explain the metamorphic systems we observe. One mechanism to overcome this, involving rapid thermal erosion of the lithosphere (Figure 11.10), has been proposed by Cloetingh et al. (2013) with a similar model proposed by Sobolev et al. (2011). The thermal time constant is then reduced to that of a thinned crust (say 30 My or less).

11.5.4 Delamination

Delamination either in the form of Rayleigh–Taylor instabilities or as 'peeling-off events' or combinations of both (Houseman and Molnar, 1997; Elkins-Tanton, 2007; Gorczyk

Homogeneous viscous lithosphere

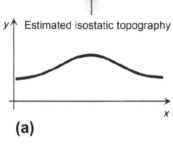

Layered elastic-plastic-viscous lithosphere

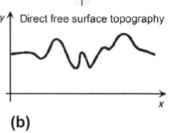

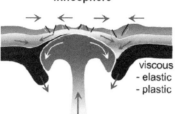

viscous
- elastic
- plastic

FIGURE 11.10 Plume impingement at the base of the lithosphere *From Cloetingh et al., 2013*. (a) Impingement at the base of a purely viscous lithosphere. (b) Impingement at the base of an elastic-plastic-viscous lithosphere with rapid erosion of the subcontinental lithosphere.

(a)

(b)

et al., 2013) is a mechanism for introducing hot asthenospheric material into a position close to the base of the crust. The time scale for heat conduction is then reduced from that associated with a relatively thick lithosphere (say 100–200 km thickness and a timescale of 316–1265.8 My) to that associated with a crustal thickness of perhaps 40 km (a timescale of 50.6 My). PTt paths are mainly clockwise but some can be anti-clockwise (Figure 11.11).

11.6 THE THERMODYNAMICS OF HEAT CONDUCTION, THERMAL EXPANSION AND ENTROPY PRODUCTION

When a change in temperature occurs in a body, the shape of the body changes due to thermal expansion. If this change in shape is expressed as a strain, ε_{ij}, then the strain is related to the change in temperature, ΔT, by

$$\varepsilon_{ij} = \alpha_{ij}\Delta T \qquad (11.18)$$

where α_{ij} is a second order tensor known as the *coefficient of thermal expansion*. Since ε_{ij} is symmetrical, so also is α_{ij} and can be represented by a quadric (see Nye, 1957, pp. 106–109).

Entropy production due to heat flow is an important example to consider since the process of heat flow is commonly thought of as a linear process and a common statement in the literature is that the entropy production at steady state is a minimum. Consider an element of an isolated rod of material with thermal conductivity, k, with the temperature on one side held fixed at T_1 and at the other side at T_2 with $T_1 > T_2$ (Figure 11.12). Let dQ be the amount of heat flow from the hotter to the cooler part of the element in time dt. We assume that the energy change is solely due to the flow of heat. Then $dQ_1 = -dQ_2 = dQ$. The total change in entropy, dS, of the system is

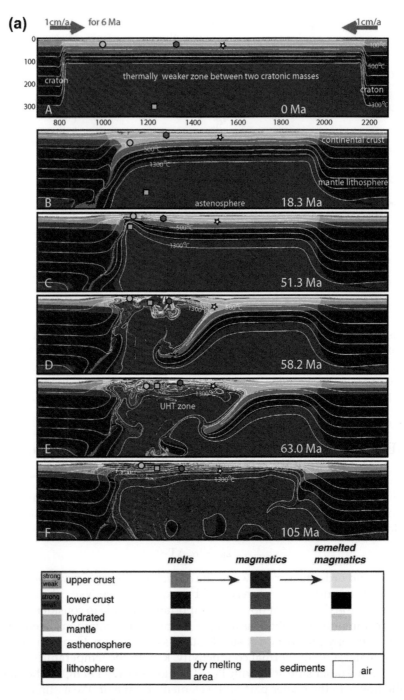

FIGURE 11.11 PT*t* paths generated by delamination. (a) A thinned lithosphere is shortened at 10 millimetres per year for 6 My and then left to evolve. Delamination and subsequent decompression melting takes place over the subsequent 100 My. PT*t* paths are recorded for four sites marked as green, yellow (with black outline), blue and red. The nucleation of delamination at about 50 My rapidly advects the green marker from deep in the lithosphere to a position close to the surface. (b) PT*t* paths. Notice that three are clockwise and one (red) is anti-clockwise. Most of the PT evolution takes place over 50 My. *Figures supplied by Weronika Gorczyk.*

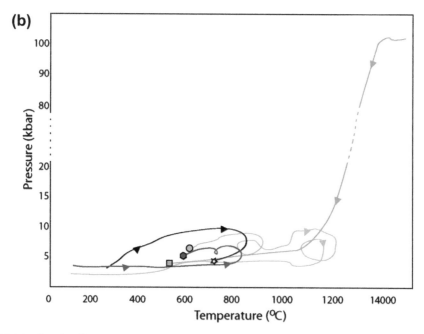

FIGURE 11.11   (*continued*)

$$dS = -\frac{dQ}{T_1} + \frac{dQ}{T_2} = \left(\frac{1}{T_2} - \frac{1}{T_1}\right)dQ$$

And so

$$\frac{dS}{dt} = \left(\frac{1}{T_2} - \frac{1}{T_1}\right)\frac{dQ}{dt} = J\left(\frac{1}{T_2} - \frac{1}{T_1}\right)$$

Thus the *thermodynamic force* that drives thermal conduction in a system with a temperature gradient is $grad\left(\frac{1}{T}\right)$ and not $grad(T)$ so that thermal conduction in an isolated system constitutes a nonlinear thermodynamic system. This has important repercussions when one

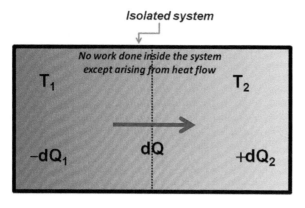

FIGURE 11.12   An isolated thermal conduction system.

B. PROCESSES INVOLVED IN THE DEVELOPMENT OF GEOLOGICAL STRUCTURES considers entropy production. The following discussion is taken from Ross. (2008, p. 115). For a general isolated system in which temperature gradients exist the entropy production of the whole system is a functional of the temperature field, which we define as $T(\mathbf{r})$ where $\mathbf{r}$ is a position vector. The entropy production at a point due to thermal conduction is

$$\dot{s} = \mathbf{J} \cdot \nabla T^{-1} \geq 0$$

where $\mathbf{J}$ is the heat flux vector at $\mathbf{r}$.

Thus the total entropy production is

$$\dot{s}(\mathbf{r}) = \int \mathbf{J} \cdot \nabla T^{-1} d\mathbf{r} = k \int \frac{[\nabla T(\mathbf{r})]^2}{[T(\mathbf{r})]^2} d\mathbf{r} \geq 0 \tag{11.19}$$

Also the functional derivative of the entropy production is:

$$\frac{\delta}{\delta T(\mathbf{r}')} \dot{s}(T(\mathbf{r})) = 2k \left[\frac{[\nabla T(\mathbf{r}')]^2}{[T(\mathbf{r}')]^3} - \frac{\nabla^2 T(\mathbf{r}')}{[T(\mathbf{r}')]^2} \right] \tag{11.20}$$

For a stationary state, $\nabla^2 T_{ss}(\mathbf{r}) = 0$, where the subscript ss refers to the stationary state and this corresponds to a stationary state heat flux, $\mathbf{J}_{ss}$. We substitute $\mathbf{J}_{ss} = -k\nabla T_{ss}(\mathbf{r})$ into (11.19) and (11.20) to obtain

$$\dot{s}(T_{ss}(\mathbf{r})) = \frac{[\mathbf{J}_{ss}]^2}{k} \int \frac{d\mathbf{r}}{[T_{ss}(\mathbf{r})]^2} = \begin{array}{l} 0 \text{ if } |\mathbf{J}_{ss}| = 0 \\ > 0 \text{ if } |\mathbf{J}_{ss}| > 0 \end{array}$$

and

$$\frac{\delta}{\delta T(\mathbf{r}')} [\dot{s}(T_{ss}(\mathbf{r}))] = \frac{2\mathbf{J}_{ss}}{k \left[(T_{ss}(\mathbf{r}'))^3 \right]} = \begin{array}{l} 0 \text{ if } |\mathbf{J}_{ss}| = 0 \\ > 0 \text{ if } |\mathbf{J}_{ss}| > 0 \end{array}$$

Thus the entropy production is an extremum only at equilibrium corresponding to $\mathbf{J}_{ss} = 0$. Contrary to many published opinions there are no non-equilibrium stationary states that represent extrema in entropy production for this system.

Recommended Additional Reading

Boyce, W. E., & DiPrima, R. C. (2005). *Elementary Differential Equations and Boundary Value Problems*. John Wiley.
  A readable treatment of the mathematics involved in the solution of heat transport problems

Carslaw, H. S., & Jaeger, J. C. (1959). *Conduction of Heat in Solids*. Oxford University Press.
  This is the definitive text on thermal transport. It contains a large number of solutions for specific boundary conditions and system geometries

Ross, J. (2008). *Thermodynamics and Fluctuations Far from Equilibrium*. Springer.
  This book is concerned mainly with nonlinear chemical systems not at equilibrium but it contains important discussions on heat flow and associated entropy production (Chapter 12). In particular Ross shows (Section 8.2.2) that the excess work is minimised for steady heat flow in an isolated system

Turcotte, D. L., & Schubert, G. (1982). *Geodynamics*.
  Chapter 4 in this book is an important treatment of heat flow in geological systems

Fluid Flow

12.1 FLUID FLOW AND DEFORMATION/METAMORPHISM

Fluids in the form of aqueous solutions, solutions rich in CO_2, NaCl or hydrocarbons and partial melts play an important role in deformation and metamorphism (Yardley, 2009; Jamtveit and Austrheim, 2010). They influence the strength of deforming materials both through reactions with chemical bonds and through the *effective stress principle*; they facilitate deformation through dissolution and deposition of minerals. Fluids directly contribute to metamorphism through the supply of H^+ and $(OH)^-$ for mineral reactions (Austrheim, 1987; John and Schenk, 2003) particularly during retrogression, metasomatism and mineralisation and have a profound effect on the rates of mineral reactions and coupled deformation rates (Rubie, 1998; Wintsch and Yi, 2002). The extreme behaviour is in metasomatic and mineralising systems where the fluids not only carry reactive ions that generate alteration assemblages and the mineral deposits themselves but also *advect* the heat necessary to initiate and drive the reacting system. The flow of melts and partial melts is an integral part of high-temperature metamorphism and these melts ultimately aggregate to form magma bodies that can ascend high into the crust (Brown, 2010; Hobbs and Ord, 2010). In this chapter we outline the mechanics of fluid generation and transport through the crust, discuss the factors that drive fluid flow and consider the coupling of fluid flow with heat transport. We discuss the thermodynamics of fluid advection in deforming rocks and complete the chapter with an overview of crustal plumbing systems, including melt systems.

The important questions are *What role does fluid flow play in the evolution of metamorphic systems? What is the time integrated volume of fluid that passes through a rock during metamorphism (Yardley, 2009)? Are there parts of the system that remain dry (Thompson, 1983; Yardley and Valley, 1997)? What are the processes that drive fluid flow during metamorphism? In particular, how are meteoric waters driven into the deep crust (Fricke et al., 1992; Yardley et al., 1993; Cartwright et al., 1994; Munz et al., 1995)? Is convection an important process in the deep crust (Etheridge et al., 1983)? Can convection operate in a system with upward flow-through (Wood and Walther, 1986; Oliver, 1996)? How important are relatively impermeable 'caps' and 'seals' in metamorphic systems (Etheridge et al., 1983; Sibson, 1994)? Is flow up, down or sideways? Is flow up or down temperature gradients? How important is fluid focussing (Ague, 2011)? Is metamorphism isochemical? Or are large amounts of material removed during metamorphism (Etheridge et al., 1983; Vernon, 1998)?* The aim of this chapter is to set out the principles (based on mechanics) that allow such questions to be addressed and to answer some of these questions.

12.1.1 Evidence for the Influx of Fluids

Etheridge et al. (1983, 1984), following Fyfe et al. (1978), presented a case for large-scale infiltration of fluids during regional metamorphism, particularly at low metamorphic grades including retrogression. Since then a large number of studies (Dipple and Ferry, 1992; Ferry, 1994; Oliver, 1996; Ague, 2011) has supported this concept of extensive infiltration although notable disagreements exist (Yardley, 2009). A classification of fluid flow regimes during deformation and metamorphism given by Oliver (1996) highlights the range of observations. We present a slightly modified version of that classification in Figure 12.1. It differs from that of Oliver in that the presence of closed systems dominated by diffusive flow has been explicitly highlighted together with a thermal overlay which we develop throughout the

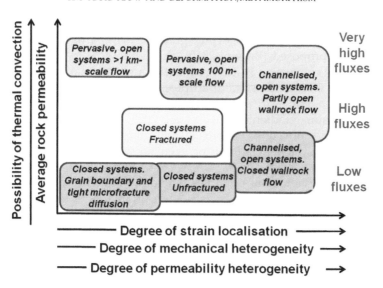

FIGURE 12.1   A classification of fluid flow regimes in metamorphic systems. *Modified after Oliver (1996).*

remainder of this chapter. Another view of this classification with respect to melt segregation and migration is presented by Rushmer (2001).

There is no doubt that some form of fluid infiltration is necessary to promote many metamorphic reactions and in some instances metamorphic rocks have been exposed to high temperatures and pressures with no reactions taking place until H_2O (or H^+) is introduced (Austrheim, 1987; White and Clarke, 1997, see Figure 12.2; Rubie, 1998; John and Schenk, 2003). There is still some debate as to the role of fluid flow in pore space in high grade rocks (Yardley, 2009) or in some mylonites (Fitzgerald et al., 2006). Some (Oliver, 1996; Ord and Oliver, 1997; Ague, 2011) have emphasised that flow can be channelized so that some parts

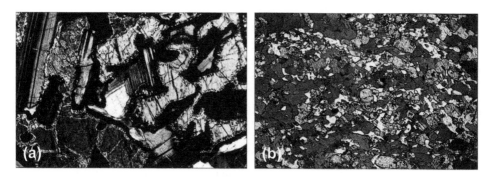

FIGURE 12.2   The influence of deformation and H_2O infiltration on mineral reactions. (a) Partially reacted eclogites with corona structures outside of shear zone, West Musgrave Ranges, Australia. (b) Fully reacted eclogites forming a hornblende gneiss inside shear zone. *See White and Clarke (1997) for discussion. Photomicrographs from Ron Vernon. Each photomicrograph approximately 4 mm across.*

of metamorphic terrains see low fluid fluxes whilst neighbouring parts are exposed to strongly focussed flow. At high metamorphic grades where partial melting occurs, flow in leucosomes has been proposed (Brown, 2010). For some workers (Etheridge et al., 1983) the concept of thermal convection is important. Although thermal convection is said to be impossible in systems with a lithostatic fluid pressure gradient (Wood and Walther, 1986), it turns out that it is possible, in principle, in some forms of open systems with super-hydrostatic fluid pressure gradients (Zhao et al., 2008; Section 4.2). The question is the following: What form do convective systems take in compartments with lithostatic pressure gradients and are such systems common or even possible under crustal conditions? We explore these concepts in Section 12.5.3.

In Figure 12.3, we present some examples where the influence of fluids is widely proposed. These include grain boundary pore structures in mylonites (Figure 12.3(a)), dissolution seams associated with differentiated crenulation cleavage at low metamorphic grades (Figure 12.3(b)), vein systems (Figure 12.3(c)) and leucosomes in migmatite complexes (Figure 12.3(d)).

Fluids associated with metamorphism arise from a number of sources including meteoric sources, connate waters, fluids released by devolatilisation (including decarbonisation,

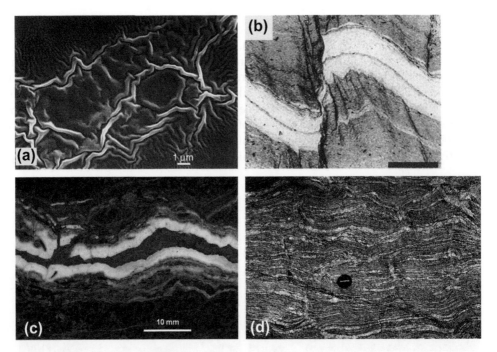

FIGURE 12.3   Some examples of fabrics associated with fluid flow in deforming rocks. (a) 'Brain' structure on a grain boundary in a mylonite. *From Mancktelow et al. (1998)*. (b) Dissolution seams associated with differentiated crenulation cleavage in low-grade metamorphic rocks. *From Worley et al. (1997)*. Scale bar: 1 mm. (c) Vein system, Sirius ore body, Western Australia. (d) Leucosomes. Port Navalo, Brittany, France. *From Brown (2008). Also figure D30 in Sawyer (2008)*.

hydrocarbon release from organic material and dehydration of (OH)-bearing minerals), and release of volatiles from crystallising melts. The chemical and isotopic characteristics of these fluids and the roles they play in metamorphic processes are discussed by Hollister and Crawford (1986) and Yardley (2009).

12.1.2 Influence of Fluids on Deformation. *Effective Stress*

A detailed discussion of the coupling between deformation and fluid flow is difficult because deformation is commonly viewed from a Lagrangian point of view (Chapter 2) whereas fluid flow is viewed from an Eulerian point of view. Integration of the two approaches is developed by Coussy (1995). We consider only one aspect of the coupling here, namely, the concept of *effective stress*. Some other forms of coupling are considered in Volume II where interesting nonlinear behaviour can occur depending on the relative rates of deformation and of fluid flow. In particular, an important coupling between fluid flow and deformation resulting in *porosity waves* (Connolly, 2010) is deferred to Volume II.

For rocks with connected pore space (including fracture networks), the concept of effective stress (Terzaghi, 1936) is commonly adopted. This concept proposes that the pore fluid pressure decreases the stress felt by the solid framework of the rock so that the deformation is controlled by an *effective stress* given by

$$\sigma_{ij}^{effective} = \sigma_{ij} + \delta_{ij}P^{fluid} \qquad (12.1)$$

The plus sign arises because we assume compressive stresses to be negative. This principle is true only if the solid framework of the porous solid undergoes negligible volume change during deformation. Notice that P^{fluid} is not in general equal to the mean stress so that $\sigma_{ij}^{effective}$ is not in general equal to the deviatoric stress. This also means (Section 12.3.3) that, in general, fluid flow is not driven by gradients in mean stress (Ridley, 1993). For discussions of the effective stress principle, see Detournay and Cheng (1993), Coussy (1995, 2004, 2010), Vardoulakis and Sulem (1995, Section 5.3) and Paterson and Wong (2005; Chapter 7).

The common interpretation of the effective stress principle is presented in Figure 12.4, where the yield criterion for a frictional material is shown in the absence of a pore pressure in Figure 12.4(a). The addition of a pore fluid pressure translates the stress σ_{ij} to $\sigma_{ij}^{effective}$ (Figure 12.4(b)) as described by (12.1) and ultimately results in failure in these materials if P^{fluid} is large enough.

Even though the deformation of frictional materials is commonly considered to be rate independent, the applicability of the effective stress principle to failure is dependent on the relative rates of deformation and of fluid flow. Above a critical strain-rate the effective stress principle is not a good approximation to the observed behaviour (Brace and Martin, 1968; Rutter, 1972; Paterson and Wong, 2005).

The concept of failure induced by a change in effective stress is used indiscriminately in the geosciences for any material. One should note that Figure 12.4(a) and (b) applies to a frictional material such as Mohr−Coulomb or Drucker−Prager (Chapter 6). For materials where the yield surface everywhere parallels the hydrostatic axis (Figure 12.4(c) and (d);

see Chapter 6), changes in the total stress arising from an increase in pore fluid pressure initiate yield only if one postulates a *tensile yield cap* on the yield surface as shown. Although this is reasonable from a mechanical point of view, and well-studied experimentally for brittle failure (Paterson and Wong, 2005), experimental data on tensile yield of non-frictional visco-plastic materials due to changes in effective stress are rare. Notice that in general the fluid pressure cannot equal the mean stress without initiating failure (Figure 12.4(d)) before that pressure is attained. Clearly (from Figure 12.4(d)) such a situation is possible if the tensile strength is large compared to the shear stress in the material.

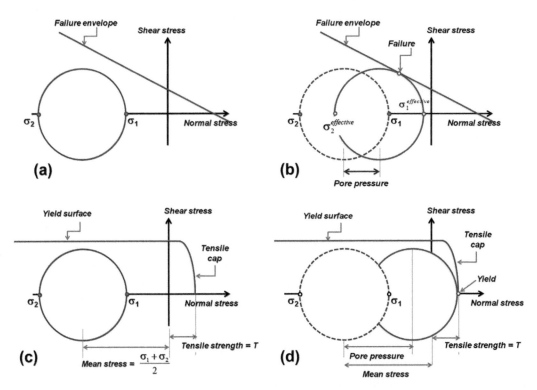

FIGURE 12.4 The effective stress principle applied to failure of frictional and non-frictional plastic materials. (a) A frictional porous material with a failure envelope shown in red. The imposed stress is (σ_1, σ_2) represented by the full blue circle. No fluid is present and the stress circle does not touch the failure envelope so that the stress state corresponds to elastic deformation. (b) A fluid pressure is added as indicated. This translates the initial stress state to a new effective stress state $\left(\sigma_1^{effective}, \sigma_2^{effective}\right)$ as given by (12.1). If the fluid pressure is high enough, the material fails. (c) A non-frictional plastic porous material with a yield surface shown in red. A tensile cap is added to describe tensile yield at a tensile yield strength, T. The imposed stress is (σ_1, σ_2) represented by the full blue circle. No fluid is present and the stress circle does not touch the yield surface so that the stress state corresponds to elastic deformation. (d) A fluid pressure is added as indicated. This translates the initial stress state to a new effective stress state given by (12.1) and represented by the full red circle. If the fluid pressure is high enough, the material yields at a stress state given by $\left(\frac{\sigma_1 + \sigma_2}{2} + T\right)$. However this does not imply (as shown here) that the pore pressure is equal to the mean stress.

12.1.3 Hydrostatic and Over-Pressured Fluid Systems: Critical Height

It is commonly proposed that during metamorphism rocks are subjected to fluid pressures that are everywhere lithostatic. This means that the *fluid pressure gradient* is lithostatic and hence there must be an upward flow of fluid throughout the system. If the drivers for fluid influx decrease, this flow tends to drive the fluid pressure gradient towards hydrostatic. When the lithostatic gradient is not maintained by the supply of new fluid, the fluid pressure gradient relaxes to near hydrostatic on a time scale of (H^2/κ^{fluid}) where H is the height of the system and κ^{fluid} is the *fluid diffusivity* given (Phillips, 1991, p. 80) by

$$\kappa^{fluid} = \frac{KV_P^2 \rho_0^{fluid}}{\mu^{fluid} \phi} \tag{12.2}$$

where K is the permeability, V_P is the P-wave velocity in the fluid, ϕ is the porosity, ρ_0^{fluid} is the fluid density at some reference state and μ^{fluid} is the fluid viscosity. This means that although the compartment is over-pressured, the fluid pressure gradient relaxes to hydrostatic. If we take the values given in Table 12.1 for water together with $V_P = 1.4 \times 10^3 \text{ m s}^{-1}$, $\phi = 0.2$ and

TABLE 12.1   Representative Numerical Values for Various Material Constants and Parameters for Water, Quartz Rich Silicate Melt and for Rock

Quantity	Symbol	Value
WATER		
Density	ρ_0^{fluid}	1000 kg m^{-3}
Specific heat	c_P^{fluid}	$4185 \text{ J kg}^{-1}\text{K}^{-1}$
Coefficient of thermal expansion	β_T^{fluid}	$2 \times 10^{-4}\text{K}^{-1}$
Dynamic viscosity	μ^{fluid}	10^{-3} Pa s (see also Figure 12.9)
Thermal conductivity	k^{fluid}	$0.6 \text{ W m}^{-1}\text{K}^{-1}$
SILICATE MELT		
Density	ρ_0^{melt}	2300 kg m^{-3}
Specific heat	c_P^{melt}	$1200 \text{ J kg}^{-1}\text{K}^{-1}$
Coefficient of thermal expansion	β_T^{melt}	10^{-3}K^{-1}
Dynamic viscosity	μ^{melt}	$10^{4.8}$ Pa s
Thermal conductivity	k^{melt}	$1.26 \text{ W m}^{-1}\text{K}^{-1}$
ROCK		
Density	ρ_0^{solid}	2700 kg m^{-3}
Specific heat	c_P^{solid}	$815 \text{ J kg}^{-1}\text{K}^{-1}$
Thermal conductivity	k^{solid}	$3.35 \text{ W m}^{-1}\text{K}^{-1}$

For discussion of the values for silicate melts see Hobbs and Ord (2010).

$K = 10^{-13}$ m^2, we obtain $\kappa^{fluid} \approx 1$ m^2 s^{-1}. Clearly this value depends strongly on the values of K and ϕ. For instance, for a tight metamorphic rock with $K = 10^{-18}$ m^2 and $\phi = 0.01$, $\kappa^{fluid} \approx 2 \times 10^{-4}$ m^2 s^{-1}. This is still about two orders of magnitude larger than the typical thermal diffusivity for rocks (Chapter 11) of $\kappa^{thermal} \approx 10^{-6}$ m^2 s^{-1}. Thus in general *fluid pressure will diffuse through rocks faster than temperature.*

If we do the same calculations (see Hobbs and Ord, 2010) for granitic melts flowing in leucosomes with $V_P = 5.6 \times 10^3$ m s^{-1}, $\phi = 10^{-2}$ and $K = 1.58 \times 10^{-9}$ m^2, we arrive at $\kappa^{melt} \approx 4 \times 10^{-6}$ m^2 s^{-1}. Again there are great uncertainties associated with the values of K and ϕ. In general, $\kappa^{fluid} \gg \kappa^{melt} > \kappa^{thermal}$.

For the values taken for water, a value of $\kappa^{fluid} \approx 1$ m^2 s^{-1} means that a lithospheric pressure gradient in an isolated compartment 1 km high will relax to a hydrostatic pressure gradient on a time scale of 10^6 s or about 11 days. A value of $\kappa^{fluid} \approx 2 \times 10^{-4}$ m^2 s^{-1} increases the relaxation time in this chamber to 160 years. This relaxation in fluid pressure gradient produces compressive stresses at the base of the compartment and effective tensile stresses at the top as discussed by Zhao et al. (2008). The maximum height, $H^{critical}$ (Figure 12.5) of a body of rock that can support a given fluid pressure gradient is:

$$H^{critical} = \frac{\overline{\sigma}^{compressive} + \overline{\sigma}^{tensile}}{\left(\rho^{rock} - \alpha\rho^{fluid}\right)g} \qquad (12.3)$$

where $\overline{\sigma}^{compressive}$ and $\overline{\sigma}^{tensile}$ are the compressive and tensile strengths of the rock and α is a factor that measures how far the fluid pressure gradient is above hydrostatic; $\alpha = 1$ for a hydrostatic fluid pressure and $\alpha = 2.7$ for a lithostatic fluid pressure gradient with an average rock density of $\rho^{rock} = 2700$ kg m^{-3}. Thus if $(\overline{\sigma}^{compressive} + \overline{\sigma}^{tensile}) = 50$ MPa, say, then $H^{critical} = 3$ km for a hydrostatic fluid pressure gradient. If the height of the hydrostatically pressured compartment is greater than $H^{critical}$, then the pore space at the base collapses and the top localises in the form of fractures in a manner considered by Connolly and Podladchikov (1998) for the generation of porosity waves (Volume II).

Although we have framed the argument above in terms of a system where an initial lithospheric pressure gradient is relaxed following the closure of the fluid supply, an identical

FIGURE 12.5   The critical height for a hydrostatic fluid pressure gradient.

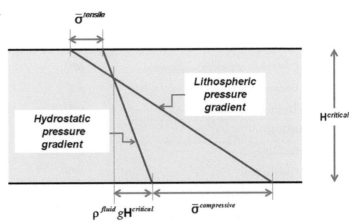

argument applies to a layered crustal system where a low permeability lithostatically pressured system is overlain by a layer with higher permeability. We will see (Section 12.2.3) that the high permeability layer must have a fluid pressure gradient that is below lithostatic and its thickness is then controlled by (12.3).

12.2 TYPES OF FLUID FLOW

Flow in metamorphic systems takes place at small spatial scales within nano-films or channels on grain boundaries up to the scale of kilometres in fluidised hydrothermal breccias (Figure 12.6(a)). Within this wide range of spatial scales, a wide range of mechanisms is involved in the transport of fluid phases. At the scale of grain interfaces, or closed microcracks within grains, transport is presumably by diffusion; some molecular modelling at this scale sheds some light on how this process works (see Figure 12.8 and Section 12.2.2). At a slightly larger length scale, there is a transition between what one would normally call *diffusion mechanisms* and *flow in nano- to micro-pore structures*. This is the regime where flow in the pore structures recognised by workers such as Mancktelow et al. (1998), Mancktelow and Pennacchioni (2004) and Putnis (2009) is relevant. At the nano-scale, flow is not described by the *Navier–Stokes equations* (Hughes and Brighton, 1999) which are the classical equations that describe the flow of linear fluids under the influence of a pressure gradient and is better described by the *Burnett equations* (Roy et al., 2003).

As one increases the length scale, there is a transition to where the Navier–Stokes equations can be used but the fluid does not stick to the walls of the flow channel; at larger length scales, the fluid does stick to the walls. At the next scale up, fluids in grain boundaries and closed fractures can aggregate to form bubbles and the bubbles can diffuse. The mechanics of this process is discussed by Burton (2001a,b). At the next scale, assuming that the fluid is Newtonian viscous, the transport process is described by the *Navier–Stokes equations* (Hughes and Brighton, 1999) and the fluids can be assumed to stick to the walls of the flow channel. This is the scale where Darcy's law begins to be applicable. At the scale where fluid transport can be considered to take place in open fractures, as is the case in many hydrothermal and breccia systems, the flow is described by Navier–Stokes equations although it may be possible in some situations (where the flow is slow enough) to describe flow through a large enough volume in terms of Darcy's law. These transitions are illustrated in Figure 12.6(d).

In Section 12.2.5, we point out that the complex topology of three dimensional fluid-filled porous networks leads to fluid stretching and folding within the flow channels and hence *chaotic advection* (Ottino, 1989a, 1990; Metcalfe, 2010; Lester et al., 2012). This is an important process for enhancing both fluid mixing and the rates of chemical reactions (Tel et al., 2005). At the scale where a porous solid can be considered homogeneous with respect to pore distribution, the chaotic advection within pores can be averaged (Phillips, 1991) so that the flow (as long as it is slow enough) is represented, macroscopically, by *Darcy's law*.

Chaotic advection is possible from the smallest to the largest scales in open pore and fracture networks so long as the networks are three dimensional and produce a flow geometry that branches and merges (Metcalfe, 2010; Lester et al., 2012). If the fluid velocity is large enough for a given length scale, the flow becomes *turbulent*. This is the case in

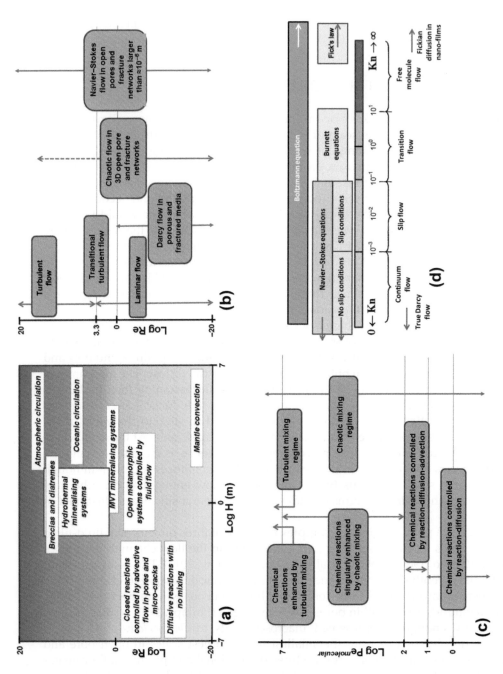

FIGURE 12.6   Characteristics and examples of various kinds of fluid flows. (a) Positions, within spatial scale – Reynolds number space, of various kinds of flow processes in the geosciences. The positions are diagrammatic and not meant to represent conditions precisely. *(Motivated by Ottino (1990) and Metcalfe (2010).)* (b). (c) Distinctions between various kinds of flows based on the Reynolds number and Peclet number related to molecular diffusion. (d) Classification of flows according to the Knudsen number. For $Kn > 10^{-1}$ flow in nano- to micron-scale pores is below the scale where the Navier–Stokes equations are applicable. Ultimately, at very fine scales, there is a transition to situations where Fick's law becomes applicable. *Adapted from Roy et al. (2003).*

some hydrothermal systems (Figure 12.6(b)). It is important to note that *chaotic advection* is different to *turbulent flow* of fluids; chaotic advection can occur in a laminar flow (Metcalfe, 2010). In non-turbulent chaotic flows, the Eulerian description of the flow at any instant is laminar, however, the flow is unsteady so that the history of the Lagrangian description is chaotic. For turbulent flows, the Eulerian description of the flow at any instant is not laminar and the streamlines define a cascade of dissipative structures over many length scales. Chaotic mixing can occur for both unsteady laminar flows and turbulent flows (Ottino, 1990; Metcalfe, 2010).

Some distinctions between various flows and examples in the geosciences are shown in Figure 12.6(a). We discuss these various processes throughout the remainder of this chapter but first we need some tools that enable us to describe the nature and geometry of these various flows and the ways in which they transport heat, dissipate energy and promote chemical reactions (Figure 12.6(b)–(d)).

12.2.1 The Dimensionless Groups: The Knudsen Number, the Reynolds Number, the Peclet Numbers and the Rayleigh Numbers

Dimensionless numbers of various kinds are useful in discussing the flow of fluids in a range of environments. Four of these dimensionless numbers are the Knudsen number, Kn, the Reynolds number, Re, the Peclet number, Pe, which can be defined both for molecular diffusion, $Pe^{molecular}$ and the diffusion of heat, $Pe^{thermal}$, and the Rayleigh number which can be defined for both temperature boundary conditions, Ra^T, and for fluid flux boundary conditions, Ra^{flux}. These are defined below.

The Knudsen number, Kn, is a number useful in defining the lower limit in length scale where the Navier–Stokes equations are relevant and the transitions to other forms of transport in micron- to nanometre- scale flow channels. Kn is defined as the ratio of the mean-free-path, π, of molecules and the macroscopic length scale of the pore space, Π:

$$Kn = \frac{Mean - free - path\ of\ molecules}{Length\ scale\ of\ pores} = \frac{\pi}{\Pi}$$

For $Kn > 10^{-1}$, the Navier–Stokes equations cease to be relevant in describing the flow (Figure 12.6(d)).

The *Reynolds number* is the ratio of inertial forces to viscous forces in a flow:

$$Re = \frac{Inertial\ forces}{Viscous\ forces} = \frac{Total\ momentum\ transfer}{Molecular\ momentum\ transfer}$$

or,

$$Re = \frac{H\rho_0^{fluid}}{\mu^{fluid}}v \tag{12.4}$$

where H is the characteristic length for the system in the direction of flow, v is the physical velocity of the fluid, ρ_0^{fluid} is the density of the fluid in some reference state and μ^{fluid} is the fluid viscosity (units: Pa s). If flow is slow and within the pores of a porous solid, v is given by

$$v = \widehat{V}/\phi \qquad (12.5)$$

where $\widehat{V}$ is the Darcy velocity (units: $m^3 \, m^{-2} \, s^{-1}$; see Section 12.2.3 and Figure 12.9) and ϕ is the porosity. For porous flow in metamorphic rocks typical values are $\widehat{V} = 1 \, m \, year^{-1}$ (or less), $\phi = 0.05$, $H = 10^{-3} \, m$, and $\mu^{fluid} = 10^{-4}$ Pa s. Thus, $Re \approx 6 \times 10^{-3}$ which means viscous forces dominate. Low Reynolds numbers are typical of fluid flow in many metamorphic systems.

The condition $Re < 1$ is necessary for Darcy's law to hold (see Section 12.2.3; Phillips, 1991, p. 28). For $Re < 1$ the flow is laminar which means that in a simple shearing flow, the fluid particles move in parallel planes. For these conditions, the effect of fluid advection upon mixing is negligible; any chemical reactions that take place in the fluid occur by diffusion across interfaces between different fluids and hence are very slow. Above the threshold $Pe = 10^2$ (if the flow is chaotic; Figure 12.6(c)), mixing rates and chemical reactions are enhanced by several orders of magnitude (Tel et al., 2005) although for such flows, Darcy's law is no longer applicable. Chemical reactions are enhanced by turbulent mixing above $Pe = 10^7$ (Villermaux, 2012). We revisit this issue in Section 12.2.5. Oliver et al. (2006) calculate fluid velocities greater than $1 \, m \, s^{-1}$ in a breccia system from Cloncurry, Australia, with length scales up to 1 km. Assuming the values given in Table 12.1 for water, the Reynolds number is of the order of 10^{10}. This is an extreme value and one could expect strong turbulent mixing with greatly enhanced chemical reaction rates. The point is that although the value of the Reynolds number is commonly <1 in metamorphic systems there are examples, particularly in hydrothermal breccia systems, where values $>10^7$ exist.

For fluid flows that transport mass and heat, the *Peclet number* is the ratio of the time scale for molecular diffusion, or the diffusion of heat, to the time scale for advection in a flow:

$$Pe = \frac{Time \; scale \; for \; diffusion}{Time \; scale \; for \; advection}$$

In particular, for Darcy flow coupled to the diffusion of heat or mass,

$$Pe^{thermal} = \frac{H \rho_0^{fluid} c_P}{k_e} \widehat{V} \quad and \quad Pe^{molecular} = \frac{H}{D} \widehat{V} \qquad (12.6)$$

where H is the length scale for the system in the direction of flow, ρ_0^{fluid} is the reference density of the fluid, c_P is the specific heat of the fluid, $\widehat{V}$ is the Darcy velocity (related to the physical fluid velocity by (12.5)) and k_e is the effective thermal conductivity of the fluid saturated solid given by

$$k_e = \phi k^{fluid} + (1 - \phi)k^{solid}$$

k^{fluid} and k^{solid} are the thermal conductivities of the fluid and unsaturated solid, respectively. D is the molecular diffusivity. Taking the values for water given in Table 12.1 and assuming $\phi = 0.1$ then $k_e = 3.08 \, W \, m^{-1} K^{-1}$. We assume this value in future calculations in this chapter.

If $\hat{V} = 1$ m year^{-1}, then given the values of ρ_0^{fluid} and c_P for water in Table 12.1 and taking $k_e = 3.08$ W m^{-1} K^{-1}, we obtain Pe$^{thermal} = 43$ for H = 1 km. Phillips (1991, pp. 213–216) gives examples of Pe$^{thermal} = 300$ for flows in sedimentary basins and of how such large values influence the possible temperature distributions in such basins. We will see later in this chapter that in metamorphic systems with a lithostatic fluid pressure gradient, thermal Peclet numbers greater than about 1 are unlikely unless large-scale advection of heat dominates the system as in some large hydrothermal systems.

The *Rayleigh number* is the ratio of the buoyancy forces to the viscous forces in a fluid:

$$\mathrm{Ra} = \frac{Buoyancy\ forces}{Viscosity\ forces}$$

For a system with given boundary conditions and geometry, a value for a *critical Rayleigh number* exists that marks the transition from pure conduction to fluid convection as the mode of heat transfer. This value of Ra is known as a *critical Rayleigh number*, Ra$_{critical}$, for that system. The boundary conditions for a system with a given geometry consist of various combinations of fixed temperature, fixed heat flux, fixed fluid pressure and fixed fluid flux. Examples of values of Ra$_{critical}$ for 10 different sets of boundary conditions are given by Nield and Bejan (2013, Table 6.1). As the temperature gradient or heat supply to a given system is increased, the convection pattern commonly undergoes transitions to new modes of behaviour at new critical Rayleigh numbers. These transitions include switches to new wavelengths and patterns of convection including oscillatory and chaotic behaviour (Nield and Bejan, 2013, Section 6.8).

Thus the precise form of the Rayleigh number depends on the boundary conditions for the problem. If the boundary conditions are fixed temperatures, T and T_0, respectively, at the base and top of the compartment (corresponding to the classical Horton–Rogers–Lapwood view of a convecting system; Nield and Bejan, 2013, Chapter 6), then the relevant Rayleigh number is

$$\mathrm{Ra}^T = \frac{\left(\rho_0^{fluid} c_P\right) \rho^{fluid} g \beta_T^{fluid} (\Delta T) K H}{\mu^{fluid} k_e} \tag{12.7}$$

where

$$\rho^{fluid} = \rho_0^{fluid}\left[1 - \beta_T^{fluid}(T - T_0)\right], \tag{12.8}$$

and $\Delta T = T - T_0$ is the temperature difference between the top and bottom of the compartment. β_T^{fluid} is the coefficient of thermal expansion of the fluid.

If on the other hand the boundary conditions comprise applied thermal and fluid fluxes at the base of the system and fixed temperature and fluid pressure at the top, then the relevant Rayleigh number is

$$\mathrm{Ra}^{flux} = \frac{\left(\rho_0^{fluid} c_P\right) \rho^{fluid} g \beta_T^{fluid} q K H^2}{\mu^{fluid} k_e^2} \tag{12.9}$$

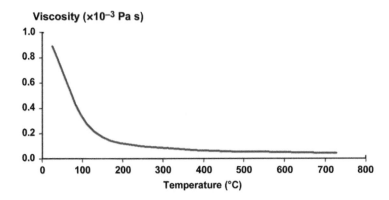

FIGURE 12.7   Variation of the viscosity of pure water with temperature.

where q is the heat flux imposed at the base of the system. This corresponds to the situation in most metamorphic/hydrothermal systems. Given the values for water in Table 12.1 and assuming $\phi = 0.1$ then numerically,

$$Pe^{thermal} = 1.359 \times 10^6 H \widehat{V} \tag{12.10}$$

$$Ra^T = 2.66 \times 10^9 (\Delta T) KH \tag{12.11}$$

$$Ra^{flux} = 8.82 \times 10^8 qKH^2 \tag{12.12}$$

As examples, if $H = 1$ km and $\widehat{V}$ is 10 mm year^{-1} then $Pe^{thermal} = 0.43$. Ra^T is 79.8 for a temperature difference between the top and bottom of 300 °C and a permeability of 10^{-13} m^2. For $H = 10$ km, a basal heat flux of 60 mW m^{-2}, a lithostatic fluid pressure gradient and $K = 10^{-18}$ m^2, $Pe^{thermal} = 2.3$ and $Ra^{flux} = 5.3$.

Note that the viscosity of water decreases rapidly with increasing temperature (Figure 12.7) at crustal pressures in the range $0-\sim 700$ °C (Abramson, 2007); the variation in viscosity, μ^{fluid}, with temperature is given approximately by $\mu^{fluid} = A[10^{B/(T-C)}]$ where $A = 2.414 \times 10^{-5}$ Pa s, $B = 247.8$ K and $C = 140$ K. For temperatures greater than about 300 °C, the viscosity of water is about 10^{-4} Pa s.

12.2.2 Grain Boundary and Micro-Fracture Flow

Transport of fluids in deforming metamorphic rocks takes place at a number of scales ranging from the grain scale, in the case of reacting metamorphic rocks at high pressures, or deformation by pressure solution, to flow in open cracks both within grains and at coarser scales (Etheridge et al., 1983, 1984). In hydrothermal systems, particularly in hydrothermal breccias, flow may be in open fractures and take place on the kilometre-scale. In this section, we consider flow at the grain boundary or intra-grain crack scales.

The detailed mechanism (or mechanisms) of transport of fluids in metamorphic rocks remains a matter of speculation although several models have been proposed that are worthy contenders. As Etheridge et al. (1983) point out, the initial porosity that is present in unmetamorphosed rocks is modified and probably largely obliterated by chemical reactions and grain boundary adjustments during metamorphism to produce a *metamorphic porosity* that comprises planar films on grain boundaries, pores ('bubbles' or 'voids') along grain boundaries (Billia et al., 2013; Mancktelow et al., 1998; Mancktelow and Pennacchioni, 2004; White and White, 1981), channels or tubes along the boundaries between three grains (Hay and Evans, 1988), and nano-porosity generated by dissolution processes (Putnis, 2009). In addition, Etheridge et al. (1983) proposed a *fracture porosity* that is envisaged to develop by hydrofracturing at the grain scale (Cox and Etheridge, 1989; Etheridge et al., 1984). We discuss below porosity structure and generation, and hence permeability, at the grain scale.

The evolving grain boundary structure. The nature of grain boundaries is important not only for controlling the transport of fluids but also for the mechanical behaviour of a polycrystalline aggregate during deformation. Thus Paterson (1995) distinguished three different models for fluid transport in grain boundaries in the context of fluid-assisted granular flow and showed that each results in a different dependence of the flow stress upon the grain size and/or other parameters such as the porosity or molar volume. Thus the details of grain boundary structure and the ways in which fluids are incorporated into that structure have wide implications not only for fluid transport mechanisms but also for the mechanical response during deformation and for the mechanisms of metamorphic reactions.

The geometrical makeup and evolution of grain boundary microstructure is discussed in Chapter 13; the configuration of grain boundaries is essentially controlled by interfacial energies. Of interest here is how fluids are incorporated into the boundaries and the controls on fluid transport. Figure 12.8 shows molecular dynamic models for some interfaces between H_2O and SiO_2. The model in the upper left of Figure 12.8 shows the structure of a water film between two quartz grains at 1000 K and 300 MPa pressure modelled by Adeagbo et al. (2008). The bonds at the water–quartz interface are hydroxylised and the modelling shows that silica is dissolved in the interfacial film in the form of $Si(OH)_4$. The calculations indicate that the mobility of $Si(OH)_4$ within the interfacial film is similar to that in pure water; the diffusion coefficient in pure water is $D^{Si(OH)_4}_{pure\ water} = 0.85 \times 10^{-8}\ m^2\ s^{-1}$, whereas that calculated for the interface is $D^{Si(OH)_4}_{interfacialfilm} = 1.01 \times 10^{-8}\ m^2\ s^{-1}$. The structure of the interface changes as the film becomes wider and develops into a bubble. A mechanism for doing this is discussed by Burton (2001a,b). The top-right panel in Figure 12.8 shows the edge of a water bubble in contact with quartz at 330 K (Ho et al., 2011). The structure of the interface is now much more complicated and the quartz surface becomes hydrophyllic with large concentration of hydrogen bonds. As the density of surface (OH) groups increases, the substrate becomes more hydrophyllic. There is also a weak, layered structure developing within the water bubble within 1.0–1.5 nm of the quartz interface. One would expect the diffusion coefficient for $Si(OH)_4$ to be different in the structured boundary layer to that in the bubble itself and to be different again to the diffusion coefficient measured in a boundary film.

Thus the detailed structure of interfaces between grains, and of 'tight' cracks within grains in contact with fluids, needs to be considered in developing models for fluid migration, species diffusion and mechanisms for mineral reactions at the grain scale. To date relatively little is available in the way of molecular modelling but one expects this field of study to grow rapidly and to supply data on diffusion coefficients and chemical reaction mechanisms that are difficult to obtain experimentally.

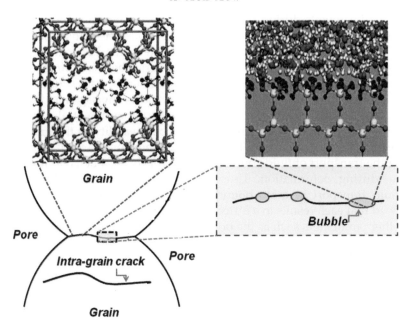

FIGURE 12.8 Molecular structure of a film of H_2O between two crystals of SiO_2 (top left) and between a bubble of H_2O and SiO_2 (top right). *Compiled from Raphanel (2011), Adeagbo et al. (2008) and Ho et al. (2011).* In the top-left panel, red, yellow, blue and white represent, respectively, quartz oxygen, silicon, hydrogen and water oxygen. In the top-right panel, blue, red, yellow and white spheres represent, respectively, surface hydrogen, oxygen, silicon and water hydrogen. Such molecular scale ordering can be applied to intra-grain cracks as well as grain boundaries.

Micro-cracking. The mechanics of micro-cracking is discussed in Chapters 8 and 10. Such processes are by far the most dramatic ways of changing permeability. The actual increase in permeability produced by arrays of cracks and/or tubes depends on the precise geometry and statistics of the defects but a feeling for the effect can be gained by considering an array of parallel cracks with apertures δ. The permeability arising from arrays of parallel cracks which generate a porosity, ϕ, is $\approx 10^{-2}\phi\delta^2$ (Phillips, 1991; pp. 29–34). An array of micro-cracks with an average number of cracks per unit area of ε gives a porosity $\phi = \delta\varepsilon$ and so (assuming all the new porosity is connected) the permeability is $\approx 10^{-2}\varepsilon\delta^3$. Thus if $\delta = 10^{-5}$ m and $\varepsilon = 10^3$ m^{-2}, then the permeability is $\approx 10^{-14}$ m^2. If the rock initially had a permeability below the percolation limit (say 10^{-22} m^2) then this micro-cracking represents a very substantial increase in permeability.

These kinds of arguments are supported experimentally. Micro-cracking induces a permeability increase that can be quite dramatic (Fredrich and Wong, 1986). Near the percolation threshold a number of workers (see Meredith et al., 2012 for a review) have proposed a relation:

$$K = K_0(\phi - \phi_c)^n$$

where K is the permeability (in m^2), ϕ is the total porosity (expressed as a %) and ϕ_c is the percolation threshold for fluid flow; K_0 is a characteristic permeability when $\phi - \phi_c$ is 1%. n is a critical parameter to be determined by experiment but values in the range 2.3–3.8

have been reported (Meredith et al., 2012). The experiments of Meredith et al. (2012) demonstrate a permeability increase from $<10^{-22}$ m^2 with a porosity of 1% increasing to about 10^{-16} m^2 at a porosity of 5%. This change in permeability has been proposed on the basis of percolation theory by Berkowitz and Balberg (1992) and Feng et al. (1987). This means, as suggested by Etheridge et al. (1983), that metamorphic rocks that are considered to be of very low permeability by writers such as Yardley (2009) and Connolly (2010) can become highly permeable with just the slightest amount of micro-cracking.

Reaction induced porosity. Many authors have proposed that permeability can be generated directly by metamorphic reactions. The mechanisms involve a spectrum of behaviours that range from an influence of released fluid or melt pressure on the effective stress thus inducing fracture (Connolly et al., 1997) to volume changes associated with the chemical reactions: positive ΔV reactions are proposed to generate local stresses that initiate fracture (Watt et al., 2000) and negative ΔV reactions generate local porosity increases directly. For instance, Rushmer (2001) describes situations where partial melting involving biotite (zero or small positive ΔV) results in local trapping of partial melt, whereas partial melting of muscovite (large positive ΔV) results in escape of partial melts.

An important part of this spectrum of behaviour involves the mechanisms by which mineral reactions take place. The issue is strongly linked to the observation that many metamorphic reactions such as kyanite replacing andalusite, serpentinite replacing olivine or albite replacing K-feldspar, are pseudomorphic (see Vernon, 2004, Section 4.13.3 and Chapter 15 in this book). Pseudomorphic reactions are classically considered, by definition, to be constant volume replacements even though the molar volumes of the initial and replaced phases may be quite different. One view of this process is presented by Merino and Canals (2011) who propose that the replacement process involves atom for atom displacements driven by a pressure solution process; the process does not involve solution and precipitation. Stresses are generated during this process (Fletcher and Merino, 2001) that drive pressure solution, fracture and vein formation. These stress induced fractures that are then responsible for increases in permeability.

A second view of mineral reactions (particularly those that involve pseudomorphism) derives from the work of Putnis and others (Putnis, 2009; Putnis and John, 2010) where a solution−dissolution-assisted phase transition is proposed whereby solutions dissolve the parent grain to form a network of nano-metre pores and/tubes (Figure 12.11(a)) and the new replacement mineral is precipitated in these pores. It is this nano-porous network that is capable of increasing the local permeability but stress-induced fracturing may be important also (Putnis, 2009).

This subject is treated in a little more detail in Chapter 15; it may be that a complete range of behaviour between the Merino- and Putnis-models exists. The subject is in its infancy and an exciting future in this area is developing. The issue is important not only for models of porosity and permeability development during metamorphic reactions, it has relevance as to how chemical reactions are written for metamorphic systems; *should the reaction be written in terms of constant pressure (as is the classical approach) or in terms of constant volume (Lindgren, 1912, 1918; Putnis, 2009; Merino and Canals, 2011)?*

In summary, it appears that there is a complete spectrum of behaviour with respect to mechanisms of fluid transport during deformation and metamorphism. At the finest of intergranular scales or within tight microcracks, the fluid comprises a nano-film with the kinds of structures illustrated in Figure 12.8. Transport is probably described by Fick's law. At a coarser scale, interfaces are sites for nano-pores derived largely by chemical reactions including dissolution. Transport is described by the Burnett equations. As the scale increases and the pores become larger (perhaps the micron-scale), the flow can be described by the Navier–Stokes equations but with slip at the solid interface. Increasing the scale further leads to non-slip conditions at the interface and a transition to true Darcy flow. At larger scales again, in open fractures, Navier–Stokes flow describes laminar flow. Ultimately at large Reynolds numbers, such as in breccias, the flow becomes turbulent. Chaotic flow leading to enhanced mixing is possible in all regimes where the flow at each instant can be described as laminar. Enhancement of chemical reactions takes place in these flows for Pe > 10^2. Chemical reaction enhancement by chaotic mixing is overwhelmed by turbulent mixing for Pe > 10^7.

12.2.3 Darcy Flow; Porosity and Permeability

Slow fluid flow through porous media with coordinates (x, y), with x horizontal and y vertical (positive down), is well described by *Darcy's law* which simply says that in the absence of gravity, the fluid flux is proportional to the pore fluid pressure gradient. In two dimensions with gravity Darcy's law is expressed as

$$\widehat{V}_x = \frac{K}{\mu^{fluid}} \left(-\frac{\partial P^{fluid}}{\partial x} \right) \qquad (12.13)$$

$$\widehat{V}_y = -\frac{K}{\mu^{fluid}} \frac{\partial}{\partial y} \left(P^{fluid} - \rho^{fluid} gy \right) = \frac{K}{\mu^{fluid}} \left(-\frac{\partial P^{fluid}}{\partial y} + \rho^{fluid} g \right) \qquad (12.14)$$

where $\widehat{V}_x$, $\widehat{V}_y$ are the horizontal and vertical components of the *Darcy fluid velocity*. These relations for Darcy's law are true for Re < 1, Kn < $\approx 10^{-3}$, and for linear viscous fluids (Phillips, 1991).

The *Darcy velocity* is the *volume of fluid* that passes through a unit area of the material in unit time and thus has the units: $(m^3 \, m^{-2} \, s^{-1})$ or $(m \, s^{-1})$. Thus although $\widehat{V}$ has the units of velocity, one should always remember that it represents a *volume of fluid flowing through a unit area* (Figure 12.9). The concept of the Darcy fluid velocity as a *flux* is fundamental in examining open flow systems. We loosely refer to the *Darcy flux* as the Darcy velocity.

In (12.13) and (12.14), K is a material quantity known as the *permeability* (with units m^2) and μ^{fluid} is the fluid viscosity (with units Pa s). P^{fluid} is the fluid pressure, ρ^{fluid} is the fluid density and g is the acceleration due to gravity. The negative sign in these equations means the fluid flows down the gradient indicated. Notice that the flow in the presence of a gravity field is not down a gradient in fluid pressure. The flow is controlled by gradients in what is variously known as the *hydraulic head*, $\mathcal{H}$, or sometimes as the *hydraulic potential*, a concept introduced by Hubbert (1940). $\mathcal{H}$ in this case is given by

$$\mathcal{H} = \left(P^{fluid} + \rho^{fluid} gy \right) \qquad (12.15)$$

where the coordinate y can be considered as the height above (or below) an arbitrary datum. The use of the term *hydraulic potential* for $\mathcal{H}$ is useful only for isothermal flows. If thermal

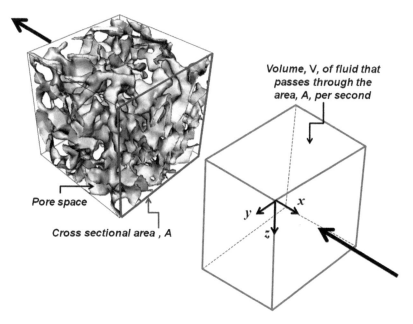

FIGURE 12.9 The definition of Darcy velocity. In this case the x-component of the Darcy velocity, $\widehat{V}_x$ (which for this flow is negative), is the volume of fluid that passes through the area, A, of the porous solid in a second. The porous solid here is constructed from a tomogram of the Castelgate sandstone and is available at http://xct.anu. edu.au/network_comparison/. Scale: cube is 2.87 mm on side. *This image is used with permission from* The Network Generation Comparison Forum *at The Australian National University, Canberra.*

effects are present, ρ^{fluid} becomes a function of position within the thermal field and $\mathcal{H}$ ceases to be a potential for the flow.

In addition, another equation is required to express the *continuity of flow* in two dimensions. The general form of this equation is (Phillips, 1991)

$$\frac{\partial\left(\phi\rho^{fluid}\right)}{\partial t} + \nabla \cdot \left(\rho^{fluid}\widehat{V}\right) = Q_s$$

where Q_s is a source term with units mass per unit volume per second. If we consider systems where fluid is not generated within the system, the source term is zero. In addition, for constant porosity and fluid density, we have for continuity in two dimensions:

$$\frac{\partial\widehat{V}_x}{\partial x} + \frac{\partial\widehat{V}_y}{\partial y} = 0 \tag{12.16}$$

We will use this form of the continuity equation frequently in what follows.

The fact that fluid flow in the presence of a gravity field is not driven by gradients in fluid pressure (in general) is illustrated in Figure 12.10 where manometers are inserted in a fluid flow system in which fluid flows from a reservoir at A at constant height above a datum X−Y to another reservoir at E at a lower height so that $\mathcal{H}_A > \mathcal{H}_E$. The fluid flows continuously from A to E under the influence of the gradient in $\mathcal{H}$. Notice that the fluid flows from C to D even though the fluid pressure at C as measured by the height of fluid in the manometer

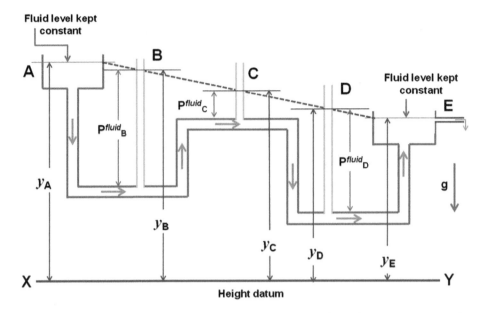

FIGURE 12.10   The concept of hydraulic head in the presence of a gravity field. Fluid reservoirs at A and E are kept at constant levels of fluid, y_A and y_E, above a height datum X–Y shown in red. The reservoirs are connected by a pipe shown in blue passing through the localities B, C and D. At these localities, manometers record the fluid pressure shown as $P^{fluid}{}_B$, $P^{fluid}{}_C$ and $P^{fluid}{}_D$. The fluid flows from high fluid pressure at B to low fluid pressure at C and then to a higher fluid pressure at D. Thus in this case, fluid flows from low fluid pressure to higher fluid pressure. The fluid flow is driven by the overall gradient in hydraulic head shown by the dotted red line and not by local differences in fluid pressure. Notice that $\mathcal{H}_A > \mathcal{H}_B > \mathcal{H}_C > \mathcal{H}_D > \mathcal{H}_E$ where the hydraulic head $\mathcal{H}$ is given by (12.15).

at C, is smaller than the fluid pressure at D so that over the path C–D the fluid is flowing *up* a fluid pressure gradient (see Mandl, 1998, figure II, 10.1).

The essential problem in applying Darcy's law to geological problems is obtaining knowledge of the permeability and its spatial distribution. As shown in Table 12.2, the permeability varies by many orders of magnitude between un-cemented sands, sandstones

TABLE 12.2   Representative Permeabilities of Various Rock Types

Rock Type	Permeability (m²)
Gravels and sands	10^{-9} to 10^{-12}
Shales	10^{-12} to 10^{-25}
Sandstones	10^{-11} to 10^{-17}
Limestones	10^{-12} to 10^{-16}
Marble	10^{-16} to 10^{-19}
Granites, gneisses, basalts	10^{-16} to 10^{-20}

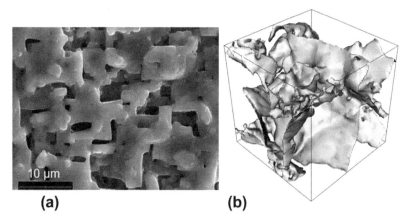

(a) **(b)**

FIGURE 12.11   Pore microstructures. (a) Pore structure developed during the dissolution of a single crystal of KBr by reaction with a KCl solution. *(From Putnis and Mezger, 2004.)* (b) Pore microstructure in Mt. Gambier limestone. Scale: cube is 155 mm on side. *This image is used with permission from* The Network Generation Comparison Forum *at The Australian National University, Canberra.*

and 'tight' metamorphic rocks with no fractures. In general, the permeability for a given rock type increases with depth (See Lyubetskaya and Ague, 2009, p. 1511, for a discussion).

The remaining parameter of importance in the flow of fluids through rocks is the *porosity*, ϕ, which is the ratio of the volume of open pore space to the total volume of the rock. The pore space may comprise the network of pores between individual grains or of fractures in the rock mass. If both pores between grains and fractures are present, one talks of *double porosity*. Typically ϕ varies from perhaps 0.3 in uncemented sandstones to 0.01 or less in metamorphic rocks. Some examples of pore structures are given in Figures 12.9 and 12.11. Since the open pore space need not be interconnected (there may be many dead end or enclosed spaces), there is no simple relationship between porosity and permeability. The subject is discussed by Walder and Nur (1984). An empirical relation known as the *Kozeny-Carman relation* which relates the porosity of a packing of regular spheres to the permeability of the array is commonly used:

$$K = K_0 \frac{\phi^3}{1 - \phi^2} \tag{12.17}$$

where K_0 is the reference permeability that depends on a reference porosity ϕ_0.

12.2.3.1 Fast Fluid Flows and Brecciation/Fluidisation

Darcy's law is derived (Phillips, 1991) for slow linearly viscous flows (Re < 1) where the dissipation arising from momentum transfer is negligible. At higher velocities, inertial effects become important and dissipation from these effects cannot be neglected. This is particularly important for fluid flow through breccias and in particular when the fluid flow is sufficiently strong to be able to lift or move individual fragments of rock, a process known as *fluidisation*. At such fluid velocities, the relation between fluid pressure gradient ∇P^{fluid} and fluid velocity, v, is no longer linear, as in Darcy's law, but includes a term with the fluid velocity squared, expressing the increased dissipation arising from transport of momentum:

$$\nabla P^{fluid} = av + bv^2 \tag{12.18}$$

Various forms of this equation are known as the *Ergun* and the *Wen and Yu equations* and are discussed by Niven (2002) and Bird et al. (1960). These equations were used by Oliver et al. (2006) in a study of breccia formation and transport.

12.2.3.2 Pressure Distribution in a Layered Crust

Most models of the crust admit of a permeability change with depth whether it be due to compaction with depth (Bethke, 1985), a transition to lower permeability metamorphic rocks (Manning and Ingebritson, 1999; Lyubetskaya and Ague, 2009) or the presence of relatively impermeable 'caps' or 'seals' at mid-crustal depths (Etheridge et al., 1983). Here we explore the effects of such permeability changes on a metamorphic system where fluid is generated at depth in the crust or is introduced into the crust from the mantle.

We first consider a homogeneous crust (Figure 12.12(a)). The fluid flow is specified directly by (12.13) and (12.14). If the vertical fluid pressure gradient is hydrostatic, that is, $\frac{\partial P^{fluid}}{\partial y} = g\rho^{fluid}$, then there is no flow. If the fluid pore pressure gradient is greater than hydrostatic, the flow is upwards and if the gradient is less than hydrostatic, the flow is downwards. Lyubetskaya and Ague (2009, Figure 3(d)) give an example of downward flow driven by retrograde metamorphic reactions at depth. Notice however that since downward flow occurs for a fluid pressure gradient that is less than hydrostatic, the permissible height of such a compartment is severely limited by (12.3).

Now consider a crust consisting of two layers (Figure 12.12(b)) in which a layer (Layer 1) with permeability, K_1, and fluid viscosity, μ_1, is overlain by another layer (Layer 2) in which the permeability is K_2 and the fluid viscosity is μ_2. We inject a fluid with upward Darcy velocity, $\widehat{V}_y$, at the base of Layer 1. For continuity, the fluid flux through both layers must be identical and hence we have

$$\widehat{V}_y = -\frac{K_1}{\mu_1^{fluid}}\left(\frac{\partial P_1}{\partial y} - \rho_1^{fluid}g\right) = -\frac{K_2}{\mu_2^{fluid}}\left(\frac{\partial P_2}{\partial y} - \rho_2^{fluid}g\right) \tag{12.19}$$

Thus if $\mu_1^{fluid} = \mu_2^{fluid}$ and $\rho_1^{fluid} = \rho_2^{fluid}$, then

$$\left(\frac{\partial P_2}{\partial y}\right) = \frac{K_1}{K_2}\left(\frac{\partial P_1}{\partial y}\right) + \rho^{fluid}g\left(1 - \frac{K_1}{K_2}\right) \tag{12.20}$$

If $\left(\frac{\partial P_1}{\partial y}\right)$ is lithostatic and $\frac{K_1}{K_2} < 1$, $\left(\frac{\partial P_2}{\partial y}\right)$ is always less than lithostatic but greater than hydrostatic. For instance, if $\frac{K_1}{K_2} = 0.1$, then

$$\left(\frac{\partial P_2}{\partial y}\right) = 0.1\left(\frac{\partial P_1}{\partial y}\right) + 0.9\rho^{fluid}g$$

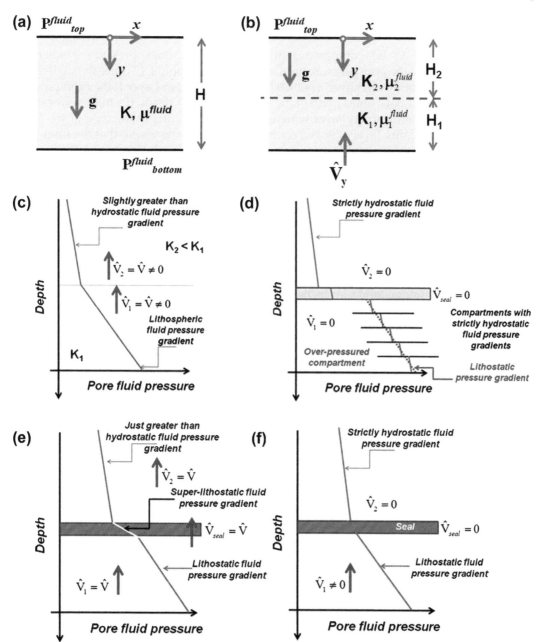

FIGURE 12.12 Simple models of fluid compartments in the crust. (a) A single compartment of height, H, permeability, K, and fluid viscosity, μ^{fluid}. (b) A two-layer system with properties as indicated. The layers are referred to as Layer 1 (bottom) and Layer 2 (top). (c) A system with continuity of fluid flux. If $K_2 < K_1$, the fluid pressure gradient in the upper compartment can be close to hydrostatic whilst the gradient in the lower

If $\left(\frac{\partial P_1}{\partial y}\right)$ is lithostatic, and hence equal to $2.646 \times 10^4 \, \text{Pa m}^{-1}$ then $\left(\frac{\partial P_2}{\partial y}\right)$ is $11.466 \times 10^3 \, \text{Pa m}^{-1}$ which is to be compared with a hydrostatic fluid pressure gradient of $9.8 \times 10^3 \, \text{Pa m}^{-1}$. Thus the pore pressure gradient in Layer 2 is only 1.17 times the hydrostatic gradient. If the permeability in Layer 2 is 1000 times larger than in Layer 1, the fluid pressure gradient in Layer 2 is 1.0017 times hydrostatic. This is a general result. If a high permeability layer overlies a low permeability layer with an imposed fluid flux at the base of the lower layer, the low permeability layer acts as a control valve for the system so that the fluid pressure gradient in the high permeability layer is always less than in the low permeability layer and commonly close to hydrostatic. Similar arguments may be developed from (12.19) if the fluid viscosity and/or density changes between compartments.

This argument is readily extended to a multilayered crust (Zhao et al. 1998, 2008). The overall conclusion is that it is not possible everywhere to have the fluid pressure gradient at lithostatic. The lowest permeability layer acts as a control valve for the system as a whole. Continuity of flux demands that in the highest permeability layers, the fluid pressure gradient must be below lithostatic, commonly close to hydrostatic, with no need for a low permeability 'seal' between the layers.

It is important to understand the mechanical and hydrological implications of proposing a low permeability 'seal' or 'cap' in a system where the fluid pressure gradient is maintained below the seal at lithostatic. The lithostatic fluid pressure gradient means there must be upward flow-through in the lower layer and continuity of flux demands this same fluid flux is maintained in the seal although at a higher fluid pressure gradient (from (12.20)). This means that the fluid pressure in the cap is above lithostatic and hence the cap is prone to yielding thus reducing the fluid pressure gradient. The cap in general cannot survive in a stressed system without failure.

Four different models for a layered crust with variations in fluid pressure gradient are shown in Figure 12.12(c)–(f). The continuity equation, (12.16), permits the models in Figures 12.12(c)–(e) but (f) is not permitted.

12.2.4 Advection of Heat in Porous Flow

Heat is said to be *advected* in porous flow when it is carried with the fluid. A special form of advection is *convection* when the flow is driven by density contrasts; the density contrasts may derive from thermal and/or chemical effects. When convection is driven

---

compartment is lithostatic. A seal between the two compartments is not necessary for this transition to exist. (d) A system with zero fluid flux. Fluid pressure gradients in all compartments are hydrostatic. No upward flow. The lower compartment is over-pressured and separated from the upper compartment by a low permeability seal. This lower compartment has an average fluid pressure gradient which is lithostatic but is divided into smaller compartments where the gradient is hydrostatic. (e) Another system with continuity of fluid flux. Upper compartment with a fluid pressure gradient just above hydrostatic. Lower compartment with a lithostatic fluid pressure gradient. Fluid pressure gradient in the 'seal' is super-lithostatic. Fluid flux is the same everywhere. (f) A system with no continuity of fluid flux and hence is physically unrealistic. Fluid pressure gradient in upper compartment is hydrostatic and lithostatic in lower compartment with a separating seal with no fluid flow. This model does not obey the fluid flux continuity (12.16).

by gradients in both temperature and density arising from chemical changes, the process is known as *double diffusion driven convection* (Zhao et al., 2008, Chapter 11). When heat is advected in a steady manner, the governing equations comprise (12.13), (12.14) and (12.16) together with

$$\rho_0^{fluid} c_P^{fluid} \left(\widehat{V}_x \frac{\partial T}{\partial x} + \widehat{V}_y \frac{\partial T}{\partial y} \right) = k_e \left(\frac{\partial^2 T}{\partial x^2} + \frac{\partial^2 T}{\partial y^2} \right) \tag{12.21}$$

where ρ_0^{fluid} and c_P^{fluid} are the reference fluid density and the fluid specific heat, respectively. The solutions to these equations are given by Zhao et al. (2008). It is convenient to introduce the following dimensionless parameters:

$$P^* = \frac{P - P_{top}}{\rho_0^{fluid} gH}, \quad T^* = -\frac{(T - T_{top}) k_e}{q_0 H} \tag{12.22}$$

$$Pe^{thermal} = \frac{\rho_0^{fluid} H c_P^{fluid}}{k_e} \widehat{V}, \quad y^* = \frac{y}{H} \tag{12.23}$$

where $Pe^{thermal}$ is the Peclet number for the hydrothermal system. For simplicity below, we write Pe for $Pe^{thermal}$. Then the solutions to the above equations become

$$T^* = \frac{1}{Pe} \exp[Pe]\{1 - \exp[-Pey^*]\} \tag{12.24}$$

and

$$P^* = y^* + \frac{q_0 \beta_T H}{k_e Pe} \exp[Pe] \left\{ y^* - \frac{1}{Pe}[1 - \exp(-Pey^*)] \right\} - \frac{\mu \widehat{V}}{K\rho_0^{fluid} g} \tag{12.25}$$

from which we obtain

$$\frac{\partial P^*}{\partial y^*} = 1 + \frac{q_0 \beta_T H}{k_e Pe} \exp[Pe]\{1 - \exp(-Pey^*)\} - \frac{\mu^{fluid} \widehat{V}}{K\rho_0^{fluid} g} \tag{12.26}$$

The expressions (12.24), (12.25) and (12.26) give the dimensionless temperature, pressure and fluid pressure gradients as functions of dimensionless depth. If we take the values of the various material constants and parameters given in Table 12.1, then the temperature distribution defined by (12.24) is shown in Figure 12.13 for $1 \leq Pe \leq 5$. The figure shows that for any value of Pe above 1, the temperature at the base of the system is unacceptably high for observed temperature distributions in the crust. For Pe = 1, 2 and 5, the temperature at the base of the system is 1.7, 3.2 and 29.5 times that predicted by the conduction solution. Thus in the absence of evidence for major discontinuities in the temperature distribution with depth in the crust together with temperature values at the Moho close to what we would expect from a conduction solution, maximum values of Pe around 1 seem to be indicated. A value of Pe ≥ 1 for the crust places severe restrictions on the physically possible values of $\widehat{V}$ for the crust and whether convection is possible for lithostatic pore pressure gradients; we discuss this issue in Section 12.7.

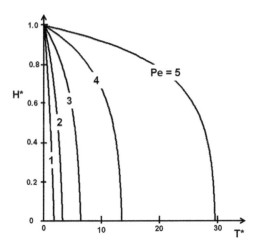

FIGURE 12.13 The influence of upward flow-through of fluids on the temperature distribution in a compartment with constant permeability. Plot of dimensionless height, H*, of the system against the dimensionless temperature, T*, for various Peclet numbers. For a Peclet number of 1, the temperature profile with depth is close to the conduction solution and blends into a conduction solution at the top of the compartment. For a Peclet number of 5, the temperature at the base of the system is close to 30 times what is expected of the conduction profile and there is a strong discontinuity in temperature profile with respect to a conduction gradient at the top of the compartment.

12.2.5 Fluid Mixing; Chaotic and Turbulent Flows

Although fluid mixing must be involved in the dissolution and reaction processes within pore spaces and at fluid—solid boundaries and is widely proposed as an important ingredient in the formation of hydrothermal systems, it remains very poorly understood (Ord et al., 2012). We describe *fluid mixing* as the dispersion of constituent species through the actions of fluid advection and/or molecular diffusion toward a spatially homogeneous distribution. Mixing of two fluids can occur in three ways: (1) two laminar flowing fluid streams may meet and mix by molecular diffusion across the interface between the fluids; (2) differences in the chemical or physical properties of the two fluid streams (for instance, chemical potential, density, surface tension, and temperature) generate flow instabilities at the interfaces between fluids which enhance fluid mixing (De Wit, 2001, 2008); (3) two fluid streams may meet in an open three dimensional network, either at the grain-scale, where the network comprises a series of interconnected pores between grains, or at larger scales where the network comprises a series of interconnected open fractures (Figure 12.14). The inherent three dimensional geometric complexity of these networks (Ord et al., 2012; Lester et al., 2012) results in a flow phenomenon known as *chaotic advection*, that is, fluid particle trajectories form a chaotic tangle, leading to efficient mixing.

The first mixing mechanism above involves diffusion alone, and is not efficient since the molecular diffusivity, D, for relevant aqueous species is typically of magnitude $10^{-10}\ \mathrm{m^2\ s^{-1}}$. The issue is discussed by Appold and Garven (2000). The latter two mechanisms involve the interplay of advection and diffusion, the relative timescales of which are characterised by the molecular Peclet number, $Pe^{molecular}$.

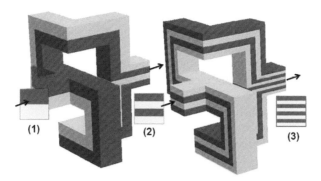

FIGURE 12.14   Chaotic advection in a three-dimensional network with branching and merging pore or fracture space. *(After Carrière (2007).)* The fluid is stretched and compressed as it traverses each set of pores from (1) to (2) to (3). The ultimate result is a Baker's transform where two initially adjacent fluids (red and yellow) are brought together as finer and finer striae.

For $Pe^{molecular} \leq 1$, chemical reactions are dominated by reaction–diffusion (RD) processes (Figure 12.6(c); see Chapters 7 and 15). Chaotic mixing involving laminar flow occurs in the range $1 < Pe^{molecular} < 10^7$ but enhanced chemical reaction rates due to chaotic flow occur only for $10^2 < Pe^{molecular} < 10^7$ (Ord et al., 2012; Tel et al., 2000, 2005). Turbulent mixing begins at $Pe^{molecular} = 10^7$ (Villermaux, 2012), corresponding to a Reynolds number of $\approx 10^3$. Turbulent flow results in enhanced chemical reaction rates (Villermaux, 2012) but turbulent flow is unlikely in crustal systems except for large hydrothermal breccia systems. Below the threshold $Pe^{molecular} = 1$, the effect of mixing arising from fluid advection upon chemical reaction rates is negligible and diffusion dominates, whereas above the threshold $Pe^{molecular} = 10^2$ advection can enhance mixing rates by several orders of magnitude. Fluid advection not only has a profound effect upon mixing rates but can also alter the stability and speciation of non-equilibrium chemical reactions (Tel et al., 2005) such as those encountered in hydrothermal systems. Hence the threshold Peclet numbers not only represent approximate thresholds in terms of mixing rate, but also the qualitative properties of mineral reaction and deposition are expected to differ across these thresholds. Although fluid advection is important below $Pe^{molecular} = 10^2$ for transporting, reacting mineralizing systems, the argument above suggests that these latter systems may be validly treated as RD systems (see Chapters 7 and 15). Above this transition, the reactions are better treated as advection–diffusion reaction (ADR) systems (Chapter 15). Although the literature on RD systems is immense, ADR systems have enjoyed much less attention, and it is the properties of these systems in the context of hydrothermal systems which are a focus of Volume II.

In order to understand the mechanisms by which fluid advection accelerates fluid mixing in hydrothermal systems, we provide a brief overview of the phenomenon of chaotic advection, involving terms and concepts which may be unfamiliar to the geoscientist. The reader is directed to several review articles (Metcalfe, 2010; Ottino, 1989b; Tel et al., 2000, 2005; Wiggins and Ottino, 2004) for further reading. The ability of turbulent flows rapidly to mix and disperse constituents is a widely recognized phenomenon (Villermaux, 2012). Turbulent flows possess a wide spectrum of eddy length scales, which act to distort concentration fields into complex spatial distributions with very fine striations and large interfacial area between constituents of the flow. In conjunction with molecular diffusion, such fluid advection leads to rapid dispersion, which is commonly termed mixing. Dispersion occurs even if the molecular diffusivity is vanishingly small. Conversely, slow, laminar flows are smooth and regular, and do not possess small length scales for the organization of fluid elements into fine striations.

As such, slow molecular diffusion usually does little to accelerate dispersion in steady laminar flows. The transition from laminar to turbulent flow typically occurs at Reynolds number Re $\sim 10^3$. We can express Re as

$$\text{Re} = \frac{\text{Pe}}{\text{Sc}} = \frac{viscous\ diffusion}{molecular\ diffusion}$$

where Sc $= \mu^{fluid}/(\rho^{fluid} D)$ is the *Schmidt number* quantifying the ratio of viscous diffusion to molecular diffusion; ρ^{fluid} and μ^{fluid} are the fluid density and viscosity, respectively. The Schmidt number is a material property, typically 10^3 to 10^4 for aqueous systems (as $\rho^{fluid} \sim 10^3$ kg m^{-3}, $\mu^{fluid} \sim 10^{-3}$ to 10^{-4} Pa s, $D \sim 10^{-10}$ m^2 s^{-1}), and so Re $= 10^{-4}$ Pe to 10^{-3} Pe.

Whilst this may be true for a wide range of laminar flows, there exist important exceptions that arise when the laminar flows are unsteady. Then the opportunity exists for the streamlines (which never intersect at any instant of the flow) to cross from one instant to the next. Although the Eulerian description of the flow is always laminar, the particle track for any material particle can become chaotic (Metcalfe, 2010; Ottino, 1990). This arises because the kinematic advection equation

$$\frac{dx}{dt} = \mathbf{v}(x, t)$$

describing the evolution with time t of the position x of a passive fluid tracer particle under the action of the fluid velocity field $\mathbf{v}$, is a nonlinear dynamical system capable of exhibiting chaotic dynamics. As such, although the velocity field $\mathbf{v}(x, t)$ itself is smooth, the fluid particle trajectories may be chaotic, leading to rapid dispersion and mixing. This phenomenon is termed *chaotic advection* (Aref, 1984; Ottino, 1990), and has been well studied over the past quarter century in the fields of fluid mechanics and dynamical systems. One familiar example of chaotic advection is expressed as the *Baker's transform*, examples of which are the kneading of bread dough or the manufacture of salt-water taffy. Kneading may be considered as a laminar flow comprising iterated stretching and folding motions. Although the flow field associated with these motions is smooth and regular, if one were to track the evolution of a dyed element of dough or taffy, it would soon be stretched to form a highly striated distribution and eventually be mixed throughout the dough or taffy form. This is chaotic advection in action. The actions of stretching and folding are in fact the fundamental mechanisms of chaotic dynamics in general: stretching acts to separate particles at an exponential rate, and folding acts to reorganize the flow to distribute these highly striated interwoven structures throughout the domain. Figure 12.14 provides an example of the Baker's transform for flow in a three dimensional pore or fracture network.

The rate of exponential stretching is quantified by the *Lyapunov exponent*, λ, which is taken as a measure of the strength of the chaotic dynamics. In the case of chaotic advection, the

Lyapunov exponent is defined as the long time limit of the exponential growth rate of the length of a material line δx between two particles initially separated by distance δX:

$$\lambda = \lim_{t \to \infty} \frac{1}{t} \ln \frac{|\delta x|}{|\delta X|}$$

Whilst the flow domain may be finite-sized, $|\delta x|$ can grow without bound due to the interwoven nature of the striations and hence of the initial material line. Although chaotic particle paths appear to be irregular and random, their underlying structures are highly organized and are often self-similar and multifractal (Muzzio et al., 1992). This organizing structure is termed the *chaotic template*, and active processes within the fluid such as diffusion and reaction play out on this template.

Although many laminar flows are engineered to exhibit chaotic advection to promote mixing and dispersion in a wide variety of applications, chaotic advection can occur in natural flows also. Systems subject to transient forcing or flow reorientation can exhibit chaotic advection, such as the evolution of plankton communities in oceanic currents (Hernandez-Garcia and Lopez, 2004; Karolyi et al., 1999) or the spreading of the gulf oil spill (Mezić et al., 2010; Thiffeault, 2010). Flows in porous media subject to transient forcing have been shown (Jones and Aref, 1988; Lester et al., 2009, 2010; Metcalfe et al., 2010a,b; Trefry et al., 2012; Zhang et al., 2009) to exhibit chaotic advection again due to the stretching and folding motions of the transient flow field. Recent studies (Carrière, 2007; Lester et al., 2012) have also established the propensity for chaotic advection to occur in porous media under steady flow, via the natural tortuosity of the pore space (Figure 12.14). Mixing due to chaotic flow has also been described in nano-porous networks (Ottino and Wiggins, 2004 and references therein) and rapid developments are happening in this area which have direct application to fluid transport and mineral reactions in metamorphic systems.

12.2.6 Multi-Phase Flow

The flow of two immiscible fluids together in a porous solid is described by a modified form of Darcy's law:

$$\hat{V}_i^{wetting} = -\frac{K_{ij}}{\mu^{wetting}} k_r^{wetting} \frac{\partial}{\partial x_j} \left(P^{wetting\ fluid} - \rho^{wetting\ fluid} g_k x_k \right)$$

$$\hat{V}_i^{non-wetting} = -\frac{K_{ij}}{\mu^{non-wetting}} k_r^{non-wetting} \frac{\partial}{\partial x_j} \left(P^{non-wetting\ fluid} - \rho^{non-wetting\ fluid} g_k x_k \right)$$

These equations define the Darcy fluxes for the two fluids (Figure 12.15(a)), one of which (the *wetting fluid*) wets the surface of the pore space more than the other (the *non-wetting fluid*). Each fluid has its own viscosity, μ, and density, ρ. The permeability of the medium is anisotropic in this case and represented by K_{ij}. As a result of the curved interface between the two fluids, the pressure in the non-wetting fluid is higher than in the wetting fluid. The difference in fluid pressure is known as the *capillary pressure*, $P^{capillary}$ and this is a function of the *wetting saturation*, $S^{wetting}$, which is the relative proportion of the wetting fluid:

$$S^{wetting} + S^{non-wetting} = 1.$$

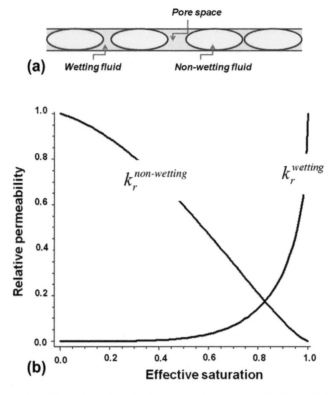

FIGURE 12.15   Principles of two-phase flow. (a) An idealised pore channel with pockets of non-wetting fluid separated by pockets of wetting fluid. (b) The relative permeabilities plotted against the effective saturation. The relative permeability of the wetting fluid remains low until inter-connections between pockets of wetting fluid can be established.

The products, $K_{ij}k_r^{wetting}$ and $K_{ij}k_r^{non-wetting}$, mean that the effective permeabilities for the two fluids are different. The *relative permeabilities*, $k_r^{wetting}$ and $k_r^{non-wetting}$, are given empirically by (van Genuchten, 1980):

$$k_r^{wetting} = S_e^b\left[1 - \left(1 - S_e^{1/a}\right)^a\right]^2 \quad \text{and} \quad k_r^{non-wetting} = (1 - S_e)^c\left[1 - S_e^{1/a}\right]^{2a}$$

where $0.1 < a < 1$, $0 < b < 1$ and $0 < c < 1$ are empirically derived constants and the *effective saturation*, S_e, is defined by

$$S_e = \frac{S^{wetting} - S_r^{wetting}}{1 - S_r^{wetting}}$$

where $S_r^{wetting}$ is a measure of the proportion of the wetting phase that remains in the system when $k_r^{non-wetting}$ is zero. Plots of the relative permeabilities against the effective saturation are given in Figure 12.15(b) for $a = b = 0.5$ and $c = 0.333$. The behaviour represented here is commonly observed: as S_e increases, that is, the proportion of wetting fluid increases, through removal of non-wetting fluid from the system, the permeability of the wetting fluid remains low until connections between pockets of wetting fluid can occur.

For examples of modelling of two phase flow with respect to the $H_2O-NaCl$ system, see Geiger et al. (2006a,b).

12.3 DRIVERS OF FLUID FLOW

Fluid flow in porous rocks is driven by one of four processes or by combinations of these processes. The one most commonly modelled arises from hydraulic head gradients induced by *topographic gradients* (Garven, 1985; Garven and Freeze, 1984a,b; ; Murphy et al., 2008; Phillips, 1991). A second process, *buoyancy* leading to convection, arises from density variations induced by temperature or chemical variations (Nield and Bejan, 2013). A third process arises from the generation of *fluid pressure gradients larger than hydrostatic* induced by compaction, devolatilisation reactions or crystallisation of volatile rich magmas (Burnham, 1979, 1985; Connolly, 2010; Fyfe et al., 1978; Phillips, 1991) and a fourth process arises from *fluid pressure gradients induced by deformation* (Cox, 1995, 1999; Ge and Garven, 1989; Ord and Oliver, 1997; Sibson, 1987, 1995). We discuss each of these processes in turn below (Table 12.3).

12.3.1 Topographically Driven Flow

A gradient in topography for saturated porous material produces a gradient in hydraulic head which drives fluid flow in exactly the same manner as is illustrated in Figure 12.11. The system is an open flow system that is controlled by the imposed hydraulic head (Figure 5.2(d)). This means that the local Darcy flow is influenced by the local permeability and fluid is focussed into the highest permeability layers, lenses or shear zones. If the local permeability changes due to clogging of pore space by mineral deposition, or to mineral dissolution or to fracturing, the local Darcy velocity changes as described by Darcy's law. In particular, if the porosity is clogged up by mineral deposition, then the fluid flow stops. This form of flow is the most commonly studied in the geological literature and derives from hydrological studies in near surface aquifers (Garven and Freeze, 1984a,b). Except in the uppermost crust, topographically driven flow is not the most important fluid driver in

TABLE 12.3 Magnitudes of Hydraulic Head Gradients Driving Fluid Flow for Various Processes

Driver	Magnitude (Pa m^{-1})	Direction of Flow
Lithostatic fluid pressure gradient	1.7×10^4	Vertical
5 km of topographic relief over 100 km	500	Downwards and horizontal
Deformation assuming 50 MPa gradient in fluid pressure over 1 km	5×10^4	Depends on scale of structures
Deformation assuming 50 MPa gradient in fluid pressure over 10 km	5×10^3	Depends on scale of structures

The fluid is assumed to be water with a density of 1000 kg m^{-3}.

metamorphic systems. Such systems are commonly flow controlled open systems (see Chapter 5; Figure 5.2(c)) where the flux into the system is the controlling factor rather than the hydrological head. However if large gradients in topography exist above the metamorphic system, as would arise from the presence of large mountain ranges, they would add a horizontal component to flow throughout the crust; this is true for both aqueous fluids and melts (see Figure 5.1(a)).

Examples of systems where the flow is controlled by topographic gradients are given by Garven (1985), Phillips (1991), Person and Baumgartner (1995), and Murphy et al. (2008). Lyubetskaya and Ague (2009) present examples where the flow of deep metamorphic fluids driven essentially by lithospheric pore pressure gradients is influenced by large topographic gradients.

12.3.2 Systems with Super-Hydrostatic Fluid Pressure Gradients. Production of Fluid and the Pressure Control Valve

The argument presented below is developed for partial melt systems by Hobbs and Ord (2010). Here we present the identical argument mainly for aqueous fluids but with some examples from melt systems. Following Phillips (1991), we consider mineral reactions of the form

$$S_1 \rightarrow S_2 + fluid$$

where the fluid may be a hydrous fluid, a melt or some other fluid. We write the following in terms of a hydrous fluid but it is equally applicable to any set of reactions that produce a fluid. S_1 is an assemblage bearing an initial volume concentration, s_0, of a hydrous phase such as muscovite, biotite or amphibole and S_2 is an anhydrous assemblage. In this reaction, s cubic metres of a cubic metre of the assemblage S_1 decrease with time,

$$\frac{\partial s}{\partial t} = -Q_s$$

as they are replaced by S_2. If T_e is the equilibrium temperature at which the reaction takes place, for $T < T_e$, $Q_s = 0$. When $T > T_e$ the rate of disappearance of S_1 is proportional to the amount per unit volume present and, near equilibrium, to $(T - T_e)$ so that Q_s can be written as

$$Q_s = \gamma s (T - T_e)/T_e$$

where γ is the reaction rate. Thus,

$$\frac{\partial s}{\partial t} = 0, \quad \text{for } T < T_e$$

$$\frac{\partial s}{\partial t} = -\gamma s (T - T_e)/T_e, \quad \text{for } T \geq T_e,$$

If the equilibrium isotherm is progressing upward through the rock mass with a speed U, then at any point near the equilibrium isotherm,

$$T - T_e = Ut gradT$$

where t is the time since the equilibrium isotherm passed through that point and $gradT$ (>0) is the negative temperature gradient in the direction of U. Thus,

$$\frac{\partial s}{\partial t} = -\frac{\gamma UtsgradT}{T_e} \quad \text{behind the equilibrium isotherm and}$$

$$= 0 \qquad \text{ahead of the equilibrium isotherm.}$$

In terms of the distance $\xi = Ut$ behind the equilibrium isotherm,

$$U\frac{\partial s}{\partial \xi} = -\left(\frac{\gamma gradT}{T_e}\right)\xi s, \quad \text{for } \xi \geq 0 \text{ (behind)}$$

$$= 0, \qquad \text{for } \xi < 0 \text{ (ahead)}.$$

Subject to the condition $s = s_0$ ahead of the reaction zone, these equations have the solution

$$s = s_0 \exp\left(-\frac{\xi^2}{l^2}\right), \quad \text{for } \xi \geq 0$$

$$= s_0, \qquad \text{for } \xi < 0$$

where l, the thickness scale of the reaction zone, is

$$l = \left(\frac{UT_e}{2\gamma gradT}\right)^{\frac{1}{2}}.$$

To gain an indication of the magnitude of l for a reaction that produces granitic melts, we take the estimates of Brown (2001) that the melting zone can be 15 km thick and forms in 10^4–10^6 years. This gives average values of U between 4.76×10^{-8} m s^{-1} and 4.76×10^{-10} m s^{-1}. If we take $T_e = 720\,°C$ which corresponds to the reaction,

muscovite + albite + quartz → K − feldspar + aluminosilicate + melt

at 1 GPa pressure (Peto, 1976), $gradT = 20\,°C\,km^{-1}$ and $\gamma = 10^{-2}\,s^{-1}$ to $10^{-3}\,s^{-1}$ (for which there is great uncertainty), then l is in the range 1 mm to 1 m which is thin enough to be neglected. However, as emphasised by Phillips (1991) for reactions where the fluid is produced over a temperature range, the physical height of the devolatilising or melting interval will be smeared out over that range. Montel and Vielzeuf (1997) for instance show that for the reaction:

biotite + plagioclase + quartz → orthopyroxene + garnet + K − feldspar + melt

melt is produced over a temperature range of 50 °C at low pressure (100 MPa) and over at least 200 °C at high pressure (1 GPa). Thus in this case, the actively melting zone is expected to be quite thick especially at high pressures. However, Phillips (1991) shows that smearing out the thickness of the devolatilising or melting zone results in the total generation of fluid, and hence the fluid pressure generated just above the reaction isotherm, being the same as when the reaction zone is thin.

Notice that if the isotherms rise only by conduction of heat, then the time scale for a thermal perturbation at the base of the crust to reach 10 km above the base of the crust is $(10 \times 10^3)^2/10^{-6}$ s or 3.2 million years. Thus if the devolatilising or melting region is to reach a thickness of 10 km in 10^4–10^6 years, as proposed by Brown (2001), the upward advection of heat by moving fluid is slightly faster than the rate at which heat would be conducted upwards.

When the reaction involves devolatilisation with 1 m^3 of the mineral assemblage releasing n^{fluid} cubic metres of fluid, the volumetric rate of fluid generation, Q^{fluid}, is $n^{fluid}Q_s$. Thus,

$$Q^{fluid} = n^{fluid}Q_s = -n^{fluid}\frac{\partial s}{\partial t}$$

And for this simple case of a well-defined T_e, the rate of fluid generation per unit volume is

$$Q^{fluid} = -n^{fluid}U\frac{\partial s}{\partial \xi}$$

where $s = s(\xi)$ is now given by $s = s_0 \exp\left(-\frac{\xi^2}{l^2}\right)$. This rate is a maximum at a distance $\frac{l}{\sqrt{2}}$ behind the moving isotherm, and decreases to zero beyond that point. The total rate of generation of fluid in this reaction zone (volume of fluid per unit time per unit area of the reacting zone) is

$$\Re^{fluid} = \int_0^\infty Q^{fluid}d\xi = n^{fluid}Us_0$$

which is independent of the reaction rate, γ, and of the temperature gradient, $gradT$.

If the porosity of the rock is constant, the vertical flux of fluid, $\widehat{V}$, at the top of the reaction zone is the volume of fluid produced per unit time per unit horizontal area or $\Re^{fluid}$. Thus if y is the vertical distance above the reaction isotherm, the vertical gradient of the total fluid pressure at the top of the reaction zone is

$$\frac{\partial P^{fluid}}{\partial y} = -\frac{\mu^{fluid}}{K}\widehat{V} - \rho^{fluid}g = -\frac{\mu^{fluid}}{K}\Re^{fluid} - \rho^{fluid}g$$

where P^{fluid} is the total fluid pressure, μ^{fluid} is the fluid viscosity, K is the vertical permeability, ρ^{fluid} is the density of the fluid and g is the acceleration due to gravity. This pressure is constant once the isotherm has risen a distance equal to the thickness of the reaction zone and the fluid pressure induced by the upward flow diffuses upwards according to

$$\frac{\partial \widehat{P}^{fluid}}{\partial t} = \kappa^{fluid}\frac{\partial^2 \widehat{P}^{fluid}}{\partial y^2} \tag{12.27}$$

where κ^{fluid} is the pressure diffusivity of the fluid and $\widehat{P}^{fluid}$ is the *reduced pressure* defined as $\widehat{P}^{fluid} = P^{fluid} + \rho_0^{fluid}gy$, where ρ_0 is the average density of the fluid and y is directed vertically upwards.

At the reaction front,

$$\frac{\partial \widehat{P}^{fluid}}{\partial y} = -\frac{\mu^{fluid}}{K}\Re^{fluid}$$

and the pressure at the reaction front increases with time thus,

$$\widehat{P}^{fluid} = -\frac{\mu^{fluid}\Re^{fluid}}{K}\left(\kappa^{fluid}t\right)^{1/2}$$

Thus the total pressure at the reaction front evolves as

$$P^{fluid} = P^{fluid}_{hydrostatic} + \frac{\mu^{fluid}n^{fluid}Us_o}{K}\left(\kappa^{fluid}t\right)^{1/2}$$

and so is always greater than the hydrostatic fluid pressure and with increase in time becomes super-lithostatic (Phillips, 1991) if the initial permeability of the rocks above the devolatilisation zone is small enough and/or $\mu^{fluid}\Re^{fluid}$ is large enough. It is easy to see however that if any or a combination of any of the following occur, then the fluid pressure at the reaction front evolves to a new value: (1) the fluid producing reaction alters thus changing the value of n^{fluid}, (2) the upward velocity, U, of the reaction isotherm changes, (3) the modal concentration of the hydrous phase changes thus altering s_o. Thus the processes occurring near the devolatilisation front control the pressure at the reaction front and hence the pressure distribution throughout the devolatilisation system. Notice also from this argument that it is the upward velocity of the fluid, $\widehat{V}$, that is controlled by these processes. The permeability is a dependent variable that has to evolve, by compaction or positive dilation, to a value that accommodates this flux given the fluid pressure gradient that develops.

The solution to (12.27) is given in Figure 12.16 where the dimensionless height, y/H, is plotted against the dimensionless value of the reduced pressure, expressed as $\frac{\widehat{P}^{fluid}}{\mu^{fluid}\Re^{fluid}}\frac{K}{H}$, for a system where a low permeability compartment is overlain by a higher permeability layer. The lithospheric minus the hydrostatic gradient is indicated by the solid red line which has a slope of $\tan^{-1}(1/1.7)$, the gradient of the reduced pressure being 1.7×10^4 Pa m^{-1}. The

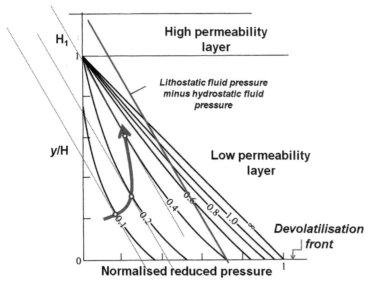

FIGURE 12.16 Diffusion of fluid pressure from a devolatilisation front. *After Phillips, 1991.* Plot of non-dimensional depth against non-dimensional reduced fluid pressure at dimensionless times (tH^2/κ^{fluid}). The red arrow shows the progressive rise of the lithostatically pressured fluid front above the devolatilisation front.

contours of dimensionless fluid pressure are given in terms of the dimensionless time (tH^2/κ^{fluid}). The positions for each time where the reduced pressure gradient passes upward from super-lithostatic to sub-lithostatic are shown as open dots and the evolution of the super-lithostatic front with time is shown by the red arrow. By dimensionless time 0.6, the whole compartment is lithostatically pressured and from then on the compartment is pressured above lithostatic.

This analysis makes two basic assumptions that need further investigation: (1) The evolution described by Figure 12.16 neglects mechanical failure of the system and so it not clear that the compartments can sustain large super-lithostatic fluid pressure gradients without deforming. (2) The flow at the reaction front is assumed to be stable so that a small perturbation in the flow relaxes back to the ground state and does not grow into a completely different flow pattern. Connolly (2010) examines these assumptions and proposes that indeed the flow is unstable so that solitary waves of porosity develop that propagate upwards and sideways; these are called *porosity waves*. We explore such effects in Volume II. An example of such a solitary wave is given in Figure 12.17.

12.3.3 Deformation Driven Flow

Fluid flow driven by deformation arises from two processes that influence the local pore pressure. This can be seen from the following argument. Formally, the fluid mass balance is

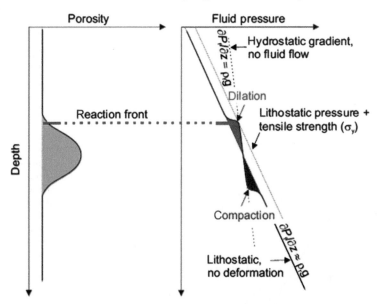

FIGURE 12.17   Nucleation of a porosity wave. *From Connolly (2010).* The fluid pressure gradient in the region of high porosity (left) below the reaction front becomes hydrostatic even though the mean pressure gradient is lithostatic. The top of this region dilates and the base compacts. This region of fluid with a hydrostatic pore pressure gradient then propagates upwards as a *porosity wave* (see Volume II). The argument in the text shows that the thickness associated with the initial region of high porosity is $l = \left(\frac{UT_e}{2\gamma grad T}\right)^{\frac{1}{2}}$ *See Phillips, 1991, Figure 4.17.*

$$\frac{\partial \zeta}{\partial t} = -\frac{\partial \widehat{V}_i}{\partial x_i} + q_V \tag{12.28}$$

where $\widehat{V}_i$ is the component of the Darcy flux, q_V is the flux of fluid per unit volume of the porous material and ζ is the *variation of fluid content* (Detournay and Cheng, 1993; Coussy, 1995) which can be thought of as the 'hydraulic strain' associated with the introduction of fluid; ζ is defined as the variation of fluid volume per unit volume of porous material and hence is dimensionless. A positive ζ means a gain in fluid content by the porous material.

As well as (12.28) which expresses the mass balance of fluid, we need a relation for the balance of momentum (see (4.21)) that is a function of the local porosity, ϕ:

$$\frac{\partial \sigma_{ij}}{\partial x_j} + \rho g_i = \rho \frac{\partial \dot{u}_i}{\partial t} \tag{12.29}$$

where $\rho = (1 - \phi)\rho^{solid} + \phi\rho^{solid}$ and ρ^{solid} and ρ^{fluid} are the densities of the solid and fluid phases, respectively.

The change in the fluid pressure, P^{fluid}, is then given by

$$\frac{\partial P^{fluid}}{\partial t} = M\left(\frac{\partial \zeta}{\partial t} - \alpha \frac{\partial \varepsilon_V}{\partial t}\right) \tag{12.30}$$

where ε_V is the volumetric strain that can arise from both deformation and chemical reactions: $\varepsilon_V = \varepsilon_V^{deformation} + \varepsilon_V^{chemical\ reactions}$. M and α (Detournay and Cheng, 1993) are material parameters known as the *Biot modulus* and the *Biot coefficient*, respectively. If the compressibility of the solid grains can be neglected compared to that of the fluid, then we can take $\alpha = 1$ and

$$M = \frac{K^{fluid}}{\phi} \tag{12.31}$$

where K^{fluid} is the bulk modulus of the fluid. Thus the change in fluid pressure is given by

$$\frac{\partial P^{fluid}}{\partial t} = \frac{K^{fluid}}{\phi}\left(\frac{\partial \zeta}{\partial t} - \frac{\partial \varepsilon_V}{\partial t}\right) \tag{12.32}$$

Expressions (12.28), (12.29) and (12.32) are sufficient to describe the influence of deformation and chemical reactions on fluid flow. Note that the mean stress on the solid does not enter into these equations. It is sometimes claimed (for example, Ridley, 1993) that fluid flow is driven by mean stress gradients the implication being that the fluid pressure is locally equal to the mean stress. This however is not true; fluid flow is driven by gradients in fluid pressure (in the absence of gravity) and so is strongly influenced by local changes in porosity which arise from changes in stress (Detournay and Cheng, 1993) and from changes in pore volume as expressed by (12.32). Figure 12.18 is presented as an example to demonstrate the effects of local dilation. In this example, a material with Mohr–Coulomb dilatant constitutive properties undergoes dextral simple shearing with an imposed fluid pressure gradient from left to right. The fluid focuses into regions of positive volumetric strain rate and shows no relation to the mean stress. These effects have been shown to be important at a regional scale by Ord and Oliver (1997). In general, the fluid pressure cannot reach the mean stress in value without yield occurring as shown in Figure 12.4 (b) and (d).

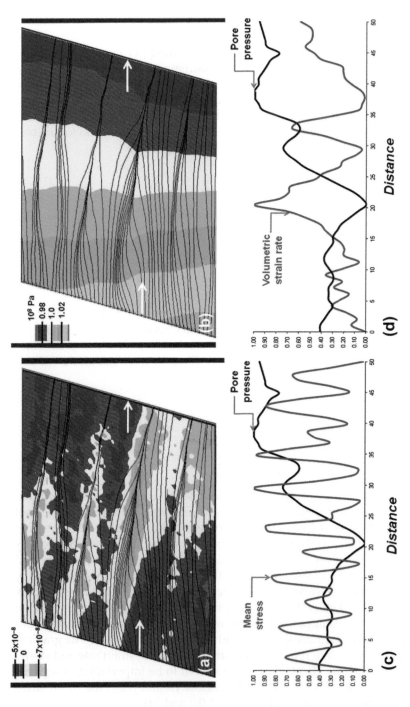

FIGURE 12.18 Control of fluid flow by pore pressure perturbations related to volumetric strains. Model deformed by simple shearing with pore pressure gradient from left to right. Mohr–Coulomb material with dilation. (a) Stream lines superimposed on contours of volumetric strain rate. The white arrow is the direction of flow. Legend shows volumetric strain rate. (b) Stream lines superimposed on contours of pore pressure. (c) Central section from base of model to top. Plots of normalised pore pressure (black) and normalised mean stress (red) against distance. No correlation exists. The mean stress correlates with high shearing rates in localised shear zones. (d) Central section from base of model to top. Plots of normalised pore pressure (black) and normalised volumetric strain rate (red) against distance. A value of 0.5 here indicates zero volumetric strain rate. Values above and below zero correspond to positive and negative volumetric strain rates, respectively. Material properties are the following: Cohesion: 10 MPa. Friction angle: 30°. Dilation angle: 10°. Pore pressure gradient: 1.08×10^4 Pa m^{-1} from left to right. Shearing strain rate: 1.08×10^{-8} s^{-1}.

The example of the influence of deformation upon fluid flow that we explore in a little detail is the *fault-valve* and *seismic pumping* models proposed by Sibson (1981, 1987, 2001, 2004) and extended by Cox (1995, 1999) and Cox et al. (2001). These models involve a fault that crosses a seal between one compartment (above) and another (below) which is over-pressured below the seal; the fault acts as a valve in which the permeability decreases during fault slip. Notice, although not commonly appreciated, that these models can only apply to the situation pictured in Figure 12.12(d) and not Figure 12.12 (c), (e) and (f). The models propose a seismic event that decreases the permeability in the fault so that fluid is driven upwards from the over-pressured compartment beneath. The fault then seals but can reactivate in subsequent seismic events thus resulting in seismic pumping from one compartment to the other. A number of workers (Matthai and Fischer, 1996; Matthai and Roberts, 1997, Braun et al., 2003; Sheldon and Ord, 2005; Weatherley and Henley, 2013) have explored the mechanics of these models with particular emphasis on the dilation associated with the faulting event. Sheldon and Ord (2005) model the fault valve process and show that if the fault zone is dilatant, fluid may be sucked into the fault zone rather than be simply transferred from the over-pressured compartment to the overlying compartment. Weatherley and Henley (2013) couple this process strongly to the magnitude of the seismic event associated with faulting and show that the increase in volume associated with such events can lead to boiling and subsequent deposition of minerals such as quartz and gold as a result of the phase change.

12.4 FOCUSSING OF FLUID FLOW

When a lens of relatively high permeability, in the form of a sedimentary lens, a fault zone or an individual fracture, is embedded in a material of lower permeability, the fluid flow is focussed into the lens as shown in Figure 12.19(a). The reason for this is that, neglecting the effect of gravity, the Darcy flow is higher in the lens than in the embedding material for the same pore pressure gradient. This initiates a pore pressure gradient around the lens that enhances the focussing effect. The mathematical formulation is identical to that in many heat flow problems. Again, neglecting the effect of gravity, the governing equations in two dimensions are the continuity equation and the two equations describing Darcy flow:

$$\frac{\partial \widehat{V}_x}{\partial x} + \frac{\partial \widehat{V}_y}{\partial y} = 0 \tag{12.33}$$

$$\widehat{V}_x = \frac{K}{\mu^{fluid}} \left(-\frac{\partial P^{fluid}}{\partial x} \right) \tag{12.34}$$

$$\widehat{V}_y = \frac{K}{\mu^{fluid}} \left(-\frac{\partial P^{fluid}}{\partial y} \right) \tag{12.35}$$

Substitution of (12.34) and (12.35) into (12.33) gives

$$\frac{\partial^2 P^{fluid}}{\partial x^2} + \frac{\partial^2 P^{fluid}}{\partial y^2} = 0 \tag{12.36}$$

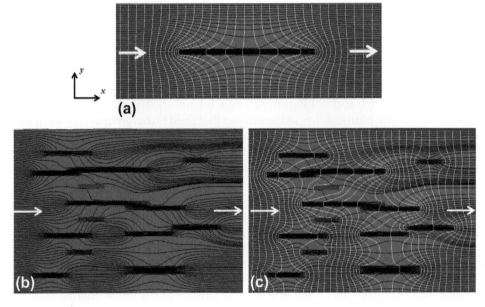

FIGURE 12.19   Focussing of fluid flow into lenses. Streamlines in black, pore fluid pressure contours in yellow. White arrows show the direction of flow.

which is the classical Laplace's equation, commonly used in heat flow problems (Carslaw and Jaeger, 1959) and in many branches of physics (Boyce and DiPrima, 2005, pp. 638—655). Solutions to this equation for complicated geometries usually require numerical methods. The analytical solution for a sphere is discussed by Phillips (1991, pp. 68—69) and Zhao et al. (2008, Chapters 6—8) present analytical solutions for an elliptical lens of low permeability embedded in a higher permeability material of any orientation with respect to the imposed flow direction and with thermal transport. Some results of modelling fluid focussing are presented in Figures 12.19.

An important measure of such fluid focussing is the degree of fluid focussing measured by the parameter Λ. This is given for an elliptical lens with major axis a and minor axis b and aspect ratio $A = \frac{a}{b}$ and permeability ratio $\Pi = \frac{K^{lens}}{K^{matrix}}$ by

$$\Lambda = \frac{A(\Pi + 1)}{\Pi + A} \tag{12.37}$$

where K^{lens} and K^{matrix} are the permeabilities inside the lens and for the matrix.

Figure 12.20 shows the degree of fluid focussing for a lens oriented with its long axis parallel to the flow. One can see that a lens with aspect ratio of 100 (a thin crack) results in fluid focussing close to 180 for a permeability ratio of 100. Other examples are given by Zhao et al. (2008). These effects have been emphasised by Lyubetskaya and Ague (2009) and Ague (2011) in discussions of high fluid fluxes in metamorphic rocks.

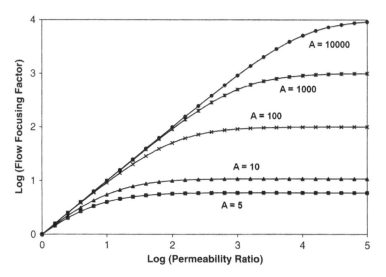

FIGURE 12.20 Focussing factor for an elliptical lens with long axis parallel to the direction of flow. *After Zhao et al. (2008).* A is the aspect ratio of the high permeability elliptical lens.

12.5 CONVECTIVE FLOW

If fluid density variations exist within a porous medium in a gravity field, then it is possible that flow instabilities arise in order to homogenise the density fluctuations. Such density variations can arise from chemical or thermal fluctuations. Here we consider only density variations arising from thermal effects. The unstable flow that arises in such systems is called *convection*. The nature of the convective flows that form depends both on the geometry of the system and its boundary conditions. If we consider a system, such as that shown in Figure 12.21, where the sides of the system stretch to infinity, then boundary conditions at the top and bottom of the system can consist of various combinations of fixed temperature and fluid pressure or fixed heat flow and fluid flow (Nield and Bejan, 2013). We consider two of these combinations below. In the first instance (Figure 12.21(a)), temperature and fluid pressure are fixed at the top and bottom of the system such that $T_{top} < T_{bottom}$ and the initial fluid pressure gradient is hydrostatic. This corresponds to the classical Horton–Rodgers–Lapwood model (Horton and Rodgers, 1945; Lapwood, 1948) and is the type of system envisaged by Etheridge et al. (1983). In the second instance (Figure 12.21(b)), fluid and heat fluxes are fixed at the bottom of the system and temperature and fluid pressure are fixed at the top. We refer to such a system as one characterised by *upward flow-through* and consider such systems in detail in Section 12.5.3. Such upward flow-through systems must be the situation for compartments with lithospheric fluid pressure gradients.

We define the variation in fluid density with temperature by what is called the *Oberbeck–Boussinesq approximation*:

$$\rho^{fluid} = \rho_0^{fluid}\left[1 - \beta_T^{fluid}(T - T_0)\right] \tag{12.38}$$

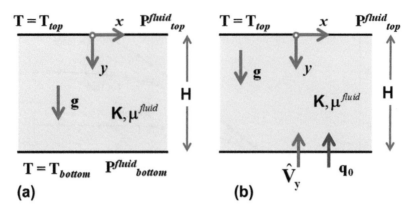

FIGURE 12.21   Models for thermal convection with two different boundary conditions. (a) A closed system with temperature and fluid pressure fixed at top and bottom of the system. (b) An open system where a fluid flux, $\hat{V}$, and heat flux, q_0, are applied to the base of the system and the temperature and fluid pressure are fixed at the top of the system.

where ρ_0^{fluid} is the fluid pressure at temperature T_0 and β_T^{fluid} is the thermal volume expansion coefficient of the fluid. The system is stable, that is, heat transfer takes place solely by conduction through the saturated solid, if the Rayleigh number, Ra_T, defined in (12.7) is less than a critical value which is labelled $Ra_T^{critical}$. If $Ra_T > Ra_T^{critical}$, then convective instabilities arise as shown in Figure 12.22. In (12.11), we point out that for the values of various parameters given in Table 12.1 for water $Ra_T = 2.66 \times 10^9 \, (\Delta T)KH$, so that the Rayleigh number is proportional to the temperature difference between the top and bottom of the system, the permeability of the porous medium and the height of the system.

The critical Rayleigh number for the two-dimensional system illustrated in Figure 12.21(a) is $4\pi^2$ or 39.48 (Nield and Bejan, 2013, Chapter 6) so that, using (12.11) and the values in Table 12.1 for water a system 1 km high with a permeability of $10^{-13} \, m^2$ becomes unstable when $\Delta T = T_{bottom} - T_{top} = 148.4 \,°C$. An example is shown in Figure 12.22 where we have taken the opportunity to include an example of some coupling with chemical processes.

12.5.1 Three-Dimensional Convective Flow

Thermal convection in three dimensional systems with temperature and fluid pressure fixed at the top and bottom of the system is treated by Zhao et al. (2009; Chapter 2). Again, the system becomes unstable at a critical Rayleigh number which is a function of the system geometry and boundary conditions. In general, the flow pattern consists of a three dimensional array of upward and downward flow domains whose geometry depends on the value of the Rayleigh number. Figure 12.23 shows the results for a three dimensional system 3 km high and 50 km in length with homogeneous permeability, an initial hydrostatic fluid pressure gradient and different salinities. For details of the simulations, see Sheldon et al. (2012).

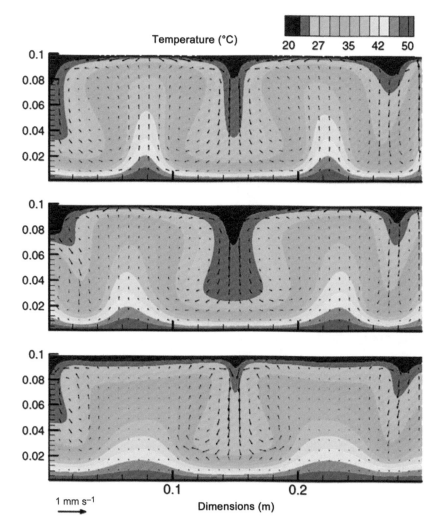

FIGURE 12.22 Two-dimensional thermal convection showing temperature and fluid flow field evolution resulting from dissolution and precipitation of anhydrite. *From Kuhn (2009). See that publication for details.*

12.5.2 Convective Flow in Faults

Many metamorphic systems provide evidence that faults or shear zones have been sites for focussed fluid flow including both aqueous fluids (Beach and Fyfe, 1972; Blenkinsop and Kadzviti, 2006) and melts (Brown and Solar, 1998). The following question arises: *What are the conditions for thermal convection and what are the fluid flow distributions resulting from convection in faults and shear zones?* The situation for fault zones heated from below has been extensively investigated by Zhao et al. (2008, Chapters 8–11). A model for the geometry of a vertical fault zone is shown in Figure 12.24, the important geometrical

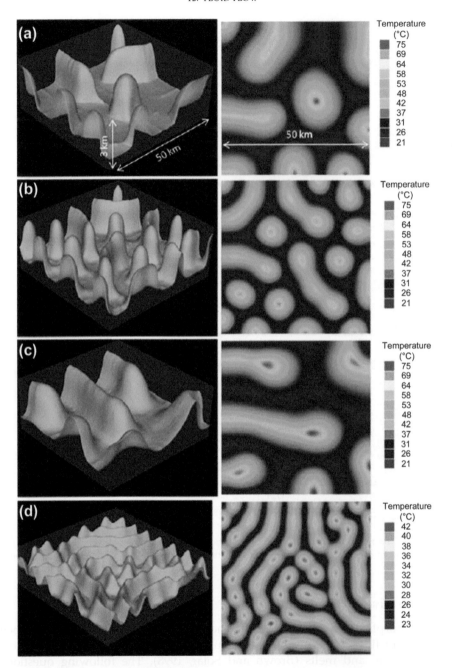

FIGURE 12.23 Fluid convection in a three dimensional system. Three dimensional view of convective system with temperature scale indicated on the right. System is 3 km high. Temperature isosurfaces of 60 °C in (a−c); 40 °C in (d). Vertical exaggeration, 5. The right-hand column shows the horizontal sections through the system at 567 m depth in (a−c) and 460 m in (d). *From Sheldon et al. (2012). See that paper for further details.*

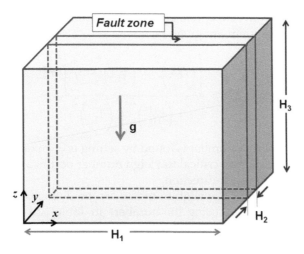

FIGURE 12.24   Geometry of the fault zone problem.

parameters being the length, H_1, the width, H_2, and the height, H_3, of the fault zone. For the coordinate system shown in Figure 12.24, with gravity acting parallel to the z-axis, the governing equations are the following:

$$\frac{\partial \widehat{V}_x}{\partial x} + \frac{\partial \widehat{V}_y}{\partial y} + \frac{\partial \widehat{V}_z}{\partial z} = 0 \tag{12.39}$$

$$\widehat{V}_x = \frac{K_x}{\mu^{fluid}} \left(-\frac{\partial P^{fluid}}{\partial x} \right), \widehat{V}_y = \frac{K_y}{\mu^{fluid}} \left(-\frac{\partial P^{fluid}}{\partial y} \right), \widehat{V}_z = \frac{K_z}{\mu^{fluid}} \left(-\frac{\partial P^{fluid}}{\partial z} + \rho^{fluid} g \right) \tag{12.40}$$

$$\rho_0^{fluid} c_P \left(\widehat{V}_x \frac{\partial T}{\partial x} + \widehat{V}_x \frac{\partial T}{\partial y} + \widehat{V}_x \frac{\partial T}{\partial z} \right) = \left(k_{ex} \frac{\partial^2 T}{\partial x^2} + k_{ey} \frac{\partial^2 T}{\partial y^2} + k_{ez} \frac{\partial^2 T}{\partial z^2} \right) \tag{12.41}$$

$$\rho^{fluid} = \rho_0^{fluid} \left[1 - \beta_T^{fluid} (T - T_0) \right] \tag{12.42}$$

$$k_{ex} = \phi k^{fluid} + (1 - \phi) k_x^{solid}, k_{ey} = \phi k^{fluid} + (1 - \phi) k_y^{solid}, k_{ez} = \phi k^{fluid} + (1 - \phi) k_z^{solid} \tag{12.43}$$

These equations are solved (Zhao et al., 2008; pp. 146–156) for an isotropic, homogeneous porous medium with no upward flow-through and temperatures fixed at top and bottom. The Rayleigh number for the fault is as given in (12.7) but now it is possible that convective instabilities can form in all of the x-, y- and z-directions. Dimensionless wave numbers are defined for these three directions as follows:

$$k_1^* = \frac{m\pi}{H_1^*}, k_2^* = \frac{n\pi}{H_2^*}, k_3^* = q\pi \tag{12.44}$$

with

$$H_1^* = \frac{H_1}{H_3} \quad \text{and} \quad H_2^* = \frac{H_2}{H_3} \tag{12.45}$$

and m, n, q take the values 1, 2, 3,....

A number of different convective instabilities are possible in the fault zone. The critical Rayleigh numbers for the system are given by

$$\mathrm{Ra}_T^{critical} = \frac{\left[\left(\frac{mH_3}{H_1}\right)^2 + \left(\frac{nH_3}{H_2}\right)^2 + q^2\right]^2 \pi^2}{\left(\frac{mH_3}{H_1}\right)^2 + \left(\frac{nH_3}{H_2}\right)^2} \tag{12.46}$$

The smallest critical Rayleigh number is found by setting $m = n = q = 1$ and allowing H_1 to vary. It is found that the minimum critical Rayleigh number occurs as $H_1 \to \infty$ when (12.44) gives a zero wave number in the x-direction.

Although a number of critical Rayleigh numbers can be defined for the problem shown in Figure 12.24 some calculations using the numbers in Table 12.1 for water (Zhao et al., 2008; Section 9.3) show that these Rayleigh numbers are quite close to each other in magnitude so that the linear stability analysis that produces (12.46) does not necessarily indicate which instability (Figure 12.25) will grow to finite size. Modelling shows that in many

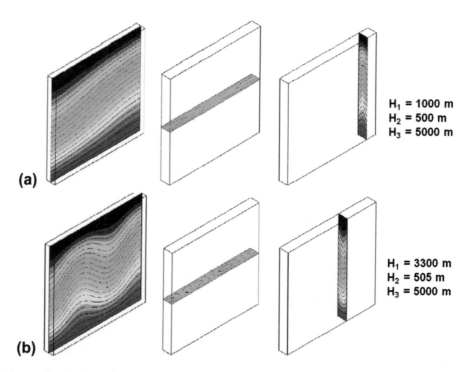

FIGURE 12.25 Sketches of temperature distribution patterns associated with convective flow patterns in vertical fault zones. (a) Fundamental mode. (b) Finger-like mode. *After Zhao et al. (2004).*

instances, the 3D-finger type instability is the mode that grows with critical Rayleigh numbers given by

$$\mathrm{Ra}_T^{critical-3D-finger} = \frac{\left[m^2 + \left(\frac{H_3}{H_2}\right)^2 + 1\right]^2 \pi^2}{m^2 + \left(\frac{H_3}{H_2}\right)^2} \tag{12.47}$$

An example is shown in Figure 12.26. Figure 12.26(d) and (e) shows the temperature and flow regime for the rocks surrounding the fault and the distribution of SiO_2 deposition that is associated with the temperature and flow regime.

If the fault is not vertical (Figure 12.27(a)) and dips at an angle, φ, then the critical Rayleigh number depends on the geometry of the fault as shown in Figure 12.27(b). Decreasing the dip angle stabilises the flow so that $\mathrm{Ra}^{critical}$ increases as the dip decreases; the effect also depends on the aspect ratio, $HR = H_3/H_1$.

12.5.3 Convective Flow in Over-Pressured and Flow-Through Systems

In this section we outline the conditions for a flow system, in which the vertical fluid pressure gradient is super-hydrostatic, to become unstable and for fluid convection to initiate. Convection in systems with upward flow-through driven by a fluid pressure gradient greater than hydrostatic is considered by Zhao et al. (2008). In contrast to the classical system where the temperatures and pressures are fixed at both the top and base where flow-through stabilises the flow (Jones and Persichetti, 1986), a system where the basal boundary conditions comprise fixed heat and mass fluxes (Figure 12.21(b)) is destabilised by flow-through so that increases in $Pe^{thermal}$ increase the possibility of convection (Figure 12.28(a) from Zhao et al. 2008). The Rayleigh number for such a system in which the temperature and fluid pressure is fixed at the top and a heat flux, q, and fluid flux are fixed at the base, is given by

$$\mathrm{Ra}^{flux} = \frac{\left(\rho^{fluid}\right)^2 c_P g \beta_T^{fluid} qKH^2}{\mu k_e^2} \tag{12.48}$$

where β_T^{fluid} is the volumetric thermal expansion coefficient of the fluid.

In Section 12.2.1, an example of a system is given where $H = 10$ km, $q = 60$ mW m^{-2}, the fluid pressure gradient is lithostatic and $K = 10^{-18}$ m^2. This gives $Pe^{thermal} = 2.3$ and $\mathrm{Ra}^{flux} = 5.3$. Figure 12.28(a) shows that these conditions are within the field where convection is possible for upward flow-through systems. Consider a situation where this 10-km-thick compartment comprises the bottom part of a 40-km-thick crust with a temperature gradient in the absence of fluid flow of 20 °C km^{-1}. This means the temperature at the top of the compartment (at 30 km depth) is 600 °C. At the base of the crust in the absence of thermal advection, the temperature would be 800 °C. Figure 12.13 and (12.24) show that for a Peclet number of 2.3, the temperature at the base of the compartment is 3.9 times the conduction solution which means a temperature at the base of the compartment of 3120 °C which is

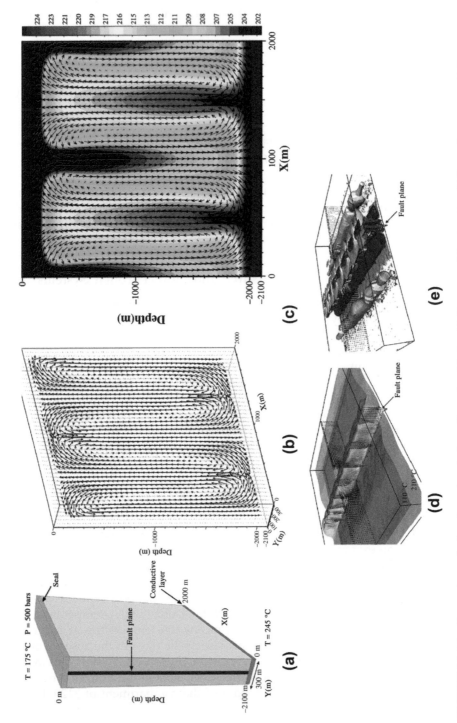

FIGURE 12.26 Convection and mineral reactions in three-dimensional faults. (a) Geometry of the situation. (b) Darcy flow vectors. (c) Temperature distribution. (d) A three-dimensional fault showing Darcy flow vectors and temperature distribution. Two temperature iso-surfaces are shown for 110 and 210 °C. (e) Pattern of quartz deposition for the geometry shown in (d). *From Alt-Epping and Zhao (2010). See that paper for details.*

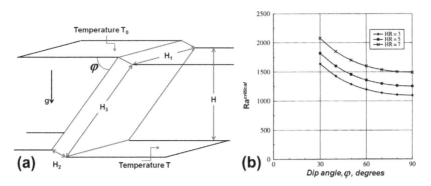

FIGURE 12.27 Thermal convection in an inclined fault. (a) Geometry of the problem. Dip angle is φ, temperature is fixed at top and bottom of a system of height, H, dimensions of the fault as shown. (b) Critical Rayleigh number for various dip angles drawn for the case of $H_2/H_3 = 0.1$. HR is the ratio H_3/H_1. For details of results for other geometries, see *Zhao et al. (2008)*.

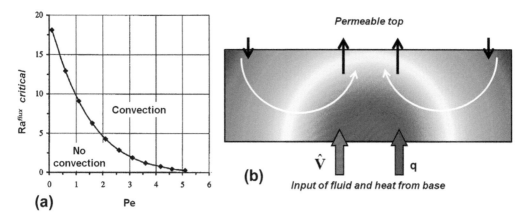

FIGURE 12.28 Convection in a system with flow-through. (a) Plot of $Ra^{flux}_{critical}$ against $Pe^{thermal}$ showing that increasing the upward flow-through in a system with open boundaries promotes instability. (b) Pattern of convection in a system with upward flow-through and input of heat at the base.

clearly unacceptable. Thus although convection is theoretically possible in this system, the resulting temperature distribution that would arise is geologically unacceptable.

Convective flow in upward flow-through systems is slightly different to that developed in systems at an ambient hydrostatic pore pressure gradient in that the convective flow tends to reinforce or compete against the upward flow generated by the non-hydrostatic pressure gradient as shown in Figure 12.28(b). Full convective rolls tend to develop in upward flow-through systems (Figure 12.29(c)) for fluid pressure gradients close to hydrostatic as discussed by Zhao et al. (2008). This results in localised high-velocity flow-through regions separated by

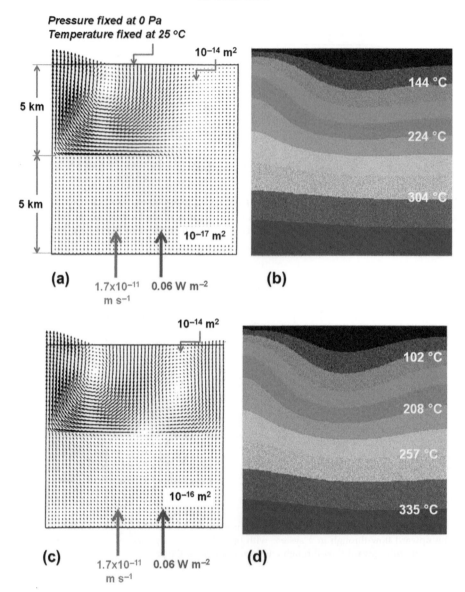

FIGURE 12.29 Two-layer flow-through systems with convection. (a) Two compartments, lower one with $K = 10^{-17}$ m^2 and upper, 10^{-14} m^2. Fluid pressure gradient in lower one lithostatic. Flow pattern is indicated in both compartments. (b) Temperature distribution corresponding to (a). (c) Two compartments, lower one with $K = 10^{-16}$ m^2 and upper, 10^{-14} m^2. Fluid pressure gradient in lower one is below lithostatic. Flow pattern is indicated in both compartments. (d) Temperature distribution corresponding to (c). *From Zhao et al. (2000).*

lower velocity flow-through regions with elevated isotherms in the higher velocity regions. At least for systems close to the critical conditions for convective instability, the pattern is periodic in space with a horizontal wave number, k, given by (Zhao et al. 2008)

$$k = \frac{2\pi}{\lambda} = 1.74(1 - 0.18Pe)^{\frac{1}{4}}/H \qquad (12.49)$$

where λ is the horizontal wavelength for the upwelling regions. For $Pe = 1.0$ and $H = 10$ km, the resulting wavelength for the upwelling regions is 37.9 km. Examples of flow and temperature distributions in convective systems with flow-through for different values of Ra^{flux} and Pe are given in Zhao et al. (1999a, b, c). For the conditions represented in Figure 12.29, convection rolls form in the high permeability compartments (where the fluid pressure gradient is close to hydrostatic) with fluid drawn down to produce colder regions and upwards to produce hotter regions. In the low permeability compartments, the flow is fairly uniform and upwards.

12.6 THE THERMODYNAMICS OF FLUID FLOW. ENTROPY PRODUCTION. BEJAN'S POSTULATE

The flow of linear viscous fluids in the form of Darcy's law constitutes a *linear thermodynamic system* (Section 5.11) in which the thermodynamic flux is the Darcy flux, $\widehat{\mathbf{V}}$, and the thermodynamic force or affinity is the pore pressure gradient, $-\nabla P^{fluid}$ (in the absence of a gravity field). Thus the dissipation density at a point is

$$\Phi^{fluid-flow} = -\widehat{\mathbf{V}} \cdot grad\left(P^{fluid}\right)$$

Using Darcy's law, this becomes

$$\Phi^{fluid-flow} = \frac{\mu^{fluid}}{K}\left|\widehat{\mathbf{V}}\right|^2$$

and so the dissipation is a quadratic function of the Darcy velocity as shown in Figure 12.30(a) or equivalently, of the pore pressure gradient as shown in Figure 12.30(b).

Since the system is thermodynamically linear one expects, based on arguments presented by Ross (2008) for chemical and thermal systems (Chapter 5), that extrema principles should exist for entropy production. In exactly the same way as we discussed various extrema for entropy production within linear thermodynamic systems in Section 5.11 under various constraints, we can recognise two situations here. One corresponds to the system being constrained solely by the power produced by the flow of fluid down the pore pressure gradient. The corresponding dissipation is a maximum (Figure 12.30(a)). A statement of the relevant extrema principle is equivalent to that of Ziegler (1963; see Chapter 5): *For a system with imposed fluid flux boundary conditions, the fluid velocity distribution that actually occurs within the system is the one that comes closest to maximising the entropy production.*

The other principle applies to systems held at steady state and corresponds to the pore pressure gradient being fixed as shown in Figure 12.30(b). Here the dissipation is a minimum. A statement of the principle is: *For a system at steady state with the pore pressure gradient fixed, the*

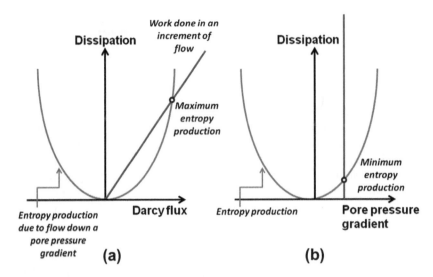

FIGURE 12.30 Entropy production during Darcy flow for different constraints. (a) Entropy production constrained by the power arising from fluid flow. Entropy production is a maximum. (b) Entropy production for a system at steady state constrained by a fixed pore pressure gradient. Entropy production is a minimum.

fluid velocity minimises the entropy production. This is the principle first proposed by Helmholtz and used extensively in the solution of ground water flow problems (Phillips, 1991, p. 60). For a discussion of these two entropy production extrema principles, see Rajagopal and Srinivasa (2004).

Another statement of a principle that is proposed to govern the evolution of flow systems is given by Bejan (see Bejan and Lorente, 2011, Nield and Bejan, 2013 and references therein) and is commonly expressed as: *For a finite sized flow system to persist in time it must evolve in such a way that it provides easier and easier access to the currents that flow through it.* This is called the *Constructal Law* by Bejan who proposes it is as a universal law; there is continuing debate concerning the status of Bejan's proposal. This proposition is presumably related to entropy production extrema principles in some manner because it implies that the system will optimise its permeability structure to accommodate an imposed flow rate or maximise its flow rate for an imposed permeability structure. We return to this proposal when we consider hydrothermal systems in Volume II.

One of the most instructive applications of extrema principles to flow-controlled systems is that of Niven (2010) for flow in single- and two (parallel)-pipe systems. He shows that for a single pipe, the Zeigler Maximum Entropy Production principle can be used as a selection criterion for the transition between laminar and turbulent flow. For a constant flow, two pipe system, a minimum entropy production principle can be used to define a stationary state for a given imposed flow and the Zeigler maximum entropy production principle still defines the transition from laminar to turbulent flow. In contrast, for a two-pipe system with constant hydraulic head as the driving force, the selection of flow regimes is based on a minimum entropy production criterion, whereas the definition of stationary states is based on a maximum entropy production principle. These results offer the potential that similar selection rules could be developed for hydrothermal systems whereby a maximum entropy production principle may select between different modes of operation of the system (for instance, fluidised versus non-fluidised flow) whereas a minimum entropy production principle may select different stationary states within each mode of operation.

12.7 OVERVIEW OF CRUSTAL PLUMBING SYSTEMS

Partly as a summary of this chapter, we develop a model for crustal plumbing systems based on the principles discussed here. We concentrate on non-melt systems but do offer some examples of melt systems. Similar arguments to those presented above have been developed for silicate melt systems by Hobbs and Ord (2010). The basic principles that govern the patterns of fluid flow and the thermal structure of the crust are as follows:

1. The continuity equation (12.16) must be honoured throughout a metamorphic system. This means that for systems in which there is no internal source of fluid and if the horizontal Darcy flux component, $\widehat{V}_x$ is zero then if the upward flux, $\widehat{V}_y$, is zero at any point in such a system, then $\widehat{V}_y = 0$ everywhere. Hence a hydrostatic fluid pressure gradient at any point in the system demands a hydrostatic fluid pressure gradient throughout the system. Conversely a non-zero upward Darcy flux at one point in the system demands a non-zero flux everywhere.
2. For a layered crust, the lowest permeability layer acts as a fluid pressure control valve for the flux of fluid through the rest of the crust. If the fluid pressure gradient is lithostatic in this layer, continuity of flux (see (12.16)) demands that the fluid pressure gradient be less than lithostatic in higher permeability layers, shear zones or lenses. A factor of 10 increase in permeability contrast is enough to reduce the fluid pressure gradient in the higher permeability layer to within 20% of hydrostatic. This fluid pressure gradient distribution is solely a result of changes in permeability and is independent of any need for low permeability 'seals' or 'caps'. Continuity of fluid flux defines the permissible distributions of fluid pressure gradients in a layered crust as shown in Figure 12.12.
3. The same principles as in (2) apply to dipping fault or shear zones so that if the permeability is high within the fault zone, the fluid pressure gradient will be close to hydrostatic even though the fluid pressure gradient in the surrounding rocks is lithostatic.
4. Fluid is focussed into high permeability layers or lenses. The focussing factor can be greater than an order of magnitude for a permeability contrast of 10 (Figure 12.20). This means that for many systems, most of the fluid is diverted from low permeability regions and focussed into high permeability faults, shear zones or fracture networks.
5. A crustal compartment where the fluid pressure gradient is close to hydrostatic is restricted in height because of the stresses produced at the base and top of the compartment. This places restrictions on the height of compartments and of high permeability faults and shear zones as defined by (12.3).
6. The patterns of convection that develop in crustal compartments with fluid flow-through depend on the boundary conditions for thermal and fluid fluxes. For such systems with impermeable upper and lower boundaries, thermal convection cannot occur. If the upper

and lower boundaries allow fluid fluxes, and the base is a heat flux boundary, fluid convection is possible, in principle, even for a lithostatic fluid pressure gradient but only if the Peclet number is less than about 5. We will see below that for a lithostatic pressure gradient, the value of the Peclet number places severe restrictions on the possibility of thermal convection in natural geological settings.

7. The patterns of convection that develop in high permeability faults and shear zones is invariably three dimensional in that both upward and downward flow regions develop forming a finger-like pattern (Figure 12.26). This means that two dimensional sections through such systems are inadequate to show the true patterns of fluid circulation and temperature distribution.

8. Topographic gradients are important in driving substantial horizontal components of fluid flow (including surface derived fluids, over-pressured fluids and melts) especially if the relief is large. Such processes are also capable of driving surface-derived fluids to at least mid-crustal levels.

There are three end-member models of the fluid flow-thermal structure of metamorphic systems. One is typified by the model presented by Etheridge et al. (1983) where convection dominates in the lower half of the crust and the permeability of the crust is relatively high (say $>10^{-16}$ m^2) controlled by fracture permeability. Another is typified by the models of Lyubetskaya and Ague (2009) where no convection occurs, the flow is essentially upwards, although influenced by topography, and the permeability of the lower crust is low ($\approx 10^{-19}$ m^2); localised fluid flow occurs in shear zones or narrow zones of lower permeability. The third type of model, typified by Oliver (1996) and summarised in Figure 12.1, is similar to that of Lyubetskaya and Ague (2009) but the role of localised flow is emphasised.

With the eight basic principles outlined above in mind, the question is: *Which of these models is more likely in metamorphic systems?* A related issue concerns the fault valve and seismic pumping models of Sibson (1994) and Cox (1995, 1999). In view of the flux continuity constraints imposed by (12.16), *what are the conditions under which the fault valve mechanism can operate?*

Each of the models has its own implications for the thermal structure of the crust. We take as a first-order observation that there are no major discontinuities in the crust where the temperature gradient suddenly increases dramatically. The temperature within metamorphic systems is clearly elevated with respect to adjacent regions in the crust but overall the temperature structure is what one would expect from elevated heat flow from the mantle or from changes in the thickness of the crust (Chapter 11). Overall the temperature distribution is similar to what one would expect from a model that involves simple conduction of heat. Above all, the temperature at the Moho is limited to perhaps 1000 °C as a maximum at about 40 km depth (corresponding to a mean temperature gradient of 25 °C km^{-1}). If one is prepared to accept, these first-order approximations to the thermal structure of metamorphic systems, then major constraints are placed on the advection of heat vertically carried by moving fluids.

In a system where the fluid pressure gradient is lithostatic, fluids flow upwards unless influenced by a topographic gradient. Figure 12.13 shows that for such a fluid pressure gradient, high values of the Peclet number (say 5) result in large temperature gradients at the top of the system and grossly elevated temperatures (≈ 30 times the conduction solution) at the base of the system. The observations point to a maximum Peclet number closer to one, that is, close to the conduction solution that one would expect from elevated heat flow of thickened crust. In fact, Pe $= 1$ is an upper limit for the physically possible Peclet number since it gives a temperature at the base of the system of 1.7 times the conduction solution. For hot crustal conditions, Pe < 1 is more compatible with observations.

If we assume Pe $= 1$ and use the values for water from Table 12.1 for a compartment with a lithostatic fluid pressure gradient, then

$$\text{Pe} = 1 = \frac{H\rho_0^{fluid} c_P}{k_e} \widehat{V} = \frac{10^4 \times 10^3 \times 4185}{3.08} \widehat{V}$$

which gives $\widehat{V} = 7.36 \times 10^{-11}$ m s^{-1}. Again this is an upper limit for the Darcy flux. Connolly (2010) using a different argument arrives at a value for the Darcy flux in metamorphic systems of $\approx 10^{-12}$ m s^{-1}. Accepting the above value of $\widehat{V}$ and using Darcy's law, we obtain

$$K = \mu^{fluid} \widehat{V} (\nabla \mathcal{H})^{-1} = \frac{10^{-4} \times 7.36 \times 10^{-11}}{1.7 \times 10^4} = 4.33 \times 10^{-19} \text{ m}^2$$

This value of the permeability agrees well with the value 10^{-19} m^2 assumed by Lyubet-skaya and Ague (2009); Connolly (2010) arrives at a permeability value of 10^{-20} m^2. We now ask: *What value of thermal flux is necessary to initiate convection in this system?* For Pe $= 1$, Figure 12.28(a) says we need a Rayleigh number >10 to initiate convection. Hence we write

$$q = \frac{Ra\mu^{fluid} k_e^2}{\left(\rho_0^{fluid}\right)^2 H^2 c_P g \beta_T^{fluid} K} = \frac{10 \times 10^{-4} \times (3.08)^2}{10^6 \times 10^8 \times 4185 \times 9.8 \times 2.01 \times 10^{-4} \times 4.33 \times 10^{-19}}$$

from which we obtain $q = 26.6$ W m^{-2}. This is about four orders of magnitude larger than the average crustal heat flow of 60 mW m^{-2}. Reducing Pe to 0.1 increases q to 266 W m^{-2} and increasing Pe to 2 gives $q = 13.3$ W m^{-2}. Hence we reach the conclusion that for realistic heat flows and a thermal structure that is near the conduction solution, thermal convection is not possible in a system with a lithostatic fluid pressure gradient. Although others have reached the same conclusion (Wood and Walther, 1986; Oliver, 1996), this argument is based on strict thermal–hydrological arguments for systems with upward fluid flow-through.

If however a more permeable compartment develops in the crust, then even though the surrounding rocks maintain a lithostatic fluid pressure gradient, the fluid pressure gradient (from (12.20)) must be less than lithostatic in order to maintain continuity of fluid flux. As we have seen even a 10-fold increase in permeability is sufficient to decrease the fluid pressure gradient in the high permeability compartment to within 20% of hydrostatic. If indeed the fluid flow regime in metamorphic systems is characterised by Peclet numbers close to one, then the permeability in such systems is close to 10^{-19} m^2 which is close to the percolation threshhold (Meredith et al., 2012). As such, even a small change in porosity (from 0.01 to 0.05) resulting from deformation or chemical reactions can induce a permeability increase of several orders of magnitude.

The limitation on the height of such high permeability compartments is that it must be less than the critical height defined by (12.3). This means that high permeability shear zones can

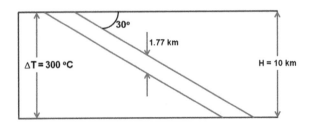

FIGURE 12.31   Model of shear zone dipping at 30° in a 10-km-thick compartment.

form and persist with a near hydrostatic fluid pressure gradient as long as the thickness of the shear zone measured vertically is less than the critical height. For instance, a shear zone with $\overline{\sigma}^{tensile} = 10$ MPa and $\overline{\sigma}^{compressive} = 30$ MPa has a critical height of 1.77 km. If the dip of the shear zone is 30°, the true thickness of a zone that persists with a hydrostatic fluid pressure gradient is 1.5 km. The geometry is shown in Figure 12.31. Figure 12.27(b) shows that the critical Rayleigh number for a zone with 30° dip and an aspect ratio of 3 is about 1650.

If we calculate the Rayleigh number for this fault from (Zhao et al., 2008, p. 178),

$$\text{Ra} = 1650 = \frac{\left(\rho_0^{fluid}\right)^2 c_P g \beta_T^{fluid} \Delta TKH}{\mu^{fluid}k_e} = \frac{10^6 \times 4185 \times 9.8 \times 2.1 \times 10^{-4} \times 300 \times 10^4}{10^{-4} \times 3.08}K$$

or,

$$K = 1.97 \times 10^{-14} \text{ m}^2$$

Thus we reach the conclusion that unless the permeability is increased in shear zones (dipping at 30°) to above 10^{-14} m^2, then convection is not possible even in such zones. If we maintain the aspect ratio and increase the dip to 60°, the true thickness for a shear zone that can maintain a hydrostatic fluid pressure becomes 0.89 km and the permeability required to initiate convection is 1.43×10^{-14} m^2.

The overall conclusion is that fluid convection systems are not possible in compartments with lithostatic fluid pressure gradients under geological conditions. If the permeability in a compartment overlying the compartment with a lithostatic gradient is about 10^{-14} m^2 or greater, then convection is possible in that compartment depending on its height which must be less than or equal to the critical height. Such high permeabilities in metamorphic systems are questionable.

Recommended Additional Reading

Coussy, O. (1995). *Mechanics of Porous Continua*. Chichester, UK: Wiley.

Coussy, O. (2004). *Poromechanics*. Chichester, UK: Wiley.

These are the definitive books on the behaviour of porous media. Both of these books treat the flow of fluids through deforming porous media. The 1995 version is more complete and thorough and includes both Eulerian and Lagrangian treatments of the fluid flow; the 2004 version is easier to read. Deformation topics include poro-elasticity and plasticity. Coupling to chemical reactions and phase changes are included.

Coussy, O. (2010). *Mechanics and Physics of Porous Solids*. Chichester, UK: Wiley.
This book treats classical thermodynamics from the viewpoint of someone interested in flow through porous media. It provides an interesting different look at classical thermodynamics.

Nield, D. A., & Bejan, A. (2013). *Convection in Porous Media*. Verlag: Springer.
This is the definitive book on thermal and mixed convection in porous media.

Phillips, O. M. (1991). *Flow and Reactions in Permeable Rocks*. Cambridge: Cambridge University Press.
An important and easily readable book on fluid flow through porous media. It is written from a geologist's viewpoint and includes equilibrium treatments of chemical reactions coupled to advection.

Phillips, O. M. (2009). *Geological Fluid Dynamics: Sub-Surface Flow and Reactions*. Cambridge: Cambridge University Press.
This is a second version of Phillips (1991) and is not as complete as the earlier version.

Zhao, C., Hobbs, B. E., & Ord, A. (2008). *Convective and Advective Heat Transfer in Geological Systems*. Springer.
This book presents a thorough treatment of the mathematical background to fluid flow in hydrothermal systems including fluid focussing, thermal convection in faults and thermal convection in systems with flow-through.

13

Microstructural Rearrangements

423

13.1 INTRODUCTION. *WHY IS MICROSTRUCTURE IMPORTANT?*

The mechanical behaviour of a polycrystalline aggregate, including the flow stress, the anisotropy of both the elasticity and the flow stress and whether the material localises during deformation are to a large extent controlled by the details of the grain microstructure. A well-known example is the *Hall–Petch effect* where dislocations generated during plastic deformation pile up at grain boundaries. The length scale defining the distance between pile-ups is the distance between grain boundaries and hence this effect is expressed at the macroscale as a dependence of flow stress on grain size and shape. The heterogeneous distribution of dislocations resulting from dislocation pile-ups can nucleate dynamic recrystallisation. The flow stress during diffusion-controlled deformation and stress-assisted dissolution ('pressure solution') depends on the grain size but in the opposite manner to the Hall–Petch effect. Hence it is important to understand the conditions in the Earth that favour a strengthening versus a weakening effect of grain size. It is widely proposed that grain size reduction during deformation is an important influence on localising deformation. The flow stress and the anisotropy of strength of a polycrystalline aggregate also depend on the grain size *distribution* and the grain shape *distribution* and on the intensity and type of crystallographic preferred orientation (CPO) in the aggregate. Thus microstructure plays a fundamental role in controlling the mechanical response of materials to deformation and its history.

It is important to emphasise the role that *microstructural anisotropy* plays in the mechanical behaviour of materials. During deformation, some grain boundaries align parallel and/ or normal to the manifolds (Chapter 3) of the flow resulting in oblique foliations and an associated anisotropy in the flow stress. This same process also operates in a multiphase polycrystalline aggregate where the progressive rearrangement of phases of different strengths (including grain boundaries) leads to the development of interconnected weak layers parallel to the unstable manifolds of the flow, resulting in macroscale weakening of the aggregate and localised deformation. Anisotropy of the flow stress arising from CPO development also leads to localisation. A related effect, the development of *elastic* anisotropy during CPO development and its interrelation with hardening (or softening) of the flow stress is an important control on localisation of the deformation. Microstructures associated with these fabric elements, including the symmetry of CPO and its relation to other microfabrics, are widely used as kinematic indicators. Hence it is important that we have an understanding of how these fabric elements develop and how they are related to the deformation.

This chapter addresses issues around the grain scale adjustments that lead to the common microstructures we see in deformed metamorphic rocks and relates these adjustments to the flow stress, its anisotropy and the controls exerted on localisation of deformation. Section 13.2 sets the basis for the chapter by examining the process of grain growth with no deformation. We examine grain size distributions and shapes in single-phase aggregates, including the significance of the so-called *foam structures* and in Section 13.3, we extend the discussion to polyphase aggregates where some effects not apparent in single phase aggregates emerge. In these sections, the *Fokker–Planck equation* is introduced as a way of understanding grain size distributions. In Section 13.4, we examine grain size growth and reduction during deformation, issues to do with compatibility of deformation throughout the aggregate and stress–grain size relations. It becomes clear in this section that it is important to distinguish between *translational* and *rotational* modes of deformation at the grain scale and hence the distinction is made between *dislocation-* and *disclination*-dominated modes of deformation. Reaction–diffusion equations are used as a way of modelling grain size and shape evolution coupled to dislocation motion. Section 13.5 is a discussion of the development of CPO during deformation. We consider the classical *Taylor–Bishop–Hill approach* together with other approaches such as *self-consistent schemes* and approaches involving *'single slip'*. We also examine the influence of *disclinations* and of diffusion on CPO development. Section 13.6 examines theories for recrystallisation, including *rotation recrystallisation*. Anisotropy of fabric and hence of elasticity and of flow stress and the influence on mechanical behaviour, including localisation of deformation, are considered in Section 13.7. Section 13.8 considers various controls on microstructure development, including the multifractal nature of microstructures, and raises the question whether metamorphic microstructures can be considered as deriving from criticality in a nonlinear system. We also examine the control that the kinematical framework exerts on microfabric. Figure 13.1 shows some examples of the grain microstructures considered in this chapter and, as an example, Figure 13.2 sketches the control of both elastic and flow stress anisotropy on patterns of localisation.

13.2 NORMAL GRAIN GROWTH. GRAIN SIZE DISTRIBUTIONS AND GRAIN SHAPES IN SINGLE PHASE AGGREGATES; NO DEFORMATION

In order to understand the microstructures that develop during deformation, we first need to understand the processes that operate in single-phase aggregates with no deformation. An important process involved is normal grain growth. *Normal grain growth* is a process where existing grains in a polycrystalline aggregate grow in response to two requirements: (1) a topological requirement whereby the grains pack together to fill space and (2) an energy requirement whereby the interfacial energy is reduced, ideally, minimised. The grain configuration ultimately reaches a stationary state whereby the interfacial tensions are balanced throughout the aggregate. This configuration is loosely referred to as an equilibrium configuration in the geoscience literature but at best it represents an unstable equilibrium where a local perturbation results in adjustments throughout the aggregate and the adoption of a new configuration. Some authors (Zollner, 2006) would argue that these configurations are never equilibrium configurations, since the grain boundaries are crystal defects. No nucleation stage exists in this process.

The literature on normal grain growth in metals begins with Smith (1948, 1964); subsequent studies in the geosciences are summarised by Vernon (1976, 2004, pp. 172–194) and concern the development of 'foam structures' (resembling those developed in soap or beer foams) where the interfacial energy of grains is minimised, or modifications of such a structure arising from the anisotropy of interfacial tensions. An example is the type of grain shape shown in Figure 13.3.

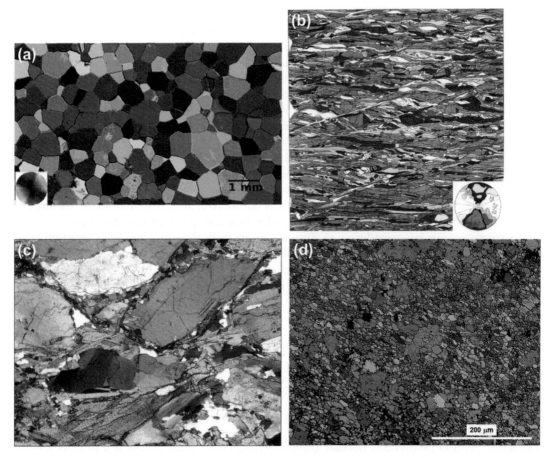

FIGURE 13.1 Grain microstructures. (a) A classical polygonal aggregate comprised almost solely of triple-junctions produced by normal grain growth in ice. The image is colour coded according to the orientation of crystallographic c-axes as in the inset. *From Montagnat et al. (2014).* (b) Oblique foliations developed in a quartzite with strong crystallographic preferred orientation as shown in the inset c-axis plot. Image is 3 mm across. *From Sander (1950).* (c) Interconnected biotite (brown) grains in a deformed granite. Image is 2 mm across. *From Ron Vernon.* (d) Two-dimensional orientation map prepared from 200,000 diffraction patterns at 1 μm spacing. Mylonitic marble from Calabria. Colours represent different orientations in Euler space: yellow lines are twins, red lines are low-angle boundaries with <10° misorientation and black lines are high-angle grain boundaries. *Supplied by Steve Reddy.*

An initial proposition was that such structures, at least for isotropic interfacial tensions, should be based on Lord Kelvin's proposal (Thomson, 1887) that a 14-sided tetrakaidecahedron with some faces slightly curved is the polygon of minimum surface area that packs together to fill space. A more optimal topology was proposed (Figure 13.4) by Weaire and Phelan (1994). The issue of specifying the three-dimensional polyhedron that fills space and has minimum surface area is discussed by Gabbrielli (2009). The subject goes back further than Smith; in 1682, Grew examined bubble shapes in beer (Grew, 1682). This means

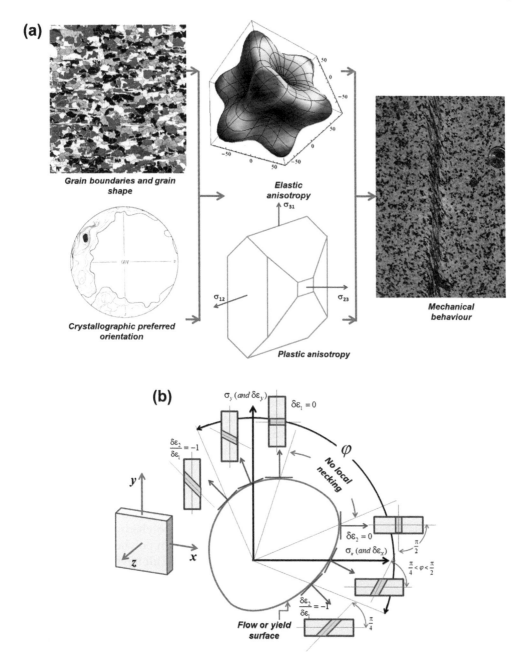

FIGURE 13.2 Controls of microstructure on deformation. (a) Grain microstructure (shape and size distribution) together with the crystallographic preferred orientation contribute to the surfaces that define both the elastic and plastic anisotropy. These two surfaces determine the mechanical properties of the material, in particular whether the material will localise or not. See Section 13.7 for a discussion. *Microstructure and CPO from Sander (1950), elastic modulus surface modified from Anisotropic Elasticity. From the Wolfram Demonstration Project: http://demonstrations. wolfram.com/AnisotropicElasticity/. Contributed by Megan Frary (Boise State University).* Plastic anisotropy surface is for quartz at high temperatures where prism slip dominates. *Shear zone from http://www.geoscienze.unipd.it/ egu-summerschool/photos/Fig06a.html.* (b) Controls on the pattern of deformation localisation by loading conditions and the anisotropy of the yield stress. See Section 13.7 for a discussion. *Modified from Backofen (1972).*

FIGURE 13.3   Computer-generated grain shapes developed during grain growth. The two images represent 180° rotations of one grain. *From Elsey et al. (2011).* Compare these images with those in Figure 13.4.

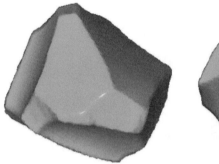

that the study of beer bubbles as an interesting scientific endeavour is not to be attributed to Vernon (1976, p. 136) but instead to Grew (1682). It is surprising that in 1945, Matzke (1945) examined soap bubbles with a Camera Lucida and reported that none of the bubbles studied by him correspond to the Kelvin model, as does Kose (1996) who studied foams using NMR. The polyhedrons observed by Matzke and Kose however do conform to the minimal surface area polyhedra discovered by Gabbrielli (2009), which have a smaller area than the classical Kelvin polygon. The distinction is somewhat academic since, at least according to the measurements taken by Kose the mean interfacial angle is 109.33° which is almost identical to the Kelvin tetrahedral angle of 109.47°. However the shapes of grains and grain faces are completely different for the models of Gabbrielli, Weaire–Phelan and Kelvin.

In order to visualise the differences between these various models, it is convenient to use the concept of *Schlegel diagrams* (Schlegel, 1883), which are projections of the different polyhedra onto a plane. Schlegel diagrams for various polyhedra are shown in Figure 13.4. The eight hexagons and six squares are shown (somewhat distorted by the projection) for the Kelvin polyhedron (Figure 13.4(a)). The Schlegel diagram does neither show the curved edges of the Kelvin polyhedron nor the curved nature of the squares, but this detail could be coded on to the diagram if required.

The Weaire–Phelan solution (Figure 13.4(b)) consists of two polyhedra which stack to fill space and together have a surface area smaller than two Kelvin polyhedra. The faces on these two polyhedra are still regular in shape, as are those of the Kelvin polyhedron. The family of polyhedra discovered by Gabbrielli all have irregular faces but all have total surface areas smaller than the Kelvin solution and some have areas smaller than the Weaire–Phelan solution. Two examples of the Gabbrielli family are shown in Figure 13.4(c) and (d). One, Figure 13.4(c), consists of a single polyhedron. Another, Figure 13.4(d), consists of three polyhedra that stack to fill space.

It would be interesting to revisit this issue in polygonal mineral aggregates and investigate the true shape of grains in these aggregates. At the present stage of development of the theory of minimal surfaces, although one can show that a given surface is a minimal surface, it is not possible to show that the surface is the surface of absolute minimum area. Hence a study of grain shapes in recrystallised mineral aggregates has much to contribute to the theory of minimal surfaces. It is also not clear that a given minimal surface in a stress-free aggregate is the

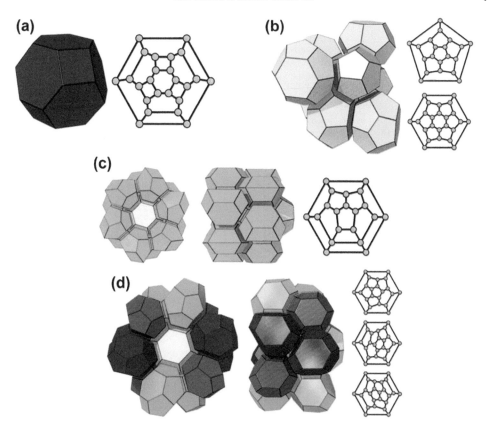

FIGURE 13.4 Schlegel diagrams for various polyhedra of minimum surface area. A Schlegel diagram is a projection of the three-dimensional configuration of a polyhedron on to a convenient plane. The figure shows projections of various polyhedra of interest. Each configuration constitutes a 'minimum surface' configuration. (a) The Kelvin polyhedron. (b) The Weaire–Phelan solution. (c) and (d) Gabbrielli polyhedra. *From Gabbrielli (2009).*

same as the one that develops under load, so detailed experimental work would be interesting as well. However at present it is still not possible to specify the polygon that fills space with a globally minimum surface area.

Kruhl and Peternell (2002) and Kuntcheva et al. (2006) have studied the relationship between crystallographic orientation and grain boundary orientation in recrystallised aggregates of quartz and show that there is commonly a strong crystallographic control. This represents a refinement on the models discussed above where the grain boundary energy is further minimised by adopting lower-energy orientations locally (See Vernon, 1976, Chapter 5 and Vernon, 2004, Chapter 4). The anisotropy of interfacial energy can be calculated using molecular dynamics modelling, and an example for olivine is shown in Figure 13.5.

It is becoming increasingly clear (Beyerlein et al., 2014; Zheng et al., 2014) that in very strongly deformed metals, grain boundaries that represent local minima in interfacial energy are fundamental in producing high ductility and stability in polyphase aggregates. The

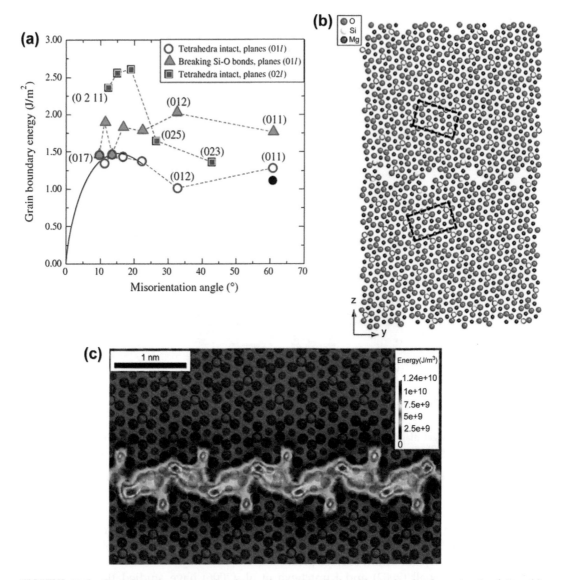

FIGURE 13.5   Grain boundary energy for olivine (forsterite) calculated by molecular dynamics simulation. (a) Grain boundary energy plotted against misorientation for various (01*l*) and (02*l*) grain boundaries and for various assumptions regarding the structure of Si−O bonds. (b) Molecular dynamics simulation of the relaxed atomic arrangement for a symmetric (012) [100] tilt boundary in olivine with misorientation 32.7°. The black rectangles are unit cells with the short side equal to **c** and the long side equal to **b**. The tilt axis is [100]. *From Adjaoud et al. (2012).* (c) Calculated grain boundary energy density (in Joules m^{-3}) for the (011) [100] 60°-tilt boundary in forsterite modelled by Adjaoud et al. (2012). *From Cordier et al. (2014).*

identification and modelling of such interfaces in rocks that have undergone grain growth and deformation represents an important area for future study.

The possible grain configurations can be understood from a different perspective in terms of a topological theorem known as the *Poincare–Euclid theorem* (Thompson, 2001):

$$N_G + N_V - N_B = 1$$

where, for an array of polygons in two dimensions, N_G is the number of grains, N_V is the number of vertices and N_B is the number of boundaries. Thompson (2001) shows that this means that the average number of boundaries is six. In three dimensions, this relation is replaced for a single simple polygon by

$$N_V + N_F - N_E = 2$$

which is the topological equivalent to the phase rule for chemical equilibrium (Chapter 14), where N_V is equivalent to the number of degrees of freedom in a chemical system, the number of faces, N_F, is equivalent to the number of chemical phases and the number of edges, N_E, is equivalent to the number of chemical components (Levin, 1946; Fink, 2009). This presents an easily visualised version of the chemical phase rule where one can see immediately, for instance, the effect of adding (or removing) an edge (equivalent to a chemical component) in the system.

13.2.1 Theories of Grain Growth

Theories of normal grain growth began with the classical work of Burke and Turnbull (1952), the proposal being that grain boundaries migrate towards their centre of curvature, by diffusion of material from one grain to an adjacent grain, to reduce the local interfacial energy. The result is a growth law expressed as

$$\langle R \rangle^{\frac{1}{n}} = \alpha t + \langle R \rangle_0^{\frac{1}{n}} \tag{13.1}$$

where $\langle R \rangle$ is the average grain size at time t, $\langle R \rangle_0$ is the average initial grain size and α is a constant. n is commonly taken to be 0.5, which is supported by many experimental, theoretical and simulation results (Evans et al., 2001). We see in Section 13.3 that for a two-phase aggregate, $n = 1/3$.

From a topological point of view, von Neumann (1952) proposed that the *two-dimensional* growth of foam structures depends on the number of sides on a bubble. If the number of sides, s, is less than six, the bubble tends to grow; if the number of sides is greater than six, the grain tends to shrink (see Figure 13.13). Mullins (1956, 1988) extended this observation to include the area, A, of a grain. The growth law that results, the *von Neumann–Mullins law*, is

$$\dot{A} = \frac{k\pi}{3}(s - 6) \tag{13.2}$$

so that $\dot{A} < 0$ if $s < 6$ and $\dot{A} > 0$ if $s > 6$. An extension of this law to three dimensions for the growth of a grain of radius R has been suggested by Streitenberger and Zollner (2006) as

$$R\dot{R} = C_0 + C_1\sqrt{s(x)} \tag{13.3}$$

where C_0 and C_1 are experimentally determined coefficients. $s(x)$ is the number of faces on a grain with *relative size*, x, given by

$$x = \frac{R}{\langle R \rangle} \tag{13.4}$$

$\langle R \rangle$ is the average grain radius in the aggregate. The term $(R\dot{R})$ is related to the volume, V, and the rate of volume change,$\dot{V}$, by

$$R\dot{R} = \left(48\pi^2\right)^{-\frac{1}{3}} V^{-\frac{1}{3}}\dot{V}$$

It seems that, in three dimensions, the growth rule becomes: *grains grow if the number of faces on a grain is less than 14 and shrink if the number of faces is greater than 14.* How such a statement needs to be modified to account for the 12-faced or 11/13-faced minimum area polyhedra reported by Gabbrielli and by Weaire–Phelan is an open question.

In addition, it has long been recognised (Burke, 1949; Mullins, 1986) that the grain size distribution that develops from normal grain growth is *self-similar*. This means that at any time during the growth process, the distribution of relative grain size, x, is always of the same form and is only modified by a function $g(t)$ that is independent of R and $\langle R \rangle$ so that the grain size distribution is given by

$$F(R,t) = g(t).f\left(\frac{R}{\langle R \rangle}\right) \tag{13.5}$$

The two observations expressed by (13.1) and (13.5) need to be captured by any reasonable model of normal grain growth.

13.2.1.1 The Fokker–Planck Equation and the Mean Field Approach

The grain growth process is *stochastic*, meaning that the evolution of the geometry and grain size distribution is not deterministic and hence probabilities need to be considered. The non-deterministic nature of the process arises because the behaviour of an individual grain is not determined solely by processes that occur at that grain, as in the theory developed by Burke and Turnbull (1952), but also by fluctuations in grain arrangements and sizes at sites far removed from that grain. In considering the behaviour of large stochastic systems, an approach is to consider a small subset of the system, in this case a single grain, and assume that the influence of the rest of the system on that grain is expressed as some average of the total system behaviour. This type of approach is called a *mean field, homogenisation* or *self-consistent* approach.

In adopting a mean field approach to normal grain growth, one assumes that the grain size distribution is the result of two competing processes (Figure 13.6). One is a process that moves (*drifts*) the grain size through grain size distribution space, an example is the classical Burke and Turnbull process where the mean grain size increases with time (Figure 13.6(a) and (b)) according to (13.1). Second is a process that diffuses, or broadens, the grain size distribution in grain size distribution space; an example is the process of grain-switching (Thompson, 2001 and references therein) whereby the local grain size can change

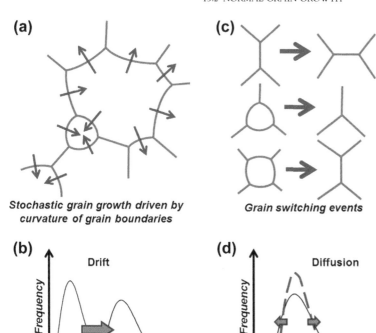

(a) Stochastic grain growth driven by curvature of grain boundaries

(c) Grain switching events

(b) Drift

(d) Diffusion

FIGURE 13.6 The competition between drift and diffusion terms during grain growth. (a) A typical drift process is the stochastic increase in grain size driven by grain boundary migration and the resulting elimination of small grains. (b) The drift in grain size distribution resulting from the process in (a). (c) A grain size distribution diffusive process resulting from grain-switching events. Some switching events preserve or increase grain size, others eliminate small grains. During or subsequent to deformation, recrystallisation reduces the grain size. Local changes in grain shape result in broadening of the grain size distribution as shown in (d).

(Figure 13.6(c) and (d)). The competition between drift and diffusion processes is expressed as the *Fokker–Planck equation*:

Total rate of change of grain size distribution

$$\frac{\partial F(R,t)}{\partial t}$$

Diffusion through grain size distribution space

$$= \qquad D\frac{\partial^2 F(R,t)}{\partial R^2}$$

Drift through grain size distribution space

$$- \qquad \frac{\partial}{\partial R}\left(\dot{R}, F(R,t)\right) \tag{13.6}$$

which says that the rate of change of the grain size distribution is equal to a diffusion term minus a drift term.

Reviews of the use of the Fokker–Planck equation are given by Pande (1987), Atkinson (1988), Thompson (2001) and Zollner (2006). Two end-member versions of this equation are those of Hillert (1965), who assumed that the drift term dominates, and Louat (1974) who assumed that the diffusion term dominates. Hillert (1965) obtained the grain size distribution:

$$f(x) = Dx_0^D \exp(D)x(x_0 - x)^{-(D+2)} \exp\left(-\frac{Dx_0}{x_0 - x}\right)$$

where $D = 2$ or 3 for two or three dimensions and x_0 is the initial value of $(R/\langle R \rangle)$. Louat (1974) obtained a *Rayleigh distribution*:

$$f(x) = \frac{\pi}{2}x \exp\left(-\frac{\pi}{4}x^2\right)$$

These two distributions are plotted in Figure 13.7. The Hillert distribution has a sharp cutoff at large grain sizes and has not been observed in experiments or computer simulations.

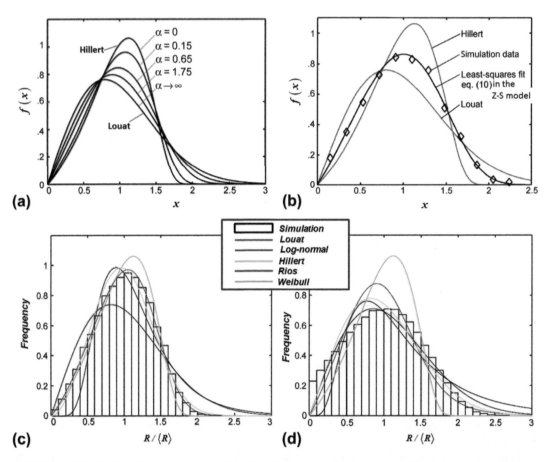

FIGURE 13.7 Grain size distributions. (a), (b) Grain size distributions obtained from the Zollner—Streitenberger (Z-S) model. (c), (d) Grain size distributions for a three-dimensional model of normal grain growth. For parameters used, see Elsey et al. (2011). *From Elsey et al. (2011).*

Slotemaker (2006) has derived expressions that approach a Hillert distribution for dynamic recrystallisation of olivine but again such distributions have not been observed in experimentally deformed olivine. The Louat distribution reproduces some experimental results, although Pande (1987) claims that many experimental distributions are log-normal. These various distributions are compared in Figure 13.7(c) and (d) which also include the results of computer simulations for three-dimensional models.

Motivated by computer simulations, Streitenberger (1998, 2001), Zollner and Streitenberger (2006) and Zollner (2006) propose a quadratic expression involving the growth rate:

$$R\dot{R} = a_2 x^2 + a_1 x + a_0, \quad x = \frac{R}{\langle R \rangle} \tag{13.7}$$

where a_0, a_1, a_2 are constants.

Then, assuming the self-similar relation, (13.5), Zollner and Streitenberger arrive at an expression (Equation (10) in Zollner and Streitenberger, 2008) that represents a self-similar two-parameter family of curves for grain size distribution. An end-member family of these distributions is given by

$$f(x) = ax_0^a \exp(a) \frac{x}{(x_0 - x)^{a+2}} \exp\left(-\frac{ax_0}{x_0 - x}\right), \ x \le x_0 \tag{13.8}$$

$$f(x) = 0, \ x > x_0$$

This distribution is plotted in Figure 13.7(a) for various values of α, which is a parameter such that $\alpha \ge 0$ and is given by

$$\alpha = \frac{a_2}{a_1} \frac{R_{stationary}}{\langle R \rangle}, \quad \text{and} \quad x_0 = 2(1 + \alpha), \quad a = D(1 + 2\alpha)^2$$

where $R_{stationary}$ is the critical grain size when $\dot{R} = 0$ and D is the spatial dimension of the system.

One can see that the Louat distribution is reproduced using this distribution for $\alpha \to \infty$ and the Hillert distribution for $\alpha = 0$. A range of distributions exists between these end members that are obtained in computer simulations. As Pande (1987) points out, many experimental grain size distributions lie close to the Louat (Rayleigh) distribution. However the precise distribution that develops in a particular circumstance depends on competition between the drift and diffusion components of the particular processes involved. Unfortunately there is very little information on grain size distributions in rocks with normal grain growth fabrics so little can be said. There is a general assumption that such distributions are log-normal, in fact Ranalli (1984) presents a case that stochastic processes should be log-normal. However such arguments are not borne out by the later work of Zollner and co-workers. Many more systematic and documented data sets are needed.

13.3 GRAIN RELATIONSHIPS IN POLYPHASE AGGREGATES

The theoretical development of single-phase normal growth is now well developed as indicated in Section 13.2. The same cannot be said for multiphase grain growth; some general principles regarding topological relations are spelt out by Vernon (1976, Chapters 5, and Vernon, 2004, Chapter 4). The foundations for a general theory of multiphase grain growth are developed by Cahn (1991). His treatment involves only two phases in which phase volume is not conserved, that is, the volume fraction of each phase can change during the evolution of the microstructure. One outcome of Cahn's theory is that in two-phase aggregates, the simple von Neumann–Mullins law, (13.2), no longer holds and the conditions that control the growth of an individual grain are much more complicated (Fan and Chen, 1997b). Cahn's work has been extended by Fan and Chen (1997a,b) who show that many of the principles outlined by Cahn (1991) are applicable to two-phase systems where phase volume is conserved. A fundamental difference from a single-phase aggregate is that n in (13.1) becomes 1/3.

The non-conserved treatment is important because it is applicable to situations where one polymorph grows into another (such as kyanite growing into andalusite) and particularly to where a single-phase material has a strong CPO comprised of two well-developed populations or where recrystallised grains grow into deformed initial grains, subgrains grow into deformed host grains or where secondary grain growth occurs. Thus the Cahn nonconservation approach is particularly relevant to recrystallisation processes and to grain growth in aggregates with strong CPO.

The two phases are labelled α and β. We distinguish boundaries between two α grains, $\alpha\alpha$-grain boundaries, boundaries between two β grains, $\beta\beta$-grain boundaries and boundaries between α and β grains, $\alpha\beta$-interphase boundaries. Associated with these boundaries are $\gamma_{\alpha\alpha}$ and $\gamma_{\beta\beta}$, the grain boundary energies and $\gamma_{\alpha\beta}$, the interphase energy. We define

$$R_\alpha = \frac{\gamma_{\alpha\alpha}}{\gamma_{\alpha\beta}} \quad \text{and} \quad R_\beta = \frac{\gamma_{\beta\beta}}{\gamma_{\alpha\beta}}$$

These two ratios are related to angles at triple-junctions (Figure 13.8(a) and (b)) by

$$R_\alpha = 2\cos(\phi_\beta/2) \quad \text{and} \quad R_\beta = 2\cos(\phi_\alpha/2)$$

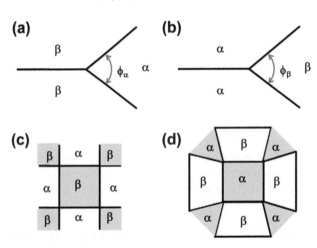

FIGURE 13.8   Grain configurations. (a) An $\alpha\beta\beta$ configuration where ϕ_α is defined. (b) An $\alpha\alpha\beta$ configuration where ϕ_β is defined. (c) A quadri-junction configuration (only stable in two-phase aggregates). (d) Another stable quadri-junction configuration.

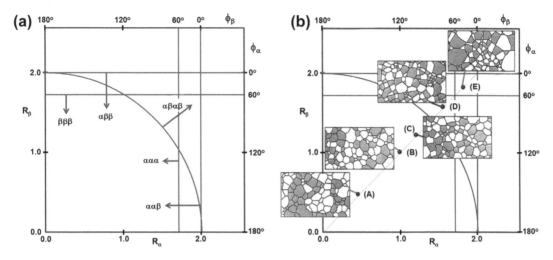

FIGURE 13.9 Control on microstructure in two-phase aggregates by grain boundary and interphase energies. (a) Phase diagram showing the stability fields of various types of grain configurations. *After Cahn (1991).* (b) The types of microstructures that develop for various ratios of grain boundary energy to interfacial energy; $R_\alpha = R_\beta$ for all examples. α is white and β is grey. (a) Clustering. (b) Normal grain growth with triple-junctions. (c) Alternating phases. (d) All systems stable including both triple- and quadri-junctions. (e) $\alpha\alpha\alpha$ and $\beta\beta\beta$ triple-junctions not present, quadri-junctions prevail. *Inspired by Holm et al. (1993).*

Grain configurations can be classified as $\alpha\alpha\alpha$, $\alpha\alpha\beta$, $\alpha\beta\beta$ and $\beta\beta\beta$, all of which Cahn (1991) shows to be stable under conditions shown in Figure 13.9(a). In addition, one other configuration, $\alpha\beta\alpha\beta$, is stable for a two-phase aggregate (Figure 13.8(c) and (d)); this configuration is not stable in a single-phase aggregate. Moreover, Cahn (1991) shows that for the $\alpha\beta\alpha\beta$ configuration, the range of angles, Φ_α, for which the configuration can exist is $\phi_\alpha \le \Phi_\alpha \le (\pi - \phi_\beta)$ and so is not restricted to a single angle as is the case of triple-junctions in isotropic two-phase aggregates.

The stability fields (from Cahn, 1991) for various configurations are shown in Figure 13.9(a); computer simulations (from Holm et al., 1993) of the microstructures for various situations, where $R_\alpha = R_\beta$ are shown in Figure 13.9(b). This shows that it is important to distinguish microstructures characterised by clustering, normal triple-junction arrays with no clustering, alternating α/β configurations, configurations with coexisting triple- and quadri-junction arrays and configurations with only quadri-junction arrays. With the images in Figure 13.9(b) in mind, perhaps it is time to revisit microstructures with a new set of eyes. For instance, in single mineral-phase aggregates with a strong CPO consisting of two maxima, α and β, one expects $\gamma_{\alpha\alpha}$ and $\gamma_{\beta\beta}$ to be small and hence clustering of grains with orientations α and β, respectively should be observed as is commonly the case.

13.4 GRAIN SIZE GROWTH AND REDUCTION DURING DEFORMATION

Grain formation and growth during deformation is commonly called *dynamic recrystallisation*. What is dynamic recrystallisation? This simple question does not have a straightforward answer. Recrystallisation is commonly used to mean the nucleation of new grains in an older, host grain and implies grain size reduction. In the case where the host grains are deforming, a substructure usually forms (but not always) and no nucleation stage may be apparent. The substructure may develop by bulging and expansion of existing grain boundaries (the so-called bulge 'nucleation') or distinct subgrains may evolve which then rotate to form distinct 'new grains' with high misorientations relative to the host grain. In the case where no substructures form, initial grains may simply increase in size. Some types of dynamic recrystallisation take place solely by grain boundary migration. All these processes are included in the term *dynamic recrystallisation*. To some, dynamic recrystallisation is viewed as a form of *recovery*. Although this is true for static recrystallisation, it is not true for dynamic recrystallisation which is actually a deformation mechanism that stores energy (Chapter 9). Thus the term dynamic recrystallisation is used very loosely and we will see that many processes may contribute to its development. Stipp et al. (2002b) have documented an example where grain boundary migration characterises dynamic recrystallisation at high temperatures ($T \geq \sim 500$ °C), subgrain rotation is the characteristic of moderate temperatures (500 °C $\geq T \geq 400$ °C) and bulge nucleation characterises low-temperature (400 °C $\geq T \geq 280$ °C) dynamic recrystallisation.

There is a clear distinction to be made between the equiaxed, triple-junction configurations of grains that arise from normal grain growth (Figure 13.1(a)) and the somewhat irregular grain configurations and broad grain size distributions developed in deformed rocks (Figures 13.1(b) and (d), 13.10(b)–(d) and 13.12(a)). In this section, we examine situations where grain adjustments take place during deformation in order to define some underlying principles. The questions of interest are: *What is the influence of stress (or deformation?) on grain size and on grain size distribution? Are the resulting grain size distributions different to those that develop in normal grain growth? How do deformation processes interact with existing grains to modify the microstructure? How do aggregates adjust to maintain compatibility during deformation? What controls grain shape in deformed aggregates?* We attempt to address these questions in what follows.

We have seen in Chapter 9 that in most crystalline materials deformed at relatively low temperatures and relatively fast strain rates, a *translation dominated mode* of deformation operates at low plastic strains. Here the deformation is achieved mainly by the motion of *dislocations*. If different populations of dislocations compete for production and annihilation, then subgrains (or cells) form with low (say <10°) misorientations across the subgrain boundaries (see Section 9.4). By contrast, at high plastic strains, the deformation is characterised by a *rotation dominated mode* whereby the deformation is achieved mainly by the motion of *disclinations* together with the cooperative motion of dislocations to relieve local stress concentrations. This rotation mode produces large rotations of subgrains (up to 60° is reported by Hughes and Hansen, 2000) and domains of subgrain rotation may develop. Kuhlmann-Wilsdorf and Hansen (1991) have christened the low angle, dislocation accommodated subgrain boundaries, *incidental dislocation boundaries (IDBs)* and the boundaries that mark domains of disclination accommodated rotation, *geometrically necessary boundaries (GNBs)* (Figure 13.11(a)).

At higher temperatures and lower strain rates, there is a transition to mechanisms where diffusion contributes to the deformation. Both dislocation- and disclination-dominated modes can still operate but *subgrain translation mechanisms* associated with grain boundary sliding and cooperative cellular dislocation motion can also contribute.

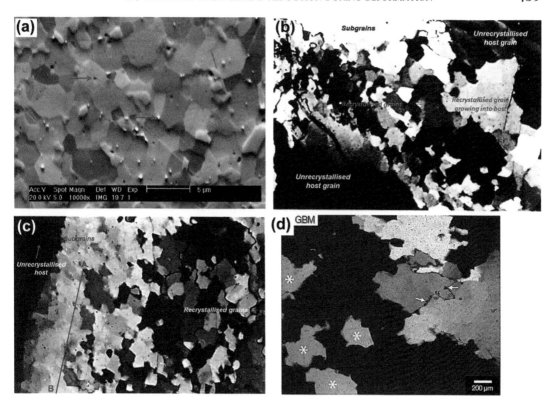

FIGURE 13.10 Dynamic recrystallisation microstructures. (a) Scanning electron microscope orientation contrast image of experimentally deformed (28% horizontal shortening) olivine. The red arrows point to two of the many voids that remain after deformation. *From Slotemaker (2006).* (b) Experimentally produced rotation recrystallisation in what was a single crystal of quartz showing patchy distribution of misorientations ranging from subgrains to recrystallised grains with one large grain growing into unrecrystallised host. *From Hobbs (1968).* Horizontal shortening 46%. Image is 0.18 mm across. (c) Experimentally produced rotation recrystallisation in what was a single crystal of quartz showing banding of misorientation from subgrains to recrystallised grains. The boundary A-B is a geometrically necessary boundary in the terminology of Kuhlmann-Wilsdorf and Hansen (1991). *From Hobbs (1968).* Horizontal shortening 51%. Image is 0.18 mm across. (d) Recrystallised grains in experimentally deformed quartzite. The arrows point to pinning sites. The asterisks are suggested to represent parts of a single large grain with a common crystallographic orientation. GBM, grain boundary migration. *From Stipp et al. (2010).*

Geologists will recognise the rotation dominated mode of deformation as *rotation recrystallisation*. In the metals literature, this mode is generally regarded as a distinct *deformation*, rather than a *recrystallisation*, mechanism. We show two examples of the rotation dominated mode for deformed single crystals of quartz in Figure 13.10(b) and (c). In these figures, the banding between low- and high-angle subgrain boundary domains can be seen as well as one grain (Figure 13.10(b)) that grows into the deformed host. These microstructures contrast with the more or less polygonal grain structures in deformed olivine shown in Figures 13.10(a) and 13.12(a) and the sutured grain boundary microstructure resulting from localised growth (bulge 'nucleation') in deformed quartzite in Figure 13.10(d).

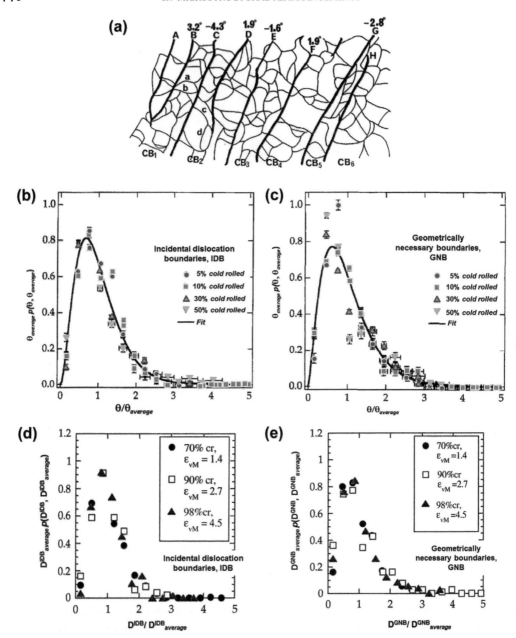

FIGURE 13.11 Scaling of misorientations and substructure size for various strains. (a) Arrays of incidental dislocation boundaries (IDBs) marked in light grey lines (*a, b, c, d*); these constitute low-angle subgrain (cell) boundaries. Also present are geometrically necessary boundaries (GNBs) marked in heavy black lines (**A, B, C, ...,** **H**); these have higher misorientations. Between the GNBs are cell blocks labelled CB_1, CB_2, ..., CB_6. (b) Plot of probability function (13.10) against $\theta/\theta_{average}$ for IDBs. (c) Plot of probability function (13.10) against $\theta/\theta_{average}$ for GNBs. (d) Plot of probability function (13.10) against $D^{IDB}/D^{IDB}_{average}$ for IDBs. (e) Plot of probability function (13.10) against $D^{GNB}/D^{GNB}_{average}$ for GNBs. *(a), (b) and (c) from Hughes et al. (1997). (d) and (e) from Hughes and Hansen (2000).*

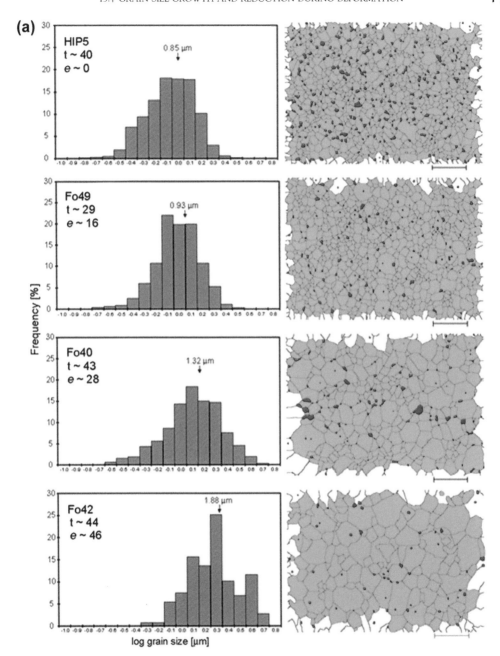

FIGURE 13.12 (a) Grain size distributions and microstructure for initially hot-pressed forsterite (HIP5) and three other specimens (Fo49, Fo40, Fo42) deformed to shortening strains of $e = 0$, $e \sim 16\%$, $e \sim 28\%$ and $e \sim 46\%$, respectively. t is the time in hours at temperature. The right-hand column shows the resulting microstructure. Green is forsterite, red is enstatite. The scale bar is 5 μm. *From Slotemaker (2006)*. (b) Stress–strain curves for the specimens Fo49, Fo40 and Fo42. *From Slotemaker (2006)*. (c) Mean grain size versus axial strain for this series of runs. (d) Defect fraction versus axial strain for this series of runs. Black arrow is value for hot pressed only specimen.

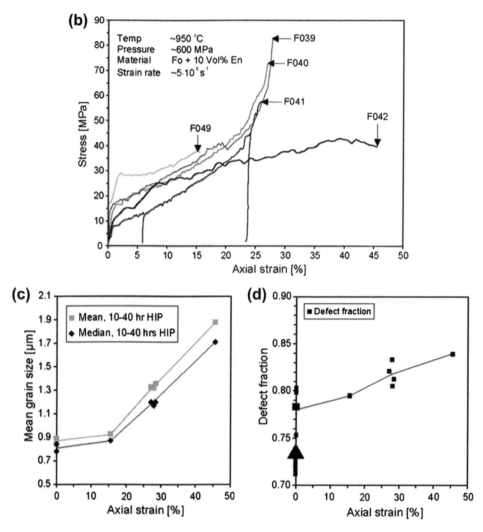

FIGURE 13.12   (Continued)

In Section 13.2, we show that normal grain growth in the absence of deformation can be explained by a stochastic process whereby grain boundary migration, driven by a reduction in interfacial energy, competes with grain size distribution diffusion processes such as grain-switching to produce a grain size distribution that is somewhere between a Rayleigh and a Hillert distribution. Many more processes potentially operate if grain size distributions undergo modification concurrently with deformation and there is still considerable uncertainty surrounding the relative importance of some of these processes for various environmental conditions and grain size distributions. Broadly, the grain size distribution that arises during deformation is influenced by the temperature, the strain rate, the current grain size distribution and the relative importance of processes such as grain growth, grain boundary sliding, coupled grain boundary migration (Section 9.5.2), cooperative grain boundary sliding, the motion of cellular dislocations and whether translational or rotational modes of deformation are dominant. The operation of some of these processes depends on the amount of strain the aggregate has experienced as opposed to the amount of time that the aggregate has been deformed or the current level of stress. It is possible for one part of a grain size distribution to be influenced by one mode of deformation whilst simultaneously another part of the same grain size distribution is influenced by another mode. Gutkin and Ovid'ko (2004, pp. 35–38), following Masumura et al. (1998), define a grain size, d*, within a single grain size distribution where the deformation mechanism switches from one characterised by the Hall–Petch relation to one dominated by Coble creep. We consider this analysis in Section 13.4.1. Four examples of the interaction of deformation processes and grain size distributions are presented below to demonstrate the range of complexity that needs to be taken into account when considering the relations between grain growth and deformation.

The first example concerns the development of *IDBs* and *GNBs* during deformation. We define θ^{IDB}, θ^{GNB} as the misorientations associated with IDBs and GNBs and D^{IDB}, D^{GNB} as the cell sizes associated with these two boundaries. A large amount of experimental work on metals deformed in the strain range measured by the equivalent von Mises strain from 0.06 to 0.8 by Hughes (Hughes and Hansen, 2000; Hughes, 2001; Hughes et al., 1997, 1998, 2003) arrives at the following scaling relations:

- IDB (see Figure 13.11(a)) misorientation *distributions*, *f(R)*, may be expressed empirically as

$$f\left(\theta^{IDB}\right) = \frac{\alpha^\alpha}{\Gamma(\alpha)}\left(\theta^{IDB}\right)^{\alpha-1}\exp\left(-\alpha\,\theta^{IDB}\right) \tag{13.9}$$

with $\alpha = 3$. $\Gamma(\alpha)$ is the *gamma function* evaluated for α: $\Gamma(\alpha) = (\alpha-1)!$ for α an integer and $\Gamma(\alpha) = \int_0^\infty e^{-t}t^{\alpha-1}dt$ for $\alpha > 0$ and α non-integer (Bronshtein et al., 2007). If $p(\theta^{IDB}, \theta^{IDB}{}_{average})\,\Delta\theta^{IDB}$ is the probability that θ^{IDB} will lie between θ^{IDB} and $\Delta\theta^{IDB}$ given that the average misorientation is $\theta^{IDB}{}_{average}$, then the distribution of θ^{IDB}, $f(\theta^{IDB})$, follows a relation,

$$f\left(\theta^{IDB}\big/\theta_{average}^{IDB}\right) = \left(\theta_{average}^{IDB}\right)^{-1}p\left(\theta^{IDB}, \theta_{average}^{IDB}\right) \tag{13.10}$$

which is independent of strain. This relation is shown in Figure 13.11(b).

- GNB (see Figure 13.11(a)) misorientation *distributions* may also be expressed by (13.9) but with θ^{GNB} replacing θ^{IDB} and $\alpha = 2.5$ for which $\Gamma(2.5) = \frac{3\sqrt{\pi}}{4}$; the fit is not as good as for IDBs. The distribution of θ^{GNB} is similar to that of θ^{IDB} as given in (13.10) with θ^{GNB} replacing θ^{IDB} as shown in Figure 13.11(c). This means that the misorientation distributions for both IDBs and GNBs are self-similar and as the plastic strain increases the misorientation distributions shift to higher misorientations with broader spreads. Hughes et al. (1998) show that the scaling law (13.9) is applicable to cold rolled Al, stainless steel, nickel and copper and seems to be a universal relationship.

- Cell *size distributions* scale in much the same way as misorientations so that D^{IDB} scales as indicated by (13.10) with D^{IDB} replacing θ^{IDB} as shown in Figure 13.10(d).
- Cell block *size distributions* scale in much the same way as misorientations so that D^{GNB} scales as indicated by (13.10) with D^{GNB} replacing θ^{IDB} as shown in Figure 13.11(e).
- *Average* IDB misorientation is proportional to $(\bar{\varepsilon}^{plastic})^{1/2}$, where $\bar{\varepsilon}^{plastic}$ is the effective plastic macroscopic strain.
- *Average* GNB misorientation is proportional to $(\bar{\varepsilon}^{plastic})^{2/3}$. This means that the average misorientation of GNBs increases with strain 1.3 times faster than average misorientation across IDBs and ultimately high-angle boundaries dominate.
- *Average* low angle subgrain size decreases as $(\bar{\varepsilon}^{plastic})^{-1/2}$.
- *Average* high angle boundary subgrain size decreases as $(\bar{\varepsilon}^{plastic})^{-2/3}$. This means that the average size of high angle boundary subgrains decreases with strain 1.3 times faster than the average size of low angle boundary subgrains and ultimately high angle boundary subgrains become the smallest.

Thus, the average misorientations and sizes are characterised as functions of the plastic strain and not of the stress. Moreover the misorientation and size distributions are self-similar across a large range of plastic strains. Unfortunately data sets of this type are lacking for geological materials, but we expect the same kind of behaviours.

As a second example: Slotemaker (2006) deformed hot-pressed synthetic forsterite (with ~10% enstatite) with an initial average grain size of 0.85 μm at 600 MPa, 950 °C and 5×10^{-6} s^{-1} (Figure 13.12). The results of these experiments provide a contrast with those of Hughes. In the Slotemaker experiments no substructure forms, the grains are relatively dislocation free and no CPO develops even for ~50% shortening. It is proposed that the mechanism of deformation is essentially dominated by diffusion processes. Hardening of the stress–strain curve characterises the mechanical behaviour with hardening still apparent at 45% overall shortening although it is not clear whether some of this hardening results from an increase in strain rate during the experiment. The average grain size increases progressively during deformation and depends on the amount of strain and not on the time at temperature. This trend, of grain size *increase* related to strain, is the opposite of that established by Hughes. Here grain growth (not grain size reduction) is driven by deformation and not by an attempt to reduce grain boundary energy. The grain size distribution does not evolve in a self-similar manner and becomes quite irregular by 45% shortening.

A grain switching mechanism is proposed by Slotemaker (Slotemaker, 2006; Slotemaker and de Bresser, 2006) to describe the grain growth process (Figure 13.13). This process is discussed by Sato et al. (1990) and by Sherwood and Hamilton (1992, 1994) as a deformation process. Basically, as a grain growth process the presence of cellular dislocation defects in a deforming aggregate means that migration of a grain with less than six sides through the aggregate increases the probability that a grain switching event will occur thus increasing the grain size. The process is illustrated in Figure 13.13. If a two dimensional aggregate undergoes static normal grain growth, one expects a defect ratio (the ratio of non-hexagonal to hexagonal grains) to be ~0.7. For dynamic grain growth following the grain switching process illustrated in Figure 13.13(e)–(h), one expects a defect ratio of ~0.8. These expectations are confirmed in two dimensional models using ELLE (Jessell et al., 2001) by Slotemaker and

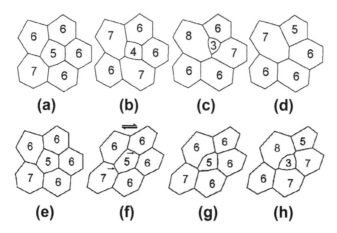

FIGURE 13.13   Grain switching. (a)–(d). Normal grain growth by progressive elimination of a defect. The defect fraction changes progressively: 0.33 → 0.5 → 0.5 → 0.4. (e)–(h) Grain evolution during cellular defect migration during a simple shearing deformation. The defect fraction can change dramatically: 0.33 → 0.33 → 0.33 → 0.83. *From Slotemaker (2006).*

de Bresser (2006). Here is another example of how the detailed topology of the deformed material can give insights into the deformation history.

As a third example, we look at the experimental simple shearing deformation of Carrara marble by Pieri et al. (2001). The microstructures developed at various shear strains for specimens deformed at 1000K, and a shear strain rate of $3 \times 10^{-4}\,\mathrm{s}^{-1}$ are shown in Figure 13.14. Original grains that are more or less equiaxed become elongate parallel to the principal axis of elongation and progressively recrystallised with a gradual decrease in grain size as deformation proceeds. After a shear strain of about five, the original grains are completely replaced by recrystallised grains. In these experiments, the grain size decreases during deformation whilst the stress first hardens, then softens and approaches steady state. This effect is the opposite of that observed by Slotemaker (2006). The grain size decreases overall as the strain increases which is similar to the trend reported by Hughes. The grain size decrease trend is still apparent between shear strains of 5 and 11 (Figure 13.14(b)) and the rate of change of grain size decreases with increasing strain as does the data reported by Hughes.

Based on these experimental observations, Pieri et al. (2001) propose the deformation map shown in Figure 13.15. A distinct field of grain size-sensitive flow is delineated with a power law exponent of 3.3 as opposed to a diffusion-dominated grain size-sensitive field with a power law exponent of 1.7. We propose that future experimental work may reveal that this kind of behaviour is common in many materials other than calcite. In other words, we expect Hall–Petch behaviour to be far more common than is revealed by many published deformation maps for minerals. The expression of such behaviour would be the delineation of grain size-sensitive regimes dominated by dislocation mechanisms where at present grain size-insensitive dislocation mechanisms are postulated.

Finally we consider the results of Halfpenny et al. (2012) who measured misorientations between host grains and subgrains/recrystallised grains, using electron back scattered diffraction (EBSD), for dynamically recrystallised quartz in rocks selected to span the three

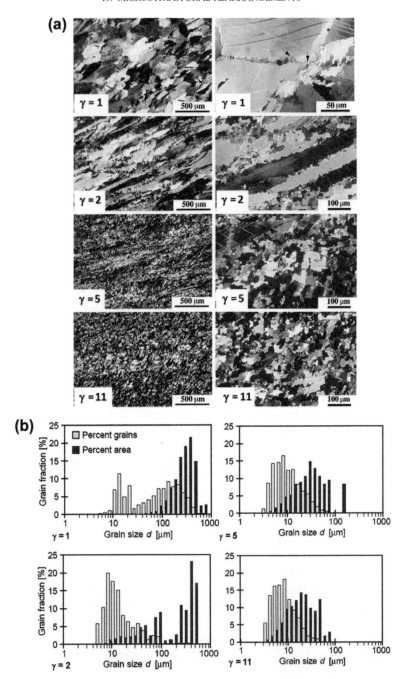

FIGURE 13.14 Experimental deformation of Carrara marble. (a) Microstructure for various shear strains. Shearing is dextral parallel to the base of the figures. Left- and right-hand columns represent low and high magnifications. Scale bars are shown. (b) Grain size distributions for various shear strains. *From Pieri et al. (2001).*

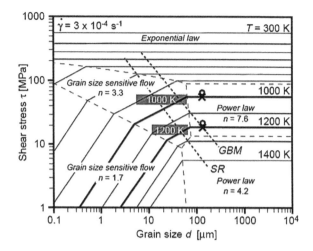

FIGURE 13.15 Deformation mechanism map for marble from Pieri et al. (2001). A grain size-sensitive flow regime is proposed with n = 3.3. Predicted and measured stresses for starting grain sizes are indicated by the crosses and circles. The ranges of recrystallised grain sizes for deformation at 1000 and 1200K are indicted by the grey bars. The calcite palaeopiezometers of Rutter (1995) for subgrain rotation (SR) and of Schmid et al. (1980) for grain boundary migration (GBM) are indicated.

recrystallisation regimes identified by Hirth and Tullis (1992). They show that the ratio of subgrain size to recrystallised grain size can be used as an indication of the recrystallisation mechanism and that dynamic grain boundary migration dominates in regimes 1 and 3 whilst rotation recrystallisation dominates in regime 2. Most importantly, the misorientations can be very large (up to 90°) and twin relationships are common in all the three regimes. They propose that grain boundary sliding is important also in all the three regimes. The observations of high misorientations and twin relationships are very important and we will return to these subjects in Section 13.6.

Hence to propose that the grain size distribution that arises from dynamic recrystallisation results simply from the competition between a process that increases grain size and one that decreases grain size as a function of stress level is somewhat misleading. A large number of processes potentially operate depending on the environmental conditions. This statement is also supported by the data presented by Stipp et al. (2010) who show that the reported average grain sizes for deformed quartzites fall into three classes with the smallest grain sizes corresponding to bulge nucleation, a medium range corresponding to subgrain rotation and the largest grain size class corresponding to grain boundary migration.

Some of the processes (such as subgrain formation and rotation) produce a grain size distribution that depends on strain. In terms of a stochastic view, these processes constitute drift terms in a Fokker—Planck formulation. The processes outlined by Hughes et al. (1997) and by Pieri et al. (2001) lead to a decrease in grain size with increasing strain, whereas the processes outlined by Sloteman (2006) have the opposite trend. Other processes, such as grain growth due to thermal activation or to contrasts in stored energy, also contribute a drift term to the evolution of the grain size distribution. Even other processes, such as cooperative cellular dislocation motion, contribute a diffusion term arising from grain-switching events. Thus it would appear that the grain size distribution evolution should be treated in terms of a Fokker—Planck equation as has been done for normal grain growth but with additional processes operating. This is a daunting task and so far has not been attempted. Such developments will depend on first gaining a more precise understanding of the processes that operate and their control by environmental factors. One important test of such models derives from the details of the grain size distributions and so it is important that efforts are made to rigorously collect and collate these data sets both from natural examples and from carefully designed experiments.

13.4.1 Stress/Grain Size Relations

Grain boundaries have properties that are different from those of the adjacent grains. They act as sources or sinks for dislocations and disclinations and as barriers to dislocation motion. They can also facilitate diffusion and commonly act as sites for recrystallisation and chemical reactions. The overall behaviour of a polycrystalline aggregate may depend on the content (total surface area) of grain boundaries and hence on the grain size and shape. The effect on the overall properties such as flow stress or fracture toughness of an aggregate depends on whether the grain boundaries impede or enhance the motion of defects such as dislocations, disclinations and diffusing species. Thus at low temperatures and/or fast strain rates, for a given grain size distribution, grain boundaries may act as obstacles to dislocation motion, generating dislocation pile-ups at grain boundaries, whereas at high temperatures and/or slow strain rates grain boundaries may act as sinks for dislocations, leading to grain boundary sliding, or the boundaries may act as channel-ways, facilitating diffusion creep. Detailed analyses of many of these processes are presented by Gutkin and Ovid'ko (2004).

The contrasting behaviour of grain boundaries at low and high temperatures is well known in industrial applications where one attempts to produce fine-grained materials for low-temperature applications to increase the strength and fracture toughness (McClintock and Argon, 1966; Myers et al., 2006) and coarse-grained materials that inhibit diffusion creep for high-temperature applications. The extreme example is the production of single-crystal turbine blades for use at high temperatures. The two types of behaviour are illustrated in Figure 13.16. We refer to situations where an increase in grain size leads to a decrease in strength as *grain size weakening* (Figure 13.16(c)). If an increase in grain size leads to an increase in strength, we refer to *grain size strengthening* (Figure 13.16(f)). In some literature (Hansen, 2004), these two contrasting effects are referred to as *grain boundary strengthening* and *grain boundary weakening*, respectively.

Although experimental data are relatively scarce for minerals and rocks (the emphasis over the past 40 years or so has been on 'steady state' creep behaviour), the early experimental work of Griggs et al. (1960, Figure 30) demonstrated the Hall—Petch effect for calcite rocks: fine-grained (10 μm) Solenhofen limestone is stronger than medium-grained (10—1000 μm) Yule marble which in turn is stronger than 10-mm sized single crystals of calcite (Figure 13.17). In these experiments, the strain rates are quite fast (10^{-4}—10^{-5} s^{-1}) and hence some fracturing may have occurred. The effect is probably due to the difficulty of propagating twinning across grain boundaries (Rutter et al., 1994). Similar grain size weakening effects are reported for calcite-rich rocks by Rutter et al. (1994) and by Renner et al. (2002) for strain rates down to 10^{-7} s^{-1}. This grain size weakening effect as a function

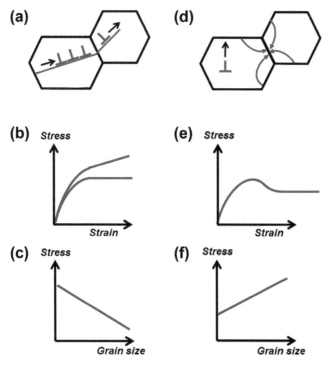

FIGURE 13.16   The two types of mechanical behaviour resulting in grain size weakening (the Hall–Petch effect) and grain size strengthening effects. (a) Dislocation pile-up at a grain boundary results in a stress concentration that nucleates dislocation motion in an adjacent grain. (b) The hardening or steady state stress–strain curves that result from the pile-up behaviour in (a). (c) The grain size weakening behaviour that results from the pile-up behaviour in (a). (d) Vacancy migration and dislocation climb characteristic of high-temperature creep. (e) The softening or steady state stress–strain curves that result from the diffusion behaviour in (d). (f) The grain size strengthening behaviour that results from the diffusion behaviour in (d).

of strain rather than stress is a much neglected topic in metamorphic geology. Drury (2005) has incorporated the effect of dislocation-dominated grain size-sensitive flow into deformation maps for olivine where he refers to the effect as *grain size-sensitive power law creep* and another example for calcite-rich rocks is given in Figure 13.15. One should note that most studies of these grain size weakening effects in the metals literature involve transient stress states and not steady creep, a concept that has dominated the geological literature over the past 40 years or so.

The low temperature grain size weakening behaviour is widely known as the *Hall–Petch effect*, named after Petch (1953) and Hall (1951) for their pioneering work on fracture toughness and plasticity of materials, respectively. The dependence of the yield stress, σ^{yield}, on the grain size, d, is generally written as a power law:

$$\sigma^{yield} \; = \; \sigma_0 + kd^{-\frac{1}{2}} \tag{13.11}$$

A simple derivation (Zhu et al., 2008) of this relation proposes that the stress required to initiate slip in a grain unfavourably oriented for slip is produced by a pile-up of dislocations

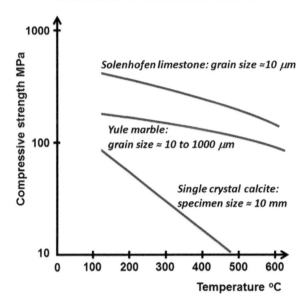

FIGURE 13.17 Experimental data on deformation of fine-grained limestone, medium-grained marble and a single crystal of calcite showing grain size weakening. *This figure is compiled from Griggs et al. (1960, figure 30).* Strain rates are 10^{-4}–10^{-5} s^{-1}.

at the boundary with a neighbouring grain where slip is active (Figure 13.16(a)). The number, n, of edge dislocations in a pile-up is

$$n = \frac{(1-v)\pi\tau}{4Gb}d$$

The stress at the tip of the pile-up is

$$\tau^{critical} = n\tau = \frac{(1-v)\pi\tau^2}{4Gb}d$$

Thus, if a frictional effect on dislocation motion such as the influence of solute atoms in inhibiting dislocation motion is added,

$$\tau = \left\{\frac{4Gb\tau^{critical}}{(1-v)\pi}\frac{1}{d}\right\}^{\frac{1}{2}} + \tau^{solute}$$

which is of the same form as (13.11). Although this is a common explanation of the Hall–Petch effect, many other explanations exist (Meyers et al., 2006).

At the high temperature end of the spectrum, the mechanical behaviour is described by some form of diffusional creep where grain boundaries facilitate, rather than impede, the deformation process. One form is Herring-Nabarro creep (Paterson, 2013) where the strain rate is given by a *grain size strengthening* relation:

$$\dot{\varepsilon} = A\left(\frac{D_{bulk}}{d^2}\right)\left(\frac{\sigma\Omega}{k^BT}\right)$$

A is a constant, D_{bulk} is a bulk diffusion coefficient, Ω is the atomic volume and k^B is Boltzmann's constant. Analogies to this grain size strengthening process can be seen in the constitutive relations (see Section 15.4) derived for the 'pressure solution' mechanisms of Paterson (1995) and Shimizu (1995), who derive expressions of the form:

$$\dot{\varepsilon} = A\sigma d^{-p}$$

where $p = 1, 2$ or 3. Here, for a given strain rate, the stress increases as the grain size increases. This regime is commonly called the *grain size sensitive diffusion creep regime* (Drury, 2005).

Although the models for grain size strengthening have been well developed over the past 60 years or so, the same cannot be said about models for grain size weakening and some controversy still exists especially concerning the role of grain size reduction in promoting localisation. Some (De Bresser et al., 2001) would claim that dynamic recrystallisation is not important in this process as far as localisation is concerned. Others (Platt and Behr, 2011) argue strongly for the importance of dynamic recrystallisation. It is important to establish a model for this effect, since it is widely proposed that grain size resulting from dynamic recrystallisation can be used as a *piezometer* (Twiss, 1977; Christie and Ord, 1980; Etheridge and Wilkie, 1981; Ord and Christie, 1984; Koch et al., 1980; Shimizu, 1998a,b, 1999; 2008, De Bresser et al., 1998, 2001; Platt and Behr, 2011) and, as indicated above, the grain size weakening effect is widely quoted as a mechanism for the initiation of shear localisation in deforming rocks.

A number of models have been developed for grain size weakening including Twiss (1977), Edward et al. (1982), Derby and Ashby (1987), Shimizu (1998a,b, 1999); De Bresser et al. (1998, 2001), Slotemaker (2006), Austin and Evans (2007, 2009), Shimizu (2008), Ricard and Bercovici (2009), Rozel et al. (2011), Platt and Behr (2011) and Herwegh et al. (2014). Some of these models are reviewed by De Bresser et al. (2001), Slotemaker (2006) and Shimizu (2008). Many models are based on simple concepts involving competition between recrystallisation and grain growth.

The issue with all models presented is to arrive at a process that explains the grain size distribution characteristics (which are poorly known) and results from a process that depends on stress leading to an average grain size that is an inverse function of the imposed stress. Ricard and Bercovici (2009) point out that for many arguments, there is an inconsistency in that if a material is initially in the grain size insensitive dislocation creep field then clearly the stress is insensitive to grain size. Thus there is no tendency to decrease the grain size solely due to the stress. In metals (Hughes and Hansen, 2000), the decrease in grain size is related to the amount of *strain*, not the *stress* although the stress is related to the grain size through the Hall–Petch effect. If the material is initially in the diffusion creep field, it is proposed that temperature effects will tend to increase the grain size; again, the effect is not dependent on stress. These issues arise because although there is a clear relationship between stress and grain size in both fields, there is no inbuilt mechanism to produce a feedback influence that results in the grain size being controlled in some manner by the stress. De Bresser et al. (2001) argue that competition between grain growth and grain reduction arising from recrystallisation leads to a stationary state located near the transition from dislocation creep to diffusion creep. They further argue that dynamic recrystallisation alone, although leading to grain size reduction, produces softening insufficient to initiate localisation.

Most arguments assume 'steady state' conditions, although it is difficult, conceptually, to fit such arguments into scenarios where softening, grain size reduction and localisation are occurring, all of which represent nonsteady state conditions. At least the evolution of grain size *distributions* towards a steady state needs to be considered in any model as has been done for normal grain growth (Section 13.2).

More of a 'helicopter' view of these processes is obtained by adopting an approach based in thermodynamics and expressing the Helmholtz energy, Ψ, in terms of grain size, represented as an internal variable (Chapter 5):

$$\Psi = \Psi(\varepsilon, T, d) \tag{13.12}$$

where ε is the strain and d is the grain size. If one wanted to emphasise the influence of stress, then the Gibbs energy would be substituted for the Helmholtz energy but that adds another level of complexity which we consider in Chapter 14. Other internal variables could be added, such as the extent (Chapter 14) of any mineral reactions, ξ, if recrystallisation involves a change in phase and/or a tensor function, $g_{\alpha\beta}$, that describes the orientation distribution function (ODF) if CPO is involved in grain size evolution (Faria et al., 2003). In particular, if there is more than one process operating to produce grain growth, perhaps a grain size reduction and a grain growth process, then the Helmholtz energy would be written as

$$\Psi = \Psi\left(\varepsilon, T, d^{reduction}, d^{growth}, \xi, g_{\alpha\beta}\right)$$

with a postulate that

$$\dot{d}^{total} = \alpha \dot{d}^{reduction} + \beta \dot{d}^{growth} \tag{13.13}$$

where α and β are constants. We emphasise that (13.13) is indeed a postulate and in fact has little experimental basis. Both $\dot{d}^{reduction}$ and $\dot{d}^{growth}$ appear to be related to strain, not stress, and it is by no means clear from the experimental data on geological materials that the concept of a field where both $\dot{d}^{reduction}$ and $\dot{d}^{growth}$ exist simultaneously actually exists for a single grain size. Nevertheless in the following we propose that (13.13) holds in order to outline the assumptions made in developing a general theory.

If we choose to neglect the influences of chemical reactions and CPO development, then it follows that

$$\dot{\Psi} = \frac{\partial \Psi}{\partial \varepsilon}\dot{\varepsilon} + \frac{\partial \Psi}{\partial T}\dot{T} + \frac{\partial \Psi}{\partial d^{reduction}}\dot{d}^{reduction} + \frac{\partial \Psi}{\partial d^{growth}}\dot{d}^{growth} \tag{13.14}$$

Otherwise the evolutionary equations involve the chemical reaction rate and the rate at which the ODF is changing. Some advances in this latter regard are reported for ice by Faria (2006a) (Faria et al., 2006). These arguments, based on thermodynamics, show that, for isothermal conditions, the thermodynamically admissible relation is not between grain size and stress but instead between the *rates* of grain size change, $\dot{d}^{reduction}$ and $\dot{d}^{growth}$, and the *dissipation* arising from plastic straining, $\sigma_{ij}\dot{\varepsilon}_{ij}^{plastic}$.

If one proceeds down the route outlined in Section 5.10, one can derive the full energy equation that couples temperature changes to deformation rates and grain size evolution terms:

$$
\underbrace{\rho c_P \frac{dT}{dt}}_{\text{Rate of heating}} = \overbrace{\left\{ \sigma_{ij}\dot{\varepsilon}_{ij} - \rho \frac{\partial \Psi}{\partial \varepsilon_{ij}^{elastic}} \dot{\varepsilon}_{ij}^{elastic} - \rho \frac{\partial \Psi}{\partial d^{reduction}} \dot{d}^{reduction} - \rho \frac{\partial \Psi}{\partial d^{growth}} \dot{d}^{growth} \right\}}^{\text{Rate of heating due to deformation assuming grain size changes contribute to deformation}}
$$

$$
+ \overbrace{\rho T \frac{\partial^2 \Psi}{\partial \varepsilon_{ij}^{elastic} \partial T} \dot{\varepsilon}_{ij}^{elastic}}^{\text{Thermo-elastic heating}} + \overbrace{\rho T \frac{\partial^2 \Psi}{\partial d^{reduction} \partial T} \dot{d}^{reduction}}^{\text{Heating due to grain size reduction}} + \overbrace{\rho T \frac{\partial^2 \Psi}{\partial d^{growth} \partial T} \dot{d}^{growth}}^{\text{Heating due to grain growth}}
$$

$$
+ \overbrace{\rho c_P \kappa^{thermal} \nabla^2 T}^{\text{Heat conduction}} \tag{13.15}
$$

For isothermal conditions, where $\frac{dT}{dt} = 0$ and $\nabla^2 T = 0$, and for constitutive behaviour where the stress lies on a yield surface that separates elastic from plastic behaviour, so that $\rho \frac{\partial \Psi}{\partial \varepsilon_{ij}^{elastic}} = \sigma_{ij}$, (13.15) reduces to

$$
0 = \overbrace{\left\{ \frac{1}{\rho} \sigma_{ij}\dot{\varepsilon}_{ij}^{plastic} - \frac{\partial \Psi}{\partial d^{reduction}} \dot{d}^{reduction} - \frac{\partial \Psi}{\partial d^{growth}} \dot{d}^{growth} \right\}}^{\text{Rate of heating due to deformation assuming grain size changes contribute to deformation}}
$$

$$
+ \overbrace{T \frac{\partial^2 \Psi}{\partial \varepsilon_{ij}^{elastic} \partial T} \dot{\varepsilon}_{ij}^{elastic}}^{\text{Thermo-elastic heating}} + \overbrace{T \frac{\partial^2 \Psi}{\partial d^{reduction} \partial T} \dot{d}^{reduction}}^{\text{Heating due to grain size reduction}} + \overbrace{T \frac{\partial^2 \Psi}{\partial d^{growth} \partial T} \dot{d}^{growth}}^{\text{Heating due to grain growth}}
$$

The common procedure is to neglect thermal effects arising from thermal expansion and from grain size growth and reduction so that one ends up with a much simplified version of the processes operating:

$$
\overbrace{\sigma_{ij}\dot{\varepsilon}_{ij}^{plastic}}^{\text{Dissipation arising from plastic deformation}} = \overbrace{\rho \left\{ \frac{\partial \Psi}{\partial d^{reduction}} \dot{d}^{reduction} + \frac{\partial \Psi}{\partial d^{growth}} \dot{d}^{growth} \right\}}^{\text{Dissipation arising from growth and reduction processes that contribute to deformation}}
$$

Thus if a steady state grain size evolves so that $\dot{d}^{total} = 0$ and $\dot{d}^{reduction} = -\frac{\beta}{\alpha}\dot{d}^{growth}$, then

$$
\overbrace{\sigma_{ij}\dot{\varepsilon}_{ij}^{plastic}}^{\text{Dissipation arising from plastic deformation}} = \overbrace{\rho \dot{d}^{growth} \left\{ -\frac{\beta}{\alpha} \frac{\partial \Psi}{\partial d^{reduction}} + \frac{\partial \Psi}{\partial d^{growth}} \right\}}^{\text{Dissipation arising from steady state grain size conditions}} \tag{13.16}
$$

Equation (13.16) expresses a balance between plastic dissipation and grain size reduction/growth process but the balance is more explicit (albeit more complicated) than that proposed by Austin and Evans (2007, 2009). One cannot proceed further unless one has knowledge of the quantities $\left(\chi^{reduction} = \frac{\partial \Psi}{\partial d^{reduction}}\right)$ and $\left(\chi^{growth} = \frac{\partial \Psi}{\partial d^{growth}}\right)$, which are general-ised thermodynamic forces or affinities that drive the grain size reduction and growth processes, respectively. The evidence seems to be, from the observations of Hughes and co-workers, from the experiments of Slotemaker and De Bresser and of Pieri and co-workers, and from the calculations we present in Section 13.6 from Ortiz and Repetto (1999), that these quantities are functions of the amount of *strain*. One suggestion from the above studies is that $\chi^{reduction}$ is related to substructure refinement arising from rotation dominated deformation modes, whereas χ^{growth} is related to grain growth processes arising from cellular dislocation migration processes, but detailed expressions for these quantities have not been established.

A relation between yield stress and grain size is presented below from Masumura et al. (1998) where a *grain size distribution* is considered rather than a single measure of grain size such as the *average*. We begin by assuming that the large grains within the grain size distribution follow the Hall–Petch relation, (13.11), which we write as

$$\tau^{Hall-Petch} = \tau - \tau_0 = kd^{-1/2} \tag{13.17}$$

where k is the *Hall–Petch constant*. The small grains in the distribution deform by Coble creep expressed as

$$\tau_c = \frac{A}{d} + Bd^3 \tag{13.18}$$

where A and B are both temperature and strain rate dependent. If d* is a critical grain size, where a switch from Hall–Petch to Coble behaviour occurs (Figure 13.18(a)), then we define a quantity, p, which measures the ratio of A to B at the switchover:

$$p = \frac{A/d^*}{B(d^*)^3} \tag{13.19}$$

Notice that p is a function of temperature and strain rate.

If we equate the stresses given by (13.17) and (13.18) at d*, then

$$k(d^*)^{-1/2} = \frac{A}{d^*} + B(d^*)^3$$

or, using (13.19),

$$k(d^*)^{-1/2} = (p+1)B(d^*)^3 \tag{13.20}$$

Masumura et al. (1998) assume a log-normal grain size distribution with the claim that the final result is not critically dependent on the precise form of the initial distribution assumed; the intent is that the finally derived expression for the stress, τ, is determined from the whole grain size distribution and not the mean, $\langle d \rangle$.

The yield stress is then written as

$$(\tau - \tau_0) = T^{Hall-Petch} + T^{Coble}$$

where $T^{Hall-Petch}$ and T^{Coble} are the values of the means of the stresses contributed to the total stress from the two parts of the grain size distribution. These means can be calculated from the assumed log-normal grain size distribution (Masumura et al., 1998). If we then define a parameter, $\xi = \langle d \rangle / d^*$, then (Masumura et al., 1998) finally arrive at an expression for the stress in terms of the grain size:

$$\frac{2(\tau - \tau_0)}{k(d^*)^{-1/2}} = f^{Hall-Petch} + \frac{pf^{Coble\_1} + f^{Coble\_2}}{1+p} \tag{13.21}$$

where $f^{Hall-Petch}$, f^{Coble_1}, f^{Coble_2} are contributions to the stress arising from the Hall−Petch effect, the first term in the Coble expression, (13.18), and the second term in (13.18),

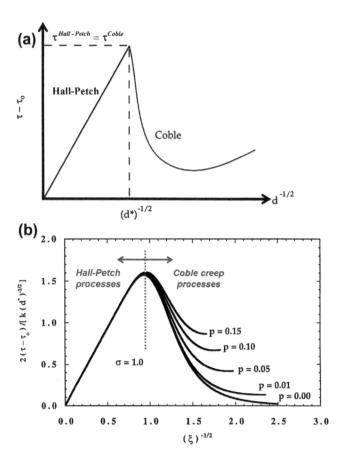

FIGURE 13.18 Grain size stress relations for a material with a grain size distribution. (a) Stress versus $d^{-1/2}$ showing a field where Hall−Petch mechanisms dominate as opposed to a field where diffusion creep dominates. The grain size, d*, represents the crossover from one field to the other. (b) Normalised yield stress versus $\xi^{-1/2}$ where $\xi = \langle d \rangle / d^*$. $\langle d \rangle$ is the average grain size assuming a log-normal distribution. The diagram is drawn for a standard deviation for the grain size distribution of $\sigma = 1$. p is the ratio defined in (13.19). Since the coordinates on both axes are normalised, this diagram is true for all materials where the Masumura argument is valid. Both (a) and (b) are drawn for constant temperature and strain rate. *From Masumura et al., 1998.* (c) The Masumura results plotted in log(grain size)−log(stress) space at a prescribed temperature and strain rate. (d) A conventional argument for grain size weakening in terms of a steady state deformation mechanism map.

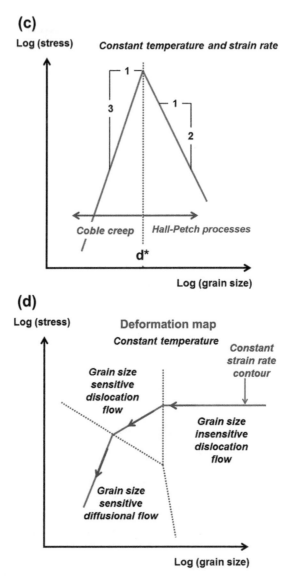

FIGURE 13.18   *(Continued)*

respectively; they are functions of the statistics of the log-normal distribution and are error functions involving the mean and standard distribution; they also involve the parameter, ξ. Clearly, if the first term in (13.18) is neglected as in most treatments in the geological literature, then (13.21) simplifies to

$$\frac{2(\tau - \tau_0)}{k(\mathbf{d}^*)^{-1/2}} = f^{Hall-Petch} + f^{Coble\_2}$$

Masumura et al. (1998) give a plot of (13.21) for various values of p as shown in Figure 13.18(b).

The contrast between the Masumura approach and that commonly adopted in the geoscience literature is illustrated in Figure 13.18(c) and (d). In Figure 13.18(c), the Masumura result is plotted in log(grain size)—log(stress) space at a prescribed temperature and strain rate. This is not a conventional deformation mechanism map since strain rate is presumed constant across the diagram. The Masumura model proposes that grain size distributions evolve on this map, driven by increasing strain. If most of the distribution is to the right of d*, then Hall—Petch processes result in grain size weakening. If most of the distribution is the left of d* then diffusional processes result in grain size strengthening. These processes need not involve steady state creep, in fact, they occur in hardening, steady or softening regimes.

Figure 13.18(d) is a conventional deformation mechanism map drawn at constant temperature but assuming steady state creep. Because of this assumption, the system is constrained to evolve along a contour of constant strain rate driven by a reduction in grain size arising from recrystallisation but in a manner that is loosely attributed to some influence of stress, not strain. Any steady state grain size is then attributed to a balance between grain size reduction and grain size increase processes.

Herwegh et al. (2014) adopt an approach similar to the thermodynamic outlined above, but simply use (13.12) so that

$$\dot{\Psi} = \frac{\partial \Psi}{\partial \varepsilon}\dot{\varepsilon} + \frac{\partial \Psi}{\partial T}\dot{T} + \frac{\partial \Psi}{\partial d}\dot{d}$$

In order to proceed to a relation between the stress and the grain size, one needs an evolution law for $\dot{d}$. Herwegh et al. (2014) incorporate the proposal of Austin and Evans (2007, 2009) that the total rate of change of grain size is given by

$$\dot{d}^{total} = \dot{d}^{reduction} + \dot{d}^{growth}$$

where $\dot{d}^{reduction}$ is the rate of grain size reduction arising from dynamic recrystallisation and $\dot{d}^{growth}$ is the rate of normal grain growth. One will appreciate from the previous discussion in this section that such an assumption is questionable. For a stationary state, they arrive at the deterministic relation:

$$d^{stationary\ state} = k\sigma^{-m} \exp\left(\frac{Q'}{RT}\right)$$

Notice that relations of this type are commonly called palaeowattmeters. This is something of a misnomer since it is the rate of change of the grain size, $\dot{d}$, that is related to the dissipation and not the grain size.

Using the above arguments, Herwegh et al. (2014) model the stress—strain response and grain size evolution (Figure 13.19(a)—(c)) for a material with hardening elastic—plastic—viscous constitutive behaviour where grain size and stress are coupled. The grain size evolves into a very sharp peak since the governing equations are completely deterministic and involve no stochastic processes, that is, they do not consider the influence of drift and diffusion terms on $\dot{d}^{growth}$ along the lines of studies of normal grain growth (Section 13.2).

The grain size evolves during the transient stages of hardening and softening before steady state deformation conditions are reached; grain growth occurs during the elastic and hardening parts of the deformation and grain reduction during the softening part (see Figure 13.19(a)–(c)); no grain size reduction occurs in the model with continued straining once the stress is steady, although the data of Pieri et al. (2001) show continued evolution of grain size to high strains. Notice that in these models the grain size *distribution* within a

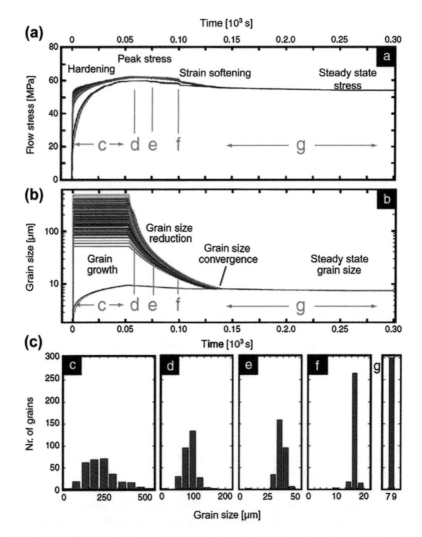

FIGURE 13.19 Modelling of stress–strain behaviour and coupled grain size evolution. (a) Stress–strain behaviour showing a hardening regime, c, peak stress at d, softening in the regime e to f and steady state for regime g. (b) Grain size evolutions for each of the stress–strain regimes. (c) Histograms of grain sizes in various subareas of the specimen. *From Herwegh et al. (2014).* (d) Map of feedback relations between energy dissipation arising from deformation, the various mechanisms of grain size reduction and growth and the influence on stress.

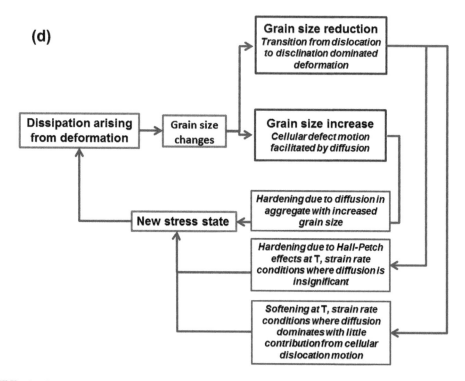

FIGURE 13.19 *(Continued)*

small volume is not modelled, rather the reported grain size 'distributions' are the sum of all the average grain sizes in each element of the computational model.

Grain size distributions have been incorporated into models of grain size weakening by Shimizu (2008) by Ricard and Bercovici (2009) and by Rozel et al. (2011) who assume log-normal distributions and self-similarity, which have not yet been demonstrated for geological materials undergoing dynamic recrystallisation although self-similarity with continued straining is well documented in the data of Hughes and co-workers reported in Figure 13.13.

In summary, we believe that the present state of affairs with respect to grain size––stress relations is quite unsatisfactory. The goal set by workers such as Twiss (1977) to establish a piezometer whereby the recrystallised grain size is related to a stress level has assumed, for the most part, steady state conditions with an emphasis on mean grain size rather than grain size distributions. Simple models have been proposed involving mainly a balance between grain size reduction processes associated with recrystallisation and grain growth processes such as those arising from a reduction of interfacial or stored strain energies. Although such a balance has to be broadly correct for a steady state to arise, the theories have not involved transient states that lead up to steady state, have not considered the evolution of these states governed by strain rather than stress, have not considered grain size distributions, or, if they have, simple log-normal distributions have been imposed and have not considered widely recognised processes such as translation- versus rotation-dominated modes of deformation and grain growth modes involving cellular dislocation motion. The theories that have been developed are essentially deterministic, whereas a stochastic approach is required that examines the evolution of the complete grain size distribution.

A new approach could be to adopt a Fokker–Planck model that incorporates these mechanisms as competitive processes involving drift and diffusion terms. We need experimental data that document the types of scaling relations illustrated in Figure 13.13. A map of how some of these processes might be brought together is presented in Figure 13.19(d). We emphasise that new experiments, which show whether a competitive environment actually exists where grain growth processes operate simultaneously with grain reduction processes, are required. The alternative approach, and the one we suggest is relevant to rocks, is a model such as that proposed by Masumura et al. (1998) where Hall–Petch mechanisms compete with diffusional mechanisms and the evolution of the grain size distribution is driven by increasing strain.

13.4.2 Grain Growth and Dislocation Interaction as a Reaction–Diffusion Process

In Chapter 9, we discuss the development of dislocation patterns, such as cell structures, in terms of reaction–diffusion equations where the generation and annihilation of two classes of dislocations, mobile and relatively immobile dislocations, compete with the diffusion of dislocation densities (Aifantis, 1986, 1987; Walgraef and Aifantis, 1985a,b,c). Hallberg and Ristinmaa (2013) have extended this model to couple the reaction–diffusion processes and interaction to grain boundaries so that dislocations pile up at the boundaries. Thus the Walgraef–Aifantis equations (see Chapter 9) are written:

$$\frac{\partial \rho_m}{\partial t} = D_m \frac{\partial^2 \rho_m}{\partial x^2} - \left[k_2 \rho_m + k_3 \sqrt{\rho_i} - k_1 \frac{\rho_i}{\rho_m} \right] \dot{\varepsilon}_{effective}^{plastic}$$

$$\frac{\partial \rho_i}{\partial t} = D_i \frac{\partial^2 \rho_i}{\partial x^2} + \left[k_2 \rho_m + k_3 \sqrt{\rho_i} - k_4 \rho_i \right] \dot{\varepsilon}_{effective}^{plastic}$$

where D_i and D_m are the diffusion coefficients for relatively immobile and mobile dislocations, ρ_i and ρ_m are the respective dislocation densities, the k's are constants and $\dot{\varepsilon}_{effective}^{plastic}$ is the effective plastic strain rate. Hallberg and Ristinmaa (2013) modify the Aifantis–Walgraef equations by proposing that the density of relatively immobile dislocations depends on a term that varies within a grain exponentially with the distance, l, from a grain boundary:

$$k_3(l) = k_3^{max} - \left(k_3^{max} - k_3^{min} \right)(1 - \exp(-wl))$$

so that the parameter k_3 varies from k_3^{max} at a grain boundary to k_3^{min} at a grain interior; w is a parameter that lies in the range 5–15 in the models investigated. Hallberg and Ristinmaa (2013) show that the Hall–Petch relation is replicated by this model but is an incidental outcome of the model. If the grain boundary energy is assumed to be a function of the dislocation density and the mobility of a new grain boundary is a function of the misorientation across the boundary, then microstructures typical of dynamic recrystallisation are produced as shown in Figure 13.20. The grain size reduction process as a function of *strain*, not *stress*, is well illustrated by this model.

13.4.3 Compatibility of Deformation across Grain Boundaries

In Figure 13.21, we illustrate the types of grain deformation that develop if just one crystallographic slip plane is present in each grain. The models are described in Zhang et al. (1994a,b). The initial grain configuration with random orientations of the slip plane is shown in Figure 13.21(a). Figure 13.21(b) and (c) shows the resulting grain configurations after a simple shearing deformation. If no grain sliding is allowed, then compatibility of deformation across grain boundaries is achieved by inhomogeneous deformation within each grain. If grain boundary sliding is allowed, then the deformation within individual grains is much more homogeneous and gaps open between grains. Deformed metamorphic rocks lack gross discontinuities in the grain structure as illustrated in Figure 13.21(c) and so some other mechanisms must operate in order to ensure deformation compatibility across grain boundaries.

In general, deformation in polycrystalline aggregates is inhomogeneous with commonly large jumps in strain across grain boundaries as illustrated in Figure 13.22(b). The question is: *How is compatibility of deformation maintained whilst deformation is progressing?* We have seen in Chapter 9 that if $\mathbf{F}$ is the deformation gradient (Chapter 2) for a particular part of a grain, then $\mathbf{F}$ can be expressed as a combination of elastic, $\mathbf{F}^{elastic}$, and plastic, $\mathbf{F}^{plastic}$, deformation gradients:

$$\mathbf{F} = \mathbf{F}^{elastic}\mathbf{F}^{plastic}$$

Moreover, for single slip on a slip system, α, with slip direction, $\mathbf{s}^{\alpha}$, and slip plane normal, $\mathbf{m}^{\alpha}$

$$\mathbf{F}^{plastic} = \mathbf{I} + \gamma^{\alpha}\mathbf{s}^{\alpha} \otimes \mathbf{m}^{\alpha}$$

Hence

$$\mathbf{F} = \mathbf{F}^{elastic}\left(\mathbf{I} + \gamma^{\alpha}\mathbf{s}^{\alpha} \otimes \mathbf{m}^{\alpha}\right)$$

But, in the absence of long-range stress fields,

$$\mathbf{F} = \mathbf{R}(\mathbf{I} + \gamma^{\alpha}\mathbf{s}^{\alpha} \otimes \mathbf{m}^{\alpha})$$

Thus for compatibility in the absence of long-range stress fields,

$$\mathbf{F}^{elastic} = \mathbf{R}$$

Although $\mathbf{F}$ is compatible across an interface between grains, there is no need for $\mathbf{F}^{plastic}$ to be compatible; any incompatibility can be accommodated by an elastic rotation.

Physically, incompatibility of the plastic deformation gradient is equivalent to an array of dislocations and/or disclinations (Figure 13.23). It is convenient to introduce the *Nye dislocation density tensor*, A_{ij} (Nye, 1953), which is a way of quantifying, in a continuum manner, the

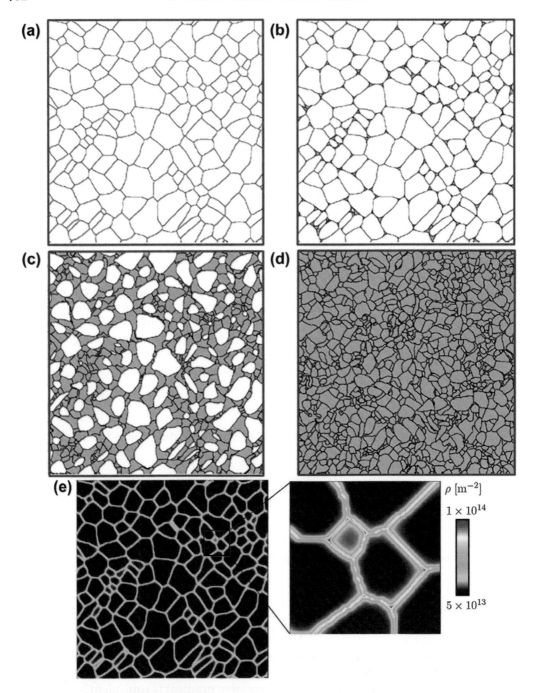

FIGURE 13.20   Results of modelling dynamic recrystallisation using reaction–diffusion equations. (a) Initial grain structure, 150 grains with an average size of 45 μm. (b) Recrystallisation (grey) nucleating at sites of the highest gradients in dislocation density, which are mostly at triple-junctions. (c) Relict grains (white) in a sea of recrystallised grains (grey) at an effective plastic strain of 0.63. (d) The final recrystallised microstructure, 977 grains with an average grain size of 16 μm at an effective plastic strain of 2. (e) Map of dislocation density corresponding to the beginning of nucleation. Inset shows detail with the highest dislocation densities at triple-junctions. *From Hallberg and Ristinmaa (2013).*

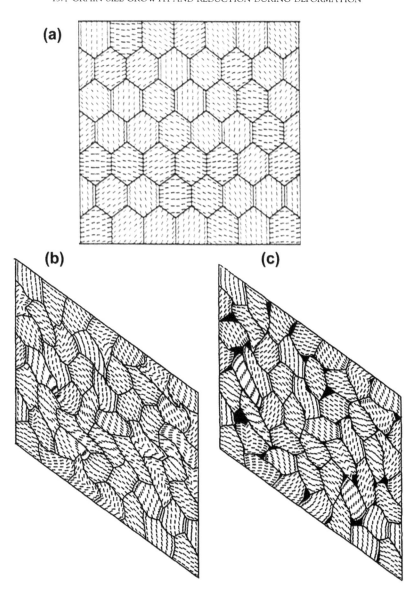

FIGURE 13.21 Two different modes of deformation arising from single slip. (a) Undeformed configuration showing orientations of a single slip plane within each grain. (b) Simple shearing with no grain boundary sliding. (c) Simple shearing with grain boundary sliding. The black areas are open spaces. *From Zhang et al. (1994a, b).*

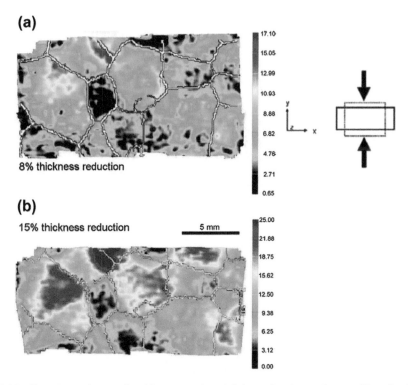

FIGURE 13.22   Experimental example of heterogeneity of deformation in a polycrystalline aluminium aggregate. Colours represent accumulated von Mises strain (see Section 6.7.4) after 8% (a) and 15% (b) overall shortening. *From Roters et al. (2010).* Strains are established by digital image correlation on the surface of the deforming specimen.

dislocation distribution in a deformed crystal. If b is a Burgers vector and t is a unit vector tangent to the dislocation line, L, then the Nye dislocation density tensor is

$$A_{ij} = \frac{1}{V} \int_L b_i t_j ds$$

so that A_{ij} is the integral along all the dislocation lines of the scalar product of the Burgers vector and the tangent to the dislocation line at each point in the crystal (Arsenlis and Parks, 1999). The components of A_{ij} can be established using high resolution EBSD (Pantleon, 2008; Hardin et al., 2011). One establishes the derivatives of the infinitesimal elastic deformation gradient tensor, β_{ij}, which measures not only the local symmetric part of the lattice strain but also local rotations of the lattice. Thus if $\mathbf{u}$ is the local (elastic) lattice displacement,

$$\beta_{ij} = \frac{\partial u_i}{\partial x_j}$$

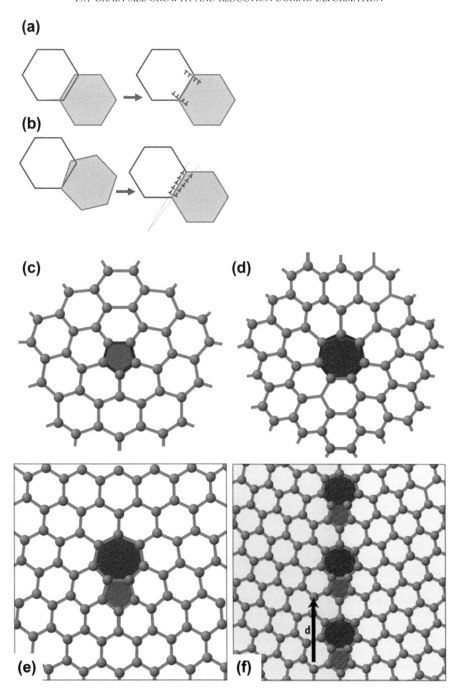

FIGURE 13.23 Mechanisms for achieving compatibility of deformation. (a) Homogeneous overlap compensated by extra half planes in dislocation dipole arrays. (b) Wedge-shaped overlap compensated by extra half planes in tilted dislocation arrays. *After Ashby (1970).* (c) A negative disclination in a hexagonal structure. (d) A positive disclination in a hexagonal structure. In both (c) and (d), the distortion of the structure extends many unit cells from the single disclination. (e) A positive—negative disclination dipole in a hexagonal structure. The distortion of the structure extends only about one unit cell from the dipole. (f) An array of disclination dipoles with spacing, d, comprising a high angle tilt boundary. Such an array can be substituted for the dislocation array in (b). (c)—(f) are models of disclinations in graphene. *From Kim (2010).*

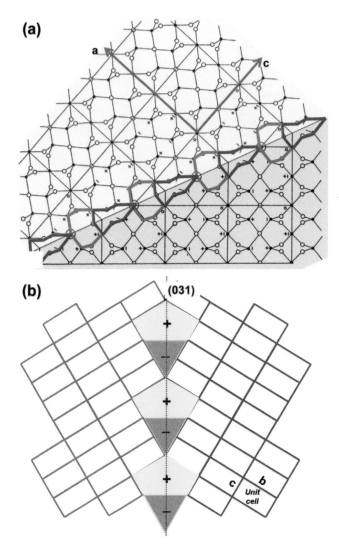

FIGURE 13.24 Proposed twin disclination dipoles configurations in quartz and olivine. (a) Quartz. The boundary represents a high coincidence boundary in β-quartz and is a Breithaupt twin. The boundary is ($\bar{2}$111) and the rotation across the boundary is 48.9°. Black circles are silicon atoms, open circles are oxygen atoms and double circles are coincident oxygen atoms in the boundary. The + and − signs indicate silicon atoms above and below the plane of the diagram. Notice that the introduction of disclinations to form the boundary reverses the chirality of the structure across the boundary. Blue and red outline positive and negative disclinations, respectively. There are two dangling bonds in the negative disclination. The two disclinations together constitute a disclination dipole. *Modified from McLaren (1986).* (b) Olivine. The twin boundary is (031) and the rotation across the boundary is 59° 16'. Red and blue outline positive and negative disclinations, respectively. *After Mikouchi et al. (1995).* The twin is referred to as a {031}-twin.

and the Nye tensor is given by

$$A = curl(\beta)$$

Ortiz and Repetto (1999) show that for two crystals undergoing single slip,

$$A(x) = (\llbracket \mathbf{R}^T \rrbracket \mathbf{F}^+) \times \mathbf{N} \delta_\Pi(x) = (\llbracket \mathbf{R}^T \rrbracket \mathbf{F}^-) \times \mathbf{N} \delta_\Pi(x)$$

which shows that the dislocation density tensor is related to the misorientation, $\llbracket \mathbf{R}^T \rrbracket$, between the two grains. $\delta_\Pi(x)$ is the Dirac-delta supported on the plane between the two grains, Π, so that $\delta_\Pi(x) = 0$ everywhere except on Π, where $\delta_\Pi(x) = 1$ and $\mathbf{F}^\pm$ are the deformation gradients in the two grains.

Just as the Nye dislocation density tensor, A, is given by the curl of the elastic deformation gradient tensor, β, the *disclination density tensor*, θ, is given by the curl of the elastic curvature tensor, κ:

$$\theta = curl(\kappa)$$

θ may be established using high resolution EBSD (Taupin et al., 2013; Cordier et al., 2014 and Chapter 9). Accommodation of deformation incompatibility by dislocation and disclination arrays is illustrated in Figure 13.23. Disclination arrays of this type have been documented in olivine by Cordier et al. (2014) and disclination dipoles such as the arrays illustrated in Figure 13.24(a) and (b) may be common in deformed quartzites and peridotites.

13.5 CPO DEVELOPMENT

When polycrystalline aggregates are plastically deformed, preferred orientations of the crystal structures that make up the individual minerals in the rock commonly develop. These preferred orientations are known as crystallographic preferred orientations (CPO) in the geoscience literature and as *textures* in the metals literature; some authors prefer the term *lattice preferred orientations*. The pattern of CPO depends on the symmetry and details of the *crystal structures* that make up the minerals in the rock (which control the slip systems potentially available for deformation), the type of deformation history (uniaxial shortening, simple shearing and so on), the degree of deformation, the temperature, the deformation rate and the presence or absence of H_2O in the system.

The mechanism of producing a CPO is generally attributed to systematic crystal lattice rotations derived from plastic slip on crystallographic planes and we focus on such mechanisms in this section. There is some evidence that CPO can develop in systems where diffusion is the only deformation mechanism (Miyazaki et al., 2013) and we consider this case briefly in Section 13.7. CPO is commonly measured using *electron back scattered diffraction* (ESBD) techniques (Prior et al., 1999) where the complete crystallography of each grain is measured together with a spatial map (Figure 13.1(d)) of the crystallographic orientation distribution. The results are expressed in the form of equal area projections of the orientations of crystallographic directions relative to geographic coordinates or deformation fabric features (Figure 13.25(a)). In the metals literature, stereographic projections are commonly used. In the geological literature, equal area projections are commonly used, but are sometimes erroneously referred to as stereographic projections and just as commonly the distinction between stereographic and equal area projections is not made. These projections are also known as *pole figures*. Also in common use are *inverse pole figures*, where the density of geographic or deformation fabric directions is plotted relative to the crystallography (Figure 13.25(b)), and orientation distribution functions (ODF) (Figure 13.25(d)), which express the volume density of a particular part of the CPO in Euler space defined (Figure 13.25(c)) by the Euler angles Ψ, φ_1 and φ_1 (Bunge, 1965; Roe, 1965; Bunge and Wenk, 1977; Kocks, 1998a,b). All pole figures and inverse pole figures can be derived from the ODF. Software for constructing these various forms of presentation and for converting from one to another can be found at https://code.google.com/p/mtex/ (Mainprice et al., 2011; Bachmann et al., 2011).

Compatibility of deformation is the key to mechanisms that produce CPO (Kocks, 1998b). The simplest model that ensures deformation compatibility is to assume that every grain in the aggregate experiences the same deformation. This is known as the *Taylor model*, proposed by Taylor (1938). For the condition that every grain experiences the

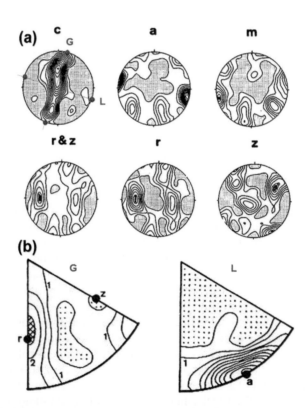

FIGURE 13.25 Methods of representing crystallographic preferred orientation data. (a) Equal area projections of enough crystallographic directions to uniquely specify the fabric. (b) Inverse pole figures corresponding to the data in (a). The distributions of points G and L in (a) are plotted on the left and right diagrams relative to the quartz crystallography. (c) Definition of Euler angles and a rectangular representation showing a single section through Euler space in red. (d) A ($\varphi_2-\Phi$) section through an orientation distribution function at $\varphi_1 = 80°$. All data are for a quartzite. *After Schmid and Casey (1986)*. See that paper for details.

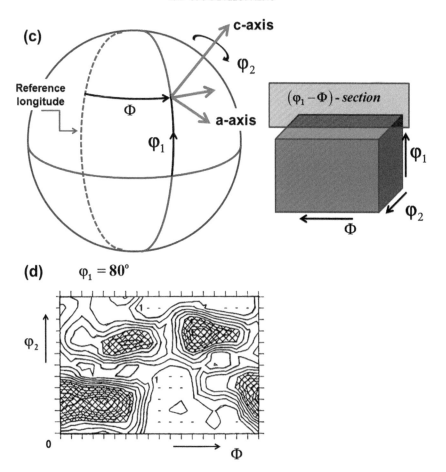

FIGURE 13.25 (Continued)

same general affine strain to be geometrically possible, every grain must have five independent slip systems (Chapter 9) if the deformation mechanism is solely by crystal slip (Groves and Kelly, 1963; Kelly and Groves, 1970; Paterson, 1969). This requirement is replaced (Chapter 9) by a need for three non-coplanar Burgers vectors if dislocations can both slip and climb although a number of other possibilities exist if climb contributes to the deformation (Groves and Kelly, 1969). In principle, the requirement for five independent slip systems is satisfied for quartz (Paterson, 1969) but not for olivine (Cordier et al., 2014). The Taylor model fulfils the requirement for deformation compatibility throughout the aggregate but does not lead to stress equilibrium at grain boundaries since the stress required to activate five independent slip systems in each grain differs from one grain to the next. However these stresses, so long as they are small compared to the yield stress, are accommodated by elastic deformations. The Taylor model is one end member in a family of models for polycrystalline deformation and we explore some of these below. At best the Taylor model is a good approximation to what happens in real aggregates of grains with five independent slip systems.

However, grain deformation is commonly observed to be inhomogeneous (Figures 13.21 and 13.22) not only from grain to grain but within individual grains as well. The challenge is to develop realistic models that maintain deformation compatibility but also allow for the lack of five independent slip systems in some cases and/or for inhomogeneous deformation to be taken into account. One way of doing so (Figure 13.23) is to allow geometrical necessary dislocations to account for incompatibility (Ashby, 1970). Another way is to introduce disclinations (Cordier et al., 2014).

13.5.1 Taylor—Bishop—Hill

Two geometrically equivalent models for CPO development with the constraint of homogeneous strain throughout the aggregate were developed by Taylor (1938) and Bishop and Hill (1951a,b). The basic assumption is that plastic deformation is achieved solely by slip on crystallographic defined slip systems, α, comprised of slip planes with normals, $\mathbf{m}^{(\alpha)}$, and slip directions, $\mathbf{s}^{(\alpha)}$. As an example, the values of the direction cosines defining some of the slip systems for α-quartz are given in Table 13.1. The deformation preserves the gross crystal lattice although dislocations are introduced into the crystal structure.

If σ_{ij} is the stress within a grain, then the resolved shear stress on the slip plane in the direction of slip is (Kelly and Groves, 1970; also Chapter 9).

$$\tau^{(\alpha)} = \sigma_{ij} s_i^{(\alpha)} m_j^{(\alpha)} \tag{13.22}$$

An additional assumption is that *Schmid's law* holds for each slip system so that slip is initiated once some critical resolved shear stress is exceeded for that system. In the strict Taylor—Bishop—Hill model, elastic strains are neglected and the deformation processes are assumed to be rate insensitive.

Taylor (1938) proposed a *principle of minimum work* which states that *of all the possible combinations of slip systems available to produce a given imposed increment of strain, only those that produce the minimum work operate.* As we have discussed in Chapter 5, this is in fact equivalent to a *principle of maximum entropy production* enunciated by Ziegler (1963). If we assume that the critical resolved shear stress is the same for all systems (true perhaps for some metals but not for silicates), Taylor's principle translates in practice into: *of the possible combinations of slip systems available to produce a given imposed incremental strain, only that combination for which the sum of the shears is a minimum actually operates.*

Bishop and Hill (1951a,b) introduced a *maximum work principle* which states that *of all the stress states that are possible within a grain subjected to a given imposed incremental strain and that can activate five independent slip systems, that stress state is selected that maximises the work done.*

Hence the Bishop—Hill approach becomes a linear optimisation problem: *maximise the work done during a prescribed increment of strain subject to the constraint that five independent slip systems with given critical resolved shear stresses must operate.* This is a *linear programming*

TABLE 13.1 Some of the Slip Systems Proposed for α-Quartz. Direction Cosines of the Directions of Slip, **s**, and the Normal to Slip Planes, **m**, for Structurally Right-handed Quartz for Slip Planes (0001), $\{10\bar{1}0\}$ and Slip Directions, $\langle \mathbf{a} \rangle$, $\langle \mathbf{c} \rangle$ and $\langle \mathbf{c+a} \rangle$. The Coordinate System Used is Left-handed with x_1 Parallel to $\mathbf{a}_1$, x_2 Normal to $\mathbf{a}_1$ but in (0001) and x_3 Parallel to $\mathbf{c}$

Slip System Number	Slip system	s_1	s_2	s_3	m_1	m_2	m_3
	Basal slip						
1	$(0001)+a_1$	0	0	1	1	0	0
2	$(0001)+a_2$	0	0	1	$-\frac{1}{2}$	$-\frac{\sqrt{3}}{2}$	0
3	$(0001)+a_3$	0	0	1	$-\frac{1}{2}$	$\frac{\sqrt{3}}{2}$	0
$\bar{1}$	$(0001)-a_1$	0	0	1	-1	0	0
$\bar{2}$	$(0001)-a_2$	0	0	1	$\frac{1}{2}$	$\frac{\sqrt{3}}{2}$	0
$\bar{3}$	$(0001)-a_3$	0	0	1	$\frac{1}{2}$	$-\frac{\sqrt{3}}{2}$	0
	Prismatic slip						
4	$(01\bar{1}0) + a_1$	0	-1	0	1	0	0
5	$(01\bar{1}0) + c$	0	-1	0	0	0	1
6	$(01\bar{1}0)[c + a_1]$	0	-1	0	0.6726	0	0.74
7	$(01\bar{1}0)[c - a_1]$	0	-1	0	-0.6726	0	0.74
$\bar{4}$	$(01\bar{1}0) - a_1$	0	-1	0	-1	0	0
$\bar{5}$	$(01\bar{1}0) - c$	0	-1	0	0	0	-1
$\bar{6}$	$(01\bar{1}0)[-c - a_1]$	0	-1	0	-0.6726	0	-0.74
$\bar{7}$	$(01\bar{1}0)[-c + a_1]$	0	-1	0	0.6726	0	-0.74
8	$(10\bar{1}0) + a_2$	$\frac{\sqrt{3}}{2}$	$-\frac{1}{2}$	0	$-\frac{1}{2}$	$-\frac{\sqrt{3}}{2}$	0
9	$(10\bar{1}0) + c$	$\frac{\sqrt{3}}{2}$	$-\frac{1}{2}$	0	0	0	1
10	$(10\bar{1}0)[c + a_2]$	$\frac{\sqrt{3}}{2}$	$-\frac{1}{2}$	0	-0.3363	-0.5825	0.74
11	$(10\bar{1}0)[c - a_2]$	$\frac{\sqrt{3}}{2}$	$-\frac{1}{2}$	0	0.3363	0.5825	0.74
$\bar{8}$	$(10\bar{1}0) - a_2$	$\frac{\sqrt{3}}{2}$	$-\frac{1}{2}$	0	$\frac{1}{2}$	$\frac{\sqrt{3}}{2}$	0
$\bar{9}$	$(10\bar{1}0) - c$	$\frac{\sqrt{3}}{2}$	$-\frac{1}{2}$	0	0	0	-1
$\overline{10}$	$(10\bar{1}0)[-c - a_2]$	$\frac{\sqrt{3}}{2}$	$-\frac{1}{2}$	0	0.3363	0.5825	-0.74
$\overline{11}$	$(10\bar{1}0)[-c + a_2]$	$\frac{\sqrt{3}}{2}$	$-\frac{1}{2}$	0	-0.3363	-0.5825	-0.74
12	$(1\bar{1}00) + a_3$	$\frac{\sqrt{3}}{2}$	$\frac{1}{2}$	0	$-\frac{1}{2}$	$\frac{\sqrt{3}}{2}$	0
13	$(1\bar{1}00) + c$	$\frac{\sqrt{3}}{2}$	$\frac{1}{2}$	0	0	0	1
14	$(1\bar{1}00)[c + a_3]$	$\frac{\sqrt{3}}{2}$	$\frac{1}{2}$	0	-0.3363	0.5825	0.74
15	$(1\bar{1}00)[c - a_3]$	$\frac{\sqrt{3}}{2}$	$\frac{1}{2}$	0	0.3363	-0.5825	0.74
$\overline{12}$	$(1\bar{1}00) - a_3$	$\frac{\sqrt{3}}{2}$	$\frac{1}{2}$	0	$\frac{1}{2}$	$-\frac{\sqrt{3}}{2}$	0
$\overline{13}$	$(1\bar{1}00) - c$	$\frac{\sqrt{3}}{2}$	$\frac{1}{2}$	0	0	0	-1
$\overline{14}$	$(1\bar{1}00)[-c - a_3]$	$\frac{\sqrt{3}}{2}$	$\frac{1}{2}$	0	0.3363	-0.5825	-0.74
$\overline{15}$	$(1\bar{1}00)[-c + a_3]$	$\frac{\sqrt{3}}{2}$	$\frac{1}{2}$	0	-0.3363	0.5825	-0.74

After Hobbs (1985).

problem and can be solved using standard software (http://www.wolfram.com/learningcenter/tutorialcollection/ConstrainedOptimization/ConstrainedOptimization.pdf). The Taylor problem, in linear programming parlance, is the *dual* of the Bishop–Hill problem.

The steps in calculating CPO development in the Bishop–Hill approach are (see Kocks, 1998b, for more details):

- Define the yield surface in six dimensional stress space. The faces on this surface are given by

$$\tau_{critical}^{(\alpha)} = \sigma_{ij} s_i^{(\alpha)} m_j^{(\alpha)} \tag{13.23}$$

Cross sections through such a surface for quartz in two dimensions are shown in Figure 13.26 assuming that $\tau_{critical}$ is the same for all slip systems. If this is not the case, then the cross sections in Figure 13.26 shrink in some directions and expand in others. Some slip systems may not operate.

- Superimpose incremental strain space upon stress space (Figure 13.26(b)).
- The incremental work done, δW, in an increment of strain, $\delta \varepsilon_{ij}$, is given by $\delta W = \sigma_{ij} \delta \varepsilon_{ij}$; this represents a plane in stress space, the *work plane*, distant δW from the origin. In order to satisfy the Bishop–Hill postulate, we need to find the maximum distance of the work plane from the origin of stress space such that the work plane touches the yield surface (Figure 13.26(b)). This in general will be at a vertex of the yield surface. Just as in two dimensions a vertex of a polygon is where two edges meet, the vertex of a polygon in six dimensional space is where six surfaces meet. Thus maximising the incremental work automatically satisfies the requirement for five independent slip systems to operate. As indicated above, the process of establishing the maximum work can be carried out using a standard linear programming package.
- Since the strain produced by the operation of the slip systems is given by

$$\varepsilon_{ij} = \begin{bmatrix} \sum_\alpha \gamma^\alpha m_1^\alpha s_1^\alpha & \sum_\alpha \frac{\gamma^\alpha}{2}(m_2^\alpha s_1^\alpha + m_1^\alpha s_2^\alpha) & \sum_\alpha \frac{\gamma^\alpha}{2}(m_3^\alpha s_1^\alpha + m_1^\alpha s_3^\alpha) \\ \sum_\alpha \frac{\gamma^\alpha}{2}(m_2^\alpha s_1^\alpha + m_1^\alpha s_2^\alpha) & \sum_\alpha \gamma^\alpha m_2^\alpha s_2^\alpha & \sum_\alpha \frac{\gamma^\alpha}{2}(m_2^\alpha s_3^\alpha + m_3^\alpha s_2^\alpha) \\ \sum_\alpha \frac{\gamma^\alpha}{2}(m_3^\alpha s_1^\alpha + m_1^\alpha s_3^\alpha) & \sum_\alpha \frac{\gamma^\alpha}{2}(m_2^\alpha s_3^\alpha + m_3^\alpha s_2^\alpha) & \sum_\alpha \gamma^\alpha m_3^\alpha s_3^\alpha \end{bmatrix}$$

and we now know the $\mathbf{m}^\alpha$ and $\mathbf{s}^\alpha$ from the linear programming exercise, since ε_{ij} is prescribed, we have five simultaneous equations for the five unknowns γ^α. Hence the shears on each slip system needed to achieve the imposed increment of strain can be calculated.

- The strain increment that corresponds to the maximum work is now known and now that the active slip systems have been identified, along with the shears necessary to achieve

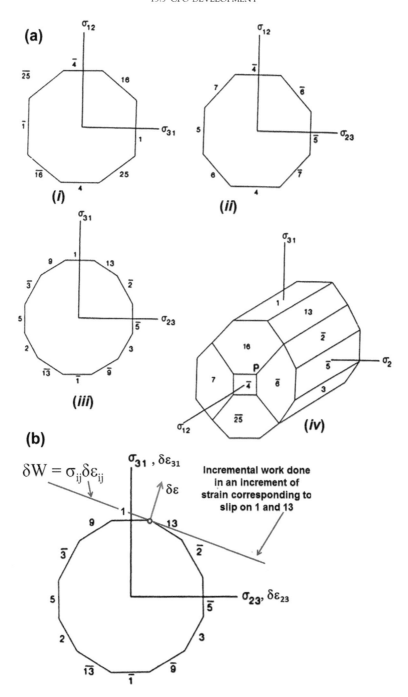

FIGURE 13.26 The yield surface for quartz. (a) Cross sections through the yield surface for quartz, assuming rate insensitive plasticity and that the critical resolved shear stresses for all slip systems are equal. The slip systems refer to the slip system numbers in Table 13.1. *From Hobbs (1985).* (b) The Bishop–Hill criterion for determining slip systems and stress state. The increment of work done in an increment of strain is δW and is represented by a plane in stress space distant δW from the origin. When this plane touches the yield surface, the slip systems corresponding to the point of contact can be activated and the stress state is identified. At that point, the work done is maximised for that strain increment.

that strain increment, the rotation associated with slip on those systems can be established from:

$$\omega_{ij} = \begin{bmatrix} 0 & \sum_\alpha \frac{\gamma^\alpha}{2}\left(m_2^\alpha s_1^\alpha - m_1^\alpha s_2^\alpha\right) & \sum_\alpha \frac{\gamma^\alpha}{2}\left(m_3^\alpha s_1^\alpha - m_1^\alpha s_3^\alpha\right) \\ -\sum_\alpha \frac{\gamma^\alpha}{2}\left(m_2^\alpha s_1^\alpha - m_1^\alpha s_2^\alpha\right) & 0 & \sum_\alpha \frac{\gamma^\alpha}{2}\left(m_2^\alpha s_3^\alpha - m_3^\alpha s_2^\alpha\right) \\ -\sum_\alpha \frac{\gamma^\alpha}{2}\left(m_3^\alpha s_1^\alpha - m_1^\alpha s_3^\alpha\right) & -\sum_\alpha \frac{\gamma^\alpha}{2}\left(m_2^\alpha s_3^\alpha - m_3^\alpha s_2^\alpha\right) & 0 \end{bmatrix} \quad (13.24)$$

- If the imposed deformation is prescribed (see Chapter 2) as $x_1 = a_{11}X_1 + a_{12}X_2 + a_{13}X_3$, $x_2 = a_{21}X_1 + a_{22}X_2 + a_{23}X_3$, $x_3 = a_{31}X_1 + a_{32}X_2 + a_{33}X_3$, then there is an imposed rotation associated with this deformation given by

$$\Omega_{ij} = \begin{bmatrix} 0 & \frac{1}{2}(a_{12} - a_{21}) & \frac{1}{2}(a_{13} - a_{31}) \\ \frac{1}{2}(a_{21} - a_{12}) & 0 & \frac{1}{2}(a_{23} - a_{32}) \\ \frac{1}{2}(a_{31} - a_{13}) & \frac{1}{2}(a_{32} - a_{23}) & 0 \end{bmatrix}$$

The rotation of the lattice relative to the coordinate frame used to define the deformation is $(\Omega_{ij} - \omega_{ij})$, where ω_{ij} is given by (13.24).
- We will have noticed that in general the choice of five independent slip systems at the active vertex of the yield surface has to be made from six possibilities and hence the rotation is not uniquely determined by the maximisation process. This ambiguity is always present for perfect rate-insensitive plasticity and one common practice is to make a random selection of five systems out of the six available. Another (realistic) solution is to introduce some rate sensitivity into the plasticity. This rounds the vertices (Section 6.7.4) and so eliminates the ambiguity (Kocks, 1998b; his Figure 15).
- The process is then repeated for another increment of deformation and the evolution of the CPO is incrementally mapped out. Since the evolution of the CPO depends critically on the relation of one increment of imposed deformation to the next increment, the resulting pattern is sensitive to the kinematics of the deformation and not the orientation of the final strain ellipsoid. Thus a simple shearing deformation history will result in a different pattern of CPO than a pure shearing deformation history even though the strain ellipsoids that result from these two different histories are identical.

In order to understand the basic principles involved in the production of CPO by crystal lattice rotations arising from slip, consider the two dimensional situation depicted in Figure 13.27(a) where a single grain with one slip system, α, is shown. The arguments

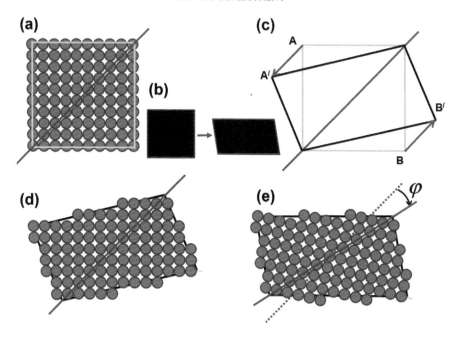

FIGURE 13.27 The principles behind the development of a CPO by crystal slip. (a) The initial crystal lattice within a square grain. The single potential slip plane is shown by the red line. (b) The imposed deformation; the black square is to become the black parallelogram (both in shape and orientation) by operation of the single slip plane. (c) The simple shearing operation parallel to the red line ($A \to A'$; $B \to B'$) that produces the required deformation but the orientation of the resulting parallelogram is wrong. (d) The crystal lattice corresponding to the deformation in (c). Although the lattice has experienced slip during simple shearing, it remains the same crystal lattice. (d) In order to arrive at the imposed deformation in the required orientation, the crystal lattice needs to be rotated through an angle, φ. This rotates the slip plane towards the shearing plane of the imposed deformation.

presented in Section 9.4 show that for an isochoric deformation, this single slip system is not sufficient to produce a general two dimensional homogeneous deformation.

Also shown is the crystal lattice defined by an array of circles. We impose a deformation given by

$$x_1 = a_{11}X_1 + a_{12}X_2$$

$$x_2 = a_{21}X_1 + a_{22}X_2$$

The rotation associated with this deformation is

$$\Omega_{ij} = \begin{bmatrix} 0 & \frac{1}{2}(a_{12} - a_{21}) \\ \frac{1}{2}(a_{21} - a_{12}) & 0 \end{bmatrix}$$

For the parameters $a_{11} = 1.28$, $a_{12} = -0.12$, $a_{21} = 0$, $a_{22} = 0.78$, the grain in Figure 13.27(a) becomes the parallelogram as shown in Figure 13.27(c). We also see from Section 9.4 that the strain that can be produced from the single slip system, α, is given by

$$
\varepsilon_{ij} = \begin{bmatrix} \gamma^\alpha m_1^\alpha s_1^\alpha & \dfrac{\gamma^\alpha}{2}\left(m_1^\alpha s_2^\alpha + m_2^\alpha s_1^\alpha\right) \\[2ex] \dfrac{\gamma^\alpha}{2}\left(m_1^\alpha s_2^\alpha + m_2^\alpha s_1^\alpha\right) & -\gamma^\alpha m_1^\alpha s_1^\alpha \end{bmatrix}
$$

The rotation that can be produced from the action of the slip system α is given by

$$
\omega_{ij} = \begin{bmatrix} 0 & \dfrac{\gamma^\alpha}{2}\left(m_2^\alpha s_1^\alpha - m_1^\alpha s_2^\alpha\right) \\[2ex] -\dfrac{\gamma^\alpha}{2}\left(m_2^\alpha s_1^\alpha - m_1^\alpha s_2^\alpha\right) & 0 \end{bmatrix}
$$

The rotations involved are illustrated in Figures 13.27 and 13.28.

13.5.2 Relaxed Constraints

If grains become flat or spindle shaped during deformation it is possible, as an approximation, to neglect some of the components of the imposed strain (Kocks and Canova, 1981). Thus if the grains become pancake shaped, then the two shear strain components oriented normal to the plane of the pancake can be neglected (Figure 13.29(b)). This means the number of strain components for an isochoric deformation is reduced from five in the general case to three and hence only three independent slip systems are needed throughout much of the grain (Figure 13.29(a)). For a spindle shaped grain Kocks and Canova (1981) argue that only one strain component can be neglected, namely the shear stress on the plane normal to the extension direction. Hence for an isochoric axial extension deformation, four independent slip systems are required. Kocks (1998b) points out that for a flattening deformation of an aggregate of initially equiaxed grains, the number of independent slip systems decreases from five at low strains to three at high strains. Thus the mechanisms of producing the CPO and hence the CPO may change during the deformation history. Figure 13.29(c) compares the CPOs that result from deformation of a quartzite for the strict Taylor–Bishop–Hill model and for a relaxed constraint model (Ord, 1988).

13.5.3 Self-Consistent Theories

The models discussed above are based on the Taylor constraint that all grains in the deforming aggregate undergo the same strain. Clearly this is not true so that the Taylor model is one bound on what actually happens in natural aggregates. Another way of calculating CPO development is based on what are referred to as *mean field* or *self-consistent* approaches (Molinari et al., 1987; Tome and Canova, 1998). These methods recognise that the stress and strain fields within a deforming polycrystal will be inhomogeneous and they consider the deformation of each grain, or of a local group of grains, in isolation and assume that the behaviour of this subset of the system is influenced by the average behaviour of

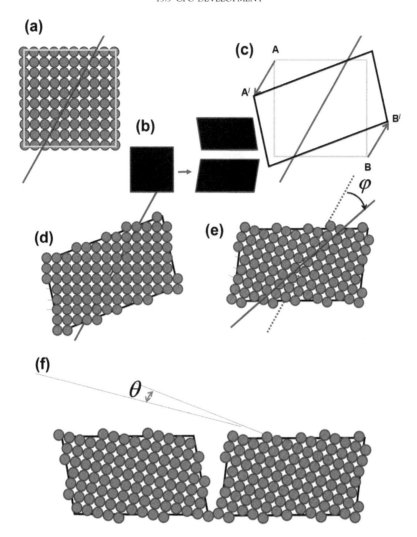

FIGURE 13.28 The example of Figure 13.27 but with a different slip system. (a) The initial lattice with slip plane marked in red. (b) The imposed deformation is represented by the upper parallelogram in (b) but cannot be achieved with the new slip system. With the same amount of simple shearing as in Figure 13.27 and in the same sense, the deformation is as indicated. (c) The deformation achieved by the simple shear indicated. (d) The lattice in the unconstrained state. (e) The lattice in a constrained state. The angle of rotation, φ, is larger than in Figure 13.27(d). (f) The results of Figures 13.27 and this figure compared. The rotated lattices are close to parallel with a misfit of θ but the deformation in each grain is entirely different and hence large incompatibility exists. This incompatibility is eliminated if two independent slip systems are available or if a disclination array in the grain boundary is introduced.

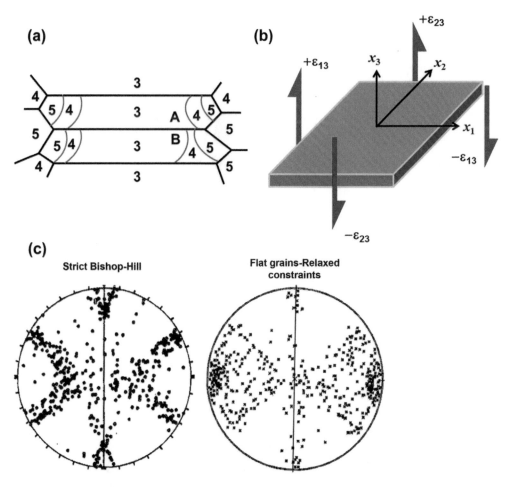

FIGURE 13.29 Relaxed constraints model. (a) Flat grains with the number of slip systems operating in each part of a grain. *Modified from Kocks and Canova (1981)*. (b) A flat grain with the strain components that are neglected in the relaxed constraint model. (c) Comparison of the strict Bishop–Hill (left) and relaxed constraint model (right) for a quartzite (Model 4 of Lister and Hobbs, 1980) deformed in pure shearing. The Bishop–Hill result is from Lister and Hobbs (1980). The relaxed constraint result is from Ord (1988). Equal area projections.

the surrounding medium. The surrounding medium is treated as though it is homogeneous and is commonly called the *homogeneous equivalent medium* (HEM). If single grains are treated, the methods are relevant to monophase polycrystals; if aggregates of grains form the subset, then polyphase materials can be treated.

The subunit is embedded in the HEM and the mechanical interaction between the subunit and the HEM is treated as though the subunit is an *Eshelby inclusion* (Eshelby, 1957). This means that the stress–strain field within the subunit is homogeneous and so the activation of slip (or twinning) systems can be calculated along with rotations arising from slip (or twinning). Deformation compatibility is ensured throughout the polycrystal since the velocity fields are continuous. The essential assumption is that the material behaviour is linear as is the case with ideal elasticity or Newtonian viscous materials. Errors arise for nonlinear materials but as long as this is appreciated, some corrections can be made (Canova, 1994). Some calculations for CPO development in quartzites are given in Figure 13.30(b) for simple shearing deformations and compared with results from the Taylor model. A number of deformation modes are explored and for details see Wenk et al. (1989). In particular, the effect of introducing visco-plasticity rather than assuming rate-independent plasticity is presented by changing the value of the stress exponent, N, in the power law constitutive relation. Here N = 99 or 31 corresponds to plasticity with low rate sensitivity, whereas N = 3 corresponds to high rate sensitivity. Examples of the use of self-consistent methods for calculating CPO development in other minerals are given in Wenk (1999), Wenk and van Houtte (2004) and for olivine by Tommasi et al. (2000).

The calculations shown in Figure 13.30 are important because they provide one of the few ways of checking on the extrapolation of the stress exponents, N, established by experimental methods to geological strain rates. In Chapter 9, we discuss the calculations of Cordier et al. (2014) that indicate experimental values of N, established at laboratory accessible strain rates may be too low when extrapolated to geological strain rates where deformation may become rate insensitive, implying large values of N. In Chapter 6, we indicate that the yield surface for large values of N becomes quite angular when compared to the corresponding yield surface for low values of N, and the calculations of Wenk et al. (1989) show the influence of such changes of N on the CPO for quartz. The self-consistent and Taylor models can give quite different CPOs for large values of N and so it is important to explore these results in greater detail to see if a test for the Cordier calculations can be developed.

13.5.4 Single Slip Theories; The Role of Disclinations

Although five independent slip systems are required (the *von Mises condition*) for a general homogeneous strain of equiaxed grains in order to preserve compatibility of deformation throughout the aggregate, we have seen that these strict conditions can be relaxed if the grain shape becomes flat or spindle shaped (the relaxed constraint conditions) or if one is prepared to accept some degree of inhomogeneous deformation as in the self-consistent approaches. However there are simple situations where the strict requirement for five independent slip systems is relaxed simply because the imposed deformation is such that five independent systems are not required to be compatible with that particular deformation. Thus for a general plane isochoric strain, the number of independent strain components is reduced to two (Hobbs, 1985). Such deformations clearly include simple shearing but also include plane deformations involving shearing on a single plane and shortening normal to that plane. In quartz, slip on the basal plane in the direction of a combination of two **a**-axes constitutes two independent slip systems (Paterson, 1969; Lister et al., 1978) so that some plane deformations can be achieved with large contribution only from basal **a** slip. Similarly slip on the first-order prisms in combinations of **a** and **c** slips constitutes four independent slip systems (Paterson, 1969) so many plane deformations can be achieved solely with prism slip.

These conditions for quartz can be seen from the sections through the yield surface for quartz shown in Figure 13.26. If the specimen axes are taken to be parallel to the x_1, x_2 and x_3 axes used to define the stress field, then for simple shearing on the $x_2 - x_3$ plane in the direction of x_2, and for the situation where all slip systems slip with equal ease, the slip systems

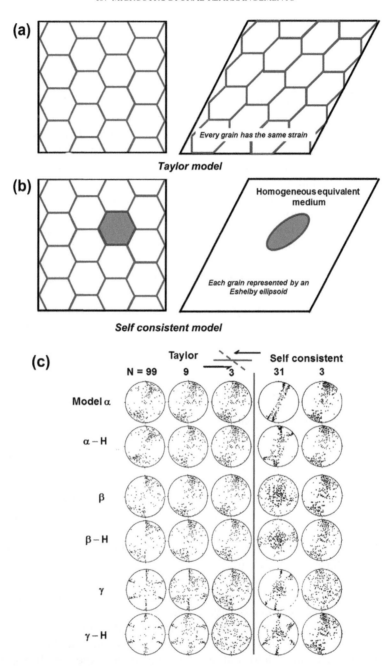

FIGURE 13.30   The self-consistent model for CPO development. (a) The Taylor model. (b) The self-consistent model. (c) Comparison of results from the Taylor and self-consistent models for a simple shearing deformation history of a quartzite. N is the assumed power law stress exponent and three models, α, β, γ, are presented. H stands for simulations with microstructural hardening. Equal area projections. *From Wenk et al. (1989).* See that paper for details.

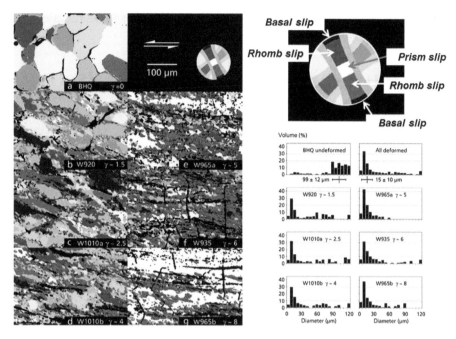

FIGURE 13.31 Microstructure developed in deformed Black Hills quartzite. *See Heilbronner and Tullis (2006) for details.* Deformation involves shearing parallel to the lower boundaries of each frame with some shortening vertically. Shear strain is marked as γ on each frame. Crystallographic preferred orientation is indicated in top right frame. Postulated dominant slip system is indicated in the right-hand inset together with estimated grain size distributions for each shear strain. *From Heilbronner and Tullis (2006).*

necessary to produce the deformation can be $(01\bar{1}0)[-\mathbf{c}]$, $(01\bar{1}0)[-\mathbf{c} - \mathbf{a}_1]$ and $(01\bar{1}0)[-\mathbf{c} + \mathbf{a}_1]$ or $(0001)[-\mathbf{a}_2]$, $(0001)[-\mathbf{a}_3]$ and $(01\bar{1}0)[-\mathbf{c}]$. In the latter case, almost all of the deformation can be achieved by basal **<a>** slip; the prism$[-\mathbf{c}]$ slip is necessary for local deformation compatibility.

An experimental example of CPO development in a deformation that approximates simple shearing is shown in Figure 13.31 from Heilbronner and Tullis (2006). The postulated dominant slip mechanism is shown for each shear strain. The slip system combinations progress from ones where basal ⟨**a**⟩ and rhomb ⟨**a**⟩ slip systems are present at low shear strains to combinations dominated by rhomb ⟨**a**⟩ slip to patterns dominated by prism ⟨**a**⟩ slip at high shear strains. The grain size decreases progressively as the shear strain increases, but grain size for a given shear strain is also controlled by crystallographic orientation. The rate of change of grain size with strain is rapid at first but becomes more or less independent of strain at high shear strains. These trends resemble those observed in metals (Figure 13.11).

An additional means of accommodating strain incompatibility in situations where five independent slip systems are not available is to introduce disclinations into the grain boundaries. Arrays of disclination dipoles in olivine grain boundaries have been documented by Cordier et al. (2014) and some aspects of disclination arrays are illustrated in Figures 13.23

and 13.24. Disclinations are considered in Chapter 9. We consider these situations under the topic of *rotation recrystallisation* in Section 13.6.

13.6 MECHANISMS OF DYNAMIC RECRYSTALLISATION

Dynamic recrystallisation in minerals has been examined by a large number of authors and for discussions and reviews, one should consult Drury et al. (1985), Hirth and Tullis (1992), Drury and Urai (1990), Stipp et al. (2002a) and Drury and Pennock (2007). Three main mechanisms have been recognised although they may be expressions of the same process operating at different scales. One involves classical nucleation and growth where a 'strain free' grain nucleates in a strongly deformed matrix and subsequent growth is driven by the stored energy of deformation in the matrix (see Section 13.4.2). A second involves *bulge nucleation* (Figure 9.22) whereby a part of an existing grain boundary (which may have been pinned by small particles of another phase such as mica) in a deforming aggregate begins to bulge into an adjacent grain driven again by the difference in energy across the interface (see Figure 13.10(d)). A third mechanism, and the one we will concentrate on here, is called *rotation recrystallisation* (Figures 13.10(b) and (c)).

13.6.1 Rotation Recrystallisation

Rotation recrystallisation is a process whereby subgrains form in a deforming host grain and progressively change orientation by rotation whilst maintaining deformation compatibility with adjacent subgrains or the host. The concept, *rotation recrystallisation*, was christened by Poirier and Nicolas (1975) although experimental observations (Figures 13.11(b) and (c)) on the process were reported earlier by Hobbs (1968). In metals, rotation of subgrains to form highly misoriented boundaries is widely recognised in strongly deformed materials, but the process is not usually referred to as a recrystallisation process but rather as a deformation process (see for example, Hughes et al., 2003). The process seems to be very widespread in deformed rocks and metals and is important because it is evidently a process whereby a slip system can become aligned in many spatially associated grains so that the aggregate develops domains characterised by the activity of one slip system. An example is presented in Figure 13.31(g) and the process is illustrated in Figure 13.33(e)–(h). The common view of this process is that subgrains form by dislocation arrays localising in subgrain walls and as the deformation proceeds, the tilt across the subgrain walls increases until finally the misorientation is large enough that the former subgrain is recognised as a 'new grain'. Although such a view is undoubtedly correct in a broad sense, it lacks any insight into why or how the subgrain walls form in the first place, the manner in which misorientation progressively increases, controls on the subgrain/grain size and orientation distributions, why and how domains characterised by different slip systems develop and the formalism that would allow CPO development to be modelled.

In order to place the role of subgrain rotation into a broader context, we need to revisit some of the discussions in Chapters 8 and 9. At each point in a deforming body, the deformation is represented by the deformation gradient tensor, **F**, which for a general deformation with volume change has *nine* independent components (Chapter 2). Hence *nine* independent types of deformation need to exist to accommodate a general imposed non-isochoric deformation at that point. This can be achieved in a number of ways. The simplest way, for the plastic deformation of a crystal, is by a combination of the operation of

- *Five* independent slip modes involving the motion of edge and/or screw *dislocations*,
- *Three* independent rotation modes involving the formation of walls of wedge and/or twist *disclination* dipoles,
- *One* volume change mode involving the formation of vacancies or climb mechanisms; these constitute *dilclinations* (see Chapter 8).

Other combinations are possible: Although a dilclination cannot contribute to a rotation, it can contribute to an extension or shortening strain if the dilclinations are arranged in some form of spatial pattern. Thus dislocation climb can reduce the number of independent slip systems needed for a general deformation (Groves and Kelly, 1969). Also the sideways migration of a wall of disclination dipoles can contribute a shear strain (Cordier et al., 2014). From this point of view, grain/subgrain boundary migration acts as a powerful deformation mechanism. Thus the three independent shear strains needed for a general deformation can be achieved by a combination of slip systems and grain/subgrain boundary migration. The three independent normal strains can be achieved by a combination of dislocation and dilclination motions.

For an isochoric plane deformation, $\mathbf{F}$ has three independent components that can be expressed as two shearing deformations and one rotation mechanism. Thus a grain can accommodate an imposed isochoric plane deformation with the operation of two independent dislocation mechanisms and one disclination mechanism.

As an imposed deformation progresses, the deformation can initially be accommodated by dislocation motion. This commonly involves the formation of subgrain boundaries as described by the reaction—diffusion mechanisms (Chapter 9) of Walgraef and Aifantis (Aifantis, 1986, 1987; Walgraef and Aifantis, 1985a,b,c). The operation of five independent dislocation systems together with three independent rotations generated by low dislocation density tilt walls is enough to accommodate the imposed deformation. Some climb or the formation of vacancy arrays may be necessary if a volume change exists.

As the dislocation density within subgrain walls increases and dislocation cores begin to overlap, the system switches to disclination-dominated subgrain walls, which is expressed as large misorientations between adjacent subgrains. At this stage, the migration of subgrain walls is a powerful deformation mechanism and the formation of smaller and smaller subgrains as the deformation progresses becomes an efficient deformation process. This reduction in subgrain size is a direct consequence of coupled grain boundary migration processes (see Section 9.5.2).

In the past 20 years, a number of apparently unrelated approaches to the subject of rotation recrystallisation have been converging and a unified view of the mechanics of the process is emerging. The final word has not yet been said. We present an overview of progress in this regard in the hope that it may guide and encourage further detailed studies of this important process in deformed rocks.

The various strands of the approach derive first from the work of Ortiz and Repetto (1999) and Ortiz et al. (2000) on the development of subgrain structures as a process that enables compatibility with an imposed deformation to be achieved without the necessity for five independent slip systems at every point in the crystal and at the same time minimises the Helmholtz energy of the system. The second strand derives from the work of many people, summarised by Romanov et al. (2009), indicating that disclinations are important in the high strain deformation of many materials. As we have seen, just as a *dislocation* is a linear lattice defect that marks the boundary between a slipped region and an unslipped region in a crystal slip plane, a *disclination* is a linear lattice defect that marks the boundary between rotated and unrotated parts of a crystal lattice. Hence disclinations become important when one seeks mechanisms for lattice rotation. The third strand extends the work of Walgraef and Aifantis (Aifantis, 1986, 1987; Walgraef and Aifantis, 1985a,b,c) on the stability of populations of mobile and immobile dislocations and considers the deforming system as comprised of populations of mobile and relatively immobile dislocations together with populations of mobile and relatively immobile disclinations (Romanov and Aifantis, 1993; Seefeldt, 1998, 2001; Seefeldt et al., 2001a,b; Seefeldt and Aifantis, 2002; Seefeldt and Klimanek, 1998). The system is then analysed as though it is a nonlinear chemical system and conditions established for dominance and stability of one population or the other or as a reaction–diffusion system where patterning of dislocations and of disclinations can emerge. Another strand arises from the observation that coupled grain boundary migration must, in general, lead to subgrain or grain rotation (Cahn and Taylor, 2004). Thus as a subgrain shrinks in size, it must rotate in order to maintain continuity (see Chapter 9). Finally these diverse strands are brought together in comprehensive nonlinear models by Clayton et al. (2006) and Clayton (2011) and extended by Fressengeas et al. (2011). We summarise aspects of this work below.

Ortiz and Repetto (1999) and Ortiz et al. (2000) build on the framework established for the deformation of materials with nonlinear elastic constitutive behaviour which has been widely developed particularly for martensitic transformations (Bhattacharya, 2003). The nonlinear elastic theory (Ball, 1977, 2004; Ball and James, 1987) leads to a non-convex Helmholtz energy (Chapter 7) and in order to minimise this energy function, a sequence of substructures form, which have a fractal, branching geometry (see Figures 7.10, 7.20 and 7.21). Ortiz and co-workers extend this framework to include elastic–plastic materials where the nonlinear behaviour is introduced either as a geometrical softening arising from single slip and/or a constitutive softening arising from dislocation interactions (Ortiz and Repetto, 1999). They point out that substructure refinement is necessary in order for an array of deformation domains to constitute a deformation that is compatible with the imposed bulk deformation. In general, an imposed deformation gradient has nine independent components (Chapter 3), or eight if the deformation is isochoric, and hence eight or nine deformation *domains* need to form to accommodate the imposed deformation. If the deformation is an isochoric plane straining, then three levels of deformation domains are required for compatibility, involving the operation of two independent slip systems. These domains develop in a hierarchical, branching fashion (Figure 13.32(d)). This compatibility requirement is the *geometrical* basis for subgrain formation. For any imposed deformation, any finite array of deformation domains cannot completely match an imposed deformation (see Figure 13.33(d)) and so a fractal array of subgrains develops (see Figure 6 of Ball and James, 1987). Ortiz et al. (2000) proceed to show that deformation by subgrain development requires less work than if subgrains do not develop (Figure 13.32(f)). They show that a Hall–Petch effect can be derived from these processes (Figure 13.32(g)).

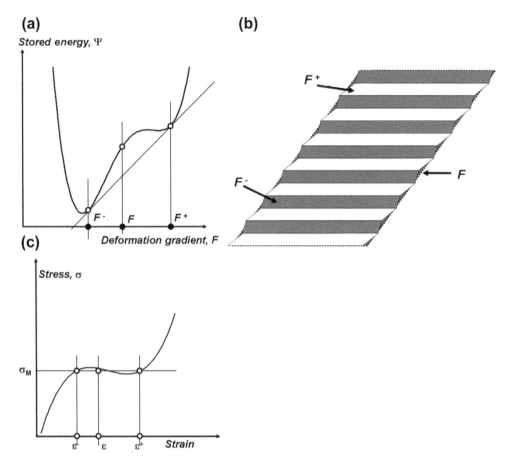

FIGURE 13.32 The formation of subgrains that minimises the stored energy of deformation expressed as a deformation compatibility problem. (a) A non-convex Helmholtz energy function. If the imposed deformation gradient is **F**, then the energy function can be minimised by combinations of domains where the deformation gradients are **F⁻** and **F⁺**. (b) A bulk simple shearing deformation, **F**, where the deformation is divided into two domains with deformation gradients **F⁻** and **F⁺**. (c) The stress—strain curve for the corresponding energy function in (a); σ_M is the Maxwell stress (see Section 7.7). (d) A branching hierarchy proposed by Ortiz and Repetto (1999) to describe the evolutionary development of subgrain structure. (e) A configuration of subgrains corresponding to the hierarchy in (d). (f) Stress—strain curves calculated by Ortiz et al. (2000) for a plane strain deformation of a single crystal of FCC material. If subgrains form, the material is weaker than without subgrains. Insets are shown the hierarchical development of subgrains. Each colour represents a different combination of slip systems within a subgrain. (g) Calculated stress—strain curves for different sizes of subgrains.

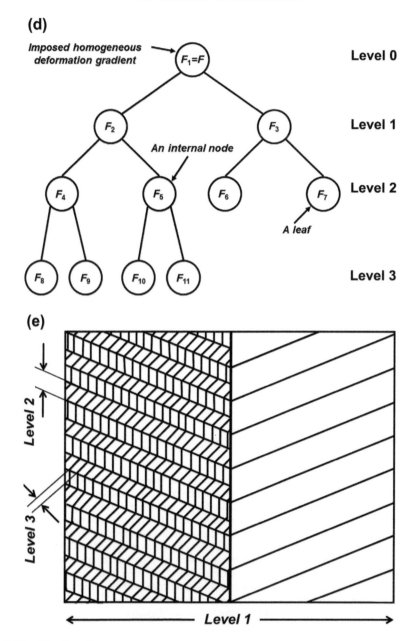

FIGURE 13.32   (Continued)

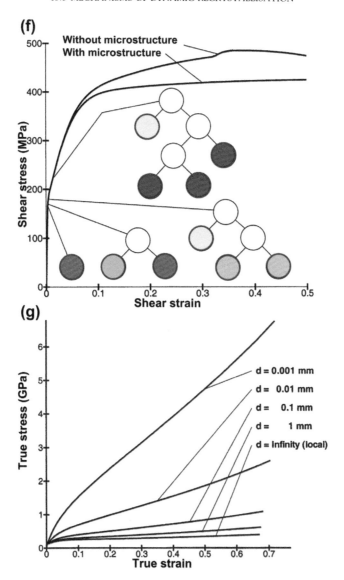

FIGURE 13.32 (Continued)

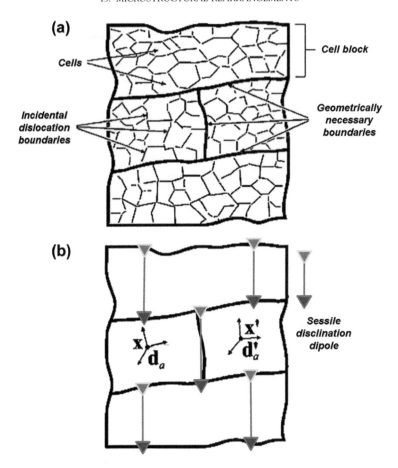

FIGURE 13.33 Subgrain formation and rotation. (a) Subgrain (cell) array showing geometry of incidental dislocation boundaries and geometrically necessary boundaries. (b) The role of disclinations. *From Clayton et al. (2006)*. (c) Three dimensional sketch of the relations between incidental dislocation boundaries and geometrically necessary boundaries. ND, TD and RD stand respectively for the normal, transverse and rolling directions for the deformation. D^{INB} and D^{GNB} are the spacings of incidental dislocation and geometrically necessary boundaries. *From Hughes and Hansen (2000)*. (d) Phase diagram. A plot of total disclination density against total dislocation density as controlled by the evolution equations, (13.25). Three deformation regimes are delineated: one with only dislocations, one dominated by dislocation motion and another dominated by disclination motion but with cooperative dislocation motion. (e) Initial array of grains with dominant slip system marked in white in each grain. Disclinations induce rotation under simple shearing as shown. (f) Rotation of dominant slip system into near parallelism. Initial connecting grain boundaries become subgrain boundaries. (g) Subgrain boundaries vanish; a group of grains becomes a ribbon of similar orientation throughout. (h) Shearing of ribbon by cooperative grain boundary sliding mechanism. Rotation of aggregate so that ribbons approach the eigenvector of the simple shearing deformation. *Inspired by Meyers et al. (2006)*.

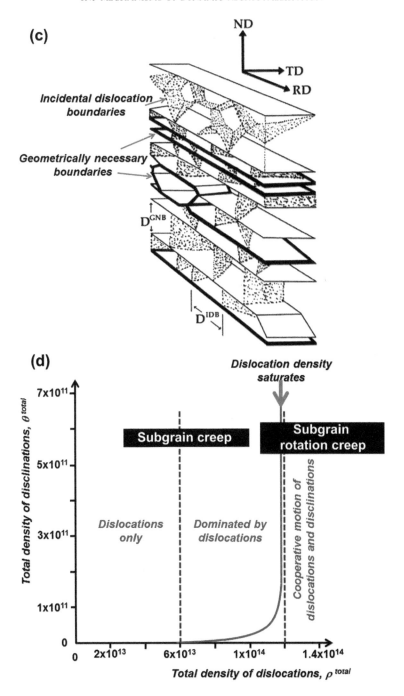

FIGURE 13.33 (Continued)

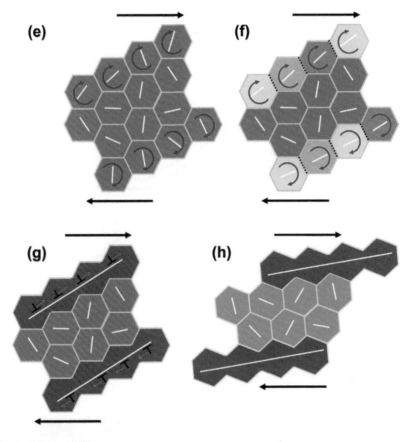

FIGURE 13.33   (Continued)

Most importantly, the arguments of Ortiz provide insight into why subgrain size depends on strain. Ortiz and Repetto (1999) express the total energy of a deforming crystal, E^{total}, as the sum of the surface energies of the crystal and the subgrains:

Surface energy of grain

| *Total surface energy* | *Surface energy of subgrains* | |
| | | |

$$E^{total} \quad = \quad C_1 \mu \gamma^2 L^2 l \quad + \quad C_2 \mu |b| \gamma \frac{L^3}{l} \quad ,$$

where C_1 and C_2 are constants, μ is the shear modulus, γ is the shear strain, $|b|$ is the magnitude of the Burgers vector, L is the grain size and l is the subgrain size. If we minimise E^{total} with respect to l, we obtain

$$l = C_3 \sqrt{\frac{|b|L}{\gamma}}$$

where $C_3 = (C_2/C_1)^{1/2}$. Thus the subgrain size decreases with the square root of the shear strain as discussed in Section 13.4. Further work in this regard is in Ortiz et al. (2000).

Although the work of Ortiz and Repetto (1999) and Ortiz et al. (2000) provides a foundation for the formation of subgrains based on the necessity for deformation compatibility and as a means of minimising the stored energy of the system, their treatment does not define the geometry of the spatial arrays of subgrains that may develop for a given deformation of a specific material. In order to do this, more detailed models of the kinetics of subgrain formation and evolution are required.

Such models are based on the experimental work of a number of workers (Bay et al., 1992; Driver et al., 1994; Hughes and Hansen, 2000; Hughes, 2001; Hughes et al., 1997, 1998, 2003) from which the model for strongly deformed materials shown in Figure 13.33(a) and (b) is developed by Seefeldt and Klimanek (1998) and Clayton et al. (2006). As indicated in Section 13.4, two classes of subgrain boundaries are recognised. One (Figure 13.33(a)) comprises of low angle boundaries between subgrains (or cells) where the misorientation between subgrains is typically less than 10°. These boundaries are made up of dislocation arrays and form at relatively low strains. Such boundaries are called *IDBs*. The cells are arranged into *cell blocks* (Figure 13.33(a)) that are commonly elongate. The boundaries of the cell blocks are *GNBs* that separate regions of completely different orientations. They are comprised of sessile disclination dipoles (Figure 13.33(b)) and form at large deformations.

Why do IDBs and GNBs develop and what controls their patterning? This question is answered in part by the work of Seefeldt and Aifantis. Seefeldt and Klimanek (1997) propose that the deformation of a material can be expressed in terms of competition between populations of mobile dislocations, relatively immobile dislocations, glissile disclinations and sessile disclinations with instantaneous densities (in units of m^{-2}) of ρ_{mobile}, $\rho_{immobile}$, $\theta_{glissile}$, $\theta_{sessile}$, respectively. The following evolutionary equations are derived to express the kinetics of competition between the production and destruction of the populations:

$$\frac{\partial \rho_{mobile}}{\bar{v}\partial t} = B + C\rho_{mobile} - D\rho_{mobile}^2 - A\sqrt{\rho_{immobile}}\rho_{mobile} - A'\rho_{immobile}\rho_{mobile}$$

$$- E_1\rho_{immobile}\rho_{mobile}\theta_{glissile} - F_1\rho_{mobile}\theta_{glissile} - J_1\rho_{mobile}$$

$$\frac{\partial \rho_{immobile}}{\bar{v}\partial t} = A\sqrt{\rho_{immobile}}\rho_{mobile}A'\rho_{immobile}\rho_{mobile}$$

$$\frac{\partial \theta_{mobile}}{\bar{v}\partial t} = E_2\rho_{immobile}\rho_{mobile}\theta_{glissile} + F_2\rho_{mobile}\theta_{glissile} + J_2\rho_{mobile} - H\theta_{glissile}\theta_{sessile}$$

$$\frac{\partial \theta_{immobile}}{\bar{v}\partial t} = H\theta_{glissile}\theta_{sessile} \tag{13.25}$$

where $\bar{v}$ is the mean dislocation velocity and the terms involving the constants A, B, C, D, E, F, H, J represent multiplication, annihilation and interactions between the various populations of dislocations and disclinations.

Solutions to these equations show that the system evolves from a dislocation dominated regime (characterised by deformations dominated by translation and low-angle cell walls) to a disclination controlled regime (characterised by deformations dominated by rotations and geometrically necessary high-angle boundaries). These solutions are illustrated in

Figure 13.33(d), which is a $\rho_{total}-\theta_{total}$ phase diagram. Two distinct fields are delineated; one is characterised by *subgrain creep* and the other by *subgrain rotation creep*. Some aspects of subgrain rotation creep are shown in Figure 13.33(e)–(h).

Rominov and Aifantis (1993) and Seefeldt and Aifantis (2002) add diffusion terms for dislocation and disclination densities to kinetic equations for dislocation and disclination evolution so that the system is described in terms of reaction–diffusion equations instead of reaction equations, such as (13.25). They show that spatial patterning can develop consisting of alternating regions dominated by either dislocations (subgrains with low angle boundaries) or disclinations (rotated subgrains). Unfortunately, the theory has not been developed to the stage where details of these spatial patterns are delineated. We have discussed subgrain rotation due to shrinking in size of the subgrains in Chapter 9. This is an area of active study at the time of writing (Trautt and Mishin, 2014) and one looks forward to future developments. The process supplies considerable insight into the mechanisms of subgrain rotation.

The models discussed above have been unified in the approach of Clayton et al. (2006) and Clayton (2011). They consider the rotation indicated in Figure 13.33(b) between two lattice direction vectors $\mathbf{d}_\alpha$ and $\mathbf{d}'_\alpha$ located at x and x', respectively. Clayton et al. (2006) express the density of disclinations in terms of this rotation which leads eventually to an expression for the non-recoverable energy function. This energy function is convex if only dislocations or only disclinations are present but becomes non-convex for mixtures of the two. Under such conditions, a spatial patterning of both dislocation and disclination densities is expected in order to minimise the energy function for a given imposed inelastic deformation and so the link back to the work of Ortiz is made. These models need to be further developed. A step in this direction is made in a continuum approach to combined dislocation/disclination motion by Fressengeas et al. (2011).

13.7 ANISOTROPY AND THE INFLUENCE OF ELASTIC ANISOTROPY AND FLOW STRESS ANISOTROPY ON LOCALISATION

As a CPO develops in an aggregate and/or as a preferred orientation of grain boundaries develops, the material becomes anisotropic with respect to both the elastic and visco-plastic properties. CPO-induced elastic anisotropy has been well studied over the past decade or so with respect to the development of anisotropy in seismic wave propagation speeds (Mainprice et al., 2011). The elastic anisotropy arises because individual crystalline minerals are intrinsically elastically anisotropic (Nye, 1957) and hence a preferred orientation of the crystal structures in an aggregate produces a bulk elastic anisotropy, which may be derived from the orientation distribution of crystal structures. Methods of calculating the elastic anisotropy function from the ODF and the elastic anisotropy of individual minerals are discussed in Mainprice et al. (2011). An example is shown in Figure 13.42. Although elastic anisotropy is important for interpretations of Earth structure using seismic methods (Mainprice et al., 2007, 2008), we do not discuss that aspect here. The importance of elastic anisotropy for deformed metamorphic rocks lies in the fact that elasticity plays a role in localising deformation within shear zones and hence any anisotropy influences both the conditions for localisation and the orientation of the shear band.

Just as a CPO induces elastic anisotropy arising from some average of the orientation distribution of the intrinsic elastic anisotropy of individual grains, a CPO induces anisotropy in the visco-plastic response of an aggregate arising from the intrinsic visco-plastic response of individual grains. The problem is that, although calculations to determine the bulk elastic anisotropy are based on well-established principles and are well bounded by models governing how to take reasonable averages (the Voigt and the Reuss models: Voigt, 1928; Reuss, 1929; Hill, 1952), no such averaging procedure is available for visco-plastic behaviour. The simplest procedure is that proposed by Hill (1948) where it is assumed that the yield surface is modified by the CPO as shown in Figure 13.34(d) where the isotropic von Mises yield surface changes shape but is still represented by an ellipse in the $(\sigma_{11}-\sigma_{33})$-plane (see Chapter 5). The yield surface is then given by a parabolic expression of the form:

$$F(\sigma_{yy} - \sigma_{zz})^2 + G(\sigma_{zz} - \sigma_{xx})^2 + H(\sigma_{xx} - \sigma_{yy})^2 + 2L\sigma_{yz}^2 + 2M\sigma_{zx}^2 + 2N\sigma_{xy}^2 = 1 \quad (13.26)$$

where F, G, H, L, M, N are parameters that characterise the current anisotropy. For (13.26) to be compatible with a von Mises yield condition, Section 6.6.2, the requirement is $F = G = H = L = M = N = 1/(2\sigma_{yield}^2)$, where σ_{yield} is the yield stress (Backofen, 1972).

Such an expression describes some relatively small deformation behaviour in metals but fails to explain some common observations at large deformations. An example is the development of 'ears' on the edges of a cup produced by punching a circular cross section die into a sheet of metal. The Hill parabolic model predicts six ears, whereas commonly eight are observed (Gambin, 2001). The approach has been to develop 'higher-order' models as the demands of large deformation forming applications have developed in industry. However as Stout and Kocks (1998) point out, it is doubtful that a unified analytical procedure for calculating yield surfaces will ever be developed since details of the CPO, which develops in individual cases, are sensitive to initial and boundary conditions.

The yield surface is also extremely difficult to establish experimentally even for ductile metals at room temperature and pressure. As far as we are aware, there are no experimental observations for rocks at high temperatures and pressures and at slow strain rates, although, as we point out in Section 6.6.4, the opportunity exists to establish yield surfaces for geologically relevant materials in high pressure, high temperature apparatus with combined torsional and shortening facilities. One experimental observation on visco-plastic anisotropy is that of Ralser et al. (1991) who deformed a quartz-mica mylonite normal, parallel and at 45° to the foliation and lineation and observed that at 700–800 °C the material was twice as strong normal to the foliation than in the 45° and parallel orientations. At 900 °C, the anisotropy is small. The fact that plastic yield surfaces are difficult to determine experimentally makes the development of rigorous and realistic (rather than simply convenient) computational models for CPO development of considerable importance.

An example of a yield surface, for an isotropic case, is given in Figure 13.34(a)–(c) together with early calculations of Bishop and Hill (1951a,b). For situations involving deformation of a layer (Figure 13.34(d)) where the anisotropy is rotationally symmetric about the normal to the layer, Backofen (1972, pp. 47–50) derives some expressions for H, G and F in (13.26) and so enables the construction of the yield surface for various degrees of

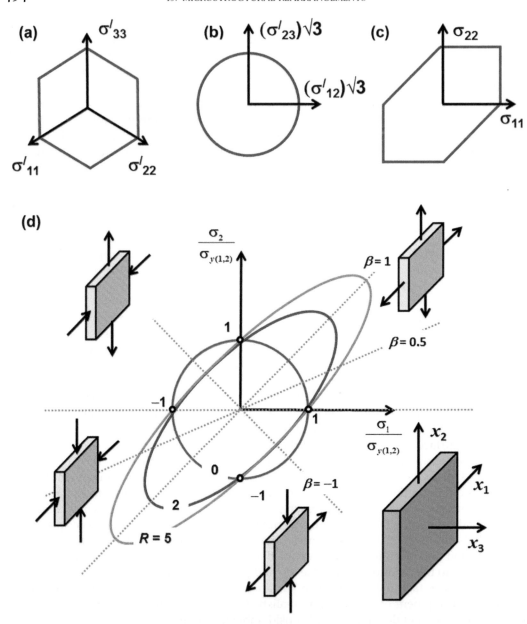

FIGURE 13.34   Sections through a yield surface for an isotropic material. (a) A section in the π-plane; σ'_{ij} are the deviatoric stresses. (b) Section in a space defined by two shear stresses. (c) Section in space defined by two normal stresses with the third one zero. *Modified from Stout and Kocks (1998).* (d) Yield loci for plane strain deformations and materials with planar anisotropy or anisotropy with rotational symmetry about x_3. R is the degree of anisotropy, $R = d\varepsilon_2/d\varepsilon_3$ and β is the stress ratio, $\beta = \sigma_2/\sigma_1$. $\sigma_{y(1,2)}$ is the yield stress in either the x_1 or x_2 directions. *Inspired by Backofen (1972).*

anisotropy, measured by a parameter, R, and for various loading paths, measured by a parameter, β:

$$R = \frac{d\varepsilon_2}{d\varepsilon_3} \ along \ X_1 \quad and \quad \beta = \frac{\sigma_2}{\sigma_1}$$

where the coordinate axes, X_1, X_2 and X_3 are shown in Figure 13.34(d). For the geometrical situation shown in Figure 13.34(d), Backofen shows that

$$R = \frac{H}{G} = \frac{H}{F} \quad and \quad H = \frac{1}{\sigma_{Y(1)}^2} \frac{R}{1+R}$$

where $\sigma_{Y(1)}$ is the yield stress in the X_1 direction.

An example of a yield surface formed where no CPO develops but grain shape anisotropy does is shown in Figure 13.35(a). Various other examples of anisotropic yield surfaces are shown in Figures 13.35 and 13.36 including the influence of twinning (Figure 13.36).

13.7.1 Anisotropy Induced by Diffusion

At first thought, one might expect that if diffusion dominates the deformation process then the mechanical properties may remain isotropic. Wheeler (2010) has shown that for hexagonal-shaped grains, strong anisotropy in the viscosity can arise associated with strong grain rotation. His results are summarised in Figure 13.37. In Figure 13.37(a) and (b), arrays of differently shaped hexagons are shown under pure shearing and simple

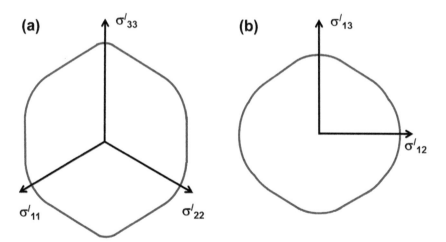

FIGURE 13.35 (a) The influence of grain shape on the yield surface of a polycrystal with random CPO. The result is for the Bishop-Hill model under relaxed constraints. (b) The effect of twinning during deformation in a shear stress section. *Modified from Stout and Kocks (1998).*

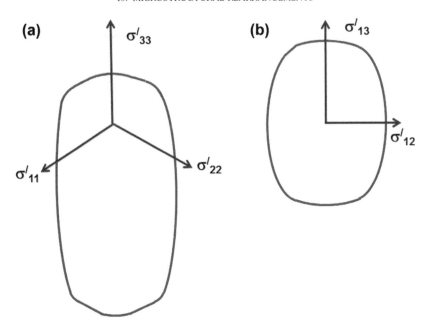

FIGURE 13.36 Sections through a yield surface where twinning is an important mechanism. Simulations. (a) and (b) 20% extension. In sections involving the normal stresses, the section lacks a centre of symmetry. These effects are expected in materials where twinning is common and include calcite, quartz and olivine. *Modified from Stout and Kocks (1998).*

shearing deformations. Grains can undergo rotation as shown, sometimes in counter-intuitive senses; the angular velocity is quoted as ω in each frame and the precipitation rates (rates of removal or addition of material during deformation) are quoted as a number on symmetrically equivalent boundaries. Figure 13.37(c) and (d) shows the anisotropy of viscosity for two different grain shapes as shown. The numbers on the contours represent values of a parameter, ζ^*, that is directly related to the effective viscosity of the grain boundaries. For low values of ζ^*, the material has low resistance to shearing in some directions. Even at relatively high values of ζ^*, the material remains anisotropic. In Figure 13.37(e) and (f), the progressive change in grain shape is shown for simple shearing and for pure shearing. The resultant normal stress—strain and shear stress—strain curves are shown in Figure 13.37(g), where various patterns of hardening and softening are shown depending on the orientation of grain boundaries relative to the deformation axes. The softening behaviour could initiate shear localisation if the softening becomes critical (see Section 13.7.1).

The rotations considered by Wheeler together with the coupled diffusive transport are controlled by grain shape so that if all grains are the same shape, no CPO develops by this mechanism. However if grain shape is controlled by crystal structure, then it is possible that a CPO can develop. An alternative mechanism that has not been explored is that the 'plating' and 'dissolution' processes operating at grain boundaries may be inhomogeneous so that some of the incompatibility arising from grain rotation is accommodated by disclinations. Then the development and motion of disclinations with preferred Frank vectors, facilitated by diffusion, could produce a CPO. The subject needs further investigation. The process discussed by Wheeler (2010) also needs to be considered in terms of the coupled grain boundary migration process (Chapter 9) considered by Cahn and Taylor (2004).

13.7.2 The Localisation of Deformation

It is commonly proposed that some form of weakening or softening is the cause of localisation of deformation in deformed rocks and that hardening is not associated with localisation. The purpose of this section is to explore such propositions in some detail. We discuss the conditions for localisation and, in particular, the influence of elastic anisotropy on the

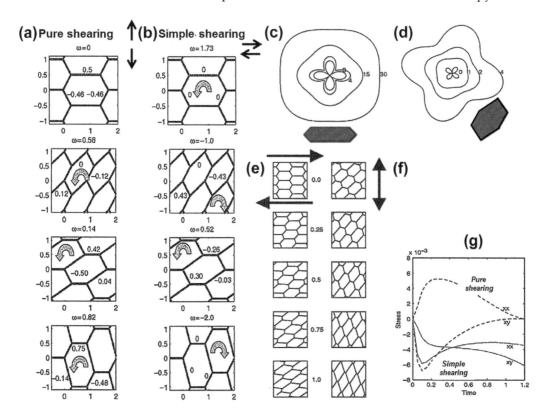

FIGURE 13.37 Rotation and anisotropy produced by diffusion. (a) and (b) Rotations induced by diffusion coupled to changes in grain shape for pure shearing and simple shearing histories. The numbers on grain boundaries indicate the rates of removal or of addition of material at grain boundaries. ω is the angular velocity. (c) and (d) Surfaces expressing the anisotropy of viscosity for the two grain shapes indicated in red. The contours are for various values of the parameter ζ^*, which is a measure of the viscosity. (e) and (f) Progressive changes in grain shape for simple shearing and pure shearing. (g) Stress time curves for the two situations shown in (e) and (f). Both hardening and softening behaviours are developed. *After Wheeler (2010).*

conditions for localisation and on the orientation of the resulting shear bands. The overall result is that *softening* is a *necessary* condition for localisation in a material with *associative* constitutive behaviour. However it is not a *sufficient* condition and the hardening modulus must reach a critical value (which depends on the elastic anisotropy) for localisation to occur. In materials with *non-associative* behaviour, the conditions for localisation are similar except that localisation can occur in the *hardening* regime.

The array of localised responses to deformation is illustrated in Figure 13.38 and some models of the resulting strain distributions are discussed by Ramsay (1980b) and Carreras (2001) who point out that the patterns of localisation can be self-similar on various spatial scales. For compatibility of deformation between the shear zone and the host material, the extension parallel to the shear zone boundary must be the same in the sheared material and in the host. Otherwise, there are two possibilities. One is that compatibility is achieved by a wide strain gradient in which case the localisation is diffuse (Figure 13.38(a)). Second, a

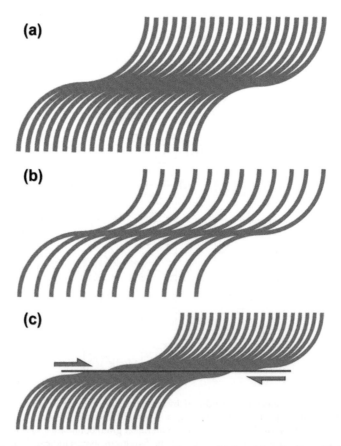

FIGURE 13.38   Styles of localisation. (a) Diffuse localisation. (b) Local localisation. (c) Fault-accommodated incompatibility.

fault is initiated at the boundary of the shear zone so that there is a jump in the deformation at the boundary or within the shear zone (Figure 13.38(c)). Compatibility can be achieved across a sharp boundary with no faulting (Figure 13.38(b)) as long as the compatibility condition (Section 2.10) is obeyed:

$$[\![\mathbf{F}]\!] = [\![x,_N]\!] \otimes \mathbf{N}$$

This condition forms the basis for a number of models of increasing complexity. The simplest model is to consider the host as deforming elastically whilst the localising zone undergoes both elastic and plastic deformation. This model is exemplified in the discussion by Backofen (1972) and is important because it brings out the fundamental features of all models. The second level of complexity is to assume the host and shear zone deform in an elastic–plastic manner but both elastic and plastic properties are isotropic. This is the model considered by Needleman and Ortiz (1991) and Rudnicki and Rice (1975). The third level of complexity is to consider elastic anisotropy. This is the model discussed by Bigoni and Loret (1999) and Bigoni et al. (2000). We consider each of these in turn below.

13.7.2.1 *The Backofen Model*

Backofen (1972, pp. 204–210) considers an anisotropic elastic–plastic model involving a deforming plate, where the yield surface is not an ellipsoid and is shown in Figures 13.2(b) and 13.39(d). Although the resulting localisation involves necking, the overall principles involving strain compatibility are just as applicable to shear localisation as to boudin development. The discussion centres around conditions for deformation compatibility between a developing zone of localisation and the elastically deforming surrounding host material. Also, since the boundaries of the shear zone are planar, compatibility demands that the plastic strain normal to the plane of the diagram (in the plane of the localisation zone and parallel to z) in the host material is zero and hence the deformation in the host is an elastic plane strain. For diffuse necking, it is only necessary that thinning occurs. For localised necking, we must also have $\delta\varepsilon_2 < 0$ (shortening positive). Because there is no necking without some form of thinning, the arrow representing $\delta\varepsilon_2$ in Figure 13.39(d) must always be shorter than the arrow representing $\delta\varepsilon_1$, which must be an extension. Thus at A in Figure 13.39(d), $\dfrac{\delta\varepsilon_2}{\delta\varepsilon_1} = -1$, and the angle between the plane of localisation and the extension axis, ϕ, is 45°. At B, $-1 < \dfrac{\delta\varepsilon_2}{\delta\varepsilon_1} < 0$ and $45° < \phi < 90°$. At C, $\dfrac{\delta\varepsilon_2}{\delta\varepsilon_1} = 0$ and $\phi = 90°$. Only diffuse necking is possible for $0 < \dfrac{\delta\varepsilon_2}{\delta\varepsilon_1} < 1$ in the range marked $\widehat{\varphi}$ in Figure 13.39(d) and either diffuse or localised necking is possible in the range φ.

13.7.2.2 *Localisation in an Elastic–Plastic Material*

In order to consider the case where localisation develops in an elastic–plastic material, we need to recall (Chapters 2 and 9) that a general deformation can be written as

$$\mathbf{F} = \mathbf{F}^{elastic}\mathbf{F}^{plastic} \tag{13.27}$$

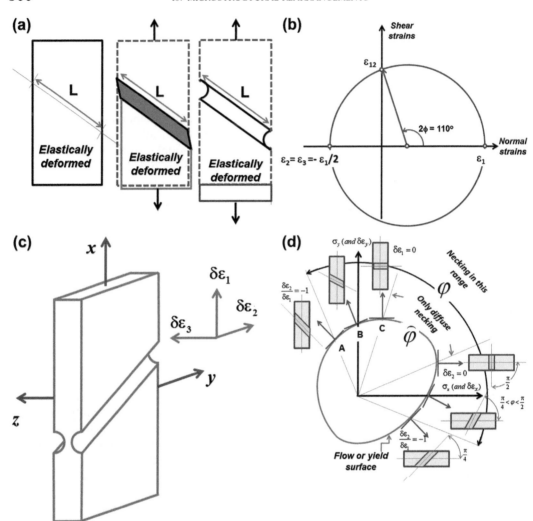

FIGURE 13.39   The Backofen (1972) model for necking localisation. The material is isotropic elastically and anisotropic plastically. (a) The basic principle is that compatibility is maintained across the interface between the elastically deformed host material and the elastic-plastically deformed zone of localisation. This means that the length L measured parallel to the incipient localisation zone in the elastically deformed host is equal to this same length in the localised zone independently of whether the mode of localisation is shearing or necking. (b) Mohr diagram showing that if the compatibility condition in (a) is fulfilled, then the normal to the plane of localisation makes an angle $\phi = 55°$ with the axis of extension. (c) Coordinate axes for localisation by necking. (d) The yield surface for the anisotropic material showing orientations of localisation zones depending on the strain state. Diffuse or localised necking can occur only in the region marked φ. In the region marked $\widehat{\varphi}$, only diffuse necking is possible.

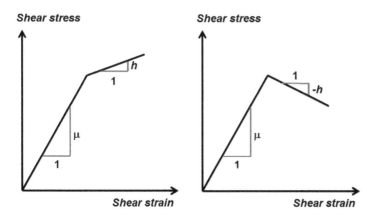

FIGURE 13.40 Definition of the elastic shear modulus, μ, and the plastic hardening modulus, h. The conditions for localisation depend on the ratio, h/μ. Localisation occurs when this ratio reaches a critical value. This critical value is commonly positive (left) for non-associative materials and negative (right) for associative materials.

This means, different to the Backofen model, that the plastic strain parallel to the shear zone boundary need no longer be zero to achieve compatibility and any incompatibility in $\mathbf{F}^{plastic}$ can be offset by an elastic deformation, $\mathbf{F}^{elastic}$, as long as the local stress arising from such deformation does not exceed the local yield or failure stress in which case the shear zone spreads laterally (becomes diffuse) or fractures appear on the shear zone boundary.

Now consider a situation where the material is deforming homogeneously in an elastic manner and, locally, yielding begins with a (positive) hardening modulus, h (Figure 13.40(a)). If the stress is σ, then the elastic strain at that stress is $\varepsilon^{elastic} = \frac{\sigma}{\mu}$, whereas the plastic strain is $\varepsilon^{plastic} = \frac{\sigma}{h}$. Thus if h is large enough, the plastic strain is comparable to the elastic strain and any incompatibilities in the deformation can be marginally accommodated by elastic strains. At that stage, it requires less work for continued yielding to begin elsewhere and the deformation remains diffuse. If the plastic strain is small when compared with the elastic strain at a given stress, then the incompatibility can be minimised by the shear band boundary adopting an orientation that as closely as possible minimises the misfit between that plastic deformation either side of the boundary. Any misfit that remains is accommodated by elastic strains that produce stresses small enough that further plastic strain does not occur. As indicated, if the stresses were to be equal to the yield stress then the plastic deformation spreads further into the host and produces a diffuse shear band. Thus one can see intuitively that any criterion for shear band formation must involve a consideration of the elastic moduli.

As indicated in Figure 13.2, both elastic and visco-plastic anisotropies play a role in controlling the development and orientation of shearing localisation during deformation. The theory behind such behaviour has been established by a large number of workers including, in particular, Needleman and Ortiz (1991) and Rudnicki and Rice (1975) for the isotropic case and Bigoni and Loret (1999) and Bigoni et al. (2000) for some forms of elastic anisotropy. We defer a detailed discussion of these theories to Volume II and instead present a qualitative discussion of localisation. The essence of the theoretical results is that localisation occurs in an elastic–plastic material with elastic shear modulus, μ, undergoing strain hardening (or softening), with a hardening modulus, h (Figure 13.40), when the ratio (h/μ) decreases to a critical value. For an isotropic material with associative plasticity, the Rudnicki–Rice condition for localisation is

$$\frac{h^{critical}}{\mu} = -\frac{1+\nu}{3} \quad \text{for which} \quad \theta_N = \tan^{-1}\sqrt{\frac{1+\nu}{2-\nu}}$$

where ν is Poisson's ratio and θ_N is the angle between the normal to the shear zone and the compression axis (Figure 13.41(b)). As an example, if $\nu = 1/3$, then for isotropic elasticity, $h^{critical} = -(4\,\mu/9)$ and $\theta_N = 41.8°$. For an associated plastic material, h is negative (strain softening) for an instability to occur but h does not need to be negative for localisation to occur in a non-associated material as shown theoretically by Rudnicki and Rice (1975) and experimentally by Ord et al. (1991).

Bigoni and Loret (1999)consider an elastic–plastic material with associated plasticity and a restricted form of elastic anisotropy described in Figure 13.41(a) and represented by an ellipsoid with b_1 the dimension in the direction of the axis of rotational symmetry and $b_2 = b_3$ normal to this axis. This form of anisotropy is quite similar to that reported for an anorthosite by Mainprice and Munch (1993) shown in Figure 13.42 and perhaps is common in deformed rocks. The anisotropy is characterised by a parameter b given by

$$b_1 = \sqrt{3}\cos b \quad \text{and} \quad b_2 = \sqrt{3}\sin b$$

so that elastic isotropy corresponds to $b_1 = b_2 = 1$ and so $b_{iso} = 54.74°$.

For this model of elastic anisotropy adopted by Bigoni and Loret (1999), the conditions for localisation are shown in Figure 13.41(c) and (d). For instance, as $b \to 90°$, or $b_2 \to 0$:

$$\text{For } b < b^{isotropic} : \quad \frac{h^{critical}}{\mu} = -\frac{\left(1+\nu\right)b_2^2}{3} \quad \text{for which} \quad \theta_N \approx \theta_N^{isotropic}$$

$$\text{For } b > b^{isotropic} \quad \text{and} \quad \theta_\sigma \geq 35.26° \quad : \quad \frac{h^{critical}}{\mu} \to 0$$

$$\text{For } b > b^{isotropic} \quad \text{and} \quad \theta_\sigma \leq 35.26° \quad : \quad \frac{h^{critical}}{\mu} = -\frac{2\tilde{Q}_1^2}{(1-\nu)}$$

In these equations, $\tilde{Q}_1 = \frac{1}{2}\sqrt{\frac{3}{2}(S + \sqrt{\Delta})}$ with $S = b_1\left(\cos^2\theta_\sigma - \frac{1}{3}\right) + b_2\left(\sin^2\theta_\sigma - \frac{1}{3}\right)$ and $\Delta = S^2 + \frac{8}{9}b_1 b_2 > 0$. Other conditions are discussed by Bigoni and Loret (1999).

The above example concerns conditions for localisation where the hardening modulus is always negative. If the material is non-associative as far as constitutive behaviour is concerned, then localisation is possible in the hardening regime as shown in Figure 13.43. The material here corresponds to a Drucker–Prager material with the equivalent of the friction angle, $\psi = 30°$ and the equivalent of the dilation angle, $\chi = 0°$. Poisson's ratio is $1/3$. The various quantities are defined in Figure 13.43(a). A material with a single plane of anisotropy (corresponding, say, to a foliation plane) is considered. This anisotropy is characterised by a parameter $\hat{b}$, similar to the parameter b discussed above. Localisation can occur for $h > 0$

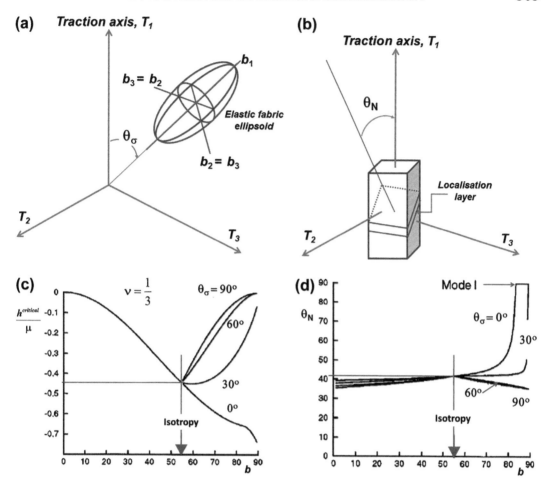

FIGURE 13.41   An example of localisation in materials with associative plasticity and anisotropic elasticity. The example has Poisson's ratio $= 1/3$. *From Bigoni and Loret (1999).* (a) Definition of terms associated with the orientation of the elastic anisotropy ellipsoid. T is the applied traction with T_1 an axis of maximum compression. θ_σ is the angle between the axis of rotation for the ellipsoid and T_1. (b) Definition of terms associated with the orientation of the shear band. θ_N is the angle between the normal to the shear band and T_1. (c) Strain localisation for an associated von Mises material. Results for Poisson's ratio, 0.33, and the type of elastic anisotropy shown in (a). Plot of the normalised critical hardening modulus, $h_{critical}/\mu$, against a measure of the anisotropy, b. Isotropy corresponds to $b = 54.74°$; for $b < 54.74°$, the elasticity ellipsoid is pancake shaped, for $b > 54.74°$ the elasticity ellipsoid is cigar shaped. For $b > 54.74°$, $h_{critical}/\mu$ can be greater than the isotropic value depending on the value of θ_σ. (d) Plot of θ_N against b for an associated von Mises material with the elastic anisotropy shown in (a). For $b > 54.74°$, this angle can be zero (Mode I orientation) for low values of θ_σ. For $b > 54.74°$ and for large values of θ_σ, θ_N is close to the isotropic value of 41.8°.

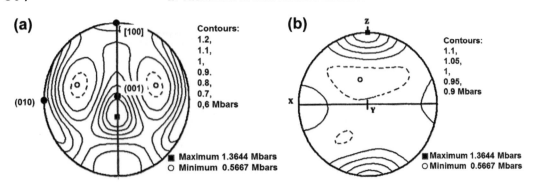

FIGURE 13.42   Anisotropy of elastic properties of anorthosite. (a) Anisotropy of Young's modulus for single crystal of An57. (b) Anisotropy of Young's modulus for Oman anorthosite. This form of anisotropy is very similar to that assumed in the Bigoni models. *From Mainprice and Munch (1993).*

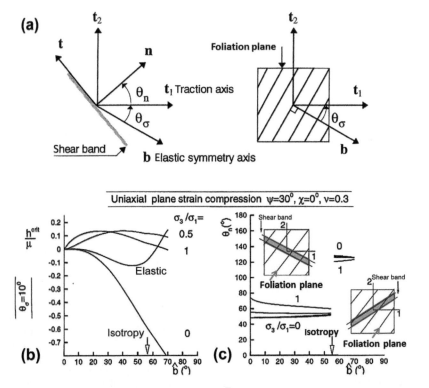

FIGURE 13.43   Conditions for localisation in a nonassociative Drucker–Prager material in plane strain. (a) Directions in physical space: **b** is the axis of elastic symmetry, t_1 and t_2 are principal stress axes, **n** is the shear band normal and **t** is the unit vector tangent to the shear zone. (b) and (c) Strain localisation for a non-associated Drucker–Prager solid with transverse isotropy described by the angle $\hat{b}$, and subjected to plane strain, uniaxial compression. Results are reported for Poisson's ratio = 0.3, the equivalent of the friction angle, 30°, zero dilatancy and for the inclination of the anisotropy axis, $\theta_\sigma = 10°$. Different values of out-of-plane stress, parameter σ_3/σ_1, are considered. The term *elastic* in the figure is a stress state calculated using a relation based on elasticity. (b) Normalised critical hardening modulus. (c) Inclination angle, θ_n, which is the angle between t_1 and the shear band normal. Since $\theta_n \neq 0°$, there is a single shear band except for the isotropic case. *From Bigoni et al. (2000).*

(Figure 13.43(b)) depending on the ratio, σ_3/σ_1. The shear band orientation relative to the plane of anisotropy depends on whether the parameter $\hat{b}$ is greater or lesser than $\hat{b}_{iso}$.

We see from the above that in non-associative materials, localisation can commonly occur in the hardening regime whilst in associative materials, softening is a necessary but not sufficient condition. Thus we expect different kinds of localisation behaviour in rocks undergoing grain size reduction by cataclasis and by chemical reactions with volume change (Figure 13.44(a) and (b)) and rocks undergoing grain size reduction by subgrain rotation and/or diffusion dominated mechanisms (Figure 13.44(c) and (d)). In the first instance, we expect that dilatancy resulting from grain sliding and chemical reactions induces non-associative plasticity and that the Hall—Petch effect results in overall hardening with increasing strain. Thus localisation is possible if the hardening becomes critical.

In the second case, diffusion accommodated plasticity at constant volume induces associative plasticity but a softening with decreasing grain size. Localisation is possible now if the softening modulus becomes critical.

FIGURE 13.44 Contrasting constitutive behaviour during grain size reduction. (a) Deformed granite with grain size reduction arising in part from chemical reactions to form a fine grained product. *(From Ron Vernon.)* (b) Cataclasis of a feldspar grain in a deformed granite. *(From Trouw et al. (2010).)* In both (a) and (b) we expect non-associative behaviour arising from dilatancy. (c) and (d) Deformation of quartz-rich materials with recrystallisation dominating the deformation mechanisms. In both (c) and (d) we expect diffusion assisted or accommodated flow with no dilatancy and associative constitutive behaviour. *(From Trouw et al. (2010).)* Localisation conditions in (a) and (b) are met by critical *hardening* behaviour, whereas localisation conditions in (c) and (d) are met by critical *softening* behaviour. Scales: widths across the base of frame are (a) 2.4 cm, (b) 4 mm, (c) 4 mm and (d) 16 mm.

13.8 CONTROLS ON MICROSTRUCTURE

In Section 3.6.1, we emphasise that the structures we see in deformed rocks result from the *movements* that have occurred during their development. The stresses involved are also the result of these movements and are related to the movements through the relevant constitutive relation. The strains are incidental in that they represent one possible geometrical measure of the accumulated incremental movements. The strains exert no control on the microstructure although they may be reflected in some aspects of the shapes of deformed objects. The essential features of the microstructure reflect the *kinematics* of the deformation history. These statements are of course reminiscent of the approach of Sander (1911, 1930, 1948, 1950, 1970): *The fabric reflects the movement picture.*

The relation between microstructure and kinematics is one important aspect of what we see in deformed metamorphic rocks but another overwhelming principle is that deformation compatibility must be established throughout the deformation history so that no material overlaps or open gaps develop. Many different processes, such as localised elastic and plastic deformation, diffusion, dissolution, grain and subgrain rotation, replacement, precipitation and, most importantly, mineral reactions, operate to ensure that such compatibility is maintained.

A common outcome of the compatibility requirement is the refinement of microstructure as an interface between two differently deformed domains is approached or, on a broader scale, in order to fit two differently deformed domains together with no overlaps or gaps. This refinement occurs in order to minimise what is commonly a non-convex Helmholtz energy of the system, the non-convexity arising because of geometrical or constitutive softening behaviour (Sections 7.7, 8.2 and 13.6; Ortiz and Repetto, 1999). The refinement process is a response to the geometrical problems involved in fitting together differently deformed regions (such as subgrains or twins), to conform to an overall imposed deformation (see Figures 7.18, 7.19, 7.20 and 7.21), unless the deformation is distributed over a number of length scales (see Section 8.2). Such a refinement process leads to a fractal distribution of length scales and the microstructure commonly has a multifractal geometry.

In this brief section, we examine some aspects of multifractal geometry, kinematics and compatibility with respect to microstructure development but we leave an in depth consideration of the application of these fundamental principles to Volume II.

13.8.1 The Multifractal Nature of Microstructure

If we are to become more quantitative about models that address the evolution of microstructure in deformed metamorphic rocks, we need ways of measuring and characterising the microfabrics we see. Such measures of microfabrics should ideally have a foundation in the physical and chemical processes that operate to produce the microfabrics rather than be of an empirical, statistical nature. We have seen that a number of processes have been proposed to describe the evolution of the microfabrics we see. One is a stochastic process, described ideally by the Fokker–Planck equation describing *competitive* processes that ultimately produce a *grain size distribution*. Other processes are better described by reaction–diffusion equations that predict spatially periodic distributions of elements of the microstructure such as metamorphic layering or porphyroblasts (see Section 15.6.1). Others, such as the models proposed by Cahn (1991) for multiphase aggregates, propose cooperative reactions within grain aggregates whereby patterning can develop because of contrasts in interfacial energies. Some other models such as those discussed by Ortiz and Repetto (1999) describe multiplicative, cascading processes that produce fractal spatial patterning of subgrains.

Some of these processes predict multifractal geometries. Some predict long range spatial correlations. Some predict local clumping or layering of individual mineral phases. Some authors (Kretz, 1969) suggest complete randomness in the distribution of mineral phases. We need easily applied measures that enable such characteristics to be quantified and used as tests for existing models and for future models that are proposed for microstructural development. This section considers some such measures. Since the processes that we deal with are nonlinear, we expect the basic geometries associated with microstructures to be fractal. The geometry of Nature is intrinsically irregular; in metamorphic rocks, various degrees of pattering are developed consisting of metamorphically produced layering or lineations or clustering. The important questions are: *Are these geometries fractal, or even multifractal? Can we use the geometrical characteristics of these microstructures to say something about the nonlinear, multiplicative or cascading processes that produced the microstructure?* An example in this regard is the work of Arneodo and co-workers (Arneodo et al., 1995, 1999; Audit et al., 2002) on the geometry of DNA and turbulence where direct links to the underlying dynamical and multiplicative processes have been made.

A number of measures that have a background in nonlinear dynamics (Sprott, 2003) may prove useful in quantifying and interpreting the geometry of microstructures. One is to clearly establish if the geometry is fractal or multifractal. Hence the establishment of singularity spectra (Section 7.8) is a first fundamental step. The demonstration of such geometry is an important advance in its own right since it implies that some form of multiplicative or iterative process was involved in the development of the microstructure. One can proceed to establish scaling laws and measures of spatial correlation for the microstructure. Examples of such measures are the *Hurst exponent* (a measure of the 'roughness' of the data) and the *lacunarity* (a measure of the heterogeneity or 'clumpiness' of the data; Plotnick et al., 1996). More advanced procedures involve the development of attractors for the system and of dynamical equations that mimic the behaviour of the attractor (Chapter 7 and Sprott, 2003). Here we consider two approaches: the measurement of the Hurst exponent and the establishment of multifractal geometry with its associated correlation dimension. Software to produce some of these measures can be found at Karperien, A., FracLac for ImageJ.

http://rsb.info.nih.gov/ij/plugins/fraclac/FLHelp/Introduction.htm.

13.8.1.1 *The Hurst Exponent*

Consider a one dimensional sequence of values, $\xi\,(d)$, representing the presence or absence of a particular mineral phase or the degree or some measure of the CPO or some other quantitative measure of the microstructure. In the example we take, ξ varies with distance, d, as shown in Figure 13.45(a). We can characterise the resulting pattern in a number of ways. One way is to use the *Hurst exponent* (Feder, 1988; Sprott, 2003), which measures the way in which the local range in variation (or roughness) scales with distance across the microstructure. In order to define the Hurst exponent we first take the mean of ξ and then (Figure 13.45(a)) the cumulative departures from the mean, $\Xi\,(d)$. If R is the range of Ξ,

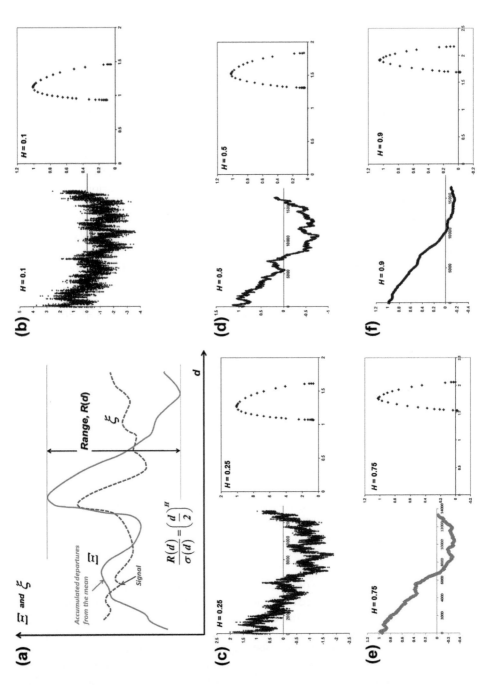

FIGURE 13.45 Hurst exponents and associated singularity spectra. (a) Definition of the Hurst exponent, H. ξ is the initial data as a function of distance, d and Ξ is derived from these data as the cumulative departure of the data from the mean. R is the range of Ξ, that is, the difference between the maximum and minimum value of Ξ. If σ is the standard deviation of ξ, then the Hurst exponent is defined by $\frac{R}{\sigma} = \left(\frac{d}{2}\right)^H$. (b)–(f): Signals with Hurst exponents 0.1, 0.25, 0.5, 0.75 and 0.9 with associated singularity spectra. (b) through to (f): Signals on the left with Hurst exponents ranging from 0.1 through to 0.9 as labelled. Singularity spectra on the right for each value of Hurst exponent

TABLE 13.2 Characteristics of the Hurst Exponent, H, for a One Dimensional Signal

Value of H	Meaning	Pattern Characteristics
$0.5 < H < 1$ Persistent	Long-range positive autocorrelations	A high (low) value tends to be followed by another high (low) value. The overall trend is to higher (lower) values. Power law decay in autocorrelations.
$H = 0.5$	Completely uncorrelated sequence	Autocorrelations at small intervals can be positive or negative. Absolute value of autocorrelations decays rapidly to zero.
$0 < H < 0.5$ Antipersistent	Long range switching between high and low values	A high value tends to be followed by a low value. Power law decay in autocorrelations.

that is, the difference between the largest and smallest value of Ξ, then the Hurst exponent, H, is defined by

$$\frac{R(d)}{\sigma(d)} = \left(\frac{d}{2}\right)^H$$

where $\sigma(d)$ is the standard deviation of ξ over the distance d. Examples of signals with different Hurst exponents are given in Figure 13.45 and characteristics of signals for various values of the Hurst exponent are given in Table 13.2.

13.8.1.2 Multifractal Geometry

The measurement of multifractal geometries has been considered in Sections 1.4 and 7.8 through the production of *scalograms* and then the use of the *Wavelet Transform Modulus Maxima* method to construct the singularity spectrum. One can then use such spectra to compare the *box counting dimension*, D_0, the *information dimension*, D_1, and the *correlation dimension*, D_2, for various geometries, which results in a rigorous means of characterising and comparing different microstructures (Chapter 7). In Figure 13.45(b)–(f), we show singularity spectra corresponding to signals with various Hurst exponents. The spectra shift progressively to higher values of α as H varies from 0.1 to 0.9. The values of D_2 and of α_{D2}, where α_{D2} is the value of α and D_2 is measured for each of the signals in Figure 13.45(b)–(f), are given in Table 13.3.

TABLE 13.3 Values of D_2 and of α_{D2} for Various Values of H

H	D_2	α_{D2}
0.1	0.974	1.053
0.25	0.951	1.201
0.5	0.966	1.461
0.75	0.974	1.724
0.9	0.998	1.884

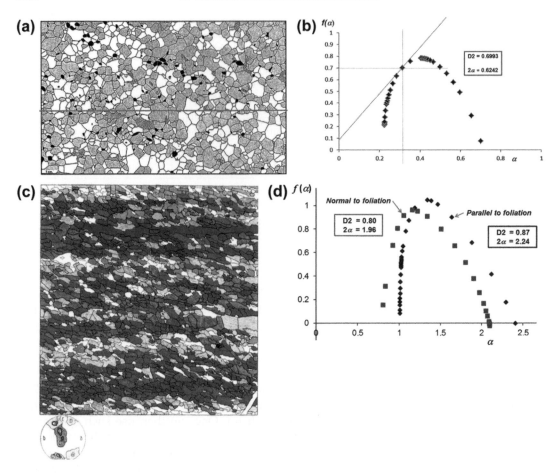

FIGURE 13.46 Microfabrics with associated singularity spectra. (a) Tracing of thin section of a pyroxene (light grey)-scapolite, (white)-sphene, (black)-amphibole, (cross hatched) rock. Section is 18.5 mm across. *(From Kretz (1969).)* (b) Singularity spectrum for the cross section marked in (a). (c) Orientation distribution map for a quartz mylonite. Orientation of quartz c-axes is shown in the inset. Map is ~1.5 mm across. *(From Sander (1950).)* (d) Singularity spectra for sections (see Figure 1.11) parallel and normal to the foliation.

In Figure 13.46(a), a figure from Kretz (1969) is presented, with the scan line marked. The singularity spectrum is shown in Figure 13.46(b). The Hurst exponent for this scan line is 0.23 indicating long range anticorrelations or 'clumpiness'. Thus although the analysis of Kretz arrives at the conclusion that the microstructure is 'random', the clumpiness of the microstructure is clearly delineated by the Hurst exponent, the fabric is multifractal and the microstructure is characterised by a correlation dimension of ~0.7.

A similar analysis is shown for the microstructure mapped by Sander (1950) in Figure 13.46(c). The Hurst exponents measured parallel and normal to the foliation are almost identical at 0.80 and 0.78 thus indicating long range positive spatial correlations, both parallel and normal to the foliation. Again the fabrics are multifractal.

13.8.2 Kinematic Controls on Microstructure Orientation

One of the most important aspects of microstructure is its use in defining the movements that have occurred during the development of the microstructure. During deformation associated with grain boundary migration and diffusion, the shapes of grains are controlled by the arrangements of the grain boundaries which in turn are controlled by two competing processes. One arises from a change in grain shape arising from crystal slip and diffusive motions within the grain that tend to distort the shape of the grain so that it approaches that of the current strain ellipsoid. The other process arises from the displacement of material markers, such as current grain boundaries, by the movements taking place locally in the polycrystal. These displacements tend to align the grain boundaries parallel to the unstable manifolds of the flow, which act as attractors for material markers. If the flow is such that a stable manifold exists, then the material tends to be instantaneously flattened and sheared normal to that manifold (Chapter 3; see also Passchier, 1997; Iacopini et al., 2007). For simple shearing deformations, there is just one unstable manifold marked by the single eigenvector of the flow (Chapter 3). Thus for simple shearing, we expect some of the grain boundaries to align parallel to the shearing plane. At shear strains of about 12, the principal axis of elongation is within $5°$ of the shearing plane (Figure 2.10 b) and so a single preferred orientation of grain boundaries, approximately parallel to the single eigenvector, is expected at high shear strains.

We have seen in Chapter 3 that in a general two dimensional affine flow, there are two eigenvectors for the velocity gradient; these define a stable and an unstable manifold for the flow. Since the manifolds are parallel everywhere to the eigenvectors of the flow, lines within the manifolds are stretched, shortened or remain of constant length. These lines are not changing direction during an increment of deformation. However shear strains generally exist parallel to the manifolds. As an example, consider a deformation history consisting of shearing parallel to a single plane together with flattening normal to that plane. Figure 13.47(a) represents one such deformation history where the deformation of an initial square (at $t = 0$) at times $t = 1$ (A), $t = 2$ (B) and $t = 3$ (C) is shown. The deformation history is given by

$$x_1 = tX_1 + t\left(X_2/\sqrt{3}\right), \quad x_2 = X_2/t \tag{13.28}$$

Following the arguments of Chapter 3, the eigenvectors of the stretching tensor in the deformed state are shown in Figure 13.47(b),(d) and (f) and the flow field in Figure 13.47(c),(e) and (g). Figure 13.47(h) shows the kinematic framework in a little more detail for $t = 6$.

The eigenvectors of the stretching tensor, which by definition must always be orthogonal, rotate very slightly in a clockwise sense during the deformation history. There are always two eigenvectors for the velocity gradient field. One, which is an unstable manifold, always remains parallel to the shearing plane, which we denote as C, whilst the other, which is a stable manifold, rotates anticlockwise towards the C-plane. The principal plane of extension is almost parallel to the shearing plane by $t = 6$. There are always large shearing displacements on the two manifolds even at high strains. Within the unstable manifold, lines are stretched whilst within the stable manifold, lines are shortened.

We suggest that oblique foliations and S-C fabrics in general have origins in this kinematic framework. We identify the unstable manifold as a C-plane whilst the line normal to the stable

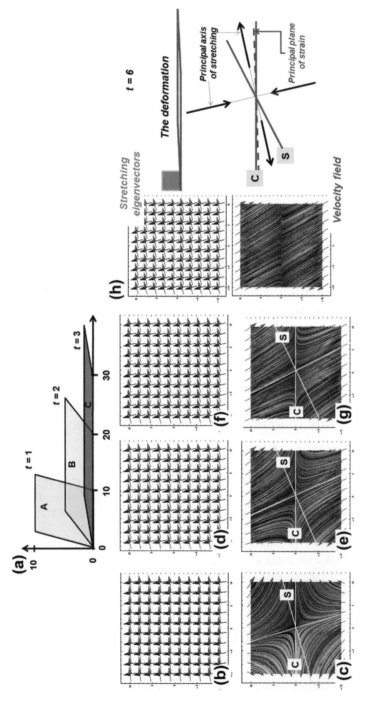

FIGURE 13.47 Stretching and velocity gradient eigenvectors for a combined shearing and flattening deformation history. (a) The deformation defined in (13.28) for $t = 1$ (A), 2 (B) and 3 (C). (c) through to (g). Stretching eigenvectors (above) and velocity gradient eigenvectors (below) for $t = 1, 2$ and 3. The stretching eigenvectors remain orthogonal and rotate slowly clockwise. One velocity gradient eigenvector (labelled C) remains constant in orientation. The other velocity gradient eigenvector (normal to the line labelled S) rotates slowly towards C. (h) The kinematic framework for $t = 6$. The eigenvector maps for the stretching and velocity gradient tensors are shown along with the deformation to the right. The relationships between the stretching eigenvectors (in black) and the velocity gradient eigenvector (in red, marked C) and the normal to the other eigenvector (in red, marked S) are shown together with the orientation of a principal plane of strain. The shear in this instance is $\gamma = 20.7$.

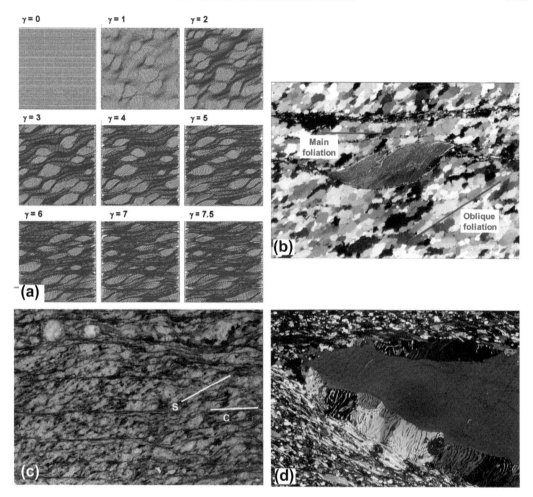

FIGURE 13.48 Kinematic controls on microstructure development. (a) Results of ELLE modelling of an aggregate with two different viscosities deformed by a simple shearing deformation. The shear strain is marked for each frame. *(From Jessell et al. (2009).)* (b) A mica 'fish' sitting in a sea of recrystallised quartz with an oblique foliation. Scale: 3 mm across. (c) S—C mylonite showing S and C foliations. Scale: 10 cm across. (d) Feldspar 'fish' with myrmekitic recrystallisation on boundaries associated with neighbouring muscovite. Scale: 5 mm across. *((b), (c) and (d) from Trouw et al. (2010).)*

manifold, which represents a plane of flattening whilst shearing takes place parallel to it, we identify with an S-plane. The angle between the S and C planes *increases* as the deformation proceeds. The angular relations shown in Figure 13.47 arise from the deformation described by (13.28). In volume II, we present a more general discussion of the eigenvectors that develop in more general deformations and the control exerted on microfabric development.

Figure 13.48 gives some examples of these effects. In Figure 13.48(a), a simple shearing deformation history is modelled (Jessell et al., 2009) with grains of two contrasting viscosities. There is one unstable manifold for this deformation history and that remains parallel to the base of the figures. The shapes and orientations of strong grains are at first controlled solely by the local strain ellipsoid but as the deformation proceeds, regions of low viscosity link up parallel to the unstable manifold and finally at a shear strain of 7.5, this unstable manifold dominates the microstructure.

In Figure 13.48(b), a mica 'fish' is shown in a sea of recrystallised quartz grains. We propose that the main foliation and the oblique foliation defined by grain boundaries of recrystallised quartz are controlled by the two eigenvectors of the flow as in Figure 13.47. The shape of the mica fish is such as to maintain deformation compatibility with the surrounding quartz. Figure 13.48(c) shows two foliations, labelled S and C, in a deformed granite gneiss. Again we propose that these foliations are controlled by the two eigenvectors of the flow. In Figure 13.48(d), a feldspar 'fish' attains compatibility with the surrounding matrix by chemically reacting on its margins to produce myrmekite.

Recommended Additional Reading

Backofen, W.A. (1972). *Deformation Processing*. Addison-Wesley.
  A clearly written text with an emphasis on plastic anisotropy.

Clayton, J.D. (2011). *Nonlinear Mechanics of Crystals*. Springer.
  An advanced text including a treatment of the mechanics of crystals in terms of generalized coordinates. Many aspects of elasticity and plasticity are treated together with a thorough treatment of dislocations and disclinations. One might look upon this book as a modern version of Nye (1957).

Gambin, W. (2001). *Plasticity and Textures*. Springer.
  This book discusses principles of deformation, plastic anisotropy and the development of CPO in an easily readable format. Little emphasis is placed on self-consistent theories of CPO development.

Gutkin, M, Yu, Ovid'ko, I.A. (2004). *Plastic Deformation in Nanocrystalline Materials*. Springer.
  Although the term nanocrystalline appears in the title of this book, much of the content is just as applicable to coarser grained materials. Topics such as the role of disclinations in deformation and the interactions with dislocations are given extensive treatment.

Kocks, U.F., Tomé, C.N., Wenk, H.,-R. (1998). *Texture & Anisotropy*. Cambridge.
  A thorough treatment of plastic anisotropy and of CPO development using the strict Taylor-Bishop-Hill theory, relaxed constraints and self-consistent approaches.

Passchier, C.W., Trouw, R.A.J. (2005). *Microtectonics*. Springer.
  An excellent book illustrating a large range of microstructures in deformed metamorphic rocks.

Sprott, J.C. (2003). *Chaos and Time-series Analysis*. Oxford University Press.
  A detailed discussion of the methods of the analysis for signals produced in chaotic nonlinear systems. The book includes an "atlas" of attractors produced by various nonlinear systems.

Trouw, R.A.J., Passchier, C.W., Wiersma, D.J. (2010). *Atlas of Mylonites - and Related Microstructures*. Springer.
  A sister book to Passchier and Trouw (2005) with an emphasis on strongly deformed rocks.

Vernon, R.H. (1976). *Metamorphic Processes*. George Allen & Unwin.
  Chapter 5 in this book is an excellent introduction to microstructures in metamorphic rocks.

Zollner, D. (2006). *Monte Carlo Potts Model Simulation and Statistical Mean-Field Theory of Normal Grain Growth*. Shaker-Verlag.
  A detailed analysis of normal grain growth simulations using mean field theory.

14

Mineral Reactions: Equilibrium and Non-Equilibrium Aspects

Structural Geology
http://dx.doi.org/10.1016/B978-0-12-407820-8.00014-X

515

14.1 INTRODUCTION

This chapter is meant as background material for chapters in Volume II where we examine the coupling between mineral reactions and other processes such as fluid flow, heat flow and deformation; it is concerned with the processes that operate during the nucleation and growth of new mineral phases in a metamorphic rock during deformation, some relationships between deformation and chemical reactions and the influence of these processes on the microfabrics and mineral assemblages we observe. Such processes are responsible for the structures that we observe at all scales in deformed metamorphic rocks and so a fundamental understanding of the principles involved is essential. Since we are concerned with processes we must, by definition, be dealing with systems not at equilibrium, although there is considerable debate as to how 'far' from equilibrium one might be during the operation of these processes.

The concept of chemical equilibrium is well defined and understood for a system where the stress is a homogeneous hydrostatic pressure, P, and the distribution of temperature, T, is homogeneous. For a deforming, chemically reacting system P and T are not homogeneous and the stress is not hydrostatic; the system as a whole is not at equilibrium. Nevertheless, it is possible in such a system to prescribe the conditions for chemical equilibrium across each interface in the system. However, these conditions depend on the constitutive behaviour of the material. If grain boundary sliding does not occur for instance, these conditions are (1) the temperature difference across the interface is zero, (2) the velocity on either side of the interface is continuous, (3) the stresses resolved normal and parallel to the interface are continuous, and (4) there exists a function that is a measure of the energy either side of the interface and this function is continuous across the interface. In a non-deforming system with a homogeneous temperature distribution these four conditions are readily met and the function mentioned is the chemical potential. In a deforming system the definition of this function has created great confusion and argument mainly because it is widely assumed that it should be a thermodynamic state function, as is the classical chemical potential. We discuss this function in this chapter and point out that it can be defined for a deforming–chemically reacting system but it is not a state function; it can only be defined at interfaces, not within grains, and depends on the orientation of the interface together with the deformation on either side of the interface. Its form also depends on whether diffusive processes operate or not and, as we have indicated above, on the constitutive behaviour of the material. Thus, the conditions for chemical equilibrium in a deforming system are quite different from those in a system with homogeneous T and hydrostatic P. After deformation has ceased the system can relax to a new state of chemical equilibrium under hydrostatic stress conditions if there is enough time but this new state is not the chemical equilibrium state that existed during deformation.

We emphasise that the problems involved have not been solved and at present there is no unified approach to the coupling between deformation and chemical reactions. We aim in this chapter to outline many of the problems yet to be addressed and some of the progress that has been made. In Chapter 15 we extend the discussion to include specific models for metamorphic reactions and some of the processes involved.

As Turner (1948) accurately observed for a deforming metamorphic rock:

> At any instant every crystal, endowed with its own surface energy and affected internally by its own system of stresses, is actively competing for growth with its neighbours (including newly initiated seed crystals), in a highly anisotropic environment, wherein such factors as availability and composition of solutions, chemical and physical nature of adjoining grains, and physical continuity of large and small fabric elements, are subject to rapid variation.

This insightful statement contains all the elements we need to address in developing a unified approach to deforming–reacting rock systems. Inherent in this statement are two fundamental concepts: the first concerns the interactions and reactions between grains while preserving geometrical continuity of the deformation field across disparate phases in a reacting–deforming system. The second is that growth of new mineral phases competes with the supply of material from adjacent mineral reactions and may also supply material for those adjacent reactions. *Metamorphic systems are networked reaction–diffusion–transport systems.*

The first concept is an intrinsically difficult concept we introduce in Chapter 13 and develop further in this chapter. However, one way of maintaining deformation compatibility during a chemical reaction is for the host and newly developing grain to undergo the same volume change throughout the deformation history. At the time Turner wrote his statement the concept of '*the force of crystallisation*' derived from the earlier work of Becker and Day (1905, 1916) was alive and well and many petrographers (Bastin, 1950; Bastin et al., 1931; Carmichael, 1987; Harker, 1950; Lindgren, 1912; Ridge, 1949; Turner, 1948) have emphasised that many mineral reactions involve constant volume (rather than constant mass) replacement of host minerals with the inevitable conclusion that stresses must be generated at the interfaces between reacting grains in order to account for the lack of volume change. One of the best ways of considering the generation of such stresses is to compare the volume per formula unit of the replacing mineral and that of the host. The volume per formula unit is the volume of the unit cell divided by the number of formula units in that unit cell (Ridge, 1949). Most minerals have more than one formula unit per unit cell (Ridge, 1949). Both quartz and coesite for instance have two SiO_2 units per unit cell but the unit cell of quartz is larger than that of coesite (Levien and Prewitt, 1981). When quartz replaces coesite pseudomorphically, the quartz is constrained to occupy the same volume as the coesite thus generating compressive stresses in the neighbourhood.

Even if the reactions involved are not pseudomorphic and are expressed as constant mass reactions there is still a ΔV associated with the reaction that must result in stresses at the reacting interface and these stresses are able to drive diffusive mass fluxes. These concepts seem to have been relegated to a category of little or no importance in most recent literature perhaps because of the conclusion of Carmichael (1987) that such processes must ultimately be relaxed by deformation (both ductile and brittle) and/or by diffusion (including pressure solution) and so the effects must be 'transient'. While such a conclusion is a truism for rocks that have reached chemical equilibrium at hydrostatic pressure conditions, the processes that guided the route to equilibrium have commonly left their mark in the microstructure and that is the aspect that concerns us. We consider the detailed processes that accompany mineral reactions to be of fundamental importance while the rock is deforming and mineral reactions are taking place; they lead to feedback relations that produce many of the microstructures observed in metamorphic rocks. To add to the complexity, the ΔH of reactions, together with the heat generated by deformation, also contributes to local stresses and temperature gradients during the reactions. Some implications of the coupling between these processes are indicated in Figure 14.1.

The second concept inherent within Turner's statement leads to considerable complexity in the rates of mineral growth especially if more than two chemical reactions are coupled and at least one is nonlinear. A deforming—chemically reacting rock mass is the same as any other system not at equilibrium. As the statement by Turner suggests, the system evolves through *competition between a number of processes*. It is forced from equilibrium by thermodynamic forces and relaxes back towards equilibrium by dissipative processes. What we observe arises from the competition between forcing and dissipation. A fundamental characteristic of most metamorphic reactions is that the various competitive processes operate at different rates and are interdependent in the sense that they are networked; *the evolution of one part of*

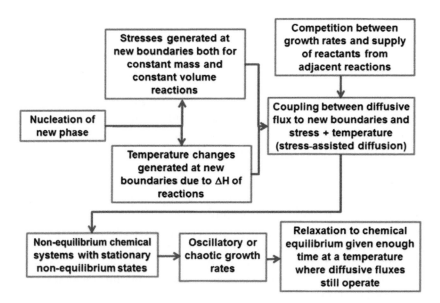

FIGURE 14.1   The progression of nucleation and mineral growth. Stresses and temperature gradients are developed at growing grain interfaces which lead to coupling between diffusive fluxes, nucleation and grain growth. This is expressed as stress-assisted diffusion or 'pressure solution'. The growth of a new grain depends on the rate of supply of reactants from adjacent reactions. There is competition between growth and the supply of reactants. This leads to the development of nonlinear chemical systems with a tendency to evolve to one or more non-equilibrium stationary states. The system behaves in this manner until the driving forces for chemical reactions and deformation dwindle when the system moves towards chemical equilibrium at a hydrostatic pressure if the temperature is high enough to allow diffusive processes to achieve this state.

the system depends on evolution in other parts of the system. Thus competition results in nonlinear behaviours that are the basis for microstructure development.

In most metamorphic systems the thermodynamic forces are gradients in deformation, in the deformation gradient that drives damage evolution, in hydraulic potential, in $1/T$, and in chemical potentials. The corresponding competing dissipative processes are fluxes of momentum, damage density, fluids, heat, and mass. If the work done by the thermodynamic forces balances the dissipation then a *non-equilibrium stationary state* develops and this may or may not be stable; loss of stability usually is expressed as a Hopf bifurcation (Chapter 7). If one or more of the thermodynamic forces are zero another stationary state called *equilibrium* with respect to that thermodynamic force develops. One important issue is the time taken for a system to relax to equilibrium once the thermodynamic forces become insignificant. The relaxation time is a function of temperature so that if the rate of decay of temperature is fast compared to the relaxation of other thermodynamic forces then non-equilibrium microstructures and mineral assemblages may be preserved.

Thus, the evolution of both microstructure and coexisting mineral phases depends on two intrinsically coupled constraints. First, deformation compatibility must be preserved during deformation that occurs synchronously with chemical reactions, meaning that only a limited number of holes (pores) and no overlaps between grains develop. To the extent that holes do develop an instantaneous porosity develops controlled by diffusive processes and such that the decrease in surface energy is balanced by the increase in grain boundary energy (Pask, 1987; for an extensive compilation of classic papers on pore development during sintering see Somiya and Moriyoshi, 1990). Deformation compatibility conditions are met by continuity of the deformation gradient across all interfaces, by grain boundary sliding and by stress-assisted diffusive fluxes; conditions for deformation compatibility control the rates of grain growth in a deforming, chemically reacting system and determine whether the growth processes are pseudomorphic, isochoric or otherwise.

The second constraint involves competition between the growth of a new grain and the supply of reactants to the growth site. If this supply depends on the production of reactants derived from a variety of local reactions all proceeding at different rates, then oscillatory or even chaotic rates of growth can emerge.

Superimposed on these two first order constraints is the influence of the local deforming–reacting environment on the nucleation sites for new grains. In a solid, new grains tend to nucleate in gradients in chemical potential and where the local deformation can accommodate the ΔV of the reaction and/or at sites of excess energy which can contribute to overstepping of the equilibrium phase boundary. We will see (Section 15.5) that nucleation is inhibited by large chemical potential gradients and so an interaction with deformation arises from the influence of stress on the chemical potential (Kamb, 1959; Paterson, 1973).

An important issue derives from additional state variables introduced in a deforming reactive system that are absent in a non-deforming system. In a non-deforming reacting system that has evolved to equilibrium the state of the system can be described solely in terms of the intensive variables, the hydrostatic pressure, P, the temperature, T, and the chemical potentials, μ_n, of the n chemical components present. In a deforming system at chemical equilibrium another intensive variable necessary to define the state of the system is the local deformation gradient, $\mathbf{F}$, which remains after the deformation ceases. This additional variable is neglected in most discussions of equilibrium states but is fundamental if deformation is synchronous with mineral reactions. The upshot is that we need a clear understanding of the detailed processes that accompany metamorphism to arrive at a clear delineation of equilibrium states under non-hydrostatic conditions. Such a requirement sets the scene for this chapter.

A number of papers have contributed to an evolving model of the behaviour of mineral reactions during deformation. These include Voll (1960), Kretz (1966, 1969, 1973, 1974, 2006), Carmichael (1969, 1987), Wintsch and Knipe (1983), Rubie and Thompson (1985), Wintsch (1985), Wintsch and Dunning (1985), Wheeler (1987, 2014), Passchier and Trouw (1996), Wintsch and Yi (2002), Ford et al. (2002), Ford and Wheeler (2004), Vernon (1976), Vernon and Clarke (2008), Merino and Canals (2011), Carlson (2011), Stokes et al. (2012), Williams and Jercinovic (2012) and Wintsch and Yeh (2013). The essential points brought out by these papers are

1. A contribution to the rates of mineral reactions comes from the energy stored in deformation products such as dislocations, sub-grain and twin boundaries.
2. Some new phases nucleate and grow preferentially in sites that reflect the ΔV of the reaction: evidence exists that if ΔV is negative the new phase grows in sites where shortening strains dominate.
3. In many metamorphic environments minerals change grain shape, and hence contribute to the macroscopic deformation by *stress induced mass transfer*. In many instances this process is accompanied by progressive changes in composition as new grains grow to reflect the changing metamorphic conditions or local changes in the supply of chemical components; these processes operate not only at low metamorphic grades where the phenomenon has been recognised for many years but also at high temperature conditions. This implies that in many metamorphic rocks dislocation flow may play a relatively minor role in the deformation process except perhaps for minerals such as quartz. Such an observation has considerable implications for the strength of the crust.
4. For many reactions the nucleation and subsequent growth of a new phase in a host is pseudomorphic. We will see in Section 14.3.2 that this is a response to the constraint of maintaining deformation compatibility. The equal volume replacement of the host induces stresses in the host that can drive stress induced mass transfer. There are other consequences of equal volume replacement that we explore in Section 14.4.
5. Examination of the Gibbs−Duhem equation for a deforming−reacting solid shows that some of these processes are responsible for increasing the rates of mineral reactions without changing the slope or position of the equilibrium mineral phase boundary in P−T space where P is taken as the hydrostatic pressure. Others are responsible for changing the slope and/or moving such a boundary while deformation and diffusion are proceeding.
6. Many, if not all, metamorphic reactions are networked in some manner in that the progress of each reaction depends on what is happening in other reactions nearby; this means that considerations of networked thermodynamics (Peusner, 1986) are useful in understanding the behaviour of such systems.

We address these points throughout this chapter and some are illustrated in Figure 14.2.

We explore many of these issues in this chapter and in Chapter 15. We first (Section 14.2) consider some preliminary concepts in physical chemistry that give us a language to use with respect to chemical reactions. Section 14.3 then explores the classical chemical equilibrium

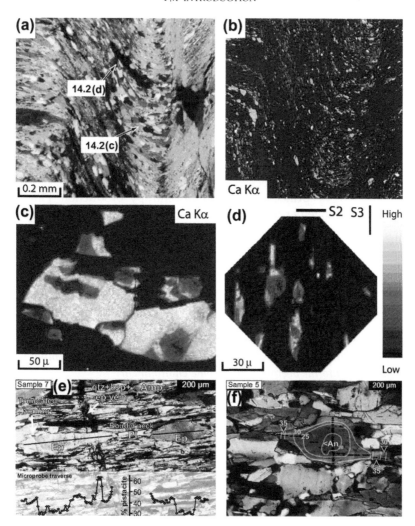

FIGURE 14.2 Dissolution and transfer of mass during deformation. (a–d): Mass transfer during crenulation cleavage development. *From Williams and Jercinovic (2012)*. (e) and (f): Mass transfer during lineation development. *From Stokes et al. (2012)*. (a) Thin section of crenulation cleavage. (b) Ca Kα wavelength dispersive spectrum (WDS) compositional map corresponding to (a). The hinge area has more high-Ca plagioclase than the crenulation limb. (c) Ca Kα WDS compositional map in the hinge area showing high Ca overgrowths on initial low-Ca plagioclase. (d) Ca Kα WDS compositional map in the mica-rich limb area showing high-Ca overgrowths elongate parallel to the new crenulation cleavage growing on initial low-Ca plagioclases. (e) Boudinaged epidote grain with microprobe chemical profile (percent pistacite). The epidote grain is zoned as indicated by shades of blue and yellow corresponding to variations in Fe^{3+} content. These zones are truncated on grain boundaries facing the shortening direction. Plagioclase grain grows in boudin neck. (f) Zoned plagioclase grain with tails of varying anorthite content.

view of metamorphic assemblages, including the roles of *tectonic overpressures* and *tectonic over-temperatures*, to highlight some aspects that, when generalised, apply to systems far from equilibrium. Also in Section 14.3 we consider deforming chemical systems at chemical equilibrium and a generalised form of the Clapeyron equation as a basis for considering the displacement of the syn-kinematic equilibrium mineral phase boundary relative to the position of such a boundary on hydrostatic P–T phase diagrams. We then (Section 14.4) discuss chemical systems not at equilibrium, the principles that govern the development of non-equilibrium stationary states and the phase rule for systems not at equilibrium. We also consider the meaning of the terms near and far from equilibrium in chemical systems. Of particular importance is the concept of autocatalytic reaction networks rather than the classical concept of autocatalytic reactions. Section 14.5 is concerned with dissipation in such systems and some comments on the Prigogine principle of minimum entropy production. Finally in Section 14.6 we attempt to bring together these various and diverse aspects of chemical-mechanical coupling.

Some of the questions we attempt to address here are: *What is equilibrium? What are the criteria for chemical equilibrium? What distinguishes a system at chemical equilibrium from one not at equilibrium? What do the terms 'near' and 'far' from chemical equilibrium mean? How do systems not at chemical equilibrium behave and what controls this behaviour? Is there any order in systems not at chemical equilibrium? What is a stationary state as opposed to an equilibrium state? What is the role of 'tectonic overpressure'? Does deformation have an influence on the position of the equilibrium mineral phase boundary defined in non-deforming systems? Does diffusion have an influence on the position of the equilibrium mineral phase boundary defined in non-deforming systems? How does pressure solution fit into the overall metamorphic environment?*

As a summary of this chapter the following important results are given:

- Chemical equilibrium between mineral phases is possible in a *deforming–reacting–transport system*. The form of the Clapeyron equation that defines the equilibrium phase boundary reflects the processes involved (deformation and diffusion) in such systems and hence differs from that in a system under hydrostatic stress with no transport. This means that the position of the equilibrium phase boundary differs in a deforming–transport system from that in a static system.
- Compatibility of deformation results in constant volume reactions leading to pseudomorphism. Such constraints are important in influencing the form of the Clapeyron equation for chemical equilibrium in deforming–transport systems. Such constraints automatically imply that dissolution/growth and dissolution/precipitation processes are intimately coupled.
- The concept of *autocatalytic systems* rather than the classical autocatalytic reactions becomes fundamental in metamorphism and is manifest as *networked mineral reaction systems*. Networked chemical reaction–transport systems lead to chemical zoning, chemical fluctuations and grain microstructures that do not occur in non-networked systems.
- The metamorphic process is a networked reaction–diffusion–transport process which controls the grain size distribution and the distribution of individual grains. This means that grain size distributions and grain microstructures are controlled by reaction–diffusion–transport processes rather than by nucleation and by growth and topological constraints on grain boundary configurations.

14.2 SOME PRELIMINARIES

In the following we use the term P to mean hydrostatic pressure. In a fluid at rest P is readily defined, as the (constant) scalar measure of the force per unit area across any plane in the fluid, but in a viscoelastic fluid that is deforming this is a quantity that is not easily defined (see Fitts, 1962, pp 159–161) especially if the behaviour is nonlinear as in a power-law material. Malvern (1969) distinguishes two 'pressures' in a deforming Newtonian fluid. One is the thermodynamic pressure, P, and the other is a quantity, $\overline{P}$, equal to the negative mean stress, $\overline{P} = -\frac{1}{3}\sigma_{ii}$. Malvern (1969, Section 6.3) shows that for an elastic-Newtonian fluid

$$P = \overline{P} + \kappa D_{kk} = \overline{P} - \kappa \frac{1}{\rho} \frac{d\rho}{dt}$$

where $\kappa = \lambda + \frac{2}{3} G$, $D_{kk} \equiv div\, \boldsymbol{v} = -\frac{1}{\rho}\frac{d\rho}{dt}$ where v is the velocity and ρ is the density; λ and G are the Lamé constants. Thus, for an incompressible flow of a Newtonian fluid $P = \overline{P}$. Otherwise *the thermodynamic pressure differs from the mean stress*. In metamorphic environments we expect $div\, \boldsymbol{v} = -\frac{1}{\rho}\frac{d\rho}{dt}$ to be negligible so that we assume that in a deforming system, $P = \overline{P}$. Nevertheless one should be aware that the commonly accepted proposal that the 'pressure' is equivalent to the mean stress is not strictly true for deforming materials.

One of the most important concepts in chemical thermodynamics is that of the *chemical potential*. We adopt the view of Callen (1960) for the definition of *chemical potential*: The internal energy, E, for a system with a volume V and entropy S is given by

$$E = E(S, V, N_1, N_2, \ldots N_n)$$

where N_i ($i = 1, \ldots, n$) are the mole numbers of the n chemical components contained in the system. Thus,

$$dE = \left(\frac{\partial E}{\partial S}\right)_{V,N_i} dS + \left(\frac{\partial E}{\partial V}\right)_{S,N_i} dV + \sum_{k=1}^{n} \left(\frac{\partial E}{\partial N_k}\right)_{S,V,\ldots N_j \ldots} dN_k \tag{14.1}$$

Following Chapter 5, the interpretations of the terms in (14.1) are

$$\left(\frac{\partial E}{\partial S}\right)_{V,N_k} \equiv T, \text{ the temperature}$$

$$-\left(\frac{\partial E}{\partial V}\right)_{S,N_k} \equiv P, \text{ the pressure}$$

$$\left(\frac{\partial E}{\partial N_j}\right)_{S,V,\ldots N_k \ldots} \equiv \mu_j, \text{ the chemical potential of the } j^{th} \text{ component}$$

With these interpretations in mind, (14.1) can be rewritten:

$$dE = TdS - PdV + \mu_1 dN_1 + \mu_2 dN_2 + \ldots \mu_r dN_r$$

so that the chemical potentials can be viewed simply as coefficients that qualify (or weight) the incremental mole numbers. Just as differences in deformation drive the transport of momentum, differences in temperature drive the transport of heat (or entropy) and differences in pressure cause changes in volume, differences in chemical potential drive the transport of mass. The incremental *chemo-mechanical* work done, $dW^{chemo-mechanical}$, by introducing increments of mole numbers of chemical components is

$$dW^{chemo-mechanical} = \sum_{k=1}^{n} \mu_k dN_k \tag{14.2}$$

It is emphasised that, although the work represented in (14.2) is commonly labelled the *chemical work* (see Callen, 1960, p. 32) it is in fact a contribution to the mechanical work in that it represents the work done in inserting increments of new chemical components into a system. For this reason we call this work the *chemo-mechanical work*. It needs to be distinguished from the quantity

$$dW^{chemical} = \mathcal{A}d\xi \tag{14.3}$$

which is the increment of true *chemical work* done arising from an increment, $d\xi$, of *reaction extent*, ξ, in a single reaction with *affinity*, $\mathcal{A}$ (see below). The extent of a reaction is a quantity $0 \le \xi \le 1$ which measures the *progress* of the chemical reaction. For the reaction $A \rightarrow B$, $\xi = 0$ at the beginning of the reaction and $\xi = 1$ at the end of the reaction when all of A is converted to B. The reaction rate in a closed system is $\dot{\xi}$ and $\dot{\xi} = 0$ at equilibrium. At equilibrium the chemical work in (14.3) is always zero and so is commonly overlooked in discussions of systems at equilibrium.

The chemical potential, $\mu_{P,T}$, of a chemical component k at a temperature, T, and pressure, P, can be calculated from (Kondepudi and Prigogine, 1998) at equilibrium, p. 139:

$$\mu_{P,T} = \frac{T}{T_0} \mu_{P_0,T_0} + \int_{P_0}^{P} V_{m(P,T)} dP + T \int_{T_0}^{T} \frac{-H_{m(P,T)}}{T^2} dT \tag{14.4}$$

where V_m and H_m are the partial molar volume and enthalpy for that chemical component and are themselves functions of P and T (Kern and Weisbrod, 1967, their Chapter VII). P_0 and T_0 are a reference pressure and temperature and μ_{P_0,T_0} is the chemical potential at the reference pressure and temperature.

The Gibbs energy, or as it was called by Gibbs, the *free enthalpy*, is given by

$$G = H - TS = E + PV - TS \tag{14.5}$$

where H is the enthalpy. If P and T are taken as constant

$$\begin{aligned} dG &= dE + PdV - TdS \\ &= \sum_{k=1}^{n} \mu_k dN_k \end{aligned} \tag{14.6}$$

The *Gibbs function for a chemical system* consisting of N_i moles of i chemical components is

$$G = \sum_i N_i \mu_i. \tag{14.7}$$

Just as each chemical component, k, is characterised by a Gibbs energy of formation from its chemical elements at a standard state, G_0^k, each chemical reaction is characterised by a *standard Gibbs energy difference of reaction*, $\Delta G_0^{reaction}$, which is the *difference* between the sum of the Gibbs energies of formation for the reactants and the sum of the Gibbs energies of the products at a standard state. For the reaction $A + B \rightarrow 2C$, $\Delta G_0^{reaction}$ is $\Delta G_0^{reaction} = G_0^A + G_0^B - 2G_0^C$.

A pragmatic form for the difference in Gibbs energy difference for a chemical reaction operating at conditions defined by P, T is (Vernon and Clarke, 2008, p. 56):

$$\Delta G_{P,T} = \Delta H_{1,298} - T\Delta S_{1,298} + \int_{298}^{T} \Delta c_P dT - T \int_{298}^{T} (\Delta c_P/T)dT + \int_{1}^{P} \Delta V dP$$

where T is measured in degrees Kelvin but P (for this example only) is measured in bars; the reference temperature and pressure are 25 °C and 1 bar respectively.

Consider the chemical reaction:

$$v_1 A_1 + v_2 A_2 + v_3 A_3 + \ldots + v_n A_n \rightleftharpoons \upsilon_1 B_1 + \upsilon_2 B_2 + \upsilon_3 B_3 + \ldots + \upsilon_m B_m \tag{14.8}$$

where A_k, B_k are sets of reactants and products respectively while v_k, υ_k are stoichiometric coefficients.

It is convenient to define a quantity called the *equilibrium constant*, $K^{equilibrium}$, for the reaction (14.8) as

$$K^{equilibrium} = \frac{b_1^{\upsilon_1} b_2^{\upsilon_2} b_3^{\upsilon_3} \ldots b_m^{\upsilon_m}}{a_1^{v_1} a_2^{v_2} a_3^{v_3} \ldots a_n^{v_n}} = \exp\left[-\Delta G_0^{reaction}/RT\right] \tag{14.9}$$

This means that

$$\Delta G_0^{reaction} = -RT \ln K^{equilibrium}$$

A chemical reaction is driven by the *affinity of the reaction*, $\mathcal{A}$ (De Donder and Van Rysselberghe, 1936, Kondepudi and Prigogine, 1998). Any *closed* system is ultimately driven to a state of chemical equilibrium in which $\mathcal{A} = 0$. The affinity for a general chemical reaction (14.8) is

$$\mathcal{A} = \sum_{k=1}^{n} \mu_{Ak} v_k - \sum_{k=1}^{m} \mu_{Bk} \upsilon_k \tag{14.10}$$

μ_{Ak}, μ_{Bk} are the sets of chemical potentials corresponding to phases A_k and B_k. If $\mathcal{A} = 0$ the system is at equilibrium, if $\mathcal{A} > 0$ the reaction proceeds to the right, if $\mathcal{A} < 0$ the reaction proceeds to the left. From (14.10), for the simple reaction $A \rightarrow B$, the affinity is

$$\mathcal{A} = (\mu_A - \mu_B)$$

As an example, for the reaction $X + Y \rightarrow 2Z$, the affinity can be interpreted as the negative change in Gibbs energy when 1 mol of X and 1 mol of Y react to form 2 mol of Z.

As we have seen the *change* in Gibbs energy (referred to standard states), associated with a chemical reaction, is called the *standard Gibbs energy of reaction*, $\Delta G_0^{reaction}$. Thus, for the reaction (14.8),

$$\Delta G_0^{reaction} = -\left\{ \sum_{k=1}^{n} \mu_{Ak}^0 v_k - \sum_{k=1}^{m} \mu_{Bk}^0 \upsilon_k \right\}$$

For this same reaction the *Gibbs energy change of the reaction* is

$$\Delta G^{reaction} = -\mathcal{A} = -\left\{ \sum_{k=1}^{n} \mu_{Ak} \nu_k - \sum_{k=1}^{m} \mu_{Bk} \upsilon_k \right\}$$

The affinity of a chemical reaction in that system is given by (14.10) but also by $\mathcal{A} = -\left(\frac{\partial G}{\partial \xi}\right)_{P,T}$ where G is given by (14.7). One can also show (Kondepudi and Prigogine, 1998, pp 148–149) that $\mathcal{A} = -\left(\frac{\partial \Psi}{\partial \xi}\right)_{V,T}$ where Ψ is the Helmholtz energy of the system.

$\Delta G^{reaction}$, and hence $(-\mathcal{A})$, are functions of ξ because if the system is not at equilibrium, the mole numbers, N_k, are constrained by the stoichiometric coefficients, ν_k, υ_k as discussed by Zemansky (1951, pp 394–399, pp 437–441); this constraint is a reflection of the phase rule written for a chemically reacting system.

In general, because of the above constraint, the behaviour of G with changes in the reaction extent, ξ, is represented in Figure 14.3(a). An argument for this general form of behaviour is given by Zemansky (1951, pp 416–410). Thus, the general form of the change in affinity with ξ, given by $\mathcal{A} = -\left(\frac{\partial G}{\partial \xi}\right)_{P,T}$, is shown in Figure 14.3(b).

As an example, if we consider the progress of the reaction $A + B \rightleftarrows 2C$ and suppose that at any instant there are N_A, N_B and N_C moles of A, B and C present, then the *Gibbs function for the system*, from (14.7), is

$$G = N_A \mu_A + N_B \mu_B + N_C \mu_C$$

The behaviour of G for this reaction is discussed by Denbigh (1968, pp 135–139), assuming the components are ideal gases, and is illustrated in Figure 14.3(c) where $(G - 2\mu_C^0)$ is plotted against a measure of the progress of the reaction, N_A. Denbigh shows that the form of the curve for G is given by

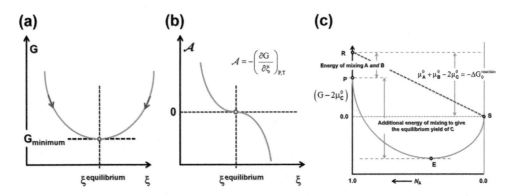

FIGURE 14.3   Gibbs energy and affinity of chemical reactions. Relation between (a) Gibbs energy of the chemical system (blue line) which is a minimum at equilibrium and (b) the affinity of the reaction (blue line) which is zero at equilibrium. If the reaction is reversible of the form $A \rightleftarrows B$ then, in general, $\xi^{equilibrium} \neq 1$. If the reaction is of the form $A \rightarrow B$ then $\xi^{equilibrium} = 1$ (see Gerhartl, 1994). (c) Form of the Gibbs function plotted against a measure of the extent of the reaction $A + B \rightleftarrows 2C$. *Adapted after Denbigh (1968), Figure 18.*

$$\left(G - 2\mu_C^0\right) = N_A\left(\mu_A^0 + \mu_B^0 - 2\mu_C^0\right) + 2RT\left(N_A \ln\frac{N_A}{2} + (1 - N_A)\ln(1 - N_A)\right)$$

This equation is plotted in Figure 14.3(c) as the blue curve, PES. Equilibrium is defined by the minimum, E. It follows that $-\frac{\partial G}{\partial N_A} = \ln(1 - N_A) - \ln\left(\frac{N_A}{2}\right)$ which is of the same form as the curve in Figure 14.3(b). The significance of other quantities is marked on Figure 14.3(c). For a complete discussion of Figure 14.3(c) see Denbigh (1968) and for a detailed discussion of diagrams such as Figure 14.3 see Gerhartl (1994).

For all closed systems, the chemical dissipation including the heat of reaction, $\mathcal{H}$, is $\Phi^{chemical} = \mathcal{A}\dot{\xi} + \dot{\mathcal{H}}$ so that the chemical dissipation (or T times the entropy production) is zero at equilibrium. As we have seen, the affinity is given by $\mathcal{A} = -\left(\frac{\partial G}{\partial \xi}\right)_{P,T}$. It is common to write $\mathcal{A}$ as equivalent to $-\Delta G^{reaction}$; this, however, obscures the fundamental difference between $\mathcal{A}$ as a function that defines the route to reaction completion and hence is related to entropy production or the dissipation (Kondepudi and Prigogine, 1998, p. 111) and $\Delta G^{reaction}$ which is used to define chemical equilibrium.

As shown in Figure 14.3, for a system specified by a given temperature, T, and an externally applied hydrostatic pressure, P, the progression of a chemical reaction towards equilibrium is defined by a function called the *Gibbs energy*, $G \leq 0$. At equilibrium, the *Gibbs energy change of the reaction*, ΔG, is zero:

$$\Delta G = 0$$

For a spontaneous reaction to progress to equilibrium a necessary and sufficient condition is

$$\Delta G \leq 0$$

An important property characteristic of ΔG (that becomes important in systems not at equilibrium) is that for any chemical system, ΔG increases *monotonically* (ΔG never oscillates) towards equilibrium from a non-equilibrium state until equilibrium is reached:

$$\frac{d\Delta G}{dt} \geq 0 \qquad (14.11)$$

where t is time. A function with the property defined in (14.11) is called a *Lyapunov function*. At equilibrium

$$\frac{d\Delta G}{dt} = 0$$

and such an equilibrium state is stable so that for a small perturbation away from that state, the system returns to that state.

14.2.1 The Gibbs–Duhem Equation; the Clapeyron Equation

It might appear that the intensive variables, P, T and μ_k can be specified independently of each other. This however is not the case and a powerful and useful relation is the

Gibbs—Duhem equation which expresses the relation between the intensive variables, P, T and μ_k (Kondepudi and Prigogine, 1998) at equilibrium:

$$SdT - VdP + \sum_{i=k}^{n} N_k d\mu_k = 0 \tag{14.12}$$

If we consider two phases, A and B, at equilibrium under hydrostatic conditions and write (14.12) in terms of molar quantities, $S_{m_j} = \frac{S}{N_j}$, $V_{m_j} = \frac{V}{N_j}$, then, since $d\mu_A = d\mu_B$, we have

$$S_{m_A} dT - V_{m_A} dP = S_{m_B} dT - V_{m_B} dP$$

Thus,

$$\frac{dP}{dT} = \frac{(S_{m_A} - S_{m_B})}{(V_{m_A} - V_{m_B})} \tag{14.13}$$

or,

$$\frac{dP}{dT} = \frac{\Delta H}{T \Delta V_m} \tag{14.14}$$

(14.14) is the *Clapeyron equation* for hydrostatic conditions and it expresses the *slope* of the equilibrium boundary between the stability fields of A and B on an equilibrium P–T mineral phase diagram where P is identified with a hydrostatic pressure. In order to fix the *position* of the boundary at a particular P and T, one needs extra information which is, in principle, obtained by minimising the Gibbs energy at that particular P and T. In Section 14.3.3 3 we discuss the form of the Clapeyron equation for situations where deformation and diffusion accompany the mineral reactions.

Consider now a mineral phase diagram for the reaction A → B shown in Figure 14.4. At the point X (Figure 14.4) A is in its equilibrium stability field and has a chemical potential which can be calculated knowing P and T and using (14.4). We increase the pressure until A

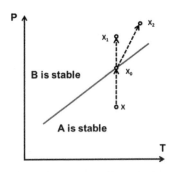

FIGURE 14.4   Phase diagram for a hydrostatic state. A temperature–pressure mineral phase diagram under hydrostatic pressure conditions with the stability fields for A and B marked. A parcel of A is taken from an equilibrium state at X to the phase boundary at X_0 where A and B are in equilibrium. No reaction of A to B takes place at X_0 because $\mu_A = \mu_B$ and there is no driving force for the reaction to take place. The parcel of A has to be taken into the stability field of B at X_1 (where the pressure is overstepped) or to X_2 (where both the temperature and the pressure are overstepped) before the affinity for the reaction, $(\mu_A - \mu_B)$, is nonzero and the reaction can proceed.

lies on the boundary of the stability fields of A and B (point X_0). At the point X_0, A is in equilibrium with B and so $(\mu_A - \mu_B) = 0$. Thus, $\mathcal{A} = 0$ and $\dot{\xi} = 0$. It is fundamental to realise that at equilibrium, there is no driving force for a chemical reaction and hence the reaction rate is zero; *there is no conversion of A into B in this case*. Thus, B is stable at X_0 but does not grow. In order for the reaction to proceed we need to increase $(\mu_A - \mu_B)$ above zero. One way of doing this is to take the parcel of A at the point X_0 (Figure 14.4) where A is in equilibrium and increase the pressure or increase both the pressure and the temperature as shown by the points X_1 and X_2, respectively. Points X_1 and X_2 correspond to states where P and/or T have been *overstepped* with respect to the equilibrium state, X_0. At these points the reaction A $\rightarrow$ B proceeds and finally reaches completion (pure B) given enough time. The reaction of A to B at points X_1 and X_2 depends on nucleation and growth processes the rates of which both depend on the value of $(\mu_A - \mu_B)$ (see Section 15.3). If the overstepped conditions are maintained phase B remains in equilibrium at the overstepped conditions. If any of A were to remain unreacted then A would be in a *metastable* state.

14.3 DEFORMING METAMORPHIC SYSTEMS: A CHEMICAL EQUILIBRIUM VIEW

14.3.1 General Statement

For a comprehensive treatment of classical chemical equilibrium thermodynamics applied to metamorphic mineral assemblages one should consult Kern and Weisbrod (1967), Wood and Fraser (1976), Powell (1978), Connolly and Kerrick (1987), Yardley (1989), Spear (1993), Powell et al. (1998, 2005) and Vernon and Clarke (2008). For a background in chemical equilibrium thermodynamics from a general point of view see Zemansky (1951), Callen (1960) and Denbigh (1968). In this section we are concerned with mineral reactions in a deforming solid and the constraints that deformation places on the growth of new phases and on the chemical equilibrium between mineral phases. We will see that three important issues arise: (1) the kinematic constraints imposed by deformation compatibility upon grain growth, (2) the contributions of both diffusive mass transfer and the deformation gradient to the Gibbs–Duhem equation and hence the Clapeyron *slope* of the mineral equilibrium phase boundary and (3) the influence of stress on the chemical potential which is responsible for the position of the mineral equilibrium phase boundary in mineral equilibrium phase space.

The classical view of metamorphism is a linear theory in which the chemical systems involved behave in a linear manner with no formal coupling between reactions that operate simultaneously. In general the kinetics are not addressed since chemical equilibrium is assumed for the whole system. *Equilibrium is a stationary state of a system characterised by zero entropy production. Chemical equilibrium* is a state of a system where no chemical processes are operating and, in particular, the entropy production arising from chemical processes is zero; on the scale of the system considered there are no gradients in temperature, pressure or chemical potentials and the system is homogeneous as far as the relevant state variables are concerned.

The classical view is that a given parcel of rocks (with a well-defined chemical composition) is placed in an environment where the phase stability boundary is overstepped resulting from changes in pressure and temperature arising from either rising or falling isotherms (see Chapter 11) in conjunction with crustal thickening or thinning (as in the models described by England and Thompson, 1984) or from advection of the rock mass through a changing temperature and pressure field (as in the models described by Jamieson, 1998) or from combinations of these two end members (as in the models described by Gorczyk et al., 2012; Cloetingh et al., 2013). The classical assumption also is that the system is closed so that all of the reactants are available within the system of interest, although some may need to be transported some small distance during the metamorphic process (Carmichael, 1969). As indicated above, an additional assumption is that all chemical reactions proceed independently of each other; although networked reactions such as those described by Carmichael (1969) exist, the reactions proceed with no feedback on each other from a thermodynamic or kinetic point of view. The neglect of all of these factors is justified because the system is at equilibrium and hence the processes involved have ceased and have no influence on the final equilibrium mineral assemblage; *the final equilibrium mineral assemblage is independent of the processes that operated in order to reach that equilibrium state*. Such an independence of processes is, however, not true for the microstructure of deformed metamorphic rocks. The microstructure is largely a leftover from the processes that operated during the chemical reactions. In this section we briefly review the salient points concerning the evolution of linear chemical systems towards equilibrium and in Section 14.4 we extend the discussion to nonlinear, non-equilibrium systems.

Although metamorphic systems are traditionally treated as closed systems (Figure 5.2), at least with respect to chemical components, most are closed *diffusive systems* in the sense of Figure 5.2(b) in that diffusion of some chemical components from a nearby reservoir is commonly postulated. For instance, Carmichael (1969) proposes that, at least for temperatures below sillimanite grade, a chemically reacting system involves migration of Al over distances of ≈ 0.2 mm while components such as Na, K, Ca, Mg, Fe^{2+} may diffuse to the reacting site over distances of $\approx 2-4$ mm; Ti and Fe^{3+} may diffuse over distances of ≈ 0.5 mm. Thus, such systems are closed diffusive systems on the scale of perhaps 1.25×10^{-7} m^3. In Figure 14.2 the systems are open to the transport of mass at least on the scale of 10^{-6} m^3 and perhaps larger. On these scales such systems are considered closed with respect to heat transport (they remain isothermal throughout the reaction) and with respect to the transport of fluids, although some inconsistency is introduced in that nominally 'closed' systems may be allowed the luxury of transport of H_2O and/or H^+ from some remote source. Some (see Figure 5.1(b)) are considered truly open with respect to the transport of H_2O or H^+ and SiO_2.

14.3.2 Mineral Reactions Constrained by Deformation Compatibility. Homochoric Reactions. Why is Pseudomorphic Replacement Common in Deformed Metamorphic Fabrics?

In a system that is not deforming a new grain can grow into a host grain or a host composed of a fine grained matrix with no constraints other than the rate of supply of reactants and heat; local stresses are generated at the interface depending on the ΔV of the reaction. There are no geometrical or kinematic constraints on the growth of the grain, the only constraint being to minimise interfacial energy. This, however, is not the situation if either or both the host and the new grain are deforming since compatibility of the deformation field must be maintained either by the growing grain deforming at the same rate as the host and/or by diffusion and/or by grain boundary sliding accommodating any discontinuities in deformation. This means that *some (at present, poorly defined) combination of coupled chemical mechanisms must operate at the reacting interface so that the rate of growth of* A *matches the rate of dissolution (or removal) of* B.

Consider a new grain of phase A that has nucleated within or at the edge of a host grain, B, and that continues to grow into B (Figure 14.5). We suppose that B is deforming and define a reference system in the undeformed state X_i. The coordinate system in the deformed state is x_i. Then the deformation of B is (see Chapter 2):

$$x_i^B = x_i^B(X_i)$$

and the deformation gradient (Chapter 2) is $\mathbf{F}^B \equiv \begin{bmatrix} \dfrac{\partial x_1^B}{\partial X_1} & \dfrac{\partial x_1^B}{\partial X_2} \\ \dfrac{\partial x_2^B}{\partial X_1} & \dfrac{\partial x_2^B}{\partial X_2} \end{bmatrix}$. Similarly we can define the

deformation gradient for A, $\mathbf{F}^A$, on the assumption that X_i can also be used as a reference frame for A (see qualifications of this assumption by Frolov and Mishin, 2012). If the deformation and growth of A are such that no holes or overlaps develop at the interface between A and B (that is the deformation of A remains compatible with that of B — see Chapter 2) then the boundary between A and B remains an invariant surface in the deformation throughout the growth of A (and dissolution of B). Thus,

$$\mathbf{F}^A = \mathbf{F}^B$$

and hence

$$J^A = J^B$$

where J represents the Jacobian (Chapter 2) of the deformation. Since the Jacobian for each grain is the dilation of that grain it follows that reactions that obey compatibility of

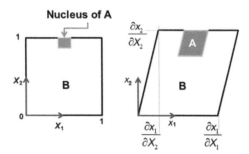

FIGURE 14.5   Nucleation and growth of a new phase A in a deforming host, B. For compatibility of deformation between A and B (that is, no holes or overlaps develop) the deformation gradient in A must be the same as that in B. This means the volume change in A is the same as that in B. It also follows that the rate of growth of A is the same as the rate of dissolution of B.

deformation must preserve the volume of the host even if the host is dilating during deformation. We call such reactions that involve identical dilations of the host and the reaction product(s), *homochoric reactions* (from the Greek, *homós*, the same and *choros*, space). An *isochoric reaction* is a subset of homochoric reactions in that the volumes both of the host and the reaction product do not change during the reaction.

In this discussion the host may be an initial grain or it may be the polycrystalline matrix of the rock (as in the example of staurolite growing in a finely layered matrix presented by Carmichael (1969) — see Figure 15.3(a), or as in the example presented in Figure 15.3(c)). If the initial material is a single crystal of phase B then an isochoric reaction produces A as a *pseudomorph* of B. The situation is relaxed if diffusion is able to assist in attaining deformation compatibility but this has other consequences as we will see in Section 14.3.3 3. The general conclusion remains however that in order to maintain deformation compatibility while a new grain is growing, in the absence of diffusion or grain boundary sliding acting to accommodate the deformation incompatibilities, the reaction must be homochoric and commonly is isochoric since the host may not be undergoing plastic deformation. This differs from the common assumption that metamorphic reactions are *constant mass* in nature. To be consistent with the Greek derivations, such reactions might be called *isomaza reactions*.

As we have indicated, homochoric (and particularly isochoric) reactions imply that the growth rate of the new phase balances the rate of dissolution of the host (so that the two rates are coupled) and that stresses are developed at the interface between the new phase and the host (see Section 15.5 and Merino and Canals, 2011, for an example). We elaborate on these issues in Chapter 15.

14.3.3 Can an Equilibrium Mineral Phase Boundary be Shifted by Deformation? the Roles of Tectonic Overpressure and of Tectonic Overtemperature

The conditions for chemical equilibrium between two mineral phases in a non-deforming system are commonly taken to be that the stress, and hence the pressure, is hydrostatic and, along with the temperature, is homogeneous throughout the system of interest. Added to this is that a convenient potential, normally the chemical potential, a scalar quantity, is continuous across all phase interfaces, and hence is also homogeneous. The equilibrium phase boundary between the two phases on a P–T diagram is then fixed by these conditions since the Clapeyron slope is fixed by (14.14) and the difference in Gibbs energy of the reaction between the two phases, $\Delta G^{reaction}$, and hence the affinity of the reaction, are both zero. In a deforming– chemically reacting system, the stress is non-hydrostatic and both the mean stress and temperature can be inhomogeneous in the region of interest. As indicated in Section 14.1, the conditions for chemical equilibrium across a boundary between two deforming mineral phases are

1. The temperature difference, ΔT, across the grain boundary is zero.
2. The difference in velocity, Δv, across the boundary is zero.
3. The differences in stress components resolved parallel and normal to the grain boundary whose normal is n, are zero, $\Delta\sigma_{ij}n_j = 0$.
4. The difference in a potential that measures the energy, including the chemical energies of the phases, is zero at the boundary of adjacent phases. It turns out that the problem that

has perplexed the metamorphic community regarding this potential is that the form of the potential depends on the orientation of the interface between the two mineral phases in the stress field. Thus, the potential is not a state function in the conventional sense. Nor can the potential be defined within a phase, only at the boundaries between phases.

Given these conditions it is possible, as we will see, to define a boundary in a suitable phase space that is the equilibrium phase boundary in that space between the two deforming phases. Since metamorphic systems are commonly characterised by mineral reactions that take place during deformation, an issue that is continuously raised is: *Is the equilibrium mineral phase boundary defined in conventional P−T space different for a syn-kinematic reaction?* In order to be clear about what this question means we need to understand that the question actually involves three quite distinct issues:

1. The *rates* of chemical reactions are increased by changes in the energy of the system introduction by defects (dislocations, twin boundaries, sub-grain boundaries and the like; Wintsch, 1985; Wintsch and Knipe, 1983; Wintsch and Dunning, 1985). Does this process change the equilibrium phase boundary? We will see that the answer is *no*. Such effects influence the *affinity* of the reaction. However, this energy, if expressed properly, can influence the Gibbs−Duhem equation as in (3) below.
2. If the pressure is equated with the mean stress then this commonly increases during deformation creating a *tectonic overpressure* (Rutland, 1965). In addition, heat is generated from plastic dissipation generating a *tectonic overtemperature*. Do these changes in pressure and temperature influence the position of the equilibrium phase boundary? We will see that the answer involves consideration of two processes. Overpressures and overtemperatures influence the *affinity* of the reaction, through overstepping of the equilibrium phase boundary, and, in particular, overpressure has an effect on the Gibbs−Duhem equation as in (3) below.
3. The form of the Gibbs−Duhem equation is changed due to the mean stress, to diffusion and to deformation including terms that involve the deformation gradient (Fletcher, 1973; Frolov and Mishin, 2012; Larche and Cahn, 1978; Plohr, 2011; Robin, 1974; Shimizu, 1992, 2001). These effects in turn are capable of altering the Clapeyron *slope*. In addition the *position* of the equilibrium phase boundary is changed by new contributions to the Gibbs function of the system. We can say in advance that these effects *do influence the position of the chemical equilibrium phase boundary* while deformation and diffusion are in progress.

14.3.3.1 *The Introduction of Deformation Defects During Mineral Reactions*

Deformation introduces defects such as dislocations, twin boundaries and sub-grain boundaries that influence the internal energy of the system. For a phase, A, sitting on the equilibrium phase boundary between A and B (point X_0 in Figure 14.4), the reaction to form B requires $\mathcal{A} = \Delta\mu = (\mu_A - \mu_B) \neq 0$ in order to proceed. The energy of deformation is capable of supplying the required energy so that $|\Delta\mu| \neq 0$. Gross (1965) and Liu et al. (1995) measured the heat stored in deformed calcite and quartz and report values of $0.8\ kJ\ mol^{-1}$ and $0.5\ kJ\ mol^{-1}$, respectively. Wintsch and Dunning (1985) calculate a maximum stored energy due to a dislocation density of 10^{12} dislocations per cubic centimetre in quartz of 1.4 kJ mol^{-1}. An example of the stored energy due to deformation in Cu is given by Williams (1965) where values up to ≈ 50 J mol^{-1} are reported. Thus, the values reported for stored energy due to deformation for calcite and quartz are much larger than recorded values for deformed metals (see also Stainier and Ortiz, 2010) and hence are expected to be important in enhancing mineral reaction rates.

The larger the $|\Delta\mu|$, the faster is the reaction. However, this does not change the position of the equilibrium phase boundary, it simply ensures that the reaction proceeds; in principle infinitesimal overstepping of the reaction boundary with respect to P and/or T is sufficient in this instance. When the energy of deformation is exhausted, either because the deformation ceases or because energy produced in a deformation is depleted, the thermodynamic state relaxes back to the equilibrium phase boundary if that corresponds to the ambient conditions.

14.3.3.2 Tectonic Overpressure and Overtemperature

Systems that are chemically isolated are not closed to the transport of momentum during plastic deformation so that a stress field $\sigma_{ij}(x, y, z, t)$ is generated within the material; strong gradients in mean stress can exist while deformation is in progress if the deformation is localised or otherwise inhomogeneous. There is sometimes a reluctance to admit that the mean stress plays any role in influencing mineral reactions. We hope to clarify this issue below through a discussion of the Gibbs energy in systems subjected to non-hydrostatic stress.

A basic postulate of chemical equilibrium thermodynamics is that the chemical equilibrium state is characterised completely by a set of state variables, namely the specific internal energy, e, the volume, V, and the mole numbers, $N_1, N_2, \ldots N_r$, of chemical components (Callen, 1960, p. 12). Such a statement assumes that the system is subjected to a uniform hydrostatic pressure, P, such that the normal (compressive) stress on every plane in the body is identical to $-P$. Callen (1960, p. 32) points out that if the stress is not a uniform pressure then other variables, completely analogous to V, are needed to characterise the equilibrium state. A requirement of any theory involving non-hydrostatic stress states is that the treatment reduces to the hydrostatic formalism when the stresses become hydrostatic.

Although we have elected to define the Gibbs energy by (14.5) there are in fact three different independent interpretations of the Gibbs energy all three of which are equivalent if the stress field is a hydrostatic pressure. The first interpretation is that the Gibbs energy is the Legendre transform (see Section 5.8) of the Helmholtz energy. The second interpretation is that the Gibbs energy is a potential that is continuous across the boundary between two phases at equilibrium. The third interpretation is that the Gibbs energy is related to the increase in total internal energy arising from insertion of a unit mass of a substance into the thermodynamic system; this is an interpretation related to the concept of a chemical potential. Although these three interpretations are equivalent for a system subjected to hydrostatic pressure this is not the case for a plastically deforming system under the influence of non-hydrostatic stress and each interpretation can be quite different. The interpretation selected for use depends on the nature of the problem involved.

The first interpretation is widely used for solving problems involving deformation (elastic, plastic, viscous and brittle) where no chemical reactions are involved. The interpretation is discussed for elastic deformations by Nye (1957) and was emphasised and extensively developed for inelastic deformations by Rice (1971, 1975). Examples are given in, and applications are discussed by, Houlsby and Puzrin (2006a) and in Chapter 5.

As indicated in Chapter 5, the definition of Gibbs energy for a deforming system with no chemical reactions means that P is replaced by $\overline{P} = -\frac{1}{3}\sigma_{ii}$ and the specific volume, $\widehat{V}$, is replaced by $\widehat{V} = \widehat{V}_0(1 + \varepsilon_{kk}) = \frac{1}{\rho}$ so that PV is replaced by $-\widehat{V}_0\left(\frac{1}{3}\sigma_{kk} + \sigma_{ij}\varepsilon_{ij}\right)$. Thus, although the pressure is replaced by (minus) the mean stress, the equivalent of the PV term contains a contribution from the total mechanical work, $\sigma_{ij}\varepsilon_{ij}$. This means that the Gibbs energy becomes

$$G = e - sT - \frac{1}{\rho_0}\sigma_{ij}\varepsilon_{ij} \tag{14.15}$$

The second interpretation is discussed by Plohr (2011). He considers the criteria for equilibrium between two solid grains that are undergoing finite, plastic deformation and the form of the relevant Gibbs energy. As we have indicated, the requirement for equilibrium is continuity of temperature, velocity, traction and of a potential that measures the energy difference across the interface between phases. The first three of these constraints are expressed by

$$\Delta T = 0, \quad \Delta v_i = 0, \quad \Delta(\sigma_{ij}n_j) = 0$$

where ΔT is the temperature difference across the grain boundary, Δv_i is the velocity difference across the grain boundary and $\Delta(\sigma_{ij}n_j)$ is the difference in the stress components parallel to the components of the grain boundary normal. Plohr (2011) shows that in order to satisfy these equilibrium conditions, the difference in Gibbs energy across an interface needs to be defined as

$$\Delta G = \Delta\left[e - Ts + \overline{P}\widehat{V} - \widehat{V}_0 J_i s_{ij} n_j\right] = 0 \tag{14.16}$$

where $\overline{P} = -\frac{1}{3}\sigma_{ii}$, s_{ij} is the deviatoric stress tensor, and $J_i \equiv \frac{F_{i\alpha}n_\alpha}{n_\gamma n_\gamma}$. F_{ij} is the deformation gradient tensor and n_α are the components of the normal to the interface between the two grains. $\widehat{V}_0, \widehat{V}$ are the specific volumes in the reference and deformed states. This means that this criterion for equilibrium between the two grains involves the mean stress, $\overline{P}$, and the orientation of the grain boundary relative to the local stress together with the nature of the deformation gradient. We return to this interpretation in Section 14.3.3 3. Clearly, since (14.16) involves the orientation of the boundary, it is defined only at the boundary and not within the body of a grain. It is not solely a function of the state of the system and hence is not a state function in the conventional sense of the concept. This potential also depends on the nature of the deformation, $\mathbf{F}$, at the boundary and hence also depends on the constitutive behaviour of the grains and issues such as: *can the grain boundary support a shear stress?* However, it does serve to define chemical equilibrium across the boundary.

The third interpretation is also considered by Plohr (2011). One arrives at

$$G = e - Ts + \overline{P}\widehat{V} \tag{14.17}$$

as an expression for G where $\overline{P}$ is minus the mean stress.

Plohr (2011) shows that if the deformation gradient involves no grain boundary sliding then the Gibbs energy is a function of $\left(\frac{1}{3}\sigma_{ii}\right)$. However if shear displacements exist, as in grain boundary sliding, then the Gibbs energy is a more complicated function of the stress. Thus the contribution to G from a non-hydrostatic stress always involves the mean stress. In addition there may be other contributions involving the shear stresses arising from the type of deformation and the orientation of the grain boundary. Similar conclusions are reached for infinitesimal deformations by Fletcher (1973), Robin (1974) and Shimizu (2001).

Thus from a thermodynamic point of view the equivalent of the pressure, the *tectonic overpressure*, is (minus) the mean stress, $\left(\frac{1}{3}\sigma_{ii}\right)$, but if one wants to understand the driving force towards equilibrium interpretations of G such as (14.16) need to be considered.

As an example of tectonic overpressures and overtemperatures in an inhomogeneous deformation, Figure 14.6 shows the distribution of mean stress, $(\sigma_{11} + \sigma_{22} + \sigma_{33})/3$, and of local dissipation due to plastic deformation in a system closed to the transport of fluid, heat and chemical components and undergoing a sinistral simple shearing. The model comprises a layered ideal elastic—plastic material with no hardening as in Figure 6.1. In Figure 14.6(a—c) plastic volume change occurs. The mean stress varies from -295 MPa above ambient to -265 MPa between the hinges of folds where plastic dilation is localised and the weaker layers on fold limbs. These pressures are in excess of the ambient pressure and constitute a *tectonic overpressure*. In Figure 14.6(d—f) no plastic volume changes accompany deformation and any volume change is elastic, $\Delta^{elastic}$, given by $\Delta^{elastic} = \sigma_{ii}/[3\lambda + 2G]$ where λ and G are Lame's parameters (Chapter 6; Jaeger, 1969). Again dilation is concentrated at fold hinges. The mean stress now varies between -4.50 MPa and -0.50 MPa through the model. The importance of the mean stress concentrations in Figure 14.6, whether they arise from elastic or plastic dilation, is that they result in an overstepping of mineral equilibrium phase boundaries and hence provide an affinity that drives chemical reactions. The high mean stress in Figure 14.6(b) relative to that in Figure 14.6(e) arises from the large plastic volume change included in the deformation in Figure 14.6(a). The overall deformation is constrained to be isochoric and the large mean stresses are a response of a swelling deforming system subjected to this constraint. Thus, the mean stress (and hence the tectonic overpressure) depends not only on the constitutive behaviour, as is commonly proposed (Rutland, 1965; Schmalholz and Podladchikov, 2013; Stuwe and Sandiford, 1994), but also on the geometrical constraints imposed on the system.

In Figure 14.6(c) and (f) the instantaneous dissipation density arising from the plastic work, $\sigma_{ij}\varepsilon_{ij}^{plastic}$, is plotted. Again the distribution is inhomogeneous with the dissipation concentrated at fold hinges. The temperature changes that arise from this dissipation (*tectonic overtemperatures*) are not plotted; they can be calculated from the energy equation, (5.44), and then integrating the incremental temperature changes from the beginning of plastic deformation up to the current deformation state. Independent of the magnitude of temperature rises due to plastic dissipation such dissipation provides a driving force for mineral reactions.

The systems discussed here are closed and once deformation ceases the temperatures relax to ambient on a thermal timescale given by $\tau^{thermal} = \frac{L^2}{\kappa^{thermal}}$ where L is a length scale associated

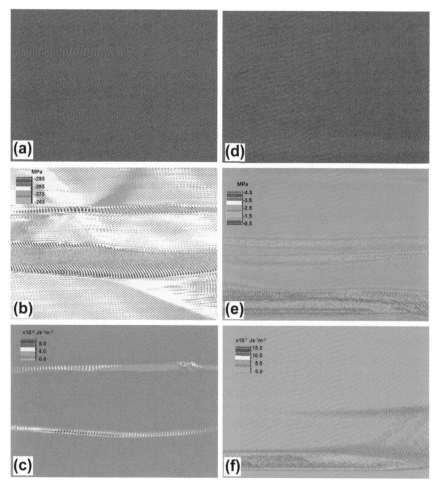

FIGURE 14.6   Sinistral simple shearing of a layered, perfectly plastic material: development of overpressures and dissipation. The initial geometry and deformation conditions are as in Figure 6.1 except that here the layers are of equal thickness. (a) Geometry with plastic volume change. (b) Mean stress with plastic volume change. (c) Dissipation density distribution with plastic volume change. (d) Geometry with no plastic volume change. (e) Mean stress with no plastic volume change. (f) Dissipation density distribution with no plastic volume change.

with the system and $\kappa^{thermal}$ is the thermal diffusivity. The stress state relaxes to the ambient hydrostatic pressure state once boundary driving forces are removed on a time scale, $\tau^{mechanical} = \frac{\eta}{E}$, where η is the viscosity associated with the relaxation process and E is the Young's modulus; for many materials a spectrum of relaxation processes exists each with its own viscosity (Phan-Thien, 2013). For a crustal scale system with $L = 20$ km as an example, and assuming $\kappa^{thermal} = 10^{-6}$ m^2 s^{-1} (Chapter 11), the time scale for thermal relaxation is approximately 13 my. There is much uncertainty about a value for η but if we take $\eta = 10^{20}$ Pa s and $E = 10^{10}$ Pa then $\tau^{mechanical} = 10^{10}$ s or ≈ 3170 y. Thus, the mechanical relaxation should track the overall tectonic relaxation of the crust as it occurs with a time lag of a few thousand years whereas the temperature lags behind the overall thermal relaxation of the crust by millions of years. The important issue is how fast the mechanical boundary conditions relax given a change in crustal scale kinematics. If these kinematic conditions relax on a timescale of 10–20 my then it is possible for stress states to persist after temperatures have relaxed. This means that non-hydrostatic equilibrium states can be preserved. Clearly until we know more about crustal scale thermal–mechanical relaxation spectra it is possible that a complete spectrum between hydrostatic and non-hydrostatic equilibrium states can be preserved. The way to explore the crustal scale thermal–mechanical spectra is to examine the record in the rock and not always assume that preserved mineral assemblages represent hydrostatic stress conditions. If the preserved mineral assemblages reflect hydrostatic conditions it is unlikely that they reflect syn-kinematic equilibrium assemblages. The challenge is to use the mineral assemblages we observe in metamorphic rocks to understand this relaxation spectrum.

Although the concept that deformation increases the mean stress in mineral reaction systems has been discussed by many workers (Rutland, 1965; Stüwe and Sandiford, 1994, and references therein) the implication commonly is that such an effect influences the equilibrium stability fields of minerals by creating a *tectonic overpressure*. Tectonic overpressures influence the affinity of a reaction and so act to drive the mineral reactions. They also influence the equilibrium phase boundary as we discuss below.

14.3.3.3 The Gibbs–Duhem and Clapeyron Equations for Deforming–Chemically Reacting Systems

It is convenient in examining deforming–chemically reacting systems to set up a virtual network, introduced by Robin (1974) and Larche and Cahn (1973, 1978), that permeates the solids that are deforming and reacting. In the case of a polycrystalline solid this could be a series of crystal lattices or it could be quite irregular and simply represent a convenient set of points that can be addressed in a fine grained matrix (Figure 14.7).

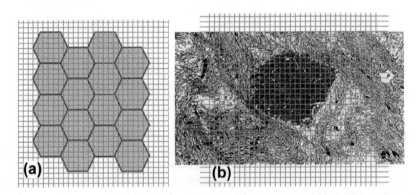

FIGURE 14.7 Examples of virtual networks that enable substitutional and interstitial sites to be identified and tracked during deformation. (a) The grid intersections could represent lattice sites in a homogeneous polycrystalline solid. (b) The grid intersections could represent both lattice sites within a single crystal and both material and compositional reference points within a fine grained matrix. *Background image from Passchier and Trouw (1996).* For presentation the grid spacing is greatly exaggerated and should be at the molecular scale.

The network is used (1) to enable a description of the deformation of the solid, (2) is capable of carrying mechanical loads and (3) provides a set of sites that can be occupied by atoms or vacancies; all chemical components can be divided into *substitutional* sites (residing on the network nodes and including vacancies) or *interstitial* (occupying sites not on the network nodes). Frolov and Mishin (2012) consider a homogeneous multicomponent solid with $\mathcal{K}$ substitutional (including vacancies) and $\mathcal{L}$ interstitial chemical components. The number of substitutional sites is N and is fixed. The number of interstitial sites is n and can vary. Diffusion of both substitutional and interstitial sites can occur.

Frolov and Mishin (2012) explore the deformation illustrated in Figure 14.8(a) for which the deformation gradient is of the form

$$F_{ij} = \begin{bmatrix} F_{11} & F_{12} & F_{13} \\ 0 & F_{22} & F_{23} \\ 0 & 0 & F_{33} \end{bmatrix} \tag{14.18}$$

with a Jacobian

$$J = F_{11}F_{22}F_{33} \tag{14.19}$$

If two adjacent grains both undergo the deformation (14.18) shown in Figure 14.8(a) then (14.18) represents a deformation where the X_1 axis is parallel to a *coherent boundary* (see Robin, 1974) between the grains (Figure 14.8(b)). This means that the interface can support a shear stress which is not necessarily the case if grain-boundary sliding is permitted. One can show that $\mathbf{F}^{-1}$ has the same form as (14.18) with the lower non-diagonal terms equal to zero and $F_{ii}^{-1} = 1/F_{ii}$ for $i = 1, 2, 3$ (no summation). All components of stress can be non-zero. The internal energy is given by

$$E = E(S, N_1, \ldots, N_{\mathcal{K}}, n_1, \ldots, n_{\mathcal{L}}, \mathbf{F})$$

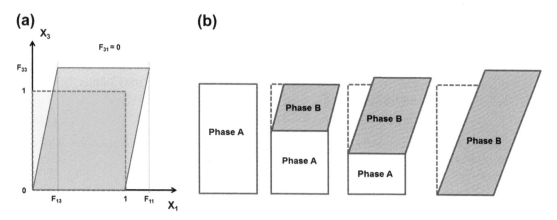

FIGURE 14.8   Coherent phase transformations. (a) The deformation explored by Frolov and Mishin (2012). (b) The progressive development of a coherent phase transformation with volume change.

with N fixed so that $\sum_{\mathcal{K}} N_{\mathcal{K}} = N$. This means that N can take on only $(\mathcal{K}-1)$ independent values. Following Larche and Cahn (1973, 1978) who define a *diffusion potential*, M_{ij}, for substitutional atoms,

$$dE = TdS + \sum_{k=2}^{\mathcal{K}} M_{k1}dN_k + \sum_{p=1}^{\mathcal{L}} \mu_p dn_p + \sum_{i,j=1,2,3} VP_{ij}dF_{ji}$$

which is more conveniently written:

$$dE = TdS + \sum_{k=2}^{\mathcal{K}} M_{k1}dN_k + \sum_{p=1}^{\mathcal{L}} \mu_p dn_p + \sum_{i,j=1,2,3} VF_{11}F_{22}\sigma_{3i}dF_{i3} + \sum_{i,j=1,2} VP_{ij}F_{22}dF_{ji}$$

$$(14.20)$$

The diffusion potential, M_{k1}, is the energy change when an atom of the substitutional component, k, is replaced by an atom of component 1 (taken as a reference component) while keeping all other components fixed:

$$M_{k1} = \frac{\partial E}{\partial N_k} - \frac{\partial E}{\partial N_1}, \quad k = 2,3,...\mathcal{K}$$

Frolov and Mishin (2012) define an intensive potential:

$$\phi_m = \frac{E}{N} - \frac{TS}{N} - \sum_{k=2}^{\mathcal{K}} \left(M_{km}\frac{N_k}{N} \right) - \sum_{p=1}^{\mathcal{L}} \left(\mu_p \frac{n_p}{N} \right) - \sum_{i=1,2,3} \left(\frac{\Omega F_{i3}}{F_{33}} \right)\sigma_{3i}$$

where E/N, S/N and Ω are the energy, entropy and volume per substitutional site. The Gibbs–Duhem relation follows as

$$0 = -SdT - \sum_{k=1}^{\mathcal{K}} N_k d\phi_k - \sum_{p=1}^{\mathcal{L}} n_p d\mu_p - \sum_{i=1,2,3} (vF_{i3}/F_{33})d\sigma_{3i} + \sum_{i,j=1,2,3} VQ_{ij}dF_{ji} \quad (14.21)$$

where

$$\mathbf{Q} \equiv J\mathbf{F}^{-1} \cdot \left(\boldsymbol{\sigma} - \sum_{m=1,2,3} (F_{m3}/F_{33})\sigma_{3m}\mathbf{I} \right) \quad (14.22)$$

In the above, V and v are the volumes in the undeformed and deformed states, so that $v = JV$, $\boldsymbol{\sigma}$ is the Cauchy stress and $(J\mathbf{F}^{-1} \cdot \boldsymbol{\sigma})$ is the first Piola–Kirchhoff stress, $\mathbf{P}$ (see Chapter 4). For the deformation gradient defined by (14.18),

$$P_{3i} = F_{11}F_{22}\sigma_{3i} = (J/F_{33})\sigma_{3i} \quad i = 1, 2, 3 \quad (14.23)$$

In (14.22) $\mathbf{Q}$ is a 3×3 matrix but only Q_{11}, Q_{12} and Q_{22} appear in the Gibbs–Duhem equation, (14.21).

In the case of reactions proceeding at hydrostatic pressure (such that $\sigma_{ij} = -P\delta_{ij}$), $Q_{ij} \equiv 0$ and $\sum_{i=1,2,3} (VF_{i3}/F_{33})d\sigma_{3i} = -VdP$, so that (14.21) reduces to the standard Gibbs–Duhem equation, (14.14).

The Gibbs–Duhem equation, (14.21) expresses a relation between $(\mathcal{K} + \mathcal{L} + 7)$ intensive variables and characterises a deforming single phase solid. In exactly the same way as is done for a system under hydrostatic stress (Section 14.2) if there are two phases, A and B, we write the Gibbs–Duhem equation for both phases and require for equilibrium that the two equations are true simultaneously. We obtain the *Clapeyron equation*:

$$0 = -[\![S]\!]dT - \sum [\![n_i]\!]d\phi_i - \sum [\![N_i]\!]d\mu_i - \sum_{i=1,2,3} [\![(vF_{i3}/F_{33})]\!]d\sigma_{3i} + \sum_{i=1,2,3} [\![VQ_{ij}]\!]dF_{ji}$$

(14.24)

where for any pair of extensive properties, Z and X,

$$[\![Z]\!]_X \equiv Z^A - Z^B \frac{X^A}{X^B}$$

Thus, $[\![Z]\!]_X$ is the difference (or *jump*) between the property, Z, of the two phases when they contain the same amount of X. Equivalents in the hydrostatic case are the molar jumps, $[\![S]\!]_N \equiv \Delta S_m$, and $[\![V]\!]_N \equiv \Delta V_m$, in the hydrostatic Clapeyron equation, (14.13, 14.14).

(14.24) is the generalised Clapeyron equation for a deforming reacting solid with diffusion and coherent grain interfaces, there being $(\mathcal{K} + \mathcal{L} + 5)$ degrees of freedom in the system (Frolov and Mishin, 2012, p. 15). In order to obtain some insight into the meaning of the generalised Clapeyron equation, (14.24), we examine the situation of an equilibrium between two coherent phases in the projection of the hyper-phase diagram into $(\sigma_1 - \sigma_2 - T)$ – space (Figure 14.9(b)). We first note (Figure 14.9(a)) that if the stress is hydrostatic then the hydrostatic pressure is represented by a line inclined at $\tan^{-1}\sqrt{2}$ to each of the σ_1, σ_2, σ_3 axes. This projects as a line at $45°$ to both σ_1 and σ_2 axes in the $(\sigma_1 - \sigma_2)$-plane. If the stress is non-hydrostatic then the line representing: $\sigma_1 + \sigma_2 + \sigma_3 = constant$ projects as a line inclined at $\tan^{-1}(\sigma_2/\sigma_1)$ to the σ_2 axis in the $(\sigma_1 - \sigma_2)$-plane and with magnitude $\tilde{P} = \sqrt{\sigma_1^2 + \sigma_2^2}$. Thus, the conventional P–T phase plane appears as the blue plane (OHIG) in Figure 14.9(b) but the hydrostatic pressure is measured in units of $\sqrt{2}P$. The (mean stress–temperature)-plane appears in red (OEFG). We add the yellow plane, ABCD, to Figure 14.9(b). This is the projection into $(\sigma_1 - \sigma_2 - T)$-space of the hyperplane that contains all equilibrium states between the two coherent grains. The line $\mathcal{L}_{hydrostatic}$ is equivalent to the conventional equilibrium phase boundary for a hydrostatic stress. The line $\mathcal{L}_{non-hydrostatic}$ is the equilibrium phase boundary for non-hydrostatic stress for a given deformation defined by (14.18). If the deformation changes for a given mean stress then so does the position of this line.

A somewhat simpler example of the effect of deformation on the position of the equilibrium phase boundary is indicated in Figure 14.10(a) and (b) where we consider the projection of the non-hydrostatic equilibrium space into the (P–T–chemical potential)-space. We assume that the pressure in the hydrostatic case is given by the overburden pressure and identify this with $-\sigma_3$. The chemical potential surfaces (see Denbigh, 1968, p. 183) for two mineral phases, α and β, are shown in (P–T–μ) space in Figure 14.10(a). The line A–B represents the equilibrium phase boundary in this space where $\mu_\alpha = \mu_\beta$. The projection of this boundary on to the P–T plane is the line A*–B* as shown in Figure 14.10(b). This line then represents the

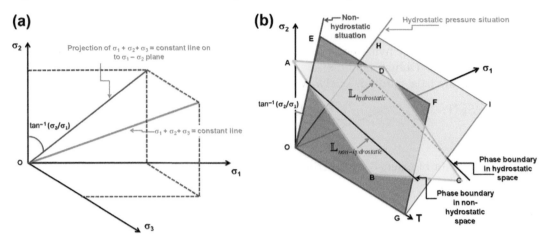

FIGURE 14.9 Aspects of the phase diagram for chemical equilibrium between two phases in a deforming solid. (a) Projection of the line (blue) $\sigma_1 + \sigma_2 + \sigma_3 = $ constant on to the $(\sigma_1 - \sigma_2)$-plane. The projection makes an angle $\tan^{-1}(\sigma_2/\sigma_1)$ with the σ_2-axis. The length of this line is $\tilde{P} = \sqrt{\sigma_1^2 + \sigma_2^2}$. (b) Projection of the hyper-phase diagram into $(\sigma_1 - \sigma_2 - T)$ space. The red plane (OEFG) is the equilibrium phase diagram in the deformed state and corresponds to the $(\tilde{P} - T)$-plane. $\mathbb{L}_{non-hydrostatic}$ is the generalised Clapeyron curve in that plane. The blue plane (OHIG) is the equilibrium phase diagram for the hydrostatic case with $\sqrt{2}P$ as the pressure axis. $\mathbb{L}_{hydrostatic}$ is the equivalent Clapeyron curve for this hydrostatic case. The yellow plane (ABCD) is the locus of the projection of all Clapeyron curves into $(\sigma_1 - \sigma_2 - T)$ space.

conventional equilibrium phase boundary. In Figure 14.10(c) we stay in $(P^* (=-\sigma_3)-T-\mu)$ space but the chemical potentials are changed in accordance with the theory of Kamb (1959) so that a chemical potential of each phase, μ_0^i, becomes $\mu^i = \mu_0^i \pm \sigma_N^i \widehat{V}^i$ (see Chapter 15) where σ_N^i is the normal stress across an interface of the ith phase and $\widehat{V}^i$ is the specific volume of that phase. The new equilibrium phase boundary in the stressed state, where $\mu_\alpha = \mu_\beta$, is the line C−D which projects on to the P^*−T plane as C^*−D^* as shown in Figure 14.10(d). Clearly the displacement of the phase boundary depends on the relative orientation of the two chemical potential surfaces and the values of $\sigma_N^i \widehat{V}^i$ but except when $\sigma_N^\alpha \widehat{V}^\alpha = \sigma_N^\beta \widehat{V}^\beta$ there is always a displacement of the equilibrium phase boundary on the P^*−T plane.

The generalised Clapeyron equation, (14.24), indicates that in a deforming−reacting−diffusing solid, for a homogeneous state of mean stress, the equilibrium phase boundary is different for grains that have experienced different deformations so that $\sum_{i=1,2,3} [\![(F_{i3}/F_{33})]\!]$ differs from grain to grain. This emphasises the need to understand the mechanisms of grain growth in a deforming aggregate. The difference between homochoric, isochoric and iso-mass reactions is fundamental. For instance, if the reactions are isochoric, as when a porphyroblast, A, replaces a fine-grained matrix, M, with no change in volume, then $F^A = F^M = F$. The outcome is that

$$[\![Q_{ij}]\!] = JF_{ij}^{-1}\left([\![\sigma_{ij}]\!] - \sum_{m=1,2,3}(F_{i3}/F_{33})[\![(\sigma_{m3})]\!]\delta_{ij}\right) \qquad (14.25)$$

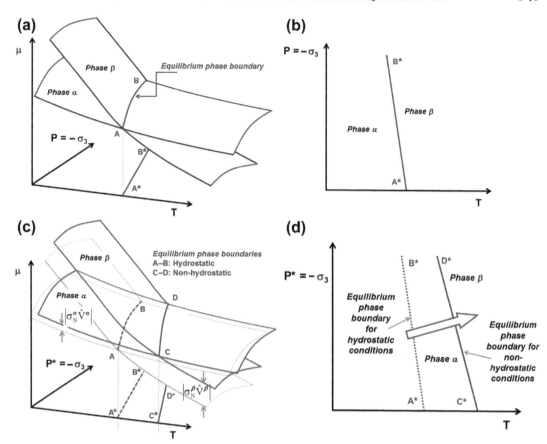

FIGURE 14.10   The chemical potential surface for two phases in a deforming solid. (a) A (P–T–μ) phase diagram for two coexisting mineral phases α and β with a hydrostatic pressure. The surfaces for the two phases intersect in the equilibrium boundary A–B which projects on to the P–T plane as the phase boundary A*–B* in (b). In this case the hydrostatic pressure is identified with the overburden stress $-\sigma_3$. (c) A section similar to (a) in the non-hydrostatic case where the hydrostatic phase surfaces for α and β have been displaced by $\sigma_N^\alpha \hat{V}^\alpha$ and by $\sigma_N^\beta \hat{V}^\beta$, respectively, for phases α and β. The non-hydrostatic equilibrium phase boundary is shifted from A–B to C–D with the result that the non-hydrostatic equilibrium phase boundary in the P*–T plane is moved from A*–B* to C*–D* as shown in (d).

This means that the equilibrium depends on the differences in stress states between the two grains (or domains). For a coherent interface the shear stress differences may be zero and so the problem reduces to the Kamb–Paterson model where only the normal stresses influence the non-hydrostatic equilibrium phase boundary. For a general deformation in each grain (or domain) this will not be the case and different grains will be in different equilibrium states. However, if the stresses relax to hydrostatic and the temperature remains high enough for diffusive fluxes to operate then the system approaches the hydrostatic pressure equilibrium state.

All of the treatments of chemical potential or Gibbs energy for non-hydrostatically stressed solids (Frolov and Mishin, 2012; Plohr, 2011; Shimizu, 1992, 2001) contain a term that depends on the grain boundary orientation as well as the deformation at the grain boundary and the stress within the grain. In the elastic case with only a normal stress on the boundary this is the Gibbs−Kamb−Paterson term, $\sigma_N \widehat{V}$ (see Section 15.4). For the treatment of Shimizu (2001), this is related to her term, $\theta_{ij} = \frac{a_{ij}N_j}{a_kN_k}$, where a is a vector that represents the crystal growth during a chemical reaction and N is a vector normal to the reacting interface (Figure 14.11(a)). Then the chemical potential for the phase change, μ^+, which we emphasise is only defined at the reacting interface, is given by

$$\mu^+ = -\sigma_{ij}\theta_{ij}\widehat{V} + \Psi$$

$$\equiv -\frac{|T|\cos\beta}{\cos\alpha}\widehat{V} + \Psi$$

where Ψ is the Helmholtz energy, α and β are defined in Figure 14.11(a) and T is the traction acting on a unit area of the interface. The term $\frac{|T|\cos\beta}{\cos\alpha}$ reaches a maximum when $\alpha = 0$ and $\beta = 0$ when the traction and growth directions are normal to the interface. This is the situation in the Gibbs−Kamb−Paterson model when μ^+ becomes $(-\sigma_N\widehat{V} + \Psi)$. If $\sigma_N = 100$ MPa and the molar volume, $\widehat{V} = 5 \times 10^{-5}$ m^3 mol^{-1} (see Fletcher and Merino, 2001), the term $\sigma_N\widehat{V}$ contributes the equivalent of 5 kJ mol^{-1} to the effective chemical potential so the effect can be significant.

Further insight into the nature of equilibrium between two grains in a deforming solid can be obtained by considering the case of coherent equilibrium between two grains in a binary solid solution. For variation of the shear stress at constant temperature and strain,

$$\frac{dM_{21}}{d\sigma_{3i}} = -\frac{[\![vF_{i3}/F_{33}]\!]_N}{[\![N_k]\!]_N} \tag{14.26}$$

Thus, changes in phase composition result from the stresses, σ_{3i}, and to maintain equilibrium, variations in the diffusion potential with respect to variations in the shear stress

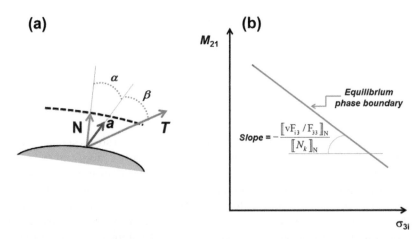

FIGURE 14.11 Aspects of non-equilibrium phase diagrams. (a) Diagram showing the significance of the quantities used by Shimizu (2001) to describe the influence of a general stress on the chemical potential. (b) Generalised Clapeyron slope in ($M_{21} − \sigma_{3i}$)-space.

must be proportional to the jump in deformation gradient and inversely proportional to the jump in phase compositions. If this is not the case then compositional differences arise due to the shearing stresses and these differences may persist after deformation ceases. (14.26) is a generalised Clapeyron equation in $(M_{21} - \sigma_{3i})$-space as shown in Figure 14.11(b). The conventional interpretation of such compositional differences (see Figure 14.2) is that changes in P and/or T were responsible. However, since such compositional differences are predicted for deforming systems, this possibility should also be explored.

The above discussion from Frolov and Mishin (2012) is strictly applicable only to coherent relations between two adjacent grains. If this condition is not met then deformation compatibility between adjacent grains has to be met by diffusive mass transfer or by grain boundary sliding or by both processes operating simultaneously. This general situation has not been analysed but one outcome is that shearing stresses parallel to grain interfaces are relaxed by these processes. We can expect results similar in principle to what is depicted in Figures 14.9–14.11(b) but the analysis is yet to be done for the general case where diffusion is in progress.

14.3.4 The Phase Rule for Closed Equilibrium System under Hydrostatic Stress

The phase rule is a statement of the number of independent degrees of freedom, $\mathcal{F}$, available to any chemical system defined by a number of mineral phases, $\mathcal{M}$, and molar fractions, $r_{\mathcal{C}}$, of $\mathcal{C}$ chemical components. $\mathcal{F}$ can be defined for any system depending on the constraints on $\mathcal{M}$ and on $r_{\mathcal{C}}$. The latter constraint is always $\sum_{n=1}^{n=\mathcal{C}} r_{\mathcal{C}} = 1$. $\mathcal{F}$ is the number of independent intensive variables necessary to define the system. For a system at equilibrium there are $r_{\mathcal{C}}(\mathcal{M} - 1)$ equations defining the relations between the chemical potentials for each chemical component in the $\mathcal{M}$ phases and a set of $[2 + r_{\mathcal{C}}(\mathcal{M} - 1)]$ independent variables corresponding to P, T and the $(r_{\mathcal{C}} - 1)$ independent mole fractions. Thus, $\mathcal{F} = r_{\mathcal{C}} - \mathcal{M} + 2$ for a system at equilibrium (Callen, 1960). For a detailed discussion of the phase rule and its application to the construction of mineral equilibrium phase diagrams see Kern and Weisbrod (1967, Chapter XII). The phase rule is used to constrain the topology of equilibrium mineral phase diagrams (Powell et al., 1998). Levin (1946) and Fink (2009) point out similarities between the phase rule and the Euler–Poincare formula relating the numbers of faces, edges and vertices on a simple polyhedron (see Chapter 13). Some readers may find this analogy useful in understanding the phase rule and the way the degrees of freedom change due to the addition or removal of a phase or chemical component.

14.4 MINERAL REACTIONS: SYSTEMS NOT AT EQUILIBRIUM

14.4.1 General Statement

We have seen in Section 14.3 that deformation compatibility places kinematic constraints on the growth of new grains such that in many instances the reactions are homochoric and commonly isochoric. In addition we see that both diffusion and deformation can influence the form of the Gibbs—Duhem equation and hence the Clapeyron slope of the *chemical equilibrium* phase boundary between reacting phases. If we want to understand something about the *processes* that result in mineral assemblages we need to explore the *kinetics* of reacting mineral systems, in particular the coupling between the rates of supply of reactants to the reacting site. Further, in this section, we explore some general features of chemical systems not at equilibrium particularly those associated with *networked chemical reactions*. We defer a discussion of the nonlinear growth rates implied by homochoric reactions to Chapter 15.

14.4.2 Kinetics of Mineral Reactions

In Chapter 7 we defined the concept of a stationary state as one where the rate of change of the mean of a particular quantity of interest is zero. For any chemical system one possible stationary state is an equilibrium state where all chemical reactions have ceased and hence the rate of change of chemical concentrations is everywhere zero. However, other, non-equilibrium stationary states can exist in chemical systems. First let us consider an isothermal chemical reaction in an isolated system of the type

$$A \overset{k}{\to} B \tag{14.27}$$

with a rate constant k and a reaction rate, $\dot{\xi}$, given in terms of the concentrations of A and B, a and b, by

$$\dot{\xi} = -\frac{da}{dt} = \frac{db}{dt} = ka \tag{14.28}$$

where the over-dot denotes differentiation with respect to time and we have used capital font Arial for chemical components and lower case italic font Arial for concentrations of those components. It is common to use the activities instead of concentrations in which case the activity of species i, a_i, is given by $a_i = \gamma_i c_i$ where γ_i the activity coefficient of species i with concentration c_i.

It follows from (14.28) that

$$a = a_0 \exp(-kt) \quad \text{and} \quad b = b_0[1 - \exp(-kt)]$$

where a_0 and b_0 are the initial concentrations of A and B. We define the *extent of the reaction* by

$$\xi = \frac{a_0 - a}{a_0}$$

ξ varies from $\xi = 0$ at the start of a reaction to $\xi = 1$ at completion when all of A has been converted to B. The evolution of the reaction rate and the concentration are shown in Figure 14.12(a) and (b). Since the system is closed, $a_0 + b_0 = a + b$ and $\xi = \frac{a_0-a}{a_0} = \frac{b-b_0}{a_0}$.

One should note that for the reaction $A \overset{k}{\to} B$, the evolution of the extent of the reaction is given by

$$\xi = 1 - \exp(-kt)$$

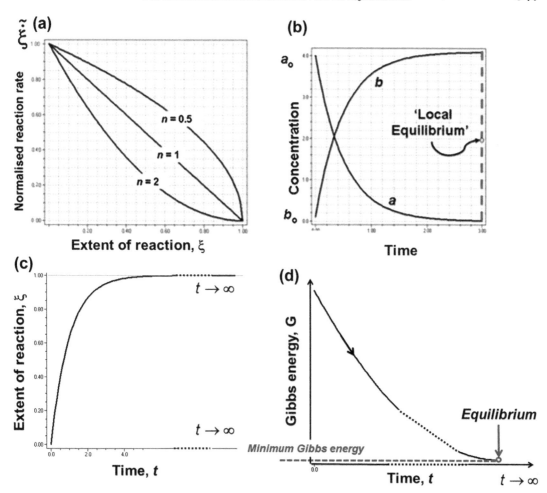

FIGURE 14.12   Uncoupled kinetics. (a) Normalised reaction rate, $\tilde{\tilde{\xi}}$, for the reactions $A \to B$ ($n = 1$), $0.5A \to B$ ($n = 0.5$) and $A \to 2B$ ($n = 2$) in an isolated, isothermal system. The reaction rate is normalised against the maximum reaction rate for these reactions. (b) Evolution of the concentrations a and b for the reaction (14.27) with initial concentrations a_0 and b_0. After dimensionless time equal to 3 the concentrations, a and b, have approached steady state and the reaction proceeds to completion. In this situation the term 'local equilibrium' is used to describe the state to the right where concentrations are near to constant. (c) Evolution of the extent of reaction, ξ, towards equilibrium for reaction (14.27). (d) Evolution of the Gibbs energy for the system described by (14.27) towards a minimum at equilibrium as $t \to \infty$.

which is plotted in Figure 14.12(c). $\xi \to 1$ as $t \to \infty$. During this evolution, the Gibbs energy, G, evolves to a minimum at $t \to \infty$ (Figure 14.12(d)). Note also that the conditions $\xi = 1$ and $G = G_{minimum}$ at $t \to \infty$ are not generally true. If, for instance, the reaction is $A \rightleftharpoons_k^k B$ so that the forward and reverse rate constants are equal, then the equilibrium, $G = G_{minimum}$, occurs at $\xi = \frac{a_0 - b_0}{2}$ with $a_{equilibrium} = b_{equilibrium} = \frac{a_0 + b_0}{2}$ where a_0 and b_0 are the initial concentrations of A and B at $t = 0$ (see Figure 14.3). Examples are given by Kondepudi and Prigogine (1998, pp 245–249) and by Gerhartl (1994).

One can see that after a period of time the rate of change of both *a* and *b* approaches zero. Thus, the system approaches a stationary state for the system and it is an equilibrium stationary state. In an isolated system this is the only stationary state for a reaction of the type (14.27); however, for an open flow system, where A is continuously input, another possibility exists (Figure 14.13) which is considered in Section 14.4.3.

Consider a chemical reaction

$$A + B \rightarrow C$$

Then the rate of consumption of A is (Epstein and Pojman, 1998):

$$\dot{\xi}_A = -\frac{da}{dt}$$

Similarly

$$-\frac{da}{dt} = -\frac{db}{dt} = +\frac{dc}{dt} = \textit{rate of reaction, } \dot{\xi}$$

For the reaction

$$A + 2B \rightarrow 3C$$

we have

$$\textit{rate of reaction, } \dot{\xi} = -\frac{da}{dt} = -\frac{1}{2}\frac{db}{dt} = +\frac{1}{3}\frac{dc}{dt}$$

In general the rate of a reaction is given by

$$\dot{\xi} = \frac{1}{\upsilon_J}\frac{dj}{dt}$$

where υ_J is the stoichiometric coefficient of component J and is positive if J is a product of the reaction and negative if it is a reactant.

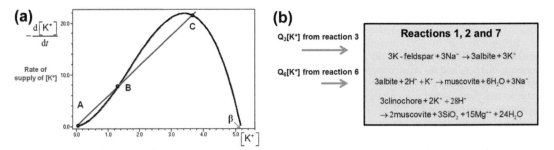

FIGURE 14.13   A sub-system from the reactions in Table 14.1. (a) The form of the curve (in black) that expresses the variation of the rate of production of [K$^+$] with changes in [K$^+$] for the coupling of three of the reactions in Table 14.1. To this curve we have added (in red) a line representing the rate supply of [K$^+$]. This line intersects the black curve in three points (A, B, C) each one of which represents the conditions for the rate of supply of [K$^+$] to equal the production of [K$^+$]. Thus, A, B, C are stationary states of the reaction–transport system. (b) The reactions 1, 2, and 7 from Table 14.1 acting as a reacting unit. These reactions cannot proceed without the supply of K$^+$ from reaction 3 and from reaction 6. These supplies presumably occur at different volumetric flow rates, Q$_3$ and Q$_6$, governed by the rates of reactions 3 and 6.

14.4.3 Linear and Nonlinear Chemical Systems

The use of the term *nonlinear* in the context of chemical systems needs clarification. Distinctions need to be made between linear and nonlinear *thermodynamic chemical reactions*, linear and nonlinear *chemical kinetics* and linear and nonlinear *chemical systems*. All chemical reactions (including the reaction A → B) are thermodynamically nonlinear (Ross, 2008) in the sense that the thermodynamic force that drives the reaction (the affinity of the reaction, $\mathcal{A}$) is never a linear function of the thermodynamic flux (the rate of the reaction, $\dot{\xi}$). To see this consider the reversible reaction

$$A \underset{k^-}{\overset{k^+}{\rightleftarrows}} B$$

where k^+ and k^- are the forward and reverse rate constants for the reaction. We write r^+ and r^- as the forward and reverse reaction rates. The affinity for the reaction is (Ross, 2008)

$$\mathcal{A} = k^B T \ln\left(\frac{r^+}{r^-}\right)$$

where k^B is Boltzmann's constant, while the reaction rate is

$$\dot{\xi} = \left[r^+ - r^-\right]$$

Clearly $\mathcal{A}$ is not a linear function of $\dot{\xi}$. This is true for all chemical reactions (Ross, 2008).

Early work on systems not at equilibrium made the distinction between chemical systems 'close to equilibrium' that were said to be linear systems, whereas systems 'far from equilibrium' were said to be nonlinear and *dissipative*. A criterion for measuring how far a chemical system is from equilibrium was given by de Groot and Mazur (1984) and by Glansdorff and Prigogine (1971): for the reaction A → B, for instance, the criterion would be $|\mu_A - \mu_B| \ll RT$. The proposed linear nature of chemical systems 'close' to equilibrium was taken as justification for using the *Onsager reciprocal relations* for such systems (Kondepudi and Prigogine, 1998). The term *dissipative* was used to distinguish systems 'close' to equilibrium where no spatial structures evolved from those 'far' from equilibrium characterised by the development of spatial 'dissipative' structures.

Most of these assertions turn out to be in error or are misleading. For a start, all chemical systems not at equilibrium dissipate energy through both the chemo-mechanical work, (14.2), and the chemical work, (14.3). This is true both 'far' and 'close' to equilibrium. Secondly, in a series of publications (Hunt et al., 1987, 1988; Ross, 2008), Ross and co-workers have pointed out that the statement that chemical systems are linear systems close to equilibrium derives from issues concerning the expansion of expressions for the thermodynamic fluxes and for the dissipation as power series. In essence the Prigogine—Kondepudi approach is to truncate the expression for the dissipation away from equilibrium at the first term and then establish a minimum for the dissipation for variations in the composition. *This automatically guarantees the system is treated as a linear one.* The approach of Hunt and others is to take the full expression for the dissipation and expand it in a power series with respect to the departure from equilibrium, $\delta^{equilibrium}$, and the departure of concentration, δ_x, from an equilibrium value. This expression is then differentiated with respect to δ_x and then truncated with respect to $\delta^{equilibrium}$. As Ross (2008) emphasises, the operations of truncation and differentiation do not commute and one needs to be careful with the order in which one performs these operations. The result is that no extremum value exists for variations in δ_x except for the equilibrium state, $\delta^{equilibrium} = 0$, when the dissipation is zero. For a discussion of these issues see the interchange between Hunt et al. (1987, 1988) and Kondepudi (1988) and also Ross (2008).

In Figure 14.12(a) the straight line labelled $n = 1$ corresponds to the reaction $A \rightarrow B$; such a reaction is known as a *first order reaction*. The curves labelled $n = 0.5$ and $n = 2$ correspond to the reactions $0.5A \rightarrow B$ and $2A \rightarrow B$, respectively (known as 'half' order and second order reactions). Only the curve for $n = 1$ represents *linear kinetics*. The other two involve *nonlinear kinetics*. All three curves approach zero monotonically at $\xi = 1$. Provided the system is able to evolve towards completion a smooth ultimate approach towards zero reaction rate at $\xi = 1$ is true for all non-reversible chemical reactions in isolated or diffusive systems or whether the reaction has had a stable or unstable history. However, we are more interested in chemical reactions where the reaction rate does not always approach zero at $\xi = 1$ in a monotonic manner.

Even though a reaction such as $A \rightarrow B$ is thermodynamically nonlinear it still has linear reaction kinetics as shown by (14.28) and the straight line plot corresponding to $n = 1$ in Figure 14.12(a). We emphasise that the plots for $n = 2$ and $n = 0.5$ in Figure 14.12(a) exhibit *nonlinear kinetics*.

In Section 7.4.2 we discuss the concept of coupled chemical reactions where the product of one or more reactions are used by another simultaneous reaction as a reactant. These are referred to as *networked* (Epstein and Pojman, 1998) or cyclic (Vernon, 2004) reactions. Such reactions are commonly *autocatalytic* in at least one chemical component and/or *autoinhibitory* in others. By *autocatalytic* we mean that the networked array of reactions produces more of a component than is used. *Autoinhibitory* means that less of a component is produced than used. Thus, in the example quoted in (7.15) from Whitmeyer and Wintsch (2005) SiO_2 is an autocatalytic component and H^+ is autoinhibitory. Another set of reactions is considered by Wintsch and Yeh (2013) and presented in Table 14.1 (see also Figure 15.4). This set of reactions is autocatalytic in K^+ and autoinhibitory in H^+.

TABLE 14.1   The Set of Networked Mineral Reactions Considered by Wintsch and Yeh (2013). See also Figure 15.4

1	K-feldspar $+ Na^+ =$ Albite $+ K^+$
2	3Albite $+ 2H^+ + K^+ =$ muscovite $+ 6SiO_2 + 3Na^+$
3	3K-feldspar $+ 2H^+ =$ muscovite $+ 6SiO_2 + 2K^+$
4	2phlogopite $+ 4H^+ + H_2O =$ clinochlore $+ 3SiO_2 + Mg^{++} + 2K^+$
5	phlogopite $+ 6H^+ =$ K-feldspar $+ 3Mg^{++} + 4H_2O$
6	2K-feldspar $+ 5Mg^{++} + 8H_2O =$ clinochlore $+ 3SiO_2 + 2K^+ + 8H^+$
7	3clinochlore $+ 2K^+ + 28H^+ = 2$muscovite $+ 3SiO_2 + 15\,Mg^{++} + 24H_2O$
8	3muscovite $+ 5$biotite $+ 9SiO_2 + 4H_2O = 8$K-feldspar $+ 3$clinochlore

The classical approach to autocatalysis and the behaviour of autocatalytic reactions is to treat such reactions as coupled reactions (as in Section 7.4.2) and express the behaviour in terms of a set of coupled differential equations (see Section 7.4.2, 3). Such a procedure becomes quite tedious or intractable if the network is complicated and/or if the reactions are highly nonlinear (as in some metamorphic systems; Figures 7.5, 15.4, 15.15). Considerable progress has been made in studying very large, complicated, nonlinear reaction networks in biology using an approach often referred to as *network thermodynamics* (Mikulecky, 2001). The approach is to treat the system as a networked flow system and analyse the transport patterns and behaviour as though the system was an electrical circuit so that Kirchhoff's laws become an important tool. This approach is appealing because it treats the reaction network as a plumbing system with competing nodes of supply, production and consumption exactly as a metamorphic geologist might view a networked reaction system. We discuss autocatalytic reactions and autocatalytic networks in Section 14.4.5. One important characteristic of *autocatalytic networks* is that, because of competition between supply and consumption of nutrients, the network as a whole can behave in an autocatalytic manner without any single autocatalytic reaction appearing in the network (Plasson et al., 2011).

Since the various reactions in Table 14.1 each have their own rate constants, some operate faster than others so that the possibility exists for competition between the supply and consumption of reactants as discussed by Wintsch and Yeh (2013). The system is highly nonlinear and so we can expect many of the behaviours discussed in Chapter 7, in particular, the development of oscillatory or even chaotic variations in the concentrations of chemical components and hence in the growth and dissolution rates of individual minerals.

One could proceed to write coupled rate equations for all of the components in Table 14.1 and then proceed to solve the coupled set of simultaneous differential equations but the task is large and we leave that as an exercise for the enthusiastic reader. The general approach is discussed in Section 7.4.2. Instead we proceed to illustrate the general principles involved. We take the simplest example from the list of reactions in Table 14.1 and, in order to preserve the mass of albite, write the first two reactions in the form:

$$3K-feldspar + 3Na^+ \xrightarrow{k_1} 3albite + 3K^+$$

$$3albite + 2H^+ + K^+ \xrightarrow{k_2} muscovite + 6SiO_2 + 3Na^+$$

In order to select the simplest sub-set of reactions from Table 14.1 that illustrates the principles involved we also select reaction 7:

$$3clinochore + 2K^+ + 28H^+ \xrightarrow{k_7} 2muscovite + 3SiO_2 + 15Mg^{++} + 24H_2O$$

Then the equation for the rate of change of the activity of K^+, $[K^+]$, is

$$\frac{d[K^+]}{dt} = -k_2^*[K^+] - k_7^*[K^+]^2[H^+]^{28} + k_1^*[K^+]^3 = [K^+]\left\{ -k_2^* - k_7^*[K^+][H^+]^{28} + k_1^*[K^+]^2 \right\}$$

where k_1^*, k_2^*, k_7^* are functions of the rate constants k_1, k_2 and k_7 and also, for the most part, of the activities of K-feldspar, albite, H^+, SiO_2 and Na^+, which themselves depend on what is being produced and consumed in the other five reactions. One can see that the rate of change of $[K^+]$ is a function of $[K^+]$ and of $[H^+]$. This rate equation is plotted in Figure 14.13 for constant $[H^+]$. Instead of the kinetics illustrated in Figure 14.12(a) where the rate is single valued, the rate of production of $[K^+]$ is a curve, which is not single valued, the rate becoming zero at a value of $[K^+]$ given by $\beta = \left\{(k_7^*)[H^+]^{28} + \sqrt{[(k_7^*)[H^+]^{28}]^2 + 4k_1^*k_2^*}\right\}/2k_1^*$ which is a strong function of $[H^+]$. This multivalued form of the rate curve is similar for a large number of networked reactions. If the three reactions 1, 2 and 7 from Table 14.1 are considered as comprising a subsystem to which K^+ is being supplied from neighbouring reactions 4 and 6 (Figure 14.13) then the subsystem can be thought of as an *isolated diffusive* system in the sense of Figure 5.2(b). The supply of K^+ from reaction 4 could be represented as the line ABC in Figure 14.13. The supply of K^+ from reaction 6 could be represented as another line which is not shown. The line ABC intersects the black curve in Figure 14.13 in three points. These points correspond to non-equilibrium stationary states for this sub-system (see Chapter 7). Figure 14.13 is an example of a *flow diagram*, a concept developed and used extensively by Gray and Scott (1994).

The type of behaviour discussed in Figure 14.13 is an example of a much wider set of behaviours. In Figure 14.14 we show the normalised reaction rates for what are called *quadratic* and *cubic autocatalytic* reactions and for a reaction of the form (14.27) but which is exothermic. All of these reactions have nonlinear kinetics (see Gray and Scott, 1994) but they differ from the nonlinear kinetics for half- and second-order reactions shown in Figure 14.12(a) in that it is possible that a given normalised reaction rate corresponds to two values of the reaction

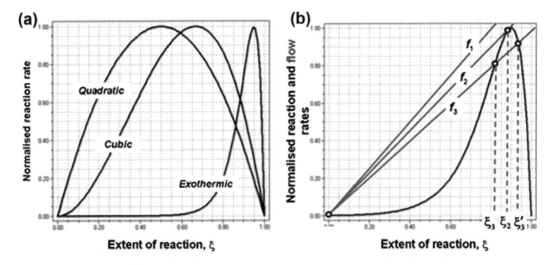

FIGURE 14.14   Nonlinear chemical kinetics. (a) Normalised reaction rate versus extent of reaction for the reactions $A + B \rightarrow 2B$ (quadratic), $A + B \rightarrow 3B$ (cubic) and an exothermic first-order reaction $A \rightarrow B$. (b) An open flow-controlled system. Normalised flow rates f_1, f_2 and f_3 and normalised reaction rate for a first order exothermic reaction $A \rightarrow B$ showing system behaviour. The behaviour is that of a *nonlinear chemical system*.

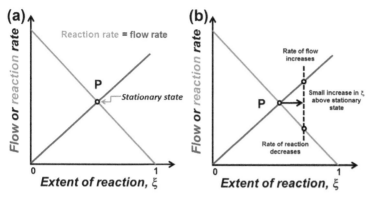

FIGURE 14.15 Flow diagrams showing the development of a non-equilibrium stationary state in an open flow system involving the reaction $A \rightarrow B$. (a) A stationary state is reached when the net flow rate equals the reaction rate. (b) This stationary state is stable since a small displacement from the stationary state results in an increase in flow rate but a decrease in reaction rate so the system returns to the stationary state. The same argument applies for a small decrease in the extent of the reaction.

extent; those in Figure 14.12(a) show a unique reaction extent corresponding to a given normalised reaction rate.

The importance of these various kinds of reactions becomes apparent when the chemical reaction is coupled with transport of chemical components by fluid flow or by diffusion. If we consider a volume of rock where fluid is flowing into and out of the volume of interest at a volumetric flow rate of q m^3 m^{-2} s^{-1} with an inflow concentration of A equal to a_0 (with units moles per cubic metre) and an outflow concentration, a, then the contribution to the production rate of A within the volume of rock is

$$ f = q\phi(a_0 - a) = q\phi a_0 \xi $$

with units [mol m^{-2} s^{-1}]; ϕ is the porosity. We call this quantity, f, the *net flow rate* of A. A plot of the *net flow rate* against reaction extent is a straight line through the origin with slope $q\phi a_0$. If the flow rate equals the reaction rate then the system is at a stationary state (Figure 14.15a). This state is stable since a small increase (or decrease) in net flow rate is accompanied by a compensating decrease (or increase) in reaction rate so that the system is returned to the stationary state (Figure 14.15b).

In Figure 14.14b we show the relation of the net flow rate of A to the reaction rate for various values of f and of ξ and for an exothermic reaction of the type $A \rightarrow B$. The intersection of the line representing the net flow rate with the reaction rate curve defines a stationary state. A small change in f (which could result from a small fluctuation in the porosity, ϕ, or the diffusion coefficient in a diffusive system) can result in a large change in behaviour. For a large net flow rate, f_1, the only stationary reaction rate corresponds to $\xi = 0$. For a slightly smaller flow rate, f_2, there result two stationary states corresponding to $\xi = 0$ and $\xi = \xi_2$. For a slightly smaller value of f, f_3, there are three stationary states corresponding to $\xi = 0$, $\xi = \xi_3$ and $\xi = \xi_3'$. In systems of this type small changes in net flow rate can result in large changes in ξ and in the qualitative behaviour of the system. In contrast, for systems such as those illustrated in Figure 14.15a, small changes in flow rate result in small changes in ξ and the system behaviour remains qualitatively the same independently of the flow rate. We refer to systems such as that illustrated in Figures 14.13 and 14.14b as nonlinear *chemical systems*.

14.4.4 Exothermic and Endothermic Reactions

Consider a first-order chemical reaction that involves an intermediate step, so that A is converted to P and then P is converted to B with rate constants k_A and k_B as shown below:

$$A \xrightarrow{k_A} P \xrightarrow{k_B} B$$

We also assume that the reaction $P \xrightarrow{k_B} B$ is exothermic with a heat of reaction, $\Delta H_B < 0$. The rate constant k_B obeys an Arrhenius temperature dependence:

$$k_B = k_B^0 \exp\left(\frac{-Q}{RT}\right)$$

where Q is the activation energy for the reaction, R is the gas constant and T is the local absolute temperature. We assume the rate constant for the reaction $A \xrightarrow{k_A} P$ is independent of temperature. We also assume the mixture sits in a heat bath at a temperature T_0. The equations for mass and heat balance are

$$\frac{da}{dt} = -k_A a \tag{14.29}$$

$$\frac{dp}{dt} = k_A a - k_B p \tag{14.30}$$

$$V\rho c_P \frac{dT}{dt} = (-\Delta H_B V k_B p) - \Sigma \chi (T - T_0) \tag{14.31}$$

where V and ρ are the volume and density of the reacting mixture, c_P is the specific heat, Σ is the surface area of the system and χ is the surface heat transfer coefficient (with units $Js^{-1}m^{-2}K^{-1}$). If $\chi = 0$ then the system is adiabatic. As $\chi \to \infty$, the system approaches isothermal. t is time. We take as the initial conditions, $a(t=0) = a_0$, $p(t=0) = 0$, $b(t=0) = 0$, and $T(t=0) = T_0$. Then the solution to (14.29) is

$$a(t) = a_0 \exp(-k_A t) \tag{14.32}$$

Substitution of (14.32) into (14.30) gives

$$\frac{dp}{dt} = k_A a_0 \exp(-k_A t) - k_B p$$

We define $\Delta T = (T - T_0)$ so that (14.31) becomes

$$V\rho c_P \frac{dT}{dt} = (-\Delta H_B V k_B p) - \Sigma \chi \Delta T$$

The solution to these equations is discussed by Gray and Scott (1994, Chapter 7). For a closed system the temperature oscillates before the system settles down to equilibrium. For open exothermic systems a diverse range of behaviours develops depending on the flow rate. In order to represent the behaviour it is convenient to show the system on a series of diagrams where the extent of the reaction is plotted against the inverse of the flow rate as shown in Figure 14.16 (see Gray and Scott, 1994 and Ord et al., 2012); some of these states are stable, others are unstable and characterised by Hopf

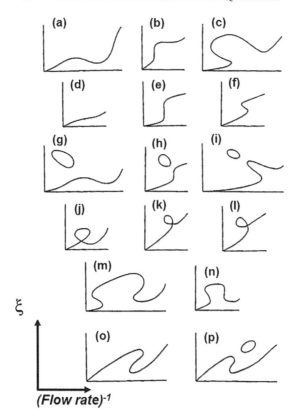

FIGURE 14.16   Behaviour of a first-order open exothermal chemical reaction of the form A → B. *From Gray and Scott (1994)*. In each diagram the extent of the reaction is plotted against (flow rate)$^{-1}$. It is clear that the behaviour of the system is extremely sensitive to the flow rate. Each diagram represents the behaviour for different fractions of the possible dimensionless temperature rise.

bifurcations. Each diagram represents a different state between isothermal and adiabatic conditions. For details of this system one should consult Gray and Scott (1994).

Thus what at first sight would appear to be one of the simplest chemical systems imaginable turns out to exhibit a diverse range of behaviours that depends *inter alia* on the rate of supply of A to the system. There is an important lesson to be learnt here. One does not need exceedingly complicated chemical systems to generate complexity in behaviour. Even the simplest of systems will exhibit the range of complexity that one observes in complicated systems so long as some feedback mechanism is in operation. Closed systems must always progress to equilibrium perhaps after some complex behaviour. Open systems can exhibit the whole range of complex behaviour indefinitely as long as the system remains open. In metamorphic chemical network systems the scale of 'openness' may be quite small (perhaps 10^{-6} m^3) but on that scale they operate as open flow systems and remain open on that scale as long as the reactions are progressing.

Coupled exothermic/endothermic reactions can also behave in chaotic manners as shown in Figure 14.17. The system illustrated is discussed by Kalhert et al. (1981) and the modelling

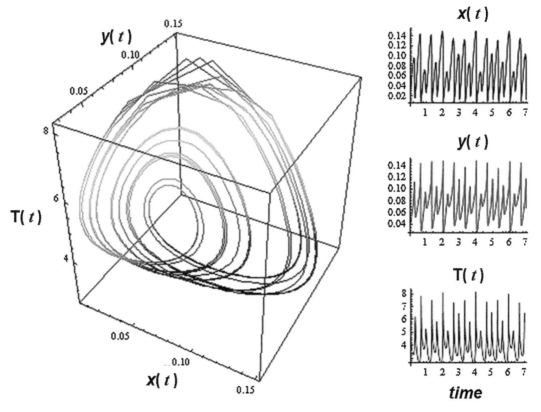

FIGURE 14.17 The behaviour of coupled exothermic/endothermic reactions. *From Gruesbeck (2013): http://demonstrations.wolfram.com/DynamicsOfAContinuousStirredTankReactorWithConsecutiveExothe/ Wolfram demonstration project.* The plot to the left is a phase plot with axes (x, y, T); the plots to the right are x, y, T against dimensionless time.

shown in Figure 14.17 comes from Gruesbeck (2013). The reaction consists of $X \rightarrow Y \rightarrow Z$ with the first reaction exothermic and the second endothermic. The system is open.

14.4.5 Autocatalytic Reactions and Autocatalytic Systems

Instabilities (transitions to oscillatory or chaotic behaviour) in isothermal chemical systems have traditionally been viewed as resulting from autocatalytic behaviour where the production of a particular chemical component, A_i, enhances the production of that component:

$$\frac{da_i}{dt} = k(a_i)a_i^n + f(\boldsymbol{b}) \tag{14.33}$$

where n is a number, $k(a_i)$ is a rate constant for the reaction that produces A_i and is a function of a_i, and $f(\boldsymbol{b})$ is a function that expresses the production of all the other components, B, in the system. (14.33) is a kinetic equation and expresses the gross rate of production of a_i; the equation may, but generally does not, represent the detailed mechanisms involved in the production of a_i (Plasson et al., 2011). A simple form of a chemical reaction that is expressed by (14.33) is

$$A + B \rightarrow nB \tag{14.34}$$

We adopt the terminology proposed by Plasson et al. (2011) in what follows. These simple autocatalytic reactions are called *template autocatalytic reactions* and the particular form of (14.34) is an example of a *direct template reaction*. Template autocatalytic reactions involve a direct association between the reactants and the products. The reaction is represented diagrammatically for $n = 2$ in Figure 14.18(a).

The lack of obvious reactions in the metamorphic literature of the form of (14.34) probably is responsible for a lack of interest in nonlinear chemical reactions. We have pointed to the fact that the networked reactions described by Carmichael (1969) and Wintsch and Yeh (2013) are autocatalytic in a gross sense but individual template autocatalytic reactions seem to be rare. Perhaps this is because such reactions are in fact rare in metamorphic systems? Perhaps the rarity arises from the tradition of expressing metamorphic reactions as constant mass reactions? It is notable that at least in mineral reactions that involve sulfides, template autocatalytic reactions are common if the reactions are expressed as constant (or near to constant) volume reactions rather than constant mass reactions. Such constant volume reactions follow from widespread observations that replacement reactions in sulfide systems are commonly isochoric (Lindgren, 1912). Ridge (1949) gives many examples of constant volume reactions involving sulfides that are template autocatalytic in (SO_4^{2-}). An example involving 0.92% volume change is (Ridge, 1949, p. 535):

$$8FeS_2 + 42Fe^{3+} + 77(SO_4^{2-}) + 14Cu^{2+} + 36H_2O \rightarrow 7Cu_2S + 50Fe^{2+} + 86(SO_4^{2-}) + 72H^+$$

This reaction, incidentally, also converts 42 mol of Fe^{3+} to 50 mol of Fe^{2+} and so resembles the reactions written in Chapter 7. Another equal volume template autocatalytic reaction that is autocatalytic in the production of Fe^{2+} given by Ridge (1949, p. 541) is

$$20Cu_5FeS_4 + 42Cu^{2+} + 10Fe^{3+} + 36H_2O \rightarrow 71Cu_2S + 30Fe^{2+} + 9(SO_4^{2-}) + 72H^+$$

This reaction involves a volume change of 0.09%.

Alternatives to template autocatalytic reactions are common. One possibility is that template autocatalytic reactions are actually expressed as *direct autocatalytic reactions* involving an intermediary:

$$A + B \rightarrow C$$

$$C \rightarrow 2B$$

This reaction is shown as a networked scheme in Figure 14.18(b). The two reactions 1 and 2 in Table 14.1 are of this type. Another possibility is that there is no direct coupling between A and B or the direct formation of 2B. This is called indirect autocatalysis. An example is

$$\begin{aligned} A + D &\rightarrow C \\ C &\rightarrow B + E \\ E &\rightarrow B \\ B &\rightarrow D \end{aligned} \tag{14.35}$$

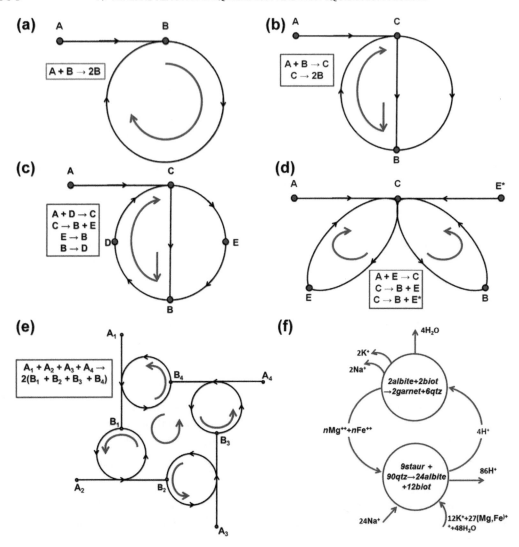

FIGURE 14.18 Classes of autocatalytic behaviour. (a) Template autocatalysis. (b) Direct autocatalysis. (c) Indirect autocatalysis. (d) Autoinductive autocatalysis. (e) Collective autocatalysis. (f) A subset of the Carmichael (1969) reaction network. *Figure inspired by Plasson et al. (2011).*

This reaction is shown in Figure 14.18(c) and can still correspond to the template reaction $A + B \rightarrow 2B$. When at least one of the two reactions $(14.35)_{1,2}$ is rate limiting the system behaves as a template reaction. When $(14.35)_4$ is rate limiting B is produced exponentially until the reaction ceases. When $(14.35)_3$ is rate limiting there is no autocatalytic effect. Thus, while these reactions can be autocatalytic there is no *templating* effect where one molecule of a component is involved directly in generating another molecule of the same component.

A slightly more complicated scheme known as *autoinductive autocatalysis* involves no direct interaction between the reactants and products and no coupling between A and B. A set of reactions can be written:

$$A + E \rightarrow C$$
$$C \rightarrow B + E \qquad (14.36)$$
$$C \rightarrow B + E^*$$

Such a system is shown in Figure 14.18(d) and can be autocatalytic if the rate of the reaction (14.36)$_2$ is equal to or slower than that of (14.36)$_1$ and non-autocatalytic otherwise. Thus the reactive network (14.36) behaves in much the same manner as an indirect autocatalytic system from a kinetic point of view but mechanistically the two are quite different.

A more general scheme, and one that probably operates in many silicate reactions, is *collective autocatalysis* whereby no chemical component influences its own formation rate. Instead, it influences the formation rate of other components which in turn influence the formation of other components. The system acts in a collective manner such that a set of reactions catalyses the generation of the system as a whole. This behaviour is exemplified in the networked reactions described by Carmichael (1969), Vernon (2004), Wintsch and Yeh (2013). Plasson et al. (2011) presents an example of four independent networked reactions as a set of four equations:

$$A_i + B_{i-1} \rightarrow B_i + B_{i-1}$$

with $i = 1, 2, 3, 4$ and $B_4 \equiv B_0$. The scheme is shown in Figure 14.18(e). If the rates of each of these reactions are equal then the system is autocatalytic and although there is no direct influence of any component upon its own growth rate, the system behaves collectively through indirect influences to produce

$$(A_1 + A_2 + A_3 + A_4) + (B_1 + B_2 + B_3 + B_4) \rightarrow 2(B_1 + B_2 + B_3 + B_4)$$

Whether networked chemical systems in metamorphic rocks behave in an autocatalytic manner or not is still an open question. It is important to address this issue because the process provides a mechanism for producing zoning in growing metamorphic minerals that is intrinsically associated with the growth history of the mineral rather than appealing to some *ad hoc* mechanism that conveniently cycles nutrient supply to the mineral from an unspecified source. Autocatalysis is also important as a mechanism of producing many microstructures through reaction–diffusion processes; an example is presented in Section 15.6. The answer probably lies in constructing molecular models of mineral reactions so that robust mathematical models can be developed to be tested using high resolution chemical images of zoned grain assemblages.

We close by pointing to a subset of the networked system proposed by Carmichael (1969) shown in Figure 14.18(f). Here the addition of H^+ to the reaction that produces garnet facilitates the reaction that produces albite and biotite and in doing so releases more H^+ than was added to the garnet-forming reaction. In the spirit of the above discussion on collective autocatalysis, if the reactions shown in the two circles are rate controlling within the total Carmichael system then autocatalytic production of H^+ results and it is therefore possible (following the discussion in Section 7.4.2,3) that zoning (both continuous and oscillatory) within growing quartz, garnet, albite and biotite grains can result.

Autocatalysis can also derive from sources other than networked reactions. In the models of pseudomorphism proposed by Merino (Merino and Canals, 2011; and see Section 15.5) the replacement process, the production of dolomite, is self-accelerating; this is a form of autocatalysis. Also, some of the models to explain the coupling between dissolution and precipitation in the Putnis model of pseudomorphism (Putnis, 2009; and see Section 15.5) also involve autocatalysis arising from interacting mechanisms at the dissolution/precipitation interface.

14.4.6 The Phase Rule for a System Not at Equilibrium

The concept of a phase rule applies equally to a closed chemical system at a non-equilibrium stationary state or an open flow system not at equilibrium (Korzhinskii, 1959, 1965, 1966, 1967; Landsberg, 1961; Zemansky, 1951) just as it does to a system at an equilibrium stationary state. If there are $\mathcal{R}$ independent chemical reactions in progress the phase rule for a nondeforming system becomes (Kondepudi and Prigogine, 1998, p. 182) $\mathcal{F} = r_C - \mathcal{M} - \mathcal{R} + 2$. Korzhinskii (1965, 1966) points out that for open flow systems there is a set of chemical components (called by Korzhinskii, mobile components) where the chemical potentials are fixed outside of the control volume. This reduces the number of independent chemical potentials and hence reduces $\mathcal{F}$ so that the resulting mineral assemblages tend to be simpler than in systems in closed metamorphic systems at equilibrium. The arguments of both Korzhinskii and Landsberg are quite general and do not depend on an appeal to *mosaic* or *local equilibrium*. In a similar fashion, the fact that the chemical potentials at a non-equilibrium stationary state are fixed by the concept of the excess work (Ross, 2008 and Section 14.4.7 below) means that $\mathcal{F}$ is again reduced from the equilibrium value and so the number of phases coexisting at a non-equilibrium stationary state is smaller than at equilibrium in a closed system. These results are a statement of the observation that systems at equilibrium are defined by states of maximum entropy (the greatest number of degrees of freedom) rather than both open flow systems and closed systems at non-equilibrium stationary states where the entropy is constrained in some manner and the number of degrees of freedom is smaller than in the same system at equilibrium. All systems however, whether at equilibrium or non-equilibrium stationary states, are characterised by ordered mineral assemblages.

14.4.7 The Ross Excess Work

For systems that approach a non-equilibrium stationary state, Ross (2008) defines a function called the *excess work*, $\phi(n_x)$, which is a function of the mole number, n_x:

$$\phi(n_x) = \int_{n_x^{stationary}}^{n_x} \left(\mu_x - \mu_x^{stationary} \right) dn_x \tag{14.37}$$

where μ_x is the chemical potential at an arbitrary state and the superscript stationary refers to that quantity at a *stationary* state.

The *excess work* represents the work, other than the contribution from PV, done in moving an isothermal system from an arbitrary state to a stationary state. At the stationary state, $\phi(x) = 0$. If we start at the stationary state and increase x then dx is positive and the integrand is greater than zero and $\phi(x) > 0$. If we start at the stationary state and decrease x then dx is negative and the integrand is also negative so that, again, $\phi(x) > 0$. Thus, $\phi(x)$ is a minimum at the stable stationary state just as the Gibbs energy is a minimum at an equilibrium stationary state. As with the Gibbs energy change, that is a Lyapunov function for systems approaching equilibrium, the excess work is a Lyapunov function for systems approaching a non-equilibrium stationary state. The properties of the integrand in the excess work, (14.37), are

$$\mu_x - \mu_x^{stationary} = 0 \ at \ a \ stationary \ state$$

$$d\left(\mu_x - \mu_x^{stationary}\right)\Big/dn_x > 0 \ at \ each \ stable \ stationary \ state$$

$$d\left(\mu_x - \mu_x^{stationary}\right)\Big/dn_x < 0 \ at \ each \ unstable \ stationary \ state$$

Thus, the behaviour of systems not at equilibrium is guided by principles that resemble the principles that guide the behaviour of systems approaching equilibrium. The excess energy behaves as a Lyapunov function and prescribes the route to a non-equilibrium stationary state. The behaviour of systems not at equilibrium is not 'random' as some may believe. The mineral assemblages at a non-equilibrium stationary state are fixed by the excess energy and are described by a relevant version of the phase rule just as the topology of an equilibrium phase diagram is described by a similar set of rules.

14.5 CHEMICAL DISSIPATION

In systems with a single non-equilibrium stationary state, the commonly quoted extremum principle is that the entropy production is minimised (Biot, 1958; Prigogine, 1955). Examples of such systems that are widely quoted are steady heat conduction in a material with constant thermal conductivity (Kondepudi and Prigogine, 1998) and simple, uncoupled chemical reactions (Kondepudi and Prigogine, 1998). However, Ross and Vlad (2005 and references therein) and Ross (2008, Chapter 12) show that, for these two classical systems, there is no extremum in the entropy production. It is only for systems where the relationship between thermodynamic fluxes and forces is linear that an extremum exists for the entropy production rate and this corresponds to a stationary state that is not an equilibrium state. If the relationship between thermodynamic fluxes and forces is nonlinear then only one extremum in the entropy production exists and that is an equilibrium state where the entropy production is zero. The linear situation is that commonly quoted and discussed by Prigogine (1955), Kondepudi and Prigogine (1998) and a host of others. However, reference to Table 5.5 shows that at least for thermal conduction and chemical reactions the thermodynamic fluxes are not proportional to the thermodynamic forces. For mass diffusive processes the thermodynamic flux is proportional to the thermodynamic force only for isothermal situations. For further discussion and clarification of what is meant by 'close' and 'far' from equilibrium and the implications for whether nonlinear systems may approximate linear behaviour 'near' equilibrium see Section 14.4.5. Reference should also be made to the interchange between Hunt, Hunt, Ross and Kondepudi (Hunt et al. 1987, 1988; Kondepudi, 1988; Ross, 2008; Ross and Vlad, 2005).

The entropy production, $\dot{s}$, for any chemical reaction is given by (Ross, 2008):

$$\dot{s} = k^B \ln\left(\frac{r^+}{r^-}\right)\left[r^+ - r^-\right]$$

where r^+ and r^- are the forward and reverse rate constants for the chemical reaction and k^B is Boltzmann's constant. If we take a simple chemical system given by

$$\nu A \underset{k_1^-}{\overset{k_1^+}{\rightleftarrows}} \nu X \quad\text{and}\quad \nu X \underset{k_2^-}{\overset{k_2^+}{\rightleftarrows}} \nu B$$

where ν is a stoichiometric coefficient, then the stationary state for the production of X is given by

$$\frac{dx}{dt} = \left(k_1^+ a^\nu - k_1^- x^\nu\right) + \left(-k_2^+ x^\nu + k_2^- b^\nu\right) = 0$$

If we now take $k_1^\pm = k_2^\pm = k$ then the stationary state concentration of X is

$$x^{ss} = \left[\frac{1}{2}(a^\nu + b^\nu)\right]^{1/\nu}$$

The variation of the entropy production about x^{ss} is

$$\delta\dot{s}(a,b,x^{ss}) = (\delta \ln x^{ss})\overline{V}k^B \nu k \frac{1}{2}(a^\nu + b^\nu)\ln\left[\frac{(a^\nu + b^\nu)^2}{4a^\nu b^\nu}\right]$$

Now,

$$\ln\left[\frac{(a^\nu + b^\nu)^2}{4a^\nu b^\nu}\right] = \begin{array}{l} 1 \text{ for } a = b \\ > 1 \text{ for } a \neq b \end{array}$$

and hence

$$\frac{\delta\dot{s}(a,b,x^{ss})}{\delta \ln x^{ss}} = \begin{array}{l} 0 \text{ for } a = b \\ > 0 \text{ for } a \neq b \end{array}$$

The condition $a = b$ corresponds to equilibrium, whereas $a \neq b$ corresponds to a non-equilibrium stationary state. Hence *the only extremum in entropy production corresponds to equilibrium*. This argument is extended to the general case by Hunt et al. (1987, 1988), Ross and Vlad (2005), Ross (2008) and Ross and Villaverde (2010). Such results hold for isolated and diffusive systems where there are no flow constraints.

As far as we are aware the general case of entropy production in an open flow reactor has not yet been examined. Ord et al. (2012) give some examples of entropy production in systems more complicated than those discussed by Ross (2008). A general principle (quite different from that discussed by Ross) seems to emerge: the actual state adopted by the system corresponds to a state of maximum entropy production whereas switches from one mode of operation to another in a system are picked by a minimum entropy production principle (not that proposed by Prigogine, 1955). Identical selection principles for open flow systems have been proposed by Niven (2010) and Niven and Noack (2014).

14.6 SYNTHESIS

One of the questions commonly posed is: *What is the influence of deformation on the position of the mineral equilibrium phase boundary defined at hydrostatic pressure?* It turns out that in order to answer this question one needs to understand many of the processes that accompany the growth of a new phase in a deforming matrix and so the answer to the question is a convenient starting point for a synthesis of processes taking place in chemically reacting–deforming materials. The short answer to the question is that both the stored energy of deformation and overstepping of phase boundaries by tectonic overpressures and overtemperatures have effects on a deforming/reacting system. First they increase the affinities that drive chemical reactions. Second they can contribute to the form of the Gibbs–Duhem equation and hence influence the position of the mineral equilibrium phase boundary that one would observe for hydrostatic pressure conditions. If the deformation ceases the system relaxes to the equilibrium state corresponding to hydrostatic conditions if enough time is available at the equilibrium P and T. However, it is important to understand that both deformation and diffusion can alter the position of the equilibrium phase boundary as summarised below.

For simplicity we consider the equilibrium between two phases A and B. The equilibrium phase diagram is obtained in part from the Gibbs–Duhem equation which expresses the relation between the intensive variables commonly used in equilibrium chemical thermodynamics, the hydrostatic pressure, P, the temperature, T, and the chemical potentials of the two components A and B, μ_A and μ_B. The conjugate extensive variables in the hydrostatic pressure case are the volume, V, the entropy, S, and the mole numbers, N_A and N_B. The slope of the phase boundary in $P - T - \mu_A - \mu_B$ hyperspace is obtained by writing the Gibbs–Duhem equation for both phases and prescribing that both equations are true simultaneously. As Korzhinskii (1959), Kern and Weisbrod (1967) and Powell et al. (2005) point out, many different sections can be drawn through this hyperspace, the common one being a $P - T$ section but a $\mu_A - \mu_B$ section may also be of use in some problems. The solution to solving the two Gibbs–Duhem equations involving the two phases is the Clapeyron equation which, for the P–T phase diagram, is

$$\frac{dP}{dT} = \frac{[\![S]\!]_N}{[\![V]\!]_N} = \frac{\Delta H_m}{T \Delta V_m} \tag{14.38}$$

where $[\![S]\!]_N, [\![V]\!]_N$ are the differences (or *jumps*) in the molar entropy and molar volume, respectively, between the two phases, A and B. (14.38) is the equation for the *slope* of the equilibrium phase boundary in P–T space for a hydrostatic pressure and a particular temperature, T. The actual *position* of the equilibrium phase boundary for a given P and T is found, in principle, by minimising the Gibbs energy at that P and T.

The preceding argument can be generalised to include situations where the two phases, A and B, are deforming while reacting. The conditions for chemical equilibrium between two grains in a deforming system are that the differences in temperature, T, velocity, v, and resolved stresses normal and parallel to the interface, $(\sigma_{ij} n_j)$, should be zero and that the jump in a potential that measures the energy of each grain should be zero. These conditions assume that there is no grain boundary sliding during deformation. Discussion, and considerable confusion, has existed in the literature about the form of the potential mentioned and whether it is equivalent to the chemical potential (or the Gibbs energy) or is something else. It turns out that this potential is defined only at the interface and not within the bodies of the grains. It is also dependent on the orientation of the interface relative to the stress on the interface together with the deformation on either side of the interface. Another (diffusion) potential is also defined which describes diffusive fluxes in the system. Thus, although the first potential has some of the characteristics of Gibbs energy, it is not a state function in the classical sense of being a function of variables that measure the state of the system. Nevertheless, this function is the potential that defines chemical equilibrium across an interface in a deforming system. For situations where the interface cannot support a shear stress, this potential reduces to the chemical potential function defined by the Gibbs–Kamb–Paterson model (Section 15.4).

In an Eulerian view of a reacting–deforming system, the intensive variables become the first Piola stress tensor, $\mathbf{P}$, the temperature, T, and the chemical potentials, μ_A and μ_B. We recall (Chapter 6) that the first Piola stress is measured by the force per unit of the undeformed area, whereas the Cauchy stress is the force per unit of the deformed area. The conjugate extensive variables are the deformation gradient, $\mathbf{F}$, the entropy, S, and the mole numbers, N_A and N_B. If diffusion is involved in the reaction between A and B and also acts as a deformation process then additional intensive variables need to be considered, namely, for atoms sitting at substitutional sites, M_{AB} and M_{BA}, and extensive variables n_A and n_B. Here M_{AB} is the energy change when an atom of A is replaced by an atom of B keeping all other variables fixed and is a diffusion potential. M_{BA} has a similar interpretation. n_A and n_B are the mole numbers of atoms sitting at interstitial sites. These potentials govern the diffusion of species in a deforming solid and are not dependent on the orientation of an interface between two phases.

The Clapeyron equation, (14.38), is

$$0 = -[\![S]\!]dT - \sum [\![n_i]\!]d\phi_i - \sum [\![N_i]\!]d\mu_i - \sum_{i=1,2,3} [\![(vF_{i3}/F_{33})]\!]d\sigma_{3i} + \sum_{i=1,2,3} [\![VQ_{ij}]\!]dF_{ji}$$

(14.39)

Thus, the slope of the equilibrium phase boundary between A and B in a deforming–diffusion environment depends not only on the temperature and mean stress but also on jumps in entropy, mole numbers, volume and deformation gradients between A and B. The repercussion of these dependencies is that the slope of the equilibrium phase boundary is different from that in the hydrostatic case and hence, in general, the position of the phase boundary is different from that in the hydrostatic case.

The shift in the phase boundary depends on the constitutive behaviour of the material under consideration. Deformation compatibility arguments show that this displacement of the phase boundary is different depending on the mechanisms involved in the chemical reactions that occur—whether they are isochoric, homochoric, iso-mass, and whether they are coherent or grain boundary sliding is involved. These processes control the differences in deformation gradients and stresses that can exist between adjacent grains and still maintain compatibility of the deformation.

An important conclusion is that in an inhomogeneous deformation, adjacent grains can experience different jumps in deformation gradient and stress and hence the equilibrium phase boundary for adjacent grains can differ depending where the grains are situated in the deformation field—whether in a fold hinge, a fold limb or within a shear zone. In addition, if A and B are part of a solid solution (as in, say, the plagioclases) variations in chemical composition can arise in order to maintain chemical equilibrium even though P and T do not change (see Figure 14.11(b))

This analysis, although based on coherent, plane interfaces between adjacent grains brings out the essential characteristics of deforming—reacting—diffusing systems, namely, that a multiplicity of equilibrium states exist each defined by local jumps in deformation gradients, in diffusion potentials, in mean stress and in entropy. Thus the equilibrium phase boundary is not even a single surface but can consist of multiple hyper-surfaces in mean stress, temperature, deformation gradient, diffusion potential space; this needs to be kept in mind in interpreting chemical variations and microstructures.

Although this multiplicity exists in principle, in any particular rock the deformation within individual mineral phases is likely to lie in a restricted range so that the equilibrium conditions are similarly restricted. The interesting examples to explore are those where the deformation is non-affine in the form of folds or shear zones.

In addition to the influence of deformation on the slope of the equilibrium phase boundary there is an influence on the position of the boundary. No general treatment of this effect is available but a simple example involves a chemical potential section through phase space (Figure 14.10). Since the chemical potential of a solid dissolved in an adjacent fluid film depends on the normal stress across the interface with the fluid, the phase boundary in chemical potential space is shifted as shown in Figure 14.10. Arguments involving reacting interfaces that can support shear stresses are developed by Shimizu (1992, 1995, 2001).

An important aspect of mineral reactions in metamorphic rocks is that they are invariably networked so that the behaviour of one reaction is dependent on other reactions that occur nearby. In many examples a subset of the network is autocatalytic and/or autoinhibitory. If such subsets are rate controlling for the whole network the possibility exists for nonlinear behaviour of the entire network including oscillatory or chaotic behaviour. Such characteristics become important in controlling microstructure development. Examples are considered in Chapter 15 and Volume II.

Much is yet to be learnt about deforming—chemically reacting systems and with the present arsenal of equilibrium thermodynamics (to supply bounds on what is possible), high resolution chemical mapping technologies (electron probe and synchrotron), transmission and scanning electron microscopy, electron back scattered diffraction (EBSD) and *in situ* geochronology combined with increased emphasis on molecular modelling of processes, some interesting and fascinating developments are on the horizon. However progress will not be made if one slavishly adheres to classical views of chemical equilibrium thermodynamics; recent theoretical developments in understanding the nature of systems in non-hydrostatic stress states, at or far from equilibrium, provide the opportunity to view these systems with new eyes and to test a wide range of new hypotheses that will enable metamorphic microstructures to provide greater insight into the physical and chemical processes involved in metamorphism.

Recommended Additional Reading

Callen, H. B. (1960). *Thermodynamics: An Introduction to the Physical Theories of Equilibrium Thermostatics and Irreversible Thermodynamics*. New York, London: John Wiley and Sons, 376 pp.
The classical text on thermodynamics with excursions into non-equilibrium thermodynamics.

Coussy, O. (1995). *Mechanics of Porous Continua*. Chichester, UK: Wiley.
The definitive text on the thermodynamics of fluid flow through deforming porous media with sections on chemical phase changes, dissolution and precipitation and chemical reactions.

Denbigh, K. (1968). *The Principles of Chemical Equilibrium*. Cambridge at The University Press.
A thorough, and readable, treatment of equilibrium chemical thermodynamics with important discussions of the route to equilibrium.

Gray, & Scott. (1994). *Chemical Oscillations and Instabilities*. Oxford: Clarendon Press.
This book is a comprehensive treatment of chemical reactions in open systems. Many simple, coupled reactions are analysed to define the conditions for bifurcation especially using the concept of a flow diagram.

Kern, R., & Weisbrod, A. (1967). *Thermodynamics for Geologists*. Freeman, Cooper and Co.
A comprehensive and in depth treatment of equilibrium chemical thermodynamics with many applications to mineral systems.

Kondepudi, D., & Prigogine, I. (1998). *Modern Thermodynamics. From Heat Engines to Dissipative Structures*. Wiley.
A treatment of chemical thermodynamics from both the equilibrium and non-equilibrium viewpoints.

Powell, R. (1978). *Equilibrium Thermodynamics in Petrology: An Introduction*. London: Harper and Row.
A clearly written text that acts as an introduction to equilibrium chemical thermodynamics applied to petrological examples.

Vernon, R. H., & Clarke, G. L. (2008). *Principles of Metamorphic Petrology*. Cambridge University Press.
This is an excellent treatment of many aspects of metamorphic geology including the use of phase diagrams, including pseudo-sections, the importance of microstructures and the influence of mineral reactions on deformation behaviour.

Zemansky, M. W. (1951). *Heat and Thermodynamics*. McGraw-Hill.
A pragmatic and detailed overview of thermodynamics with chemical thermodynamics as a subset. Important details concerning the extent and affinity of a chemical reaction are considered and the phase rule for a chemically reacting system.

15

Models for Mineral Phase Nucleation and Growth

15.1 INTRODUCTION

The final microstructure we observe in deformed metamorphic rocks depends on a number of poorly understood processes some of which are associated with the phase transformations involved. These processes include the nucleation rate of new grains, the rates at which new boundaries move, whether the transformation is isochemical (in which case diffusion of material is limited to the scale at which local adjustments of initially present atoms need to move) or consists of the growth of grains different in chemical composition to the host and involves the migration of chemical components on the scale of many grains, whether the growth rate is diffusion controlled or controlled by the grain boundary migration rate, whether the new phase replaces the host and preserves the initial volume of the host (isochoric replacement/pseudomorphism) or whether aggregates of new grains form not necessarily preserving initial volume (recrystallisation). An added complication arises if fluid or diffusive flow is involved and the system can be considered open on the scale of interest. Some processes may involve dissolution and precipitation, whereas others may involve various forms of pressure solution or solid state reactions involving coupled replacement and dissolution. Some relations between these processes are given in Figure 15.1.

The nucleation and subsequent growth of new mineral phases in a metamorphic rock have been considered by a large number of workers including Lindgren (1912), Bastin et al. (1931), Voll (1960), Thompson (1970), Carmichael (1969, 1987), Wintsch (1985), Passchier and Trouw (1996), Kretz (2006), Vernon (1976, 2004), Vernon and Clarke (2008), Carlson (2011), Putnis and Austrheim (2010), Merino and Canals (2011) and Paterson (2013). The range of models probably reflects the wide range of metamorphic environments delineated in the classification of Oliver (1996) and presented in Figure 12.1. In closed metamorphic systems the processes may be very similar to those proposed for metals where new phases nucleate at energetically favourable sites such as dislocations, grain/twin boundaries and the like. Transport of chemical components (if necessary) is akin to Fickian diffusion and the process involves the growth of the new phase into the host essentially as a solid state process. If a fluid phase is involved then transport of chemical components is in thin grain boundary and fracture films such as those depicted in Figure 12.8.

FIGURE 15.1 Some of the constraints placed on the nucleation and growth of new grains in solid state reactions. The red stream is iso-chemical with no large-scale diffusion. These reactions may comprise recrystallisation or replacement. The replacement reactions can be pseudomorphic or polymorphic. The blue stream involves large-scale transport, and comprises both closed and open (orange stream) systems.

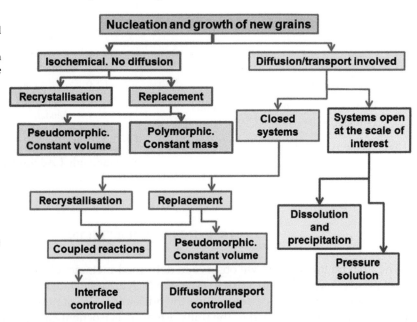

At the other extreme of metamorphic environments where open fluid flow networks exist, the process may involve dissolution of the host phase and precipitation of the new phase in the porosity that is initially present or that is newly created by dissolution as discussed by Putnis and Austrheim (2010). In between these two extremes another process seems to operate that involves transport of chemical components in a fluid phase but is better described as a replacement–dissolution sequence (Merino and Canals, 2011) rather than a dissolution–precipitation process. We consider examples of these models below. The literature tends to be polarised with respect to particular models presumably reflecting the experience and emphasis of particular workers.

One aspect that is neglected by many is that commonly phase transformations during metamorphism are replacement in character and usually are pseudomorphic. This means that the process preserves the volume (or, except for elastic strains, comes close to preserving the volume) of the host material meaning that the new phase ultimately occupies the volume initially occupied by the host even though the specific volumes of the new phase and of the host are completely different. This has consequences for the local stress field that is generated by the replacement process. The range of proposed models is illustrated in Figure 15.2.

Below we explore some models that have been proposed for the nucleation and growth of new mineral phases. Many authors, including Lindgren (1912), Bastin et al. (1931), Carmichael (1969, 1987) and Merino and Canals (2011) have pointed out that the growth of new minerals during metamorphism is close to constant volume in nature so that intimate initial detail of the host material is preserved during the new growth (Figure 15.3(a) and (c)). In order for one mineral to replace a host mineral and preserve the initial volume of the host, the rate of dissolution of the host must balance the rate of growth of the new mineral as shown in Figure 15.3(b). One model for this process, which depends on pressure solution, is given by Nahon and Merino (1997) and Merino and Canals (2011): Consider a host grain, B, exposed to a flowing fluid supersaturated with respect to a phase, A. If A is in its stability field then we consider a situation where a nucleus of A begins to grow in B through the reaction B $\rightarrow$ A. If the replacement is isochoric and the molar volume of B is smaller than the molar volume of A, pressure is exerted on B by the growing A grain causing B to be pressure dissolved. We have the following coupled reactions:

$$\text{A in solution} \xrightarrow{k_A} \text{A as a solid}$$

$$\text{B as a solid} \xrightarrow{k_B} \text{B in solution}$$

where k_A and k_B are the rate constants for these two reactions. The growth rate of A as a solid is

$$R_A \approx k_A \left[\left(Q_A / K_A^{equilibrium} \right) - 1 \right]$$

whereas the dissolution rate of B is

$$R_B \approx k_B \left[1 - \left(Q_B / K_B^{equilibrium} \right) \right] \tag{15.1}$$

In (15.1) $Q_A = \Omega_A K_A^{equilibrium}$ and $Q_B = \Omega_B K_B^{equilibrium}$ where Ω_A, Ω_B are the saturation indices of A, B, respectively. As the stress between A and B increases, the equilibrium

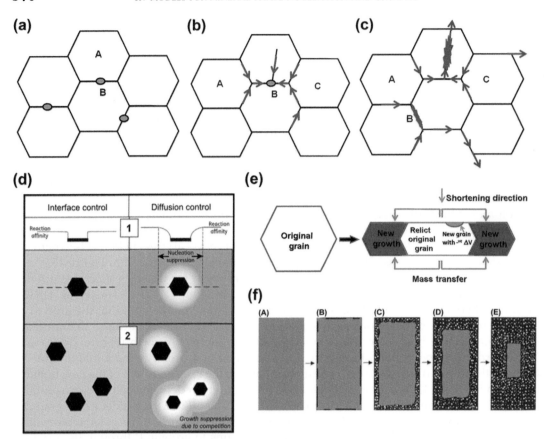

FIGURE 15.2 Models for mineral reactions at the grain scale. (a) Nucleation of polymorphic reactions at grain boundaries; no diffusive processes are involved. The replacement is isochoric. (b) Replacement reactions at grain boundaries. Diffusion may need to take place over distances comparable with the grain size. The replacement is isochoric and the system is closed on the scale of several grain sizes. *This is the process discussed by Carmichael (1969).* (c) Nucleation of pseudomorphic replacement along micro-cracks and grain boundaries. Diffusion of components on the scale of many grain sizes is proposed. The system may be open on larger scales. *This is the process discussed by Carmichael (1969) and Merino and Canals (2011).* (d) Nucleation of grains and growth controlled by interface migration or by diffusion to the growing grain *After Carlson (2011).* (e) Deformation controlled by mineral reactions. The original grain (white) undergoes dissolution on faces normal to the shortening direction and is deposited (red) at the ends of the initial grain in regions of extensional deformation. *Inspired by Stokes et al. (2012).* (f) Dissolution of the host grain with precipitation of the new phase to form a porous network of the new phase. Fluid continues to flow through this network to finally dissolve all of the host. The host is replaced by a porous network of the new phase. *From Putnis and Austrheim (2010).*

constants, $K_A^{equilibrium}$, $K_B^{equilibrium}$ both increase so that R_A decreases while R_B increases. When $R_A = R_B$ a stable non-equilibrium stationary state develops as shown in Figure 15.3(b) and the volume of **B** is preserved from then on. The examples of isochoric replacement, or pseudo-morphism, shown in Figure 15.3(c) and (d) illustrate replacement preserving initial host detail and idiomorphic shape.

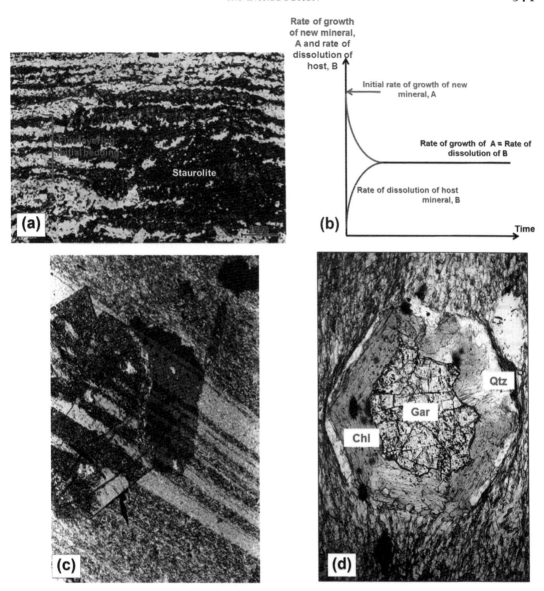

FIGURE 15.3 Constant volume (isochoric) replacement. (a) Staurolite replacing and preserving initial fine-scale layering *From Carmichael (1969)*. (b) Plot showing evolution of growth and dissolution rates with time. Ultimately a stationary state is reached where the growth rate equals the dissolution rate. *Inspired by Merino and Canals (2011)*. (c) Biotite porphyroblast (left) and staurolite porphyroblast (right) overgrowing fine-grained schist and preserving initial microstructure. Height 5.5 mm. *From Passchier and Trouw (1996)*. (d) Chlorite and quartz pseudomorphically replacing garnet. *From Guidotti and Johnson (2002)*.

15.2 SOME DETAILED MODELS OF METAMORPHIC REACTIONS

We distinguish four classes of reaction-microstructure models that have been proposed. We call these

- network models
- replacement–dissolution models
- nucleation and growth models
- dissolution and precipitation models.

These models are not distinct and some overlap with each other but in distinguishing these four models we can illustrate many of the principles involved in metamorphic reactions.

15.2.1 Network Models

Network models recognise that mineral reactions do not proceed independently of each other during metamorphism but are coupled so that the products of one reaction become reactants for another neighbouring reaction. An individual reaction cannot proceed without input from a neighbouring reaction and the reacting system as a whole cannot react unless communication between all the reactions in the system is well established. The behaviour of the whole system is dictated by the rate-controlling reaction in the network.

The concept of networked reactions was introduced by Carmichael (1969) and other examples have been presented by Vernon (1976, 2004, and references therein) and by Wintsch and co-workers (see Wintsch and Yeh, 2013 and references therein). We have presented one example in Section 7.4.2 and the Carmichael example is illustrated in Figure 15.16. In Figure 15.4 we present another example from Wintsch and Yeh (2013). In this example three

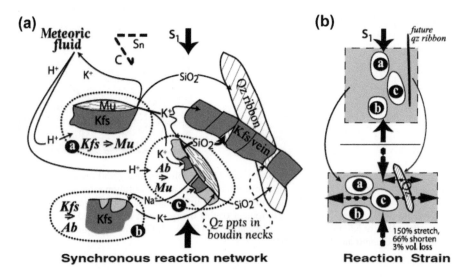

FIGURE 15.4   Networked mineral reaction system. *From Wintsch and Yeh (2013).* (a) Network relations for the three reactions **a**, **b** and **c**, showing details of microstructure associated with each reaction. Some fabric elements define S-planes while others define C-planes. (b) Contributions to the strain in the rock arising from the chemical reactions in (a). It is estimated that there is 150% stretch parallel to the S-plane and 66% shortening. Volume loss is 3%.

reactions are involved: K-feldspar $\rightarrow$ muscovite, K-feldspar $\rightarrow$ albite and albite $\rightarrow$ muscovite. The system is driven by the input of H^+ and the output is SiO_2.

These reactions could be analysed by writing the differential equations that describe the time rate of change of each chemical component in the system and solving the coupled set of nonlinear equations as discussed in Chapter 7. This is a formidable task and to date has not been done. The application of *network thermodynamics* (Mikulecky, 2001 and references therein) is an opportunity here. So far these analyses have not been conducted for metamorphic reaction networks, although such procedures are now commonplace in biological systems (Hill, 1989; Oster et al., 1971).

15.2.2 Replacement–Dissolution Models

Replacement–dissolution models have been developed by Merino and co-workers (see Merino and Canals, 2011, and references therein) mainly to describe isochoric replacement reactions. The important point made in these models is that a precipitating process is not involved. The reaction takes place first by a constant volume atom-by-atom replacement of the host. This generates a stress at the reacting interface (see Section 15.5). These stresses induce pressure solution of the host. The process is described as a coupled replacement–dissolution reaction rather than a dissolution–precipitation reaction. Such reactions are important because the constraint of constant volume can mean that the reactions are *self-accelerating* (or autocatalytic) when coupled with simultaneous dissolution of the host material.

As an example we consider the reaction which is commonly written to express the replacement of calcite by dolomite:

$$2CaCO_3 + Mg^{2+} \rightarrow CaMg(CO_3)_2 + Ca^{2+} \tag{15.2}$$

This equation says that 2 moles of calcite become 1 mole of dolomite. However, (15.2) does not express a pseudomorphic replacement where volume is preserved since the calcite formula volume is 3.69×10^{-5} m^3 and the dolomite formula volume is 6.43×10^{-5} m^3, the ratio being 1.743. Hence in order to preserve volume, (15.2) needs to be rewritten as

$$1.743CaCO_3 + Mg^{2+} + 0.26CO_3^{2-} \rightarrow CaMg(CO_3)_2 + 0.74Ca^{2+} \tag{15.3}$$

In addition the following reactions take place:

$$\textit{Dissolution of calcite}: \ CaCO_{3(solid)} + H^+ \rightarrow Ca^{2+} + HCO_3^- \tag{15.4}$$

$$\textit{Precipitation of dolomite}: \ Ca^{2+} + Mg^{2+} + 2CO_3^{2-} \rightarrow CaMg(CO_3)_{2(solid)} \tag{15.5}$$

Initially the reaction is described by (15.5), the reaction driven by excess Ca^{2+} derived from (15.4). As the replacement proceeds, the replacement–dissolution reaction is described by the constant volume reaction, (15.3). The aqueous Ca^{2+} released in (15.4) in an increment of replacement increases the saturation index, Ω, through

$$\Omega = \left(Q/K_{dolomite}^{equilibrium} \right) = \left(a_{Ca^{2+}} \right) \left(a_{Mg^{2+}} \right) \left(a_{CO_3} \right)^2 / K_{dolomite}^{equilibrium}$$

$$= \left(m_{Ca^{2+}} \gamma_{Ca^{2+}} \right) \left(a_{Mg^{2+}} \right) \left(a_{CO_3} \right)^2 / K_{dolomite}^{equilibrium}$$

which in turn accelerates the growth rate of dolomite:

$$R_{dolomite} = k_{dolomite}S_0(\Omega - 1)$$

where S_0 is the surface area of dolomite per unit volume and $k_{dolomite}$ is the rate constant for reaction (15.5). Here we have written $a_{Ca^{2+}}$ for the activity of Ca^{2+}, $m_{Ca^{2+}}$ for the mass of Ca^{2+}, and $\gamma_{Ca^{2+}}$ for the activity coefficient of Ca^{2+}.

Now, to express the constant volume replacement process we have as a statement of Figure 15.3(b):

$$hR_{dolomite} = \frac{dm_{Ca^{2+}}}{dt}$$

where h in this case is the stoichiometric constant 0.74 mol. Thus,

$$\frac{dm_{Ca^{2+}}}{m_{Ca^{2+}}} = hk_{dolomite}S_0Hdt \tag{15.6}$$

where $H = (\gamma_{Ca^{2+}})(a_{Mg^{2+}})(a_{CO_3})^2/K_{dolomite}^{equilibrium}$. Integration of (15.6):

$$\int_{m_{Ca^{2+}},initial}^{m_{Ca^{2+}}} \left[\frac{dm_{Ca^{2+}}}{m_{Ca^{2+}}}\right] = \int_0^t hk_{dolomite}S_0Hdt$$

gives

$$m_{Ca^{2+}} = m_{Ca^{2+},initial}\,\exp(t/\tau)$$

where

$$\tau = (1000hk_{dolomite}S_0H)^{-1}$$

We also have

$$R_{dolomite} = \left(k_{dolomite}S_0Hm_{Ca^{2+},initial}\right)\exp(t/\tau) \tag{15.7}$$

Thus, because of the constraint that volume be preserved, the growth rate of dolomite and the concentration of Ca^{2+} in the pore fluid increase exponentially as shown in Figure 15.5(a). This means that the growth rate of dolomite can outstrip the supply of Mg^{++} as shown in Figure 15.5(a) with the subsequent termination of dolomite growth because of the dependence of the growth rate on H in (15.6). Following this hiatus in growth the concentration of Mg^{++} grows again and the process repeats as shown in Figure 15.5(b).

In this model the interfacial stress leads not only to pressure solution as discussed in Section 15.5 but also to strain rate softening as discussed by Merino and Canals (2011); we consider this effect in greater detail in Volume II.

15.2.3 Nucleation and Growth Models

Following insightful work by Kretz (1966, 1969, 1973, 1974, 2006), Carlson and co-workers (Carlson, 1989, 1991, 2011; Carlson et al., 1995) have led the way in modelling the

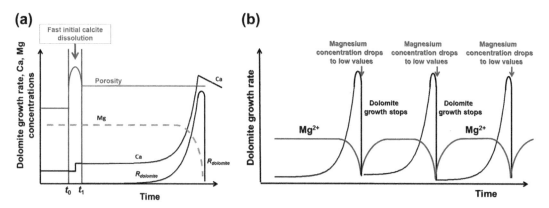

FIGURE 15.5 The Merino model for pseudomorphic replacement. (a) Evolution of the dolomite replacement system over one cycle of replacement. Mg^{2+} is supplied at a constant rate. Porosity is increased during the initial dissolution of Ca^{2+} between t_0 and t_1. Coupling of Ca^{2+} concentration with dolomite growth arising from constant volume replacement accelerates both dolomite growth and Ca^{2+} dissolution until dolomite growth outstrips the supply of Mg^{2+}. Dolomite growth then dramatically ceases until the supply of Mg^{2+} can build up again. (b) Repetition of (a) due to competition between accelerated dolomite growth and the constant input of Mg^{2+}. *After Merino and Canals (2011).*

development of porphyroblast microstructures. The influence of deformation is not considered in these models. In Chapter 9 we point out that grain growth during deformation can be treated using the concept of disconnections (Howe et al. 2000) with implication for coupled grain boundary migration and consequences for preferred mineral grain shape and the development of curved inclusion trails. The basic proposition in the Carlson models is that the affinity that drives the porphyroblast reaction arises from temperature overstepping so that the history of porphyroblast growth reflects the thermal history of the rock. The procedure involves, first, the assumption of a nucleation rate, $\dot{N}$, per unit volume of rock in the form

$$\dot{N} = \dot{N}_0 \exp[\kappa(T - T_0)]\left\{\frac{1}{V}\iiint \hat{\rho}(x,y,z)dxdydz\right\}$$

where $\dot{N}_0$ is the nucleation rate at time $t = 0$, T_0 is the temperature at $t = 0$, κ is an acceleration factor related to the activation energy for nucleation, V is the volume of the model and $\hat{\rho}(x, y, z)$ is the relative probability that a nucleation event occurs at the spatial position (x, y, z). In the models developed by Carlson this nucleation law is expressed as an exponential dependence on temperature rather than an Arrhenius law with constant activation energy.

A second step involves postulating a growth law of a form that is diffusion controlled:

$$R_i(t) = \left\{\int_{t_i}^{t} D_{T(t)}dt\right\}^{\frac{1}{2}} \tag{15.8}$$

$R_i(t)$ is the radius of the ith grain as a function of time, $D_{T(t)}$ is the diffusion coefficient at a temperature T which is a function of time and t_i is the time that the ith grain is nucleated. The diffusion coefficient depends on temperature according to

$$D_{T(t)} = D_\infty \exp\left(\frac{-Q_D}{RT(t)}\right)$$

where D_∞ is a reference diffusion coefficient, Q_D is the activation energy for diffusion and an explicit dependence of temperature on time is expressed by the function T(t). The model depends on specification of a particular thermal history. The temperature provides for overstepping of the reaction and the system evolves as the temperature increases above the equilibrium value. Notice that, although most porphyroblasts preserve initial detail of the matrix within them and hence are the result of constant volume replacement reactions, such a constraint is not incorporated into the Carlson models. The proposal (Kretz, 1974) is that the rate of growth for a particular radius can be obtained by measuring the normalised distance, c^*, between two growth zones at a particular radius, C^*, in a given grain. Such measurements are normalised with respect to measurements for the largest grain in the system (Figure 15.6(c)). The assumption is that zonal patterns in porphyroblasts are growth zones and not a result of intermittent supply of nutrients to the growing grain.

Adopting Carlson's model, the relation between c^* and C^* is

$$c^* = \frac{C^{*2}(\gamma\tau_i - 1) + 1}{C^*(\gamma\tau_i + 1 - C^{*2})} \tag{15.9}$$

where γ is a parameter that appears in the rate law for growth and τ_i is the dimensionless time when the ith grain nucleates. C^* is taken as a measure of the dimensionless radius corresponding to the dimensionless growth rate, c^*. Plots of c^* against C^* are shown in Figure 15.6(d) and a data set from natural porphyroblasts is shown in Figure 15.6(d).

This model produces grain microstructures that are very similar to natural examples. A particular aim of such studies is to gain some information on grain growth kinetics. Following Kretz (1974), the approach has been to compare observations with curves for specific kinetic laws on a plot of normalised growth rate, $\bar{\bar{R}} \equiv c^*$, against normalised radius, $\bar{R} \equiv C^*$. Such plots for natural examples are shown in Figure 15.6. Curves 1, 2 and 3 in Figure 15.6(a) and (b) correspond, respectively, to interface-controlled, isothermal diffusion-controlled and heat flow-controlled growth kinetics. The observation that none of these laws is strictly followed suggests that other factors are at play. Since some of the data follow approximately linear relations in Figure 15.6(a) and (b) we assume a general (linear) relation between $\bar{\bar{R}}$ and $\bar{R}$ of the form

$$\bar{\bar{R}} = -\alpha\bar{R} + \beta; \quad \alpha \geq 0$$

Then, for initial conditions given by $\bar{R} = 0$ when $t = 0$, $\bar{R}$ is given by

$$\bar{R} = \frac{\beta}{\alpha}\{1 - \exp(-\alpha t)\}$$

Since a series expansion gives $\exp(-\alpha t) \approx 1 - \alpha t + \frac{\alpha^2 t^2}{2} - \ldots\ldots, \bar{R} \to \beta t$ as $\alpha \to 0$. Thus, for $\alpha \to 0$ and $\beta = 1$ one obtains curve 1 in Figure 15.6(a) and (b) and the growth rate is

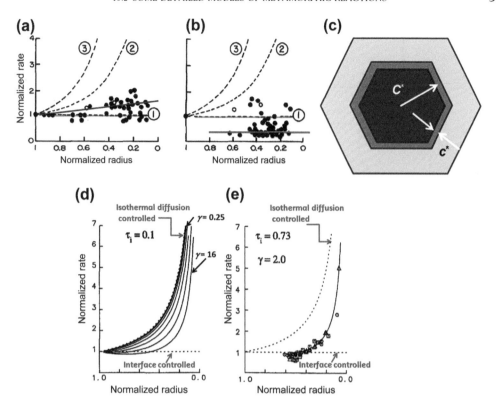

FIGURE 15.6   Garnet growth laws. (a) and (b): Plots of normalised growth rate against normalised radius for (a) case where MnO content at the centre of a grain correlates with grain radius and (b) case where garnets of intermediate size possess the highest MnO content. In both plots the two open circles are from Kretz (1974). *For details see Carlson (1989).* Curves 1, 2 and 3 correspond to interface-controlled growth, isothermal diffusion-controlled growth and heat flow-controlled growth, respectively. Brown, blue and red lines represent three different normalised growth laws, the blue one being an exponential growth law that decreases with time. (c) Measurements made on growth zones in garnet. C^* is the radial distance from the grain centre to the middle of a growth zone, c^* is the width of this zone. Both measurements are normalised with respect to similar measurements on the largest garnet in the population. (d) Plot of (15.9) for $\tau_i = 0.1$ and for various values of γ. (e). Data from natural garnet population. *After Carlson (1989).*

independent of time. Otherwise, for $\alpha \neq 0$, $\overline{R} = \beta \exp(-\alpha t)$. If $\alpha \sim 0.5$ and $\beta \sim 1.5$ one obtains the blue line in Figure 15.6(a). As $\alpha \rightarrow 0$, one obtains the red line in Figure 15.6(b) with $\beta \sim 0.33$. This suggests that an exploration of exponential growth laws would be of interest. However, all of the growth laws shown in Figure 15.6 are either independent of time or are decreasing functions of time (see Kretz, 1974). This contrasts with the models of Merino (and as we shall see, of Putnis) where the growth rate increases with time. As shown by Merino and Canals (2011), one way of generating such accelerating laws is for replacement of the matrix to be coupled with dissolution in an isochoric reaction. Clearly, more work needs to be done with respect to these models. As an indication of the zonal microstructures to be expected from various growth laws (assuming that zonation patterns are growth zones) some models are shown in Figure 15.7.

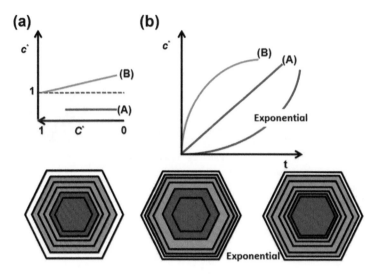

FIGURE 15.7 Growth laws and resulting zonation patterns. Top left is a sketch of a (c^*-C^*) plot from Figure 15.6(a) showing the red line (a) and blue line (b) from that figure. Top right are the resulting growth laws (c^* versus time) plots for the lines (a) and (b). Also included is a growth curve for exponentially increasing growth. The bottom row shows the zone patterns that result from these growth laws.

Schwarz et al. (2011) obtain a range of grain size distributions but their measurements do not fit the growth laws shown in Figure 15.6(a) and (b). They have extended porphyroblast modelling to include the influence of nutrient supply, and hence the degree of supersaturation, on the nucleation kinetics. Such modelling is important because it considers other mechanisms for generating an affinity for reaction rather than temperature overstepping.

15.2.4 Dissolution—Precipitation Models

Many reactions in metamorphic rocks are expressed as dissolution—precipitation reactions. Thus, one common model is that material is dissolved at one place in a rock to be precipitated at another place; in the simplest and most common models, the dissolution and precipitation processes are not coupled. Clearly this kind of behaviour demands a relatively high porosity. There is an extensive literature concerned with modelling fluid transport in porous rocks with uncoupled chemical reactions, although fluid transport is commonly coupled with dissolution and precipitation through changes in porosity. An important review with an historical summary is Steefel and Maher (2009) and the overall approach is considered by Steefel and Lasaga (1994). Another important approach is that of Bethke (1996), Reed (1997, 1998) and Grichuk and Shvarov (2002) where the extent of a reaction is simulated by a series of steps each one of which is taken to be in local equilibrium. This attempts to reproduce the approach discussed by Thompson (1959, 1970). The governing equations linking the rate of change of chemical composition with changes in composition arising from advection and reaction rate are commonly written as:

$$\frac{\partial(\phi c_i)}{\partial t} + \mathbf{J} \cdot \nabla c_i = \dot{R}_i \tag{15.10}$$

where c_i is the concentration of the ith chemical species in solution (with units moles per unit volume of the fluid), ϕ is the instantaneous porosity, $\mathbf{J}$ is the sum of the fluxes (moles per unit area of the rock per unit time) that arise from dispersive, diffusive and advective processes, ∇ is the gradient operator, $\cdot$ represents the scalar product and $\dot{R}_i$ is the total reaction rate (moles per unit volume of the rock per unit time) for the reaction involving the ith chemical species and may be expressed as

$$\dot{R}_i = \dot{R}_i^{homogeneous} + \dot{R}_i^{heterogeneous}$$

$\dot{R}_i^{homogeneous}$ is the reaction rate of the ith reaction occurring in solution and $\dot{R}_i^{heterogeneous}$ is the reaction rate of the ith reaction occurring at solid/liquid interfaces. Typically $\dot{R}_i^{heterogeneous}$ is a function of the surface area of the reacting grain and of pH (Lasaga, 1981, 1984).

A common approach (Phillips, 1991, 2009; Wood and Hewett, 1982) to using (15.10) is to assume chemical equilibrium, so that one can neglect the term $\dot{R}_i$ in (15.10), and c_i then represents the equilibrium concentration of the species C_i in solution. The porosity is assumed to remain constant and one writes

$$c_i^{equilibrium} = c_i\left(P^{fluid}, T, c_\aleph\right)$$

where $\aleph$ is the number of species and P^{fluid} is the fluid pressure. Using the chain rule of differentiation it follows that

$$\phi\frac{\partial c_i}{\partial t} = -\mathbf{J}\cdot\left\{\frac{\partial c_i}{\partial T}\nabla T + \frac{\partial c_i}{\partial P^{fluid}}\nabla P^{fluid} + \sum_{k=1}^{k=\aleph-1}\frac{\partial c_i}{\partial c_k}\nabla c_k\right\} \tag{15.11}$$

Clearly the assumptions involved here (equilibrium and constant porosity) are extreme but nevertheless the approach has a wide following (Murphy et al., 2008; Phillips, 1991, 2009; Steefel et al., 2005; Wood and Hewett, 1982; Zhang et al., 2003; Zhang et al., 2008; Zhao et al., 2009). (15.11) assumes that the processes involved are (1) dissolution of various species so that equilibrium concentrations of these species are carried in solution, and (2) a particular species is precipitated in existing open spaces according to the way in which the Darcy flux is resolved along the gradients in temperature, in fluid pressure and in the other chemical species in solution. The quantities $\frac{\partial c_i}{\partial T}, \frac{\partial c_i}{\partial P^{fluid}}, \frac{\partial c_i}{\partial c_k}$, (which are all functions of T, P^{fluid} and $c_\aleph$), act as qualifiers on these gradients.

In these classical models the processes of dissolution and precipitation proceed independently of each other and are uncoupled. Putnis (Putnis, 2002, 2009; Putnis and Austrheim, 2010; Putnis and Mezger, 2004; Putnis and Putnis, 2007; Putnis et al., 2005, 2008) has proposed a *coupled* model for dissolution–precipitation replacement reactions whereby the initial host is replaced through a porous dissolution network that progressively migrates into the host, dissolving the host and transporting the dissolved material into an adjacent fluid reservoir. The fluid-saturated porous network acts as a conduit to precipitate new material. The process is illustrated in Figure 15.2(f) and an example of an experimentally produced porous network is shown in Figure 12.11(a).

The characteristics of this model are (Putnis, 2009)

1. The dissolution and precipitation processes are coupled at the host/new phase interface. This coupling preserves the initial morphology of the host.

2. The reaction front between the host and the new phase is sharp with no obvious diffusion profile in the host.
3. The new phase develops in the porosity network that allows fluid to maintain contact with the reaction front.
4. In cases where there is a large change in molar volume between the host and the new phase a system of fractures develops ahead of the reaction front.
5. If the structures of the host and new phase allow, there is an epitaxial relation between the two.

Although the model is described as a coupled dissolution/precipitation process, the mechanism of coupling is so far not completely understood. Experiments by Putnis and Mezger (2004) and by Putnis et al. (2005) indicate that the layer of fluid next to the dissolving host plays a fundamental role; the composition of this thin film may be quite different to that of the bulk fluid.

Reactions between the solid host and this thin fluid layer may be the basis of the proposed coupling. This is supported by Xia et al. (2009) who show that the length scale of pseudomorphism can be manipulated by changing the pH of the fluid. Xia et al. (2009) examined the pseudomorphism of pentlandite by violarite experimentally between 80 and 210 °C and room pressure with a range of Eh-pH conditions, grain sizes and other variables. The conclusion, based on microstructures and the non-Arrhenius temperature behaviour of the kinetics, is that the process consists of dissolution and precipitation and is not a solid state diffusive process. The dissolution and precipitation are coupled presumably because the dissolution of pentlandite increases the supersaturation of violarite thus enhancing nucleation rates. Pentlandite dissolution is rate limiting for $1 < pH < 6$ and results in nanoscale pseudomorphism

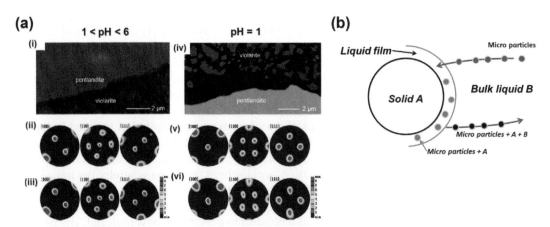

FIGURE 15.8 Aspects of pseudomorphic models involving dissolution and precipitation. (a) Influence of fluid pH on replacement morphology. (i) An image of violarite replacing pentlandite in the pH range $1 < pH < 6$. The pseudomorphism is regular and pole figures from pentlandite (ii) and violarite (iii) are sharp and demonstrate strong epitaxis. (iv) An image of irregular replacement structure for $pH = 1$. The pole figures, (v) for pentlandite and (vi) from violarite, are diffuse demonstrating weaker epitaxis. *From Xia et al. (2009).* (b) The Anderson et al. (1998a) model for coupling dissolution and precipitation within a thin liquid layer next to the dissolving solid.

at the <20 nm scale. Violarite precipitation is rate limiting for pH = 1 and results in microscale pseudomorphism at the scale of ~10 μm. Details of these microstructures are shown in Figure 15.8(a).

Another way of achieving this coupling lies in the work of Anderson et al. (1998a, 1998b) where *microphases* assist in enhancing reaction rates at interfaces between a solid and a reacting fluid. A microphase is a dispersed phase (particles, droplets, bubbles) that is smaller than the diffusion length of the solute. Thus, for instance, if the diffusion length is 50 μm, an effective microphase size is 10 μm. The basic mechanism consists of (1) diffusion of the microphase to the reacting interface; (2) the microphase reacts with dissolved solute near the interface thus decreasing the concentration of solute and increasing the dissolution rate; (3) the microphase, now carrying some solute, diffuses into the bulk fluid thus facilitating the transport of solute into the bulk fluid (Figure 15.8). We are interested in instances where the microphase consists of particles of the product which is to be precipitated on the solid host. This means that the final precipitated phase enhances the dissolution of the solid as precipitation occurs. This enhancement of reaction rate by the product-microphase is a form of *autocatalysis* or *accelerated reaction rate* where dissolution and precipitation are coupled.

Anderson et al. (1998a) develop a model for these processes that is expressed as a reaction–diffusion equation:

$$\frac{\partial a}{\partial t} = D\frac{\partial^2 a}{\partial x^2} - k_m a^m - l_0 K_0 \left\{ a - \frac{a_{mp}}{m_A} \right\} \tag{15.12}$$

where m is the reaction order for the reaction involved, k_m is a rate constant, l_0 is the volumetric fraction of microphase in the liquid, K_0 is the rate at which A is taken up by the microphase, a_{mp} is the concentration of A on the microphase and m_A is the distribution coefficient of A between microphase and continuous phase. Anderson et al. (1998a, 1998b) show that these microphase processes result in acceleration of reactions by 20- to 30-fold depending on growth rates and initial microphase size distributions. The process is important as an example of a dissolution–precipitation process that is coupled, autocatalytic and takes place in the immediate vicinity of the dissolving/precipitation interface.

15.3 HOMOGENEOUS AND HETEROGENEOUS NUCLEATION

The traditional approach to nucleation involves making a distinction between *homogeneous nucleation* and *heterogeneous nucleation*. The former is strictly applicable to liquids where a new phase nucleates within a host phase independently of position. The latter applies to nucleation where the nucleation site depends on position and typically is represented as the new phase nucleating at a surface. The theory is discussed by Vernon (2006, pp 46–51) and involves a spherical nucleus. Nucleation is controlled by the balance between a surface energy term that grows as the square of the radius of the sphere and a volume energy term that involves the difference in Gibbs energy per unit volume, ΔG_V, between the host and new grains and decreases with the cube of the radius. The driving force for growth is the difference in the total Gibbs energy, $\Delta G(R)$, as a function of radius, R, between the host and new grain. The balance between the increasing surface energy and decreasing Gibbs energy per unit volume, ΔG_V, defines a critical radius, $R_{critical}$, and a critical Gibbs energy difference, $\Delta G(R)_{critical}$, above which the new grain can grow:

$$R_{critical} = -2\frac{\sigma \widehat{V}}{\Delta G_V}, \quad \Delta G_{critical} = \frac{16\pi}{3}\frac{\sigma^3}{(\Delta G_V)^2}$$

where σ is the surface energy. Clearly such a model predicts that the larger ΔG_V, the smaller both $\Delta G_{critical}$ and $R_{critical}$. The relation between the Gibbs energy difference and the radius of the nucleus is

$$\Delta G(R) = \Delta G_{critical}\left[3\left(\frac{R}{R_{critical}}\right)^2 - 2\left(\frac{R}{R_{critical}}\right)^3\right] \tag{15.13}$$

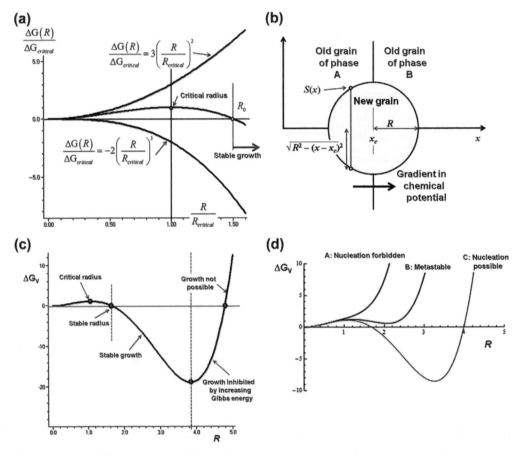

FIGURE 15.9 Nucleation of new grains. (a) The classical view of nucleation with no gradients in chemical potential. Stable growth occurs once the Gibbs energy becomes negative at a radius R_0. (b) A model of nucleation of a new grain in a chemical potential gradient. (c) Change of Gibbs energy with radius of a nucleus in a chemical potential gradient for conditions where nucleation is possible. (d) Three types of nucleation behaviour depending on the magnitude of the chemical potential gradient. Values of the parameters in (15.14) are $\alpha = 3$, $\beta = 2$, $\gamma = 1$; $\nabla\mu = 0.6$ for curve A, $\nabla\mu = 0.38$ for curve B, $\nabla\mu = 0.28$ for curve C.

This function is plotted in Figure 15.9(a). For embryos smaller than the critical radius, local fluctuations mean that the embryonic grain may dissolve. At the critical radius an increase in surface energy just balances the decrease in difference in (volumetric) Gibbs energy. Above the critical radius the nucleus can grow but the difference in Gibbs energy is positive until the radius R_0 is reached when the nucleus can undergo thermodynamically stable growth.

The above treatment, while satisfactory for nucleation in liquids, neglects some important issues that are relevant to nucleation in solids. First, the traditional model assumes that there is no volume change associated with the formation of the new nucleus and hence no stresses are generated between the new grain and the host. Second, spherical nuclei are assumed. Third, and probably the most important, in solids subcritical embryonic nuclei grow in chemical potential gradients that change with time (see Figure 15.10). New embryos interact through diffusive mass transfer with the host phases and neighbouring embryos. Heterogeneous kinetics governed by these diffusive processes control the process and not solely a balance between surface energy formation and decrease in Gibbs energy difference (Gusak, 2010).

For a situation that includes nucleation in a chemical potential gradient (Gusak, 2010) we assume a one-dimensional model with again a spherical nucleus and no volume change between the host and new grain as shown in Figure 15.9(b).

$$\Delta G = n \int \left(g^{new}(\mu(x)) - g^{old}(\mu(x)) \right) S(x) \mathrm{d}x + \sigma S$$

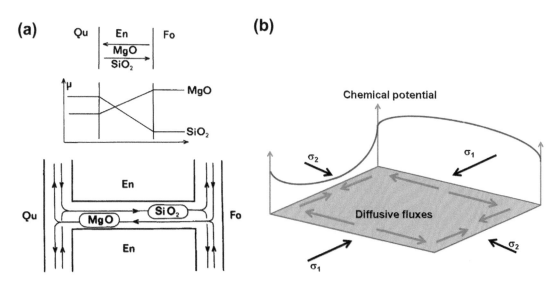

FIGURE 15.10 Chemical potential gradients in reacting and stressed solids. (a) Chemical potential gradients between quartz and forsterite reacting to produce enstatite. The gradients set up fluxes in grain boundaries as shown. (b) Chemical potential gradients and associated fluxes induced by a stress field. *From Wheeler (1987).*

where n is the atomic density (assumed the same in the nucleus and the host so that no stresses are set up), g is the Gibbs energy per atom, S is the surface area of the nucleus and σ is the surface energy. Gusak (2010) shows that the equation describing nucleation, (15.13), is replaced by an equation of the form:

$$\Delta G(R) = \alpha R^2 - \beta R^3 + \gamma(\nabla\mu)^2 R^5 \tag{15.14}$$

where α, β and γ are parameters which are functions of the chemical potential gradient and are given by Gusak (2010, Chapter 4). Nucleation is forbidden for

$$(\nabla\mu) > (\nabla\mu)_1^{critical} = \frac{\beta}{\alpha}\sqrt{\frac{\beta}{5\gamma}}$$

Nucleation is metastable for

$$(\nabla\mu)_1^{critical} > (\nabla\mu) > (\nabla\mu)_2^{critical} = \frac{\beta}{\alpha}\sqrt{\frac{4\beta}{27\gamma}}$$

Nucleation can occur for

$$(\nabla\mu) < \frac{\beta}{\alpha}\sqrt{\frac{4\beta}{27\gamma}}$$

(15.14) is plotted for this last condition in Figure 15.9(c) and the other conditions are shown in Figure 15.9(d).

Gusak (2010; Chapter 4) discusses the influence of a volume change associated with nucleation on the above argument and shows that elastic strains induced by change in volume have only a small effect. However, chemical potential gradients do influence the shape of the nucleus with flattened ellipsoidal nuclei favoured over spherical nuclei at large gradients.

A complete analysis of the influence of deformation on the shape and orientation of new nuclei depends on arguments presented by Kamb (1959) on the influence of elastic energy and by Wulff (see Kelly and Groves, 1970) on the influence of interfacial energy. Preferred growth of nuclei controlled by the deformation can either enhance or obliterate most orientations and shapes of initial nuclei and hence is the dominating factor in controlling the final microstructure. These issues are considered in Volume II.

15.4 STRESS-ASSISTED TRANSPORT

In order for mineral reactions to proceed, chemical components must be transported from one part of the reacting system to another. For the most part the mechanism for achieving such transport is unclear and depends on the details of the grain microstructure. In Chapter 12 we discuss mechanisms for mass transport that depend on the structure and size of apertures between grains. This in turn is probably a function of where the reacting−deforming system sits relative to the metamorphic regimes outlined by Oliver (1996; see Figure 12.1). As indicated in Figure 12.6, in regimes where the porosity is low (strictly, where the Knudsen number approaches ∞) the transport is presumably Fickian in nature and takes place in films similar to those illustrated in Figure 12.8. At smaller Knudsen numbers the transport is governed by the Burnett equations. This is a transport regime that is fundamental to metamorphic systems but is poorly studied; it is currently an area of active research in the nanofluidics industry and we expect important developments in this field in the future. At Knudsen numbers <0.1 (Figure 12.6) one moves into conditions where the complete Navier–Stokes equations need to be considered and then into regimes governed by Darcy flow. In this section we consider only Fickian transport and in particular, concentrate on the influence of stress on mass transport. This regime is commonly called the *diffusive transport regime*.

Details of diffusive transport processes, including diffusion in multicomponent materials, are considered by Paterson (2013) and we do not repeat that material here. We consider here only processes governed by *Fick's first and second laws*, namely (in one dimension):

$$\text{Fick's first law}: J = -\frac{cD}{RT}\frac{\partial\mu}{\partial x} \tag{15.15}$$

and

$$\text{Fick's second law}: \frac{\partial\mu}{\partial t} = D\frac{\partial^2\mu}{\partial x^2} \tag{15.16}$$

where J is the diffusive flux (units: amount of diffusing component per unit area per unit time, e.g. moles per square metres per second), c is the concentration of the diffusing component (units: moles per unit volume), D is the diffusion coefficient or diffusivity (units: square metres per second), μ is the chemical potential density of the chemical component that is diffusing (units: joules per mole) and x is distance; R is the universal gas constant (units: joules per mole per kelvin) and T is the absolute temperature.

In two or more dimensions Fick's second law becomes

$$\frac{\partial\mu}{\partial t} = D\nabla^2\mu$$

which is the same form as the heat diffusion equation (11.5). Hence the results of heat diffusion problems (Carslaw and Jaeger, 1959) can be used directly for the equivalent mass diffusion problems. In particular for a one dimensional problem where $c(x, t)$ is the concentration of a chemical component that has a concentration c_0 at $t = 0$ and $x = 0$, the concentration profile at time t is given by

$$c(x, t) = c_0 erfc\left(\frac{x}{2\sqrt{Dt}}\right)$$

which is identical in form to (11.13) and the solution is shown in Figure 11.4.

It has long been known that rocks subjected to stress undergo preferred dissolution at highly stressed interfaces (see Cox and Paterson, 1991; Lehner, 1995; Paterson, 1995; Rutter, 1976; Schutjens and Spiers, 1999 for examples and reviews). The mechanical background for this process is largely based on Gibbs' treatment (Gibbs, 1906) of the dependence of the solubility, in an adjacent fluid, of a solid with a normal stress, σ_N, on the solid/fluid interface.

The concentration of the solid, c, dissolved in the adjacent fluid is given in terms of the local chemical potential, μ, of the solid dissolved in the fluid by

$$\mu = \mu_0\left(P^{fluid}, T\right) + RT \ln(\gamma c) \tag{15.17}$$

where μ_0 is a reference chemical potential and γ is an activity coefficient (Paterson, 1973). The chemical potential is given in terms of σ_N as (Kamb, 1959; Paterson, 1973)

$$\mu = \Psi + \sigma_N \widehat{V} \tag{15.18}$$

where Ψ is the specific Helmholtz energy of the solid phase, $\widehat{V}$ in this case is the molar volume and σ_N is the normal stress across the surface whose normal in $\mathbf{N}$. The Helmholtz energy is commonly considered to be only a function of the elastic energy of the solid. Notice that Kamb (1959) and Paterson (1973) assume compressive stresses to be positive, whereas Shimizu (1992) assumes compressive stresses are negative. This means (for the Shimizu assumption) that $+\sigma_N$ is replaced by $-\sigma_N$ in (15.18).

The expressions (15.17) and (15.18) mean that for a stressed solid in contact with a fluid, the concentration of the solid dissolved in the fluid is greater than that for an unstressed solid and so the local fluid is supersaturated. As Gibbs noted, if an unstressed piece of the solid is placed in the fluid then precipitation will occur on that piece. The usual treatments then involve diffusion of the dissolved material driven by stress gradients, which induce gradients in chemical potential, with the presumption that material is transferred from highly stressed regions in the aggregate to low-stressed regions (Figure 15.10(b)). Such a process has now been recognised at all grades of metamorphism from essentially diagenetic conditions to upper amphibolite grades. This process may in fact dominate as a deformation mechanism in many metamorphic environments and operate instead of plastic deformation by dislocation processes (Wintsch and Yeh, 2013).

Most treatments of the subject consider a closed system. Paterson (1973), for instance, derives expressions for (γc) in a uniaxial stress condition: for the face normal to the stress, σ, the concentration of dissolved material, c, in the fluid adjacent to the interface is given by

$$\gamma c = \gamma_0 c_0 \exp\left[\widehat{V}\sigma/RT\right]$$

where c_0 is a reference concentration in an unstressed situation and γ_0 is the activity coefficient in the reference state. For the face parallel to the uniaxial stress,

$$\gamma c = \gamma_0 c_0 \exp\left[\widehat{V}\sigma^2/2ERT\right]$$

where E is the Young's modulus of the solid. This means that chemical potential gradients are set up on the surfaces of a stressed material and hence mass moves from one interface in the aggregate to another under the influence of these gradients. Such a process is variously called *pressure solution* or *stress-assisted diffusion*.

This process of stress-assisted dissolution and transport can clearly be dominated by one of three grain scale processes: (1) chemical reactions (dissolution and replacement/precipitation) at the stressed grain interface, (2) transfer of chemical components in the interstitial fluid, and (3) rate of transfer of dissolved material at the reaction site. The constitutive relations for uniaxial straining with strain rate, $\dot{\varepsilon}_1$, for these three situations considered by Paterson (1995) are

$$\text{fluid diffusion controlled: } \dot{\varepsilon}_1 = \frac{\widehat{V}^2 c D_f \phi^m}{RT} \frac{\left(\sigma_3 - P^{fluid}\right) + 3(\sigma_1 - \sigma_3)}{d^2}$$

$$\text{reaction controlled: } \dot{\varepsilon}_1 = \frac{\widehat{V}^2 k}{RT} \frac{\delta}{d_i} \frac{\left(\sigma_3 - P^{fluid}\right) + 3(\sigma_1 - \sigma_3)}{d}$$

$$\text{source/sink diffusion controlled: } \dot{\varepsilon}_1 = \frac{\widehat{V} \delta D_{gb}}{RT d_i^2} \frac{\left(\sigma_3 - P^{fluid}\right) + 3(\sigma_1 - \sigma_3)}{d}$$

In these expressions, c is the solubility, D_f is the diffusion coefficient in the fluid, k is the rate constant for dissolution or precipitation, D_{gb} is the diffusion factor in a grain boundary, δ is the grain boundary thickness, m is the Archie exponent (Paterson, 1995), ϕ is the porosity, d is the grain diameter, d_i is the diameter of the grain boundary island structure, P^{fluid} is the fluid pressure and σ_1, σ_2, σ_3 are the principal stresses. In all three cases the dependence of strain rate on stress is linear but different grain size and island size dependencies arise.

Reacting–deforming systems that involve stress driven transport can be various combinations of open or closed (see classification of Oliver: Figure 12.1), diffusion controlled, transport controlled, reaction controlled, or, controlled by dissolution processes that take place at the reacting interface. These various models are illustrated in Figure 15.11.

If the interface can support a shear stress (which is not the case at a fluid/solid interface), Shimizu (1992, 1995) shows that the chemical potential, μ^+, for a phase transition at a stressed interface where the growth direction is a, is given by

$$\mu^+ = \Psi - \varsigma \widehat{V}$$

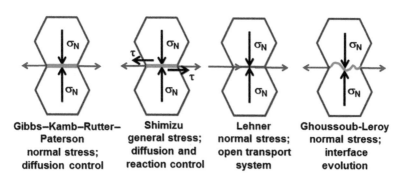

| Gibbs–Kamb–Rutter–Paterson normal stress; diffusion control | Shimizu general stress; diffusion and reaction control | Lehner normal stress; open transport system | Ghoussoub-Leroy normal stress; interface evolution |

FIGURE 15.11 Pressure solution models. From left to right: The classical Gibbs (1906), Kamb (1959), Paterson (1973), Rutter (1976) model involving only normal stress and diffusion; next the Shimizu (1992, 1995, 1997, 2001) model involving a general stress at the boundary and both diffusion and reaction at the interface; next the Lehner (1995) model involving a normal stress at the interface and open flow; last the Ghoussoub–Leroy model (Ghoussoub and Leroy, 2001; Raphanel, 2011) which discusses the evolution of the interface under a normal stress.

where $\varsigma = \frac{\sigma_{ij}N_i a_j}{N_k a_k}$ and $\mathbf{N}$ is the normal to the interface (Figure 14.11(a)). If the shear stress on the boundary is zero then the above reduces to the Kamb–Paterson expression.

The cases considered by Shimizu (1995) involve diffusion-controlled and reaction-controlled situations:

$$\text{Diffusion controlled}: \quad \dot{\varepsilon} = \left(\alpha \widehat{V}^2_{SiO_2} K D_w \right) \left(\widehat{V}_{H_2O} RT d^3 \right)^{-1} \sigma$$

$$\text{Reaction controlled}: \quad \dot{\varepsilon} = \left(\beta \widehat{V}^2_{SiO_2} k_+ \right) \left(\widehat{V}_{H_2O} RT d \right)^{-1} \sigma$$

where α is a geometrical factor that is a function of grain shape and the stress distribution, β is a geometrical factor describing the roughness of the stressed interface, K is the equilibrium constant for the dissolution of SiO_2 in H_2O, D_w is the diffusion coefficient for diffusion of the solute through channels at the interface, d is the grain diameter and k_+ is the rate constant for the dissolution reaction. Again, the strain rate is a linear function of the stress but different dependencies on grain diameter are present depending on the process.

If the system is open (Lehner, 1995) a distinction between intergranular dissolution and grain boundary diffusion (the Kamb–Paterson model) as rate limiting processes can be made on the basis of a complicated dimensionless number that involves among other quantities, the area of contact between grains, the grain diameter and the chemical potentials. The strain rates are now complicated functions of these quantities (see Lehner, 1995, Eqn 23). Evolution of the interfacial microstructure has been considered by Ghoussoub and Leroy (2001) and by Raphanel (2011) but developments in this arena are still very preliminary.

15.5 STRESSES GENERATED BY MINERAL GROWTH

We have seen that isochoric reactions (of which pseudomorphic replacement reactions are a special but common subset) generate stresses at the interface between the new mineral and the crystal or matrix that is being replaced. Since such stresses are involved in the conditions for chemical equilibrium (Section 14.3.3c) and control dissolution/diffusion rates and pathways it is important to understand their origins and magnitudes. These stresses are expressed in various ways. In some cases the surrounding material is fractured as in the case of quartz polymorphically replacing coesite (Figure 15.12(a)); in other cases plastic deformation may be clearly expressed as in the case of brucite replacing periclase (Figure 15.12(b)) or pressure solution may occur at the interface as documented by Merino and Canals (2011) and Stokes et al. (2012; see Figure 14.2(e)).

15.5.1 Stresses Induced by Constant Volume Replacement

The following discussion is taken from Fletcher and Merino (2001) where compressive stresses are taken as negative. Consider a spherical volume of rock, with radius R, in which the replacement reaction B → A is progressing (Figure 15.13). We consider a uniform array of

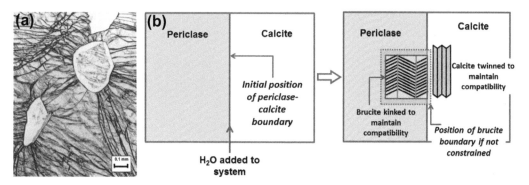

FIGURE 15.12 Examples of microstructures arising from replacement. (a) Constant mass and constant volume polymorphic replacement. Quartz (clear, low relief) and coesite (cloudy, high relief) embedded in garnet with microcracks radiating outwards from the quartz. *From Chopin (1984).* (b) Replacement of periclase by brucite. Kinks arise in brucite and twins in calcite from stresses generated by isochoric replacement. *Inspired by Turner and Weiss (1965) and Carmichael (1987).*

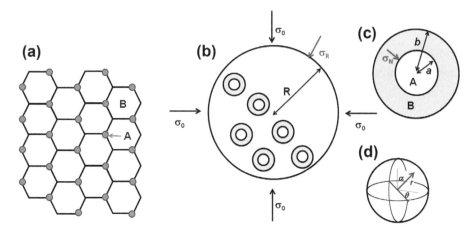

FIGURE 15.13 The model of Fletcher and Merino (2001). (a) Nucleation of a phase A shown, for discussion, at triple junctions within a polygonal array of phase B. (b) A spherical volume, radius R, of distributed mineral growth selected from the aggregate in (a) embedded in an infinite aggregate under a far field hydrostatic stress, σ_0. The normal stress on the sphere is σ_R. (c) A microscopic representative volume element (RVE); a, b are the radii of phases made of A and B; the normal stress on the interface between A and B is σ_N. (d) Spherical coordinate system r, θ and α.

spherical grains of A, with radii, a, surrounded by shells of B with radii, b. This unit, comprising A surrounded by B, is treated as a representative volume element (RVE). The regular array of RVEs means that transport of material is not a rate-controlling process. The large spherical volume is under a far field hydrostatic stress, σ_0. Since A is replacing B and the molar volume of A, $\widehat{V}_A$, is greater than that of B, $\widehat{V}_B$, dilational strains mean that the stress in the vicinity of the mineral reactions, σ_R, is different from σ_0 but it is still hydrostatic since the growth of the grains of A is assumed to be uniform.

In order to calculate the stress in an RVE we adopt a spherical coordinate system, r, θ and α as shown in Figure 15.13(d). The nonvanishing stress components are σ_{rr} and $\sigma_{\theta\theta} = \sigma_{\alpha\alpha}$; the radial displacement is u_r with $u_\theta = u_\alpha = 0$. Following Kamb (1959) the chemical potentials of A and B at their interfaces are

$$\mu^{A(interface)} = \mu_0^A - \sigma_0 \widehat{V}_A + \Delta\mu^A$$

$$\mu^{B(interface)} = \mu_0^B - \sigma_0 \widehat{V}_B \tag{15.19}$$

where component A is supersaturated by an amount $\Delta\mu^A = RT \ln(\Omega^A)$ relative to a hydrostatic value where the far field stress is σ_0 and where Ω^A is the saturation index for A. Component B is saturated at the hydrostatic value, $\Omega^B = 1$. $\widehat{V}_A$ and $\widehat{V}_B$ are the specific volumes of A and B in an unstressed state. The chemical potentials of A and B at the A/B interface at equilibrium are

$$\mu^{A(equilibrium)} = \mu_0^A - \sigma_N \widehat{V}_A$$

$$\mu^{B(equilibrium)} = \mu_0^B - \sigma_N \widehat{V}_B \tag{15.20}$$

The rates of replacement and dissolution are given by

$$\frac{da}{dt} = k_A \left[\mu_A^{interface} - \mu_A^{equilibrium} \right]$$

$$\frac{da'}{dt} = k_B \left[\mu_B^{interface} - \mu_B^{equilibrium} \right]$$

where k_A and k_B are rate constants and $\frac{da'}{dt}$ is the rate of change in the radial position of the interface due to growth of B. Since the only mechanism of growth in this model is constant volume replacement,

$$\frac{da}{dt} + \frac{da'}{dt} = 0 \tag{15.21}$$

From (15.19) to (15.21) we have

$$\frac{da}{dt} = k_A \left[\Delta\mu_A - (\sigma_0 - \sigma_N)\widehat{V}_A \right] = k_B(\sigma_0 - \sigma_N)\widehat{V}_B$$

so that

$$(\sigma_0 - \sigma_N) = k_A\Delta\mu_A / \left[k_A\widehat{V}_A + k_B\widehat{V}_B \right] = (k_A RT \ln \Omega_A) / \left[k_A\widehat{V}_A + k_B\widehat{V}_B \right] \tag{15.22}$$

which is positive since $\sigma_0 > \sigma_N$.

If $\frac{k_A}{k_B} \gg 1$ and we assume $\widehat{V}_A \approx \widehat{V}_B$ then the maximum stress difference is

$$(\sigma_0 - \sigma_N)_{maximum} = \Delta\mu_A / \widehat{V}_A$$

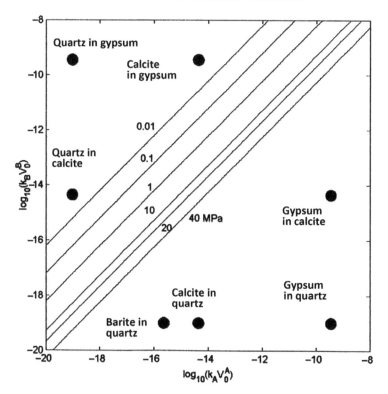

FIGURE 15.14   Contours of the interfacial normal stress, σ_N, generated by growth with a saturation $\Omega_A = 2$ at 100 °C assuming $\widehat{V}_A = 50 \times 10^{-6} \text{ m}^3 \text{ mol}^{-1}$ for several mineral pairs. *Based on data from Table 2 in Fletcher and Merino (2001).* Note that quartz replacing calcite generates negligible interfacial stress, whereas calcite replacing quartz generates ∼100 MPa interfacial stress. *From Fletcher and Merino (2001).*

If $\frac{k_A}{k_B} \ll 1$ then $(\sigma_0 - \sigma_N)$ is small and the growth rate is similar to that in a fluid. As an example, the interfacial normal stresses generated at 100 °C for various combinations of A and B are shown in Figure 15.14 assuming $\widehat{V}_A = 50 \times 10^{-6} \text{ m}^3 \text{ mol}^{-1}$ and $\Omega_A = 2$. Clearly the very high stresses generated by some mineral combinations (such as calcite replacing quartz) cannot be supported by elastic distortions and some form of dissolution, plastic flow or brittle fracturing must ensue.

15.6 AN EXAMPLE: PORPHYROBLAST FORMATION

We present below a discussion of porphyroblast growth and microstructure evolution partly as a summary of the concepts developed in this chapter and partly to emphasise the differences between what is essentially a linear approach to microstructure evolution and an approach built around nonlinear dynamics. In Chapter 13 we discuss the development of microstructures in polycrystalline aggregates from the point of view of stochastic processes described by the Fokker–Planck equation and from the point of view of deterministic processes described by reaction–diffusion equations. We repeat some of that discussion here but specifically for the development of porphyroblasts.

Discussions of porphyroblast development in the literature treat the problem from a nucleation/growth/transport viewpoint with an emphasis on models that develop an affinity for the garnet forming reaction(s) derived from temperature overstepping (Carlson, 2011) or from chemical potential changes induced by supersaturation of nutrients (Schwarz et al. 2011). Such effects are undoubtedly present in metamorphic systems; however, since the formation of porphyroblasts is an example of a reaction–diffusion–transport process we aim to emphasise that aspect here by employing reaction–diffusion–transport equations. This is essentially a nonlinear approach to porphyroblast formation where the results arise from competition between diffusion (or the rates of supply of reactants) and the rates of chemical reactions. We consider only isothermal situations but the discussion could be readily extended to nonisothermal situations by including temperature dependent diffusion coefficients.

Understanding the development of porphyroblasts is fundamental to understanding the formation of metamorphic microstructures and associated processes. The work of Kretz and Carlson and co-workers (Carlson, 1989, 1991, 2011; Carlson et al., 1995; Kretz 1966, 1969, 1973, 1974, 2006) has emphasised the potential to establish grain growth laws and to quantify values for diffusion coefficients, activation energies and thermal histories from aspects of the microstructure. Various authors have proposed that quantitative measures of the microstructure (in particular grain size frequency distributions and correlation coefficients) can be used to distinguish between various rate-controlling processes such as diffusion-controlled growth, interface-controlled growth and mass/thermal flux-controlled growth (see Carlson, 2011; Kretz, 2006; Schwarz et al., 2011). In particular, chemical mapping using X-ray, electron and synchrotron sources enables finer resolution of chemical zoning (Figure 15.15(a)). The development of three-dimensional X-ray tomography now enables

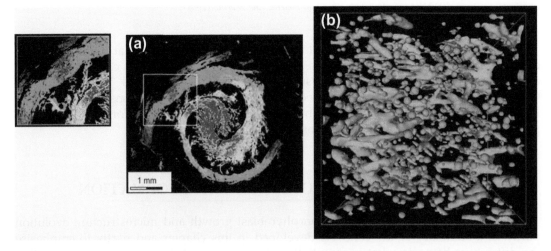

FIGURE 15.15   (a) X-ray Mn-compositional image of spiral garnet porphyroblast from Switzerland with detailed inset. *From Robyr et al. (2009).* (b) Three-dimensional X-ray tomographic image of staurolite (orange) and garnet (pink) porphyroblasts from the Picurus Range, New Mexico. *From Ketcham and Carlson (2001).*

detailed quantitative description of microstructures (Figure 15.15(b)). With this arsenal of equipment we are now in an ideal situation to compare the geometrical and chemical details of microstructure with computer-generated microstructure and hence to test various proposed models for microstructure development (Carlson, 2011). Although Carlson has shown good fits between natural and simulated microstructures, the fits rely on matching certain parameters in the computer models to observed features of natural microstructures, depend on temperature overstepping to drive the reaction and on associated models for the thermal history of the rock.

However, there is still much we do not understand. Schwarz et al. (2011) have reported a range of grain size frequency distributions that do not agree with models for growth driven solely by temperature overstepping and have proposed that the rate of supply of nutrients is the dominating control on growth. The models proposed by Kretz and by Carlson as typified by Figure 15.6(a–e) all involve growth laws where the growth rate is constant with time or a decreasing function of time. The models of Merino and of Putnis as applied to the growth of pseudomorphs predict growth laws that are increasing functions of time. There is the additional observation by Bell et al. (1986) and Bell and Johnson (1989) that porphyroblast growth and periods of localised deformation alternate. And, of course, there remains the beleaguered issue of the origin of 'spiral' garnets (Figure 15.15(a)). Thus, the influence of deformation on porphyroblast growth remains an outstanding issue. In what follows we first outline the linear models for porphyroblast formation and growth and then proceed to some nonlinear models. We defer a discussion of the influence of deformation on porphyroblast development and of the origin of spiral garnets to Volume II, although a model involving coupled grain boundary migration (Cahn et al. 2006) is discussed in Chapter 9.

The development of a porphyroblast differs from models of recrystallisation and grain growth discussed in Chapter 13 in that

- Grain growth must satisfy topological space filling requirements; porphyroblast growth does not fulfil this requirement in that there is no constraint that individual porphyroblasts need to stack together to fill space but commonly the porphyroblasts preserve initial fine-scale structure from the matrix and hence the growth is at least homochoric and commonly (perhaps always?) isochoric in order to maintain deformation compatibility with the matrix.
- Grain growth obeys topological requirements that are constrained by the need to minimise interfacial energy; porphyroblast growth is also constrained by a requirement to minimise interfacial energy but does so by adopting idiomorphic forms as opposed to polyhedral forms (with curved interfaces) that must pack together to fill space.
- Grain growth in a monomineralic aggregate is essentially controlled by interfacial energies and grain boundary diffusion and is driven by processes that reduce the interfacial energy; porphyroblast growth is an example of reaction–diffusion–transport processes where reaction rates compete with diffusion rates and interfacial energies may not be important.

Any realistic model of porphyroblast growth should incorporate these three growth constraints.

15.6.1 Stochastic Models of Porphyroblast Growth; the Fokker–Planck Equation

The work of Carlson represents one of the best developed models of porphyroblast growth. It is based on concepts of nucleation and of growth developed essentially for metals. Examples include the classical work of Avrami (1939, 1940, 1941), Burke and Turnbull (1952) and Christian (1975). Nucleation is heterogeneous and is presumably governed by the heterogeneous distribution of sites where the magnitude of the chemical potential differences between the new and host grains is greater than zero due to local arrangements of mineral phases that produce favourable chemical potential gradients, local fluctuations in chemistry, stress, temperature or stored energy due to deformation (see Kretz, 2006). The nucleation stage is described by a probability distribution in the Carlson model (see Section 15.2.3). This is essentially a stochastic description of part of the process but we should note that the essential features of stochastic processes are not captured in existing models in the geological literature. By a *stochastic process* we mean *any process where the evolution of the system is non-deterministic so that future states of the system must be based on probabilities*. We have seen in Chapter 13 that for microstructure development this involves the use of the Fokker–Plank equation. We return to this aspect later in the discussion.

The second step in the Carlson model is the proposal that the growth of nuclei is controlled by the diffusion of nutrients to the growing grain (Carlson, 1989, Eqn (A1)):

$$\boxed{\text{Radius}} = \boxed{k(\text{Diffusion coefficient} \times \text{time})^{1/2}}$$

$$R_i(t) = k\sqrt{D(t - t_i)} \tag{15.23}$$

where R_i is the radius of a grain that began growing at time t_i, k is a dimensionless constant and D is a diffusivity (a slightly different relation is quoted in (15.8)). In proposing such a control the problem is posed as a system where the isothermal growth rate is $\frac{\partial R_i}{\partial t} = \frac{1}{2}k\sqrt{\frac{D}{t}}$ and so is a decreasing function of time. The model is linear where the form of the growth law is fixed by the diffusion law and the rate of supply of heat to the system. We have discussed this model in greater detail in Section 15.2.3.

The nucleation of porphyroblasts is a problem that needs greater attention. Porphyroblast growth occurs in a matrix of continuously reacting (and deforming) grains and represents a phase transformation in a (locally) open heterogeneous system in which gradients in chemical potentials exist and in which mass is being transported both to and from the growing porphyroblast and perhaps through the system as a whole. Thus, the situation is different from the classical Avrami–Christian model of heterogeneous nucleation in an environment with no gradients in chemical potential. As an example, consider the system described by Carmichael (1969) and reproduced in Figure 15.16(a).

If we consider the four circles in Figure 15.16(a) to be neighbouring reaction sites as envisaged by Carmichael (1969) then garnet and quartz nucleate and grow in Box A at the interface between albite and biotite with an influx of H^+, Fe^{2+} and an outflow of Na^+, Mg^{++}, H_2O and K^+ (Figure 15.16(b)). Clearly the nucleation and growth processes occur in an environment involving gradients in chemical potentials and so are subject to the inhibition effects of large

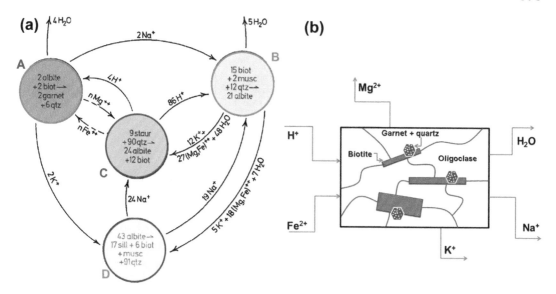

FIGURE 15.16   Metamorphic reactions as isolated diffusive systems. (a) Networked chemical reaction scheme for the formation of garnet described by Carmichael (1969). (b) Diagram of microstructure and transport input/output systems associated with the red box, A, in figure (a).

gradients in chemical potentials, $|\nabla\mu|$, described in Section 15.3. However, while this process is taking place, albite and biotite are being produced in Box C, thus creating sites where new garnet can nucleate and grow. In this scheme there is continuous opportunity for new garnet nucleation sites to be generated. Gusak (2010) refers to such processes as sequential growth processes and proposes that they are characteristic of solid state growth processes that are diffusion controlled.

In such circumstances the simplistic concepts of homogeneous and heterogeneous nucleation discussed in Section 15.3 do not apply, although the basic idea of competition between surface energy increase and Gibbs energy decrease still applies. Nucleation occurs in gradients of chemical potentials that can both enhance and suppress nucleation, depending on $|\nabla\mu|$ (Gusak, 2010; see also Section 15.3), and in an environment that is continually evolving. Some nuclei are in environments where sustained growth is possible; others grow in an oscillatory manner; some grow only to be resorbed. In order to model such a process at the microscopic level one would need to track the paths of individual atoms. One more realistic way is to model the process as a stochastic one where probabilities play a fundamental role. This leads us to a nonlinear approach.

The nonlinear approach to the problem is to explore the phase portrait associated with microstructure development by not constraining the growth rate of a single grain by (15.8), (15.23) but to rewrite (15.8), (15.23) in a rate form, (15.24), where the grain size distribution, defined by $F(R)$, evolves with time in a stochastic manner (Pande, 1987; Thompson, 2001). This evolution is envisaged to take place in a field of fluctuating macroscopic variables such as chemical potentials, temperature and grain sizes. The evolution of grain growth is then expressed as a balance between a deterministic process expressed as a function, F, of grain size, R, and time, t, that tends to increase grain size (grain growth) and some stochastic noisy process, $N(t)$, that tends to homogenise the process (in this case, diffusion):

Growth rate	=	Coarsening or grain size reduction rate	−	Diffusion of nutrients to or from the grain

$$\frac{dR_i(t)}{dt} = F\{R_i(t)\} - N(t) \qquad (15.24)$$

(15.24) describes a general case where the evolution of the grain size distribution can fluctuate (including becoming narrower corresponding to resorption); the form of the growth rate, $\dot{R}_i(t)$, is not constrained solely by temperature-dependent diffusion and needs to be established by other reasoning.

The evolution of the grain size distribution, $F(R, t)$, is then expressed as a balance between the *diffusion* of the grain size distribution through grain size distribution space and coarsening/reduction during grain growth or grain size reduction during deformation. Such a change in the grain size distribution is called *drift*. The evolution of the final grain size distribution is then expressed as

Change in grain size distribution with time	=	Diffusion of grain size distribution through grain-size distribution space	−	Coarsening or grain-size reduction with time (drift)

$$\frac{\partial F\{R(t)\}}{\partial t} = D\frac{\partial^2 F\{R(t)\}}{\partial R^2} - \frac{\partial}{\partial R}\left(\dot{R}F\{R(t)\}\right) \qquad (15.25)$$

(15.25) is a *Fokker–Planck equation* (Risken, 1996) which we consider in some detail in Chapter 13. Various authors have explored aspects of the Fokker–Planck equation including Hillert (1965), Louat (1974), Mullins (1986), Atkinson (1988), Thompson (2001) and Streitenberger and Zollner (2006). Many studies lead to a log-normal or Rayleigh distribution of grain sizes with little difference between the two distributions (Figure 15.17(b)). Sometimes a Weibull distribution is a better fit. It is important to note that this stochastic approach does not necessarily lead to a diagnostic or unique grain size distribution such as log-normal or Weibull. The distribution that does evolve is the result of competition between the diffusion and growth (drift) terms in (15.25) and so the resultant distribution depends on the balance between these two terms.

The two end members of the Fokker–Planck equation, namely, a situation where the drift term dominates as opposed to a situation where the diffusion term dominates, have been explored by Hillert (1965) and Louat (1974), respectively. The assumption that the diffusion term is insignificant and that the drift term dominates (Hillert, 1965) leads to the Hillert distribution (Figure 15.17(b)) which has not so far been observed in metals. The Hillert distribution (Figure 15.17(b)) is skewed to larger grain sizes and has a cut-off at a large grain size limit. Louat

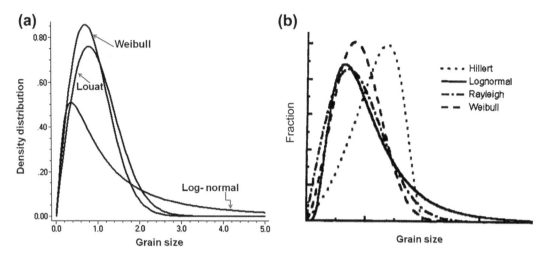

FIGURE 15.17 Comparisons of different grain size distributions. (a) A comparison of log-normal, Louat (Rayleigh) and Weibull distributions. (b) A comparison of log-normal, Hillert, Weibull and Rayleigh distributions. *From Thompson (2001).*

(1974) proposed that the drift term in the Fokker–Planck equation was insignificant compared to the diffusion term and arrived at a Rayleigh distribution of grain size, $F(R)$, given by

$$F(R) = \frac{\pi R}{2}\exp\left(-\frac{\pi}{4}R^2\right)$$

This distribution is plotted in Figure 15.17(a) and is similar to the Weibull distribution. It differs from the log-normal distribution which has a longer tail at large grain sizes. Since the Louat assumption considers only the diffusion term it may be representative of diffusion-controlled grain size evolution (see Streitenberger, 1998). We can perhaps expect grain size distributions for porphyroblasts that lie within a spectrum somewhere between Louat and Hillert distributions depending on the relative contributions from diffusion and drift terms. Some natural grain size distributions are close to Weibull distributions (see Figure 15.17(b) as an example) and such distributions do in fact lie between Louat and Hillert distributions (cf., Figure 15.17(a) and (b)). For a comparison of these distributions for models of grain growth see Atkinson (1988). Expressions for various distributions can be found in Section 16.2.4 of Bronshtein et al. (2007).

15.6.2 A Reaction–Diffusion Approach, the Gray–Scott Model

The Fokker–Planck equation represents a *stochastic* approach to modelling porphyroblast growth whereby the evolution of *grain size distribution* is tracked. Another way to model this problem but with an emphasis on the physical position of grains is to use an approach whereby *spatial patterns*, in this case, localised growth of grains, arise from *deterministic competition* between at least two coupled chemical reactions and diffusion. This approach utilises reaction–diffusion equations which we considered in some detail in Chapter 7. Reaction–diffusion equations (Chapter 7) express the coupling between diffusion processes (that tend to homogenise the system) and chemical reactions (that tend to localise the system); the chemical reactions are commonly networked as described by Carmichael (1969), Vernon (2004) and Wintsch and Yeh (2013). In addition, if the system is open (even on a local scale) then a transport process may be added to the governing equations.

The balance between the two (or three) processes can lead to complicated spatial patterns. Although one might suspect that only complicated chemical reaction schemes would lead to complicated spatial patterns, Gray and Scott (1994) show that even the simplest of reactions of the form $A \rightarrow B$, provided the reaction is exothermic, can produce the complete array of behaviours seen in complicated reaction systems depending on whether the system is open to A or to A and B (see Section 14.4.4). We will not consider the details of the reaction behaviour (see Gray and Scott, 1983, 1984, 1985; Ord et al., 2012) but note that a similar array of behaviours results from the simplest of autocatalytic reaction schemes: $A + B \rightarrow X$, $X + Y \rightarrow 3B$ even if the system is isothermal. One does not need complicated reaction schemes to produce complex behaviour.

Except for Carmichael (1969) there is no detailed discussion of the networked chemical reactions that develop garnet porphyroblasts; unfortunately there is no data available on the grain size distributions and spatial patterning developed for the Carmichael example. Hence, as an example, we explore the simplest of systems that produce spotlike spatial patterns as an example of a deterministic route to modelling porphyroblast systems. This reaction is the so called *Gray–Scott reaction* (Pearson, 1993).

The Gray–Scott chemical reaction model (Gray and Scott, 1983, 1984, 1985; Lee et al., 1993; Pearson, 1993) is the simplest model that incorporates nonlinear coupled chemical reactions and diffusive transport that holds the system away from equilibrium. Of course, real networked reaction systems in metamorphic rocks are far more complicated than in the Gray–Scott reaction and involve many more degrees of freedom. Here, we are interested in the minimal reaction–diffusion–transport system that will produce patterns similar to natural porphyroblast microstructures. It is amazing that a system as simple as the Gray–Scott reaction can produce many of these characteristics. The chemical reactions involved are written as

$$U + 2V \xrightarrow{k_1} X$$

$$X + Y \xrightarrow{k_2} 3V$$

$$V \xrightarrow{k_3} P$$

where X and Y are intermediaries that enable U to react with V to produce P. The equations that describe the evolution of this system are

$$\frac{\partial u}{\partial t} = D_U \frac{\partial^2 u}{\partial x^2} - uv^2 + F(1 - u)$$

$$\frac{\partial v}{\partial t} = D_V \frac{\partial^2 v}{\partial x^2} + uv^2 - (F + k_3)v \qquad (15.26)$$

where u and v are the concentrations of U and V and D_U, D_V are the diffusion coefficients of U and V. The term $+F(1 - u)$ is a *replenishment term* that supplies U to the system. F is the *feed rate* of U and the maximum rate of replenishment is 1. The term $-(F + k_3)v$ is a *depletion term* and represents the net supply (or extraction) of V to (from) the system. A necessary condition for spatial patterns to develop from (15.26) is that the two diffusion coefficients, D_U and D_V, are not equal. This corresponds to situations in silicate reactions where it is commonly proposed that the diffusivity of Al is much less than that of other elements.

The results of the Gray–Scott reaction for the parameters $F = 0.02$, $k_3 = 0.056$, $D_U = 0.18$, $D_V = 0.09$ are shown in Figure 15.18 (d, e, f). For comparison the results for a natural example

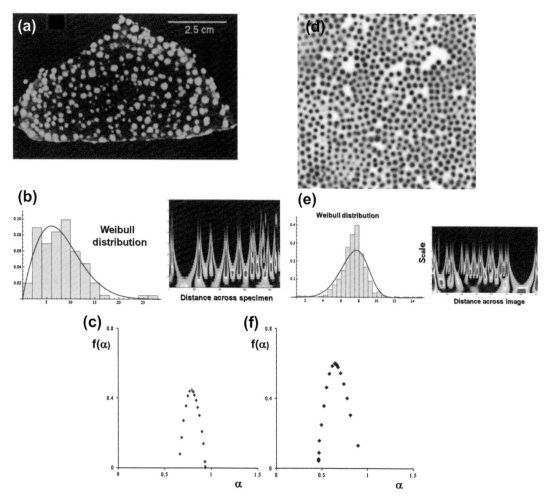

FIGURE 15.18   Natural porphyroblast population and population of reaction centres arising from a Gray–Scott reaction. (a) Garnet porphyroblasts. Specimen WR, garnet amphibolite from Carlson and Denison (1992). (b) Grain size distribution for image in (a) with Weibull distribution for comparison (left); scalogram (right) for one-dimensional scan across (a). (c) Singularity spectrum derived from scalogram in (b); $D_0 = 0.45$, $D_1 = 0.44$, $D_2 = 0.41$ at $\alpha = 0.75$. (d) Reaction centres, Gray–Scott reaction. (e) Grain size distribution for image in (d) with Weibull distribution for comparison (left); scalogram (right) for one-dimensional scan across (d). (f) Singularity spectrum derived from scalogram in (e); $D_0 = 0.60$, $D_1 = 0.59$, $D_2 = 0.54$ at $\alpha = 0.58$. For the Gray–Scott model the parameters in (15.26) are $D_U = 0.18$, $D_V = 0.09$, $F = 0.02$, $k_3 = 0.056$. Model has been constructed using software at http://www.aliensaint.com/uo/java/rd/.

(specimen WR, garnet amphibolite of Carlson and Denison, 1992) are shown in Figure 15.18 (a, b, c). The spatial distributions of both the natural example and of the Gray–Scott reaction centres are characterised by clustering so that large grains can be adjacent to other large grains and relatively large vacant areas exist. Both the natural and Gray–Scott scalograms (see Chapter 7) have near to periodic intensity distributions at mid-scales (Figure 15.18(b) and (e)). Both the natural example and the Gray–Scott model are multifractal in the spatial distribution of porphyroblasts. The grain size distribution for the natural example is near to Weibull whereas the Gray–Scott model gives a sharper distribution. The fractal dimension, D_0, of the natural example is 0.45, whereas that of the Gray–Scott distribution is 0.60 which means that the Gray–Scott distribution tends to fill space slightly more than the natural example. The information dimension for the Carlson specimen is $D_1 = 0.44$, while the correlation dimension $D_2 = 0.41$ at $\alpha = 0.75$. D_1 for the Gray–Scott reaction is 0.59 with $D_2 = 0.54$ at $\alpha = 0.58$. Both the Carlson specimen and the Gray–Scott reaction show long-range spatial correlations.

Thus, simple reaction diffusion transport models are capable of producing many features of natural porphyroblast microstructures including, in particular, multifractal spatial distributions with long-range spatial correlations. Such models do not require temperature overstepping to supply an affinity for the reaction. The affinity exists simply because the system is not at equilibrium during the reaction and the system evolves to a non-equilibrium stationary state; the (Ross) excess work drives the reaction to non-equilibrium stationary states and derives from competition between the diffusive and reaction terms in (15.26). As an additional observation, the reaction–diffusion–transport models automatically produce zoned porphyroblasts (Figure 15.19(a)) and these zonation patterns are similar to those that one would expect from a growth law that decreases with time as shown by the blue lines in Figures 15.6(a) and

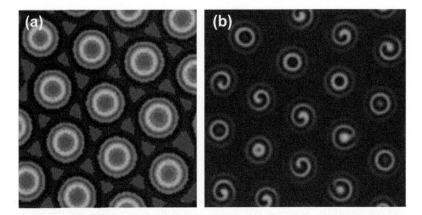

FIGURE 15.19   Microstructures arising from reaction–diffusion systems. (a) Zoning developed within grains formed by the Gray–Scott reaction; shown are contours of concentration of V with red maximum. For the Gray–Scott model the parameters in (15.26) are $D_U = 0.18$, $D_V = 0.09$, $F = 0.02$, $k_3 = 0.056$. Model has been constructed using software at http://www.aliensaint.com/uo/java/rd/ (b) Spiral travelling waves formed from a reaction–diffusion system (not the Gray–Scott model). *From Yang and Epstein (2003).* For a movie showing the development of this structure, see http://www.youtube.com/watch?v=-71t565s1b4&list=PL35FD96A5F8236109 and click on *Isolated Spirals in Yang 2003.*

15.7. As a contentious additional observation, these reaction–diffusion models are also capable of producing spiral compositional growth patterns (Figure 15.19(b)). We return to reaction–diffusion–transport models for microstructure formation in Volume II.

In order to understand the influence of deformation on the patterns developed by reaction-diffusion-transport equations we need to introduce new terms into the governing equations that describe the coupling between deformation and grain growth. This involves the preferred motion of disconnections controlled by the imposed deformation and some aspects of this process have been considered in Chapter 9. Such a coupling introduces new nonlinear terms and gives a preferred sense of rotation to the spirals in Figure 15.19(b). We return to reaction-diffusion-transport models coupled to deformation as generators of microstructure in Volume II.

15.7 SYNTHESIS

Chapter 14 (in part) concerns the chemical equilibrium state of a system undergoing deformation and diffusive processes. In common with the equilibrium mineral assemblages that develop in a syn-kinematic-metamorphic environment that may not reflect the equilibrium state under hydrostatic conditions, microstructures in deformed metamorphic rocks are the result of coupled chemical/deformation/transport processes and hence reflect the non-equilibrium history of the rocks. These processes can be expressed as reaction–diffusion–transport equations characterised by non-equilibrium stationary states.

We have seen in Chapter 14 that whether a mineral assemblage is 'close' or 'far' from an equilibrium state, as defined by non-hydrostatic pressure conditions at a particular temperature, depends on the stress distribution within the assemblage and the nature of deformation gradients between adjacent grains. For instance, these parameters appear in the term $[\![vF_{i3}/F_{33}]\!]d\sigma_{3i}$ in the Clapeyron equation for a coherent interface. If the phase transitions are such that the interface cannot sustain a shear stress, then deformation compatibility is achieved by grain boundary sliding coupled with diffusive fluxes which also enter into the Clapeyron equation, (14.65).

Thus, the mechanism(s) by which deformation compatibility is achieved in a deforming–reacting mineral assemblage, as described in the present chapter, is of fundamental importance. Commonly such mechanisms involve pseudomorphic or isochoric replacement of existing grains or matrix. However, such microstructures demand that the rate of replacement matches the rate of reaction or dissolution of the material so that the replacement/dissolution processes must be coupled at a very fine scale. Thus, the common pseudomorphic or isochoric microstructures are a geometrical expression not only of deformation compatibility but of the metamorphic reaction process itself. As indicated, these processes must be tightly coupled at the chemically reacting interface. The detailed mechanisms are not well established at present but the work of Merino, Putnis and associated workers provides some important insights.

Presumably these reaction processes take place in a thin layer (probably fluid bearing such as illustrated in Figure 12.8) adjacent to the reacting interface where the addition of material to the grain being replaced is strongly influenced by the removal of material from that grain and vice versa. In the Merino model this is a self-accelerating (autocatalytic) process that is driven ultimately by the coupling between replacement dissolution, the dissolution in turn being driven by stresses arising from isochoric replacement. The process operates at a very fine scale (nanometre to micrometre scales). In the Putnis model, as exemplified by the experimental work of Xia et al. (2009), the process involves coupling of dissolution and precipitation from solution at the interface. This is perhaps driven by the influence of the replacing material on the saturation of the solution at the interface with respect to the host, and facilitated by epitaxial relations between the host and precipitate. This could perhaps involve the microphase processes discussed by Anderson et al. (1998a, 1998b).

Clearly there is room for both the Merino (quasi solid state replacement–dissolution, stress driven) and the Putnis (fluid dominated dissolution–precipitation, epitaxial facilitated) models within the geological spectrum and perhaps these two models represent end members in a spectrum of behaviours expressed by the Oliver classification in Figure 12.1. Perhaps the Merino model is relevant at relatively high confining pressure and temperatures at low permeabilities, whereas the Putnis model is relevant at low pressures and temperatures in fluid dominated conditions? We need detailed microstructural studies to make a decision and to gain a better understanding of these processes.

Independently of whether the Merino or Putnis model operates, the interface between host and the reacting medium in a deforming metamorphic rock is a dynamic environment. Strong gradients in chemical potential exist as pointed out by Wheeler (1987, 2014) and these influence and control nucleation sites and rates (Figures 15.9 and 15.10). The reactions that produce a new grain are commonly networked (in the sense of Carmichael, Vernon and Wintsch) and each reaction in the chemical network proceeds at different rates so that the supply of reactants to the reacting sites may be episodic as growth/reaction/diffusion rates compete. Such systems are open flow chemical reactors at least at the scale of a few grains. In addition, because these systems are driven by coupled diffusive fluxes and chemical reactions they are archetypical examples of reaction–diffusion systems.

It is important to note that diffusive fluxes are driven by gradients in chemical potentials which in turn are functions of the stress at the interface. This may be simply the normal stress if the interface cannot support a shear stress (the Kamb–Paterson model), or more general stress states otherwise (the Shimizu model). These stress-driven diffusive fluxes (and hence mass transfer processes) are an integral part of the chemically reacting–deforming system and the evidence seems to be accumulating (see the Wintsch network reactions and the microstructural work of Williams in Figure 14.2) that such processes dominate as deformation mechanisms in many metamorphic environments instead of more conventional dislocation creep mechanisms. The stresses involved here arise not only from the imposed tectonic boundary conditions but also from grain scale stresses generated by chemical replacement reactions.

Major progress has been made in understanding metamorphic microstructures, particularly porphyroblast microstructures, by the quantitative work of Kretz (1969, 2006), Cashman and Ferry (1988), Daniel and Spear (1999) and Carlson (1989, 2011) and co-workers. In particular it is evident that a range of grain size distributions exist (not simply or specifically lognormal or Weibull), the microstructures are characterised by clustering (not other patterns such as grain depleted haloes around larger grains as in the model proposed by Ortoleva, 1994, pp 196–199) and multifractal spatial distributions with long-range spatial correlations (see Figure 15.18); growth rates can depend exponentially on grain size and evidently these growth laws can be both increasing and decreasing functions of time. To date such systems have been modelled in terms of classical (Christian, 1975) nucleation-growth models where the chemical affinity derives from temperature overstepping (Carlson models) together with a temperature history or from supersaturation derived from supply of nutrients (Schwarz model).

An alternative approach is to consider these systems as reaction–diffusion systems. Then the affinities of the reactions derive simply from the fact that the reacting/diffusing system is not at equilibrium and is driven to one or more non-equilibrium stationary states by the (Ross) excess work. The system is characterised by competition between diffusive (and fluid transport) fluxes, that act to homogenise the system, and chemical reactions, that tend to localise the system. The result, at a stationary state, is a microstructure, in this case an array of porphyroblasts whose positions and grain size distribution depend on these competitive processes. The precise equations describing the behaviour of natural networked reacting/diffusing systems involved in porphyroblast development have not yet been written but experience from other systems (Gray and Scott, 1994; Pearson, 1993) shows that even the simplest of reaction–diffusion systems is capable of displaying many of the complexities developed in complicated networked reacting systems. We have explored one example in Figure 15.18. All of the microstructural features observed in natural examples are developed by such a simple reaction–diffusion system including multifractal spatial patterns with long-range correlations, clustering and zoning (Figures 15.18 and 15.19). As a recognised contentious observation, spiral compositional patterns, as described by Yang and Rivers (2001), develop as well!

A reaction–diffusion approach to metamorphic microstructures also emphasises that we do not expect a specific grain size distribution (Gaussian, log-normal, Weibull) to develop from metamorphic reactions. We do expect competition between diffusive and drift terms in the Fokker–Planck equation and hence distributions that lie somewhere between Louat and Hillert distributions.

In summary, we see an exciting future in detailed microstructural and microchemical studies that examine and test models for coupled reaction diffusion systems and the influence of such processes on the position of syn-kinematic mineral equilibrium phase boundaries. In addition, a range of new questions arises to do with the multifractal nature of metamorphic microstructures. For instance, *are mineral growth laws decreasing or increasing functions of time? And what are the detailed processes that lead to these two contrasted growth laws? Are metamorphic microstructures an expression of criticality in the metamorphic process? Do metamorphic microstructures develop the characteristics of classical critical systems (multifractal behaviour, long-range correlations) only as an isograd is approached and passed?*

Finally, we emphasise that much of the material in this chapter does not consider coupling grain growth to deformation. In order to do so a number of new factors need to be considered. These are:

(i) The influence of elastic strain energy on the orientation of newly growing grains (Kamb, 1959; MacDonald, 1960).

(ii) The influence of anisotropy of the surface energy of the newly growing grain and of the difference in interfacial energy between the newly growing grain and its matrix. This has an influence on the standard Wulff construction governing grain shape (Kelly and Groves, 1970) as modified by Winterbottom (1967).

(iii) The influence of anisotropic supply of nutrients, controlled by deformation, on the growing grain.
(iv) The preferred motion of disconnections controlled by the deformation.
(v) The effects of coupling between grain boundary migration and internal grain deformation (Cahn et al. (2006)).

Some of these aspects are considered in Chapter 9. A complete theory of grain nucleation and growth during deformation involves a synthesis of these concepts and those discussed in this chapter. Such a synthesis is deferred to Volume II.

Recommended Additional Reading

Gray, & Scott. (1994). *Chemical Oscillations and Instabilities*. Oxford: Clarendon Press.
   This book is a comprehensive treatment of chemical reactions in open systems. Many simple, coupled reactions are analysed to define the conditions for bifurcation especially using the concept of a flow diagram.

Murray. (2002). *Mathematical Biology*. Springer.
   An important early rigorous and readable work on reaction–diffusion systems with an emphasis on pattern formation.

Passchier, C. W., & Trouw, R. A. J. (1996). *Microtectonics*. Springer.
   An important resource for deformation and metamorphic microstructures and models for microstructure development.

Paterson, M. S. (2013). *Material Science for Structural Geology*. Springer.
   A concise discussion of many aspects of material science with direct application to metamorphic geology.

Vernon, R. H. (2004). *A Practical Guide to Rock Microstructure*. Cambridge University Press.
   A powerful presentation and discussion of metamorphic microstructures with many examples of coupled chemical-deformation processes.

Epilogue

This volume has been concerned with the principles involved in the deformation and metamorphism of rocks in the crust of the earth from the point of view of modern solid mechanics. In doing so our goal has been to lay down the framework and a vocabulary for applications that we consider in Volume II. Metamorphic systems are considered to be giant chemical reactors that operate under non-equilibrium conditions and are driven by the kinematic boundary conditions arising from motions in the mantle of the earth, the dead weight load exerted by overlying rocks and by the influx of heat and fluids such as H_2O and CO_2. In addition energy is dissipated throughout the history of the system arising from brittle deformations, visco-plastic deformations and chemical reactions.

The evolution of these systems is governed by the following first-order principles which are five in number.

1. *Deforming metamorphic rocks are solids not fluids.*

Although there is a complete gradation between elastic—viscous—plastic materials that we refer to as *solids* and inviscid rate-insensitive materials we call *fluids*, our discussion has been concerned with rocks as solids. The principle here is that deforming rocks are characterised by a *yield surface* or in the rate-sensitive regime, a *flow surface*, whose size and shape is strongly dependent on the temperature, the rate and the amount of deformation. The existence of a yield surface means that elasticity is always important in the deformation and is fundamental in controlling the conditions for localisation. When the elasticity becomes anisotropic, as in strongly deformed rocks, it not only controls the conditions for localisation but also the orientation of the localised zone of deformation. Treatment of rocks as solids also means that constitutive laws such as power law descriptions alone are inadequate since in most conditions of interest, the deformation is not steady, elasticity is important as also is anisotropy of both elasticity and plasticity.

The stress state in any material is governed by the history of the motions that the material has experienced. This statement has far reaching implications for a solid. It means that the stress state is not only a product of the boundary conditions, which in general in rocks will be a combination of a dead weight vertically and velocity boundary conditions horizontally, but most importantly, also a product of the internal microstructural rearrangements that have contributed to the motions.

This control of the local stress state by the history of microstructural rearrangements is by far the most important feature that characterises a plastic solid from an inviscid fluid. It means that any constitutive law must be framed in terms of these microstructural adjustments; this is fundamental for any thermodynamic treatment of the deformation (Principle 5 below).

2. *The kinematic framework of the deformation is fundamental in determining the structures we see in deformed metamorphic rocks and not the strain.*

The quantity that appears in the constitutive equations that describe the deformation of materials is the stretching tensor and not the strain tensor. The exception is the constitutive equations that describe linear elasticity. This means that the kinematics assume a dominant position in the mechanics of deformation. The strain is simply a quantity that accumulates during deformation and, except for linear elasticity, it does not enter directly into any constitutive relation. Moreover, fabric elements in a deforming material are attracted to the eigenvectors of the velocity gradient tensor which in general are not coincident with the eigenvectors of the stretching tensor. In particular, the manifolds of the deformation (the planes that are tangent to the eigenvectors) ultimately become the dominant features of the fabric. Thus the axial planes of folds and the orientations of many foliations and lineations are a representation of the manifolds of the velocity gradient tensor. These fabric elements are not parallel to principal planes of strain and always have shearing deformations parallel to their orientations. The strain may be expressed ultimately in the deformed shape of inherited objects but at large deformations the manifolds of the velocity gradient tensor prevail as the dominant features.

In a general affine deformation there are always two such manifolds both of which can be parallel to surfaces of shear strain. Hence kinematic indicators always show shear strains parallel to dominant fabric elements. The two manifolds are represented in deformed rocks as arrays of fractures (joints and veins) and they control the orientations of oblique foliations and S-C fabrics.

3. *The processes that occur during deformation and metamorphism invariably lead to some form of geometrical or constitutive softening and hence non-convex energy functionals. This is the basis for all structure formation.*

Many processes that occur during deformation and metamorphism lead to softening of either a physical property such as the elastic or plastic hardening moduli or to a decrease in the stress bearing capacity of the material with continuing deformation. Such softening behaviour can arise from geometrical factors such as the rotation of a single slip plane and any plane of anisotropy or from the geometrical percolation of a weak mineral phase throughout the deforming aggregate. Softening also arises from constitutive behaviour involving damage, chemical reaction or heat generated from dissipation arising from any of the processes involved in metamorphism or in deformation.

This softening behaviour is reflected in a Helmholtz energy which is non-convex. The implications of a non-convex energy functional are profound. In general the energy of the system cannot be minimised by a single homogeneous deformation and still maintain compatibility with an imposed deformation. However it is possible to come arbitrarily close to minimising the energy, and simultaneously satisfy the constraint imposed by the deformation, by developing an inhomogeneous deformation consisting of two or more deformations at finer and finer spatial scales. A lower limit to the size of these structures is set by competition with the energy required to produce surfaces between the differently deformed regions. The continuous refinement of structure also leads to fractal geometries.

The non-convex energy is also the hallmark of critical behaviour in systems marked by cascades of energy dissipation with multifractal characteristics in both space and time. Such behaviour seems to be universal and is represented in both earthquake statistics and in the fine-scale plastic deformation of crystals.

The development of non-convex energy functionals, and the resulting development of continuously refining minimising structures, is the basis for the development of most fabrics we see in deformed metamorphic rocks.

4. *The motion of defects is the fundamental mode of deformation and of mineral reactions.*

Although the motion of dislocations (line defects associated with lattice translation) and the migration of point defects have been widely explored in the deformation of rocks over the past 50 years, other crystal/grain defects can be more important and especially provide a link to solid-state mineral reactions. These other defects are disclinations (line defects associated with lattice rotation), and disconnections (line defects associated with a change in phase). At large deformations, disclinations and disconnections, with the motion of dislocations and point defects, are important deformation mechanisms that can be the dominant processes in subgrain formation, grain size reduction and rotation, recrystallisation by grain-boundary migration and the formation of foliations and lineations defined by mineral shape. These processes arise through mechanisms of *coupled grain boundary migration* whereby the motion of a grain boundary, driven by stress, curvature or the affinity of a chemical reaction, induces a shear strain parallel to the moving boundary. The process is completely coupled so that a shear strain parallel to a grain boundary induces a movement in the grain boundary normal to the boundary. An important repercussion of this process is that material markers in a grain undergo a rotation (due to simple shearing parallel to the grain boundary) as the boundary moves. This is proposed as the dominant mechanism for subgrain rotation and for the development of curved (even spiral) inclusion trails in porphyroblasts.

An actively deforming, chemically reacting, metamorphic rock needs to be viewed as a fully coupled system wherein the motion of grain boundaries driven by the affinity of the chemical reactions (itself a function of temperature and the stored energy of deformation) is closely coupled to the deformation and is in fact a deformation mechanism in its own right so that the motion of these grain boundaries induces shear strains and rotations, which are fundamental to the development of foliations and lineations defined by mineral shape.

5. *The thermodynamics of stressed, reacting solids provides the fundamental links between the various processes that occur during deformation and metamorphism. A thermodynamic treatment is essential to ensure that hypotheses are internally consistent and thermodynamically admissible.*

As with all systems not at equilibrium, the evolution of metamorphic systems (and also the various stages that such systems might evolve through) is the result of competition between the power (work done per unit time) associated with thermodynamic forces that tends to move the system away from equilibrium and the dissipation arising from thermodynamic fluxes that tends to move the system towards equilibrium. For most systems of interest the thermodynamic forces are gradients in the deformation, in $(1/T)$, in hydraulic potential and in chemical potentials. The corresponding thermodynamic fluxes are the momentum per unit area (or the stress), heat flux, fluid flux, and chemical transport/reaction rate.

The various processes that operate during the deformation and metamorphism of rocks can be incorporated into a thermodynamic view of metamorphic systems by expressing the rearrangements involved as internal variables such as grain size, damage tensors, fabric tensors or the extents of chemical reactions. It is the presence of these internal variables that distinguishes a solid from an inviscid fluid. The evolution of these variables, with time, is associated with thermodynamic fluxes and for each flux there corresponds a generalised thermodynamic force. In the case of chemical reactions this force is known as the affinity of the reaction. The energy balance between the thermodynamic forces and the fluxes enables stationary states to be defined; one such stationary state is equilibrium but other non-equilibrium stationary states are far more interesting and are responsible for most fabric development.

The thermodynamics of deforming metamorphic rocks is in its infancy but a number of points are clear. Although the concepts of Gibbs energy and chemical potential have a clear and unambiguous meaning when a chemical system is close to equilibrium under hydrostatic pressure, the same statement is not true when the stress state is non-hydrostatic. Chemical equilibrium can exist in a deforming solid but the conditions that define equilibrium are not as straightforward as in a hydrostatic system. The quantity called the Gibbs energy has three different interpretations for a non-hydrostatic system and this emphasises the difficulty in generalising from the equilibrium to the non-equilibrium theory. It is better to proceed from the general to the specific theory and we attempt the beginnings of such an approach in this volume.

Appendix A

Commonly Used Symbols

In an interdisciplinary book such as this, we have found it too demanding to stick to a unique mathematical notation throughout. We have attempted to standardise on commonly used quantities as defined in this appendix but for individual examples we have mainly used the notation of the original authors of the example. Such terms are always defined where they are first used.

Symbol	Quantity	Units
$\mathbf{D}$, D_{ij}	Stretching tensor	s^{-1}
e	Specific internal energy	$J\,kg^{-1}$
$\mathbf{F}$, F_{ij}	Deformation gradient	Dimensionless
J	Jacobian of a matrix	Dimensionless
P	Hydrostatic pressure	Pa
$\bar{P}$	−(Mean stress)/3	Pa
$\mathbf{R}$	Rotation matrix	Dimensionless
s	Specific entropy	$J\,kg^{-1}\,K^{-1}$
T	Temperature	K
t	Time	s
V	Volume	m^3
$\mathbf{X}$, X_i	Cartesian coordinates in a reference state	m
$\mathbf{x}$, x_i	Cartesian coordinates in the current or deformed state	m
x, y, z	Cartesian coordinates	m
$\boldsymbol{\varepsilon}$, ε_{ij}	Small strain tensor	Dimensionless
Φ	Specific dissipation function	$J\,kg^{-1}\,s^{-1}$
ρ	Mass density	$kg\,m^{-3}$

(Continued)

Symbol	Quantity	Units
σ, σ_{ij}	Cauchy stress	Pa
Ψ	Specific Helmholtz energy	$J\,kg^{-1}$
MATHEMATICAL SYMBOLS		
$a \bullet b$	Scalar product of vectors a and b	
$a \times b$	Vector product of vectors a and b	
$a \otimes b$	Diadic product of vectors a and b	
$A : B$	Scalar product of tensors A and B	
$[\![A]\!]$	Jump in A	
CHEMICAL SYMBOLS		
A; SiO_2	The chemical species, A; silica	
a; $[SiO_2]$	The concentration (or activity) of A; the concentration (or activity) of SiO_2	mol^{-1}, kg^{-1}

Appendix B
Vectors, Tensors and Matrices

For a useful introduction to vectors, tensors and matrices see Tadmor et al. (2012). The following is a brief summary of useful relations. For detailed proofs see Tadmor et al. (2012). The Einstein summation convention is nicely discussed by Nye (1957).

1. **Kronecker delta.**

   The Kronecker delta is defined as

   $$\delta_{ij} = 1 \quad \text{if } i = j$$
   $$= 0 \quad \text{if } i \neq j$$

   A useful identity is $a_i \delta_{ij} = a_j$.

2. **Permutation symbol.**

   The permutation symbol, ε_{ijk} for three dimensions is defined as.

   $$\varepsilon_{ijk} = 1 \quad \text{if } i, j, k \text{ form an even permutation of } 1, 2, 3$$
   $$= -1 \quad \text{if } i, j, k \text{ form an odd permutation of } 1, 2, 3$$
   $$= 0 \quad \text{if } i, j, k \text{ do not form a permutation of } 1, 2, 3$$

   Some useful identities are

   $$\varepsilon_{ijk}\delta_{ij} = \varepsilon_{iik} = 0, \quad \varepsilon_{ijk}\varepsilon_{mjk} = 2\delta_{im}, \quad \varepsilon_{ijk}\varepsilon_{ijk} = 6$$

3. **The cofactor of a matrix.**

   The cofactor of a matrix, $\mathbf{A} = \begin{bmatrix} A_{11} & A_{12} \\ A_{21} & A_{22} \end{bmatrix}$, is

   $$cof\ \mathbf{A} = \begin{bmatrix} A_{22} & -A_{12} \\ -A_{21} & A_{11} \end{bmatrix}$$

4. **The determinant of a matrix.**

   The determinant of a matrix, $\mathbf{A} = \begin{bmatrix} A_{11} & A_{12} & A_{13} \\ A_{21} & A_{22} & A_{23} \\ A_{31} & A_{32} & A_{33} \end{bmatrix}$ is

   $$det\ \mathbf{A} = A_{11}\begin{vmatrix} A_{22} & A_{23} \\ A_{32} & A_{33} \end{vmatrix} - A_{12}\begin{vmatrix} A_{21} & A_{23} \\ A_{31} & A_{33} \end{vmatrix} + A_{13}\begin{vmatrix} A_{21} & A_{22} \\ A_{31} & A_{32} \end{vmatrix}$$

Also

$$\varepsilon_{mnp} det\ \mathbf{A} = \varepsilon_{ijk} A_{im} A_{jn} A_{kp} = \varepsilon_{ijk} A_{mi} A_{nj} A_{pk}$$

5. The scalar (or dot) product of two vectors.

The scalar product of two vectors, a and b, is the scalar

$$a \bullet b = ab\cos\theta = a_i b_i = a_1 b_1 + a_2 b_2 + a_3 b_3$$

where θ is the angle between a and b.

6. The vector (or cross) product of two vectors.

The vector product of two vectors, a and b, is the vector, c

$$c = a \times b$$

where c is orthogonal to both a and b and has a magnitude $|ab\sin\theta|$. θ is the angle between a and b and c forms a right handed triplet with a and b. If i, j, k are unit vectors parallel respectively to the x_1, x_2, x_3 coordinate axes,

$$c = a \times b = \begin{bmatrix} a_2 & a_3 \\ b_2 & b_3 \end{bmatrix} i - \begin{bmatrix} a_1 & a_3 \\ b_1 & b_3 \end{bmatrix} j + \begin{bmatrix} a_1 & a_2 \\ b_1 & b_2 \end{bmatrix} k$$

Note that $a \times b = -b \times a$.

7. The scalar product of two tensors.

The scalar product of two tensors, $\mathbf{A}$ and $\mathbf{B}$, is the scalar

$$\mathbf{A} : \mathbf{B} = A_{ij} B_{ij}$$

8. The tensor (or dyadic) product of two vectors.

The tensor (or dyadic) product of two vectors, a and b, is a tensor, $\mathbf{T}$, given by

$$\mathbf{T} = a \otimes b = \begin{bmatrix} a_1 b_1 & a_1 b_2 & a_1 b_3 \\ a_2 b_1 & a_2 b_2 & a_2 b_3 \\ a_3 b_1 & a_3 b_2 & a_3 b_3 \end{bmatrix}$$

9. Polar decomposition theorem.

A *pure rotation* is a deformation where material vectors locally undergo changes in orientation (rotations) but not changes in length. A *pure stretch* is a deformation where three initially orthogonal material lines change lengths during the deformation but not their orientations. A general affine deformation, expressed by the deformation gradient, $\mathbf{F}$, includes both rotations and stretches of local lines but there are local rotations that do not involve a stretch. The polar decomposition theorem states that the local deformation of a material vector can be expressed as a local pure rotation, $\mathbf{R}$, together with a local pure stretch. The nature of the pure stretch depends on whether the total deformation, $\mathbf{F}$, is expressed as a pure stretch followed by a pure rotation, or as a pure stretch followed by a pure rotation. In the first case the pure stretch is expressed as the tensor, $\mathbf{U}$, and $\mathbf{F} = \mathbf{RU}$. In the second case the pure stretch is expressed as the tensor, $\mathbf{V}$, and $\mathbf{F} = \mathbf{VR}$.

The polar decomposition theorem says that one obtains the same final result regardless of whether you rotate first or rotate last. The key equations involved are

$$\mathbf{U} = \left(\mathbf{F}^T\mathbf{F}\right)^{1/2}, \ \mathbf{V} = \left(\mathbf{F}\mathbf{F}^T\right)^{1/2} = \mathbf{R}\mathbf{U}\mathbf{R}^T, \ \mathbf{R} = \mathbf{F}\mathbf{U}^{-1}, \ \mathbf{R} = \mathbf{V}^{-1}\mathbf{F}$$

10. Rotation matrix in terms of components of the deformation gradient.

If a deformation gradient, $\mathbf{F}$, is given in two dimensions as

$$\mathbf{F} = \begin{bmatrix} F_{11} & F_{12} \\ F_{21} & F_{22} \end{bmatrix}$$

then the rotation matrix, $\mathbf{R}$, associated with this deformation is (see (2.14))

$$\mathbf{R} = \begin{bmatrix} \cos\varphi & -\sin\varphi \\ \sin\varphi & \cos\varphi \end{bmatrix}$$

The components of $\mathbf{R}$ can be calculated directly from the components of $\mathbf{F}$ as indicated by the equations above but the process is tedious. The main issue concerns the calculation of the square root of the matrix for $\mathbf{C} = \mathbf{F}^T\mathbf{F}$.

For a matrix given by

$$\mathbf{C} = \begin{bmatrix} C_{11} & C_{12} \\ C_{12} & C_{22} \end{bmatrix}$$

The two square roots of $\mathbf{C}$, $\mathbf{U}^{\pm}$, are given by (Levinger, 1980):

$$\mathbf{U}^{\pm} = (C_{11} + C_{22} \pm 2\Delta)^{-1/2} \begin{bmatrix} C_{11} \pm \Delta & C_{12} \\ C_{12} & C_{22} \pm \Delta \end{bmatrix}$$

where $\Delta = (det\ \mathbf{C})^{1/2} = \sqrt{C_{11}C_{22} - C_{12}^2}$. Having calculated $\mathbf{U}$, one then needs to calculate $\mathbf{U}^{-1} = \frac{1}{det\ \mathbf{U}} cof\ \mathbf{U}$ and then $\mathbf{R} = \mathbf{F}\mathbf{U}^{-1}$. These calculations can be carried out in Mathematica. The result is $(2.15)_1$:

$$\mathbf{R} = \frac{1}{\sqrt{(F_{11} + F_{22})^2 + (F_{12} - F_{21})^2}} \begin{bmatrix} F_{11} + F_{22} & F_{12} - F_{21} \\ F_{21} - F_{12} & F_{11} + F_{22} \end{bmatrix}$$

References

Levinger, B.W., 1980. The square root of a 2 × 2 matrix. Mathematics Magazine 53, 222–224.

Nye, J.F., 1957. Physical Properties of Crystals. Oxford.

Tadmor, E.B., Miller, R.E., Elliot, R.S., 2012. *Continuum Mechanics and Thermodynamics*. Cambridge University Press.

Appendix C

Some Useful Mathematical Concepts and Relations

A useful book to have is: Bronshtein, I. N., Semendyayev, K. A., Musiol, G., and Muehlig, H. 2007. *Handbook of Mathematics*. Springer. This contains a wealth of information concerning mathematical relations, formulae, and concepts. In particular it contains a series of tables involving mathematical and natural constants, units, series expansions, derivatives, integrals, and commonly used mathematical symbols.

C.1   THE CHAIN RULE OF DIFFERENTIATION

If $\xi = F(x, y)$, $\eta = G(x, y)$, ... are all differentiable functions of x and y and a compound function, u, is given by

$$u = f(\xi, \eta, ...)$$
$$= f\{F(x,y), G(x,y), ...\}$$

Then

$$\frac{\partial u}{\partial x} = \frac{\partial f}{\partial \xi}\frac{\partial \xi}{\partial x} + \frac{\partial f}{\partial \eta}\frac{\partial \mu}{\partial x} + ...$$

and

$$\frac{\partial u}{\partial y} = \frac{\partial f}{\partial \xi}\frac{\partial \xi}{\partial y} + \frac{\partial f}{\partial \eta}\frac{\partial \mu}{\partial y} + ...$$

Thus, if $\Psi = \Psi(\varepsilon, T, \alpha)$ where ε, T and α are all functions of t then

$$\frac{\partial \Psi}{\partial t} = \frac{\partial \Psi}{\partial \varepsilon}\frac{\partial \varepsilon}{\partial t} + \frac{\partial \Psi}{\partial T}\frac{\partial T}{\partial t} + \frac{\partial \Psi}{\partial \alpha}\frac{\partial \alpha}{\partial t}$$

C.2   ELLIPTIC, PARABOLIC, AND HYPERBOLIC PARTIAL DIFFERENTIAL EQUATIONS

The classification of partial differential equations into elliptic, parabolic, and hyperbolic is important in mechanics because the transition from elliptic to hyperbolic is commonly associated with a bifurcation that signals a transition from homogeneous deformation in the elliptic regime to localization of deformation in the hyperbolic regime (Hill, 1962; Rice, 1976). This transition is referred to as *loss of ellipticity* in the governing equations.

A general partial differential equation of the form

$$A\frac{\partial^2 u}{\partial x^2} + 2B\frac{\partial^2 u}{\partial x \partial y} + C\frac{\partial^2 u}{\partial y^2} + D\frac{\partial u}{\partial x} + E\frac{\partial u}{\partial y} + F = 0$$

can be classified into three types:

If $\Delta = \begin{vmatrix} A & B \\ B & C \end{vmatrix} = AC - B^2 > 0$, the equation is said to be *elliptic*. Laplace's equation, $\nabla^2 u = 0$, is an example of such an equation.

If $\Delta = \begin{vmatrix} A & B \\ B & C \end{vmatrix} = AC - B^2 = 0$, the equation is said to be *parabolic*. Diffusion equations, $\frac{\partial u}{\partial t} = D\frac{\partial^2 u}{\partial x^2}$, are of this form.

If $\Delta = \begin{vmatrix} A & B \\ B & C \end{vmatrix} = AC - B^2 < 0$, the equation is said to be *hyperbolic*. The wave equation, $\frac{\partial^2 u}{\partial t} - c\frac{\partial^2 u}{\partial x^2} = 0$, is an example of such an equation.

The differences between elliptic and hyperbolic equations are summarized in Table C1 where ϕ is a function of interest and $\mathbf{n}$ is the normal to a surface.

An example of the transition from elliptic to hyperbolic behavior is the flow of a perfect fluid governed by the equation (Malvern, 1969)

$$A\frac{\partial^2 \phi}{\partial x^2} + B\frac{\partial^2 \phi}{\partial x \partial y} + C\frac{\partial^2 \phi}{\partial y^2} = 0$$

TABLE C1   The Differences between Elliptic and Hyperbolic Functions

Elliptic Equations	Hyperbolic Equations
To avoid singularities either ϕ or $\frac{\partial \phi}{\partial n}$ must be prescribed on a closed boundary.	ϕ or $\frac{\partial \phi}{\partial n}$ must be specified on an open boundary.
A change in boundary conditions affects the whole region.	A change in boundary conditions only affects a limited domain.
The solution must be analytic and the flow is smooth and continuous.	Solutions need not be analytic and shocks or discontinuities in the flow are permitted across curves known as *characteristic curves*.
Perturbations are felt simultaneously everywhere in the flow.	Perturbations are propagated along characteristics of the flow (see Figure C1).

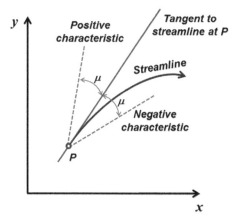

FIGURE C1   Positive and negative characteristics oriented at an angle μ either side of the tangent to the streamline of a flow at P.

where ϕ is the stream function. One can show that the character of this differential equation is governed by the sign of the discriminant $(B^2 - 4AC)$ which can be expressed as

$$B^2 - 4AC = 4c^2 \left[\left(\frac{v}{c} \right)^2 - 1 \right]$$

where v is the velocity of the fluid and c is the speed of sound. Thus the equation is

elliptic for $\left(\frac{v}{c} \right) < 1$ (subsonic).

parabolic for $\left(\frac{v}{c} \right) = 1$ (transonic).

hyperbolic for $\left(\frac{v}{c} \right) > 1$ (supersonic).

C3   THE TAYLOR EXPANSION

A Taylor expansion of a function, $f(x)$, is a representation of that function, at a point $x = a$, as an infinite series of terms that involves the derivatives of $f(x)$ with respect to x evaluated at the point $x = a$. It is common to truncate the series after a small number of terms in order to approximate the function. Sometimes this means that a nonlinear function can be approximated as a linear function.

If $f^{(n)}$ is written for the nth derivative of $f(x)$ with respect to x: $f^{(n)} = \frac{d^n f(x)}{dx^n}$, and $n!$ is the factorial of n then the Taylor series for a function, $f(x)$, at the point a is given by

$$f(a) + \frac{f^{(1)}(x)}{1!}(x - a) + \frac{f^{(2)}(x)}{2!}(x - a)^2 + \frac{f^{(3)}(x)}{3!}(x - a)^3 + \ldots\ldots$$

$$= \sum_{n=0}^{\infty} \frac{f^{(n)}(x)}{n!}(x - a)^n$$

If $a = 0$ then the series is called a MacLaurin Series. Thus the MacLaurin Series for $\ln(1-x)$ is

$$0 - x - \frac{x^2}{2} - \frac{x^3}{3} - \frac{x^4}{4} - \ldots\ldots$$

The Taylor Series for $\ln(x)$ at $a = 1$ is

$$(x-1) - \frac{1}{2}(x-1)^2 - \frac{1}{3}(x-1)^3 - \frac{1}{4}(x-1)^4 - \ldots\ldots$$

Similarly, the expansion of $\sin(x)$, where x is in radians, is

$$0 + x - \frac{x^3}{3} + \frac{x^5}{5}\ldots\ldots + (-1)^n\frac{x^{2n+1}}{n!} \pm \ldots\ldots$$

Thus if x is small, $\sin(x) \approx x$. Approximations such as this are used in linear stability analyses. The accuracy of such an approximation is discussed in Bronshtein et al. (2007, Table 7.3). Thus the approximation for $\sin(x)$ given above is good to within 1% for x measured in degrees between $-14°$ and $+14°$. Any analysis based on such truncations of a Taylor series is clearly only applicable whilst such approximations hold. Once the range of applicability is exceeded one expects nonlinear effects to appear. In nonlinear systems this commonly appears as a slowing of the growth rate for instabilities and is known as *nonlinear saturation*. The series expansions for a large number of functions are given by Bronshtein et al. (2007, Table 21.5).

C4   SOME USEFUL MATRIX RELATIONS

1. $\frac{\partial(\det A)}{\partial A} = A^{-T}\det A$

Proof: We start with the identity for the expansion of the determinant of a matrix:

$$\det A = \frac{1}{6}\varepsilon_{ijk}\varepsilon_{mnp}A_{mi}A_{nj}A_{pk} = \varepsilon_{ijk}A_{1i}A_{2j}A_{3k}$$

where ε_{ijk} is the permutation symbol (Appendix B).
Then

$$\frac{\partial(\det A)}{\partial A_{rs}} = \frac{1}{6}\varepsilon_{ijk}\varepsilon_{mnp}\left[\delta_{rm}\delta_{is}A_{nj}A_{pk} + \delta_{rn}\delta_{js}A_{mi}A_{pk} + \delta_{rp}\delta_{ks}A_{mi}A_{nj}\right]$$

$$= \frac{1}{2}\varepsilon_{sjk}\varepsilon_{rnp}A_{nj}A_{pk}$$

(A3.1)

If we replace ε_{rnp} in (A3.1) by $\varepsilon_{qnp}\delta_{qr}$ then assuming that A^{-1} exists, so that,

$$\delta_{qr} = A_{qi}A_{ir}^{-1}$$

then

$$\frac{\partial(\det\mathbf{A})}{\partial A_{rs}} = \left(\frac{1}{2}\varepsilon_{sjk}A_{ir}^{-1}\right)\left(\varepsilon_{qnp}A_{qi}A_{nj}A_{pk}\right)$$

or

$$\frac{\partial(\det\mathbf{A})}{\partial\mathbf{A}} = \mathbf{A}^{-T}\det\mathbf{A}$$

2. $\mathbf{A}: (\mathbf{BC}) = (\mathbf{B}^{T}\mathbf{A}): \mathbf{C}$

Proof: $\mathbf{A}: (\mathbf{BC}) = A_{ji}B_{ik}B_{kj}$
Also, $(\mathbf{B}^{T}\,\mathbf{A}): \mathbf{C} = B_{ki}A_{kj}C_{ij} \equiv A_{ji}B_{ik}C_{kj} = \mathbf{A}: (\mathbf{BC})$

3. $(\mathbf{a}\otimes\mathbf{N})l = \mathbf{a}(l\bullet\mathbf{N})$

Proof: $(\mathbf{a}\otimes\mathbf{N})l = (a_{i}N_{j})l_{j} = a_{i}(N_{j}l_{j}) = \mathbf{a}(l\bullet\mathbf{N})$

References

Hill, R., 1962. Acceleration waves in solids. Journal of the Mechanics and Physics Solids 10, 1–16.

Malvern, L.E., 1969. Introduction to the Mechanics of a Continuous Medium. Prentice-Hall, Inc., Englewood Cliffs, New Jersey, 713 pp.

Rice, J.R., 1976. The localization of plastic deformation. In: Theoretical and Applied mechanics. Koiter, W.T. (Ed), Proceedings of the 14th International Congress on Theoretical and Applied mechanics, Delft. vol. 1, 207–220

References

Aaronson, H.I., 2002. Mechanisms of the massive transformation. Metallurgical and Materials Transactions: A-Physical Metallurgy and Materials Science 33, 2285–2297.

Abdulagatov, I.M., Emirov, S.N., Abdulagatova, Z.Z., Askerov, S.Y., 2006. Effect of pressure and temperature on the thermal conductivity of rocks. Journal of Chemical and Engineering Data 51, 22–33.

Abraham, F.F., 2003. How fast can cracks move? A research adventure in materials failure using millions of atoms and big computers. Advances in Physics 52, 727–790.

Abramson, E.H., 2007. Viscosity of water measured to pressures of 6 GPa and temperatures of 300 degrees C. Physical Review E 76.

Adeagbo, W.A., Doltsinis, N.L., Klevakina, K., Renner, J., 2008. Transport processes at α-quartz–water interfaces: insights from first-principles molecular dynamics simulations. ChemPhysChem 9, 994–1002.

Adjaoud, O., Marquardt, K., Jahn, S., 2012. Atomic structures and energies of grain boundaries in Mg_2SiO_4 forsterite from atomistic modeling. Physics and Chemistry of Minerals 39, 749–760.

Ague, J.J., 2011. Extreme channelization of fluid and the problem of element mobility during Barrovian metamorphism. American Mineralogist 96, 333–352.

Aifantis, E.C., 1984. On the microstructural origin of certain inelastic models. Journal of Engineering Materials and Technology-Transactions of the ASME 106, 326–330.

Aifantis, E.C., 1986. On the dynamical origin of dislocation patterns. Materials Science and Engineering 81, 563–574.

Aifantis, E.C., 1987. The physics of plastic-deformation. International Journal of Plasticity 3, 211–247.

Aifantis, E.C., 1999. Gradient deformation models at nano, micro, and macro scales. Journal of Engineering Materials and Technology-Transactions of the ASME 121, 189–202.

Akai, K., Mori, H., 1967. Study on the failure mechanism of a sandstone under combined compressive stress. In: Proc. 2nd Congr. ISRM, Belgrade, Yugoslavia, pp. 3–30.

Alt-Epping, P., Zhao, C., 2010. Reactive mass transport modelling of a three-dimensional vertical fault zone with a finger-like convective flow regime. Journal of Geochemical Exploration 106, 8–23.

Amodeo, J., Carrez, P., Devincre, B., Cordier, P., 2011. Multiscale modelling of MgO plasticity. Acta Materialia 59, 2291–2301.

Anderson, G.M., Garven, G., 1987. Sulfate-sulfide-carbonate associations in Mississippi Valley-type lead-zinc deposits. Economic Geology 82, 482–488.

Anderson, J.G., Doraiswamy, L.K., Larson, M.A., 1998a. Microphase-assisted 'autocatalysis' in a solid-liquid reaction with a precipitating product – I. Theory. Chemical Engineering Science 53, 2451–2458.

Anderson, J.G., Larson, M.A., Doraiswamy, L.K., 1998b. Microphase-assisted "autocatalysis" in a solid-liquid reaction with a precipitating product – II. Experimental. Chemical Engineering Science 53, 2459–2468.

Andronov, A.A., Vitt, A.A., Khaikin, S.E., 1966. Theory of Oscillations. Pergamon, New York.

Appold, M.S., Garven, G., 2000. Reactive flow models of ore formation in the Southeast Missouri district. Economic Geology 95, 1605–1626.

Appold, M.S., Garven, G., Boles, J.R., Eichhubl, P., 2007. Numerical modeling of the origin of calcite mineralization in the Refugio-Carneros fault, Santa Barbara Basin, California. Geofluids 7, 79–95.

Aref, H., 1984. Stirring by chaotic advection. Journal of Fluid Mechanics 143, 1–21.

Arias, M., Gumiel, P., Sanderson, D.J., Martin-Izard, A., 2011. A multifractal simulation model for the distribution of VMS deposits in the Spanish segment of the Iberian Pyrite Belt. Computers & Geosciences 37, 1917–1927.

Arneodo, A., Grasseau, G., Kostelich, E.J., 1987. Fractal dimensions and F(α) spectrum of the Henon attractor. Physics Letters A 124, 426–432.

Arneodo, A., Bacry, E., Muzy, J.F., 1995. The thermodynamics of fractals revisited with wavelets. Physica A 213, 232–275.

Arneodo, A., Manneville, S., Muzy, J.F., Roux, S.G., 1999. Experimental evidence for anomalous scale dependent cascading process in turbulent velocity statistics. Applied and Computational Harmonic Analysis 6, 374–381.

Arneodo, A., Decoster, N., Kestener, P., Roux, S.G., 2003. A wavelet-based method for multifractal image analysis: from theoretical concepts to experimental applications. In: Hawkes, P.W. (Ed.), Advances in Imaging and Electron Physics, vol. 126, pp. 1–92.

Arneodo, A., Vaillant, C., Audit, B., Argoul, F., d'Aubenton-Carafa, Y., Thermes, C., 2011. Multi-scale coding of genomic information: from DNA sequence to genome structure and function. Physics Reports-Review Section of Physics Letters 498, 45–188.

Arsenlis, A., Parks, D.M., 1999. Crystallographic aspects of geometrically-necessary and statistically-stored dislocation density. Acta Materialia 47, 1597–1611.

Asaro, R.J., Needleman, A., 1985. Overview 42. Texture development and strain-hardening in rate dependent polycrystals. Acta Metallurgica 33, 923–953.

Asaro, R.J., Rice, J.R., 1977. Strain localization in ductile single crystals. Journal of the Mechanics and Physics of Solids 25, 309–338.

Asaro, R.J., 1979. Geometrical effects in the homogeneous deformation of ductile single crystals. Acta Metallurgica 27, 445–453.

Asaro, R.J., 1983. Crystal plasticity. Journal of Applied Mechanics-Transactions of the ASME 50, 921–934.

Ashby, M.F., Gandhi, C., Taplin, D.M.R., 1979. Fracture-mechanism maps and their construction for F.C.C. metals and alloys. Acta Metallurgica 27, 699–729.

Ashby, M.F., 1970. Deformation of plastically non-homogeneous materials. Philosophical Magazine 21, 399–424.

Astarita, G., 1979. Objective and generally applicable criteria for flow classification. Journal of Non-Newtonian Fluid Mechanics 6, 69–76.

Atkinson, B.K., 1979. A fracture mechanics study of subcritical tensile cracking of quartz in wet environments. Pure and Applied Geophysics 117, 1011–1024.

Atkinson, B.K., 1982. Subcritical crack propagation in rocks: theory, experimental results and applications. Journal of Structural Geology 4, 41–56.

Atkinson, B.K., 1987. Fracture Mechanics of Rock. Academic Press, London, p. 534.

Atkinson, H.V., 1988. Theories of normal grain-growth in pure single-phase systems. Acta Metallurgica 36, 469–491.

Audit, B., Vaillant, C., Arneodo, A., d'Aubenton-Carafa, Y., Thermes, C., 2002. Long-range correlations between DNA bending sites: relation to the structure and dynamics of nucleosomes. Journal of Molecular Biology 316, 903–918.

Austin, N.J., Evans, B., 2007. Paleowattmeters: a scaling relation for dynamically recrystallized grain size. Geology 35, 343–346.

Austin, N., Evans, B., 2009. The kinetics of microstructural evolution during deformation of calcite. Journal of Geophysical Research-Solid Earth 114, B09402.

Austrheim, H., 1987. Eclogitization of lower crustal granulites by fluid migration through shear zones. Earth and Planetary Science Letters 81, 221–232.

Avrami, M., 1939. Kinetics of phase change I – general theory. Journal of Chemical Physics 7, 1103–1112.

Avrami, M., 1940. Kinetics of phase change. II Transformation-time relations for random distribution of nuclei. The Journal of Chemical Physics 8, 212–224.

Avrami, M., 1941. Granulation, phase change, and microstructure – kinetics of phase change. III. Journal of Chemical Physics 9, 177–184.

Bachmann, F., Hielscher, R., Schaeben, H., 2011. Grain detection from 2d and 3d EBSD data-Specification of the MTEX algorithm. Ultramicroscopy 111, 1720–1733.

Backofen, W.A., 1972. Deformation Processing. Addison-Wesley Pub. Co.

Bai, T.X., Pollard, D.D., 2000. Closely spaced fractures in layered rocks: initiation mechanism and propagation kinematics. Journal of Structural Geology 22, 1409–1425.

Bak, P., 1996. How Nature Works: The Science of Self-Organized Criticality. Copernicus, Springer-Verlag, New York.

Ball, J.M., James, R.D., 1987. Fine phase mixtures as minimizers of energy. Archive for Rational Mechanics and Analysis 100, 13–52.

Ball, J.M., Holmes, P.J., James, R.D., Pego, R.L., Swart, P.J., 1991. On the dynamics of fine structure. Journal of Nonlinear Science 1, 17–70.

Ball, J.M., 1977. Convexity conditions and existence theorems in nonlinear elasticity. Archive for Rational Mechanics and Analysis 63, 337–403.

Ball, J.M., 2004. Mathematical models of martensitic microstructure. Materials Science and Engineering: A 378, 61–69.

Baram, R.M., Herrmann, H.J., 2004. Self-similar space-filling packings in three dimensions. Fractals-Complex Geometry Patterns and Scaling in Nature and Society 12, 293–301.

Baram, R.M., Herrmann, H.J., 2005. Random bearings and their stability. Physical Review Letters 95, 224303(4).

Baram, R.M., Herrmann, H.J., Rivier, N., 2004. Space-filling bearings in three dimensions. Physical Review Letters 92.

Baram, R.M., Lind, P.G., Andrade Jr., J.S., Herrmann, H.J., 2010. Superdiffusion of massive particles induced by multi-scale velocity fields. EPL 91.

Barenblatt, G.I., 1959. Concerning equilibrium cracks forming during brittle fracture. Journal of Applied Mathematics and Mechanics (PMM) 23, 1273–1282.

Barenblatt, G.I., 1962. The mathematical theory of equilibrium cracks in brittle fracture. In: Dryden, H.L., Kármán, T.v., Kuerti, G., Dungen, F. H. v. d., Howarth, L. (Eds.), Advances in Applied Mechanics Volume 7. Elsevier, pp. 55–129.

Barnes, H.A., 1999. The yield stress—a review or 'παντα ρει'—everything flows? Journal of Non-Newtonian Fluid Mechanics 81, 133–178.

Barnsley, M.F., 1988. Fractals Everywhere. Academic Press/Harcourt-Brace-Jovanovitch.

Basinski, Z.S., Basinski, S.J., 2004. Quantitative determination of secondary slip in copper single crystals deformed in tension. Philosophical Magazine 84, 213–251.

Bastin, E.S., Graton, L.C., Lindgren, W., Newhouse, W.H., Schwartz, G.M., Short, M.N., 1931. Criteria of age relations of minerals, with especial reference to polished sections of ores. Economic Geology 26, 561–610.

Bastin, E.S., 1950. Interpretation of Ore Textures. Geological Society of America Memoir 45, 1–101.

Bastrakov, E.N., Skirrow, R.G., Didson, G.J., 2007. Fluid evolution and origins of iron oxide Cu-Au prospects in the Olympic Dam district, Gawler craton, south Australia. Economic Geology 102, 1415–1440.

Bay, B., Hansen, N., Kuhlmannwilsdorf, D., 1992. Microstructural evolution in rolled aluminum. Materials Science and Engineering: A-Structural Materials Properties Microstructure and Processing 158, 139–146.

Beach, A., Fyfe, W.S., 1972. Fluid transport and shear zones at Scourie, Sutherland — evidence of overthrusting. Contributions to Mineralogy and Petrology 36, 175–180.

Beausir, B., Fressengeas, C., 2013. Disclination densities from EBSD orientation mapping. International Journal of Solids and Structures 50, 137–146.

Becker, G.F., Day, A.L., 1905. The linear force of growing crystals. Proceedings of the Washington Academy of Sciences VII, 283–288.

Becker, G.F., Day, A.L., 1916. Note on the linear force of growing crystals. The Journal of Geology 24, 313–333.

Becker, G.F., 1893. Finite homogeneous strain, flow and rupture in rocks. Bulletin of the Geological Society of America 4, 13–90.

Bejan, A., Lorente, S., 2011. The constructal law and the evolution of design in nature. Physics of Life Reviews 8, 209–240.

Bell, T.H., Johnson, S.E., 1989. Porphyroblast inclusion trails — the key to orogenesis. Journal of Metamorphic Geology 7, 279–310.

Bell, T.H., Johnson, S.E., 1992. Shear sense — a new approach that resolves conflicts between criteria in metamorphic rocks. Journal of Metamorphic Geology 10, 99–124.

Bell, T.H., Rubenach, M.J., Fleming, P.D., 1986. Porphyroblast nucleation, growth and dissolution in regional metamorphic rocks as a function of deformation partitioning during foliation development. Journal of Metamorphic Geology 4, 37–67.

Bell, T.H., Johnson, S.E., Davis, B., Forde, A., Hayward, N., Wilkins, C., 1992. Porphyroblast inclusion-trail orientation data — eppure-non-son-girate. Journal of Metamorphic Geology 10, 295–307.

Bell, T.H., Hickey, K.A., Upton, G.J.G., 1998. Distinguishing and correlating multiple phases of metamorphism across a multiply deformed region using the axes of spiral, staircase and sigmoidal inclusion trails in garnet. Journal of Metamorphic Geology 16, 767–794.

Bell, T.H., 1985. Deformation partitioning and porphyroblast rotation in metamorphic rocks — a radical reinterpretation. Journal of Metamorphic Geology 3, 109–118.

Ben-Zion, Y., Sammis, C., 2003. Characterization of fault zones. Pure and Applied Geophysics 160, 677–715.

Ben-Zion, Y., Sammis, C.G., 2013. Shear heating during distributed fracturing and pulverization of rocks. Geology 41, 139–142.

Ben-Zion, Y., Dahmen, K., Lyakhovsky, V., Ertas, D., Agnon, A., 1999. Self-driven mode switching of earthquake activity on a fault system. Earth and Planetary Science Letters 172, 11–21.

Ben-Zion, Y., 2008. Collective behavior of earthquakes and faults: continuum-discrete transitions, progressive evolutionary changes, and different dynamic regimes. Reviews of Geophysics 46 (RG4006 / 2008), 1–70.

Berkowitz, B., Balberg, I., 1992. Percolation approach to the problem of hydraulic conductivity in porous-media. Transport in Porous Media 9, 275–286.

Bésuelle, P., Rudnicki, J.W., 2004. Localization: shear bands and compaction bands. In: Guéguen, Y., Boutéca, M. (Eds.), Mechanics of Fluid-Saturated Rock. Academic Press, London, pp. 219–321.

Bethke, C.M., 1985. A numerical-model of compaction-driven groundwater-flow and heat-transfer and its application to the paleohydrology of intracratonic sedimentary basins. Journal of Geophysical Research-Solid Earth and Planets 90, 6817–6828.

Bethke, C.M., 1996. Geochemical Reaction Modeling. Oxford University Press, New York.

Beyerlein, I.J., Mayeur, J.R., Zheng, S., Mara, N.A., Wang, J., Misra, A., 2014. Emergence of stable interfaces under extreme plastic deformation. Proceedings of the National Academy of Sciences of the United States of America 111, 4386–4390.

Bhattacharya, K., 2003. Microstructure of Martensite: Why It Forms and How It Gives Rise to the Shape-Memory Effect. Oxford University Press.

Bigoni, D., Loret, B., 1999. Effects of elastic anisotropy on strain localization and flutter instability in plastic solids. Journal of the Mechanics and Physics of Solids 47, 1409–1436.

Bigoni, D., Loret, B., Radi, E., 2000. Localization of deformation in plane elastic-plastic solids with anisotropic elasticity. Journal of the Mechanics and Physics of Solids 48, 1441–1466.

Billia, M.A., Timms, N.E., Toy, V.G., Hart, R.D., Prior, D.J., 2013. Grain boundary dissolution porosity in quartz-ofeldspathic ultramylonites: Implications for permeability enhancement and weakening of mid-crustal shear zones. Journal of Structural Geology 53, 2–14.

Biot, M.A., 1955. Variational principles in irreversible thermodynamics with application to viscoelasticity. The Physical Review 97, 1463–1469.

Biot, M.A., 1958. Linear thermodynamics and the mechanics of solids. In: Third U.S. National Congress of Applied Mechanics. ASME, New York, Brown University, Providence, RI, pp. 1–18.

Biot, M.A., 1965a. Theory of similar folding of the first and second kind. Geological Society of America Bulletin 76, 251–258.

Biot, M.A., 1965b. Mechanics of Incremental Deformations. John Wiley, New York.

Biot, M.A., 1984. New variational-Lagrangian irreversible thermodynamics with application to viscous flow, reaction-diffusion, and solid mechanics. Advances in Applied Mechanics 24, 1–91.

Birch, F., Schairer, J.F., Spicer, H.C., 1942. Handbook of Physical Constants. Special Paper Number 36, Geological Society of America.

Bird, R.B., Stewart, W.E., Lightfoot, E.N., 1960. Transport Phenomena. Wiley, New York.

Bishop, J.F.W., Hill, R., 1951a. CXXVIII. A theoretical derivation of the plastic properties of a polycrystalline face-centred metal. Philosophical Magazine Series 7 42, 1298–1307.

Bishop, J.F.W., Hill, R., 1951b. XLVI. A theory of the plastic distortion of a polycrystalline aggregate under combined stresses. Philosophical Magazine Series 7 42, 414–427.

Bishop, A.W., 1966. The strength of soils as engineering materials – 6th Rankine Lecture. Geotechnique 16, 91–130.

Blenkinsop, T.G., Kadzviti, S., 2006. Fluid flow in shear zones: insights from the geometry and evolution of ore bodies at Renco gold mine, Zimbabwe. Geofluids 6, 334–345.

Blenkinsop, T.G., 1991. Cataclasis and processes of particle-size reduction. Pure and Applied Geophysics 136, 59–86.

Bohr, T., Tel, T., 1988. The thermodynamics of fractals. In: Hao, B.-L. (Ed.), Directions in Chaos. World Scientific, Singapore, pp. 194–237.

Bons, P.D., Elburg, M.A., Gomez-Rivas, E., 2012. A review of the formation of tectonic veins and their microstructures. Journal of Structural Geology 43, 33–62.

Borja, R., Aydin, A., 2004. Computational modeling of deformation bands in granular media. I. Geological and mathematical framework. Computer Methods in Applied Mechanics and Engineering 193, 2667–2698.

Bourdin, B., Francfort, G.A., Marigo, J.-J., 2008. The variational approach to fracture. Journal of Elasticity 91, 5–148.

Boyce, W.E., DiPrima, R.C., 2005. Elementary Differential Equations and Boundary Value Problems, eighth ed. John Wiley & Sons, Inc.

Brace, W.F., Martin, R.J., 1968. A test of the law of effective stress for crystalline rocks of low porosity. International Journal of Rock Mechanics and Mining Sciences 5, 415–426.

Brace, W.F., 1960. Chapter 2: Orientation of anisotropic minerals in a stress field: discussion. Geological Society of America Memoirs 79, 9–20.

Brady, B.H.G., Brown, E.T., 1993. Rock Mechanics for Underground Mining, first ed. Chapman & Hall, London.

Brantut, N., Heap, M.J., Meredith, P.G., Baud, P., 2013. Time-dependent cracking and brittle creep in crustal rocks: a review. Journal of Structural Geology 52, 17–43.

Braun, J., Munroe, S.M., Cox, S.F., 2003. Transient fluid flow in and around a fault. Geofluids 3, 81–87.

Bridgman, P.W., 1943. The Nature of Thermodynamics. Harvard University Press, Cambridge, MA.

Bridgman, P.W., 1950. The thermodynamics of plastic deformation and generalized entropy. Reviews of Modern Physics 22, 56–63.

Bronshtein, I.N., Semendyayev, K.A., Musiol, G., Muehlig, H., 2007. Handbook of Mathematics, fifth ed. Springer-Verlag.

Brown, M., Solar, G.S., 1998. Shear-zone systems and melts: feedback relations and self-organization in orogenic belts. Journal of Structural Geology 20, 211−227.

Brown, M., 2001. Orogeny, migmatites and leucogranites: a review. Proceedings of the Indian Academy of Sciences 110, 313−336.

Brown, M., 2004. The mechanism of melt extraction from lower continental crust of orogens. Transactions of the Royal Society of Edinburgh: Earth Sciences 95, 35−48.

Brown, M., 2005. Synergistic effects of melting and deformation: an example from the Variscan belt, western France. In: Gapais, D., Brun, J.P., Cobbold, P.R. (Eds.), Deformation Mechanisms, Rheology and Tectonics: From Minerals to the Lithosphere, pp. 205−226.

Brown, M., 2008. Granites, migmatites and residual granulits: relationships and processes. In: Sawyer, E.W., Brown, M. (Eds.), Working with Migmatites. Mineralogical Association of Canada Short Course Series 38. Mineralogical Association of Canada, pp. 97−144. Chapter 6.

Brown, M., 2010. The spatial and temporal patterning of the deep crust and implications for the process of melt extraction. Philosophical Transactions of the Royal Society A-Mathematical Physical and Engineering Sciences 368, 11−51.

Budd, C.J., Hunt, G.W., Kuske, R., 2001. Asymptotics of cellular buckling close to the Maxwell load. Proceedings of the Royal Society of London Series A: Mathematical Physical and Engineering Sciences 457, 2935−2964.

Budiansky, B., Hutchinson, J.W., Slutsky, S., 1982. Void growth and collapse in viscous solids. In: Hopkins, H.G., Sewell, M.J. (Eds.), Mechanics of Solids, The Rodney Hill 60th Anniversary Volume. Pergamon Press, Oxford, pp. 13−45.

Bunge, H.J., Wenk, H.R., 1977. 3-Dimensional texture analysis of 3 quartzites (trigonal crystal and triclinic specimen symmetry). Tectonophysics 40, 257−285.

Bunge, H.J., 1965. Zur darstellung allgemeiner texturen. Zeitschrift fur Metallkunde 56, 872−874.

Burg, J.P., Gerya, T.V., 2005. The role of viscous heating in Barrovian metamorphism of collisional orogens: thermomechanical models and application to the Lepontine Dome in the Central Alps. Journal of Metamorphic Geology 23, 75−95.

Burke, J., Knobloch, E., 2007. Homoclinic snaking: structure and stability. Chaos 17, 037102-1−037102-15.

Burke, J.E., Turnbull, D., 1952. Recrystallization and grain growth. Progress in Metal Physics 3, 220−292.

Burke, J.E., 1949. Some factors affecting the rate of grain growth in metals. Transactions of the American Institute of Mining and Metallurgical Engineers 180, 73−91.

Burnham, C.W., 1979. Magmas and hydrothermal fluids. In: Barnes, H.L. (Ed.), Geochemistry of Hydrothermal Ore Deposits. Wiley, New York, pp. 71−136.

Burnham, C.W., 1985. Energy-release in subvolcanic environments − implications for breccia formation. Economic Geology 80, 1515−1522.

Burton, B., 2001a. Theory of bubble growth in crystals and implications concerning nuclear fuel. Materials Science and Technology 17, 389−398.

Burton, B., 2001b. Theory of bubble growth in crystals under stress. Materials Science and Technology 17, 399−402.

Byerlee, J.D., 1967. Frictional characteristics of granite under high confining pressure. Journal of Geophysical Research 72, 3639−3648.

Byerlee, J., 1978. Friction of rocks. Pure and Applied Geophysics 116, 615−626.

Cabral, B., Leedom, L.C., 1993. Imaging vector fields using line integral convolution. In: Proceedings of the 20th Annual Conference on Computer Graphics and Interactive Techniques. ACM, Anaheim, CA, pp. 263−270.

Cahn, J.W., Larche, F., 1984. A simple-model for coherent equilibrium. Acta Metallurgica 32, 1915−1923.

Cahn, J.W., Mishin, Y., 2009. Recrystallization initiated by low-temperature grain boundary motion coupled to stress. International Journal of Materials Research 100, 510−515.

Cahn, J.W., Taylor, J.E., 2004. A unified approach to motion of grain boundaries, relative tangential translation along grain boundaries, and grain rotation. Acta Materialia 52, 4887−4898.

Cahn, J.W., Mishin, Y., Suzuki, A., 2006. Duality of dislocation content of grain boundaries. Philosophical Magazine 86, 3965−3980.

Cahn, J.W., 1991. Stability, microstructural evolution, grain-growth, and coarsening in a 2-dimensional 2-phase microstructure. Acta Metallurgica et Materialia 39, 2189−2199.

Callen, H.B., 1960. Thermodynamics: An Introduction to the Physical Theories of Equilibrium Thermostatics and Irreversible Thermodynamics. John Wiley and Sons, New York.

Canova, G., Kocks, U.F., 1984. The development of deformation textures and resulting properties of FCC metals. In: Seventh International Conference on Textures of Materials: ICOTOM 7, Holland 1984, September 17−21, Noordwijkerhout, The Netherlands. Netherlands Society for Materials Science, Zwijndrecht, Netherlands, pp. 573−579.

Canova, G.R., 1994. Self-consistent methods — application to the prediction of the deformation texture of polyphase materials. Materials Science and Engineering: A-Structural Materials Properties Microstructure and Processing 175, 37—42.

Carlson, W.D., Denison, C., 1992. Mechanisms of porphyroblast crystallization — results from high-resolution computed X-ray tomography. Science 257, 1236—1239.

Carlson, W.D., Denison, C., Ketcham, R.A., 1995. Controls on the nucleation and growth of porphyroblasts: kinetics from natural textures and numerical models. Geological Journal 30, 207—225.

Carlson, W.D., 1989. The significance of intergranular diffusion to the mechanisms and kinetics of porphyroblast crystallization. Contributions to Mineralogy and Petrology 103, 1—24.

Carlson, W.D., 1991. Competitive diffusion-controlled growth of porphyroblasts. Mineralogical Magazine 55, 317—330.

Carlson, W.D., 2011. Porphyroblast crystallization: linking processes, kinetics, and microstructures. International Geology Review 53, 406—445.

Carmichael, D.M., 1969. On the mechanism of prograde metamorphic reactions in quartz-bearing pelitic rocks. Contributions to Mineralogy and Petrology 20, 244—267.

Carmichael, D.M., 1987. Induced stress and secondary mass transfer: thermodynamic basis for the tendency toward constant-volume constraint in diffusion metasomatism. In: Helgeson, H.C. (Ed.), Chemical Transport in Metasomatic Processes. D. Reidel Publishing Co., Dordrecht, pp. 239—264.

Carreras, J., 2001. Zooming on Northern Cap de Creus shear zones. Journal of Structural Geology 23, 1457—1486.

Carrez, P., Ferre, D., Cordier, P., 2009. Peierls-Nabarro modelling of dislocations in MgO from ambient pressure to 100 GPa. Modelling and Simulation in Materials Science and Engineering 17, 035010 (11pp).

Carriere, P., 2007. On a three-dimensional implementation of the baker's transformation. Physics of Fluids 19, 118110 (4pp).

Carslaw, H.S., Jaeger, J.C., 1959. Conduction of Heat in Solids, second ed. Oxford University Press, London.

Cartwright, I., Power, W.L., Oliver, N.H.S., Valenta, R.K., McLatchie, G.S., 1994. Fluid migration and vein formation during deformation and greenschist facies metamorphism at ormiston gorge, central Australia. Journal of Metamorphic Geology 12, 373—386.

Cartwright, I., Buick, I.S., Maas, R., 1997. Fluid flow in marbles at Jervois, central Australia: oxygen isotope disequilibrium and zoning produced by decoupling of mineralogical and isotopic resetting. Contributions to Mineralogy and Petrology 128, 335—351.

Cashman, K.V., Ferry, J.M., 1988. Crystal size distribution (csd) in rocks and the kinetics and dynamics of crystallization. 3. Metamorphic crystallization. Contributions to Mineralogy and Petrology 99, 401—415.

Castelnau, O., Cordier, P., Lebensohn, R.A., Merkel, S., Raterron, P., 2010. Microstructures and rheology of the Earth's upper mantle inferred from a multiscale approach. Comptes Rendus Physique 11, 304—315.

Champneys, A.R., Toland, J.F., 1993. Bifurcation of a plethora of multi-modal homoclinic orbits for autonomous Hamiltonian systems. Nonlinearity 6, 665—721.

Chau, K.T., Muhlhaus, H.B., Ord, A., 1998. Bifurcation in growth patterns for arrays of parallel Griffith, edge and sliding cracks. In: Tong, P., Zhang, T.Y., Kim, J.K. (Eds.), Fracture and Strength of Solids III. Key Engineering Materials Vols. 145—149, 71—76.

Chella, R., Ottino, J.M., 1985. Stretching in some classes of fluid motions and asymptotic mixing efficiencies as a measure of flow classification. Archive for Rational Mechanics and Analysis 90, 15—42.

Chen, W.F., Saleeb, A.F., 1982. Constitutive Equations for Engineering Materials. Wiley Inter-science, New York.

Chester, F.M., Chester, J.S., 1998. Ultracataclasite structure and friction processes of the Punchbowl Fault, San Andreas System, California. Tectonophysics 295, 199—221.

Choksi, R., Piero, G.D., Fonseca, I., Owen, D., 1999. Structured deformations as energy minimizers in models of fracture and hysteresis. Mathematics and Mechanics of Solids 4, 321—356.

Chopin, C., 1984. Coesite and pure pyrope in high-grade blueschists of the Western Alps — a 1st record and some consequences. Contributions to Mineralogy and Petrology 86, 107—118.

Christian, J.W., 1975. Theory of Transformations in Metals and Alloys. Pergamon Press, Oxford.

Christie, J.M., Ord, A., 1980. Flow-stress from microstructures of mylonites — example and current assessment. Journal of Geophysical Research 85, 6253—6262.

Clarke, B.L., 1976. Stability of bromate-cerium-malonic acid network. 1. Theoretical formulation. Journal of Chemical Physics 64, 4165—4178.

Clarke, B.L., 1980. Stability of complex reaction networks. In: Prigogine, I., Rice, S. (Eds.), Advances in Chemical Physics. Wiley, New York, 43, 1–216.

Clark Jr., S.P. (Ed.), 1966. Handbook of Physical Constants. Revised Edition. Geological Society of America Memoir, 97, 587 pp.

Clayton, J.D., McDowell, D.L., Bammann, D.J., 2006. Modeling dislocations and disclinations with finite micropolar elastoplasticity. International Journal of Plasticity 22, 210–256.

Clayton, J.D., 2011. Nonlinear Mechanics of Crystals. Springer, Dordrecht.

Cloetingh, S., Burov, E., Francois, T., 2013. Thermo-mechanical controls on intra-plate deformation and the role of plume-folding interactions in continental topography. Gondwana Research 24, 815–837.

Coleman, B.D., Noll, W., 1963. The thermodynamics of elastic materials with heat conduction and viscosity. Archive for Rational Mechanics and Analysis 13, 167–178.

Collins, I.F., Hilder, T., 2002. A theoretical framework for constructing elastic/plastic constitutive models of triaxial tests. International Journal for Numerical and Analytical Methods in Geomechanics 26, 1313–1347.

Collins, I.F., Houlsby, G.T., 1997. Application of thermomechanical principles to the modelling of geotechnical materials. Proceedings: Mathematical, Physical and Engineering Sciences 453, 1975–2001.

Collins, I.F., Kelly, P.A., 2002. A thermomechanical analysis of a family of soil models. Geotechnique 52, 507–518.

Connolly, J.A.D., Kerrick, D.M., 1987. An algorithm and computer-program for calculating composition phase-diagrams. CALPHAD: Computer Coupling of Phase Diagrams and Thermochemistry 11, 1–55.

Connolly, J.A.D., Podladchikov, Y.Y., 1998. Compaction-driven fluid flow in viscoelastic rock. Geodinamica Acta 11, 55–84.

Connolly, J.A.D., Holness, M.B., Rubie, D.C., Rushmer, T., 1997. Reaction-induced microcracking: an experimental investigation of a mechanism for enhancing anatectic melt extraction. Geology 25, 591–594.

Connolly, J.A.D., 1990. Multivariable phase diagrams; an algorithm based on generalized thermodynamics. American Journal of Science 290, 666–718.

Connolly, J.A.D., 2005. Computation of phase equilibria by linear programming: a tool for geodynamic modeling and its application to subduction zone decarbonation. Earth and Planetary Science Letters 236, 524–541.

Connolly, J.A.D., 2010. The mechanics of metamorphic fluid expulsion. Elements 6, 165–172.

Cordier, P., Amodeo, J., Carrez, P., 2012. Modelling the rheology of MgO under Earth's mantle pressure, temperature and strain rates. Nature 481, 177–180.

Cordier, P., Demouchy, S., Beausir, B., Taupin, V., Barou, F., Fressengeas, C., 2014. Disclinations provide the missing mechanism for deforming olivine-rich rocks in the mantle. Nature 507, 51–56.

Cosgrove, J., 2007. The use of shear zones and related structrues as kinematic indicators: a review. Geological Society, London, Special Publications 272, 61–76.

Coussy, O., 1995. Mechanics of Porous Continua. Wiley, Chichester, UK.

Coussy, O., 2004. Poromechanics. Wiley, Chichester, UK.

Coussy, O., 2010. Mechanics and Physics of Porous Solids. Wiley, Chichester, UK.

Cowie, P.A., Vanneste, C., Sornette, D., 1993. Statistical physics model for the spatiotemporal evolution of faults. Journal of Geophysical Research-Solid Earth 98, 21809–21821.

Cowie, P.A., Sornette, D., Vanneste, C., 1995. Multifractal scaling properties of a growing fault population. Geophysical Journal International 122, 457–469.

Cox, S.F., Etheridge, M.A., 1989. Coupled grain-scale dilatancy and mass transfer during deformation at high fluid pressures: Examples from Mount Lyell, Tasmania. Journal of Structural Geology 11, 147–162.

Cox, S.F., Paterson, M.S., 1991. Experimental dissolution-precipitation creep in quartz aggregates at high-temperatures. Geophysical Research Letters 18, 1401–1404.

Cox, S.F., Braun, J., Knackstedt, M.A., 2001. Principles of structural control on permeability and fluid flow in hydrothermal systems. Reviews in Economic Geology 14, 1–24.

Cox, S.F., 1995. Faulting processes at high fluid pressures — an example of fault valve behavior from the Wattle Gully fault, Victoria, Australia. Journal of Geophysical Research-Solid Earth 100, 12841–12859.

Cox, S.F., 1999. Deformational controls on the dynamics of fluid flow in mesothermal gold systems. In: McCaffrey, K., Lonergan, L., Wilkinson, J. (Eds.), Fractures, Fluid Flow and Mineralization. Geological Society of London, Special Publications 155, 123–139.

Cross, M., Greenside, H., 2009. Pattern Formation and Dynamics in Nonequilibrium Systems. Cambridge University Press, Cambridge, UK.

Cross, M.C., Hohenberg, P.C., 1993. Pattern formation outside of equilibrium. Reviews of Modern Physics 65, 851–1112.

Crutchfield, J.P., Farmer, J.D., Packard, N.H., Shaw, R.S., 1986. Chaos. Scientific American 255, 46–57.

Curie, P., 1894. Sur la symétrie dans les phénomènes physiques, symétrie d'un champ électrique et d'un champ magnétique. Journal de Physique 3, 393–415.

Curie, P., 1908. Oeuvres de Pierre Curie. Gauthier-Villars, Paris.

Dahmen, K., Ben-Zion, Y., 2009. Jerky motion in slowly driven magnetic and earthquake fault systems, physics of. In: Meyers, R.A. (Ed.), Encyclopedia of Complexity and Systems Science. Springer, New York, pp. 5021–5037.

Daniel, C.G., Spear, F.S., 1999. The clustered nucleation and growth processes of garnet in regional metamorphic rocks from north-west Connecticut, USA. Journal of Metamorphic Geology 17, 503–520.

Davies, G.F., 2011. Mantle Convection for Geologists. Cambridge University Press, Cambridge.

De Bresser, J.H.P., Peach, C.J., Reijs, J.P.J., Spiers, C.J., 1998. On dynamic recrystallization during solid state flow: effects of stress and temperature. Geophysical Research Letters 25, 3457–3460.

De Bresser, J.H.P., Ter Heege, J.H., Spiers, C.J., 2001. Grain size reduction by dynamic recrystallization: can it result in major theological weakening? International Journal of Earth Sciences 90, 28–45.

Decoster, N., Roux, S.G., Arneodo, A., 2000. A wavelet-based method for multifractal image analysis. II Applications to synthetic multifractal rough surfaces. European Physical Journal B 15, 739–764.

de Donder, T., Van Rysselberghe, P., 1936. Thermodynamic Theory of Affinity. Stanford University Press.

de Giorgi, E., 1975. Sulla convergenza di alcune successioni d'integrali del tipo dell'area. Rendiconti di Matematica 8, 277–294.

De Groot, S.R., Mazur, P., 1984. Non-Equilibrium Thermodynamics. Dover Publications, Inc., New York.

De Groot, S.R., 1952. Thermodynamics of Irreversible Processes. North-Holland Publishing Co., Amsterdam.

De Wit, A., Borckmans, P., Dewel, G., 1997. Twist grain boundaries in three-dimensional lamellar Turing structures. Proceedings of the National Academy of Sciences of the United States of America 94, 12765–12768.

De Wit, A., 1999. Spatial patterns and spatiotemporal dynamics in chemical systems. Advances in Chemical Physics 109, 435–513.

De Wit, A., 2001. Fingering of chemical fronts in porous media. Physical Review Letters 87.

De Wit, A., 2008. Chemi-hydrodynamic patterns and instabilities. Chimie Nouvelle 9, 1–7.

Del Piero, G., Owen, D., 1993. Structured deformations of continua. Archive for Rational Mechanics and Analysis 124, 99–155.

Del Piero, G., Truskinovsky, L., 2001. Macro- and micro-cracking in one-dimensional elasticity. International Journal of Solids and Structures 38, 1135–1148.

Denbigh, K., 1968. The Principles of Chemical Equilibrium. Cambridge University Press, London, UK.

Derby, B., Ashby, M.F., 1987. On dynamic recrystallization. Scripta Metallurgica 21, 879–884.

Detournay, E., Cheng, A.H.-D., 1993. Fundamentals of poroelasticity. In: Hudson, J.A., Fairhurst, C. (Eds.), Comprehensive Rock Engineering. Pergamon Press, Oxford, pp. 113–171.

Dimiduk, D.M., Nadgorny, E.M., Woodward, C., Uchic, M.D., Shade, P.A., 2010. An experimental investigation of intermittent flow and strain burst scaling behavior in LiF crystals during microcompression testing. Philosophical Magazine 90, 3621–3649.

Dimiduk, D.M., 2006. Scale-free intermittent flow in crystal plasticity. Science 312, 1188–1190.

Dipple, G.M., Ferry, J.M., 1992. Metasomatism and fluid-flow in ductile fault zones. Contributions to Mineralogy and Petrology 112, 149–164.

Drazin, P.G., Reid, W.H., 1981. Hydrodynamic Stability. Cambridge University Press, Cambridge, UK.

Driver, J.H., Jensen, D.J., Hansen, N., 1994. Large-strain deformation structures in aluminum crystals with rolling texture orientations. Acta Metallurgica et Materialia 42, 3105–3114.

Drucker, D.C., Prager, W., 1952. Soil mechanics and plastic analysis or limit design. Quarterly of Applied Mathematics 10, 157–165.

Drucker, D.C., 1959. A definition of a stable inelastic material. Journal of Applied Mechanics-Transactions of the ASME 26, 101–106.

Drury, M.R., Pennock, G.M., 2007. Subgrain rotation recrystallization in minerals. In: Prangnell, P.B., Bate, P.S. (Eds.), Fundamentals of Deformation and Annealing, pp. 95–104.

Drury, M.R., Urai, J.L., 1990. Deformation-related recrystallization processes. Tectonophysics 172, 235–253.

Drury, M.R., Humphreys, F.J., White, S.H., 1985. Large strain deformation studies using polycrystalline magnesium as a rock analogue. Part II: dynamic recrystallization mechanisms at high-temperatures. Physics of the Earth and Planetary Interiors 40, 208–222.

Drury, M.R., 2005. Dynamic recrystallization and strain softening of olivine aggregates in the laboratory and the lithosphere. In: Gapais, D., Brun, J.P., Cobbold, P.R. (Eds.), Deformation Mechanisms, Rheology and Tectonics: From Minerals to the Lithosphere, pp. 143–158.

Edward, G.H., Etheridge, M.A., Hobbs, B.E., 1982. On the stress dependence of subgrain size. Textures and Micro-structures 5, 127–152.

Eichhubl, P., 2004. Growth of ductile opening-mode fractures in geomaterials. In: Cosgrove, J.W., Engelder, T. (Eds.), The Initiation, Propagation, and Arrest of Joints and Other Fractures: Interpretations Based on Field Observations. Geological Society, London, Special Publications 231, 11–24.

Einav, I., Houlsby, G., Nguyen, G., 2007. Coupled damage and plasticity models derived from energy and dissipation potentials. International Journal of Solids and Structures 44, 2487–2508.

Einav, I., 2007a. Breakage mechanics – part I: theory. Journal of the Mechanics and Physics of Solids 55, 1274–1297.

Einav, I., 2007b. Breakage mechanics – part II: modelling granular materials. Journal of the Mechanics and Physics of Solids 55, 1298–1320.

Einav, I., 2007c. Fracture propagation in brittle granular matter. Proceedings of the Royal Society A: Mathematical Physical and Engineering Sciences 463, 3021–3035.

Einstein, A., 1903. Eine Theorie der Grundlagen der Thermodynamik. Annalen der Physik 11, 170–187.

Elkins-Tanton, L.T., 2007. Continental magmatism, volatile recycling, and a heterogeneous mantle caused by litho-spheric gravitational instabilities. Journal of Geophysical Research 112, B03405.

Elsey, M., Esedoglu, S., Smereka, P., 2011. Large-scale simulation of normal grain growth via diffusion-generated motion. Proceedings of the Royal Society A: Mathematical Physical and Engineering Sciences 467, 381–401.

England, P.C., Thompson, A.B., 1984. Pressure temperature time paths of regional metamorphism. 1. Heat-transfer during the evolution of regions of thickened continental-crust. Journal of Petrology 25, 894–928.

Epstein, I.R., Pojman, J.A., 1998. An Introduction to Nonlinear Chemical Dynamics: Oscillations, Waves, Patterns, and Chaos. Oxford University Press, Oxford.

Ericksen, J.L., 1975. Equilibrium of bars. Journal of Elasticity 5, 191–201.

Ericksen, J.L., 1998. Introduction to the Thermodynamics of Solids, revised ed. Springer-Verlag, New York.

Eringen, A.C., 1962. Nonlinear Theory of Continuous Media. McGraw-Hill, New York.

Eshelby, J.D., 1957. The determination of the elastic field of an ellipsoidal inclusion, and related problems. Proceedings of the Royal Society of London Series A: Mathematical and Physical Sciences 241, 376–396.

Eshelby, J.D., 1975. Elastic energy-momentum tensor. Journal of Elasticity 5, 321–335.

Etheridge, M.A., Wilkie, J.C., 1981. An assessment of dynamically recrystallized grain-size as a paleopiezometer in quartz-bearing mylonite zones. Tectonophysics 78, 475–508.

Etheridge, M.A., Wall, V.J., Vernon, R.H., 1983. The role of the fluid phase during regional metamorphism and deformation. Journal of Metamorphic Geology 1, 205–226.

Etheridge, M.A., Wall, V.J., Cox, S.F., Vernon, R.H., 1984. High fluid pressures during regional metamorphism and deformation – implications for mass-transport and deformation mechanisms. Journal of Geophysical Research 89, 4344–4358.

Evans, B., Kohlstedt, D.L., 1995. Rheology of rocks. In: Ahrens, T.J. (Ed.), Rock Physics and Phase Relations: A Handbook of Physical Constants. American Geophysical Union, Washington, DC, pp. 148–165.

Evans, B.E., Renner, J.R., Hirth, G.H., 2001. A few remarks on the kinetics of static grain growth in rocks. International Journal of Earth Sciences 90, 88–103.

Everall, P.R., Hunt, G.W., 1999. Arnold tongue predictions of secondary buckling in thin elastic plates. Journal of the Mechanics and Physics of Solids 47, 2187–2206.

Fan, D., Chen, L.Q., 1997a. Computer simulation of grain growth using a continuum field model. Acta Materialia 45, 611–622.

Fan, D.N., Chen, L.Q., 1997b. Diffusion-controlled grain growth in two-phase solids. Acta Materialia 45, 3297–3310.

Faria, S.H., Kremer, G.M., Hutter, K., 2003. On the inclusion of recrystallization processes in the modeling of induced anisotropy in ice sheets: a thermodynamicist's point of view. In: Duval, P. (Ed.), Annals of Glaciology, vol. 37, pp. 29–34.

Faria, S.H., Kremer, G.M., Hutter, K., 2006. Creep and recrystallization of large polycrystalline masses. II. Constitutive theory for crystalline media with transversely isotropic grains. Proceedings of the Royal Society A: Mathematical, Physical and Engineering Science 462, 1699–1720.

Faria, S.H., 2006a. Creep and recrystallization of large polycrystalline masses. I. General continuum theory. Proceedings of the Royal Society A: Mathematical, Physical and Engineering Science 462, 1493–1514.

Faria, S.H., 2006b. Creep and recrystallization of large polycrystalline masses. III. Continuum theory of ice sheets. Proceedings of the Royal Society A: Mathematical, Physical and Engineering Science 462, 2797–2816.

Feder, J., 1988. Fractals. Plenum, New York.

Feng, S.C., Halperin, B.I., Sen, P.N., 1987. Transport-properties of continuum-systems near the percolation-threshold. Physical Review B 35, 197–214.

Ferry, J.M., 1994. Overview of the petrologic record of fluid-flow during regional metamorphism in northern New England. American Journal of Science 294, 905–988.

Fink, J.K., 2009. Physical Chemistry in Depth. Springer-Verlag, Berlin.

Fisher, G.W., Lasaga, A.C., 1981. Irreversible thermodynamics in petrology. Reviews in Mineralogy and Geochemistry 8, 171–207.

Fitts, D.D., 1962. Nonequilibrium Thermodynamics. McGraw-Hill.

Fitz Gerald, J.D., Mancktelow, N.S., Pennacchioni, G., Kunze, K., 2006. Ultrafine-grained quartz mylonites from high-grade shear zones: evidence for strong dry middle to lower crust. Geology 34, 369–372.

Fletcher, R.C., Merino, E., 2001. Mineral growth in rocks: kinetic-rheological models of replacement, vein formation, and syntectonic crystallization. Geochimica et Cosmochimica Acta 65, 3733–3748.

Fletcher, R.C., 1973. Propagation of a coherent interface between 2 nonhydrostatically stressed crystals. Journal of Geophysical Research 78, 7661–7666.

Fletcher, R.C., 1974. Wavelength selection in folding of a single layer with power-law rheology. American Journal of Science 274, 1029–1043.

Ford, A., Blenkinsop, T.G., 2009. An expanded de Wijs model for multifractal analysis of mineral production data. Mineralium Deposita 44, 233–240.

Ford, J.M., Wheeler, J., 2004. Modelling interface diffusion creep in two-phase materials. Acta Materialia 52, 2365–2376.

Ford, J.M., Wheeler, J., Movchan, A.B., 2002. Computer simulation of grain-boundary diffusion creep. Acta Materialia 50, 3941–3955.

Fossen, H., Tikoff, B., 1993. The deformation matrix for simultaneous simple shearing, pure shearing and volume change, and its application to transpression-transtension tectonics. Journal of Structural Geology 15, 413–422.

Francfort, G.A., Marigo, J.J., 1998. Revisiting brittle fracture as an energy minimization problem. Journal of the Mechanics and Physics of Solids 46, 1319–1342.

Francfort, G.A., 2006. Quasistatic brittle fracture seen as an energy minimizing movement. GAMM-Mitteilungen 29, 172–191.

Fredrich, J.T., Wong, T.F., 1986. Micromechanics of thermally induced cracking in 3 crustal rocks. Journal of Geophysical Research-Solid Earth and Planets 91, 2743–2764.

Fressengeas, C., Taupin, V., Capolungo, L., 2011. An elasto-plastic theory of dislocation and disclination fields. International Journal of Solids and Structures 48, 3499–3509.

Fricke, H.C., Wickham, S.M., O'Neil, J.R., 1992. Oxygen and hydrogen isotope evidence for meteoric water infiltration during mylonitization and uplift in the Ruby Mountains-East Humboldt Range core complex, Nevada. Contributions to Mineralogy and Petrology 111, 203–221.

Friedel, J., 1967. Dislocations. Pergamon Press, Oxford.

Frisch, U., 1995. Turbulence. The Legacy of A.N. Kolmogorov. Cambridge University Press.

Frolov, T., Mishin, Y., 2012. Thermodynamics of coherent interfaces under mechanical stresses. I. Theory. Physical Review B 85, 46 pp.

Fung, Y.C., 1965. Foundations of Solid Mechanics. Prentice-Hall, Inc., Englewood Cliffs, NJ, USA.

Fyfe, W.S., Price, N.J., Thompson, A.B., 1978. Fluids in the Earth's Crust. Elsevier, Amsterdam.

Gabbrielli, R., 2009. A new counter-example to Kelvin's conjecture on minimal surfaces. Philosophical Magazine Letters 89, 483–491.

Gale, J.D., Wright, K., Hudson-Edwards, K.A., 2010. A first-principles determination of the orientation of H_3O^+ in hydronium alunite. American Mineralogist 95, 1109–1112.

Gambin, W., 2001. Plasticity and Textures. Kluwer Academic, Dordrecht.

Garven, G., Freeze, R.A., 1984a. Theoretical analysis of the role of groundwater flow in the genesis of stratabound ore deposits. 1. Mathematical and numerical model. American Journal of Science 284, 1085−1124.

Garven, G., Freeze, R.A., 1984b. Theoretical-analysis of the role of groundwater-flow in the genesis of stratabound ore-deposits. 2. Quantitative results. American Journal of Science 284, 1125−1174.

Garven, G., 1985. The role of regional fluid-flow in the genesis of the Pine Point deposit, Western Canada Sedimentary Basin. Economic Geology 80, 307−324.

Ge, S., Garven, G., 1989. Tectonically induced transient groundwater flow in foreland basins. In: Price, R.A. (Ed.), Origin and Evolution of Sedimentary Basins and Their Energy and Mineral Resources. American Geophysical Union Monograph Series, pp. 145−157.

Geiger, S., Driesner, T., Heinrich, C.A., Matthai, S.K., 2006a. Multiphase thermohaline convection in the earth's crust: I. A new finite element − finite volume solution technique combined with a new equation of state for NaCl−H$_2$O. Transport in Porous Media 63, 399−434.

Geiger, S., Driesner, T., Heinrich, C.A., Matthai, S.L., 2006b. Multiphase thermohaline convection in the earth's crust: II. Benchmarking and application of a finite element − finite volume solution technique with a NaCl−H$_2$O equation of state. Transport in Porous Media 63, 435−461.

van Genuchten, M.T., 1980. A closed-form equation for predicting the hydraulic conductivity of unsaturated soils. Soil Science Society of America 44, 892−898.

Gerhartl, F.J., 1994. The A+B$\rightleftharpoons$C of chemical thermodynamics. Journal of Chemical Education 71, 539−548.

Ghoussoub, J., Leroy, Y.M., 2001. Solid-fluid phase transformation within grain boundaries during compaction by pressure solution. Journal of the Mechanics and Physics of Solids 49, 2385−2430.

Gibbs, J.W., 1906. The Collected Works of J. Willard Gibbs. Yale University Press, New Haven.

Girard, L., Amitrano, D., Weiss, J., 2010. Failure as a critical phenomenon in a progressive damage model. Journal of Statistical Mechanics: Theory and Experiment P01013.

Glansdorff, P., Prigogine, I., 1971. Thermodynamic Theory of Structure, Stability and Fluctuations. Wiley-Interscience, New York.

Gorczyk, W., Hobbs, B., Gerya, T., 2012. Initiation of Rayleigh−Taylor instabilities in intra-cratonic settings. Tectonophysics 514−517, 146−155.

Gorczyk, W., Hobbs, B., Gessner, K., Gerya, T., 2013. Intracratonic geodynamics. Gondwana Research 24, 838−848.

Goscombe, B., Trouw, R., 1999. The geometry of folded tectonic shear sense indicators. Journal of Structural Geology 21, 123−127.

Goscombe, B.D., Passchier, C.W., Hand, M., 2004. Boudinage classification: end-member boudin types and modified boudin structures. Journal of Structural Geology 26, 739−763.

Gray, P., Scott, S.K., 1983. Autocatalytic reactions in the isothermal, continuous stirred tank reactor − isolas and other forms of multistability. Chemical Engineering Science 38, 29−43.

Gray, P., Scott, S.K., 1984. Autocatalytic reactions in the isothermal, continuous stirred tank reactor: oscillations and instabilities in the system $A + 2B \rightarrow 3B$; $B \rightarrow C$. Chemical Engineering Science 39, 1087−1097.

Gray, P., Scott, S.K., 1985. Sustained oscillations and other exotic patterns of behavior in isothermal reactions. Journal of Physical Chemistry 89, 22−32.

Gray, P., Scott, S.K., 1994. Chemical Oscillations and Instabilities. Clarendon Press, Oxford.

Grew, N., 1682. The Anatomy of Plants with an Idea of a Philosophical History of Plants, second ed. Printed by W. Rawlins, for the author, London.

Grichuk, D.V., Shvarov, Y.V., 2002. Comparative analysis of techniques used in the equilibrium-dynamic simulations of infiltration metasomatic zoning. Petrology 10, 580−592.

Griffith, A.A., 1921. The phenomena of rupture and flow in solids. Philosophical Transactions of the Royal Society of London. Series A: Containing Papers of a Mathematical or Physical Character 221, 163−198.

Griggs, D.T., Turner, F.J., Heard, H.C., 1960. Deformation of rocks at 500° to 800° C. In: Griggs, D.T., Handin, J. (Eds.), Symposium on Rock Deformation. Geological Society of America Memoir 79, pp. 39−104.

Gross, K.A., 1965. X-ray line broadening and stored energy in deformed and annealed calcite. Philosophical Magazine 12, 801−813.

Groves, G.W., Kelly, A., 1963. Independent slip systems in crystals. Philosophical Magazine 8, 877−887.

Groves, G.W., Kelly, A., 1969. Change of shape due to dislocation climb. Philosophical Magazine 19, 977−986.

Gruesbeck, C., 2013. Dynamics of a Continuous Stirred-Tank Reactor with Consecutive Exothermic and Endothermic Reactions. Wolfram Demonstrations Project.

Gu, J.C., Rice, J.R., Ruina, A.L., Tse, S.T., 1984. Slip motion and stability of a single degree of freedom elastic system withr ate and state dependent friction. Journal of the Mechanics and Physics of Solids 32, 167–196.

Guckenheimer, J., Holmes, P., 1986. Nonlinear Oscillations, Dynamical Systems and Bifurcations of Vector Fields. Springer-Verlag, New York.

Guidotti, C.V., Johnson, S.E., 2002. Pseudomorphs and associated microstructures of western Maine, USA. Journal of Structural Geology 24, 1139–1156.

Gurtin, M.E., Fried, E., Anand, L., 2010. The Mechanics and Thermodynamics of Continua. Cambridge University Press.

Gusak, A.M., 2010. Nucleation in a concentration gradient. In: Gusak, A.M. (Ed.), Diffusion-Controlled Solid State Reactions. Wiley-VCH, Weinheim, pp. 61–98.

Gutkin, M.Y., Ovid'ko, I.A., 2004. Plastic Deformation in Nanocrystalline Materials. Springer-Verlag, Berlin.

Guyer, R.A., Johnson, P.A., 2009. Nonlinear Mesoscopic Elasticity. Wiley-VCH, Berlin.

Hahner, P., Bay, K., Zaiser, M., 1998. Fractal dislocation patterning during plastic deformation. Physical Review Letters 81, 2470–2473.

Halfpenny, A., Prior, D.J., Wheeler, J., 2012. Electron backscatter diffraction analysis to determine the mechanisms that operated during dynamic recrystallisation of quartz-rich rocks. Journal of Structural Geology 36, 2–15.

Hall, E.O., 1951. The deformation and ageing of mild steel. 3. Discussion of results. Proceedings of the Physical Society of London Section B 64, 747–753.

Hallberg, H., Ristinmaa, M., 2013. Microstructure evolution influenced by dislocation density gradients modeled in a reaction-diffusion system. Computational Materials Science 67, 373–383.

Halsey, T.C., Jensen, M.H., Kadanoff, L.P., Procaccia, I., Shraiman, B.I., 1986. Fractal measures and their singularities: The characterization of strange sets. Physical Review A 33, 1141–1151.

Hamiel, Y., Liu, Y., Lyakhovsky, V., Ben-Zion, Y., Lockner, D., 2004. A viscoelastic damage model with applications to stable and unstable fracturing. Geophysical Journal International 159, 1155–1165.

Hammer, S., Passchier, C., 1991. Shear-Sense Indicators: A Review. Geological Survey of Canada, 72 pp.

Hansen, N., 2004. Hall-Petch relation and boundary strengthening. Scripta Materialia 51, 801–806.

Hardin, T., Adams, B.L., Fullwood, D.T., 2011. Recovering the full dislocation tensor from high-resolution EBSD microscopy. In: Fan, J.H., Zhang, J.Q., Chen, H.B., Jin, Z.H. (Eds.), Advances in Heterogeneous Material Mechanics, 2011, pp. 169–177.

Harker, A., 1950. Metamorphism. Methuen, London.

Haslach, H.W., 2011. Maximum Dissipation Non-Equilibrium Thermodynamics and Its Geometric Structure. Springer, New York.

Hay, R.S., Evans, B., 1988. Intergranular distribution of pore fluid and the nature of high-angle grain-boundaries in limestone and marble. Journal of Geophysical Research-Solid Earth and Planets 93, 8959–8974.

Heard, H.C., 1960. Chapter 7: transition from brittle fracture to ductile flow in Solenhofen Limestone as a function of temperature, confining pressure, and interstitial fluid pressure. In: Griggs, D.T., Handin, J. (Eds.), Symposium on Rock Deformation. Geological Society of America Memoir 79, 193–226.

Heilbronner, R., Keulen, N., 2006. Grain size and grain shape analysis of fault rocks. Tectonophysics 427, 199–216.

Heilbronner, R., Tullis, J., 2006. Evolution of c axis pole figures and grain size during dynamic recrystallization: results from experimentally sheared quartzite. Journal of Geophysical Research 111. B10202, 19 pages.

Hernandez-Garcia, E., Lopez, C., 2004. Sustained plankton blooms under open chaotic flows. Ecological Complexity 1, 253–259.

Herrmann, H.J., Mantica, G., Bessis, D., 1990. Space-filling bearings. Physical Review Letters 65, 3223–3226.

Herwegh, M., Poulet, T., Karrech, A., Regenauer-Lieb, K., 2014. From transient to steady state deformation and grain size: a thermodynamic approach using elasto-visco-plastic numerical modeling. Journal of Geophysical Research-Solid Earth 119, 900–918.

Hill, R., 1948. A theory of the yielding and plastic flow of anisotropic metals. Proceedings of the Royal Society of London Series A: Mathematical and Physical Sciences 193, 281–297.

Hill, R., 1950. The Mathematical Theory of Plasticity. Clarendon Press, Oxford.

Hill, R., 1952. The elastic behaviour of a crystalline aggregate. Proceedings of the Physical Society of London Section A 65, 349–355.

Hill, R., 1958. A general theory of uniqueness and stability in elastic-plastic solids. Journal of the Mechanics and Physics of Solids 6, 236—249.

Hill, R., 1962. Acceleration waves in solids. Journal of the Mechanics and Physics Solids 10, 1—16.

Hill, R., 1998. The Mathematical Theory of Plasticity, second ed. Clarendon Press, Oxford.

Hill, T.L., 1989. Free Energy Transduction and Biochemical Cycle Kinetics. Springer-Verlag, New York.

Hillert, M., 1965. On theory of normal and abnormal grain growth. Acta Metallurgica 13, 227—238.

Hinch, E.J., Leal, L.G., 1975. Constitutive equations in suspension mechanics. Part 1. General formulation. Journal of Fluid Mechanics 71, 481—495.

Hinch, E.J., Leal, L.G., 1976. Constitutive equations in suspension mechanics. Part 2. Approximate forms for a suspension of rigid particles affected by Brownian rotations. Journal of Fluid Mechanics 76, 187—208.

Hirsinger, V., Hobbs, B.E., 1983. A general harmonic coordinate transformation to simulate the states of strain in inhomogeneously deformed rocks. Journal of Structural Geology 5, 307—320.

Hirth, G., Tullis, J., 1992. Dislocation creep regimes in quartz aggregates. Journal of Structural Geology 14, 145—159.

Ho, T.A., Argyris, D., Papavassiliou, D.V., Striolo, A., Lee, L.L., Cole, D.R., 2011. Interfacial water on crystalline silica: a comparative molecular dynamics simulation study. Molecular Simulation 37, 172—195.

Hobbs, B.E., Ord, A., 2010. The mechanics of granitoid systems and maximum entropy production rates. Philosophical Transactions of the Royal Society A: Mathematical, Physical and Engineering Sciences 368, 53—93.

Hobbs, B.E., Ord, A., 2011. Microstructures in deforming-reactive systems. In: Prior, D.J., Rutter, E.H., Tatham, D.J. (Eds.), Deformation Mechanisms, Rheology and Tectonics: Microstructures, Mechanics and Anisotropy, Geological Society, London, Special Publications 360, 273—299.

Hobbs, B.E., Means, W.D., Williams, P.F., 1976. An Outline of Structural Geology. John Wiley & Sons, Inc., New York.

Hobbs, B.E., Muhlhaus, H.B., Ord, A., 1990. Instability, softening and localization of deformation. In: Knipe, R.J., Rutter, E.H. (Eds.), Deformation Mechanisms, Rheology and Tectonics, Geological Society Special Publication 54, 143—165.

Hobbs, B., Regenauer-Lieb, K., Ord, A., 2008. Folding with thermal—mechanical feedback. Journal of Structural Geology 30, 1572—1592.

Hobbs, B.E., Regenauer-Lieb, K., Ord, A., 2009. Folding with thermal-mechanical feedback: a reply. Journal of Structural Geology 31, 752—755.

Hobbs, B., Regenauer-Lieb, K., Ord, A., 2010. Folding with thermal mechanical feedback: another reply. Journal of Structural Geology 32, 131—134.

Hobbs, B.E., Ord, A., Regenauer-Lieb, K., 2011. The thermodynamics of deformed metamorphic rocks: a review. Journal of Structural Geology 33, 758—818.

Hobbs, B.E., 1968. Recrystallization of single crystals of quartz. Tectonophysics 6, 353—401.

Hobbs, B.E., 1971. The analysis of strain in folded layers. Tectonophysics 11, 329—375.

Hobbs, B.E., 1985. The geological significance of microfabric analysis. In: Wenk, H.-R. (Ed.), Preferred Orientation in Deformed Metals and Rocks: An Introduction to Modern Texture Analysis. Academic, San Diego, CA, pp. 463—484.

Hobbs, D.W., 1967. The formation of tension joints in sedimentary rocks: an explanation. Geological Magazine 104, 550—556.

Hoek, E., Brown, E.T., 1980a. Empirical strength criterion for rock masses. Journal of the Geotechnical Engineering Division, ASCE 106, 1013—1035.

Hoek, E., Brown, E.T., 1980b. Underground Excavations in Rock. Instn. Min. Metall, London.

Holcomb, D., Rudnicki, J.W., Issen, K.A., Sternlof, K., 2007. Compaction localization in the Earth and the laboratory: state of the research and research directions. Acta Geotechnica 2, 1—15.

Holdsworth, A.M., Kevlahan, N.K.R., Earn, D.J.D., 2012. Multifractal signatures of infectious diseases. Journal of the Royal Society Interface 9, 2167—2180.

Hollister, L.S., Crawford, M.L., 1986. Melt-enhanced deformation — a major tectonic process. Geology 14, 558—561.

Holm, E.A., Srolovitz, D.J., Cahn, J.W., 1993. Microstructural evolution in 2-dimensional 2-phase polycrystals. Acta Metallurgica et Materialia 41, 1119—1136.

Horton, C.W., Rogers, F.T., 1945. Convection currents in a porous medium. Journal of Applied Physics 16, 367—370.

Houlsby, G.T., Puzrin, A.M., 2000. An approach to plasticity based on generalised thermodynamics. In: Kolymbas, D. (Ed.), Constitutive Modelling of Granular Materials. Springer Berlin Heidelberg, pp. 319—331.

Houlsby, G.T., Puzrin, A.M., 2006a. Principles of Hyperplasticity. Springer-Verlag, London.

Houlsby, G.T., Puzrin, A.M., 2006b. Thermodynamics of porous continua. In: Wu, W., Yu, H.S. (Eds.), Modern Trends in Geomechanics, pp. 39—60.

Houseman, G.A., Molnar, P., 1997. Gravitational (Rayleigh-Taylor) instability of a layer with non-linear viscosity and convective thinning of continental lithosphere. Geophysical Journal International 128, 125–150.

Howe, J.M., Aaronson, H.I., Hirth, J.P., 2000. Aspects of interphase boundary structure in diffusional phase transformations. Acta Materialia 48, 3977–3984.

Howe, J.M., Pond, R.C., Hirth, J.P., 2009. The role of disconnections in phase transformations. Progress in Materials Science 54, 792–838.

Hubbert, M.K., 1940. The theory of ground-water motion. Journal of Geology 48, 785–944.

Hughes, W.F., Brighton, J.A., 1999. Schaum's Outline of Fluid Dynamics, third ed. McGraw-Hill.

Hughes, D.A., Hansen, N., 2000. Microstructure and strength of nickel at large strains. Acta Materialia 48, 2985–3004.

Hughes, D.A., Liu, Q., Chrzan, D.C., Hansen, N., 1997. Scaling of microstructural parameters: misorientations of deformation induced boundaries. Acta Materialia 45, 105–112.

Hughes, D.A., Chrzan, D.C., Liu, Q., Hansen, N., 1998. Scaling of misorientation angle distributions. Physical Review Letters 81, 4664–4667.

Hughes, D.A., Hansen, N., Bammann, D.J., 2003. Geometrically necessary boundaries, incidental dislocation boundaries and geometrically necessary dislocations. Scripta Materialia 48, 147–153.

Hughes, D.A., 2001. Microstructure evolution, slip patterns and flow stress. Materials Science and Engineering: A-Structural Materials Properties Microstructure and Processing 319, 46–54.

Hull, D., 1999. Fractography: Observing, Measuring and Interpreting Fracture Surface Topography. Cambridge University Press.

Hunt, G., 2006. Buckling in space and time. Nonlinear Dynamics 43, 29–46.

Hunt, G.W., Everall, P.R., 1999. Arnold tongues and mode-jumping in the supercritical post-buckling of an archetypal elastic structure. Proceedings of the Royal Society of London Series A 455, 125–140.

Hunt, G.W., Hammond, J., 2012. Mechanics of shear banding in a regularized two-dimensional model of a granular medium. Philosophical Magazine 1–18.

Hunt, G.W., Wadee, M.K., 1991. Comparative Lagrangian formulations for localized buckling. Proceedings: Mathematical and Physical Sciences 434, 485–502.

Hunt, G.W., Lawther, R., Costa, P.P.E., 1997a. Finite element modelling of spatially chaotic structures. International Journal for Numerical Methods in Engineering 40, 2237–2256.

Hunt, G.W., Muhlhaus, H.B., Whiting, A.I.M., 1997b. Folding processes and solitary waves in structural geology. Philosophical Transactions of the Royal Society of London Series, A: Mathematical Physical and Engineering Sciences 355, 2197–2213.

Hunt, G.W., Peletier, M.A., Champneys, A.R., Woods, P.D., Wadee, M.A., Budd, C.J., Lord, G.J., 2000a. Cellular buckling in long structures. Nonlinear Dynamics 21, 3–29.

Hunt, G.W., Peletier, M.A., Wadee, M.A., 2000b. The Maxwell stability criterion in pseudo-energy models of kink banding. Journal of Structural Geology 22, 669–681.

Hunt, J.A., Baker, T., Thorkelson, D.J., 2007. A review of iron oxide copper-gold deposits, with focus on the Wernecke Breccias, Yukon, Canada, as an example of a non-magmatic end member and implications for IOCG genesis and classification. Exploration and Mining Geology 16, 209–232.

Hunt, K.L.C., Hunt, P.M., Ross, J., 1987. Dissipation in steady states of chemical systems and deviations from minimum entropy production. Physica A 147, 48–60.

Hunt, K.L.C., Hunt, P.M., Ross, J., 1988. Deviations from minimum entropy production at steady states of reacting chemical systems arbitrarily close to equilibrium. Physica A 154, 207–211.

Iacopini, D., Passchier, C.W., Koehn, D., Carosi, R., 2007. Fabric attractors in general triclinic flow systems and their application to high strain shear zones: a dynamical system approach. Journal of Structural Geology 29, 298–317.

Iacopini, D., Carosi, R., Xypolias, P., 2010. Implications of complex eigenvalues in homogeneous flow: a three-dimensional kinematic analysis. Journal of Structural Geology 32, 93–106.

Imanaka, T., Sano, K., Shimizu, M., 1973. Dislocation attenuation and acoustic emission during deformation in copper single crystal. Crystal Lattice Defects 4, 57–64.

Irwin, G.R., 1957. Analysis of stresses and strains near the end of a crack traversing a plate. Journal of Applied Mechanics 24, 361–364.

Issen, K.A., Rudnicki, J.W., 2000. Conditions for compaction bands in porous rock. Journal of Geophysical Research-Solid Earth 105, 21529–21536.

Issen, K.A., Rudnicki, J.W., 2001. Theory of compaction bands in porous rock. Physics and Chemistry of the Earth Part A-Solid Earth and Geodesy 26, 95–100.

Jaeger, J.C., Cook, N.G.W., 1969. Fundamentals of Rock Mechanics. Methuen, London.

Jaeger, J.C., 1969. Elasticity, Fracture and Flow with Engineering and Geological Applications, third ed. Chapman & Hall, London.

Jamieson, R.A., Beaumont, C., Fullsack, P., Lee, B., 1998. Barrovian regional metamorphism: where's the heat? In: Treloar, P.J., O'Brien, P.J. (Eds.), What Drives Metamorphism and Metamorphic Reactions? Geological Society, London, Special Publications 138, 23–51.

Jamtveit, B., Austrheim, H., 2010. Metamorphism: the role of fluids. Elements 6, 153–158.

Jaynes, E.T., 1963. Information theory and statistical mechanics. In: Ford, K.W. (Ed.), Brandeis University Summer Institute, Lectures in Theoretical Physics. Benjamin-Cummings Publ. Co., pp. 181–218.

Jaynes, E.T., 2003. Probability Theory: The Logic of Science. Cambridge University Press.

Jeffery, G.B., 1922. The motion of ellipsoidal particles in a viscous fluid. Proceedings of the Royal Society of London Series A: Containing Papers of a Mathematical and Physical Character 102, 161–179.

Jessell, M., Bons, P., Evans, L., Barr, T., Stuwe, K., 2001. Elle: the numerical simulation of metamorphic and deformation microstructures. Computers and Geosciences 27, 17–30.

Jessell, M.W., Bons, P.D., Griera, A., Evans, L.A., Wilson, C.J.L., 2009. A tale of two viscosities. Journal of Structural Geology 31, 719–736.

Jiang, D., Williams, P.F., 1998. High-strain zones: a unified model. Journal of Structural Geology 20, 1105–1120.

Jiang, D., 2010. Flow and finite deformation of surface elements in three dimensional homogeneous progressive deformations. Tectonophysics 487, 85–99.

John, T., Schenk, V., 2003. Partial eclogitisation of gabbroic rocks in a late precambrian subduction zone (Zambia): prograde metamorphism triggered by fluid infiltration. Contributions to Mineralogy and Petrology 146, 174–191.

Johnson, A.M., Fletcher, R.C., 1994. Folding of Viscous Layers. Columbia University Press, New York.

Johnson, S.E., 1993a. Testing models for the development of spiral-shaped inclusion trails in garnet porphyroblasts - to rotate or not to rotate, that is the question. Journal of Metamorphic Geology 11, 635–659. Wiley.

Johnson, S.E., 1993b. Unraveling the spirals — a serial thin-section study and 3-dimensional computer-aided reconstruction of spiral-shaped inclusion trails in garnet porphyroblasts. Journal of Metamorphic Geology 11, 621–634.

Jones, S.W., Aref, H., 1988. Chaotic advection in pulsed-source sink systems. Physics of Fluids 31, 469–485.

Jones, M.C., Persichetti, J.M., 1986. Convective instability in packed-beds with throughflow. AIChE Journal 32, 1555–1557.

Kahlert, C., Rossler, O.E., Varma, A., 1981. Chaos in a continuous stirred tank reactor with two consecutive first-order reactions, one exo-, one endothermic. In: Ebert, W.H., Deuflhard, P., Jager, W. (Eds.), Modelling of Chemical Reactions Systems. Springer Verlag, New York, pp. 355–365.

Kamb, W.B., 1959. Theory of preferred crystal orientation developed by crystallization under stress. Journal of Geology 67, 153–170.

Kamb, W.B., 1961. Thermodynamic theory of nonhydrostatically stressed solids. Journal of Geophysical Research 66, 259–271.

Karato, S., Wenk, H.-R. (Eds.), 2002. Plastic Deformation of Minerals and Rocks. In: Ribbe, P.H. (Ed.), Reviews in Mineralogy and Geochemistry. Mineralogical Society of America, vol. 51.

Karato, S., 2008. Deformation of Earth Materials: An Introduction to the Rheology of Solid Earth. Cambridge University Press, New York.

Karolyi, G., Pentek, A., Toroczkai, Z., Tel, T., Grebogi, C., 1999. Chemical or biological activity in open chaotic flows. Physical Review E 59, 5468–5481.

Karrech, A., Regenauer-Lieb, K., Poulet, T., 2011a. Continuum damage mechanics for the lithosphere. Journal of Geophysical Research-Solid Earth 116, B04205.

Karrech, A., Regenauer-Lieb, K., Poulet, T., 2011b. A damaged visco-plasticity model for pressure and temperature sensitive geomaterials. International Journal of Engineering Science 49, 1141–1150.

Karrech, A., Regenauer-Lieb, K., Poulet, T., 2011c. Frame indifferent elastoplasticity of frictional materials at finite strain. International Journal of Solids and Structures 48, 397–407.

Karrech, A., Poulet, T., Regenauer-Lieb, K., 2012. A limit analysis approach to derive a thermodynamic damage potential for non-linear geomaterials. Philosophical Magazine 92, 3439–3450.

Kelly, A., Groves, G.W., 1970. Crystallography and Crystal Defects. Addison-Wesley Publishing Company, Reading, MA.

Kern, R., Weisbrod, A., 1967. Thermodynamics for Geologists. Freeman, Cooper and Co.

Kestener, P., Arneodo, A., 2003. Three-dimensional wavelet-based multifractal method: the need for revisiting the multifractal description of turbulence dissipation data. Physical Review Letters 91.

Kestener, P., Arneodo, A., 2004. Generalizing the wavelet-based multifractal formalism to random vector fields: application to three-dimensional turbulence velocity and vorticity data. Physical Review Letters 93.

Kestener, P., Arneodo, A., 2008. A multifractal formalism for vector-valued random fields based on wavelet analysis: application to turbulent velocity and vorticity 3D numerical data. Stochastic Environmental Research and Risk Assessment 22, 421–435.

Kestin, J., Rice, J.R., 1970. Paradoxes in the application of thermodynamics to strained solids. In: Stuart, E.G., Gal-Or, B., Brainard, A.J. (Eds.), A Critical Review of Thermodynamics. Mono Book Corp., Baltimore, pp. 275–298.

Kestin, J., 1968. On the application of the principles of thermodynamics to strained solid materials. In: Parkus, H., Sedov, L.I. (Eds.), Irreversible Aspects of Continuum Mechanics and Transfer of Physical Characteristics in Moving Fluids. Springer, Vienna, pp. 177–212.

Kestin, J., 1969. A Course in Thermodynamics. Vol. II. Chapter 24. Blaisdell.

Kestin, J., 1990. A note on the relation between the hypothesis of local equilibrium and the Clausius–Duhem inequality. Journal of Non-Equilibrium Thermodynamics, p. 193.

Kestin, J., 1992. Local-equilibrium formalism applied to mechanics of solids. International Journal of Solids and Structures 29, 1827–1836.

Ketcham, R.A., Carlson, W.D., 2001. Acquisition, optimization and interpretation of X-ray computed tomographic imagery: applications to the geosciences. Computers and Geosciences 27, 381–400.

Kilian, R., Heilbronner, R., Stunitz, H., 2011. Quartz grain size reduction in a granitoid rock and the transition from dislocation to diffusion creep. Journal of Structural Geology 33, 1265–1284.

Kim, M.K., Lade, P.V., 1984. Modeling rock strength in 3 dimensions. International Journal of Rock Mechanics and Mining Sciences 21, 21–33.

Kim, P., 2010. Graphene across the border. Nature Materials 9, 792–793.

Knobloch, E., 2008. Spatially localized structures in dissipative systems: open problems. Nonlinearity 21, T45–T60.

Koch, P.S., Christie, J.M., George, R.P., 1980. Flow law of 'wet' quartzite in the a-quartz field. EOS, Transactions of the American Geophysical Union 63, 376.

Kocks, U.F., Canova, G., 1981. Mechanisms and microstructures. In: Hansen, N., Horsewell, A., Leffers, T., Lilholt, H. (Eds.), Deformation of Polycrystals. Riso National Laboratory, Roskilde, Denmark, pp. 35–44.

Kocks, U.F., Argon, A.S., Ashby, M.F., 1975. Thermodynamics and Kinetics of Slip. Pergamon Press, Oxford.

Kocks, U.F., Tome, C.N., Wenk, H.-R., 1998. Texture and Anisotropy. Cambridge University Press, Cambridge, UK.

Kocks, U.F., 1987. Constitutive behavior based on crystal plasticity. In: Miller, A.K. (Ed.), Unified Constitutive Equations for Creep and Plasticity. Elsevier, pp. 1–88.

Kocks, U.F., 1998a. Kinematics and kinetics of plasticity. In: Kocks, U.F., Tome, C.N., Wenk, H.-R. (Eds.), Texture and Anisotropy. Cambridge University Press, Cambridge, pp. 326–389.

Kocks, U.F., 1998b. The representation of orientations and textures. In: Kocks, U.F., Tome, C.N., Wenk, H.-R. (Eds.), Texture and Anisotropy. Cambridge University Press, Cambridge, pp. 44–101.

Kohlstedt, D.L., Mackwell, S.J., 2009. Strength and deformation of planetary lithospheres. In: Watters, T.R., Schultz, R.A. (Eds.), Planetary Tectonics. Cambridge University Press, Cambridge, pp. 397–456.

Kohn, R.V., 1991. The relaxation of a double-well energy. Continuum Mechanics and Thermodynamics 3, 193–236.

Kondepudi, D., Prigogine, I., 1998. Modern Thermodynamics. John Wiley, Chichester.

Kondepudi, D.K., 1988. Remarks on the validity of the theorem of minimum entropy production. Physica A 154, 204–206.

Korzhinskii, D.S., 1959. Physicochemical Basis of the Analysis of the Paragenesis of Minerals. Consultants Bureau, Inc., New York.

Korzhinskii, D.S., 1965. The theory of systems with perfectly mobile components and processes of mineral formation. American Journal of Science 263, 193–205.

Korzhinskii, D.S., 1966. On thermodynamics of open systems and the phase rule (A reply to Weill, D.F., Fyfe, W.S.). Geochimica et Cosmochimica Acta 30, 829–835.

Korzhinskii, D.S., 1967. On thermodynamics of open systems and the phase rule (A reply to the second critical paper of Weill, D.F., Fyfe, W.S.). Geochimica et Cosmochimica Acta 31, 1177–1180.

Kose, K., 1996. 3D NMR imaging of foam structures. Journal of Magnetic Resonance Series A 118, 195–201.

Kretz, R., 1966. Grain-size distribution for certain metamorphic minerals in relation to nucleation and growth. Journal of Geology 74, 147–173.

Kretz, R., 1969. On the spatial distribution of crystals in rocks. Lithos 2, 39–66.

Kretz, R., 1973. Kinetics of the crystallization of garnet at two localities near Yellowknife. Canadian Mineralogist 12, 1–20.

Kretz, R., 1974. Some models for the rate of crystallization of garnet in metamorphic rocks. Lithos 7, 123–131.

Kretz, R., 2006. Shape, size, spatial distribution and composition of garnet crystals in highly deformed gneiss of the Otter Lake area, Quebec, and a model for garnet crystallization. Journal of Metamorphic Geology 24, 431–449.

Kroner, E., Anthony, K.-H., 1975. Dislocations and disclinations in material structures: the basic topological concepts. Annual Review of Materials Science 5, 43–72.

Kruhl, J.H., Nega, M., 1996. The fractal shape of sutured quartz grain boundaries: application as a geothermometer. Geologische Rundschau 85, 38–43.

Kruhl, J.H., Peternell, M., 2002. The equilibration of high-angle grain boundaries in dynamically recrystallized quartz: the effect of crystallography and temperature. Journal of Structural Geology 24, 1125–1137.

Kruhl, J.H., 1994. The formation of extensional veins: an application of the Cantor-dust model. In: Kruhl, J.H. (Ed.), Fractals and Dynamic Systems in Geoscience. Springer-Verlag, Berlin, pp. 95–104.

Kuhlman-Wilsdorf, D., Hansen, N., 1991. Geometrically necessary, incidental and subgrain boundaries. Scripta Metallurgica et Materialia 25, 1557–1562.

Kuhn, M., 2009. Reactive Flow Modeling of Hydrothermal Systems. Lecture Notes in Earth Sciences. Springer Science & Business Media, 261 pp.

Kukkonen, I.T., Jokinen, J., Seipold, U., 1999. Temperature and pressure dependencies of thermal transport properties of rocks: implications for uncertainties in thermal lithosphere models and new laboratory measurements of high-grade rocks in the Central Fennoscandian Shield. Surveys in Geophysics 20, 33–59.

Kuntcheva, B., Kruhl, J.H., Kunze, K., 2006. Crystallographic orientations of high-angle grain boundaries in dynamically recrystallized quartz: first results. Tectonophysics 421, 331–346.

Lade, P.V., Duncan, J.M., 1975. Elastoplastic stress-strain theory for cohesionless soil. Journal of the Geotechnical Engineering Division, ASCE 101, 1037–1053.

Lade, P.V., 1977. Elastoplastic stress-strain theory for cohesionless soil with curved yield surfaces. International Journal of Solids and Structures 13, 1019–1035.

Lade, P.V., 1993. Rock strength criteria: the theories and the evidence. In: Hudson, J.A. (Ed.), Comprehensive Rock Engineering: Principles, Practice & Projects, vol. 1. Fundamentals, Pergamon, pp. 255–284.

Landsberg, P.T., 1961. Thermodynamics with Quantum Statistical Illustration. Interscience Publishers, New York.

Lapwood, E.R., 1948. Convection of a fluid in a porous medium. Proceedings of the Cambridge Philosophical Society 44, 508–521.

Larche, F., Cahn, J.W., 1973. Linear theory of thermochemical equilibrium of solids under stress. Acta Metallurgica 21, 1051–1063.

Larche, F.C., Cahn, J.W., 1978. Thermochemical equilibrium of multiphase solids under stress. Acta Metallurgica 26, 1579–1589.

Lasaga, A.C., Gibbs, G.V., 1990. Ab-initio quantum mechanical calculations of water-rock interactions; adsorption and hydrolysis reactions. American Journal of Science 290, 263–295.

Lasaga, A.C., 1981. Rate laws of chemical reactions. In: Lasaga, A.C., Kirkpatrick, R.J. (Eds.), Kinetics of Geochemical Processes. Rev. Mineral. Mineralogical Society of America, Washington, DC, pp. 1–67.

Lasaga, A.C., 1984. Chemical kinetics of water-rock interactions. Journal of Geophysical Research 89, 4009–4025.

Law, C.K., 2006. Combustion Physics. Cambridge University Press, Cambridge.

Leal, L.G., 2007. Advanced Transport Phenomena: Fluid Mechanics and Convective Transport Processes. Cambridge University Press, New York.

Lee, K.J., McCormick, W.D., Ouyang, Q., Swinney, H.L., 1993. Pattern-formation by interacting chemical fronts. Science 261, 192–194.

Lehner, F.K., 1995. A model for intergranular pressure solution in open systems. Tectonophysics 245, 153–170.

Lester, D.R., Metcalfe, G., Trefry, M.G., Ord, A., Hobbs, B., Rudman, M., 2009. Lagrangian topology of a periodically reoriented potential flow: symmetry, optimization, and mixing. Physical Review E 80.

Lester, D.R., Rudman, M., Metcalfe, G., Trefry, M.G., Ord, A., Hobbs, B., 2010. Scalar dispersion in a periodically reoriented potential flow: acceleration via Lagrangian chaos. Physical Review E 81.

Lester, D.R., Ord, A., Hobbs, B.E., 2012. The mechanics of hydrothermal systems: II. Fluid mixing and chemical reactions. Ore Geology Reviews 49, 45–71.

Levien, L., Prewitt, C.T., 1981. High-pressure crystal-structure and compressibility of coesite. American Mineralogist 66, 324–333.

Levin, I., 1946. The phase rule and topology. Journal of Chemical Education 23, 183–185.

Levinger, B.W., 1980. The square root of a 2×2 matrix. Mathematics Magazine 53, 222–224.

Li, J.C.M., 1972. Disclination model of high angle grain-boundaries. Surface Science 31, 12–26.

Li, W., Jun, D., Qingfei, W., 2009. Multifractal modeling and evaluation of inhomogeneity of metallogenic elements distribution in the Shangzhuang Deposit, Shandong Province, China. Computational Intelligence and Software Engineering (CiSE). International Conference, 4 pp.

Liebowitz, H., 1968. Fracture, an Advanced Treatise. Academic Press, New York, p. 759.

Liljenroth, F.G., 1918. Starting and stability phenomena of ammonia-oxidation and similar reactions. Chemical and Metallurgical Engineering 19, 287–293.

Lin, T.H., Ito, M., 1966. Theoretical plastic stress-strain relationship of a polycrystal and the comparisons with the von Mises and the Tresca plasticity theories. International Journal of Engineering Science 4, 543–561.

Lind, P.G., Baram, R.M., Herrmann, H.J., 2008. Obtaining the size distribution of fault gouges with polydisperse bearings. Physical Review E 77.

Lindgren, W., 1912. The nature of replacement. Economic Geology 7, 521–535.

Lindgren, W., 1918. Volume changes in metamorphism. Journal of Geology 26, 542–554.

Lister, G.S., Hobbs, B.E., 1980. The simulation of fabric development during plastic-deformation and its application to quartzite — the influence of deformation history. Journal of Structural Geology 2, 355–370.

Lister, G.S., Snoke, A.W., 1984. S-C mylonites. Journal of Structural Geology 6, 617–638.

Lister, G.S., Paterson, M.S., Hobbs, B.E., 1978. The simulation of fabric development in plastic deformation and its application to quartzite: the model. Tectonophysics 45, 107–158.

Liu, M., Yund, R.A., Tullis, J., Topor, L., Navrotsky, A., 1995. Energy associated with dislocations — a calorimetric study using synthetic quartz. Physics and Chemistry of Minerals 22, 67–73.

Lockner, D.A., Byerlee, J., 1980. Development of fracture planes during creep in granite. In: Hardy, H.R., Leighton, F.W. (Eds.), 2nd Conference on Acoustic Emission/Microseismic Activity in Geological Structures and Materials. Trans-Tech. Publications, Clausthal-Zellerfeld, pp. 11–25.

Lockner, D.A., Byerlee, J.D., Kuksenko, V., Ponomarev, A., Sidorin, A., 1991. Quasi-static fault growth and shear fracture energy in granite. Nature 350, 39–42.

Louat, N.P., 1974. Theory of normal grain-growth. Acta Metallurgica 22, 721–724.

Lyakhovsky, V., Myasnikov, V., 1984. On the behavior of elastic cracked solid. Physics of the Solid Earth 10, 71–75.

Lyakhovsky, V., Ben-Zion, Y., Agnon, A., 1997. Distributed damage, faulting, and friction. Journal of Geophysical Research-Solid Earth 102, 27635–27649.

Lyakhovsky, V., Hamiel, Y., Ben-Zion, Y., 2011. A non-local visco-elastic damage model and dynamic fracturing. Journal of the Mechanics and Physics of Solids 59, 1752–1776.

Lynch, D.T., Rogers, T.D., Wanke, S.E., 1982. Chaos in a continuous stirred tank reactor. Mathematical Modelling 3, 103–116.

Lynch, S., 2007. Dynamical Systems with Applications Using Mathematica®. Birkhauser.

Lyubetskaya, T., Ague, J.J., 2009. Modeling the magnitudes and directions of regional metamorphic fluid flow in collisional orogens. Journal of Petrology 50, 1505–1531.

MacDonald, G.J.F., 1960. Chapter 1: Orientation of anisotropic minerals in a stress field. Geological Society of America Memoirs 79, 1–8.

Mainprice, D., Munch, P., 1993. Quantitative texture analysis of an anorthosite — application to thermal expansion, Young's modulus and thermal stresses. Textures and Microstructures 21, 79–92.

Mainprice, D., Le Page, Y., Rodgers, J., Jouanna, P., 2007. Predicted elastic properties of the hydrous D phase at mantle pressures: Implications for the anisotropy of subducted slabs near 670-km discontinuity and in the lower mantle. Earth and Planetary Science Letters 259, 283–296.

Mainprice, D., Tommasi, A., Ferre, D., Carrez, P., Cordier, P., 2008. Predicted glide systems and crystal preferred orientations of polycrystalline silicate Mg-perovskite at high pressure: Implications for the seismic anisotropy in the lower mantle. Earth and Planetary Science Letters 271, 135–144.

Mainprice, D., Hielscher, R., Schaeben, H., 2011. Calculating anisotropic physical properties from texture data using the MTEX open-source package. Geological Society, London, Special Publications 360, 175–192.

Mallat, S., 1991. Zero-crossings of a wavelet transform. IEEE Transactions on Information Theory 37, 1019–1033.

Mallat, S., Hwang, W.L., 1992. Singularity detection and processing with wavelets. IEEE Transactions on Information Theory 38, 617–643.

Malvern, L.E., 1969. Introduction to the Mechanics of a Continuous Medium. Prentice-Hall, Inc., Englewood Cliffs, NJ.

Mamtani, M.A., Greiling, R.O., 2010. Serrated quartz grain boundaries, temperature and strain rate: testing fractal techniques in a syntectonic granite. In: Spalla, M.I., Marotta, A.M., Gosso, G. (Eds.), Advances in Interpretation of Geological Processes: Refinement of Multi-scale Data and Integration in Numerical Modelling, pp. 35–48.

Mamtani, M.A., 2010. Strain-rate estimation using fractal analysis of quartz grains in naturally deformed rocks. Journal of the Geological Society of India 75, 202–209.

Mancktelow, N.S., Pennacchioni, G., 2004. The influence of grain boundary fluids on the microstructure of quartz-feldspar mylonites. Journal of Structural Geology 26, 47–69.

Mancktelow, N.S., Grujic, D., Johnson, E.L., 1998. An SEM study of porosity and grain boundary microstructure in quartz mylonites, Simplon fault zone, central Alps. Contributions to Mineralogy and Petrology 131, 71–85.

Mancktelow, N., 2002. Finite element modelling of shear zone development in viscoelastic materials and its implications for localisation of partial melting. Journal of Structural Geology 24, 1045–1053.

Mandelbrot, B.B., 1974. Multiplications aleatoires iterees et distributions invariantes par moyenne ponderee aleatoire. Comptes Rendus des Seances de l'Academie des Sciences Paris 278A, pp. 289–292 & 355–358.

Mandelbrot, B., 1982. The Fractal Geometry of Nature. W.H. Freeman & co.

Mandl, G., 1988. Mechanics of Tectonic Faulting: Models and Basic Concepts. Elsevier.

Mandl, G., 2005. Rock Joints: The Mechanical Genesis. Springer, Berlin.

Mangasarian, O.L., 1994. Non-linear Programming. Society for Industrial and Applied Mathematics, Philadelphia.

Manning, C.E., Ingebritsen, S.E., 1999. Permeability of the continental crust: Implications of geothermal data and metamorphic systems. Reviews of Geophysics 37, 127–150.

Marder, M., Fineberg, J., 1996. How things break. Physics Today 49, 24–29.

Marigo, J.J., Truskinovsky, L., 2004. Initiation and propagation of fracture in the models of Griffith and Barenblatt. Continuum Mechanics and Thermodynamics 16, 391–409.

Massalski, T.B., Soffa, W.A., Laughlin, D.E., 2006. The nature and role of incoherent interphase interfaces in diffusional solid-solid phase transformations. Metallurgical and Materials Transactions: A-Physical Metallurgy and Materials Science 37A, 825–831.

Masumura, R.A., Hazzledine, P.M., Pande, C.S., 1998. Yield stress of fine grained materials. Acta Materialia 46, 4527–4534.

Matsuoka, H., Nakai, T., 1974. Stress-deformation and strength characteristics of soil under three different principal stresses. Proceedings of JSCE 232, 59–70.

Matthai, S.K., Fischer, G., 1996. Quantitative modeling of fault-fluid-discharge and fault-dilation-induced fluid-pressure variations in the seismogenic zone. Geology 24, 183–186.

Matthai, S.K., Roberts, S.G., 1997. Transient versus continuous fluid flow in seismically active faults: An investigation by electric analogue and numerical modelling. In: Jamtveit, B., Yardley, B.W.D. (Eds.), Fluid Flow and Transport in Rocks: Mechanisms and Effects. Springer, Netherlands, pp. 263–295.

Matzke, E.B., 1945. The three-dimensional shapes of bubbles in foams. Proceedings of the National Academy of Sciences of the United States of America 31, 281–289.

Maugin, G.A., 1999. Thermomechanics of Nonlinear Irreversible Behaviors: An Introduction. World Scientific Publ. Co., Singapore.

McClintock, F.A., Argon, A.S., 1966. Mechanical Behavior of Materials. Addison-Wesley Pub. Co.

McKenzie, D., 1979. Finite deformation during fluid flow. Geophysical Journal of the Royal Astronomical Society 58, 689–715.

McLaren, A.C., Phakey, P.P., 1965. Dislocations in quartz observed by transmission electron microscopy. Journal of Applied Physics 36, 3244–3246.

McLaren, A.C., 1986. Some speculations on the nature of high-angle grain boundaries in quartz rocks. In: Hobbs, B.E., Heard, H.C. (Eds.), Mineral and Rock Deformation: Laboratory Studies: The Paterson Volume. Geophysical Monograph Series 36, AGU, Washington, DC, pp. 233–245.

McLellan, A.G., 1980. The Classical Thermodynamics of Deformable Materials. Cambridge University Press, Cambridge.

Means, W.D., Hobbs, B.E., Lister, G.S., Williams, P.F., 1980. Vorticity and non-coaxiality in progressive deformations. Journal of Structural Geology 2, 371–378.

Means, W.D., 1976. Stress and Strain: Basic Concepts of Continuum Mechanics for Geologists. Springer-Verlag, New York.

Means, W.D., 1982. An unfamiliar Mohr circle construction for finite strain. Tectonophysics 89, T1–T6.

Means, W.D., 1983. Application of the Mohr-circle construction to problems of inhomogeneous deformation. Journal of Structural Geology 5, 279–286.

Mei, S., Suzuki, A.M., Kohlstedt, D.L., Dixon, N.A., Durham, W.B., 2010. Experimental constraints on the strength of the lithospheric mantle. Journal of Geophysical Research - Solid Earth 115, B08204, 9 pages.

Meredith, P.G., Main, I.G., Clint, O.C., Li, L., 2012. On the threshold of flow in a tight natural rock. Geophysical Research Letters 39, L04307.

Merino, E., Canals, A., 2011. Self-accelerating dolomite-for-calcite replacement: self-organized dynamics of burial dolomitization and associated mineralization. American Journal of Science 311, 573–607.

Metcalfe, G., Lester, D., Ord, A., Kulkarni, P., Rudman, M., Trefry, M., Hobbs, B., Regenauer-Lieb, K., Morris, J., 2010a. An experimental and theoretical study of the mixing characteristics of a periodically reoriented irrotational flow. Philosophical Transactions of the Royal Society A-Mathematical Physical and Engineering Sciences 368, 2147–2162.

Metcalfe, G., Lester, D., Ord, A., Kulkarni, P., Trefry, M., Hobbs, B.E., Regenauer-Lieb, K., Morris, J., 2010b. A partially open porous media flow with chaotic advection: towards a model of coupled fields. Philosophical Transactions of the Royal Society A: Mathematical, Physical and Engineering Sciences 368, 217–230.

Metcalfe, G., 2010. Applied fluid chaos: designing advection with periodically reoriented flows for micro to geophysical mixing and transport enhancement. In: Dewar, R.L., Detering, F. (Eds.), Complex Physical, Biophysical and Econophysical Systems. Proceedings of the 22nd Canberra International Physics Summer School. World Scientific, Canberra, pp. 187–239.

Meyers, M.A., Mishra, A., Benson, D.J., 2006. Mechanical properties of nanocrystalline materials. Progress in Materials Science 51, 427–556.

Mezic, I., Loire, S., Fonoberov, V.A., Hogan, P., 2010. A new mixing diagnostic and gulf oil spill movement. Science 330, 486–489.

Mikouchi, T., Takeda, H., Miyamoto, M., Ohsumi, K., McKay, G.A., 1995. Exsolution lamellae of kirschsteinite in magnesium-iron olivine from an angrite meteorite. American Mineralogist 80, 585–592.

Mikulecky, D.C., 2001. Network thermodynamics and complexity: a transition to relational systems theory. Computers and Chemistry 25, 369–391.

Miyazaki, T., Sueyoshi, K., Hiraga, T., 2013. Olivine crystals align during diffusion creep of Earth's upper mantle. Nature 502, 321–326.

Molinari, A., Canova, G.R., Ahzi, S., 1987. A self-consistent approach of the large deformation polycrystal viscoplasticity. Acta Metallurgica 35, 2983–2994.

Monnet, G., Devincre, B., Kubin, L.P., 2004. Dislocation study of prismatic slip systems and their interactions in hexagonal close packed metals: application to zirconium. Acta Materialia 52, 4317–4328.

Montagnat, M., Castelnau, O., Bons, P.D., Faria, S.H., Gagliardini, O., Gillet-Chaulet, F., Grennerat, F., Griera, A., Lebensohn, R.A., Moulinec, H., Roessiger, J., Suquet, P., 2014. Multiscale modeling of ice deformation behavior. Journal of Structural Geology 61, 78–108.

Montel, J.-M., Vielzeuf, D., 1997. Partial melting of metagreywackes, part II. Compositions of minerals and melts. Contributions to Mineralogy and Petrology 128, 176–196.

Morrey, C.B., 1952. Quasi-convexity and the lower semicontinuity of multiple integrals, Pacific Journal of Mathematics 2, pp. 25–53.

Mottaghy, D., Vosteen, H.-D., Schellschmidt, R., 2008. Temperature dependence of the relationship of thermal diffusivity versus thermal conductivity for crystalline rocks. International Journal of Earth Sciences 97, 435–442.

Muhlhaus, H.B., Chau, K.T., Ord, A., 1996. Bifurcation of crack pattern in arrays of two-dimensional cracks. International Journal of Fracture 77, 1–14.

Muller, I., Villaggio, P., 1977. Model for an elastic-plastic body. Archive for Rational Mechanics and Analysis 65, 25–46.

Mullins, W.W., 1956. 2-Dimensional motion of idealized grain boundaries. Journal of Applied Physics 27, 900–904.

Mullins, W.W., 1986. The statistical self-similarity hypothesis in grain-growth and particle coarsening. Journal of Applied Physics 59, 1341–1349.

Mullins, W.W., 1988. On idealized 2 dimensional grain-growth. Scripta Metallurgica 22, 1441–1444.

Munz, I.A., Yardley, B.W.D., Banks, D.A., Wayne, D., 1995. Deep penetration of sedimentary fluids in basement rocks from Southern Norway – evidence from hydrocarbon and brine inclusions in quartz veins. Geochimica et Cosmochimica Acta 59, 239–254.

Murphy, F.C., Ord, A., Hobbs, B.E., Willetts, G., Barnicoat, A.C., 2008. Targeting stratiform Zn–Pb–Ag massive sulfide deposits in Ireland through numerical modeling of coupled deformation, thermal transport, and fluid flow. Economic Geology 103, 1437–1458.

Murray, J.D., 1989. Mathematical Biology. Springer-Verlag, Berlin.

Murrell, S.A.F., 1963. A criterion for brittle fracture of rocks and concrete under triaxial stress and the effect of pore pressure on the criterion. In: Fairhurst, C. (Ed.), Rock Mechanics, Proceedings of the 5th U.S. Symposium on Rock Mechanics. Minneapolis, MN, Pergamon Press, Oxford, pp. 563–577.

Muzzio, F.J., Meneveau, C., Swanson, P.D., Ottino, J.M., 1992. Scaling and multifractal properties of mixing in chaotic flows. Physics of Fluids A-Fluid Dynamics 4, 1439–1456.

Naamane, S., Monnet, G., Devincre, B., 2010. Low temperature deformation in iron studied with dislocation dynamics simulations. International Journal of Plasticity 26, 84–92.

Nabarro, F.R.N., 1967. Theory of Crystal Dislocations. Clarendon Press, Oxford.

Nabarro, F.R.N., 2003. One-dimensional models of thermal activation under shear stress. Philosophical Magazine 83, 3047–3054.

Nadai, A., 1950. Theory of Flow and Fracture of Solids. McGraw-Hill, New York.

Nahon, D., Merino, E., 1997. Pseudomorphic replacement in tropical weathering: evidence, geochemical consequences, and kinetic-rheological origin. American Journal of Science 297, 393–417.

Naus-Thijssen, F.M.J., Goupee, A.J., Johnson, S.E., Vel, S.S., Gerbi, C., 2011. The influence of crenulation cleavage development on the bulk elastic and seismic properties of phyllosilicate-rich rocks. Earth and Planetary Science Letters 311, 212–224.

Needleman, A., Ortiz, M., 1991. Effect of boundaries and interfaces on shear-band localization. International Journal of Solids and Structures 28, 859–877.

Needleman, A., Tvergaard, V., Van Der Giessen, E., 1995. Evolution of void shape and size in creeping solids. International Journal of Damage Mechanics 4, 134–152.

von Neumann, J., 1952. Discussion. In: Herring, C. (Ed.), Metal Interfaces. American Society for Metals, Cleveland, pp. 108–110.

Nguyen, G.D., Einav, I., 2009. The energetics of cataclasis based on breakage mechanics. Pure and Applied Geophysics 166, 1693–1724.

Nield, D.A., Bejan, A., 2013. Convection in Porous Media, fourth ed. Springer, New York.

Niven, R.K., Andresen, B., 2010. Jaynes' maximum entropy principle, Riemannian metrics and generalised least action bound. In: Dewar, R.L., Detering, F. (Eds.), Complex Physical, Biophysical and Econophysical Systems. Proceedings of the 22nd Canberra International Physics Summer School. World Scientific Publishing Co. Pte. Ltd., Canberra, pp. 283–317.

Niven, R., Noack, B., 2014. Control volume analysis, entropy balance and the entropy production in flow systems. In: Dewar, R.C., Lineweaver, C.H., Niven, R.K., Regenauer-Lieb, K. (Eds.), Beyond the Second Law. Springer Berlin Heidelberg, pp. 129–162.

Niven, R.K., 2002. Physical insight into the Ergun and Wen & Yu equations for fluid flow in packed and fluidised beds. Chemical Engineering Science 57, 527–534.

Niven, R.K., 2009. Steady state of a dissipative flow-controlled system and the maximum entropy production principle. Physical Review E 80, 021113-1–02113-15.

Niven, R.K., 2010a. Minimization of a free-energy-like potential for non-equilibrium flow systems at steady state. Philosophical Transactions of the Royal Society B-Biological Sciences 365, 1323–1331.

Niven, R.K., 2010b. Simultaneous extrema in the entropy production for steady-state fluid flow in parallel pipes. Journal of Non-Equilibrium Thermodynamics 35, 347–378.

Nye, J.F., 1953. Some geometrical relations in dislocated crystals. Acta Metallurgica 1, 153–162.

Nye, J.F., 1957. Physical Properties of Crystals. Oxford Press.

Olbricht, W.L., Rallison, J.M., Leal, L.G., 1982. Strong flow criteria based on microstructure deformation. Journal of Non-Newtonian Fluid Mechanics 10, 291–318.

Oliver, N.H.S., Rubenach, M.J., Fu, B., Baker, T., Blenkinsop, T.G., Cleverley, J.S., Marshall, L.J., Ridd, P.J., 2006. Granite-related overpressure and volatile release in the mid crust: fluidized breccias from the Cloncurry District, Australia. Geofluids 6, 346–358.

Oliver, N.H.S., 1996. Review and classification of structural controls on fluid flow during regional metamorphism. Journal of Metamorphic Geology 14, 477–492.

Olsen, E.T., Bernstein, B., 1984. A class of hypoelastic non-elastic materials and their thermodynamics. Archive for Rational Mechanics and Analysis 86, 291–303.

Olson, J.E., Pollard, D.D., 1991. The initiation and growth of en échelon veins. Journal of Structural Geology 13, 595–608.

Onsager, L., Machlup, S., 1953. Fluctuations and irreversible processes. Physical Review 91, 1505–1512.

Onsager, L., 1931a. Reciprocal relations in irreversible processes. I. Physical Review 37, 405–426.

Onsager, L., 1931b. Reciprocal relations in irreversible processes. II. Physical Review 38, 2265–2279.

Ord, A., Christie, J.M., 1984. Flow stresses from microstructures in mylonitic quartzites of the moine thrust zone, assynt area, Scotland. Journal of Structural Geology 6, 639–654.

Ord, A., Oliver, N.H.S., 1997. Mechanical controls on fluid flow during regional metamorphism: some numerical models. Journal of Metamorphic Geology 15, 345–359.

Ord, A., Vardoulakis, I., Kajewski, R., 1991. Shear band formation in gosford sandstone. International Journal of Rock Mechanics and Mining Sciences and Geomechanics Abstracts 28, 397–409.

Ord, A., Hobbs, B.E., Lester, D.R., 2012. The mechanics of hydrothermal systems: I. Ore systems as chemical reactors. Ore Geology Reviews 49, 1–44.

Ord, A., 1988. Deformation texture development in geological materials. In: Kallend, J.S., Gottstein, G. (Eds.), Eighth International Conference on Textures of Materials (ICOTOM 8). The Metallurgical Society, Santa Fe, pp. 765–776.

Ord, A., 1994. The fractal geometry of patterned structures in numerical models for rock deformation. In: Kruhl, J.H. (Ed.), Fractals and Dynamic Systems in Geoscience. Springer-Verlag, Berlin, pp. 131–155.

Oreskes, N., Einaudi, M.T., 1990. Origin of rare earth element-enriched hematite breccias at the Olympic Dam Cu-U-Au-Ag deposit, Roxby Downs, South Australia. Economic Geology 85, 1–28.

Orowan, E., 1949. Fracture and strength of solids. Reports on Progress in Physics 12, 185.

Ortiz, M., Repetto, E.A., 1999. Nonconvex energy minimization and dislocation structures in ductile single crystals. Journal of the Mechanics and Physics of Solids 47, 397–462.

Ortiz, M., Repetto, E.A., Stainier, L., 2000. A theory of subgrain dislocation structures. Journal of the Mechanics and Physics of Solids 48, 2077–2114.

Ortiz, M., 1999. Plastic yielding as a phase transition. Journal of Applied Mechanics 66, 289–298.

Ortoleva, P., Chadam, J., Merino, E., Sen, A., 1987a. Geochemical self-organization II: the reactive-infiltration instability. American Journal of Science 287, 1008–1040.

Ortoleva, P., Merino, E., Moore, C., Chadam, J., 1987b. Geochemical self-organization I: reaction-transport feedbacks and modelling approach. American Journal of Science 287, 979–1007.

Ortoleva, P.J., 1994. Geochemical Self-Organisation. Oxford University Press, Oxford.

Oster, G., Perelson, A., Katchals, A., 1971. Network thermodynamics. Nature 234, 393–399.

Ott, E., 1993. Chaos in Dynamical Systems. Cambridge University Press, Cambridge.

Ottino, J.M., Wiggins, S., 2004. Introduction: mixing in microfluidics. Philosophical Transactions of the Royal Society of London Series, A: Mathematical Physical and Engineering Sciences 362, 923–935.

Ottino, J.M., 1989a. The Kinematics of Mixing: Stretching, Chaos, and Transport. Cambridge University Press, Cambridge, UK.

Ottino, J.M., 1989b. The mixing of fluids. Scientific American 260, 56–67.

Ottino, J.M., 1990. Mixing, chaotic advection, and turbulence. Annual Review of Fluid Mechanics 22, 207–253.

Ozawa, H., Ohmura, A., 1997. Thermodynamics of a global-mean state of the atmosphere — a state of maximum entropy increase. Journal of Climate 10, 441–445.

Ozawa, H., Ohmura, A., Lorenz, R.D., Pujol, T., 2003. The second law of thermodynamics and the global climate system: a review of the maximum entropy production principle. Reviews of Geophysics 41, 1018.

Packard, N.H., Crutchfield, J.P., Farmer, J.D., Shaw, R.S., 1980. Geometry from a time-series. Physical Review Letters 45, 712–716.

Paltridge, G.W., Farquhar, G.D., Cuntz, M., 2007. Maximum entropy production, cloud feedback, and climate change. Geophysical Research Letters 34, L14708.

Paltridge, G.W., 1975. Global dynamics and climate — system of minimum entropy exchange. Quarterly Journal of the Royal Meteorological Society 101, 475—484.

Paltridge, G.W., 1978. The steady-state format of global climate. Quarterly Journal of the Royal Meteorological Society 104, 927—945.

Paltridge, G.W., 1981. Thermodynamic dissipation and the global climate system. Quarterly Journal of the Royal Meteorological Society 107, 531—547.

Paltridge, G.W., 2001. A physical basis for a maximum of thermodynamic dissipation of the climate system. Quarterly Journal of the Royal Meteorological Society 127, 305—313.

Pande, C.S., 1987. On a stochastic-theory of grain-growth. Acta Metallurgica 35, 2671—2678.

Pantleon, W., 2008. Resolving the geometrically necessary dislocation content by conventional electron backscattering diffraction. Scripta Materialia 58, 994—997.

de Paor, D.G., 1983. Orthographic analysis of geological structures—I. Deformation theory. Journal of Structural Geology 5, 255—277.

Pask, J.A., 1987. From technology to the science of glass metal and ceramic metal sealing. American Ceramic Society Bulletin 66, 1587—1592.

Passchier, C.W., Simpson, C., 1986. Porphyroclast systems as kinematic indicators. Journal of Structural Geology 8, 831—843.

Passchier, C.W., Trouw, R.A.J., 1996. Microtectonics. Springer-Verlag, Berlin.

Passchier, C.W., Trouw, R.A.J., 2005. Microtectonics, second ed. Springer, Berlin.

Passchier, C.W., Urai, J.L., 1988. Vorticity and strain analysis using Mohr diagrams. Journal of Structural Geology 10, 755—763.

Passchier, C., 1988. Analysis of deformation paths in shear zones. Geologische Rundschau 77, 309—318.

Passchier, C., 1990. Reconstruction of deformation and flow parameters from deformed vein sets. Tectonophysics 180, 185—199.

Passchier, C.W., 1991. The classification of dilatant flow types. Journal of Structural Geology 13, 101—104.

Passchier, C.W., 1994. Mixing in flow perturbations: a model for development of mantled porphyroclasts in mylonites. Journal of Structural Geology 16, 733—736.

Passchier, C.W., 1997. The fabric attractor. Journal of Structural Geology 19, 113—127.

Paterson, M.S., Weiss, L.E., 1961. Symmetry concepts in the structural analysis of deformed rocks. Geological Society of America Bulletin 72, 841—882.

Paterson, M.S., Wong, T.-F., 2005. Experimental Rock Deformation — The Brittle Field, second ed. Springer, Berlin.

Paterson, M.S., 1969. The ductility of rocks. In: Argon, A.S. (Ed.), Physics of Strength and Plasticity. M.I.T. Press, Cambridge, MA, pp. 377—392.

Paterson, M.S., 1973. Nonhydrostatic thermodynamics and its geologic applications. Reviews of Geophysics and Space Physics 11, 355—389.

Paterson, M.S., 1978. Experimental Rock Deformation — The Brittle Field. Springer-Verlag, Berlin.

Paterson, M.S., 1987. Problems in the extrapolation of laboratory rheological data. Tectonophysics 133, 33—43.

Paterson, M.S., 1995. A theory for granular flow accommodated by material transfer via an intergranular fluid. Tectonophysics 245, 135—151.

Paterson, M.S., 2001. Relating experimental and geological theology. International Journal of Earth Sciences 90, 157—167.

Paterson, M.S., 2013. Materials Science for Structural Geology. Springer, Dordrecht.

Patton, R., Watkinson, A., 2005. A viscoelastic strain energy principle expressed in fold—thrust belts and other compressional regimes. Journal of Structural Geology 27, 1143—1154.

Patton, R.L., Watkinson, A.J., 2010. Shear localization in solids: insights for mountain building processes from a frame-indifferent ideal material model. Geological Society, London, Special Publications 335, 739—766.

Patton, R.L., Watkinson, A.J., 2013. Deformation localization in orogens: spatiotemporal expression and thermodynamic constraint. Journal of Structural Geology 50, 221—236.

Pearson, J.E., 1993. Complex patterns in a simple system. Science 261, 189—192.

Peletier, L.A., Troy, W.C., 2001. Spatial Patterns: Higher Order Models in Physics and Mechanics. In: Progress in NonLinear Differential Equations and Their Applications, vol. 45. Birkhauser, 320 pp.

Perez-Reche, F.-J., Truskinovsky, L., Zanzotto, G., 2008. Driving-induced crossover: from classical criticality to self-organized criticality. Physical Review Letters 101, 230601.

Person, M., Baumgartner, L., 1995. New evidence for long-distance fluid migration within the earth's crust. Reviews of Geophysics 33, 1083–1091.

Petch, N.J., 1953. The cleavage strength of polycrystals. Journal of the Iron and Steel Institute 174, 25–28.

Petersen, K.B., Pedersen, M.S., 2008. The Matrix Cookbook. http://matrixcookbook.com.

Peto, P., 1976. An experimental investigation of melting relations involving muscovite and paragonite in the silica-saturated portion of the system $K_2O–Na_2O–Al_2O_3–SiO_2–H_2O$ to 15 kb total pressure. Progress in Experimental Petrology 3, 41–45.

Peusner, L., 1986. Hierarchies of energy-conversion processes. 3. Why are Onsager equations reciprocal – the euclidean geometry of fluctuation-dissipation space. Journal of Theoretical Biology 122, 125–155.

Phan-Thien, N., 2013. Understanding Viscoelasticity, second ed. Springer, New York.

Phillips, O.M., 1990. Flow-controlled reactions in rock fabrics. Journal of Fluid Mechanics 212, 263–278.

Phillips, O.M., 1991. Flow and Reactions in Permeable Rocks. Cambridge University Press, Cambridge.

Phillips, O.M., 2009. Geological Fluid Dynamics: Sub-surface Flow and Reactions. Cambridge University Press, Cambridge.

Pieri, M., Burlini, L., Kunze, K., Stretton, I., Olgaard, D.L., 2001. Rheological and microstructural evolution of Carrara marble with high shear strain: results from high temperature torsion experiments. Journal of Structural Geology 23, 1393–1413.

Pipkin, A.C., 1993. Convexity conditions for strain-dependent energy functions for membranes. Archive for Rational Mechanics and Analysis 121, 361–376.

Plasson, R., Brandenburg, A., Jullien, L., Bersini, H., 2011. Autocatalyses. Journal of Physical Chemistry A 115, 8073–8085.

Platt, J.P., Behr, W.M., 2011. Grainsize evolution in ductile shear zones: Implications for strain localization and the strength of the lithosphere. Journal of Structural Geology 33, 537–550.

Plohr, J.N., 2011. Equilibrium conditions at a solid-solid interface. Hindawi Publishing Corporation. Journal of Thermodynamics. 940385. 10 pp.

Plotnick, R.E., Gardner, R.H., Hargrove, W.W., Prestegaard, K., Perlmutter, M., 1996. Lacunarity analysis: a general technique for the analysis of spatial patterns. Physical Review E 53, 5461–5468.

Poirier, J.P., Nicolas, A., 1975. Deformation-induced recrystallization due to progressive misorientation of subgrains, with special reference to Mantle peridotites. Journal of Geology 83, 707–720.

Poirier, J.-P., 1985. Creep of Crystals. Cambridge University Press, Cambridge, UK.

Pollard, D.D., Aydin, A., 1988. Progress in understanding jointing over the past century. Geological Society of America Bulletin 100, 1181–1204.

Pontes, J., Walgraef, D., Aifantis, E., 2006. On dislocation patterning: multiple slip effects in the rate equation approach. International Journal of Plasticity 22, 1486–1505.

Powell, R., Downes, J., 1990. Garnet porphyroblast-bearing leucosomes in metapelites: mechanisms, phase diagrams, and an example from Broken Hill, Australia. In: Ashworth, J.R., Brown, M. (Eds.), High-Temperature Metamorphism and Crustal Anatexis. Springer, Netherlands, pp. 105–123.

Powell, R., Holland, T., Worley, B., 1998. Calculating phase diagrams involving solid solutions via non-linear equations, with examples using THERMOCALC. Journal of Metamorphic Geology 16, 577–588.

Powell, R., Guiraud, M., White, R.W., 2005. Truth and beauty in metamorphic phase-equilibria: conjugate variables and phase diagrams. The Canadian Mineralogist 43, 21–33.

Powell, R., 1978. Equilibrium Thermodynamics in Petrology. Harper and Row, New York.

Prigogine, I., 1955. Introduction to Thermodynamics of Irreversible Processes. John Wiley & Sons, New York.

Prior, D.J., Boyle, A.P., Brenker, F., Cheadle, M.C., Day, A., Lopez, G., Peruzzo, L., Potts, G.J., Reddy, S., Spiess, R., Timms, N.E., Trimby, P., Wheeler, J., Zetterstrom, L., 1999. The application of electron backscatter diffraction and orientation contrast imaging in the SEM to textural problems in rocks. American Mineralogist 84, 1741–1759.

Puglisi, G., Truskinovsky, L., 2000. Mechanics of a discrete chain with bi-stable elements. Journal of the Mechanics and Physics of Solids 48, 1–27.

Puglisi, G., Truskinovsky, L., 2002. A mechanism of transformational plasticity. Continuum Mechanics and Thermodynamics 14, 437–457.

Puglisi, G., Truskinovsky, L., 2005. Thermodynamics of rate-independent plasticity. Journal of the Mechanics and Physics of Solids 53, 655—679.

Putnis, A., Austrheim, H., 2010. Fluid-induced processes: metasomatism and metamorphism. Geofluids 10, 254—269.

Putnis, A., John, T., 2010. Replacement processes in the Earth's crust. Elements 6, 159—164.

Putnis, C.V., Mezger, K., 2004. A mechanism of mineral replacement: Isotope tracing in the model system KCl—KBr—H$_2$O. Geochimica et Cosmochimica Acta 68, 2839—2848.

Putnis, A., Putnis, C.V., 2007. The mechanism of reequilibration of solids in the presence of a fluid phase. Journal of Solid State Chemistry 180, 1783—1786.

Putnis, C.V., Tsukamoto, K., Nishimura, Y., 2005. Direct observations of pseudomorphism: compositional and textural evolution at a fluid-solid interface. American Mineralogist 90, 1909—1912.

Putnis, C.V., Kowacz, M., Putnis, A., 2008. The mechanism and kinetics of DTPA-promoted dissolution of barite. Applied Geochemistry 23, 2778—2788.

Putnis, A., 2002. Mineral replacement reactions: from macroscopic observations to microscopic mechanisms. Mineralogical Magazine 66, 689—708.

Putnis, A., 2009. Mineral replacement reactions. In: Oelkers, E.H., Schott, J. (Eds.), Thermodynamics and Kinetics of Water-Rock Interaction. Mineralogical Soc Amer, Chantilly, pp. 87—124.

Qingfei, W., Jun, D., Li, W.A.N., Jie, Z., Qingjie, G., Liqiang, Y., Lei, Z., Zhijun, Z., 2008. Multifractal analysis of element distribution in skarn-type deposits in the Shizishan Orefield, Tongling Area, Anhui Province, China. Acta Geologica Sinica - English Edition - 82, 896—905.

Raffensperger, J.P., Garven, G., 1995a. The formation of unconformity-type uranium ore deposits. 1. Coupled groundwater flow and heat transport modeling. American Journal of Science 295, 581—636.

Raffensperger, J.P., Garven, G., 1995b. The formation of unconformity-type uranium ore deposits. 2. Coupled hydrochemical modeling. American Journal of Science 295, 639—696.

Raffler, N., Howe, J.M., 2006. A high-resolution time-resolved study of incoherent interface motion during the massive transformation in TiAl alloy. Metallurgical and Materials Transactions: A-Physical Metallurgy and Materials Science 37A, 873—878.

Rajagopal, K.R., Srinivasa, A.R., 2004. On thermomechanical restrictions of continua. Proceedings of the Royal Society of London A: Mathematical, Physical and Engineering Science 460, 631—651.

Rajagopal, K.R., Srinivasa, A.R., 2007. On the response of non-dissipative solids. Proceedings of the Royal Society A: Mathematical Physical and Engineering Sciences 463, 357—367.

Rajagopal, K.R., Srinivasa, A.R., 2009. On a class of non-dissipative materials that are not hyperelastic. Proceedings of the Royal Society A: Mathematical Physical and Engineering Sciences 465, 493—500.

Rajagopal, K.R., 1995. Multiple Configurations in Continuum Mechanics. Computational and Applied Mechanics. University of Pittsburgh, Pittsburgh, PA. Report 6.

Ralser, S., Hobbs, B.E., Ord, A., 1991. Experimental deformation of a quartz mylonite. Journal of Structural Geology 13, 837—850.

Ramberg, H., 1963. Fluid dynamics of viscous buckling applicable to folding of layered rocks. Bulletin of the American Association of Petroleum Geologists 47, 484—505.

Ramberg, H., 1975. Particle paths, displacement and progressive strain applicable to rocks. Tectonophysics 28, 1—37.

Ramsay, J.G., Graham, R.H., 1970. Strain variation in shear belts. Canadian Journal of Earth Science 7, 786—813.

Ramsay, J.G., Huber, M.I., 1983. The Techniques of Modern Structural Geology. In: Strain Analysis, vol. 1. Academic Press Limited, London, UK. Reprinted 2003.

Ramsay, J.G., 1962. The geometry and mechanics of formation of "similar" type folds. The Journal of Geology 70, 309—327.

Ramsay, J.G., 1967. Folding and Fracturing of Rocks. McGraw-Hill, New York.

Ramsay, J.G., 1969. The measurement of strain and displacement in orogenic belts. In: Kent, P.E., Salter-Thwaite, G.E., Spencer, A.M. (Eds.), Time and Place in Orogeny, pp. 43—79.

Ramsay, J.G., 1980a. The crack-seal mechanism of rock deformation. Nature 284, 135—139.

Ramsay, J.G., 1980b. Shear zone geometry — a review. Journal of Structural Geology 2, 83—99.

Ranalli, G., 1984. Grain-size distribution and flow-stress in tectonites. Journal of Structural Geology 6, 443—447.

Ranalli, G., 1995. Rheology of the Earth, second ed. Chapman & Hall, London.

Raphanel, J., 2011. Three-dimensional morphology evolution of solid-fluid interfaces by pressure solution. In: Leroy, Y., Lehner, F. (Eds.), Mechanics of Crustal Rocks. Springer, Vienna, pp. 127—155.

Razumovskiy, V.I., Ruban, A.V., Korzhavyi, P.A., 2011. Effect of temperature on the elastic anisotropy of pure Fe and $Fe_{0.9} Cr_{0.1}$ random Alloy. Physical Review Letters 107, 205504.

Reches, Z., Lockner, D.A., 1994. Nucleation and growth of faults in brittle rocks. Journal of Geophysical Research 99, 18159–18173.

Reed, M.H., 1997. Hydrothermal alteration and its relationship to ore fluid compositions. In: Barnes, H.L. (Ed.), Geochemistry of Hydrothermal Ore Deposits. John Wiley & Sons, Inc., New York, pp. 303–365.

Reed, M.H., 1998. Calculation of simultaneous chemical equilibrium in aqueous-mineral-gas systems and its application to modeling hydrothermal processes. In: Richard, J.P., Larson, P.B. (Eds.), Techniques in Hydrothermal Ore Deposits Geology. Society of Economic Geologists, Colorado, pp. 109–123.

Regenauer-Leib, K., Yuen, D.A., 2003. Modeling shear zones in geological and planetary sciences: solid- and fluid-thermal-mechanical approaches. Earth-Science Reviews 63, 295–349.

Regenauer-Lieb, K., Yuen, D., 2004. Positive feedback of interacting ductile faults from coupling of equation of state, rheology and thermal-mechanics. Physics of the Earth and Planetary Interiors 142, 113–135.

Regenauer-Lieb, K., Hobbs, B., Ord, A., Gaede, O., Vernon, R., 2009. Deformation with coupled chemical diffusion. Physics of the Earth and Planetary Interiors 172, 43–54.

Regenauer-Lieb, K., 1998. Dilatant plasticity applied to Alpine collision: ductile void growth in the intraplate area beneath the Eifel volcanic field. Journal of Geodynamics 27, 1–21.

Renner, J., Evans, B., Siddiqi, G., 2002. Dislocation creep of calcite. Journal of Geophysical Research 107 (B12), 2364.

Reuss, A., 1929. Account of the liquid limit of mixed crystals on the basis of the plasticity condition for single crystal. Zeitschrift fur Angewandte Mathematik und Mechanik 9, 49–58.

Ricard, Y., Bercovici, D., 2009. A continuum theory of grain size evolution and damage. Journal of Geophysical Research-Solid Earth 114, B01204.

Rice, J.R., 1968. A path independent integral and approximate analysis of strain concentration by notches and cracks. Journal of Applied Mechanics 35, 379–386.

Rice, J.R., 1970. On the structure of stress-strain relations for time-dependent plastic deformation in metals. Journal of Applied Mechanics 37, 728–737.

Rice, J.R., 1971. Inelastic constitutive relations for solids: an internal-variable theory and its application to metal plasticity. Journal of the Mechanics and Physics of Solids 19, 433–455.

Rice, J.R., 1975. Continuum mechanics and thermodynamics of plasticity in relation to microscale deformation mechanisms. In: Argon, A.S. (Ed.), Constitutive Equations in Plasticity. MIT Press, Cambridge, MA, pp. 23–79.

Rice, J.R., 1976. The localization of plastic deformation. In: Koiter, W.T. (Ed.), 14th International Congress of Theoretical and Applied Mechanics. North-Holland Publishing Co., Delft, pp. 207–220.

Rice, J.R., 1993. Mechanics of solids. In: Encyclopaedia Britannica, fifteenth ed., pp. 734–747, 773.

Richardson, L.F., 1922. Weather Prediction by Numerical Process. Cambridge University Press.

Ridge, J., 1949. Replacement and the equating of volume and weight. Journal of Geology 57, 522–550.

Ridley, J., 1993. The relations between mean rock stress and fluid flow in the crust: with reference to vein- and lode-style gold deposits. Ore Geology Reviews 8, 23–37.

Riedi, R.H., 1998. Multifractals and wavelets: A potential tool in geophysics. In: Proceedings of the 68th Annual International Meeting of the Society of Exploration Geophysicists. Society of Exploration Geophysicists. New Orleans, LA, 4 pp.

Risken, H., 1996. The Fokker–Planck Equation, second ed. Springer-Verlag, Berlin. 3rd Printing ed.

Robin, P.Y.F., Cruden, A.R., 1994. Strain and vorticity patterns in ideally ductile transpression zones. Journal of Structural Geology 8, 831–844.

Robin, P.Y.F., 1974. Thermodynamic-equilibrium across a coherent interface in a stressed crystal. American Mineralogist 59, 1286–1298.

Robyr, M., Carlson, W.D., Passchier, C., Vonlanthen, P., 2009. Microstructural, chemical and textural records during growth of snowball garnet. Journal of Metamorphic Geology 27, 423–437.

Roe, R.J., 1965. Description of crystallite orientation in polycrystalline materials. III. General solution to pole figure inversion. Journal of Applied Physics 36, 2024–2031.

Romanov, A.E., Aifantis, E.C., 1993. On the kinetic and diffusional nature of linear defects. Scripta Metallurgica et Materialia 29, 707–712.

Romanov, A.E., Kolesnikova, A.L., 2009. Application of disclination concept to solid structures. Progress in Materials Science 54, 740–769.

Romanov, A.E., Kolesnikova, A.L., Ovid'ko, I.A., Aifantis, E.C., 2009. Disclinations in nanocrystalline materials: manifestation of the relay mechanism of plastic deformation. Materials Science and Engineering: A-Structural Materials Properties Microstructure and Processing 503, 62–67.

Ross, J., Villaverde, A.F., 2010. Thermodynamics and fluctuations far from equilibrium. Entropy 12, 2199–2243.

Ross, J., Vlad, M.O., 2005. Exact solutions for the entropy production rate of several irreversible processes. Journal of Physical Chemistry A 109, 10607–10612.

Ross, J., Hunt, K.L.C., Hunt, P.M., 1988. Thermodynamics far from equilibrium – reactions with multiple stationary states. Journal of Chemical Physics 88, 2719–2729.

Ross, J., Corlan, A.D., Müller, S.C., 2012. Proposed principles of maximum local entropy production. The Journal of Physical Chemistry B 116, 7858–7865.

Ross, J., 2008. Thermodynamics and Fluctuations Far from Equilibrium. Springer-Verlag, Berlin.

Roters, F., Eisenlohr, P., Hantcherli, L., Tjahjanto, D.D., Bieler, T.R., Raabe, D., 2010. Overview of constitutive laws, kinematics, homogenization and multiscale methods in crystal plasticity finite-element modeling: theory, experiments, applications. Acta Materialia 58, 1152–1211.

Roux, S.G., Arneodo, A., Decoster, N., 2000. A wavelet-based method for multifractal image analysis. III Applications to high-resolution satellite images of cloud structure. European Physical Journal B 15, 765–786.

Roy, S., Raju, R., Chuang, H.F., Cruden, B.A., Meyyappan, M., 2003. Modeling gas flow through microchannels and nanopores. Journal of Applied Physics 93, 4870–4879.

Rozel, A., Ricard, Y., Bercovici, D., 2011. A thermodynamically self-consistent damage equation for grain size evolution during dynamic recrystallization. Geophysical Journal International 184, 719–728.

Rubie, D.C., Thompson, A.B., 1985. Kinetics of metamorphic reactions at elevated temperatures and pressures: an appraisal of available experimental data. In: Thompson, A., Rubie, D. (Eds.), Metamorphic Reactions. Springer, New York, pp. 27–79.

Rubie, D.C., 1998. Disequilibrium during metamorphism: the role of nucleation kinetics. In: Taylor, P.J., O'Brien, P.J. (Eds.), What Drives Metamorphism and Metamorphic Reactions? pp. 199–214.

Rudnicki, J.W., Rice, J.R., 1975. Conditions for the localization of deformation in pressure-sensitive dilatant materials. Journal of the Mechanics and Physics of Solids 23, 371–394.

Rushmer, T., 2001. Volume change during partial melting reactions: implications for melt extraction, melt geochemistry and crustal rheology. Tectonophysics 342, 389–405.

Rutland, R.W.R., 1965. Tectonic overpressures. In: Pitcher, W.S., Flynn, G.W. (Eds.), Controls of Metamorphism. Oliver and Boyd, Edinburgh, pp. 119–139.

Rutter, E.H., Brodie, K.H., 1988. The role of tectonic grain-size reduction in the rheological stratification of the lithosphere. Geologische Rundschau 77, 295–308.

Rutter, E.H., Casey, M., Burlini, L., 1994. Preferred crystallographic orientation development during the plastic and superplastic flow of calcite rocks. Journal of Structural Geology 16, 1431–1446.

Rutter, E.H., 1972. Influence of interstitial water on rheological behavior of calcite rocks. Tectonophysics 14, 13–33.

Rutter, E.H., 1976. Kinetics of rock deformation by pressure solution. Philosophical Transactions of the Royal Society A-Mathematical Physical and Engineering Sciences 283, 203–219.

Rutter, E.H., 1995. Experimental study of the influence of stress, temperature, and strain on the dynamic recrystallization of Carrara marble. Journal of Geophysical Research-Solid Earth 100, 24651–24663.

Salman, O.U., Truskinovsky, L., 2012. On the critical nature of plastic flow: one and two dimensional models. International Journal of Engineering Science 59, 219–254.

Sammis, C., King, G., Biegel, R., 1987. The kinematics of gouge deformation. Pure and Applied Geophysics 125, 777–812.

Sander, B., 1911. Uber Zusammenhange zwischen Teilbewegung und Gefuge in Gesteinen. Tschermaks mineralogische und petrographische Mitteilungen 30, 281–315.

Sander, B., 1930. Gefügekunde der Gesteine mit besonderer Berücksichtigung der Tektonite. Springer, Vienna.

Sander, B., 1948. Einfuhrung in die Gefugekunde der geologischen Korper. 1. Allgemeine Gefugekunde und Arbeiten im Bereich Handstuck bis Profil. Springer, Vienna.

Sander, B., 1950. Einfuhrung in die Gefugekunde geologischer Korper, II: Die Korngefuge. Springer, Vienna.

Sander, B., 1970. An Introduction to the Study of Fabrics of Geological Bodies (Phillips, F.C., Windsor, G., trans.). Pergamon Press, Oxford.

Sanderson, D.J., Roberts, S., Gumiel, P., Greenfield, C., 2008. Quantitative analysis of tin- and tungsten-bearing sheeted vein systems. Economic Geology 103, 1043–1056.

Sandiford, M., McLaren, S., 2002. Tectonic feedback and the ordering of heat producing elements within the continental lithosphere. Earth and Planetary Science Letters 204, 133–150.

Sass, J.H., Lachenbruch, A.H., Moses, T.H., Morgan, P., 1992. Heat-flow from a scientific-research well at Cajon Pass, California. Journal of Geophysical Research-Solid Earth 97, 5017–5030.

Sato, E., Kuribayashi, K., Horiuchi, R., 1990. A mechanism of superplastic deformation and deformation induced grain growth based on grain switching. MRS Online Proceedings Library 196. In: Mayo, M.J., Kobayashi, M., Wadsworth, J. (Eds.), Superplasticity in Metals, Ceramics and Intermetallics, 196, 27–32.

Sawyer, E.W., 2008. Atlas of Migmatites. NRC Research Press.

Schiller, C., Walgraef, D., 1988. Numerical-simulation of persistent slip band formation. Acta Metallurgica 36, 563–574.

Schlegel, V., 1883. Theorie der homogen zusammengesetzten Raumgebilde. Nova Acta, der Ksl. Leop.-Carol. Deutschen Akademie der Naturforscher XLIV, 337–459.

Schmalholz, S.M., Podladchikov, Y.Y., 2013. Tectonic overpressure in weak crustal-scale shear zones and implications for the exhumation of high-pressure rocks. Geophysical Research Letters 40, 1984–1988.

Schmid, S.M., Casey, M., 1986. Complete fabric analysis of some commonly observed quartz c-axis patterns. In: Hobbs, B.E., Heard, H.C. (Eds.), Mineral and Rock Deformation: Laboratory Studies: The Paterson Volume. Geophysical Monograph Series 36, AGU, Washington, DC, pp. 263–286.

Schmid, S.M., Paterson, M.S., Boland, J.N., 1980. High temperature flow and dynamic recrystallization in Carrara marble. Tectonophysics 65, 245–280.

Schodde, R.C., Hronsky, J.M.A., 2006. The role of world-class mines in wealth creation. In: Doggett, M.D., Parry, J.R. (Eds.), Wealth Creation in the Minerals Industry: Integrating Science, Business, and Education, vol. 12. Society of Economic Geologists Special Publications Series, pp. 71–90.

Schreiber, I., Ross, J., 2003. Mechanisms of oscillatory reactions deduced from bifurcation diagrams. Journal of Physical Chemistry A 107, 9846–9859.

Schroeder, M., 1991. Fractals, Chaos, Power Laws. W.H. Freeman, New York.

Schutjens, P., Spiers, C.J., 1999. Intergranular pressure solution in NaCl: grain-to-grain contact experiments under the optical microscope. Oil and Gas Science and Technology − Revue d'IFP Energies Nouvelles 54, 729–750.

Schwarz, J.O., Engi, M., Berger, A., 2011. Porphyroblast crystallization kinetics: the role of the nutrient production rate. Journal of Metamorphic Geology 29, 497–512.

Scott, S.K., 1994. Oscillations, Waves, and Chaos in Chemical Kinetics. Oxford University Press, Oxford.

Seefeldt, M., Aifantis, E.C., 2002. Disclination patterning under steady-state creep at intermediate temperatures. In: Klimanek, P., Romanov, A.E., Seefeldt, M. (Eds.), Local Lattice Rotations and Disclinations in Microstructures of Distorted Crystalline Materials, pp. 221–226.

Seefeldt, M., Klimanek, P., 1997. Interpretation of plastic deformation by means of dislocation-disclination reaction kinetics. Materials Science and Engineering: A-Structural Materials Properties Microstructure and Processing 234, 758–761.

Seefeldt, M., Klimanek, P., 1998. Modelling of flow behaviour of metals by means of a dislocation-disclination reaction kinetics. Modelling and Simulation in Materials Science and Engineering 6, 349–360.

Seefeldt, M., Delannay, L., Peeters, B., Aernoudt, E., Van Houtte, P., 2001a. Modelling the initial stage of grain subdivision with the help of a coupled substructure and texture evolution algorithm. Acta Materialia 49, 2129–2143.

Seefeldt, M., Delannay, L., Peeters, B., Kalidindi, S.R., Van Houtte, P., 2001b. A disclination-based model for grain subdivision. Materials Science and Engineering: A-Structural Materials Properties Microstructure and Processing 319, 192–196.

Seefeldt, M., 2001. Disclinations in large-strain plastic deformation and work-hardening. Reviews on Advanced Materials Science 2, 44–79.

Sheldon, H.A., Ord, A., 2005. Evolution of porosity, permeability and fluid pressure in dilatant faults post-failure: implications for fluid flow and mineralization. Geofluids 5, 272–288.

Sheldon, H., Florio, B., Trefry, M., Reid, L., Ricard, L., Ghori, K.A., 2012. The potential for convection and implications for geothermal energy in the Perth Basin, Western Australia. Hydrogeology Journal 20, 1251–1268.

Sherwood, D.J., Hamilton, C.H., 1991. A mechanism for deformation-enhanced grain-growth in single-phase materials. Scripta Metallurgica et Materialia 25, 2873–2878.

Sherwood, D.J., Hamilton, C.H., 1992. Production of cellular defects contributing to deformation-enhanced grain-growth. Scripta Metallurgica et Materialia 27, 1771–1776.

Sherwood, D.J., Hamilton, C.H., 1994. The neighbor-switching mechanism of superplastic deformation — the constitutive relationship and deformation-enhanced grain-growth. Philosophical Magazine A-Physics of Condensed Matter Structure Defects and Mechanical Properties 70, 109—143.

Shimizu, I., 1992. Nonhydrostatic and nonequilibrium thermodynamics of deformable materials. Journal of Geophysical Research-Solid Earth 97, 4587—4597.

Shimizu, I., 1995. Kinetics of pressure solution creep in quartz — theoretical considerations. Tectonophysics 245, 121—134.

Shimizu, I., 1997. The non-equilibrium thermodynamics of intracrystalline diffusion under non-hydrostatic stress. Philosophical Magazine A 75, 1221—1235.

Shimizu, I., 1998a. Lognormality in crystal size distribution in dynamic recrystallization. Forma 13, 1—11.

Shimizu, I., 1998b. Stress and temperature dependence of recrystallized grain size: a subgrain misorientation model. Geophysical Research Letters 25, 4237—4240.

Shimizu, I., 1999. A stochastic model of grain size distribution during dynamic recrystallization. Philosophical Magazine A-Physics of Condensed Matter Structure Defects and Mechanical Properties 79, 1217—1231.

Shimizu, I., 2001. Nonequilibrium thermodynamics of nonhydrostatically stressed solids. In: Teisseyre, R., Majewski, E. (Eds.), Earthquake Thermodynamics and Phase Transformations in the Earth's Interior. Academic Press, pp. 81—102.

Shimizu, I., 2008. Theories and applicability of grain size piezometers: the role of dynamic recrystallization mechanisms. Journal of Structural Geology 30, 899—917.

Shimokawa, S., Ozawa, H., 2001. On the thermodynamics of the oceanic general circulation: entropy increase rate of an open dissipative system and its surroundings. Tellus Series A-Dynamic Meteorology and Oceanography 53, 266—277.

Shimokawa, S., Ozawa, H., 2002. On the thermodynamics of the oceanic general circulation: Irreversible transition to a state with higher rate of entropy production. Quarterly Journal of the Royal Meteorological Society 128, 2115—2128.

Shizawa, K., Zbib, H.M., 1999. A thermodynamical theory of gradient elastoplasticity with dislocation density tensor. I: fundamentals. International Journal of Plasticity 15, 899—938.

Shizawa, K., Kikuchi, K., Zbib, H.M., 2001. A strain-gradient thermodynamic theory of plasticity based on dislocation density and incompatibility tensors. Materials Science and Engineering: A-Structural Materials Properties Microstructure and Processing 309, 416—419.

Sibson, R.H., 1981. Controls on low-stress hydro-fracture dilatancy in thrust, wrench and normal-fault terrains. Nature 289, 665—667.

Sibson, R.H., 1987. Earthquake rupturing as a mineralizing agent in hydrothermal systems. Geology 15, 701—704.

Sibson, R.H., 1994. Crustal stress, faulting and fluid flow. Geological Society, London, Special Publications 78, 69—84.

Sibson, R.H., 1995. Selective fault reactivation during basin inversion: potential for fluid redistribution through fault-valve action. In: Buchanan, J.G., Buchanan, P.G. (Eds.), Basin Inversion. Geol. Soc. Lond. Spec. Publ, pp. 3—19.

Sibson, R.H., 2001. Seismogenic framework for hydrothermal transport and ore deposition. Reviews in Economic Geology 14, 25—47.

Sibson, R.H., 2004. Controls on maximum fluid overpressure defining conditions for mesozonal mineralisation. Journal of Structural Geology 26, 1127—1136.

Silhavy, M., 1997. The Mechanics and Thermodynamics of Continuous Media. Springer, Berlin.

Simpson, C., Schmid, S.M., 1983. An evaluation of criteria to deduce the sense of movement in sheared rocks. Geological Society of America Bulletin 94, 1281—1288.

Singleton, J.S., Mosher, S., 2012. Mylonitization in the lower plate of the Buckskin-Rawhide detachment fault, west-central Arizona: Implications for the geometric evolution of metamorphic core complexes. Journal of Structural Geology 39, 180—198.

Slotemaker, A.K., de Bresser, J.H.P., 2006. On the role of grain topology in dynamic grain growth — 2D micro-structural modeling. Tectonophysics 427, 73—93.

Slotemaker, A.K., 2006. Dynamic Recrystallization and Grain Growth in Olivine Rocks. Department of Earth Sciences. Utrecht University, Utrecht, p. 187.

Smith, C.S., 1948. Grains, phases, and interfaces: an interpretation of microstructure. Transactions of the American Institute of Mining and Metallurgical Engineers 175, 15—43.

Smith, C.S., 1964. Some elementary principles of polycrystalline microstructure. Metallurgical Reviews 9, 1—48.

Smith, R.B., 1977. Formation of folds, boudinage, and mullions in non-Newtonian materials. Geological Society of America Bulletin 88, 312–320.

Smithies, R.H., Howard, H.M., Evins, P.M., Kirkland, C.L., Kelsey, D.E., Hand, M., Wingate, M.T.D., Collins, A.S., Belousova, E., 2011. High-temperature granite magmatism, crust-mantle interaction and the Mesoproterozoic intracontinental evolution of the Musgrave Province, Central Australia. Journal of Petrology 52, 931–958.

Sobolev, S.V., Sobolev, A.V., Kuzmin, D.V., Krivolutskaya, N.A., Petrunin, A.G., Arndt, N.T., Radko, V.A., Vasiliev, Y.R., 2011. Linking mantle plumes, large igneous provinces and environmental catastrophes. Nature 477, 312–318.

Somiya, S., Moriyoshi, Y., 1990. Sintering Key Papers. Elsevier Applied Science, London, p. 801.

Sornette, D., 2000. Critical Phenomena in Natural Sciences; Chaos, Fractals, Self-Organization and Disorder: Concepts and Tools. Springer-Verlag, Berlin.

Spear, F.S., 1993. Metamorphic Phase Equilibria and Pressure-Temperature-Time Paths. Mineralogical Society of America, Washington, DC.

Sprott, J.C., 2003. Chaos and Time-Series Analysis. Oxford University Press, Oxford.

Srinivasan, S.G., Cahn, J.W., 2002. Challenging some free-energy reduction criteria for grain growth. In: Ankem, S., Pande, C.S., Ovid'ko, I., Ranganathan, S. (Eds.), Science and Technology of Interfaces: International Symposium Honoring the Contributions of Dr. Bhakta Rath. 3–14.

Stainier, L., Ortiz, M., 2010. Study and validation of a variational theory of thermo-mechanical coupling in finite visco-plasticity. International Journal of Solids and Structures 47, 705–715.

Stallard, A., Ikei, H., Masuda, T., 2002. Numerical simulations of spiral-shaped inclusion trails: can 3D geometry distinguish between end-member models of spiral formation? Journal of Metamorphic Geology 20, 801–812.

Steefel, C.I., Lasaga, A.C., 1994. A coupled model for transport of multiple chemical species and kinetic precipitation/dissolution reactions with application to reactive flow in single phase hydrothermal systems. American Journal of Science 294, 529–592.

Steefel, C.I., Maher, K., 2009. Fluid-rock interaction: a reactive transport approach. In: Oelkers, E.H., Schott, J. (Eds.), Thermodynamics and Kinetics of Water-Rock Interaction. Mineralogical Soc Amer, Chantilly, pp. 485–532.

Steefel, C.I., DePaolo, D.J., Lichtner, P.C., 2005. Reactive transport modeling: an essential tool and a new research approach for the Earth sciences. Earth and Planetary Science Letters 240, 539–558.

Stefanou, I., Sulem, J., 2014. Chemically induced compaction bands: triggering conditions and band thickness. Journal of Geophysical Research-Solid Earth 119, 880–899.

Sternlof, K.R., PhD thesis, 2006. Structural Geology, Propagation Mechanics and Hydraulic Effects of Compaction Bands in Sandstone, Geological and Environmental Sciences. Stanford University, Stanford, p. 214.

Stesky, R.M., 1978. Mechanisms of high-temperature frictional sliding in westerly granite. Canadian Journal of Earth Sciences 15, 361–375.

Stipp, M., Stunitz, H., Heilbronner, R., Schmid, S.M., 2002a. Dynamic recrystallization of quartz: correlation between natural and experimental conditions. Geological Society, London, Special Publications 200, 171–190.

Stipp, M., Stunitz, H., Heilbronner, R., Schmid, S.M., 2002b. The eastern Tonale fault zone: a 'natural laboratory' for crystal plastic deformation of quartz over a temperature range from 250 to 700 degrees C. Journal of Structural Geology 24, 1861–1884.

Stipp, M., Tullis, J., Scherwath, M., Behrmann, J.H., 2010. A new perspective on paleopiezometry: dynamically recrystallized grain size distributions indicate mechanism changes. Geology 38, 759–762.

Stokes, M.R., Wintsch, R.P., Southworth, C.S., 2012. Deformation of amphibolites via dissolution-precipitation creep in the middle and lower crust. Journal of Metamorphic Geology 30, 723–737.

Storti, F., Billi, A., Salvini, F., 2003. Particle size distributions in natural carbonate fault rocks: insights for non-self-similar cataclasis. Earth and Planetary Science Letters 206, 173–186.

Storti, F., Balsamo, F., Salvini, F., 2007. Particle shape evolution in natural carbonate granular wear material. Terra Nova 19, 344–352.

Stout, M.G., Kocks, U.F., 1998. Effects of texture on plasticity. In: Kocks, U.F., Tome, C.N., Wenk, H.-R. (Eds.), Texture and Anisotropy. Cambridge University Press, Cambridge, pp. 420–465.

Streitenberger, P., Zollner, D., 2006. Effective growth law from three-dimensional grain growth simulations and new analytical grain size distribution. Scripta Materialia 55, 461–464.

Streitenberger, P., Forster, D., Kolbe, G., Veit, P., 1995. The fractal geometry of grain-boundaries in deformed and recovered zinc. Scripta Metallurgica et Materialia 33, 541–546.

Streitenberger, P., 1998. Generalized Lifshitz-Slyozov theory of grain and particle coarsening for arbitrary cut-off parameter. Scripta Materialia 39, 1719—1724.

Streitenberger, P., 2001. Analytic model of grain growth based on a generalized LS stability argument and topological relationships. In: Gottstein, G., Molodov, D.A. (Eds.), Recrystallization and Grain Growth. Springer-Verlag, Berlin, pp. 257—262.

Strogatz, S.H., 1994. Nonlinear Dynamics and Chaos: With Applications to Physics, Biology, Chemistry, and Engineering. Addison-Wesley, Reading, MA.

Strogatz, S.H., 2001. Exploring complex networks. Nature 410, 268—276.

Stunitz, H., Keulen, N., Hirose, T., Heilbronner, R., 2010. Grain size distribution and microstructures of experimentally sheared granitoid gouge at coseismic slip rates — criteria to distinguish seismic and aseismic faults? Journal of Structural Geology 32, 59—69.

Stuwe, K., Sandiford, M., 1994. Contribution of deviatoric stresses to metamorphic P-T paths: an example appropriate to low-P, high-T metamorphism. Journal of Metamorphic Geology 12, 445—454.

Tadmor, E.B., Miller, R.E., 2011. Modeling Materials: Continuum, Atomistic and Multiscale Techniques. Cambridge University Press.

Tadmor, E.B., Miller, R.E., Elliott, R.S., 2012. Continuum Mechanics and Thermodynamics: From Fundamental Concepts to Governing Equations. Cambridge University Press.

Takahashi, M., 1998. Fractal analysis of experimentally, dynamically recrystallized quartz grains and its possible application as a strain rate meter. Journal of Structural Geology 20, 269—275.

Takens, F., 1981. Detecting strange attractors in turbulence. In: Rand, D.A., Young, L.-S. (Eds.), Dynamical Systems and Turbulence. Lecture Notes in Mathematics. Springer, New York, pp. 366—381.

Tang, M., Kubin, L.P., Canova, G.R., 1998. Dislocation mobility and the mechanical response of BCC single crystals: a mesoscopic approach. Acta Materialia 46, 3221—3235.

Tanner, R.I., Huilgol, R.R., 1975. On a classification scheme for flow fields. Rheologica Acta 14, 959—962.

Tanner, R.I., 1976. A test particle approach to flow classification for viscoelastic fluids. AIChE Journal 22, 910—918.

Taupin, V., Capolungo, L., Fressengeas, C., Das, A., Upadhyay, M., 2013. Grain boundary modeling using an elastoplastic theory of dislocation and disclination fields. Journal of the Mechanics and Physics of Solids 61, 370—384.

Taylor, J.E., Cahn, J.W., 1988. Theory of orientation textures due to surface-energy anisotropies. Journal of Electronic Materials 17, 443—445.

Taylor, J.E., Cahn, J.W., 2007. Shape accommodation of a rotating embedded crystal via a new variational formulation. Interfaces and Free Boundaries 9, 493—512.

Taylor, G.I., Quinney, H., 1934. The latent energy remaining in a metal after cold working. Proceedings of the Royal Society of London Series A 143, 307—326.

Taylor, G.I., 1938. Plastic strain in metals. Journal of the Institute of Metals 62, 307—324.

Tel, T., Karolyi, G., Pentek, A., Scheuring, I., Toroczkai, Z., Grebogi, C., Kadtke, J., 2000. Chaotic advection, diffusion, and reactions in open flows. Chaos 10, 89—98.

Tel, T., de Moura, A., Grebogi, C., Karolyi, G., 2005. Chemical and biological activity in open flows: a dynamical system approach. Physics Reports-Review Section of Physics Letters 413, 91—196.

ten Grotenhuis, S.M., Trouw, R.A.J., Passchier, C.W., 2003. Evolution of mica fish in mylonitic rocks. Tectonophysics 372, 1—21.

Terzaghi, K., 1936. The shear resistance of saturated soils. In: 1st International Conference on Soil Mechanics and Foundation Engineering, Cambridge, MA, pp. 54—56.

Thiffeault, J.-L., 2010. Chaos in the Gulf. Science 330, 458—459.

Thompson, A.B., 1983. Fluid-absent metamorphism. Journal of the Geological Society 140, 533—547.

Thompson, A.B., England, P.C., 1984. Pressure temperature time paths of regional metamorphism. 2. Their inference and interpretation using mineral assemblages in metamorphic rocks. Journal of Petrology 25, 929—955.

Thompson, C.V., 2001. Grain growth and evolution of other cellular structures. Solid State Physics: Advances in Research and Applications 55, 269—314.

Thompson Jr, J.B., 1959. Local equilibrium in metasomatic processes. In: Abelson, P.H. (Ed.), Researches in Geochemistry. John Wiley & Sons, New York, pp. 427—457.

Thompson, J.B., 1970. Geochemical reaction and open systems. Geochimica et Cosmochimica Acta 34, 529—551.

Thompson, J.M.T., Sieber, J., 2010. Predicting climate tipping as a noisy bifurcation: a review. International Journal of Bifurcation and Chaos 1—28.

Thompson, J.M.T., Stewart, H.B., 2002. Nonlinear Dynamics and Chaos, second ed. Wiley, Chichester, UK.

Thompson, J.M.T., 1982. Instabilities and Catastrophes in Science and Engineering. Wiley, Chichester, UK.

Thomson, W., Tait, P.G., 1872. Elements of Natural Philosophy, second ed. Cambridge University Press.

Thomson, W., 1887. LXIII. On the division of space with minimum partitional area. Philosophical Magazine Series 5 (24), 503–514.

Tome, C.N., Canova, G.R., 1998. Self-consistent modelling of heterogeneous plasticity. In: Kocks, U.F., Tome, C.N., Wenk, H.-R. (Eds.), Texture and Anisotropy. Cambridge University Press, Cambridge, pp. 466–511.

Tommasi, A., Mainprice, D., Canova, G., Chastel, Y., 2000. Viscoplastic self-consistent and equilibrium-based modeling of olivine lattice preferred orientations: Implications for the upper mantle seismic anisotropy. Journal of Geophysical Research-Solid Earth 105, 7893–7908.

Tonelli, L., 1921. Fondamenti di Calcolo delle Variazioni, vol. I. Zanichelli, Bologna.

Tonelli, L., 1923. Fondamenti di Calcolo delle Variazioni, vol. II. Zanichelli, Bologna.

Trautt, Z.T., Mishin, Y., 2012. Grain boundary migration and grain rotation studied by molecular dynamics. Acta Materialia 60, 2407–2424.

Trautt, Z.T., Mishin, Y., 2014. Capillary-driven grain boundary motion and grain rotation in a tricrystal: a molecular dynamics study. Acta Materialia 65, 19–31.

Trefry, M.G., Lester, D.R., Metcalfe, G., Ord, A., Regenauer-Lieb, K., 2012. Toward enhanced subsurface intervention methods using chaotic advection. Journal of Contaminant Hydrology 127, 15–29.

Tresca, H.E., 1864. Mémoire sur l'écoulement des corps solides soumis à de fortes pressions. Comptes Rendus des Seances de l'Academie des Sciences 59, 754–758.

Tresca, H.E., 1865. Rapport sur un Memoire presente par M. H. Tresca dans la seance du 7 novembre 1864, et intitule : de l'ecoulement des corps solides. Comptes Rendus des Seances de l'Academie des Sciences 60, 1226–1232.

Tresca, H.E., 1867. Mémoire sur l'écoulement des corps solides soumis à de fortes pressions. Comptes Rendus des Seances de l'Academie des Sciences 64, 809–812.

Tresca, H.E., 1868. Rapport sur un troisieme et un quatrieme Memoires relatifs a l'ecoulement des solides, presentes par M. H. Tresca. Comptes Rendus des Seances de l'Academie des Sciences 66, 263–270.

Tresca, H.E., 1869. Memoire sur le poinconnage et la theorie mecanique de la deformation des metaux. Comptes Rendus des Seances de l'Academie des Sciences 68, 1197–1201.

Tresca, H.E., 1870a. Memoire sur le poinconnage et la theorie mecanique de la deformation des metaux. Comptes Rendus des Seances de l'Academie des Sciences 70, 27–31.

Tresca, H.E., 1870b. Rapport sur le Memoire presente a l'Academie le 29 mai 1869 par M. Tresca, sur le poinconnage et sur la theorie mecanique de la deformation des corps solides. Comptes Rendus des Seances de l'Academie des Sciences 70, 288–308.

Trouw, R.A.J., Passchier, C.W., Wiersma, D.J., 2010. Atlas of Mylonites — And Related Microstructructures. Springer, Heidelberg.

Truesdell, C., Toupin, R.A., 1960. The classical field theories. In: Flugge, S. (Ed.), Encyclopaedia of Physics. Springer-Verlag, Berlin, pp. 226–793.

Truesdell, C., 1952. The mechanical foundations of elasticity and fluid mechanics. Journal of Rational Mechanics and Analysis 1, pp. 125–171, 173–300.

Truesdell, C., 1953. Corrections and additions to 'The mechanical foundations of elasticity and fluid dynamics'. Journal of Rational Mechanics and Analysis 2, 593–616.

Truesdell, C., 1954. The Kinematics of Vorticity. Indiana University Press, Bloomington, 232 pp.

Truesdell, C.A., 1966. Six Lectures on Modern Natural Philosophy. Springer-Verlag, Berlin.

Truesdell III, C.A., 1969. Rational Thermodynamics, second ed. Springer-Verlag, New York.

Truesdell III, C.A., 1977. A First Course in Rational Continuum Mechanics, Part I. Academic Press, New York.

Truskinovsky, L., Zanzotto, G., 1995. Finite-scale microstructures and metastability in one-dimensional elasticity. Meccanica 30, 577–589.

Truskinovsky, L., Zanzotto, G., 1996. Ericksen's bar revisited: energy wiggles. Journal of the Mechanics and Physics of Solids 44, 1371–1408.

Truskinovsky, L., 1996. Fracture as a phase transition. In: Batra, R.C., Beatty, M.F. (Eds.), Contemporary Research in the Mechanics and Mathematics of Materials. Symposium on Recent Developments in Elasticity. John Hopkins University, Baltimore, Maryland. June 12–15, pp. 322–332.

Turcotte, D.L., Schubert, G., 1982. Geodynamics: Applications of Continuum Physics to Geological Problems. John Wiley and Sons Inc.

Turiel, A., Perez-Vicente, C.J., Grazzini, J., 2006. Numerical methods for the estimation of multifractal singularity spectra on sampled data: a comparative study. Journal of Computational Physics 216, 362–390.

Turing, A.M., 1952. The chemical basis of morphogenesis. Philosophical Transactions of the Royal Society of London Series, B: Biological Sciences 237, 37–72.

Turner, F.J., Weiss, L.E., 1963. Structural Analysis of Metamorphic Tectonites. McGraw-Hill Book Company, New York.

Turner, F.J., Weiss, L.E., 1965. Deformational kinks in brucite and gypsum. Proceedings of the National Academy of Sciences of the United States of America 54, 359–364.

Turner, F.J., 1948. Mineralogical and structural evolution of the metamorphic rocks. Geological Society of America Memoir 30, 1–332.

Tvergaard, V., Needleman, A., 1980. On the localisation of buckle patterns. Journal of Applied Mechanics 47, 613–619.

Twiss, R.J., 1977. Theory and applicability of a recrystallized grain size paleopiezometer. Pure and Applied Geophysics 115, 227–244.

Ulm, F.-J., Coussy, O., 2003. Mechanics and Durability of Solids. Prentice Hall, Upper Saddle River, NJ.

Upmanyu, M., Srolovitz, D.J., Lobkovsky, A.E., Warren, J.A., Carter, W.C., 2006. Simultaneous grain boundary migration and grain rotation. Acta Materialia 54, 1707–1719.

Van Heerden, C., 1953. Autothermic processes: properties and reactor design. Industrial and Engineering Chemistry 45, 1242–1247.

Vardoulakis, I., Sulem, J., 1995. Bifurcation Analysis in Geomechanics. Blackie Academic and Professional.

Velde, B., Dubois, J., Moore, D., Touchard, G., 1991. Fractal patterns of fractures in granites. Earth and Planetary Science Letters 104, 25–35.

Venugopal, V., Roux, S.G., Foufoula-Georgiou, E., Arneodo, A., 2006. Revisiting multifractality of high-resolution temporal rainfall using a wavelet-based formalism. Water Resources Research 42, W06D14, 20 pages.

Vernon, R.H., Clarke, G.L., 2008. Principles of Metamorphic Petrology. Cambridge University Press.

Vernon, R.H., 1976. Metamorphic Processes, Reactions and Microstructure Development. Murby, London.

Vernon, R.H., 1998. Chemical and volume changes during deformation and prograde metamorphism of sediments. In: Taylor, P.J., O'Brien, P.J. (Eds.), What Drives Metamorphism and Metamorphic Reactions? pp. 215–246.

Vernon, R.H., 2004. A Practical Guide to Rock Microstructure. Cambridge University Press.

Veveakis, E., Alevizos, S., Vardoulakis, I., 2010. Chemical reaction capping of thermal instabilities during shear of frictional faults. Journal of the Mechanics and Physics of Solids 58, 1175–1194.

Veveakis, M., Stefanou, I., Sulem, J., 2013. Failure in shear bands for granular materials: thermo-hydro-chemo-mechanical effects. Geotechnique Letters 3, 31–36.

Villaverde, A.F., Ross, J., Moran, F., Balsa-Canto, E., Banga, J.R., 2011. Use of a generalized Fisher equation for global optimization in chemical kinetics. Journal of Physical Chemistry A 115, 8426–8436.

Villermaux, E., 2012. On dissipation in stirred mixtures. In: Erik van der, G., Hassan, A. (Eds.), Advances in Applied Mechanics. Elsevier, pp. 91–107.

Voigt, W., 1928. Lehrbuch der Kristallphysik. Teubner, Leipzig.

Voll, G., 1960. New work on petrofabrics. Liverpool and Manchester Geological Journal 2, 503–567.

Volterra, V., 1907. L'equilibre des corps élastique multiplement connexes. Annales Scientifiques de l'Ecole Normale Superieure, Paris 24, 401–517.

Vosteen, H.D., Schellschmidt, R., 2003. Influence of temperature on thermal conductivity, thermal capacity and thermal diffusivity for different types of rock. Physics and Chemistry of the Earth 28, 499–509.

Voyiadjis, G.Z., Kattan, P.I., 2006. Damage mechanics with fabric tensors. Mechanics of Advanced Materials and Structures 13, 285–301.

Voyiadjis, G.Z., 2011. Damage Mechanics and Micromechanics of Localized Fracture Phenomena in Inelastic Solids. In: Maier, G., Rammerstorfer, F.G., Salencon, J., Schrefler, B., Serafini, P. (Eds.), International Centre for Mechanical Sciences Courses and Lectures. Springer, Vienna, p. 411.

Wagner, T., Lee, J., Hacker, B.R., Seward, G., 2010. Kinematics and vorticity in Kangmar Dome, Southern Tibet: testing midcrustal channel flow models for the Himalaya. Tectonics 29, TC6011, 1–26.

Walder, J., Nur, A., 1984. Porosity reduction and crustal pore pressure development. Journal of Geophysical Research 89, 1539–1548.

Walgraef, D., Aifantis, E.C., 1985a. On the formation and stability of dislocation patterns − I: one-dimensional considerations. International Journal of Engineering Science 23, 1351–1358.

Walgraef, D., Aifantis, E.C., 1985b. On the formation and stability of dislocation patterns — II: two-dimensional considerations. International Journal of Engineering Science 23, 1359—1364.

Walgraef, D., Aifantis, E.C., 1985c. On the formation and stability of dislocation patterns — III: three-dimensional considerations. International Journal of Engineering Science 23, 1365—1372.

Wang, Y.F., Merino, E., 1992. Dynamic model of oscillatory zoning of trace elements in calcite: double layer, inhibition, and self-organization. Geochimica et Cosmochimica Acta 56, 587—596.

Watt, G.R., Oliver, N.H.S., Griffin, B.J., 2000. Evidence for reaction-induced microfracturing in granulite facies migmatites. Geology 28, 327—330.

Weaire, D., Phelan, R., 1994. A counter-example to Kelvin's conjecture on minimal surfaces. Philosophical Magazine Letters 69, 107—110.

Weatherley, D.K., Henley, R.W., 2013. Flash vaporization during earthquakes evidenced by gold deposits. Nature Geoscience 6, 294—298.

Wei, Y., Wu, J., Yin, H., Shi, X., Yang, R., Dresselhaus, M., 2012. The nature of strength enhancement and weakening by pentagon-heptagon defects in graphene. Nature Materials 11, 759—763.

Weijermars, R., Poliakov, A., 1993. Stream functions and complex potentials — implications for development of rock fabric and the continuum assumption. Tectonophysics 220, 33—50.

Weijermars, R., 1991. The role of stress in ductile deformation. Journal of Structural Geology 13, 1061—1078.

Weijermars, R., 1993. Pulsating strains. Tectonophysics 220, 51—67.

Weinhold, F., 1975a. Metric geometry of equilibrium thermodynamics. Journal of Chemical Physics 63, 2479—2483.

Weinhold, F., 1975b. Metric geometry of equilibrium thermodynamics. II. Scaling, homogeneity, and generalized Gibbs—Duhem relations. Journal of Chemical Physics 63, 2484—2487.

Weinhold, F., 1975c. Metric geometry of equilibrium thermodynamics. III. Elementary formal structure of a vector-algebraic representation of equilibrium thermodynamics. Journal of Chemical Physics 63, 2488—2495.

Weinhold, F., 1975d. Metric geometry of equilibrium thermodynamics. IV. Vector-algebraic evaluation of thermodynamic derivatives. Journal of Chemical Physics 63, 2496—2501.

Weinhold, F., 1976. Metric geometry of equilibrium thermodynamics. V. Aspects of heterogeneous equilibrium. Journal of Chemical Physics 65, 559—564.

Weiss, J., Miguel, M.C., 2004. Dislocation avalanche correlations. Materials Science and Engineering: A 387—389, 292—296.

Weiss, J., Lahaie, F., Grasso, J.R., 2000. Statistical analysis of dislocation dynamics during viscoplastic deformation from acoustic emission. Journal of Geophysical Research 105, 433—442.

Wenk, H.R., van Houtte, P., 2004. Texture and anisotropy. Reports on Progress in Physics 67, 1367—1428.

Wenk, H.-R., Canova, G., Molinari, A., Kocks, U.F., 1989. Viscoplastic modeling of texture development in quartzite. Journal of Geophysical Research 94, 17895—17906.

Wenk, H.R., 1999. A voyage through the deformed Earth with the self-consistent model. Modelling and Simulation in Materials Science and Engineering 7, 699—722.

Wheeler, J., 1987. The significance of grain-scale stresses in the kinetics of metamorphism. Contributions to Mineralogy and Petrology 97, 397—404.

Wheeler, J., 2010. Anisotropic rheology during grain boundary diffusion creep and its relation to grain rotation, grain bound sliding and superplasticity. Philosophical Magazine 90, 2841—2864.

Wheeler, J., 2014. Dramatic effects of stress on metamorphic reactions. Geology 42, 647—650.

White, R.W., Clarke, G.L., 1997. The role of deformation in aiding recrystallization: an example from a high-pressure shear zone, central Australia. Journal of Petrology 38, 1307—1329.

White, J.C., White, S.H., 1981. On the structure of grain-boundaries in tectonites. Tectonophysics 78, 613—628.

Whiting, A.I.M., Hunt, G.W., 1997. Evolution of nonperiodic forms in geological folds. Mathematical Geology 29, 705—723.

Whitmeyer, S.J., Wintsch, R.P., 2005. Reaction localization and softening of texturally hardened mylonites in a reactivated fault zone, central Argentina. Journal of Metamorphic Geology 23, 411—424.

Wiggins, S., Ottino, J.M., 2004. Foundations of chaotic mixing. Philosophical Transactions: Mathematical, Physical and Engineering Sciences 362, 937—970.

Wiggins, S., 2003. Introduction to Applied Nonlinear Dynamical Systems and Chaos. Springer, New York.

Williams, R.O., 1965. The stored energy of copper deformed at 24 C. Acta Metallurgica 13, 163—168.

Williams, M.L., Jercinovic, M.J., 2012. Tectonic interpretation of metamorphic tectonites: integrating compositional mapping, microstructural analysis and in situ monazite dating. Journal of Metamorphic Geology 30, 739—752.

Willis, J.R., 1967. A comparison of fracture criteria of Griffith and Barenblatt. Journal of the Mechanics and Physics of Solids 15, 151–162.

Winkler, K., Nur, A., Gladwin, M., 1979. Friction and seismic attenuation in rocks. Nature 277, 528–531.

Winterbottom, W.L., 1967. Equilibrium shape of a small particle in contact with a foreign substrate. Acta Metallurgica 15, 303–310.

Wintsch, R.P., Dunning, J., 1985. The effect of dislocation density on the aqueous solubility of quartz and some geologic implications – a theoretical approach. Journal of Geophysical Research-Solid Earth and Planets 90, 3649–3657.

Wintsch, R.P., Knipe, R.J., 1983. Growth of a zoned plagioclase porphyroblast in a mylonite. Geology 11, 360–363.

Wintsch, R.P., Yeh, M.-W., 2013. Oscillating brittle and viscous behavior through the earthquake cycle in the Red River Shear Zone: monitoring flips between reaction and textural softening and hardening. Tectonophysics 587, 46–62.

Wintsch, R.P., Yi, K., 2002. Dissolution and replacement creep: a significant deformation mechanism in mid-crustal rocks. Journal of Structural Geology 24, 1179–1193.

Wintsch, R.P., 1985. The possible effects of deformation on chemical processes in metamorphic fault zones. In: Thompson, A., Rubie, D. (Eds.), Metamorphic Reactions. Springer, New York, pp. 251–268.

Wolfram Research, Inc., 2012. Mathematica, Version 9.0, Champaign, IL. Available at http://support.wolfram.com/kb/472

Wong, T.-f., Baud, P., 2012. The brittle-ductile transition in porous rock: a review. Journal of Structural Geology 44, 25–53.

Wood, B.J., Fraser, D.C., 1976. Elementary Thermodynamics for Geologists. Oxford University Press.

Wood, J.R., Hewett, T.A., 1982. Fluid convection and mass-transfer in porous sandstones – a theoretical model. Geochimica et Cosmochimica Acta 46, 1707–1713.

Wood, B.J., Walther, J.V., 1986. Fluid flow during metamorphism and its implications for fluid-rock ratios. In: Walther, J.V., Wood, B.J. (Eds.), Fluid-Rock Interactions During Metamorphism. Springer-Verlag, New York, pp. 89–108.

Worley, B., Powell, R., Wilson, C.J.L., 1997. Crenulation cleavage formation: Evolving diffusion, deformation and equilibration mechanisms with increaseing metamorphic grade. Journal of Structural Geology 19, 1121–1135.

Xia, F., Brugger, J., Chen, G., Ngothai, Y., O'Neill, B., Putnis, A., Pring, A., 2009. Mechanism and kinetics of pseudomorphic mineral replacement reactions: a case study of the replacement of pentlandite by violarite. Geochimica et Cosmochimica Acta 73, 1945–1969.

Xiao, Y., Lasaga, A.C., 1996. Ab initio quantum mechanical studies of the kinetics and mechanisms of quartz dissolution: OH^- catalysis. Geochimica et Cosmochimica Acta 60, 2283–2295.

Yang, L.F., Epstein, I.R., 2003. Oscillatory Turing patterns in reaction-diffusion systems with two coupled layers. Physical Review Letters 90, 178303.

Yang, P., Rivers, T., 2001. Chromium and manganese zoning in pelitic garnet and kyanite: spiral, overprint, and oscillatory (?) zoning patterns and the role of growth rate. Journal of Metamorphic Geology 19, 455–474.

Yardley, B.W.D., Valley, J.W., 1997. The petrologic case for a dry lower crust. Journal of Geophysical Research-Solid Earth 102, 12173–12185.

Yardley, B.W.D., Banks, D.A., Bottrell, S.H., Diamond, L.W., 1993. Post-metamorphic gold quartz veins from N.W. Italy – the composition and origin of the ore fluid. Mineralogical Magazine 57, 407–422.

Yardley, B.W.D., 1989. An Introduction to Metamorphic Petrology. Longman, Harlow, UK.

Yardley, B.W.D., 2009. The role of water in the evolution of the continental crust. Journal of the Geological Society 166, 585–600.

Yuen, D.A., Schubert, G., 1979. Stability of frictionally heated shear flows in the asthenosphere. Geophysical Journal of the Royal Astronomical Society 57, 189–207.

Zaiser, M., 2006. Scale invariance in plastic flow of crystalline solids. Advances in Physics 55, 185–245.

Zbib, H.M., Rhee, M., Hirth, J.P., 1996. 3-D simulation of curved dislocations: discretization and long range interactions, advances in engineering plasticity and its applications. In: Proceedings of the 3rd Asia-Pacific Symposium on Advances in Engineering Plasticity and Its Applications. AEPA '96, August 12–24, 1996. Pergamon, Amsterdam, Hiroshima, Japan, pp. 15–20.

Zemansky, M.W., 1951. Heat and Thermodynamics, third ed. McGraw-Hill, New York.

Zhang, P., DeVries, S.L., Dathe, A., Bagtzoglou, A.C., 2009. Enhanced mixing and plume containment in porous media under time-dependent oscillatory flow. Environmental Science and Technology 43, 6283–6288.

Zhang, Y., Hobbs, B.E., Jessell, M.W., 1994a. The effect of grain-boundary sliding on fabric development in polycrystalline aggregates. Journal of Structural Geology 16, 1315—1325.

Zhang, Y., Hobbs, B.E., Ord, A., 1994b. A numerical-simulation of fabric development in polycrystalline aggregates with one slip system. Journal of Structural Geology 16, 1297—1313.

Zhang, Y., Hobbs, B.E., Ord, A., Barnicoat, A., Zhao, C., Walshe, J.L., Lin, G., 2003. The influence of faulting on host-rock permeability, fluid flow and ore genesis of gold deposits: a theoretical 2D numerical model. Journal of Geochemical Exploration 78-9, 279—284.

Zhang, Y., Schaubs, P.M., Zhao, C., Ord, A., Hobbs, B.E., Barnicoat, A.C., 2008. Fault-related dilation, permeability enhancement, fluid flow and mineral precipitation patterns: numerical models. Geological Society, London, Special Publications 299, 239—255.

Zhang, Y., Schaubs, P.M., Sheldon, H.A., Poulet, T., Karrech, A., 2013. Modelling fault reactivation and fluid flow around a fault restraining step-over structure in the Laverton gold region, Yilgarn Craton, Western Australia. Geofluids 13, 127—139.

Zhao, C.B., Hobbs, B.E., Muhlhaus, H.B., 1998. Analysis of pore-fluid pressure gradient and effective vertical-stress gradient distribution in layered hydrodynamic systems. Geophysical Journal International 134, 519—526.

Zhao, C.B., Hobbs, B.E., Baxter, K., Muhlhaus, H.B., Ord, A., 1999a. A numerical study of pore-fluid, thermal and mass flow iu fluid-saturated porous rock basins. Engineering Computations 16, 202—214.

Zhao, C.B., Hobbs, B.E., Muhlhaus, H.B., 1999b. Effects of medium thermoelasticity on high Rayleigh number steady-state heat transfer and mineralization in deformable fluid-saturated porous media heated from below. Computer Methods in Applied Mechanics and Engineering 173, 41—54.

Zhao, C.B., Hobbs, B.E., Muhlhaus, H.B., 1999c. Theoretical and numerical analyses of convective instability in porous media with upward throughflow. International Journal for Numerical and Analytical Methods in Geomechanics 23, 629—646.

Zhao, C.B., Hobbs, B.E., Muhlhaus, H.B., 2000. Finite element analysis of heat transfer and mineralization in layered hydrothermal systems with upward throughflow. Computer Methods in Applied Mechanics and Engineering 186, 49—64.

Zhao, C., Hobbs, B., Ord, A., Peng, S., Muhlhaus, H., Liu, L., 2004. Theoretical investigation of convective instability in inclined and fluid-saturated three-dimensional fault zones. Tectonophysics 387, 47—64.

Zhao, C., Hobbs, B.E., Ord, A., 2008. Convective and Advective Heat Transfer in Geological Systems. Springer-Verlag, Berlin.

Zhao, C., Hobbs, B.E., Ord, A., 2009. Fundamentals of Computational Geoscience: Numerical Methods and Algorithms. Springer-Verlag, Berlin.

Zheng, S., Carpenter, J.S., McCabe, R.J., Beyerlein, I.J., Mara, N.A., 2014. Engineering interface structures and thermal stabilities via SPD processing in bulk nanostructured metals. Scientific Reports 4, 4226, 1—6.

Zhu, T., Li, J., Lin, X., Yip, S., 2005. Stress-dependent molecular pathways of silica-water reaction. Journal of the Mechanics and Physics of Solids 53, 1597—1623.

Zhu, T.T., Bushby, A.J., Dunstan, D.J., 2008. Materials mechanical size effects: a review. Materials Technology 23, 193—209.

Ziegler, H., 1963. Some extremum principles in irreversible thermodynamics with application to continuum mechanics. In: Sneddon, L.N., Hill, R. (Eds.), Progress in Solid Mechanics, pp. 91—193.

Ziegler, H., 1983a. Chemical reactions and the principle of maximal rate of entropy production. Journal of Applied Mathematics and Physics 34, 832—844.

Ziegler, H., 1983b. An Introduction to Thermomechanics, 2nd, revised ed. North-Holland Publishing Company, Amsterdam.

Zollner, D., Streitenberger, P., 2006. Three-dimensional normal grain growth: Monte Carlo Potts model simulation and analytical mean field theory. Scripta Materialia 54, 1697—1702.

Zollner, D., 2006. Monte Carlo Potts Model Simulation and Statistical Mean-Field Theory of Normal Grain Growth. Shaker-Verlag, Aachen.

Zollner, D., Streitenberger, P., 2008. Normal grain growth: Monte Carlo Potts model simulation and mean-field theory. In: Bertram, A., Tomas, J. (Eds.), Micro-Macro-Interactions in Structured Media and Particle Systems. Springer Verlag, pp. 3—18.

Index

Printed and bound by CPI Group (UK) Ltd, Croydon, CR0 4YY

08/05/2025

01864871-0004